AMERICAN MATHEMATICAL CONTESTS
A Guide to Success

Harold Reiter
University of North Carolina Charlotte

Jonathan Kane
University of Wisconsin

Yunzhi Zou
Sichuan University

Kendall Hunt
publishing company

Cover image by Harold Reiter

Illustrations provided by Yunzhi Zou

Kendall Hunt
publishing company

www.kendallhunt.com
Send all inquiries to:
4050 Westmark Drive
Dubuque, IA 52004-1840

CONTENTS IN BRIEF

PART V   SOLUTION TO PART II PRACTICE PROBLEMS

PART VI   SOLUTIONS TO PART III PAST CONTEST PROBLEMS

CONTENTS

PART II   ESSAYS AND PRACTISE PROBLEMS

PART VI   SOLUTIONS TO PART III PAST CONTEST PROBLEMS

xii CONTENTS

INTRODUCTION

Doing mathematics in general and creating and solving problems in particular are social endeavors. My American Mathematics Competitions companions included dear friends Walter Mientka and Leo Schneider, both deceased, and Richard 'Dick' Gibbs. My university life has been enriched by the enthusiasm and curiosity of three gifted undergraduates, David Wieland, James Rudzinski, and Juan Vargas. Early Charlotte Mathematics Club students Nathan Bronson, Akira Negi, Ashley Reiter Ahlin, Scott Harrington and later ones Brian Dean, Sarah Dean, Tung Tran, Garrett Mitchener, Ryan Vinroot, Paul Rupe, Joseph Schaeffer, Anders Kaseorg, Zack Lee, Drew Boyuka, Parker Garrison and Brendan Fletcher all turned out to be world class problem solvers.

And current ones are Griffin Cherniss, Oliver Lippard, Aman Singh, Timmy Deng, Dev Chheda, Olivia Yang, Sreyas Adiraju, Anthony Yang, Bhushan Mohanraj, and Soumyadeep Bhattacharjee.

During the years 2005 to 2016 I have enjoyed teaching at several wonderful summer camps including Math Path, Math Works, Math Zoom, and Epsilon Camp. Wonderful students John Ahlin, Connor Mooney, Daniel Liu, Kalman Strauss, Catherine Phillips, Angela Deng (MathPath), Vaughan McInerney (MathWorks), and Tyson Lin (Punahou School) have taught me new ways to look at old problems.

My friends Richard Rusczyk, David Patrick and Jason Batterson at Art of Problem Solving have always been encouraging and helpful. I am also grateful to friends at other competitions, Paul Dreyer (ARML), Zvezdelina Stankova and Paul Zeitz (BAMO), and Jon Kane (PurpleComet), and State Math Contest friends John Goebel, Archie Benton, Philip Rash and Randy Harter.

MATHCOUNTS friends John Jensen, Kristen Chandler, Patrick Vennebush, and John Benson have been great companions. My life is also enriched by nearly daily contact with two local problem solving friends Arthur Holshouser and Ben Klein.

And, of course, there is my new and dear friend Yunzhi Zou, whose wonderful hospitality and company I have enjoyed on my two trips to Chengdu, and friends there Zengbao Wu, Jian Mei and Binhao Fan.

Finally, I want to thank my wonderful wife Betty who has encouraged my travels and problem puzzling times, and even helped me solve some of them.

Harold B. Reiter

I taught mathematics, statistics, computer science, and actuarial mathematics at the university level for my entire career, but aside from annually proctoring the collegiate William Lowell Putnam Mathematical Competition at the University of Wisconsin - Whitewater, I had no involvement with mathematics competitions before my son, Daniel Kane, began participating in mathematics competitions in 1999. That spring, to the surprise of my wife and me, Daniel did exceptionally well in the AMC 12, the AIME, and the USA Math Olympiad. To allow Daniel to participate in MATH-COUNTS, I coached a team from his school and went on to assist in coaching the Wisconsin State MATHCOUNTS team. This experience not only introduced me to these competitions, it allowed me to meet some of the key players in the mathematics competition community including Titu Andreescu, who was the director of the American Mathematics Competitions (AMC) at the time, Susan Wild-strom, a high school teacher active in the AMC, and Steve Dunbar, who succeeded Titu Andreescu as AMC director. I began to do problem writing and review for the AMC, and continued to assist with the Wisconsin MATHCOUNTS teams.

In 2003-2004 Titu Andreescu spent a year as a visiting scholar at UW - Whitewater. He and I worked closely on several projects and became close friends. Our collaboration resulted in what became known as the Purple Comet! Math Meet, now in its fifteenth year. In 2005 I joined the AIME Committee which has the responsibility of creating the two sets of AIME contest problems each year. This quickly became my most rewarding professional activity, and I always look forward to the annual AIME Committee meetings. I am currently chair of the AIME Committee, and as chair am automatically a member of the AMC 10/12 Committee and the USAMO Committee. I have enjoyed and learned from working with many talented people on these committees including David Hankin, Zuming Feng, Steve Blasberg, Elgin Johnston, Jerry Grossman, Mark Saul, and Chris Jeuell.

In 2006 Titu Andreescu, now working at the University of Texas at Dallas, created the Awe-someMath Summer Program, a summer camp experience for middle and high school students want-ing to improve their mathematics competition skills. From the beginning, I have enjoyed teaching at this camp, and, through it, have worked with some great teachers of mathematics including Harold Reiter, Richard Rusczyk, Razvan Gelca, Walter Stromquist, Natalya St. Clair, and Richard Newcomb.

When I retired from university teaching in 2012, I immediately joined the Wisconsin Math Talent Search Committee, a natural extension of the work I was already doing for the AMC. My work there with Benedek Valko and Melanie Matchett-Wood has been very rewarding.

I am greatly indebted to my coauthors, Harold and Yunzhi, who, less than a year ago, invited me to teach mathematics problem solving in China. In January, 2017 I made my first ever trip to Asia with a busy two-week working vacation in Chengdu, Nanjing, and Shanghai where I met many nice people and was thrilled to work with Chinese mathematics students, to learn about their history, and to experience their culture. I was glad to contribute what I could to this book and hope to have many years of continued collaboration with my new friends in China.

Jonathan M. Kane

During the past decade, I have witnessed a rising number of students going overseas to further their education. Popular destinations are English speaking countries such as the United States of America, United Kingdom, Canada, and Australia. I have been very pleased to work with Dipont for more than ten years on its international curricula programs, and therefore have had chances to know and teach overseas math courses and standards such as British A-level, Collegeboard AP, and IBDP. Teaching mathematics has become more fun for me. However, for many years, I have not had the chance to work in the field of mathematics competitions and problem solving. In 2015 I met my dear friend Dr. Harold Reiter. We had a good time traveling around Chengdu as well as talking mathematics. Then, I got to know more about math competitions especially the American Mathematics Competitions (AMC). Later that year, Harold had the initiative to offer his University of North Carolina Charlotte (UNCC) contest problems and publish them in China for students preparing for AMC. That was the beginning of this work. I was excited about this idea. Routine problems can be solved by machines but creative problem solving is becoming ever so important for future global talents and leaders. Math is such an important subject for students to develop their thinking and problem solving skills and succeed in future careers. Diversity enhances education. I am happy that more and more students are interested in global perspectives. Some of them even decide to go abroad to broaden their horizons or for their higher education. But there is still no such problem book published in China. I therefore decided to work with Harold on this interesting project. Also, I proposed to write a part with a list of key points in each major branch of math: algebra, geometry, number theory, counting, logarithm, trigonometry, and complex numbers. Last October, introduced by Harold, I invited Jon to visit me and help our math camps in January. We enjoyed very much our trips in Chengdu, Nanjing and Shanghai. More importantly, we found very similar visions and interests in many math projects. Later, Harold and I sent Jon our request for an essay on geometry.

We were very pleased that Jon promptly replied with his interest to join this project. This work wouldn't have achieved this level without Jon's contributions on vectors, the geometry essay and, very importantly, his thorough proofreading. This book consists of three parts. In the first part, instead of giving a full proof for each theorem, we just list theorems and follow them with one or two demonstrative examples. We hope they serve as a quick checklist helping readers identify what they know and what they don't know. The second part of this book consists of essays and exercises. The essays are on combinatorics, place value, counting and geometry. The problems in the exercises are all from the UNCC contest. The third part has past problems from renowned competitions in USA. Besides AMC and UNCC, Harold also secured permissions from some other popular contests in the states to publish some of their problems, usually, an entire year, for our readers to know what those contests look like. We thank them for their help in making this work more productive. The following is a brief introduction to each of them.

The American Mathematics Competitions (AMC) The American Mathematics Competitions (AMC), sponsored by the Mathematical Association of America (MAA), is the biggest math competition in USA. It is a series of six pre-college contests, the AMC 8, AMC 10, AMC 12, American Invitational Mathematics Exam (AIME), the United States Math Olympiad (USAMO), and USAJMO, for students in grades 10 and below. The United States team for the International Mathematical Olympiad is selected based on the results of USAMO. International students can take AMC 8, AMC 10, AMC 12 and AIME in their own countries. Before the year 2000, the AMC 12 was called the American High School Mathematics Exam (AHSME), and before 1999 the AMC 8 was called the American Junior High School Mathematics Exam (AJHSME). For more information, please refer to http://www.maa.org/math-competitions/about-amc.

MATHCOUNTS MATHCOUNTS is another popular national mathematics program for middle school students. The competition includes a competition series, a club package, and a Math Video Challenge. The competition series has four levels of contests: school, chapter, state, and national. A team from each of the states takes part in the national competition. The winners of MATHCOUNTS may be invited to visit the White House. For more information, please refer to https://www.mathcounts.org.

United States of America Mathematical Talent Search (USAMTS) USAMTS was founded in 1989. It is now a program run by the Art of Problem Solving Foundation. The competition allows students a full month to work out their solutions. USAMTS is open to all United States middle and high school students. Contestants may use any material such as books, calculators and computers; however, they must work on their own. For more information, please refer to http://www.usamts.org.

Wisconsin Mathematical Talent Search (WMTS) WMTS has five sets of five problems in each school year. Also, contestants can make use of any material that wish to use, but they have to do the work on their own. Those problems are open to the state of Wisconsin and the world as well. Top participants are invited to take a proctored exam, and the top participants from Wisconsin are eligible to win the prize: a scholarship to UW-Madison. For more information, please refer to http://www.math.wisc.edu/talent.

American Regions Mathematics League (ARML) ARML is a team mathematics competition. The competition consists of several events, which include a team round, a power question, an individual round, two relay rounds, and a super relay. Power questions are proof-oriented. In relay rounds, a contestant solves a problem and passes his/her answer to another team member, who uses this answer to solve another problem. For more information, please refer to http://www.arm.com

Bay Area Math Olympiad (BAMO) BAMO is an annual competition. It has two four-hour exams both consisting of five proof-type math problems. One exam is for students in grade 8 or below and the other is for students in grade 12 or below. The competition takes place on the last Tuesday of every February. For more information, please refer to http://www.bamo.org.

Purple Comet! Math Meet (PCMM) The Purple Comet! Math Meet was founded in 2003 by Professor Jonathan Kane and Professor Titu Andreescu. It is an annual, international team mathematics competition. It has two exams, one for the middle school students, and the other for the high school students. The middle school exam has 20 questions to be solved in 60 minutes and the high school one has 30 problems to be solved in 90 minutes. The contest has been increasing in popularity. The 2016 contest attracted more than 3000 teams from more than 50 countries. Teams are composed by up to six students. Teams can either represent a school or be made up of students who attend different schools. Each team must have an adult supervisor. Students can take the completion at any time during the ten-day exam period. For more information, please refer to http://purplecomet.org.

State of North Carolina Math Competition (NC-SMC) The NC-SMC began in 1979 to provide state level competition in comprehensive mathematics. Regional site winners are invited to take part in this competition. There are currently twelve regional test sites. One of the sites is in University of North Carolina Charlotte, whose problems are featured in this book. Winners of the contest make up the North Carolina ARML teams. For more information, please refer to https://sites.google.com/site/statemathcontest/

Harvard-MIT Mathematics Tournament (HMMT) HMMT has two tournaments; one is in November and the other is in February. Each tournament includes individual tests, the team round, and the guts round. Participants are not allowed to use books, note or calculators or other computational or drawing aids. All members of a team must attend school within 150 miles of each other. Not every team that applies to HMMT can be accommodated due to the space limitation. For more information, please refer to Http://www.hmmt.co.

I would like to thank people who encourage and help me during the past two years which make this work come to pass. First, I would like to thank the two coauthors, Harold and Jon for their professional inspiration. Working with them is truly enjoyable. I also thank Dipont CEO Benson

Zhang for his vision, insight and generosity, my colleagues Binhao Fan, Yali Xiong, Yan Gou and Kun Tang for their enthusiasm in promoting math education, which impresses and inspires me much, and my students, Zengbao Wu and Jian Mei, for their help on Latex and typesetting. Also, I thank my friend John Jensen, and students Hao Liu and Aileen Luo, my dear wife Xinrong Shui for her patience and companionship.

We worked hard on this first version. However, there might still be typos and even mistakes. The responsibility for those errors in this book lie entirely with me, as I am the one who finalizes the edit.

Yunzhi Zou

PART I

BASIC KNOWLEDGE

CHAPTER 1

ALGEBRA

Algebra is an important part of mathematics. It deals with mathematical symbols and the rules for manipulating these symbols. The more basic parts of algebra are called elementary algebra, while the more abstract parts are called abstract algebra. In this section, we list some important terminology and results involving elementary arithmetic and algebra. In Part III of this book, you will have algebra practice exercises. These exercises include problems about polynomials and their roots, exponentials and radicals, equations and inequalities, sequences and series.

✂ ✂

☞ A **linear expression** has the form

$$ax + b.$$

☞ A **quadratic expression** has the form

$$ax^2 + bx + c, \text{ or, } a(x - h)^2 + k, \text{ or } \quad a(x - x_1)(x - x_2), \text{where} \quad a \neq 0.$$

☞ For a **quadratic equation** $ax^2 + bx + c = 0$, complete the square to obtain

$$\left(x + \frac{b}{2a}\right)^2 = \frac{b^2 - 4ac}{4a^2}.$$

Then, the **quadratic formula**

$$x_{1,2} = \frac{-b \pm \sqrt{b^2 - 4ac}}{2a}$$

gives two roots of the quadratic equation. The **discriminant** $D = b^2 - 4ac$. If a, b, and c are real, the two solutions are both real if $D > 0$, both **complex** if $D < 0$, or repeated and real if $D = 0$.

☞ Linear and quadratic expressions are special cases of **polynomials**. A polynomial is of the form

$$a_n x^n + a_{n-1} x^{n-1} + a_{n-2} x^{n-2} + \cdots + a_1 x + a_0,$$

where $a_n \neq 0$ is the **leading coefficient**. If $a_n = 1$, then the polynomial is called **monic**. a_0 is the constant term, and n is the **degree** of this polynomial. When $n = 3$, it is **cubic**. When $n = 4$, it is **quartic**.

A polynomial can be added, subtracted, multiplied and divided by another polynomial.

☞ For a polynomial function

$$p(x) = a_n x^n + a_{n-1} x^{n-1} + a_{n-2} x^{n-2} + \cdots + a_1 x + a_0,$$

it is easy to see

$$
\begin{aligned}
p(0) &= a_0, \\
p(1) &= a_n + a_{n-1} + \cdots + a_0, \\
p(-1) &= a_0 - a_1 + a_2 - \cdots + (-1)^n a_n.
\end{aligned}
$$

☞ If a number c satisfies $p(c) = 0$, then c is a **root** or **zero** of $p(x)$.

Theorem 1.1 Vieta's Theorem for quadratic equations

If x_1 and x_2 are two roots of the quadratic equation $ax^2 + bx + c = 0$, then

$$x_1 + x_2 = -\frac{b}{a}, \quad \text{and} \quad x_1 x_2 = \frac{c}{a}.$$

The **Vieta's Theorem** reveals the relationship between the coefficients of a quadratic equation and its roots x_1 and x_2. This theorem is named after the French mathematician Franciscus Vieta (1540-1603), whose innovative use of letters as parameters in equations was an important step towards modern algebra.

Example 1.1 Evaluating Expressions Involving Roots

Assume α and β are two roots of the equation $2x^2 + x - 7 = 0$. Find

$$(1)\, \alpha^2 + \beta^2, \quad (2)\, \frac{1}{\alpha} + \frac{1}{\beta}, \quad (3)\, \alpha^3 + \beta^3.$$

✎ **Solution** By the Vieta's Theorem, we have

$$\alpha + \beta = -\frac{1}{2} \text{ and } \alpha\beta = \frac{-7}{2}.$$

Therefore

(1) $\alpha^2 + \beta^2 = (\alpha + \beta)^2 - 2\alpha\beta = \left(-\dfrac{1}{2}\right)^2 - 2\left(-\dfrac{7}{2}\right) = \dfrac{29}{4}.$

(2) $\dfrac{1}{\alpha} + \dfrac{1}{\beta} = \dfrac{\alpha + \beta}{\alpha\beta} = \dfrac{-\frac{1}{2}}{-\frac{7}{2}} = \dfrac{1}{7}.$

(3) Since $2\alpha^3 + \alpha^2 - 7\alpha = 0$, and $2\beta^3 + \beta^2 - 7\beta = 0$, we add up the two equations to obtain

$$2\left(a^3 + \beta^3\right) + \left(\alpha^2 + \beta^2\right) - 7\left(\alpha + \beta\right) = 0.$$

And so

$$a^3 + \beta^3 = -\dfrac{\left(\alpha^2 + \beta^2\right) - 7\left(\alpha + \beta\right)}{2} = -\dfrac{\frac{29}{4} - 7\frac{-1}{2}}{2} = -\dfrac{43}{8}.$$

Theorem 1.2 Remainder Theorem

If a polynomial $p\left(x\right)$ is divided by a linear factor $x - c$, then the remainder is $p\left(c\right)$.

Example 1.2 Finding Remainder

Find the remainder when $x + 2$ is divided into the polynomial $p\left(x\right) = 2x^7 - 6x^5 + 3x + 1$.

✎ **Solution** By the Remainder Theorem，the remainder is given by

$$p\left(-2\right) = 2\left(-2\right)^7 - 6\left(-2\right)^5 + 3\left(-2\right) + 1 = -69.$$

In fact, $2x^7 - 6x^5 + 3x + 1 = (2x^6 - 4x^5 + 2x^4 - 4x^3 + 8x^2 - 16x + 35)(x + 2) - 69.$

Example 1.3 1999 AHSME Problem 17

Let $P(x)$ be a polynomial such that when $P(x)$ is divided by $x - 19$, the remainder is 99, and when $P(x)$ is divided by $x - 99$, the remainder is 19. What is the remainder when $P(x)$ is divided by $(x - 19)(x - 99)$?
(A) $-x + 80$ **(B)** $x + 80$ **(C)** $-x + 118$ **(D)** $x + 118$ **(E)** 0

✎ **Solution** (C). From the hypothesis, $P(19) = 99$ and $P(99) = 19$. Let

$$P(x) = (x - 19)(x - 99)Q(x) + ax + b,$$

where a and b are constants and $Q(x)$ is a polynomial. Then

$$99 = P(19) = 19a + b \quad \text{and} \quad 19 = P(99) = 99a + b.$$

It follows that $99a - 19a = 19 - 99$, hence $a = -1$ and $b = 99 + 19 = 118$. Thus the remainder is $-x + 118$.

> **Theorem 1.3 Factor Theorem**
>
> c is a root of a polynomial $p(x)$, that is $p(c) = 0$, if and only if $(x - c)$ is factor of $p(x)$.

> **Example 1.4**
>
> Suppose a, b, c are integers such that
>
> 1. $0 < a < b$,
>
> 2. The polynomial $x(x - a)(x - b) - 17$ is divisible by $(x - c)$.
>
> What is $a + b + c$?
>
> (**A**) 14   (**B**) 17   (**C**) 21   (**D**) 24   (**E**) 27

Solution (C). Since the polynomial is divisible by $(x - c)$, we have

$$c(c - a)(c - b) = 17.$$

Since $c(c - a)(c - b) = 17 > 0$, it follows that $c > 0$, and we have the following two cases:

Case 1: $0 < (c - b) < (c - a) < c$,

Case 2: $(c - b) < (c - a) < 0 < c$.

Since 17 is a prime number, case 1 does not occur. In case 2, $c = 1, c - a = -1, c - b = -17$. Hence $a = 2, b = 18, c = 1$. Thus, $a + b + c = 21$.

> **Example 1.5 1988 AHSME Problem 15**
>
> If a and b are integers such that $x^2 - x - 1$ is a factor of $ax^3 + bx^2 + 1$, then b is
> (**A**) -2 (**B**) -1 (**C**) 0   (**D**) 1   (**E**) 2

Solution (A). By long division, one finds that $x^2 - x - 1$ divides $ax^3 + bx^2 + 1$ with quotient $ax + (a + b)$ and remainder $(2a + b)x + (a + b + 1)$. But $x^2 - x - 1$ is a factor of $ax^3 + bx^2 + 1$, so the remainder is 0. In other words,

$$2a + b = 0$$
$$a + b = -1.$$

Solving, one obtains $a = 1$, $b = -2$.

OR

Since $x^2 - x - 1$ is a factor of $ax^3 + bx^2 + 1$, the quotient must be $ax - 1$ (why?). Thus

$$\begin{aligned} ax^3 + bx^2 + 1 &= (ax - 1)(x^2 - x - 1) \\ &= ax^3 + (-a - 1)x^2 + (1 - a)x + 1. \end{aligned}$$

Equating the coefficients of x^2 on the left and right, and then the coefficients of x, we obtain

$$b = -a - 1, \quad 0 = 1 - a.$$

Hence $a = 1$ and $b = -2$.

Theorem 1.4 The Rational Zero Test

If $a_n, a_{n-1}, ..., a_0$ are integers, and if p/q is a solution of

$$a_n x^n + a_{n-1} x^{n-1} + a_{n-2} x^{n-2} + \cdots + a_1 x + a_0 = 0,$$

and p and q relatively prime, then, $p \mid a_0$ and $q \mid a_n$.

Example 1.6

The number of rational solutions to $x^4 - 3x^3 - 20x^2 + 30x + 100 = 0$ is

(**A**) 0   (**B**) 1   (**C**) 2   (**D**) 3   (**E**) 4

Solution (C). All rational solutions must be factors of 100. Neither ± 1 nor 2 are solutions, but $x = -2$ is a solution, and we find

$$x^4 - 3x^3 - 20x^2 + 30x + 100 = (x + 2)(x^3 - 5x^2 - 10x + 50).$$

It is easy to see that $x = 5$ is a solution to

$$x^3 - 5x^2 - 10x + 50 = 0,$$

so by dividing, we find that

$$x^4 - 3x^3 - 20x^2 + 30x + 100 = (x + 2)(x - 5)(x^2 - 10).$$

The solutions to $x^2 - 10 = 0$ are $\pm\sqrt{10}$, and these are not rational, so there are just two rational solutions, $x = -2$ and $x = 5$.

The Fundamental Theorem of Algebra, which is explained in more detail later, states that a polynomial equation of degree n has exactly n roots. The following is the extended version of the Vieta's theorem discussed before.

Theorem 1.5 Vieta's Theorem/Zero-coefficient Relationship

Assume $x_1, x_2, ..., x_n$ are the roots of the polynomial

$$P(x) = a_n x^n + a_{n-1} x^{n-1} + a_{n-2} x^{n-2} + \cdots + a_1 x + a_0,$$

then

$$x_1 + x_2 + x_3 + \cdots + x_n = -\frac{a_{n-1}}{a_n}.$$

$$x_1 x_2 x_3 \cdots x_n = (-1)^n \frac{a_0}{a_n}.$$

$$x_1 x_2 + x_1 x_3 + \cdots + x_n x_{n-1} = \frac{a_{n-2}}{a_n}.$$

Example 1.7

Let

$$p(x) = (x - 7)(x^3 + 5x^2 + 7x - 11) + (x - 9)(x^3 + 5x^2 + 7x - 11).$$

What is the sum of the roots of $p(x) = 0$?

(**A**) -11 (**B**) -3 (**C**) 3 (**D**) 11 (**E**) 13

✎ **Solution** (C). Factor $p(x)$ to get

$$p(x) = (2x - 16)(x^3 + 5x^2 + 7x - 11),$$

which has the sum of the roots $8 + (-5) = 3$.

Example 1.8 2010 AMC 10A Problem 21

The polynomial $x^3 - ax^2 + bx - 2010$ has three positive integer roots. What is the smallest possible value of a?

(**A**) 78   (**B**) 88   (**C**) 98   (**D**) 108   (**E**) 118

✎ **Solution** (A). Let the polynomial be $(x - r)(x - s)(x - t)$ with $0 < r \le s \le t$. Then $rst = 2010 = 2 \cdot 3 \cdot 5 \cdot 67$, and $r + s + t = a$. If $t = 67$, then $rs = 30$, and $r + s$ is minimized when $r = 5$ and $s = 6$.

In that case $a = 67 + 5 + 6 = 78$. If $t \neq 67$, then $a > t \geq 2 \cdot 67 = 134$, so the minimum value of a is 78.

Sometimes, we may not be able to work out the exact roots of a complicated polynomial. However, we may get some ideas about the number of roots that it may have. Descartes' Rule of Sign is helpful in this case. René Descartes, a French mathematician and philosopher, is credited as the father of analytical geometry, the bridge between algebra and geometry, which enables algebraic equations to be expressed as geometric shapes in a two- or three-dimensional coordinate system. This was influential in the discovery of calculus. He was also called the father of modern western philosophy. Much of his writing has been closely studied to this day.

Theorem 1.6 Descartes' Rule of Sign

If $p(x)$ is a polynomial with real coefficients, then

1. the number of positive real zeros of $p(x)$ is either equal to the number of variations in sign of $p(x)$ or less than this by an even number.

2. the number of negative real zeros of $p(x)$ is either equal to the number of variations in sign of $p(-x)$ or less than this by an even number.

Example 1.9 Number of Positive or Negative Real Roots

For the polynomial function

$$p(x) = x^6 + 3x^5 - 4x^4 + 6x^3 + 7x - 10,$$

find the number of positive real roots and the number of negative real roots.

Solution Since the number of variations in sign of $p(x)$ is 3, the number of positive real zeros of $p(x)$ is either 3 or 1. Because

$$p(-x) = x^6 - 3x^5 - 4x^4 - 6x^3 - 7x - 10$$

and the number of variations in sign of $p(x)$ is 1, the number of real negative zeros of $p(x)$ must be 1. Graphing this function with a graphing device, we can easily see that $p(x)$ indeed has one positive real root and one negative real root.

Theorem 1.7 Properties of Exponents and Radicals

Assume m and n are two integers, a and b are two real numbers, and none of the numbers is equal to 0 in case that it appears as a denominator. Then,

- $a^0 = 1, a \neq 0$
- $\dfrac{a^m}{a^n} = a^{m-n}$
- $a^{-m} = \dfrac{1}{a^m}$
- $\sqrt[n]{a} = a^{\frac{1}{n}}, a \geq 0$
- $\sqrt[n]{ab} = \sqrt[n]{a} \times \sqrt[n]{b}, a \geq 0, b \geq 0$

- $a^m \times a^n = a^{m+n}$
- $(a^m)^n = a^{mn}$
- $\left(\dfrac{a}{b}\right)^m = \dfrac{a^m}{b^m}$
- $\sqrt[n]{a^m} = a^{\frac{m}{n}}$
- $\sqrt[n]{\dfrac{a}{b}} = \dfrac{\sqrt[n]{a}}{\sqrt[n]{b}}$

Example 1.10

Let $x = \dfrac{1}{2}^{\left(\frac{1}{2}^{-\frac{1}{2}}\right)}$. To which of the following intervals does x belong?

(A) $(0, 1/8]$ (B) $(1/8, 1/4]$ (C) $(1/4, 1/2]$ (D) $(1/2, 1]$ (E) $(1, \infty)$

✎ **Solution** (C). Let $y = \frac{1}{2}^{\frac{-1}{2}}$. Then $y = \sqrt{2}$. Because $1 < y < 2$ and $0 < \frac{1}{2} < 1$, we have $\left(\frac{1}{2}\right)^2 < \left(\frac{1}{2}\right)^y < \left(\frac{1}{2}\right)^1$.

Example 1.11 1974 AHSME Problem 3

The coefficient of x^7 in the polynomial expansion of

$$(1 + 2x - x^2)^4$$

is

(A) -8 (B) 12 (C) 6 (D) -12 (E) none of these

✎ **Solution** (A). The coefficient of x^7 in $(1 + 2x - x^2)^4$ is the coefficient of the sum of four identical terms $2x\left(-x^2\right)^3$, which sum is $-8x^7$.

☞ The sequence $a_1, a_2, a_3, \cdots, a_n, \cdots$ is **arithmetic** if there is a constant d such that $a_n - a_{n-1} = d$ for $n = 2, 3, 4, \cdots$.

☞ The sequence $b_1, b_2, b_3, \cdots, b_n, \cdots$ is **geometric** if there is a constant r such that $\frac{b_n}{b_{n-1}} = r$ for $n = 2, 3, 4, \cdots$.

> **Theorem 1.8 Partial Sums of Arithmetic and Geometric Sequences**
>
> If $a_1, a_2, a_3, ..., a_n, ...$ is an arithmetic sequence with the common difference d, then
>
> $$a_n = a_1 + (n-1)d, \text{ where } a_n \text{ is the n-th term}$$
> $$s_n = \frac{(a_1 + a_n)n}{2}, \text{ where } s_n \text{ is the partial sum of its first } n \text{ terms.}$$
>
> If $a_1, a_2, a_3, ..., a_n, ...$ is a geometric sequence with the common ratio r, then
>
> $$a_n = a_1 r^{n-1}, \text{ where } a_n \text{ is the n-th term}$$
> $$s_n = \frac{a_1(1-r^n)}{1-r}, \text{ where } s_n \text{ is the partial sum of its first } n \text{ terms.}$$

Example 1.12

Laura jogs seven blocks the first day of her training program. She increases her distance by two blocks each day. On the last day, she jogs 25 blocks. How many days was she in training?

(A) 5   (B) 8   (C) 9   (D) 10   (E) 15

Solution (D). Note that $d_1 = 7, d_2 = 7 + 1 \cdot 2, d_3 = 7 + 2 \cdot 2, \ldots, d_n = 7 + (n-1) \cdot 2$. Hence, $d_n = 25 = 7 + (n-1) \cdot 2 \implies n = 10$.

Example 1.13 1981 AHSME Problem 26

Alice, Bob, Carol repeatedly take turns tossing regular six-sided die. Alice begins; Bob always follows Alice; Carol always following Bob; and Alice always follows Carol. Find the probability that Carol will be the first to toss a six.

(A) $\frac{1}{3}$ (B) $\frac{2}{9}$ (C) $\frac{5}{18}$ (D) $\frac{25}{91}$ (E) $\frac{36}{91}$

Solution (D). The probability that the first 6 is tossed on the k-th toss is the product

$$\left(\begin{array}{c} \text{probability that never a 6 was} \\ \text{tossed in the previous } k-1 \text{ tosses} \end{array} \right) \left(\begin{array}{c} \text{probability that a 6 is} \\ \text{tossed on the } k^{\text{th}} \text{ toss} \end{array} \right) = (5/6)^{k-1}(1/6).$$

The probability that Carol will toss the first 6 is the sum of the probabilities that she will toss the first 6 on her first turn (3rd toss of the game), on her second turn (6th toss of the game), on her third turn, etc. This sum is

$$\left(\frac{5}{6}\right)^2 \frac{1}{6} + \left(\frac{5}{6}\right)^5 \frac{1}{6} + \cdots + \left(\frac{5}{6}\right)^{3n-1} \frac{1}{6} + \cdots,$$

an infinite geometric series with first term $a = \left(\frac{5}{6}\right)^2 \frac{1}{6}$ and common ratio $r = \left(\frac{5}{6}\right)^3$. This sum is

$$\frac{a}{1-r} = \frac{5^2/6^3}{1-(5^3/6^3)} = \frac{5^2}{6^3-5^3} = \frac{25}{91}.$$

Note *A nicer solution not involving sums of a geometric series results from* $p + p \cdot \frac{6}{5} + p \cdot \frac{36}{25} = 1$.

☞ The long sum $a_1 + a_2 + a_3 + \cdots + a_n + \cdots$, is called a **series**. Using **sigma notation**, this series can be written as $\sum_{k=0}^{n} a_k$. That is

$$\sum_{k=1}^{n} a_k = a_1 + a_2 + a_3 + \cdots + a_n.$$

☞ **Linearity properties** of the sigma notation

$$\sum_{i=1}^{n} (ka_i + lb_i) = k \sum_{i=1}^{n} a_i + l \sum_{i=1}^{n} b_i.$$

☞ Some useful summations of series

$$\sum_{i=1}^{n} (f(i) - f(i+1)) = f(1) - f(2) + f(2) - f(3) + \cdots + f(n) - f(n+1)$$
$$= f(1) - f(n+1), \text{ Telescoping.}$$
$$\sum_{i=1}^{n} i = 1 + 2 + 3 + \cdots + n = \frac{n(n+1)}{2}.$$
$$\sum_{i=1}^{n} i^2 = 1^2 + 2^2 + 3^2 + \cdots + n^2 = \frac{n(n+1)(2n+1)}{6}.$$
$$\sum_{i=1}^{n} i^3 = 1^3 + 2^3 + 3^3 + \cdots + n^3 = \left(\frac{n(n+1)}{2}\right)^2.$$

These results can be proved by the **method of mathematical induction**.

☞ **Composition** of two functions: let f and g be two functions such that the domain of f intersects the range of g. The composition of f and g is $f(g(x))$.

Example 1.14 2000 AMC 10 Problem 24

Let f be a function for which $f(x/3) = x^2 + x + 1$. Find the sum of all values of z for which $f(3z) = 7$.
(A) $-1/3$ (B) $-1/9$ (C) 0 (D) $5/9$ (E) $5/3$

Solution (B). Let $x = 9z$. Then $f(9z/3) = f(3z) = 81z^2 + 9z + 1 = 7$. Simplifying and solving the equation for z yields $81z^2 + 9z - 6 = 0$, so $3(3z+1)(9z-2) = 0$. Thus $z = -1/3$ or $z = 2/9$. The sum of these values is $-1/9$. (The answer could also be obtained by using the sum-of-roots formula on $81z^2 + 9z - 6 = 0$. The sum of the roots is $-9/81 = -1/9$).

CHAPTER 2

GEOMETRY

Geometry is another important branch of mathematics. It concerns questions of shape, size (length, area, or volume), and relative position of figures. In this section, we list important results that you should know in **plane geometry**, **solid geometry** and **coordinate geometry**. Contest problems usually involve **triangles**, **circles**, **polygons** and three dimensional geometry. These problems vary from very easy to extremely challenging. Some problems even require some geometric constructions. Schools usually offer one semester or one year geometry courses, however, the subject matter covered in those courses may not be sufficient to solve competition problems.

✂ ✂

☞ The three angles of any triangle sum to $180°$ or equivalently π radians. This is sometimes called the **Triangle Sum Theorem**.

☞ Two triangles are **similar** if and only if two angles of a triangle are equal to two angles of the other triangle. Similar triangles have proportional sides.

☞ The **Side-splitter Theorem**: Suppose in $\triangle ABC$, points D and E are on sides $\overline{AB}$ and $\overline{BC}$ respectively, then $\overline{DE} \parallel \overline{AC}$ if and only if

$$\frac{AD}{AB} = \frac{CE}{CB}.$$

☞ **SSS:** Two triangles are **congruent** if the three sides of one are equal to the three sides of the other.

☞ **SAS**: Two triangles are congruent if two sides and the included angle of one are equal to two sides and the included angle of the other.

☞ **ASA**: Two triangles are congruent if two angles and an included side of one are equal to two angles and the included side of the other.

☞ A polygon with n-sides has a sum of **interior angles** $(n-2) \cdot 180$ degrees. A 4-sided polygon is a **quadrilateral**. A 5-sided polygon is a **pentagon**. A 6-sided polygon is a **hexagon**.

☞ A polygon whose sides have equal lengths and angles have equal measures is **regular**. Each interior angle of a regular polygon with n-sides is $\frac{(n-2) \cdot 180}{n}$.

Theorem 2.1 Triangle Inequalities

If a, b, and c are the lengths of the sides of a triangle ABC, then the sum of lengths of any two sides is greater than the length of the third side, that is

$$a + b > c, \quad b + c > a \quad \text{and} \quad a + c > b.$$

Example 2.1

The sides of a triangle are in the ratio $3 : 5 : 9$. Which of the following words best describes the triangle?
(A) obtuse   **(B)** scalene   **(C)** right   **(D)** isosceles   **(E)** impossible

---

✎ **Solution (E).** The triangle inequality, which states that the sum of the lengths of any two sides of a triangle is greater than the length of the third side, can be invoked. There are no such triangles.

Example 2.2 2012 AMC 12B Problem 20

A trapezoid has side lengths 3, 5, 7, and 11. The sums of all the possible areas of the trapezoid can be written in the form of $r_1\sqrt{n_1} + r_2\sqrt{n_2} + r_3$, where r_1, r_2, and r_3 are rational numbers and n_1 and n_2 are positive integers not divisible by the square of any prime. What is the greatest integer less than or equal to $r_1 + r_2 + r_3 + n_1 + n_2$?
(A) 57   **(B)** 59   **(C)** 61   **(D)** 63   **(E)** 65

✎ **Solution (D).** Let $ABCD$ be a trapezoid with $\overline{AB} \parallel \overline{CD}$ and $AB < CD$. Let E be the point on $\overline{CD}$ such that $CE = AB$. Then $ABCE$ is a parallelogram. Set $AB = a$, $BC = b$, $CD = c$, and $DA = d$. Then the side lengths of $\triangle ADE$ are b, d, and $c - a$. If one of b or d is equal to 11, say $b = 11$ by symmetry, then $d + (c - a) \leq 7 + (5 - 3) < 11 = d$, which contradicts the triangle inequality. Thus $c = 11$. There are three cases to consider, namely, $a = 3$, $a = 5$, and $a = 7$. If $a = 3$, the $\triangle ADE$ has side lengths $5, 7,$ and 8 and by Heron's formula its area is

$$\frac{1}{4}\sqrt{(5+7+8)(7+8-5)(8+5-7)(5+7-8)} = 10\sqrt{3}.$$

The area of $\triangle AEC$ is $\frac{3}{8}$ of the area of $\triangle ADE$, and triangles ABC and AEC have the same area. It follows that the area of the trapezoid is $\frac{1}{2}(35\sqrt{3})$.

If $a = 5$, then $\triangle ADE$ has side lengths $3, 6,$ and 7, and area

$$\frac{1}{4}\sqrt{(3+6+7)(6+7-3)(7+3-6)(3+6-7)} = 4\sqrt{5}.$$

The area of $\triangle AEC$ is $\frac{5}{6}$ of the area of $\triangle ADE$, and triangles ABC and AEC have the same area. It follows that the area of the trapezoid is $\frac{1}{3}(32\sqrt{5})$.

If $a = 7$, then $\triangle ADE$ has side lengths $3, 4,$ and 5. Hence this is a right trapezoid with height 3 and base lengths 7 and 11. This trapezoid has area $\frac{1}{2}(3(7+11)) = 27$.

The sum of the three possible areas is $\frac{35}{2}\sqrt{3} + \frac{32}{3}\sqrt{5} + 27$. Hence $r_1 = \frac{35}{2}$, $r_2 = \frac{32}{3}$, $r_3 = 27$, $n_1 = 3$, $n_2 = 5$, and $r_1 + r_2 + r_3 + n_1 + n_2 = \frac{35}{2} + \frac{32}{3} + 27 + 3 + 5 = 63 + \frac{1}{6}$. Thus the required integer is 63.

Theorem 2.2 Pythagorean Theorem

If a right triangle has legs with lengths a and b and hypotenuse with length c, then

$$c^2 = a^2 + b^2.$$

The Pythagorean Theorem, also known as Pythagoras' Theorem, is a fundamental relation in Euclidean geometry among the three sides of a right triangle. It states that the square of the hypotenuse (the side opposite the right angle) is equal to the sum of the squares of the other two sides. This theorem is credited to ancient Greek mathematician Pythagoras who was also credited with the first proof. However, there is evidence that ancient Indian and Chinese also discovered the theorem independently, and proofs were given in some special cases. Contest problems in geometry almost certainly involve questions that require the Pythagorean Theorem.

Example 2.3

A ladder is leaning against a wall with the top of the ladder 8 feet above the ground. If the bottom of the ladder is moved 2 feet farther from the wall, the top of the ladder slides all the way down the wall and rests against the foot of the wall. How long is the ladder?

(A) 14 feet   **(B)** 15 feet   **(C)** 16 feet   **(D)** 17 feet   **(E)** 18 feet

✎ **Solution** (D). If the base is x feet from the wall in the initial position, then the ladder must be $x + 2$ feet long. The Pythagoras Theorem gives $x^2 + 8^2 = (x + 2)^2$ which yields $x = 15$. The length of the ladder is 17 feet.

Theorem 2.3 Heron's Formula

If s is the semiperimeter of a triangle ABC, $s = \frac{1}{2}(a + b + c)$, then the area of the triangle ABC is

$$[ABC] = \sqrt{s\,(s - a)\,(s - b)\,(s - c)}.$$

This formula is named after Heron of Alexandria. The formula gives the area of triangle by requiring no arbitrary choice of side as base or vertex. A normal formula that gives the area of a triangle is one half the product of a base and the altitude corresponding to that base. Note that for a cyclic quadrilateral, there is a similar result called Brahmagupta's Formula.

Example 2.4

A triangle with sides $a = 15$, $b = 28$ and $c = 41$ has an altitude of integer length. What is the length of this altitude?

(A) 6   **(B)** 7   **(C)** 9   **(D)** 16   **(E)** 17

✎ **Solution** (C). With $s = (a + b + c)/2 = 42$, use Heron's formula for the area

$$F = \sqrt{s(s - a)(s - b)(s - c)} = \sqrt{42 \cdot 27 \cdot 14 \cdot 1} = \sqrt{2^2 \cdot 3^4 \cdot 7^2} = 2 \cdot 3^2 \cdot 7.$$

Since the three heights are

$$h_a = 2F/a = 4 \cdot 9 \cdot 7/(3 \cdot 5); \quad h_b = 2F/b = 4 \cdot 9 \cdot 7/(4 \cdot 7) = 9;$$

and

$$h_c = 2F/c = 4 \cdot 9 \cdot 7/41,$$

so only $h_b = 9$ is an integer.

Alternatively, notice that to the triangle with sides 15, 28, and 41, we can append a triangle with sides 9, 12 and 15 to get a right triangle with sides 9, 40, and 41 as shown in the diagram.

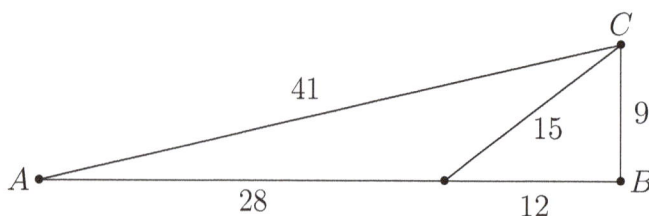

Alternatively, let θ be the angle measure of the angle opposite the side of length 41 and let ϕ be its supplement. Note that $41^2 > 28^2 + 15^2$, so the angle is obtuse. Next use the Law of Cosines to get the cosine of θ and thus

$$-\cos(\phi) = \cos(\theta) = (28^2 + 15^2 - 41^2)/(2 \cdot 28 \cdot 15) = -4/5.$$

Thus $\sin(\phi) = 3/5$, and $9 = 15 \cdot \sin(\phi)$ is the altitude for the side of length 28 ($84/5 = 28 \cdot \sin(\phi)$ is the altitude for the side of length 15, and $9 \cdot 28/41$ is the altitude for the side of length 41).

Example 2.5 2011 AMC 12A Problem 24

Consider all quadrilaterals $ABCD$ such that $AB = 14$, $BC = 9$, $CD = 7$, and $DA = 12$. What is the radius of the largest possible circle that fits inside or on the boundary of such a quadrilateral?

(A) $\sqrt{15}$ **(B)** $\sqrt{21}$ **(C)** $2\sqrt{6}$ **(D)** 5 **(E)** $2\sqrt{7}$

Solution (C). Because $AB+CD = 21 = BC+DA$, it follows that $ABCD$ always has an inscribed circle tangent to its four sides. Let r be the radius of the inscribed circle. Note that $[ABCD] = \frac{1}{2}r(AB + BC + CD + DA) = 21r$. Thus the radius is maximum when the area is maximized. Note that $[ABC] = \frac{1}{2} \cdot 14 \cdot 9 \sin B = 63 \sin B$ and $[ACD] = \frac{1}{2} \cdot 12 \cdot 7 \sin D = 42 \sin D$. On the one hand,

$$\begin{aligned}[ABCD]^2 &= ([ABC] + [ACD])^2 \\ &= 63^2 \sin^2 B + 42^2 \sin^2 D + 2 \cdot 42 \cdot 63 \sin B \sin D.\end{aligned}$$

On the other hand, by the Law of Cosines,

$$AC^2 = 12^2 + 7^2 - 2 \cdot 7 \cdot 12 \cos D = 14^2 + 9^2 - 2 \cdot 9 \cdot 14 \cos B.$$

Thus

$$\begin{aligned}21^2 &= \left(\frac{2 \cdot 26 + 2 \cdot 16}{4}\right)^2 = \left(\frac{14^2 - 12^2 + 9^2 - 7^2}{4}\right)^2 \\ &= (63 \cos B - 42 \cos D)^2 \\ &= 63^2 \cos^2 B + 42^2 \cos^2 D - 2 \cdot 42 \cdot 63 \cos B \cos D.\end{aligned}$$

Adding these two identities yields

$$\begin{aligned}[ABCD]^2 + 21^2 &= 63^2 + 42^2 - 2 \cdot 42 \cdot 63 \cos(B + D) \\ &\leq 63^2 + 42^2 + 2 \cdot 42 \cdot 63 = (63 + 42)^2 = 105^2,\end{aligned}$$

with equality if and only if $B+D = \pi$ (that is, $ABCD$ is cyclic). Therefore $[ABCD]^2 \leq 105^2 - 21^2 = 21^2(5^2 - 1) = 42^2 \cdot 6$, and the required maximum $r = \frac{1}{21}[ABCD] = 2\sqrt{6}$.

OR

Establish as in the first solution that r is maximized when the area is maximized. Bretschneider's formula, which generalizes Brahmagupta's formula, states that the area of an arbitrary quadrilateral with side lengths $a, b, c,$ and d, is given by

$$\sqrt{(s-a)(s-b)(s-c)(s-d) - abcd\cos^2\theta},$$

where $s = \frac{1}{2}(a+b+c+d)$ and θ is half the sum of either pair of opposite angles. For $a, b, c,$ and d fixed, the area is maximized when $\cos\theta = 0$. Thus the area is maximized when $\theta = \frac{1}{2}\pi$, that is, when the quadrilateral is cyclic. In this case, the area equals $\sqrt{7 \cdot 12 \cdot 14 \cdot 9} = 42\sqrt{6}$, and the required maximum radius $r = \frac{1}{21} \cdot 42\sqrt{6} = 2\sqrt{6}$.

Theorem 2.4 Angle Bisector Theorem

If D is a point on the side $\overline{AC}$ of a triangle ABC, then BD bisects angle ABC if and only if

$$\frac{AD}{CD} = \frac{AB}{BC}.$$

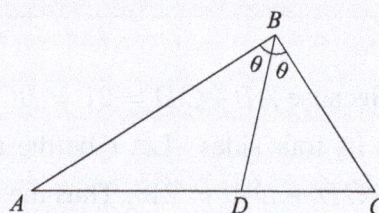

Example 2.6 2010 AMC 10A Problem 16

Nondegenerate $\triangle ABC$ has integer side lengths, $\overline{BD}$ is an angle bisector, $AD = 3$, and $DC = 8$. What is the smallest possible value of the perimeter?

(**A**) 30   (**B**) 33   (**C**) 35   (**D**) 36   (**E**) 37

Solution (**B**). By the Angle Bisector Theorem, $8 \cdot BA = 3 \cdot BC$. Thus BA must be a multiple of 3. If $BA = 3$, the triangle is degenerate. If $BA = 6$, then $BC = 16$, and the perimeter is $6 + 16 + 11 = 33$.

Theorem 2.5 The Exterior Angle Theorem

The measure of an exterior angle of a triangle is the sum of measures of the other two non-adjacent interior angles of the triangle.

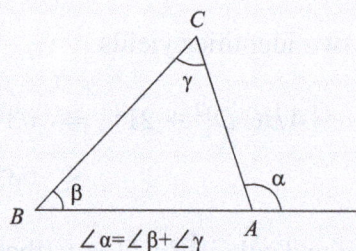

$\angle\alpha = \angle\beta + \angle\gamma$

Example 2.7

Consider a general convex 7-gon with vertices A_1, A_2, ..., A_7 (marked in the order as they appear on the polygon). Connecting the first vertex A_1 with the third vertex A_3, the second vertex A_2 with the fourth vertex A_4, and so on, we get a "star" as shown in the figure below with the vertices A_1, A_2, ..., A_7. Find the sum of the angles $\angle A_1$ to $\angle A_7$.

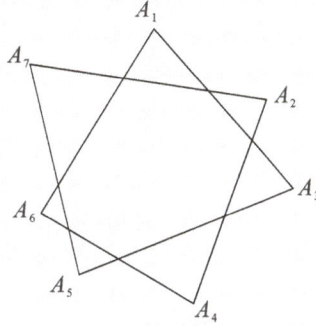

(A) $360°$ (B) $480°$ (C) $540°$ (D) $600°$ (E) $620°$

✎ **Solution** (C). For a triangle $\triangle XYZ$, the sum of the angles with vertices at X and Y is equal to the exterior angle at Z. Since $2\angle X + \angle Y + \angle Z = \angle X + 180°$, the sum of the exterior angles at Y and Z is equal to $180°$ plus $\angle X$. Thus the sum of the angles at A_1 through A_7 is twice the angle sum of the interior heptagon minus $7 \cdot 180°$. For the heptagon the angle sum is $7 \cdot 180° - 360° = 5 \cdot 180°$. So the desired sum is $10 \cdot 180° - 7 \cdot 180° = 540°$.

Note *The angle $\angle A_1$ may be expressed as*

$$\angle A_1 = 2 \cdot \angle A_7 A_1 A_2 - \angle A_7 A_1 A_2 - \angle A_7 A_1 A_6 - \angle A_3 A_1 A_2.$$

By rotating the indices we may write six other similar equations for $\angle A_2, \ldots, \angle A_7$. If we take the sum of these equations, on the left hand side we get the sum of the angle sought. On the right hand side we get twice the sum of the interior angles of the convex 7-gon, minus the sum of the angles of the triangles

$$A_7 A_1 A_2, A_1 A_2 A_3, \ldots, A_6 A_7 A_1.$$

The sum of the inner angles of the polygon is $(7-2)180° = 900°$. The sum of the inner angles of a triangle is $180°$. Thus we obtain $2 \cdot 900° - 7 \cdot 180° = 540°$.

Example 2.8   2010 AMC 12A Problem 8

Triangle ABC has $AB = 2 \cdot AC$. Let D and E be on $\overline{AB}$ and $\overline{BC}$, respectively, such that $\angle BAE = \angle ACD$. Let F be the intersection of segments $\overline{AE}$ and $\overline{CD}$, and suppose that $\triangle CFE$ is equilateral. What is $\angle ACB$?

(A) $60°$ **(B)** $75°$ **(C)** $90°$ **(D)** $105°$ **(E)** $120°$

✎ **Solution (C).** Let

$$\alpha = \angle BAE = \angle ACD = \angle ACF.$$

Because $\triangle CFE$ is equilateral, it follows that $\angle CFA = 120°$ and then

$$\angle FAC = 180° - 120° - \angle ACF = 60° - \alpha.$$

Therefore

$$\angle BAC = \angle BAE + \angle FAC = \alpha + (60° - \alpha) = 60°.$$

Because $AB = 2 \cdot AC$, it follows that $\triangle BAC$ is a $30° - 60° - 90°$ triangle, and thus $\angle ACB = 90°$.

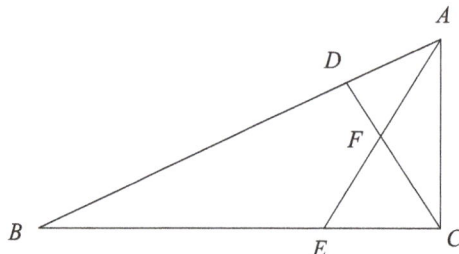

Giovanni Ceva was an Italian mathematician widely known for proving Ceva's Theorem, which is about triangles in Euclidean plane geometry. Given a triangle ABC, let the lines AO, BO and CO be drawn from the vertices to a common point O (not on one of the sides of ABC), to meet opposite sides at E, F and D respectively. The segments $\overline{AE}$, $\overline{BF}$, and $\overline{CD}$ are called **cevians**.

Theorem 2.6 Ceva's Theorem

If D, E and F are three points on the sides of $\overline{AB}$, $\overline{BC}$ and $\overline{AC}$ of a triangle ABC, respectively, then $\overline{AE}$, $\overline{BF}$ and $\overline{CD}$ are concurrent if and only if

$$\frac{AF}{FC} \cdot \frac{CE}{EB} \cdot \frac{BD}{DA} = 1.$$

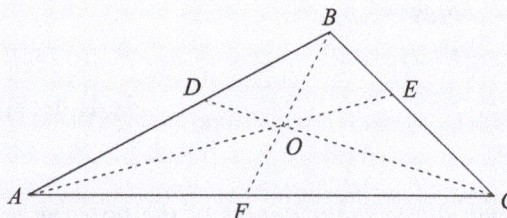

Example 2.9

In $\triangle ABC$, some important facts that can be derived by Ceva's Theorem are

☞ the three medians of $\triangle ABC$ are concurrent. They meet at the point called **centroid** of $\triangle ABC$.

☞ the three angle bisectors of $\triangle ABC$ are concurrent. They meet at the point called **incenter** of $\triangle ABC$.

☞ the three altitudes of $\triangle ABC$ are concurrent. They meet at the point called **orthocenter** of $\triangle ABC$.

Also, the three perpendicular bisectors of three sides of $\triangle ABC$ are concurrent. They meet at the point called **circumcenter** of $\triangle ABC$. (However, this doesn't follow from Ceva's Theorem).

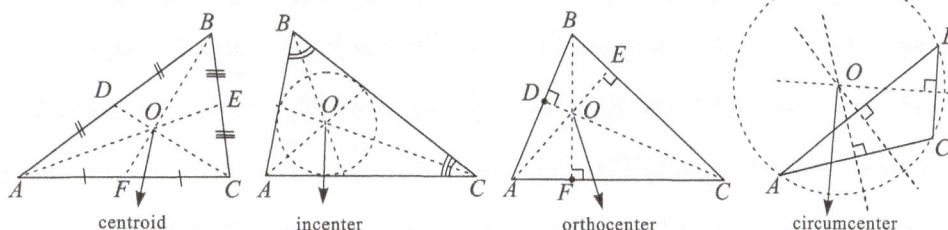

centroid incenter orthocenter circumcenter

Theorem 2.7 Stewart Theorem

If a, b and c are lengths of sides $\overline{BC}$, $\overline{AC}$ and $\overline{AB}$ of a triangle ABC, respectively, and $\overline{AD}$ is a cevian so that $BD = m$, $DC = n$ and $AD = d$, then

$$b^2m + c^2n = amn + d^2a.$$

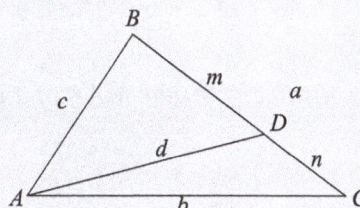

Note *If r is the inradius, R is the circumradius and s is the semi-perimeter, then $[ABC] = \frac{1}{2}ab\sin C = rs = \frac{abc}{4R}$.*

Example 2.10 2009 AIME II Problem 10

Four lighthouses are located at points A, B, C, and D. The lighthouse at A is 5 kilometers from the lighthouse at B, the lighthouse at B is 12 kilometers from the lighthouse at C, and the lighthouse at A is 13 kilometers from the lighthouse at C. To an observer at A, the angle determined by the lights at B and D and the angle determined by the lights at C and D are equal. To an observer at C, the angle determined by the lights at A and B and the angle determined by the lights at D and B are equal. The number of kilometers from A to D is given by $\frac{p\sqrt{q}}{r}$, where p, q, and r are relatively prime positive integers, and r is not divisible by the square of any prime. Find $p + q + r$.

✎ **Solution 096.**

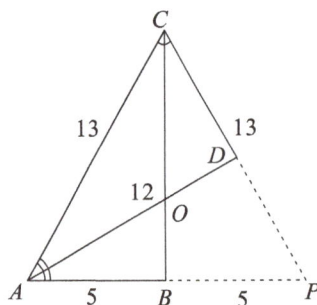

Extend $\overline{AB}$ and $\overline{CD}$ to intersect at P. Since $\triangle ABC$ is a right triangle, we conclude that $\triangle ABC \cong \triangle PBC$ by ASA congruency. Thus $AC = PC = 13$.

By the Angle Bisector Theorem, $\frac{AC}{AP} = \frac{CD}{DP}$ and $CD + DP = 13$. Note that $AC = 13$ and $AP = 10$, solving for CD and DP gives $DP = \frac{130}{23}$ and $CD = \frac{169}{23}$. Now we can apply Stewart's Theorem to find AD:

$$PC \times CD \times DP + PC \times AD^2 = AC^2 \times PD + AP^2 \times CD$$

$$13 \cdot \frac{130}{23} \cdot \frac{169}{23} + 13 \cdot AD^2 = 13 \cdot 13 \cdot \frac{130}{23} + 10 \cdot 10 \cdot \frac{169}{23}.$$

Solve for AD to obtain $AD = \frac{60\sqrt{13}}{23}$. So the final answer is $60 + 13 + 23 = 096$.

Note *The above solution is not the MAA official solution. It was provided by the authors.*

Note

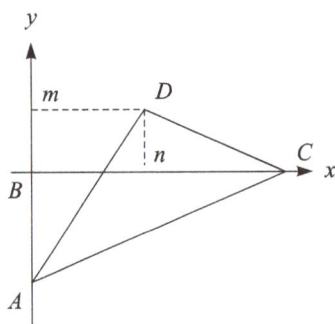

The following solution was provided by **Boyang Gu** *from Nanjing Foreign Language School. As seen from the diagram shown above, B is the origin. The line passing through $A(0, -5)$ and $C(12, 0)$ has equation $5x - 12y - 60 = 0$. Since $\overline{AD}$ is the angle bisector of $\angle BAC$, the distance from $D(m, n)$ to the $y-$axis is equal to the distance from $D(m, n)$ to the line $\overline{AC}$. Thus*

$$m = \frac{|5m - 12n - 60|}{\sqrt{5^2 + 12^2}}.$$

Let P be the projection of D onto the x-axis. Since $\overline{CB}$ is the angle bisector of $\angle ACD$, the right triangle CDP is similar to the right triangle ACB. Thus $\frac{n}{5} = \frac{12-m}{12}$. Now, solving for m and n gives $m = \frac{120}{23}$ and $n = \frac{65}{23}$. Therefore

$$AD = \sqrt{\left(\frac{120}{23} - 0\right)^2 + \left(\frac{65}{23} - (-5)\right)^2} = \frac{60}{23}\sqrt{13}.$$

Our answer is $60 + 23 + 13 = 96$.

Note *If O is the intersection of $\overline{BC}$ and $\overline{AD}$ and P is the foot of perpendicular from D to $\overline{OC}$, one can show that $\triangle AOB$ is similar to $\triangle DOP$ and $\triangle DPC$ is similar to $\triangle ABC$. Together with the Angle Bisector Theorem, one can derive another nice solution to this problem.*

Theorem 2.8 Menelaus Theorem

Suppose P, Q, and R are three points on the respective sides $\overline{BC}$, $\overline{AC}$, and $\overline{AB}$ (or their extensions) of a triangle ABC, then P, Q, and R are collinear if and only if

$$\frac{AR}{BR} \cdot \frac{BP}{CP} \cdot \frac{CQ}{AQ} = 1.$$

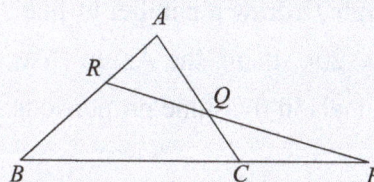

Example 2.11 2011 AIME II Problem 4

In triangle ABC, $AB = \frac{20}{11}AC$. The angle bisector of $\angle A$ intersects $\overline{BC}$ at point D, and point M is the midpoint of $\overline{AD}$. Let P be the point of the intersection of $\overline{AC}$ and $\overline{BM}$. The ratio of CP to PA can be expressed in the form $\frac{m}{n}$, where m and n are relatively prime positive integers. Find $m + n$.

Solution 51. Let $[AMP] = x$, $[CMP] = y$, $[CMD] = z$, and $[BMD] = t$ because M is the midpoint of AD, if follows that $[AMB] = [BMD] = t$.

The Angle Bisector Theorem yields

$$\frac{z}{t} = \frac{CD}{BD} = \frac{AC}{AB} = \frac{11}{20}.$$

Also,

$$\frac{[CPM]}{[CBM]} = \frac{PM}{MB} = \frac{[APM]}{[ABM]},$$

or

$$\frac{y}{z+t} = \frac{x}{t}.$$

Thus

$$\frac{CP}{PA} = \frac{y}{x} = \frac{z+t}{t} = \frac{z}{t} + 1 = \frac{11}{20} + 1 = \frac{31}{20}.$$

Hence $m + n = 51$.

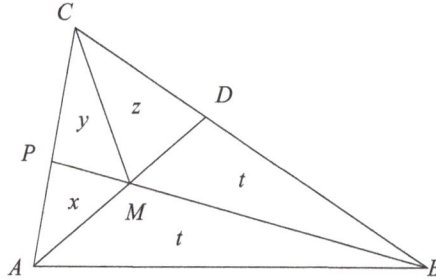

OR

Through D draw a parallel to line BP intersecting line AC at Q. Then $PQ = 20k$, $QC = 11k$ and $PA = 20k$, using the Angle Bisector Theorem and the fact that 3 or more parallel lines divide all tranversals in the same proportions. Thus $\frac{CP}{PA} = \frac{20k+11k}{20k} = \frac{31}{20}$ as in the previous solution.

Note This problem can be solved by using Menelaus Theorem.

By Menelaus' Theorem on $\triangle ACD$ with transversal $\overline{PB}$, $1 = \frac{CP}{PA} \cdot \frac{AM}{MD} \cdot \frac{DB}{CB}$. Now that $\frac{AM}{MD} = 1$ and $\frac{DC}{DB} = \frac{AC}{AB}$ by the Angle Bisector Theorem. Thus, we have

$$1 = \frac{CP}{PA} \cdot \frac{AM}{MD} \cdot \frac{DB}{CB} = \frac{CP}{PA} \cdot \frac{DB}{CB} = \frac{CP}{PA} \cdot \frac{DB}{DB+DC} = \frac{CP}{PA} \cdot \frac{1}{1+\frac{DC}{DB}}$$

And so

$$\frac{CP}{PA} = 1 + \frac{DC}{DB} = 1 + \frac{AC}{AB} = 1 + \frac{11}{20} = \frac{31}{20}.$$

Therefore our answer is $m + n = 051$.

There are many important results in **circle geometry**. Some of them you should already know from a normal school geometry course. For example

☞ The **Tangent-chord Theorem**: The **central angle** determined by a chord to a circle is twice the angle formed by the chord and the tangent with one endpoint in common with the chord.

☞ The **Inscribed Angle Theorem**: Any inscribed angle has measure half of the central angle with the same arc.

☞ **Inscribed Angle and Diameter**: An inscribed angle is a right angle if and only if the associated chord is a diameter.

Now we talk about some results that might not be covered in your geometry textbook.

Theorem 2.9 The Power of a Point

There are three cases as shown in the following figures

Case I Case II Case III

Case I: $AE \times EC = DE \times BE$ (E is a point inside the circle)

Case II: $AB^2 = BC \times BD$ (AB is tangent to the circle at the point B)

Case III: $CB \times CA = CD \times CE$ (C is point outside of the circle)

Example 2.12 2013 AMC 12A Problem 19

In $\triangle ABC$, $AB = 86$ and $AC = 97$. A circle with center A and radius $\overline{AB}$ intersects $\overline{BC}$ at points B and X. Moreover, $\overline{BX}$ and $\overline{CX}$ have integer lengths. What is BC?

(A) 11   **(B)** 28   **(C)** 33   **(D)** 61   **(E)** 72

Solution (D). By the Power of a Point Theorem, $BC \cdot CX = AC^2 - r^2$ where $r = AB$ is the radius of the circle. Thus $BC \cdot CX = 97^2 - 86^2 = 2013$. Since $BC = BX + CX$ and CX are both integers, they are complementary factors of 2013. Note that $2013 = 3 \cdot 11 \cdot 61$, and $CX < BC < AB + AC = 183$. Thus the only possibility is $CX = 33$ and $BC = 61$.

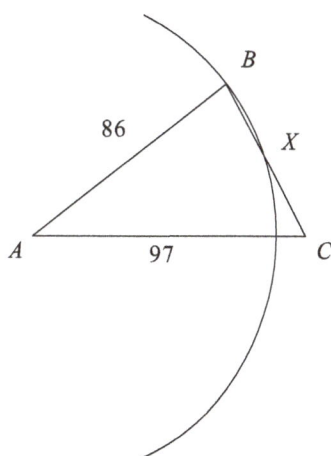

Example 2.13 2015 AMC 12B Problem 24

Four circles, no two of which are congruent, have centers at A, B, C, and D, and points P and Q lie on all four circles. The radius of circle A is $\frac{5}{8}$ times the radius of circle B, and the radius of circle C is $\frac{5}{8}$ times the radius of circle D. Furthermore, $AB = CD = 39$ and $PQ = 48$. Let R be the midpoint of $\overline{PQ}$. What is $AR + BR + CR + DR$?

(A) 180   **(B)** 184   **(C)** 188   **(D)** 192   **(E)** 196

Solution (D). Points A, B, C, D, and R all lie on the perpendicular bisector of $\overline{PQ}$. Assume R lies between A and B. Let $y = AR$ and $x = \frac{AP}{5}$. Then $BR = 39 - y$ and $BP = 8x$, so $y^2 + 24^2 = 25x^2$ and $(39 - y)^2 + 24^2 = 64x^2$. Subtracting the two equations gives $x^2 = 39 - 2y$, from which $y^2 + 50y - 399 = 0$, and the only positive solution is $y = 7$. Thus $AR = 7$, and $BR = 32$.

Note that circles A and B are determined by the assumption that R lies between A and B. Thus because the four circles are noncongruent, R does not lie between C and D. Let $w = CR$ and $z = \frac{CP}{5}$. Then $DR = 39 + w$ and $DP = 8z$, so $w^2 + 24^2 = 25z^2$ and $(39 + w)^2 + 24^2 = 64z^2$. Subtracting the two equations gives $z^2 = 39 + 2w$, from which $w^2 - 50w - 399 = 0$, and the only positive solution is $w = 57$. Thus $CR = 57$ and $DR = 96$. Again, the uniqueness of the solution implies that R must indeed lie between A and B.

The requested sum is $7 + 32 + 57 + 96 = 192$.

Note *This problem can also be solved by using the Power of a Point.*

Theorem 2.10 Tangent-Tangent Intersection Theorem

Angle of intersection between two tangents dividing a circle into arc length A and arc length B is $\frac{1}{2}$ (Arc $A°$ − Arc $B°$).

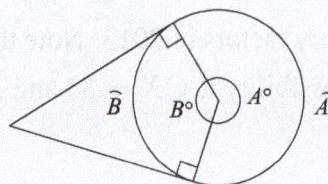

Example 2.14 2011 AMC 12B Problem 6

Two tangents to a circle are drawn from a point A. The points of contact B and C divide the circle into arcs with lengths in the ratio $2 : 3$. What is the degree measure of $\angle BAC$?

(A) 24   **(B)** 30   **(C)** 36   **(D)** 48   **(E)** 60

✎ **Solution** (C). Let O be the center of the circle, and let the degree measures of the minor and major arcs be $2x$ and $3x$, respectively. Because $2x + 3x = 360°$, it follows that $x = 72°$ and $\angle BOC = 2x = 144°$. In quadrilateral $ABOC$, the segments AB and AC are tangent to the circle, thus $\angle ABO = \angle ACO = 90°$ and $\angle BAC = 360° - (144° + 90° + 90°) = 36°$.

Theorem 2.11 Ptolemy's Theorem

The sum of the products of the opposite sides of a cyclic quadrilateral is equal to the product of the diagonals.

$$AC \cdot BD = AD \cdot BC + AB \cdot CD.$$

Example 2.15 2013 AMC 12B Problem 19

In triangle ABC, $AB = 13$, $BC = 14$, and $CA = 15$. Distinct points D, E, and F lie on segments $\overline{BC}$, $\overline{CA}$, and $\overline{DE}$, respectively, such that $\overline{AD} \perp \overline{BC}$, $\overline{DE} \perp \overline{AC}$, and $\overline{AF} \perp \overline{BF}$. The length of segment $\overline{DF}$ can be written as $\frac{m}{n}$, where m and n are relatively prime positive integers. What is $m + n$?

(**A**) 18   (**B**) 21   (**C**) 24   (**D**) 27   (**E**) 30

✎ **Solution** (B). The Pythagorean Theorem applied to right triangle ABD and ACD gives $AB^2 - BD^2 = AD^2 = AC^2 - CD^2$; that is, $13^2 - BD^2 = 15^2 - (14 - BD)^2$, from which it follows that $BD = 5$, $CD = 9$, and $AD = 12$. Because triangles AED and ADC are similar,

$$\frac{AE}{12} = \frac{DE}{9} = \frac{12}{15},$$

implying that $ED = \frac{36}{5}$ and $AE = \frac{48}{5}$.

Because $\angle AFB = \angle ADB = 90°$, it follows that $ABDF$ is cyclic. Thus $\angle ABD + \angle AFD = 180°$ from which $\angle ABD = \angle AFE$. Therefore right triangles ABD and AFE are similar. Hence

$$\frac{FE}{5} = \frac{\frac{48}{5}}{12},$$

from which it follows that $FE = 4$. Consequently $DF = DE - FE = \frac{36}{5} - 4 = \frac{16}{5}$.

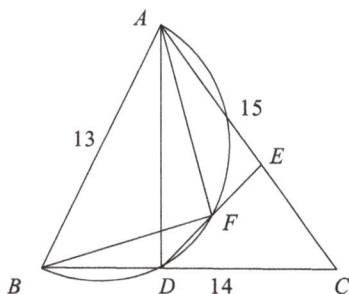

Note This problem can also be solved by using the Ptolemy Theorem.

As seen in the diagram, by Heron's Formula or otherwise, we can have $AD = 12$ and so $BD = 5$ and $DC = 9$. By evaluating the area of triangle ADC, we have $\frac{1}{2}AC \times DE = \frac{1}{2}AD \times DC$. Thus, $DE = \frac{12 \times 9}{15} = \frac{36}{5}$. By Pythagorean theorem, $AE = \sqrt{AD^2 - DE^2} = \sqrt{12^2 - \left(\frac{36}{5}\right)^2} = \frac{48}{5}$. Since quadrilateral $ABDF$ is cyclic, $\angle ABF = \angle ADF$, and the two right-angled triangles ABF and ADE are similar. Therefore, $\frac{BF}{DE} = \frac{AB}{AD} = \frac{AF}{AE}$. We have

$$
\begin{aligned}
BF &= DE \times \frac{AB}{AD} = \frac{36}{5} \times \frac{13}{12} = \frac{39}{5} \\
AF &= AE \times \frac{AB}{AD} = \frac{48}{5} \times \frac{13}{12} = \frac{52}{5}.
\end{aligned}
$$

Applying Ptolemy, we have

$$
\begin{aligned}
AB \times DF + BD \times AF &= BF \times AD \\
13 \times DF + 5 \times \frac{52}{5} &= \frac{39}{5} \times 12
\end{aligned}
$$

Solve for DF to obtain $DF = \frac{16}{5}$. So $m + n = 21$.

When dealing with mutually tangent circles, Descartes' Circle Formula can be useful. We first give the definitions of curvatures of mutually tangent circles.

Definition 2.1 Curvatures of mutually tangent circles

☞ If two circles O_1 and O_2 with radii r_1 and r_2 are externally tangent to each other, then their curvatures are defined to be $\frac{1}{r_1}$ and $\frac{1}{r_2}$, respectively.

☞ If circle O_1 with radius r_1 is internally tangent to circle O_2 with radius r_2, then the curvature of circle O_1 is defined to be $\frac{1}{r_1}$ and the curvature of the circle O_2 is defined to be $-\frac{1}{r_2}$.

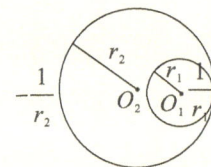

Comment A straight line can be regarded as a degenerate circle with infinite radius, thus its curvature is 0. Now we can state the formula.

Theorem 2.12 Descartes' Circle Formula

When four circles O_1, O_2, O_3 and O_4 are pairwise tangent, with respective curvatures a, b, c and d, then we have

$$(a+b+c+d)^2 = 2\left(a^2+b^2+c^2+d^2\right).$$

And

$$d = a+b+c \pm 2\sqrt{ab+ac+bc}.$$

Comment If one of the four circles is replaced by a straight line, say, $d = 0$, then the theorem becomes

$$(a+b+c)^2 = 2\left(a^2+b^2+c^2\right).$$

and

$$c = a+b \pm 2\sqrt{ab}.$$

This also gives, in case that all curvatures are positive

$$\sqrt{c} = \sqrt{a} \pm \sqrt{b}.$$

Example 2.16 2015 AMC 12A Problem 25

A collection of circles in the upper half-plane, all tangent to the x-axis, is constructed in layers as follows. Layer L_0 consists of two circles of radii 70^2 and 73^2 that are externally tangent. For $k \geq 1$, the circles in $\bigcup_{j=0}^{k-1} L_j$ are ordered according to their points of tangency with the x-axis. For every pair of consecutive circles in this order, a new circle is constructed externally tangent to each of the two circles in the pair. Layer L_k consists of the 2^{k-1} circles constructed in this way. Let $S = \bigcup_{j=0}^{6} L_j$, and for every circle C denote by $r(C)$ its radius. What is

$$\sum_{C \in S} \frac{1}{\sqrt{r(C)}}?$$

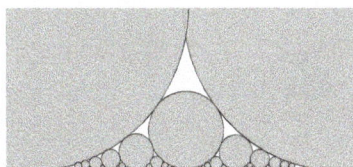

(A) $\dfrac{286}{35}$ (B) $\dfrac{583}{70}$ (C) $\dfrac{715}{73}$ (D) $\dfrac{143}{14}$ (E) $\dfrac{1573}{146}$

Solution (D). Suppose that circles C_1 and C_2 in the upper half-plane have centers O_1 and O_2 and radii r_1 and r_2, respectively. Assume that C_1 and C_2 are externally tangent and tangent to the x-axis at X_1 and X_2, respectively. Let C with center O and radius r be the circle externally tangent to C_1 and C_2 and tangent to the x-axis. Let X be the point of tangency of C with the x-axis, and let T_1 and T_2 be the points of tangency of C with C_1 and C_2, respectively. Let M_1 and M_2 be the points on the x-axis such that $\overline{M_1T_1} \perp \overline{O_1T_1}$ and $\overline{M_2T_2} \perp \overline{O_2T_2}$.

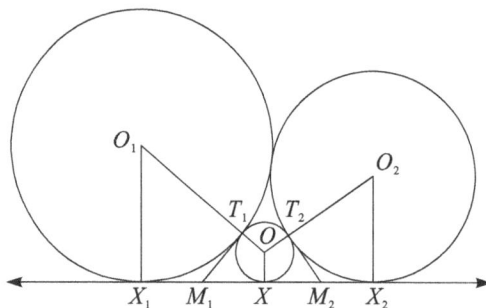

Because $\overline{M_1X_1}$ and $\overline{M_1T_1}$ are both tangent to C_1, it follows that $X_1M_1 = M_1T_1$. Similarly, $\overline{M_1T_1}$ and $\overline{M_1X}$ are both tangent to C, and thus $M_1T_1 = M_1X$. Because $\angle OT_1M_1$, $\angle M_1X_1O_1$, $\angle M_1T_1O$, and $\angle OXM_1$ are all right angles and $\angle T_1M_1X = \pi - \angle X_1M_1T_1$, it follows that quadrilaterals $O_1X_1M_1T_1$ and M_1XOT_1 are similar. Thus

$$\frac{r_1}{X_1M_1} = \frac{O_1X_1}{X_1M_1} = \frac{M_1X}{XO} = \frac{X_1M_1}{r}.$$

Therefore $X_1M_1 = \sqrt{rr_1}$, and similarly $M_2X_2 = \sqrt{rr_2}$. By the distance formula,

$$(r_1 + r_2)^2 = (O_1O_2)^2 = (X_1X_2)^2 + (r_1 - r_2)^2.$$

Thus

$$2\sqrt{r_1r_2} = X_1X_2 = X_1M_1 + M_1X + XM_2 + M_2X_2$$
$$= 2(X_1M_1 + M_2X_2) = 2\sqrt{r}(\sqrt{r_1} + \sqrt{r_2});$$

that is,

$$\frac{1}{\sqrt{r}} = \frac{1}{\sqrt{r_1}} + \frac{1}{\sqrt{r_2}}. \tag{2.1}$$

It follows that

$$\sum_{C \in L_k} \frac{1}{\sqrt{r(C)}} = \sum_{C \in L_k} \left(\frac{1}{\sqrt{r(C_1)}} + \frac{1}{\sqrt{r(C_2)}} \right),$$

where C_1 and C_2 are the consecutive circles in $\bigcup_{j=0}^{k-1} L_j$ that are tangent to C. Note that every circle in $\bigcup_{j=0}^{k-1} L_j$ appears twice in the sum on the right-hand side, except for the two circles in L_0, which appear only once. Thus

$$\sum_{C \in L_k} \frac{1}{\sqrt{r(C)}} = 2 \sum_{j=1}^{k-1} \sum_{C \in L_j} \frac{1}{\sqrt{r(C)}} + \sum_{C \in L_0} \frac{1}{\sqrt{r(C)}}.$$

In particular, if $k = 1$, then

$$\sum_{C \in L_1} \frac{1}{\sqrt{r(C)}} = \sum_{C \in L_0} \frac{1}{\sqrt{r(C)}} = \frac{1}{70} + \frac{1}{73}.$$

For simplicity let $x = \frac{1}{70} + \frac{1}{73}$. Let $k \geq 2$, and suppose by induction that for $1 \leq j \leq k - 1$,

$$\sum_{C \in L_j} \frac{1}{\sqrt{r(C)}} = 3^{j-1} x.$$

It follows that

$$\sum_{C \in L_k} \frac{1}{\sqrt{r(C)}} = 2 \left(\sum_{j=1}^{k-1} 3^{j-1} x \right) + x = 2x \left(\frac{3^{k-1} - 1}{2} \right) + x = x 3^{k-1}.$$

Therefore

$$\sum_{C \in S} \frac{1}{\sqrt{r(C)}} = \sum_{k=0}^{6} \sum_{C \in L_k} \frac{1}{\sqrt{r(C)}} = x + \sum_{k=1}^{6} x 3^{k-1} = x \left(1 + \frac{3^6 - 1}{2} \right)$$

$$= x \left(\frac{3^6 + 1}{2} \right) = \frac{143}{70 \cdot 73} \left(\frac{730}{2} \right) = \frac{143}{14}.$$

Note *Equation (2.1) is a special case of the Descartes' Kissing Circles Theorem.*

We now list some important facts that you should know in **solid geometry**.

☞ The volume of a **prism** with base area A and height h is $V = Ah$. Cubes and rectangular boxes are examples of prisms.

☞ The volume of a **pyramid** with base area A and height h is $V = \frac{1}{3}Ah$.

☞ The volume of a **circular cylinder** with base radius r and height h is $V = \pi r^2 h$.

☞ The volume of a **sphere** with radius R is $V = \frac{4}{3}\pi R^3$.

☞ The **surface area** of a sphere with radius R is $S = 4\pi R^2$.

☞ The volume of **circular cone** with base radius r and height h is $V = \frac{1}{3}\pi r^2 h$.

☞ The **lateral surface area** of a right circular cone with base radius r and height h is $S = \pi r l$, where l is the slant height.

☞ Spheres, cylinders, cones, planes, and **polyhedrons** are figures in three dimensional space. Rectangular boxes, prisms, pyramids, tetrahedron, octahedron, dodecahedrons are all examples of polyhedrons.

> **Theorem 2.13 Euler's Polyhedron Formula**
>
> If a polyhedron has F faces, E edges, and V vertices, then
>
> $$F + V - E = 2.$$

> **Example 2.17 2009 AMC 12B Problem 20**
>
> A convex polyhedron Q has vertices $V_1, V_2, \ldots, V_n$, and 100 edges. The polyhedron is cut by planes $P_1, P_2, \ldots, P_n$ in such a way that plane P_k cuts only those edges that meet at vertex V_k. In addition, no two planes intersect inside or on Q. The cuts produce n pyramids and a new polyhedron R. How many edges does R have?
>
> **(A)** 200 **(B)** $2n$ **(C)** 300 **(D)** 400 **(E)** $4n$

Solution (C). Each edge of Q is cut by two planes, so R has 200 vertices. Three edges of R meet at each vertex, so R has $\frac{1}{2} \cdot 3 \cdot 200 = 300$ edges.

OR

At each vertex, as many new edges are created by this process as there are original edges meeting that vertex. Thus the total number of new edges is the total number of endpoints of the original edges, which is 200. A middle portion of each original edge is also present in R, so R has $100 + 200 = 300$ edges.

OR

Euler's Polyhedron Formula applied to Q gives $n - 100 + F = 2$, where F is the number of faces of Q. Each edge of Q is cut by two planes, so R has 200 vertices. Each cut by a plane P_k creates an additional face on R, so Euler's Polyhedron Formula applied to R gives $200 - E + (F + n) = 2$, where E is the number of edges of R. Subtracting the first equation from the second gives $300 - E = 0$, so $E = 300$.

Analytical geometry combines algebra and geometry. It allows an algebraic equation to be expressed in terms of a graph in a coordinate plane.

☞ The graph of a **linear function**

$$f(x) = ax + b \ \text{ or } \ y = ax + b$$

is a straight line with slope a and y-intercept b.

☞ The graph of a **quadratic function**

$$f\left(x\right) = ax^2 + bx + c \ \text{ or } \ y = ax^2 + bx + c$$

is a **parabola**. Completing the square, we have

$$f\left(x\right) = a\left(x - h\right)^2 + k, \ \text{ where } h = \frac{b}{2a} \ \text{ and } k = \frac{-b^2 + 4ac}{4a}.$$

The parabola has **vertex** (h, k) and **axis of symmetry** $x = h$.

- If $a > 0$, then the function has its minimum value k when $x = h$,

- if $a < 0$, then the function has its maximum value k when $x = h$.

☞ The **focus** $(h, k + 1/4a)$ of a parabola is on its axis of symmetry. If P is a point on a parabola, then the distance between P and the focus of the parabola is same as the distance between P and the **directrix** $(y = k - 1/4a)$ of this parabola.

☞ The circle with center (a, b) and radius R has an equation

$$(x - a)^2 + (y - b)^2 = R^2.$$

☞ The sphere with center (a, b, c) and radius R has an equation

$$(x - a)^2 + (y - b)^2 + (z - c)^2 = R^2.$$

Example 2.18 2013 AMC 12B Problem 21

Consider the set of 30 parabolas defined as follows: all parabolas have as focus the point $(0,0)$ and the directrix lines have the form $y = ax + b$ with a and b integers such that $a \in \{-2, -1, 0, 1, 2\}$ and $b \in \{-3, -2, -1, 1, 2, 3\}$. No three of these parabolas have a common point. How many points in the plane are on two of these parabolas?

(A) 720   **(B)** 760   **(C)** 810   **(D)** 840   **(E)** 870

✎ **Solution** (C). If the directrices of two parabolas with the same focus intersect, then the corresponding parabolas intersect in exactly two points. The same conclusion holds if the directrices are parallel and the focus is between the two lines. Moreover, if the directrices are parallel and the focus is not between the two lines, then the corresponding parabolas do not intersect. Indeed, a point C belongs to the intersection of the parabolas with focus O and directrices l_1 and l_2, if and only if, $d(C, l_1) = OC = d(C, l_2)$. That is, the circle with center C and radius OC is tangent to both l_1 and l_2. If l_1 and l_2 are parallel and O is not between them, then clearly such circle does not exist. If l_1 and

l_2 intersect and O is not on them, then there are exactly two circles tangent to both l_1 and l_2 that go through O. The same is true if l_1 and l_2 are parallel and O is between them.

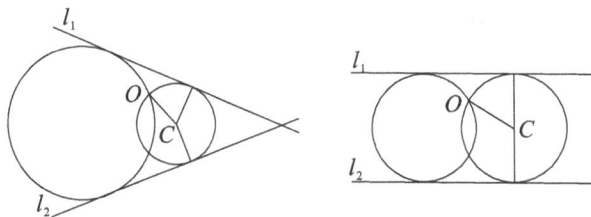

Thus there are $\binom{30}{2}$ pairs of parabolas and the pairs that do not intersect are exactly those whose directrices have the same slope and whose y-intercepts have the same sign. There are 5 different slopes and $2 \cdot \binom{3}{2} = 6$ pairs of y-intercepts with the same sign taken from $\{-3, -2, -1, 1, 2, 3\}$. Because the pairs of parabolas that intersect do so at exactly two points and no point is in three parabolas, it follows that the total number of intersection point is

$$2\left(\binom{30}{2} - 5 \cdot 6\right) = 810.$$

Note *It is possible to construct the two circles through O and tangent to the lines l_1 and l_2 as follows: Let l' be the bisector of the angle determined by the angular sector spanned by l_1 and l_2 that contains O (or the midline of l_1 and l_2 if these lines are parallel and O is between them). Let Q be the symmetric point of O with respect to l' and let P be the intersection of l_1 and the line OQ (if $O = Q$ then let P be the intersection of l_1 and a perpendicular line to l' by O). If C is one of the desired circles, then C passes through O and Q and is tangent to l_1. Let T be the point of tangency of C and l_1. By the power of a Point Theorem, $PT^2 = PO \cdot PQ$. The circle with center P and radius $\sqrt{PO \cdot PQ}$ intersects l_1 in two points T_1 and T_2. The circumcircles of OQT_1 and OQT_2 are the desired circles.*

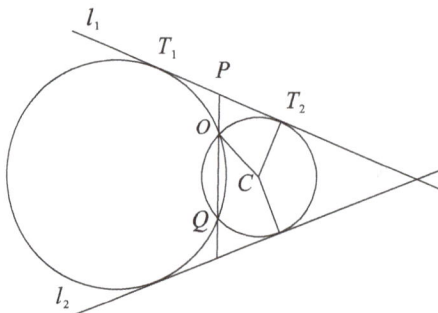

Theorem 2.14 Shoelace Theorem

If the polygon P has vertices $(a_1, b_1), (a_2, b_2), \cdots, (a_n, b_n)$, listed in clockwise order, then the area of P is given by

$$\frac{1}{2} \left| (a_1 b_2 + a_2 b_3 + \cdots + a_n b_1) - (b_1 a_2 + b_2 a_3 + \cdots + b_n a_1) \right|.$$

Example 2.19 2013 AMC 10A Problem 18

Let points $A = (0,0)$, $B = (1,2)$, $C = (3,3)$, and $D = (4,0)$. Quadrilateral $ABCD$ is cut into equal area pieces by a line passing through A. This line intersects $\overline{CD}$ at point $\left(\frac{p}{q}, \frac{r}{s} \right)$, where these fractions are in lowest terms. What is $p + q + r + s$?

(**A**) 54   (**B**) 58   (**C**) 62   (**D**) 70   (**E**) 75

Solution (**B**). Let line AG be the required line, with G on $\overline{CD}$. Divide $ABCD$ into triangle ABF, trapezoid $BCFE$, and triangle CDE, as shown. Their areas are 1, 5 and $\frac{3}{2}$, respectively. Hence the area of $ABCD = \frac{15}{2}$, and the area of triangle $ADG = \frac{15}{4}$, Because $AD = 4$, it follows that $GH = \frac{15}{8} = \frac{r}{s}$. The equation of $\overline{CD}$ is $y = -3(x - 4)$, so when $y = \frac{15}{8}, x = \frac{p}{q} = \frac{27}{8}$. Therefore $p + q + r + s = 58$.

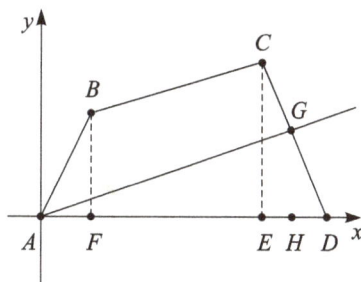

Note *This Problem can also be solved by using the Shoelace Theorem. First, we use Shoelace Theorem to find the area of the quadrilateral $ABCD$:*

$$[ABCD] = \frac{1}{2} |4 \times 3 - 0 \times 3 + 3 \times 2 - 3 \times 1 + 1 \times 0 - 2 \times 0| = \frac{15}{2}.$$

Therefore each piece of the cut quadrilateral has an area of $\frac{15}{4}$. Let the line intersect line CD at P, then

$$[ADP] = \frac{1}{2} 4 \times \frac{r}{s} = \frac{1}{2} \frac{15}{2} \rightarrow \frac{r}{s} = \frac{15}{8}.$$

So P has coordinates $\left(\frac{p}{q}, \frac{15}{8}\right)$. Use the Shoelace Theorem again to find the area of the quadrilateral APCB:

$$[APCB] = \frac{1}{2}\left|0 \times \frac{15}{8} - 0 \times \frac{p}{q} + \frac{p}{q} \times 3 - \frac{15}{8} \times 3 + 3 \times 2 - 3 \times 1\right| = \frac{1}{2}\frac{15}{2}$$

This simplies to $\left|\frac{p}{q} - \frac{7}{8}\right| = \frac{5}{2}$. Thus $\frac{p}{q} = \frac{27}{8}$ or $\frac{p}{q} = -\frac{13}{8}$ (rejected). Hence, we know that $E = \left(\frac{27}{8}, \frac{15}{8}\right)$. And then $27 + 15 + 8 + 8 = 58$ is the answer.

When finding areas of polygons involving lattice points, Pick's theorem can also be very helpful.

Theorem 2.15 Pick's Theorem

In a coordinate plane, all vertices of a polygon P are lattice points, then the area of the polygon is

$$\text{Area of } P = I + \frac{B}{2} - 1,$$

where I is the number of lattice points in the interior of P and B is the number of lattice points on the boundary of P.

Example 2.20 1998 AJHSME Problem 6

Dots are spaced one unit apart, horizontally and vertically. The number of square units enclosed by the polygon is

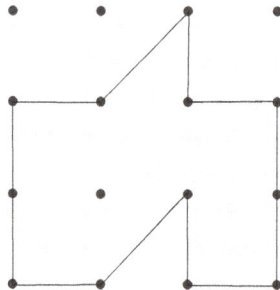

(A) 5   (B) 6   (C) 7   (D) 8   (E) 9

Solution (B). Since $I = 1$, $B = 12$, by Pick's Theorem, the area $A = I + \frac{B}{2} - 1 = 1 + \frac{12}{2} - 1 = 6$.

Comment This solution demonstrates how to use Pick's theorem to find area of a polygon. In fact, for this question, by inspection, you may see directly that the area is $2 \times 3 = 6$ by noticing that the triangle on the top row matches the hole in the bottom row.

There is another elegant theorem in plane geometry, although it rarely appears in AMC exams. This theorem is called **Simson's Theorem**, and we just state it without further discussion.

CHAPTER 3

NUMBER THEORY

Number theory is a branch of pure mathematics devoted primarily to the study of the natural numbers and the integers at large. It was deemed as the "crown of mathematics" by Karl Gauss who was widely believed as one of the greatest mathematicians in history. Problems involving number theory have been consistently popular ones in almost all types of math competitions. However, number theory is not taught in a normal school math curriculum in the depth required for contest problems. We, therefore, discuss this topic in detail in a more accessible way.

✂ ✂

Important terminology and results

☞ **Integers** are the positive and negative whole numbers and zero.

☞ The integer n is a **divisor** of the integer m if there is an integer k such that $m = kn$.

☞ An integer $n > 1$ is a **prime number** if its only positive divisors are 1 and n. Otherwise, n is call **composite**.

☞ Two integers m and n are **coprime** or **relatively prime**, denoted by $\gcd(m, n) = 1$, if their **greatest common divisor** is 1.

☞ A **Diophantine equation** is an equation that asks for integer solutions. The positive integer solutions to Diophantine equation $a^2 + b^2 = c^2$ are called Pythagorean triples which have the form $a = m^2 - n^2, b = 2mn$ and $c = m^2 + n^2$, where m and n are two integers. $(3, 4, 5)$, $(5, 12, 13)$, $(7, 24, 25)$, $(8, 15, 17)$ are all Pythagorean triples.

Theorem 3.1 Division Algorithm: Quotient and Remainder

For any integers m and $n \neq 0$, there are exactly two integers q and r such that

$$m = qn + r, \quad \text{where } 0 \leq r < |n| \text{ is the remainder and } q \text{ is the quotient.}$$

Example 3.1

For example, if you divide 5 into 37, then the quotient is 7 and the remainder is 2. We write $37 = 7 \times 5 + 2$.

- In case that the remainder $r = 0$, we say m is **divisible** by n, or n divides m, and we write $n|m$.

Theorem 3.2 Divisibility Rules

If $n = a_m \times 10^m + a_{m-1} \times 10^{m-1} + \cdots + a_1 \times 10 + a_0$, each a_i is a single digit and $a_m \neq 0$, then

- n is divisible by 2 if and only if $2|a_0$.

- n is divisible by 3 if and only if $3| (a_0 + a_1 + \cdots + a_m)$.

- n is divisible by 5 if and only if $5|a_0$.

- n is divisible by 9 if and only if $9| (a_0 + a_1 + \cdots + a_m)$.

- n is divisible by 11 if and only if $11| (a_0 - a_1 + a_2 - \cdots + (-1)^m a_m)$.

- n is divisible by 2^k if and only if $2^k| (a_{k-1} \times 10^{k-1} + a_{k-2} \times 10^{k-2} + \cdots + a_0)$.

Example 3.2

Test the divisibility of the number $m = 21482256$ by 2,3,5,9 and 11. And find the largest n such that $2^n|m$.

Solution Since $2 \mid 6$, m is divisible by 2. Since $5 \nmid 6$, it is not divisible by 5. Since $2+1+4+8+2+2+5+6 = 30$ and $3|30$ but $9 \nmid 30$, it is divisible by 3 but not 9. Since $6-5+2-2+8-4+1-2 = 4$, and $11 \nmid 4$, it is not divisible by 11. Since $2^4|256$ but $2^5 \nmid 2256$, then 4 is the largest possible value for n. The factorization of 21482256 given by computer software is $2^4 \cdot 3 \cdot 31 \cdot 14437$.

> **Example 3.3 1992 AHSME Problem 17**
>
> The 2-digit integers from 19 to 92 are written consecutively to form the integer $N = 192021\cdots9192$. Suppose that 3^k is the highest power of 3 that is a factor of N. What is k?
>
> **(A)** 0 **(B)** 1 **(C)** 2 **(D)** 3 **(E)** more than 3

Solution (B). Since $0 + 1 + 2 + \cdots + 9 = 45$ and

$$N = 19\underbrace{2021\cdots29}_{10\cdot2+45=65}\underbrace{3031\cdots39}_{10\cdot3+45=75}\cdots\underbrace{8081\cdots89}_{10\cdot8+45=125}909192,$$

the sum of the digits of N is

$$
\begin{aligned}
S &= (1+9) + 65 + 75 + \cdots + 125 + (3\cdot9+3)\\
&= (10) + \left(7\cdot\frac{65+125}{2}\right) + (30)\\
&= 705 = 9(78) + 3.
\end{aligned}
$$

Thus S has a factor of 3 but not 9, so the highest power of 3 that is a divisor of N is 3^1 and $k = 1$.

OR

The integer 3 [or 9] will divide N if and only if it divides the sum of $19, 20, \ldots, 92$. (Why? See note below.) Since
$$19 + 20 + \cdots + 92 = 74\cdot\frac{19+92}{2} = 37\cdot111 = 37^2\cdot3,$$
it follows that $k = 1$.

Note Consider N as a 74-digit base-100 number with digits $19, 20, \ldots, 92$. The sum of these digits is $3\cdot37^2$. The "casting out nines" procedure says:

- *"The greatest common factor of a positive integer and 9 is the same as the greatest common factor of the sum of the base 10 digits of the integer and 9."*

The proof of this procedure can be generalized to:

- *"The greatest common factor of a positive integer and $b-1$ is the same as the greatest common factor of the sum of the base b digits of the integer and $b-1$."*

Therefore, the greatest common factor of N and 99 equals the greatest common factor of $37^2\cdot3$ and 99.

The **Prime Factorization Theorem** is often called the the Fundamental Theorem of Arithmetic, which is useful for problem solving.

Theorem 3.3 Prime Factorization

Every natural number n can be factored into a product of prime numbers in a unique way, apart from the order of factors. That is

$$n = p_1^{k_1} p_2^{k_2} \cdots p_m^{k_m},$$

where each p_i is prime, each k_i is a positive integer, and $p_1 < p_2 < \cdots < p_m$.

Example 3.4 Prime Factorization

The prime factorization of the number $2016 = 2^5 3^2 7^1$.

Example 3.5 2001 AMC 12 Problem 21

Four positive integers a, b, c, and d have a product of $8!$ and satisfy:

$$ab + a + b = 524$$

$$bc + b + c = 146$$

$$cd + c + d = 104.$$

What is $a - d$?

(A) 4 **(B)** 6 **(C)** 8 **(D)** 10 **(E)** 12

✎ **Solution (D).** Note that

$$(a+1)(b+1) = ab + a + b + 1 = 524 + 1 = 525 = 3 \cdot 5^2 \cdot 7,$$

and

$$(b+1)(c+1) = bc + b + c + 1 = 146 + 1 = 147 = 3 \cdot 7^2.$$

Since $(a+1)(b+1)$ is a multiple of 25 and $(b+1)(c+1)$ is not a multiple of 5, it follows that $a+1$ must be a multiple of 25. Since $a+1$ divides 525, a is one of the 24, 74, 174, or 524. Among these only 24 is a divisor of $8!$, so $a = 24$. This implies that $b + 1 = 21$ and $b = 20$. From this it follows that $c + 1 = 7$ and $c = 6$. Finally, $(c+1)(d+1) = 105 = 3 \cdot 5 \cdot 7$, so $d + 1 = 15$ and $d = 14$. Therefore $a - d = 24 - 14 = 10$.

Theorem 3.4 Greatest Common Divisor and Least Common Multiple

Suppose n and m have the following prime factorizations

$$n = p_1^{k_1} p_2^{k_2} \cdots p_m^{k_m}, \quad \text{and} \quad m = p_1^{e_1} p_2^{e_2} \cdots p_m^{e_m}.$$

Then,

$$\gcd(n, m) = p_1^{\min(k_1, e_1)} p_2^{\min(k_2, e_2)} \cdots p_m^{\min(k_m, e_m)}$$

and

$$\text{lcm}(n, m) = p_1^{\max(k_1, e_1)} p_2^{\max(k_2, e_2)} \cdots p_m^{\max(k_m, e_m)}.$$

Example 3.6

Find the greatest common divisor and least common multiple of the two numbers $m = 2^3 5^7 11^2$ and $n = 3^3 5^4 7^1 11^3$.

✎ Solution Rewrite the prime factorizations as

$$m = 2^3 3^0 5^7 7^0 11^2 \quad \text{and} \quad n = 2^0 3^3 5^4 7^1 11^3.$$

Thus, $\gcd(m, n) = 2^0 3^0 5^4 7^0 11^2 = 5^4 11^2$ and $\text{lcm}(m, n) = 2^3 3^3 5^7 7^1 11^3$.

Comment Besides using the theorem above, one can also use **Euclidean Algorithm** to find the greatest common divisor for two integers. The algorithm is demonstrated below: to find $\gcd(5340, 2355)$, we do the following algorithm

$$
\begin{aligned}
5340 &= 2355 \times 2 + 630 \\
2355 &= 630 \times 3 + 465 \\
630 &= 465 \times 1 + 165 \\
465 &= 165 \times 2 + 135 \\
165 &= 135 \times 1 + 30 \\
135 &= 30 \times 4 + 15 \\
30 &= 15 \times 2, \text{ we stop here since the remainder is } 0.
\end{aligned}
$$

The remainder of the last step is 0, and $\gcd(5340, 2355) = 15$, which is the second to last remainder. In general, one can deduce from the Euclidean algorithm that: for any integers m and n, there are integers x and y such that $xm + yn = \gcd(m, n)$.

Theorem 3.5 Number of Factors (divisors)

If the positive integer n has the prime decomposition $n = p_1^{k_1} p_2^{k_2} \cdots p_m^{k_m}$, then the number of positive divisors of n is given by $\tau(n) = (k_1 + 1)(k_2 + 1) \cdots (k_m + 1)$.

Example 3.7 Number of Factors

Find the number of divisors of the number 2016.

✎ **Solution** Since $2016 = 2^5 3^2 7^1$, then the number of divisors of 2016 is $(5 + 1)(2 + 1)(1 + 1) = 36$. In fact, all the divisors of 2016 are:

$$1, 2, 3, 4, 6, 7, 8, 9, 12, 14, 16, 18, 21, 24, 28, 32, 36, 42, 48,$$

$$54, 56, 63, 72, 84, 96, 126, 144, 168, 224, 252, 288, 336, 504, 672, 1008, 2016.$$

These divisors are in the form $2^m 3^n 7^p$, where there are 6 choices for m since $m \in \{0, 1, 2, 3, 4, 5\}$, and 3 choices for n since $n \in \{0, 1, 2\}$ and 2 choices for p since $p \in \{0, 1\}$.

Example 3.8 2005 AMC 10A Problem 15

How many positive cubes divide $3! \cdot 5! \cdot 7!$?
(A) 2   **(B)** 3   **(C)** 4   **(D)** 5   **(E)** 6

✎ **Solution** **(E).** Written as a product of primes, we have

$$3! \cdot 5! \cdot 7! = 2^8 \cdot 3^4 \cdot 5^2 \cdot 7.$$

A cube that is a factor has a prime factorization of the form $2^p \cdot 3^q \cdot 5^r \cdot 7^s$, where p, q, r and s are all multiples of 3. There are 3 possible values for p, which are 0, 3 and 6. There are 2 possible values for q, which are 0 and 3. The only value for r and for s is 0. Hence there are $6 = 3 \cdot 2 \cdot 1 \cdot 1$ distinct cubes that divide $3! \cdot 5! \cdot 7!$. They are

$$1 = 2^0 \cdot 3^0 \cdot 5^0 \cdot 7^0, \quad 8 = 2^3 \cdot 3^0 \cdot 5^0 \cdot 7^0, \quad 27 = 2^0 \cdot 3^3 \cdot 5^0 \cdot 7^0,$$

$$64 = 2^6 \cdot 3^0 \cdot 5^0 \cdot 7^0, \quad 216 = 2^3 \cdot 3^3 \cdot 5^0 \cdot 7^0, \quad \text{and} \quad 1728 = 2^6 \cdot 3^3 \cdot 5^0 \cdot 7^0.$$

Definition 3.1

Two integers a and b are said to be equal modulo n if and only if n divides $a - b$. We write

$$a \equiv b \pmod{n}.$$

For example, $7 \equiv 1 \pmod 3$, and $11 \equiv 2 \pmod 3$. $2016 \equiv 0 \pmod 7$. Also, from the definition, we have $n \mid m$ if and only if $m \equiv 0 \pmod m$.

☞ For any integer n, n^2 mod 3 or 4 can only be 0 or 1.

☞ For any integer n, n^2 mod 8 can only be 0 or 1 or 4.

☞ For any positive integer n, $n \equiv$ sum of digits of n mod 9.

Theorem 3.6 Modular Arithmetic

☞ If n is a positive integer and $a \equiv b \pmod n$, $c \equiv d \pmod n$, then

$$\begin{aligned} a \pm c &\equiv b \pm d \pmod n, \\ ac &\equiv bd \pmod n, \\ a^k &\equiv b^k \pmod n. \end{aligned}$$

☞ If $am \equiv bm \pmod n$ and $\gcd(m, n) = 1$ (m and n are relatively prime), then $a \equiv b \pmod n$.

☞ If d is a common factor of a, b and n, and $a \equiv b \pmod n$, then $\dfrac{a}{d} \equiv \dfrac{b}{d} \left(\bmod \dfrac{n}{\gcd(n, d)} \right)$.

Example 3.9 1999 AMC 8 Problem 24

When 1999^{2000} is divided by 5, the remainder is

(A) 4 **(B)** 3 **(C)** 2 **(D)** 1 **(E)** 0

✎ **Solution (D).** Write 1999 as $2000 - 1$. We are taking $(2000 - 1)^{2000}$ mod 5. Since $2000 - 1 \equiv -1$ mod 5, $(2000 - 1)^{2000} \equiv (-1)^{2000} \equiv 1$ mod 5. So the answer is (D).

Note *Notice that the pattern for the units digit of the power of 9: $9^1 = 9, 9^2 = 81, 9^3 = 729, \cdots$. Since 2000 is even, the units digit of 1999^{2000} is 1.*

Example 3.10 1972 AHSME Problem 31

The number 2^{1000} is divided by 13. What is the remainder?

(A) 1   **(B)** 2   **(C)** 3   **(D)** 7   **(E)** 11

Solution (C). We make use of the following fact: If N_1, N_2 are integers whose remainders, upon division by D, are R_1 and R_2, then the products $N_1 N_2$ and $R_1 R_2$ have the same remainder upon division by D. In symbols: If $N_1 = Q_1 D + R_1$ and $N_2 = Q_2 D + R_2$, then

$$\begin{aligned} N_1 N_2 &= (Q_1 D + R_1)(Q_2 D + R_2) \\ &= (Q_1 Q_2 D + Q_1 R_2 + Q_2 R_1) D + R_1 R_2, \end{aligned}$$

and the last expression clearly has the same remainder as $R_1 R_2$.

Among the first few powers of two, we find that 2^6 has the convenient remainder 12 [or -1] upon division by 13, so $2^{12} = 2^6 \cdot 2^6$ has the same remainder as $12 \cdot 12$ [or $(-1)^2$], namely 1. We now write $2^{1000} = (2^{12})^{83} \cdot 2^4$, and conclude that the remainder upon division by 13 is $(1)^{83} \cdot 3 = 3$, since $2^4 = 16 = 1 \cdot 13 + 3$.

Using the notation of congruences, we have

$$2^6 = 64 \equiv -1 (\text{mod } 13)$$

$$\begin{aligned} 2^{1000} = \left(2^6\right)^{166} \cdot 2^4 &\equiv (-1)^{166} \cdot 16 (\text{mod } 13) \\ &\equiv 1 \cdot 3 (\text{mod } 13) \equiv 3 (\text{mod } 13). \end{aligned}$$

Theorem 3.7 Chinese Remainder Theorem

Suppose $n_1, \ldots, n_k$ are positive integers that are pairwise coprime. Then, for any given integers $a_1, \ldots, a_k$, there exists a unique solution x (modulo $n_1 n_2 \cdots n_k$) that solves the following system of simultaneous congruences

$$x \equiv a_1 (\text{mod } n_1), \ x \equiv a_2 (\text{mod } n_2), \ \ldots, \ x \equiv a_k (\text{mod } n_k).$$

Example 3.11

In an open school assembly, if students stand in rows of 10, there are 3 students left out. If students stand in rows of 13, there are 5 students left out. If students stand in rows of 17, there is only 1 student left out. What is the minimum possible number of students in this school?

✎ **Solution** This is equivalent to solve the equations

$$x \equiv 3 \quad (\bmod\ 10)$$
$$x \equiv 5 \quad (\bmod\ 13)$$
$$x \equiv 1 \quad (\bmod\ 17)$$

From the equation, we know $x = 10m + 3$ for some integer m. Then, by the second equation

$$
\begin{aligned}
10m + 3 &\equiv 5 \quad (\bmod\ 13) \\
10m &\equiv 2 \quad (\bmod\ 13) \\
5m &\equiv 1 \quad (\bmod\ 13), \quad \text{since } 8 \cdot 5 = 40 \equiv 1 \quad (\bmod\ 13) \\
m &\equiv 8 \quad (\bmod\ 13) \\
m &= 13n + 8 \qquad\qquad \text{for some integer } n.
\end{aligned}
$$

So, $x = 10m + 3 = 10\,(13n + 8) + 3 = 130n + 83$. From the last mod equation, we obtain

$$
\begin{aligned}
130n + 83 &\equiv 1 \quad (\bmod\ 17) \\
11n &\equiv 3 \quad (\bmod\ 17), \quad \text{since } 3 \cdot 11 = 33 \equiv -1 \quad (\bmod\ 17) \\
-n &\equiv 3 \cdot 3 \equiv 9 \quad (\bmod\ 17) \\
n &\equiv 8 \quad (\bmod\ 17)
\end{aligned}
$$

And so, $x = 130n + 83 = 130\,(17t + 8) + 83 = 130 \times 17t + 1123$ for some integer t. This means $x \equiv 1123 \pmod{10 \times 13 \times 17}$. The minimum positive value of x is therefore 1123. There are at least 1123 students.

Note: $nk \equiv 1 \pmod{m}$, n is called the **modular multiplicative inverse** of $k \pmod{m}$.

Definition 3.2 The Euler Function

The Euler function $\phi(n)$ is defined to be the number of positive integers that are less than n and relatively prime to n.

☞ If $n = p_1^{k_1} p_2^{k_2} \cdots p_m^{k_m}$, where p_i are distinct prime numbers, then

$$\phi(n) = n \left(1 - \frac{1}{p_1}\right)\left(1 - \frac{1}{p_2}\right) \cdots \left(1 - \frac{1}{p_m}\right).$$

For example, $\phi(7) = 6$, $\phi(10) = 4$ and $\phi(p) = p - 1$ if p is prime.

Theorem 3.8 Euler's Theorem

☞ If $\gcd(a, n) = 1$, then $a^{\phi(n)} \equiv 1 \pmod{n}$.

Example 3.12

Find the remainder when 2011^{2011} is divided by 125.

✎ **Solution** Since $125 = 5^3$, then $\phi(125) = 125\left(1 - \frac{1}{5}\right) = 100$. In modulo 125, we have $2011^{100} \equiv 1$ (mod 125) by Euler's Theorem, and also $2011 \equiv 11$ (mod 125). Therefore, we have

$$2011^{2011} = (2011^{100})^{20} \cdot 2011^{11}$$
$$\equiv 1^{20} \cdot 11^{11}$$
$$= 121^5 \cdot 11$$
$$= (-4)^5 \cdot 11 = -1024 \cdot 11$$
$$\equiv -24 \cdot 11 = -264$$
$$\equiv 111 \quad (\text{mod } 125).$$

Definition 3.3 Number bases

If there is a set of integers $a_0, a_1, \ldots, a_k$ such that $0 \leq a_i < m$, for $i = 0, \ldots, k$ and $a_k \neq 0$, and

$$n = a_k m^k + a_{k-1} m^{k-1} + \cdots + a_1 m + a_0,$$

then the number n has a base m expression, usually denoted by $n = (a_k a_{k-1} \cdots a_1 a_0)_m$.

Theorem 3.9 Uniqueness of Base Number Expression

Every number n has a unique base m expression.

Example 3.13

Convert the decimal number 2016 to a (i) binary (base 2) number, (ii) octal (base 8) number (iii) hexadecimal (base 16) number.   (note: in a hexadecimal expression, the letters A-F denote 10-15, respectively).

✎ **Solution** Divide 2016 by 2 to get 1008 with remainder 0. Divide 1008 by 2 to get 504 with remainder 0. Divide 504 by 2 to get 252 with remainder 0. Divide 252 by 2 to get 126 with remainder 0. Divide 126 by 2 to get 63 with remainder 0. Divide 63 by 2 to get 31 with remainder 1. Divide 31 by 2 to get

15 with remainder 1. Divide 15 by 2 to get 7 with remainder 1. Divide 7 by 2 to get 3 with remainder 1. Divide 3 by 2 to get 1 with remainder 1. The last quotient is less than 2, so write the last quotient followed by the remainders in the opposite order to get $(11111100000)_2$.

Similarly, divide 2016 by 8 to get 252 with remainder 0. Divide 252 by 8 to get 31 with remainder 4. Divide 31 by 8 to get 3 with remainder 7. Divide 3 by 8 to get 0 with remainder 3. Write the remainders in the opposite order to get $(3740)_8$.

Similarly, $2016 = (7E0)_{16}$.

Example 3.14 1981 AHSME Problem 16

The base-3 representation of x is
$$121122111222111112222_3.$$
What is the first digit (on the left) of the base-9 representation of x?
(A) 1 (B) 2 (C) 3 (D) 4 (E) 5

Solution (E). Grouping the base three digits of x in pairs yields
$$
\begin{aligned}
x &= \left(1 \cdot 3^{19} + 2 \cdot 3^{18}\right) + \left(1 \cdot 3^{17} + 1 \cdot 3^{16}\right)\\
&\quad + \cdots + (2 \cdot 3 + 2)\\
&= (1 \cdot 3 + 2)\left(3^2\right)^9 + (1 \cdot 3 + 1)\left(3^2\right)^8\\
&\quad + \cdots + (2 \cdot 3 + 2).
\end{aligned}
$$

Therefore, the first base nine digit of x is $1 \cdot 3 + 2 = 5$.

☞ Occasionally, contest problems involve palindromes.

Example 3.15 2010 AMC 10A Problem 9

A palindrome, such as 83438, is a number that remains the same when its digits are reversed. The numbers x and $x + 32$ are three-digit and four-digit palindromes, respectively. What is the sum of the digits of x?
(A) 20 (B) 21 (C) 22 (D) 23 (E) 24

Solution (E). Let $x+32$ be written in the form $CDDC$. Because x has three digits, $1000 < x+32 < 1032$, and so $C = 1$ and $D = 0$. Hence $x = 1001 - 32 = 969$, and the sum of the digits of x is $9 + 6 + 9 = 24$.

Note The following two theorems may be beyond the AMC 10 or AMC 12 level, but can be useful sometimes.

Theorem 3.10 Fermat's Theorem

☞ If p is prime, then $a^p \equiv a \pmod{p}$.

☞ If furthermore, $\gcd(a, p) = 1$, then $a^{p-1} \equiv 1 \pmod{p}$.

Fermat's Theorem is sometimes called Fermat's Little Theorem to distinguish it from Fermat's Last Theorem. In fact, Fermat's Theorem is a special case of Euler's Theorem, since $\phi(p) = p - 1$ if p is prime.

Theorem 3.11 Wilson's Theorem

If p is prime, then $(p-1)! + 1 \equiv 0 \pmod{p}$.

Here comes more examples in this chapter.

Example 3.16

Find the units digit of the Fibonacci number F_{2008}.

✎ **Solution** Build the table of units digits of Fibonacci numbers.

u	0	1	2	3	4	5	6	7	8	9
u	0	1	1	2	3	5	8	3	1	4
$1u$	5	9	4	3	7	0	7	7	4	1
$2u$	5	6	1	7	8	5	3	8	1	9
$3u$	0	9	9	8	7	5	2	7	9	6
$4u$	5	1	6	7	3	0	3	3	6	9
$5u$	5	4	9	3	2	5	7	2	9	1

The next row is the same as the first row. So the 2008th Fibonacci number ends in the same digit as the $2008 - 33 \cdot 60 = 28$th, which is 1.

Example 3.17 2010 AMC 12A Problem 23

The number obtained from the last two nonzero digits of $90!$ is equal to n. What is n?

(A) 12   **(B)** 32   **(C)** 48   **(D)** 52   **(E)** 68

✎ **Solution** (A). There are 18 factors of 90! that are multiples of 5, 3 factors that are multiples of 25, and no factors that are multiples of the higher powers of 5. Also, there are more than 45 factors of 2 in 90!. Thus $90! = 10^{21}N$ where N is an integer not divisible by 10, and if $N \equiv n \pmod{100}$ with $0 < n \le 99$, then n is a multiple of 4.

Let $90! = AB$ where A consists of the factors that are relatively prime to 5 and B consists of the factors that are divisible by 5. Note that

$$\prod_{j=1}^{4}(5k+j) \equiv 5k(1+2+3+4) + 1 \cdot 2 \cdot 3 \cdot 4 \equiv 24 \pmod{25},$$

thus

$$A = (1 \cdot 2 \cdot 3 \cdot 4) \cdot (6 \cdot 7 \cdot 8 \cdot 9) \cdot \cdots \cdot (86 \cdot 87 \cdot 88 \cdot 89)$$
$$\equiv 24^{18} \equiv (-1)^{18} \equiv 1 \pmod{25}.$$

Similarly,

$$B = (5 \cdot 10 \cdot 15 \cdot 20) \cdot (30 \cdot 35 \cdot 40 \cdot 45) \cdot (55 \cdot 60 \cdot 65 \cdot 70) \cdot (80 \cdot 85 \cdot 90) \cdot (25 \cdot 50 \cdot 75),$$

thus

$$\frac{B}{5^{21}} = (1 \cdot 2 \cdot 3 \cdot 4) \cdot (6 \cdot 7 \cdot 8 \cdot 9) \cdot (11 \cdot 12 \cdot 13 \cdot 14) \cdot (16 \cdot 17 \cdot 18) \cdot (1 \cdot 2 \cdot 3)$$
$$\equiv 24^3 \cdot (-9) \cdot (-8) \cdot (-7) \cdot 6 \equiv (-1)^3 \cdot 1 \equiv -1 \pmod{25}.$$

Finally,

$$2^{21} = 2 \cdot (2^{10})^2 = 2 \cdot (1024)^2 \equiv 2 \cdot (-1)^2 \equiv 2 \pmod{25},$$

so

$$13 \cdot 2^{21} \equiv 13 \cdot 2 \equiv 1 \pmod{25}.$$

Therefore

$$N \equiv (13 \cdot 2^{21})N = 13 \cdot \frac{90!}{5^{21}} = 13 \cdot A \cdot \frac{B}{5^{21}} \equiv 13 \cdot 1 \cdot (-1) \pmod{25}$$
$$\equiv -13 \equiv 12 \pmod{25}.$$

Thus n is equal to 12, 37, 62, or 87 and because n is a multiple of 4, it follows that $n = 12$.

CHAPTER 4

COUNTING

Problems involving counting can be very tricky. Many challenging contest problems involve sophisticated techniques for counting. These usually combine several branches of mathematics, for example, geometry, algebra and number theory. In this section, we list fundamental counting principals as well as other important counting techniques involving permutation and combination. These skills should enable you to solve the problems later in this book.

✂ ✂

We begin with two fundamental principles.

> **Theorem 4.1 Fundamental Counting Principles (Multiplication Rule)**
>
> If there are m ways of doing some task and n ways of doing another task, then there are mn ways of performing both tasks.

> **Example 4.1**
>
> How many 3-digit positive numbers are there?

Solution For the hundreds digit, we have 9 choices, say 1, 2, 3, 4, 5, 6, 7, 8, and 9; all the digits except 0. For the tens digits, we have 10 choices, and for the units digits we have 10 choices. Therefore, all together, we have $9 \cdot 10 \cdot 10 = 900$ 3-digit positive numbers.

Example 4.2 2003 AMC 10B Problem 10

Nebraska, the home of the AMC, changed its license plate scheme. Each old license plate consisted of a letter followed by four digits. Each new license plate consists of the three letters followed by three digits. By how many times is the number of possible license plates increased?

(A) $\frac{26}{10}$ (B) $\frac{26^2}{10^2}$ (C) $\frac{26^2}{10}$ (D) $\frac{26^3}{10^3}$ (E) $\frac{26^3}{10^2}$

Solution (C). In the old scheme 26×10^4 different plates could be constructed. In the new scheme $26^3 \times 10^3$ different plates can be constructed. There are

$$\frac{26^3 \times 10^3}{26 \times 10^4} = \frac{26^2}{10}$$

times as many possible plates with the new schemes.

Theorem 4.2 Fundamental Counting Principle (Sum Rule)

If we have m ways of doing something and n ways of doing another thing and we can not do both at the same time, then there are $m + n$ ways to choose one of the actions.

Example 4.3

If you have offers from three universities to go in the United States of American, and five offers from Canadian universities, then you have $3 + 5 = 8$ ways to choose a university to attend.

Theorem 4.3 Permutation Without Repetition

The number of distinct permutations of N distinguishable objects is given by

$$N! = N \times (N-1) \times (N-2) \times \cdots \times 2 \times 1.$$

Example 4.4

If you are going to put three letters, A,B and C in a row, then there are $3 \cdot 2 \cdot 1 = 6$ ways of doing so. The 6 outcomes are

$$ABC, \quad ACB, \quad BAC, \quad BCA, \quad CAB, \quad \text{and} \quad CBA.$$

> **Theorem 4.4 Permutation with Repetition**
>
> Suppose there are N objects consisting of m distinguishable classes, and n_1 objects of class 1 are indistinguishable, ..., n_m objects of class m are indistinguishable. Then the number of distinct permutations of the N objects is
>
> $$\frac{N!}{n_1! n_2! \cdots n_m!}.$$

> **Example 4.5**
>
> There are 1 red ball, 2 indistinguishable blue balls and 3 indistinguishable green balls in a row. The number of ways of arranging these balls in this row is $\dfrac{6!}{2! 3!} = 60$.

☞ A **combination** is a selection of a certain number of objects from a given collection, where the order in which the objects are chosen does not matter; in other words, a combination is a subset. Also, see Chapter 6.

> **Theorem 4.5**
>
> The number of different combinations of k objects chosen from n objects is given by
>
> $$\binom{n}{k} = \frac{n!}{k! (n-k)!}.$$

> **Example 4.6**
>
> Three delegates to a conference are to be picked from a class of 50 students. How many ways are there to form the team of three delegates.

✎ **Solution** There are $\dbinom{50}{3} = \dfrac{50!}{3! (50-3)!} = \dfrac{50!}{3! (47)!} = \dfrac{50 \cdot 49 \cdot 48}{3 \cdot 2 \cdot 1} = 19600$ ways, quite a large number.

> **Theorem 4.6 Stars and Bars (Balls and Urns)**
>
> The number of ways to drop n balls into k urns, or equivalently to drop n balls amongst $k-1$ dividers is $\binom{n+k-1}{n}$, given there can be 0 balls in an urn.
>
> However, if there can only be a positive number of balls in an urn, then the number of ways to do such is $\binom{n-1}{k-1}$.

Example 4.7 2008 AMC 10A Problem 23

Two subsets of the set $S = \{a, b, c, d, e\}$ are to be chosen so that their union is S and their intersection contains exactly two elements. In how many ways can this be done, assuming that the order in which the subsets are chosen does not matter?

(A) 20 (B) 40 (C) 60 (D) 160 (E) 320

✎ **Solution (B).** Let the two subsets be A and B. There are $\binom{5}{2} = 10$ ways to choose the two elements common to A and B. There are then $2^3 = 8$ ways to assign the remaining three elements to A or B, so there are 80 ordered pairs (A, B) that meet the required conditions. However, the ordered pairs (A, B) and (B, A) represent the same pair $\{A, B\}$ of subsets, so the conditions can be met in $\frac{80}{2} = 40$ ways.

Example 4.8 2003 AMC 10A Problem 21

Pat is to select six cookies from a tray containing only chocolate chip, oatmeal, and peanut butter cookies. There are at least six of each of these three kinds of cookies on the tray. How many different assortments of six cookies can be selected?

(A) 22 (B) 25 (C) 27 (D) 28 (E) 29

✎ **Solution (D).** The numbers of the three types of cookies must have a sum of six. Possible sets of whole numbers whose sum is six are

$$0, 0, 6; 0, 1, 5; 0, 2, 4; 0, 3, 3; 1, 1, 4; 1, 2, 3; \text{ and } 2, 2, 2.$$

Every ordering of each of these sets determines a different assortment of cookies. There are 3 orders for each of the sets

$$0, 0, 6; 0, 3, 3; \text{ and } 1, 1, 4.$$

There are 6 orders for each of the sets

$$0, 1, 5; 0, 2, 4; \text{ and } 1, 2, 3.$$

There is only one order for 2,2,2. Therefore the total number of assortments of six cookie is $3 \cdot 3 + 3 \cdot 6 + 1 = 28$.

OR

Construct eight slots, six to place the cookies in and two to divide the cookies by type. Let the number of chocolate chip cookies be the number of slots to the left of the first divider, the number of oatmeal cookies be the number of slots between the two dividers, and the number of peanut butter cookies be the number of slots to the right of the second divider. For example, $111|11|1$ represents three chocolate chip cookies, two oatmeal cookies, and one peanut butter cookies. There are $\binom{8}{2} = 28$ ways to place the dividers, so there are 28 ways to select the six cookies.

Theorem 4.7 Binomial Theorem

The expansion of $(x+y)^n$ for positive integer n is

$$(x+y)^n = \binom{n}{0}x^n y^0 + \binom{n}{1}x^{n-1}y + \binom{n}{2}x^{n-2}y^2 + \cdots + \binom{n}{n}x^0 y^n$$

where, for each $k = 0, 1, 2, \cdots, n$, the coefficient $\binom{n}{k}$ is the k-th binomial coefficient equal to the number of ways of choosing k objects from n distinguishable objects.

Note that, if we use sigma notation, the Binomial Theorem could be rewritten as

$$(x+y)^n = \sum_{k=0}^{n} \binom{n}{k} x^n y^{n-k} = \sum_{k=0}^{n} \frac{n!}{k!\,(n-k)!} x^n y^{n-k}.$$

☞ The **binomial coefficients** are the coefficients obtained by expanding the polynomial $(x+y)^n$. The k-th binomial coefficient is exactly $\binom{n}{k}$. These coefficients have a pattern called **Pascal's Triangle:**

$$\binom{0}{0}$$

$$\binom{1}{0} \quad \binom{1}{1}$$

$$\binom{2}{0} \quad \binom{2}{1} \quad \binom{2}{2}$$

$$\binom{3}{0} \quad \binom{3}{1} \quad \binom{3}{2} \quad \binom{3}{3}$$

$$\binom{4}{0} \quad \binom{4}{1} \quad \binom{4}{2} \quad \binom{4}{3} \quad \binom{4}{4}$$

$$\binom{5}{0} \quad \binom{5}{1} \quad \binom{5}{2} \quad \binom{5}{3} \quad \binom{5}{4} \quad \binom{5}{5}$$

Example 4.9

What is the constant term in the expansion of $(x - 2/x)^{10}$?

✎ **Solution** (**D**). Expand $(x - 2/x)^{10}$ using the Binomial Theorem

$$\left(x - \frac{2}{x}\right)^{10} = \sum_{k=0}^{10} \binom{10}{k} x^{10} \left(\frac{2}{x}\right)^{10-k}$$

$$= \binom{10}{0} x^{10} \left(\frac{2}{x}\right)^0 + \binom{10}{1} x^{10-1} \left(\frac{2}{x}\right) + \binom{10}{2} x^{n-2} \left(\frac{2}{x}\right)^2 + \cdots + \binom{10}{10} x^0 \left(\frac{2}{x}\right)^{10}.$$

We find the constant terms by noticing that $k = 10 - k$ and $k = 5$. Thus, the constant term is $\binom{10}{5} x^5 (-2/x)^{10-5} = \binom{10}{5} \cdot (-2)^5 = -8064$.

Example 4.10 2011 AMC 10B Problem 23

What is the hundreds digit of 2011^{2011}?

(**A**) 1   (**B**) 4   (**C**) 5   (**D**) 6   (**E**) 9

✎ **Solution** (**D**). In the expansion of $(2000+11)^{2011}$, all terms except 11^{2011} are divisible by 1000, so the hundreds digit of 2011^{2011} is equal to that of 11^{2011}. Furthermore, in the expansion of $(10+1)^{2011}$, all terms except 1^{2011}, $\binom{2011}{1}(10)(1^{2010})$, and $\binom{2011}{2}(10)^2(1^{2009})$ are divisible by 1000. Thus the hundreds digit of 2011^{2011} is equal to that of

$$1 + \binom{2011}{1}(10)(1^{2010}) + \binom{2011}{2}(10)^2(1^{2009})$$

$$= 1 + 2011 \cdot 10 + 2011 \cdot 1005 \cdot 100$$

$$= 1 + 2011 \cdot 100510.$$

Finally, the hundreds digit of this number is equal to that of $1 + 11 \cdot 510 = 5611$, so the requested hundreds digit is 6.

Example 4.11 1986 AHSME Problem 23

Let $N = 69^5 + 5 \cdot 69^4 + 10 \cdot 69^3 + 10 \cdot 69^2 + 5 \cdot 69 + 1$. How many positive integers are factors of N?

(**A**) 3   (**B**) 5   (**C**) 69   (**D**) 125   (**E**) 216

✎ **Solution** (**E**). By the Binomial Theorem, $N = (69 + 1)^5 = (2 \cdot 5 \cdot 7)^5$. Thus a positive integer d is a factor of N iff $d = 2^p 5^q 7^r$, where p, q, r are each one of the 6 integers 0, 1, 2, 3, 4, 5. Therefore there are $6^3 = 216$ choices for d.

Theorem 4.8 Properties of Binomial Coefficients

If k and n are two integers and $0 \leq k \leq n$, then

- $\binom{n}{k} = \binom{n}{n-k}$ Symmetry

- $k\binom{n}{k} = n\binom{n-1}{k-1}$

- $\binom{k}{k} + \binom{k+1}{k} + \cdots + \binom{n}{k} = \binom{n+1}{k+1}$ Hockey Stick Theorem.

- $\binom{n}{0} + \binom{n}{1} + \cdots + \binom{n}{n} = 2^n$

- $\binom{n}{k} + \binom{n}{k+1} = \binom{n+1}{k+1}$

Example 4.12 2012 AMC 10B Problem 22

Let $(a_1, a_2, \ldots a_{10})$ be a list of the first 10 positive integers such that for each $2 \leq i \leq 10$ either $a_i + 1$ or $a_i - 1$ or both appear somewhere before a_i in the list. How many such lists are there?

(**A**) 120   (**B**) 512   (**C**) 1024   (**D**) 181,440   (**E**) 362,880

Solution (B). If $a_1 = 1$, then the list must be an increasing sequence. Otherwise let $k = a_1$. Then the numbers 1 through $k-1$ must appear in increasing order from right to left, and the numbers from k through 10 must appear in increasing order from left to right. For $2 \leq k \leq 10$ there are $\binom{9}{k-1}$ ways to choose positions in the list for the numbers from 1 through $k-1$, and the positions of the remaining numbers are then determined. The number of lists is therefore

$$1 + \sum_{k=2}^{10} \binom{9}{k-1} = \sum_{k=0}^{9} \binom{9}{k} = 2^9 = 512.$$

OR

If a_{10} is not 1 or 10, then numbers larger than a_{10} must appear in reverse order in the list, and numbers smaller than a_{10} must appear in order. However, 1 and 10 cannot both appear first in the list, so the placement of either 1 or 10 would violate the given conditions. Hence $a_{10} = 1$ or 10. By similar reasoning, when reading the list from right to left the number that appears next must be the smallest or largest unused integer from 1 to 10. This gives 2 choices for each term until there is one number left. Hence there are $2^9 = 512$ choices.

There is an extension of the Binomial theorem, sometimes called the Trinomial Theorem.

> **Theorem 4.9 The Trinomial Theorem**
>
> If a, b and c are nonegative integers and n is a positive integer, then
> $$(x+y+z)^n = \sum_{a+b+c=n} \frac{n!}{a!b!c!} x^a y^b z^c.$$

> **Example 4.13 2011 AMC 10B Problem 23**
>
> What is the hundreds digit of 2011^{2011}?
> **(A)** 1 **(B)** 4 **(C)** 5 **(D)** 6 **(E)** 9

✎ **Solution (D).** Since

$$2011^{2011} = (2000 + 10 + 1)^{2011} = \sum_{a+b+c=2011} \frac{2011!}{a!b!c!} 2000^a 10^b 1^c$$

and we only care about the last three digits, those terms where $a > 0$ or $b > 2$ will be ignored. Thus, we only consider the three terms with

$$a = 0, b = 0 \text{ and } c = 2011,$$
$$a = 0, b = 1 \text{ and } c = 2010,$$
$$a = 0, b = 2 \text{ and } c = 2009,$$

The sum of these terms are

$$\frac{2011!}{0!0!2011!} + \frac{2011!}{0!1!2010!}10 + \frac{2011!}{0!2!2009!}10^2 = 1 + 20110 + 2011 \cdot 2010 \cdot 50$$

The last three digits must be $1 + 110 + 500 = 611$. Hence, the hundreds digit is 6.

Note *The above solution is not the MAA official solution. It was provided by the authors.*

☞ The **floor function** $\lfloor x \rfloor$ is defined to be the greatest integer that is less than or equal to x. This function is useful for counting multiples and factors as well. The floor function is sometimes denoted by $[x]$ or $\text{int}[x]$, and is also called the greatest integer function or Gaussian Function. It is easy to see that $\lfloor x \rfloor \leq x < \lfloor x \rfloor + 1$.

☞ The **ceiling function** $\lceil x \rceil$ is defined to be the least integer that is greater than or equal to x.

Example 4.14

- The number of positive multiples of 17 that are less than or equal to 1000 is $\left\lfloor \dfrac{1000}{17} \right\rfloor = 58$.

- The greatest power of 10 that divides $n!$ is given by the formula $\left\lfloor \dfrac{n}{5} \right\rfloor + \left\lfloor \dfrac{n}{25} \right\rfloor + \left\lfloor \dfrac{n}{125} \right\rfloor + \cdots$.

Theorem 4.10 Inclusion-Exclusion Principle

Let $|S|$ be the cardinality of a finite set S. A, B, and C are three finite sets. Then

$$|A \cup B| = |A| + |B| - |A \cap B|.$$

$$|A \cup B \cup C| = |A| + |B| + |C| - (|A \cap B| + |A \cap C| + |B \cap C|) + |A \cap B \cap C|.$$

Theorem 4.11 Inclusion-Exclusion Principle: In General

Let $|S|$ be the cardinality of a finite set S and $S = \bigcup\limits_{i=1}^{n} A_i$. Then

$$\begin{aligned}
|S| &= \left| \bigcup_{i=1}^{n} A_i \right| \\
&= \sum_{i=1}^{n} |A_i| - \sum_{1 \le i < j \le n} |A_i \cap A_j| + \sum_{1 \le i < j < k \le n} |A_i \cap A_j \cap A_k| - \cdots + (-1)^{n-1} |A_1 \cap \cdots \cap A_n|.
\end{aligned}$$

Example 4.15 2010 AMC 8 Problem 20

In a room, 2/5 of the people are wearing gloves, and 3/4 of the people are wearing hats. What is the minimum number of people in the room wearing both a hat and a glove?

(A) 3   (B) 5   (C) 8   (D) 15   (E) 20

✎ **Solution (A).** Because $\frac{2}{5}$ and $\frac{3}{4}$ of the people in the room are whole numbers, the number of people in the room is a multiple of both 5 and 4. The least common multiple of 4 and 5 is 20, so the minimum number of people in the room is 20. If $\frac{2}{5}$ of 20 people are wearing gloves, then 8 people are wearing gloves. If $\frac{3}{4}$ of 20 people are wearing hats, then 15 are wearing hats. The minimum number wearing gloves and hats occurs if the 5 not wearing hats are each wearing gloves. This leaves $8 - 5 = 3$ people wearing both gloves and hats.

OR

If 8 are wearing gloves and 15 are wearing hats, then $8 + 15$ are wearing gloves and/or hats. There is a minimum of 20 people in the room, so $23 - 20 = 3$ people are wearing both a hat and gloves.

Example 4.16

How many positive integers that are less than or equal 1000 and are multiples of either 2, 3 or 5?

Solution The number of positive multiples of 2 that are less than or equal 1000 is $\left\lfloor \dfrac{1000}{2} \right\rfloor = 500$, and the number of positive multiples of 3 that are less than or equal to 1000 is $\left\lfloor \dfrac{1000}{3} \right\rfloor = 333$. The number of positive multiples of 5 that are less than or equal to 1000 is $\left\lfloor \dfrac{1000}{5} \right\rfloor = 200$. The number of positiver integers that are less than or equal to 1000 and are multiples of both 2 and 3 is $\left\lfloor \dfrac{1000}{6} \right\rfloor$. The number of positive integers that are less than or equal to 1000 and are multiples of both 2 and 5 is $\left\lfloor \dfrac{1000}{10} \right\rfloor$. The number of positive integers that are less than or equal to 1000 and are multiples of both 3 and 5 is $\left\lfloor \dfrac{1000}{15} \right\rfloor$. The number of positive integers that are less than or equal to 1000 and are multiples of 2, 3 and 5 is $\left\lfloor \dfrac{1000}{30} \right\rfloor$. Therefore, by the principle of inclusion-exclusion, the number of positive integers that are less than or equal to 1000 and are multiples of either 2, 3 or 5 is given by

$$500 + 333 + 200 - \left\lfloor \dfrac{1000}{6} \right\rfloor - \left\lfloor \dfrac{1000}{10} \right\rfloor - \left\lfloor \dfrac{1000}{15} \right\rfloor + \left\lfloor \dfrac{1000}{30} \right\rfloor = 734.$$

Theorem 4.12 Pigeon Hole Principle/The Dirichlet's Box Principle

☞ If there are n holes to host $n + 1$ pigeons, then there is at least one hole that will contain at least two pigeons.

☞ (The extended version) If m objects are placed in n boxes, then at least one box must hold at least $\left\lceil \dfrac{m}{n} \right\rceil$ objects.

Example 4.17 2010 AMC 10B Problem 3

A drawer contains red, green, blue, and white socks with at least 2 of each color. What is the minimum number of socks that must be pulled from the drawer to guarantee a matching pair?

(A) 3　(B) 4　(C) 5　(D) 8　(E) 9

✎ **Solution** (C). If a set of 4 socks does not contain a pair, there must be one of each color. The fifth sock must match one of the others and guarantee a matching pair.

Example 4.18 2011 AMC 10B Problem 11

There are 52 people in a room. what is the largest value of n such that the statement "At least n people in this room have birthdays falling in the same month" is always true?

(**A**) 2   (**B**) 3   (**C**) 4   (**D**) 5   (**E**) 12

✎ **Solution** (D). If no more than 4 people have birthdays in any month, then at most 48 people would be accounted for. Therefore the statement is true for $n = 5$. The statement is false for $n \geq 6$ if, for example, 5 people have birthdays in each of the first 4 months of the year, and 4 people have birthdays in each of the last 8 months, for a total of $5 \cdot 4 + 4 \cdot 8 = 52$ people.

The average number of birthdays per month is $\frac{52}{12}$, which is strictly between 4 and 5. Therefore at least one month must contain at least 5 birthdays, and, as above, it is possible to distribute the birthdays so that all months contain 4 or 5 birthdays.

CHAPTER 5

LOGARITHMS, TRIGONOMETRY, COMPLEX NUMBERS, AND VECTORS

We put these four topics in one chapter since they are typically taught in senior high school. AMC 10 does not involve them, but AMC 12 does.

5.1 Logarithms

☞ The logarithms are inverse to exponentials. For $a > 0$

$$a^y = x \leftrightarrow y = \log_a x, \quad (y \text{ is the logarithm of } x \text{ to the base } a)$$

$$a^{\log_a x} = x.$$

Theorem 5.1 Properties of Logarithms

- $\log_a 1 = 0$
- $\log_a a = 1$
- $\log_a x^n = n \log_a x$
- $\log_a xy = \log_a x + \log_a y$
- $\log_a \dfrac{x}{y} = \log_a x - \log_a y$
- $\log_a x = \dfrac{\log_b x}{\log_b a}$ (base change formula)

Example 5.1

The number $\log_{\frac{1}{4}} \sqrt[3]{1024}$ is equal to

(A) 5 (B) $\dfrac{20}{3}$ (C) $\dfrac{-5}{3}$ (D) $\dfrac{5}{3}$ (E) $\dfrac{-20}{3}$

✎ **Solution (C).** The log in question is the solution of $\sqrt[3]{1024} = (\frac{1}{4})^x$ or $\sqrt[3]{2^{10}} = \left[(\frac{1}{2})^2\right]^x$ so $2^{\frac{10}{3}} = (\frac{1}{2})^{2x}$, from which it follows that $x = -\frac{5}{3}$.

Example 5.2 2015 AMC 12A Problem 14

What is the value of a for which $\frac{1}{\log_2 a} + \frac{1}{\log_3 a} + \frac{1}{\log_4 a} = 1$?

(A) 9   **(B)** 12   **(C)** 18   **(D)** 24   **(E)** 36

✎ **Solution (D).** By the change of base formula, $\frac{1}{\log_m n} = \log_n m$. Thus,

$$1 = \frac{1}{\log_2 a} + \frac{1}{\log_3 a} + \frac{1}{\log_4 a} = \log_a 2 + \log_a 3 + \log_a 4 = \log_a 24.$$

It follows that $a = 24$.

5.2 Trigonometry

Trigonometry is a branch of mathematics that studies relationships involving lengths and angles of triangles.

Theorem 5.2 The Law of Sines

In any triangle ABC,

$$\frac{a}{\sin A} = \frac{b}{\sin B} = \frac{c}{\sin C} = 2R \text{, where } R \text{ is the circumradius.}$$

Example 5.3

Triangle ABC below is equilateral and the length of each side is x. Angle BCD is a right angle and angle DAC is 100 degrees. The side DC has length 10. Find x. Round your answer to 2 decimal places.

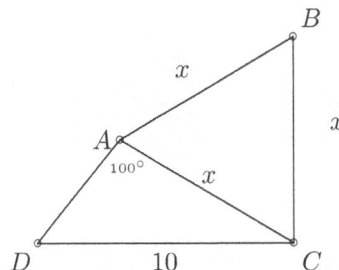

(A) 7.16   **(B)** 7.32   **(C)** 7.51   **(D)** 7.78   **(E)** 7.95

✎ **Solution** (D). Because triangle ABC is equilateral, angle ACB is 60 degrees. Because angle BCD is 90 degrees, angle ACD is 30 degrees. Therefore angle ADC is 50 degrees. Using the Law of Sines,

$$\frac{x}{\sin(50)} = \frac{10}{\sin(100)}.$$

Consequently, $x \approx 7.78$.

Theorem 5.3 The Law of Cosines

In any triangle ABC,

$$a^2 = b^2 + c^2 - 2bc \cos A,$$
$$b^2 = a^2 + c^2 - 2ac \cos B,$$
$$c^2 = a^2 + b^2 - 2ab \cos C.$$

Example 5.4

An interesting fact about circles is that if a line AB is tangent to a circle at the point B and a different line through A intersects the circle in points C and D as in the diagram below, then $AB^2 = AC \cdot AD$. If $AC = 4$, $AD = 3$ and $BC = 2$, what is the measure of the angle DAB to the nearest degree? Note that the chord $\overline{BC}$ does not pass through the center of the circle.

(A) 15   **(B)** 22.5   **(C)** 28   **(D)** 30   **(E)** 33

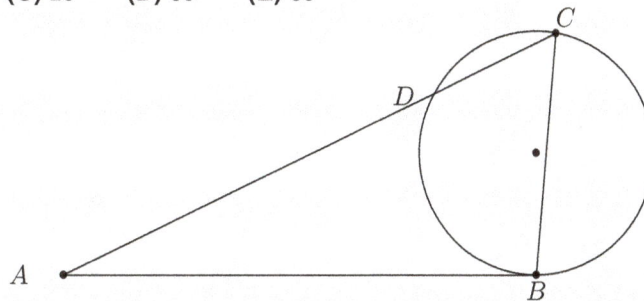

✎ **Solution** (D). First note that

$$AB^2 = AC \cdot AD = 4 \cdot 3 = 12.$$

Then by the law of cosines,

$$BC^2 = AC^2 + AB^2 - 2AC \cdot AB \cdot \cos \angle DAB.$$

Thus

$$2^2 = 4^2 + 12 - 2 \cdot 4 \cdot \sqrt{12} \cos \angle DAB.$$

We can solve

$$\cos \angle DAB = 24/(16\sqrt{3})$$

to get $\angle DAB = \arccos \sqrt{3}/2 = 30.0°$.

☞ **Important trigonometric identities**

- $\sin(-x) = -\sin x$
- $(\sin x)^2 + (\cos x)^2 = 1$
- $\sin\left(\dfrac{\pi}{2} - x\right) = \cos x$
- $\sin(\pi - x) = \sin x$
- $\tan x = \dfrac{\sin x}{\cos x}$
- $\sec x = \dfrac{1}{\cos x}$

- $\cos(-x) = \cos x$
- $\sin 2x = 2 \sin x \cos x$
- $\cos\left(\dfrac{\pi}{2} - x\right) = \sin x$
- $\cos(\pi - x) = -\cos x$
- $\cot x = \dfrac{\cos x}{\sin x}$
- $\csc x = \dfrac{1}{\sin x}$

- $\sin(2n\pi + x) = \sin x$, for each integer n.
- $\cos(2n\pi + x) = \cos x$, for each integer n.
- $\tan(n\pi + x) = \tan x$, for each integer n.
- $\cot(n\pi + x) = \cot x$, for each integer n.
- $\cos 2x = 1 - 2\sin^2 x = 2\cos^2 x - 1$
- $\sin(x + y) = \sin x \cos y + \cos x \sin y$
- $\sin(x - y) = \sin x \cos y - \cos x \sin y$
- $\cos(x + y) = \cos x \cos y - \sin x \sin y$
- $\cos(x - y) = \cos x \cos y + \sin x \sin y$
- $\tan(x - y) = \dfrac{\tan x - \tan y}{1 + \tan x \cdot \tan y}$

☞ **Product to Sum Formulas**

$$\sin x \cos y = \frac{1}{2}\left(\sin(x + y) + \sin(x - y)\right)$$

$$\cos x \cos y = \frac{1}{2}\left(\cos(x + y) + \cos(x - y)\right)$$

$$\sin x \sin y = -\frac{1}{2}\left(\cos(x + y) - \cos(x - y)\right)$$

☞ **Sum to Product Formulas**

$$\sin x - \sin y = 2\cos\left(\frac{x + y}{2}\right)\sin\left(\frac{x - y}{2}\right)$$

$$\cos x + \cos y = 2\cos\left(\frac{x + y}{2}\right)\cos\left(\frac{x - y}{2}\right)$$

$$\cos x - \cos y = -2\sin\left(\frac{x + y}{2}\right)\sin\left(\frac{x - y}{2}\right)$$

Example 5.5

What is the **exact** value of $\frac{\sin(3\alpha)+\sin(\alpha)}{\sin(2\alpha)\cos(\alpha)}$ (for all values of α such that the above expression is defined)?

(A) -2 **(B)** -1 **(C)** 0 **(D)** 1 **(E)** 2

Solution (E). We have

$$\frac{\sin(3\alpha)+\sin(\alpha)}{\sin(2\alpha)\cos(\alpha)}$$

$$=\frac{\sin(2\alpha+\alpha)+\sin(2\alpha-\alpha)}{\sin(2\alpha)\cos(\alpha)}$$

$$=\frac{\sin(2\alpha)\cos(\alpha)+\cos(2\alpha)\sin(\alpha)+\sin(2\alpha)\cos(\alpha)-\cos(2\alpha)\sin(\alpha)}{\sin(2\alpha)\cos(\alpha)}$$

$$=\frac{2\sin(2\alpha)\cos(\alpha)}{\sin(2\alpha)\cos(\alpha)}=2.$$

5.3 Complex Numbers

☞ We define $\sqrt{-1}=i$, where i is called the **imaginary unit**. Also we have $i^2=-1$.

☞ With the help of the imaginary unit i, we define the complex number as

$$a+bi, \quad \text{where } a, \text{ and } b \text{ are real numbers.}$$

☞ Complex numbers can be added, subtracted, multiplied and divided in a natural way.

$$(a+bi)+(c+di) = (a+c)+(b+d)i$$
$$(a+bi)-(c+di) = (a-c)+(b-d)i$$
$$(a+bi)(c+di) = ac-bd+(bc+ad)i$$
$$\frac{a+bi}{c+di} = \frac{(a+bi)(c-di)}{(c+di)(c-di)} = \frac{ac+bd}{c^2+d^2}+\frac{bc-ad}{c^2+d^2}i$$

☞ $a+bi$ and $a-bi$ are called **complex conjugates**.

☞ The **modulus** of the complex number $a+bi$ is $|a+bi|=\sqrt{a^2+b^2}$.

☞ A complex number $a+bi$ can also be written in a polar form $r(\cos\theta+i\sin\theta)$, where r is the modulus and θ is the anticlockwise angle measured from the positive x-axis to the line segment connecting the origin and the point (a,b). The angle θ is called the **argument** of the complex number.

☞ **Euler's Formula** is a nice connection between exponentiation, sine and cosine function:

$$e^{ix} = \cos(x) + i\sin(x).$$

In particular, if $x = \pi$, one has $e^{i\pi} + 1 = 0$ an expression connecting the five most important mathematics constant, $0, 1, i, e,$ and π.

Theorem 5.4 Demoive's Theorem

If n is an integer and $z = r(\cos\theta + i\sin\theta)$, then

$$z^n = r^n (\cos n\theta + i\sin n\theta).$$

Example 5.6

If $z = 1 + \sqrt{3}i$, find z^{100}.

✎ **Solution** Since $z = 2\left(\cos\frac{\pi}{3} + i\sin\frac{\pi}{3}\right)$, by Demoive's Theorem we have

$$
\begin{aligned}
z^{100} &= 2^{100}\left(\cos\frac{100\pi}{3} + i\sin\frac{100\pi}{3}\right) \\
&= 2^{100}\left(\cos\left(33\pi + \frac{\pi}{3}\right) + i\sin\left(33\pi + \frac{\pi}{3}\right)\right) \\
&= 2^{100}\left(-\cos\frac{\pi}{3} - i\sin\frac{\pi}{3}\right) \\
&= -2^{100}\left(\frac{1}{2} + \frac{\sqrt{3}}{2}i\right).
\end{aligned}
$$

Theorem 5.5 n-th Roots of Unity

The equation $z^n = 1$ has n roots called the n-th roots of unity. These roots are

$$z = \cos\left(\frac{2k\pi}{n}\right) + i\sin\left(\frac{2k\pi}{n}\right), k = 0, 1, ..., n - 1.$$

If you plot the n-th roots of unity in the 2-D plane, you will see that all the points are equally spaced on the unit circle.

Example 5.7

Find all the solutions to the equation $z^3 = 1$.

✎ **Solution** By the theorem above, the 3 roots are given by

$$z_1 = \cos\left(\frac{0\pi}{3}\right) + i\sin\left(\frac{0\pi}{3}\right), z_2 = \cos\left(\frac{2\pi}{3}\right) + i\sin\left(\frac{2\pi}{3}\right), z_3 = \cos\left(\frac{4\pi}{3}\right) + i\sin\left(\frac{4\pi}{3}\right),$$

$$z_1 = 1, z_2 = -\frac{1}{2} + i\frac{\sqrt{3}}{2}, z_3 = -\frac{1}{2} - i\frac{\sqrt{3}}{2}.$$

We can verify this by doing factorization. Since

$$z^3 = 1 \iff z^3 - 1 = 0 \iff (z-1)\left(z^2 + z + 1\right) = 0,$$

so $z = 1$ or $z^2 + z + 1 = 0$. Solving the quadratic gives

$$z = \frac{-1 \pm \sqrt{3}i}{2}.$$

Therefore the three roots are

$$z_1 = 1, z_2 = \frac{-1 + \sqrt{3}i}{2}, \text{ and } z_3 = \frac{-1 - \sqrt{3}i}{2}.$$

Now, we can state the Fundamental Theorem of Algebra.

Theorem 5.6 Fundamental Theorem of Algebra

A polynomial equation

$$a_n x^n + a_{n-1}x^{n-1} + a_{n-2}x^{n-2} + \cdots + a_1 x + a_0 = 0,$$

where $n > 0$ and $a_n \neq 0$, has n roots. These roots are either real or complex, and are either single or repeated. A root of multiplicity m is counted as m roots.

The Fundamental Theorem of Algebra was originally proposed by Peter Roth in about 1608. Proofs were attempted by many famous mathematicians such as d'Alembert, Euler and Bernoulli. But all of their proofs have some gaps. A rigorous proof was published by Argand in 1806 and Gauss produced two other proofs in 1816.

Example 5.8 Number of Roots of a Polynomial

Solve the polynomial equation and state the number of its roots.

$$x^9 - 7x^8 + 6x^7 + 38x^6 - 287x^5 - 1319x^4 - 3320x^3 - 5064x^2 - 4528x - 2800 = 0.$$

✏️ **Solution** With the help the computer software, Maple, we are able to obtain all the solutions: $\{x = -2\}, \{x = -2\}, \{x = 3 - 4i\}, \{x = 3 + 4i\}, \{x = 7\},$ and

$$\left\{x = -\frac{1}{2} + \frac{1}{2}i\sqrt{7}\right\}, \left\{x = -\frac{1}{2} - \frac{1}{2}i\sqrt{7}\right\}, \left\{x = -\frac{1}{2} + \frac{1}{2}i\sqrt{7}\right\}, \left\{x = -\frac{1}{2} - \frac{1}{2}i\sqrt{7}\right\}.$$

All together, it has 9 roots with 3 single roots and 3 roots of multiplicity 2. Notice that all the commplex roots appear as conjugate pairs.

5.4 Vectors

☞ Vectors are quantities that have a length and a direction. They can be thought of as segments pointing from their tail to their head.

☞ On the real line, a real number x can be thought of as a vector with length $|x|$ that points in either the positive or the negative direction.

☞ In the 2-D plane, an ordered pair $<x, y>$ can be thought of as a vector $\boldsymbol{v}$ that points from its tail at the origin $(0, 0)$ to its head at the point (x, y). This vector has length given by the Pythagorean Theorem as $|\boldsymbol{v}| = \sqrt{x^2 + y^2}$. The vector $\boldsymbol{v} = <3, 7>$ which points from the origin $(0, 0)$ to the point $(3, 7)$ is the same vector as one that points from the point $(-4, 8)$ to the point $(-1, 15)$ because a vector has length and direction but not position.

☞ In n-dimensional space, a vector can be represented by an n-tuple $<x_1, x_2, x_3, \ldots, x_n>$ and can be thought of as pointing from its tail at the origin $(0, 0, 0, \ldots, 0)$ to its head at the point $(x_1, x_2, x_3, \ldots, x_n)$. It has length $\sqrt{x_1^2 + x_2^2 + x_3^2 + \cdots + x_n^2}$.

☞ Two vectors $\boldsymbol{u}$ and $\boldsymbol{v}$ can be added to get $\boldsymbol{u} + \boldsymbol{v}$. The vector $\boldsymbol{u} + \boldsymbol{v}$ is the vector obtained by first moving the distance and direction given by vector $\boldsymbol{u}$ followed by moving the distance and direction given by the vector $\boldsymbol{v}$. If you place the tail of vector $\boldsymbol{v}$ at the head of vector $\boldsymbol{u}$, then the vector $\boldsymbol{u} + \boldsymbol{v}$ points from the tail of vector $\boldsymbol{u}$ to the head of vector $\boldsymbol{v}$ as shown in the diagram. Thus, $<5, 4> + <3, -2> = <5 + 3, 4 - 2> = <8, 2>$.

☞ Two vectors u and v can be or subtracted to get $u - v$. The vector $u - v$ is the vector obtained by first moving the distance and direction given by vector u followed by moving the distance given by vector v but in the opposite direction of vector v. If you place the head of vector v at the head of vector u, then the vector $u - v$ points from the tail of vector u to the tail of v. Thus, $<5, 4> - <3, -2> = <5 - 3, 4 - (-2)> = <2, 6>$.

Theorem 5.7 Addition of Vectors is Commutative and Associative

For any vectors u and v, it follows that $v + u = u + v$.
For any vectors u, v, and w, it follows that $(u + v) + w = u + (v + w)$.

☞ The zero vector 0 is a vector that has length 0. Its direction is not defined. Any vector v added to the zero vector results in the vector v. That is, $v + 0 = 0 + v = v$.

☞ Any vector v can be multiplied by a real number s to obtain a vector sv to get a vector that points in the same direction as v but has length $|s| \cdot |v|$. The product is called a scalar multiple of the vector v.

Theorem 5.8 Distributive Law for Scalar Multiples

For any real numbers s and t and any vectors u and v, it follows that

$$(s + t)v = sv + tv$$

$$s(u + v) = su + sv$$

☞ A linear combination of vectors $v_1, v_2, v_3, \ldots, v_k$ is a sum of scalar multiples of these vectors as in $s_1 v_1 + s_2 v_2 + s_3 v_3 + \cdots + s_k v_k$. The set of all linear combinations of the vectors $v_1, v_2, v_3, \ldots, v_k$ is called the span of these vectors.

Example 5.9 Linear Combinations

Given $u = <5, 4>$ and $v = <3, -2>$, find $5u + 2v$ and $2u - 3v$.

✎ **Solution**

$$5u + 2v = 5<5, 4> + 2v = <5 \cdot 5 + 2 \cdot 3, 5 \cdot 4 + 2(-2)> = <31, 16>.$$

$$2u - 32v = 2<5, 4> - 3v = <2 \cdot 5 - 3 \cdot 3, 2 \cdot 4 - 3(-2)> = <1, 12>.$$

☞ Vectors $v_1, v_2, v_3, \ldots, v_k$ are called linearly independent if $s_1v_1 + s_2v_2 + s_3v_3 + \cdots + s_kv_k = 0$ implies that $s_1 = s_2 = s_3 = \cdots = s_k = 0$.

Example 5.10

Show that $u = {<}5, 4{>}$ and $v = {<}3, -2{>}$ are linearly independent.

✎ **Solution** Suppose there are real numbers s and t so that $s \cdot u + t \cdot v = 0$. Then $s \cdot u + t \cdot v = s{<}5, 4{>} + t{<}3, -2{>} = {<}5s + 3t, 4s - 2t{>} = {<}0, 0{>}$. This is equivalent to the system of equations

$$5s + 3t = 0$$
$$4s - 2t = 0$$

which has the unique solution $s = 0$ and $t = 0$. This shows that u and v are linearly independent.

☞ In the 2-D plane any two vectors that point in different directions are linearly independent, but two vectors that point in the same direction have the property that one of the vectors is a scalar multiple of the other, so they are not linearly independent.

Theorem 5.9 Two Vectors Span the Plane

If u and v are 2-D vectors that are linearly independent, then the span of u and v includes every 2-D vector. Moreover, for any 2-D vector w, there are unique real numbers s and t so that $w = su + tv$.

☞ If you place the tail of vector u and the tail of vector v at the same point, let θ be the angle formed by these two vectors. The dot product (sometimes called the scalar product) of two vectors u and v is

$$u \cdot v = |u| \cdot |v| \cdot \cos \theta.$$

Theorem 5.10 Properties of Dot Products

Let $u = {<}u_1, u_2, u_3, \ldots, u_n{>}$ and $v = {<}v_1, v_2, v_3, \ldots, v_n{>}$ be two n-dimensional vectors. Then

$$u \cdot v = u_1 \cdot v_1 + u_2 \cdot v_2 + u_3 \cdot v_3 + \cdots + u_n \cdot v_n$$

$$|u|^2 = u \cdot u \qquad \text{and} \qquad u \cdot v = v \cdot u.$$

The dot product gives a way to find the angle between two vectors and, therefore, how to tell if two vectors are parallel or perpendicular to each other.

Theorem 5.11 Angles Between Vectors

The angle between two vectors u and v is given by the angle θ such that $\cos\theta = \frac{u \cdot v}{|u| \cdot |v|}$. The two vectors are parallel if $|u \cdot v| = |u| \cdot |v|$. The two vectors are perpendicular (sometimes called orthogonal) if $u \cdot v = 0$.

Example 5.11

Which of the following vectors are perpendicular to each other?

$<4, 6>$ $<-3, 6>$ $<10, 5>$ $<15, -10>$.

Solution Calculating the dot products of the six pairs of vectors gives $<4,6> \cdot <-3,6> = 24$, $<4,6> \cdot <10,5> = 70$, $<4,6> \cdot <15,-10> = 0$, $<-3,6> \cdot <10,5> = 0$, $<-3,6> \cdot <15,-10> = -105$, $<10,5> \cdot <15,-10> = 100$, so $<4,6>$ is perpendicular to $<15,-10>$ and $<-3,6>$ is perpendicular to $<10,5>$.

☞ If u is a nonzero vector, then $\frac{u}{|u|}$ is a vector parallel to u that has length equal to 1. A vector with length 1 is called a unit vector.

☞ Given two vectors $u \neq 0$ and v, there is a unique vector $\mathbf{Proj}_u(v)$ called the projection of v onto u that is parallel to u such that $v - \mathbf{Proj}_u(v)$ is perpendicular to u as seen in the diagram.

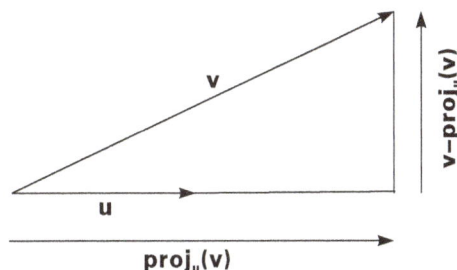

Theorem 5.12

Given two vectors $u \neq 0$ and v, the length of the projection of v onto u is given by $\frac{u \cdot v}{|u|}$ with the projection vector given by $\mathbf{Proj}_u(v) = \frac{u \cdot v}{u \cdot u} u$.

Example 5.12

A regular tetrahedron has four equilateral triangles for faces. Find the angle between one face of the regular tetrahedron and an edge that is not an edge of the given face.

Solution Without loss of generality, assume that all of the edges in the tetrahedron have length 6, and suppose that two edges of the given face of the tetrahedron are given by the vectors $<6, 0, 0>$ and $<3, 3\sqrt{3}, 0>$. Then if the tetrahedron has height h, another edge of the tetrahedron is $<3, \sqrt{3}, h>$. This vector must have length 6, so $6^2 = 3^2 + \sqrt{3}^2 + h^2$, and thus, $h = 2\sqrt{6}$. The projection of the edge $<3, \sqrt{3}, 2\sqrt{6}>$ onto the given face is then $<3, \sqrt{3}, 0>$, so the requested angle is the angle θ given by

$$\cos \theta = \frac{<3, \sqrt{3}, 2\sqrt{6}> \cdot <3, \sqrt{3}, 0>}{|<3, \sqrt{3}, 6>| \cdot |<3, \sqrt{3}, 0>|} = \frac{3^2 + \sqrt{3}^2}{\sqrt{3^2 + \sqrt{3}^2 + 24} \cdot \sqrt{3^2 + \sqrt{3}^2}} = \frac{1}{\sqrt{3}}.$$

The desired angle is $\theta \approx 54.74°$.

CHAPTER 6

EXERCISES

6.1 Algebra

1. 2013 AMC 10A Problem 21

A group of 12 pirates agree to divide a treasure chest of gold coins among themselves as follows. The kth pirate to take a share takes $\frac{k}{12}$ of the coins that remain in the chest. The number of coins initially in the chest is the smallest number for which this arrangement will allow each pirate to receive a positive whole number of coins. How many coins does the 12th pirate receive?

(A) 720 (B) 1296 (C) 1728 (D) 1925 (E) 3850

2. 1974 AHSME Problem 4

What is the remainder when $x^{51} + 51$ is divided by $x + 1$?

(A) 0 (B) 1 (C) 49 (D) 50 (E) 51

3. 2001 AMC 12 Problem 13

The parabola with equation $p(x) = ax^2 + bx + c$ and vertex (h, k) is reflected about the line $y = k$. This results in the parabola with equation $q(x) = dx^2 + ex + f$. Which of the following equals $a + b + c + d + e + f$?

(A) $2b$ (B) $2c$ (C) $2a + 2b$ (D) $2h$ (E) $2k$

4. 2013 AMC 10B Problem 19

The real numbers c, b, a form an arithmetic sequence with $a \geq b \geq c \geq 0$. The quadratic $ax^2 + bx + c$ has exactly one root. What is this root?

(A) $-7-4\sqrt{3}$ **(B)** $-2-\sqrt{3}$ **(C)** -1 **(D)** $-2+\sqrt{3}$ **(E)** $-7+4\sqrt{3}$

5. 2008 AMC 12A Problem 19

In the expansion of

$$\left(1+x+x^2+\cdots+x^{27}\right)\left(1+x+x^2+\cdots+x^{14}\right)^2,$$

what is the coefficient of x^{28}?

(A) 195 **(B)** 196 **(C)** 224 **(D)** 378 **(E)** 405

6. 1983 AHSME Problem 25

Problem: If $60^a = 3$ and $60^b = 5$, then $12^{(1-a-b)/[2(1-b)]}$ is

(A) $\sqrt{3}$ **(B)** 2 **(C)** $\sqrt{5}$ **(D)** 3 **(E)** $\sqrt{12}$

7. 2013 AMC 10B Problem 14

Define $a\clubsuit b = a^2b - ab^2$. Which of the following describes the set of points (x,y) for which $x\clubsuit y = y\clubsuit x$?

(A) a finite set of points **(B)** one line **(C)** two parallel lines **(D)** two intersecting lines

(E) three lines

8. 2006 AMC 12A Problem 18

The function f has the property that for each real number x in its domain, $1/x$ is also in its domain and $f(x) + f\left(\frac{1}{x}\right) = x$. What is the largest set of real numbers that can be in the domain of f?

(A) $\{x|x \neq 0\}$ **(B)** $\{x|x < 0\}$ **(C)** $\{x|x > 0\}$

(D) $\{x|x \neq -1 \text{ and } x \neq 0 \text{ and } x \neq 1\}$ **(E)** $\{-1, 1\}$

9. 2009 AMC 12A Problem 17

Let $a + ar_1 + ar_1^2 + ar_1^3 + \cdots$ and $a + ar_2 + ar_2^2 + ar_2^3 + \cdots$ be two different infinite geometric series of positive numbers with the same first term. The sum of the first series is r_1, and the sum of the second series is r_2. What is $r_1 + r_2$?

(A) 0 **(B)** $\frac{1}{2}$ **(C)** 1 **(D)** $\frac{1+\sqrt{5}}{2}$ **(E)** 2

10. 1992 AHSME Problem 18

The increasing sequence of positive integers $a_1, a_2, a_3, \cdots$ has the property that

$$a_{n+2} = a_n + a_{n+1} \text{ for all } n \geq 1.$$

If $a_7 = 120$, then a_8 is

(A) 128   (B) 168   (C) 193   (D) 194   (E) 210

11. 2010 AMC 10B Problem 25

Let $a > 0$, and let $P(x)$ be a polynomial with integer coefficients such that

$P(1) = P(3) = P(5) = P(7) = a$, and $P(2) = P(4) = P(6) = P(8) = -a$.

What is the smallest possible value of a?

(A) 105   (B) 315   (C) 945   (D) 7!   (E) 8!

12. 2010 AMC 12B Problem 23

Monic quadratic polynomial $P(x)$ and $Q(x)$ have the property that $P(Q(x))$ has zeros at $x = -23, -21, -17$, and -15, and $Q(P(x))$ has zeros at $x = -59, -57, -51$ and -49. What is the sum of the minimum values of $P(x)$ and $Q(x)$?

(A) -100 (B) -82 (C) -73 (D) -64 (E) 0

13. 2004 AMC 10B problem 21

Let $1, 4, \ldots$ and $9, 16, \ldots$ be two arithmetic progressions. The set S is the union of the first 2004 terms of each sequence. How many distinct numbers are in S?

(A) 3722   (B) 3732   (C) 3914   (D) 3924   (E) 4007

6.2 Geometry

6.2.1 Triangle and Geometry

1. 1997 AHSME Problem 5

A rectangle with perimeter 176 is divided into five congruent rectangles as shown in the diagram. What is the perimeter of one of the five congruent rectangles?

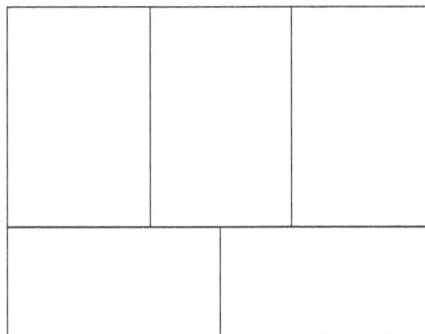

(A) 35.2   (B) 76   (C) 80   (D) 84   (E) 86

2. 2007 AMC 12B Problem 14

Point P is inside equilateral $\triangle ABC$. Points Q, R, and S are the feet of the perpendiculars from P to $\overline{AB}$, $\overline{BC}$, and $\overline{CA}$, respectively. Given that $PQ = 1$, $PR = 2$, and $PS = 3$, what is AB?

(A) 4 **(B)** $3\sqrt{3}$ **(C)** 6 **(D)** $4\sqrt{3}$ **(E)** 9

3. 2015 AMC 10A Problem 19

The isosceles right triangle ABC has right angle at C and area 12.5. The rays trisecting $\angle ACB$ intersect AB at D and E. What is the area of $\triangle CDE$?

(A) $\dfrac{5\sqrt{2}}{3}$ **(B)** $\dfrac{50\sqrt{3} - 75}{4}$ **(C)** $\dfrac{15\sqrt{3}}{8}$ **(D)** $\dfrac{50 - 25\sqrt{3}}{2}$ **(E)** $\dfrac{25}{6}$

4. 1983 AHSME Problem 19

Point D is on side BC of $\triangle ABC$ with $AB = 3$, $AC = 6$, and $\angle CAD = \angle DAB = 60°$. What is the length AD ?

(A) 2 **(B)** 2.5 **(C)** 3 **(D)** 3.5 **(E)** 4

5. 1990 AHSME Problem 14

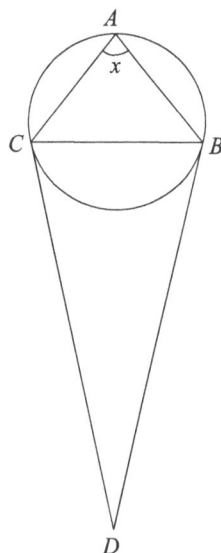

An acute isosceles triangle, ABC, is inscribed in a circle. Through B and C, tangents to the circle are drawn, meeting at point D. If $\angle ABC = \angle ACB = 2\angle D$ and x is the radian measure of $\angle A$, then $x =$

(A) $\dfrac{3\pi}{7}$ **(B)** $\dfrac{4\pi}{9}$ **(C)** $\dfrac{5\pi}{11}$ **(D)** $\dfrac{6\pi}{13}$ **(E)** $\dfrac{7\pi}{15}$

6. 1997 AHSME Problem 9

In the figure, $ABCD$ is a 2×2 square, E is the midpoint of $\overline{AD}$, and F is on $\overline{BE}$. If $\overline{CF}$ is perpendicular to $\overline{BE}$, then the area of quadrilateral $CDEF$ is

(A) 2 (B) $3 - \dfrac{\sqrt{3}}{2}$ (C) $\dfrac{11}{5}$ (D) $\sqrt{5}$ (E) $\dfrac{9}{4}$

7. 2008 AMC 10A Problem 16 or 2008 AMC 12A Problem 13

Points A and B lie on a circle centered at O, and $\angle AOB = 60°$. A second circle is internally tangent to the first and tangent to both $\overline{OA}$ and $\overline{OB}$. What is the ratio of the area of the smaller circle to that of the larger circle?

(A) $\dfrac{1}{16}$ (B) $\dfrac{1}{9}$ (C) $\dfrac{1}{8}$ (D) $\dfrac{1}{6}$ (E) $\dfrac{1}{4}$

8. 2003 AMC 12A Problem 15

A semicircle of diameter 1 sits at the top of a semicircle of diameter 2, as shown. The shaded area inside the smaller semicircle and outside the larger semicircle is called a lune. Determine the area of this lune.

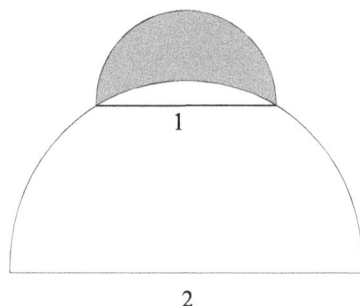

(A) $\dfrac{1}{6}\pi - \dfrac{\sqrt{3}}{4}$ (B) $\dfrac{\sqrt{3}}{4} - \dfrac{1}{12}\pi$ (C) $\dfrac{\sqrt{3}}{4} - \dfrac{1}{24}\pi$ (D) $\dfrac{\sqrt{3}}{4} + \dfrac{1}{24}\pi$ (E) $\dfrac{\sqrt{3}}{4} + \dfrac{1}{12}\pi$

9. 1997 AHSME Problem 26

Triangle ABC and point P in the same plane are given. Point P is equidistant from A and B, angle APB is twice angle ACB, and $\overline{AC}$ intersects $\overline{BP}$ at point D. If $PB = 3$ and $PD = 2$, then $AD \cdot CD =$

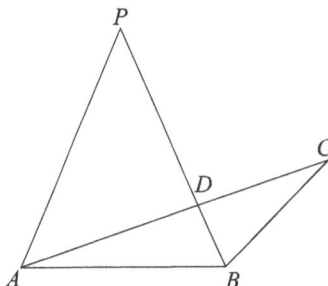

(A) 5   (B) 6   (C) 7   (D) 8   (E) 9

10. 2015 AMC 12A Problem 11

On a sheet of paper, Isabella draws a circle of radius 2, a circle of radius 3, and all possible lines simultaneously tangent to both circles. Isabella notices that she has drawn exactly $k \geq 0$ lines. How many different values of k are possible?

(A) 2   (B) 3   (C) 4   (D) 5   (E) 6

11. 2008 AMC 12A Problem 24

Triangle ABC has $\angle C = 60°$ and $BC = 4$. Point D is the midpoint of BC. What is the largest possible value of $\tan \angle BAD$?

(A) $\dfrac{\sqrt{3}}{6}$ (B) $\dfrac{\sqrt{3}}{3}$ (C) $\dfrac{\sqrt{3}}{2\sqrt{2}}$ (D) $\dfrac{\sqrt{3}}{4\sqrt{2}-3}$ (E) 1

12. 2000 AMC 12 Problem 24

If circular arcs $\overset{\frown}{AC}$ and $\overset{\frown}{BC}$ have centers at B and A, respectively, then there exists a circle tangent to both $\overset{\frown}{AC}$ and $\overset{\frown}{BC}$, and to $\overline{AB}$. If the length of $\overset{\frown}{BC}$ is 12, then the circumference of the circle is

(A) 24   (B) 25   (C) 26   (D) 27   (E) 28

6.2.2 Polygons and Solid Geometry

1. 1993 AHSME Problem 14

The convex pentagon $ABCDE$ has $\angle A = \angle B = 120°$, $EA = AB = BC = 2$ and $CD = DE = 4$. What is the area of ABCDE?

(A) 10   (B) $7\sqrt{3}$ (C) 15   (D) $9\sqrt{3}$ (E) $12\sqrt{5}$

2. 1995 AHSME Problem 17

Given regular pentagon $ABCDE$, a circle can be drawn that is tangent to $\overline{DC}$ at D and to $\overline{AB}$ at A. The number of degrees in minor arc AD is

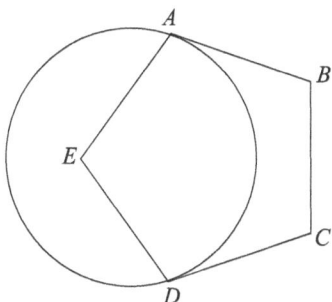

(A) 72   (B) 108   (C) 120   (D) 135   (E) 144

3. 1992 AHSME Problem 20

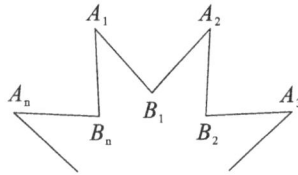

Part of an "n-pointed regular star" is shown. It is a simple closed polygon in which all $2n$ edges are congruent, angles $A_1, A_2, \cdots, A_n$ are congruent, and angles $B_1, B_2, \cdots, B_n$ are congruent. If the acute angle at A_1 is $10°$ less than the acute angle at B_1, then $n =$

(A) 12   (B) 18   (C) 24   (D) 36   (E) 60

4. 2014 AMC 10A Problem 23

A rectangular piece of paper whose length is $\sqrt{3}$ times the width has area A. The paper is divided into three equal sections along the opposite lengths, and then a dotted line is drawn from the first divider to the second divider on the opposite side as shown. The paper is then folded flat along this dotted line to create a new shape with area B. What is the ratio $B : A$?

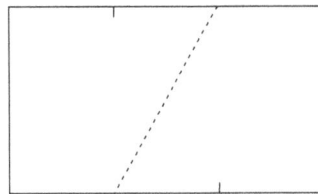

(A) $1 : 2$ (B) $3 : 5$ (C) $2 : 3$ (D) $3 : 4$ (E) $4 : 5$

5. 2015 AMC 10B Problem 22

In the figure shown below, $ABCDE$ is a regular pentagon and $AG = 1$. What is $FG + JH + CD$?

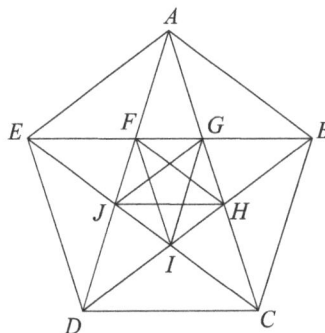

(A) 3   (B) $12 - 4\sqrt{5}$ (C) $\dfrac{5 + 2\sqrt{5}}{3}$ (D) $1 + \sqrt{5}$ (E) $\dfrac{11 + 11\sqrt{5}}{10}$

6. 2002 AMC 12B Problem 24

A convex quadrilateral $ABCD$ with area 2002 contains a point P in its interior such that $PA = 24, PB = 32, PC = 28, PD = 45$. Find the perimeter of $ABCD$.

(A) $4\sqrt{2002}$ **(B)** $2\sqrt{8465}$ **(C)** $2\left(48 + \sqrt{2002}\right)$ **(D)** $2\sqrt{8633}$

(E) $4(36 + \sqrt{113})$

7. 2006 AMC 10B Problem 23

A triangle is partitioned into three triangles and a quadrilateral by drawing two lines from vertices to their opposite sides. The areas of the three triangles are 3, 7, and 7 as shown. What is the area of the shaded quadrilateral?

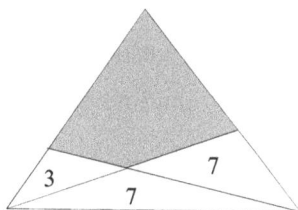

(A) 15 **(B)** 17 **(C)** $\dfrac{35}{2}$ **(D)** 18 **(E)** $\dfrac{55}{3}$

8. 1994 AHSME Problem 11

Three cubes of volume $1, 8$ and 27 are glued together at their faces. The smallest possible surface area of the resulting configuration is

(A) 36 **(B)** 56 **(C)** 70 **(D)** 72 **(E)** 74

9. 1996 AHSME Problem 28

On a $4 \times 4 \times 3$ rectangular parallelepiped, vertices A, B, and C are adjacent to vertex D. The perpendicular distance from D to the plane containing A, B, and C is closest to

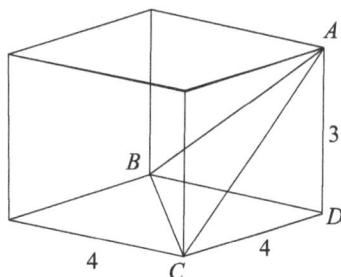

(A) 1.6 **(B)** 1.9 **(C)** 2.1 **(D)** 2.7 **(E)** 2.9

10. 2004 AMC 10A Problem 25 or 2004 AMC 12A Problem 22

Three mutually tangent spheres of radius 1 rest on a horizontal plane. A sphere of radius 2 rests on them. What is the distance from the plane to the top of the larger sphere?

(A) $3 + \dfrac{\sqrt{30}}{2}$ (B) $3 + \dfrac{\sqrt{69}}{3}$ (C) $3 + \dfrac{\sqrt{123}}{4}$ (D) $\dfrac{52}{9}$ (E) $3 + 2\sqrt{2}$

11. 2008 AMC 10B Problem 24

Quadrilateral $ABCD$ has $AB = BC = CD$, angle $ABC = 70$ and angle $BCD = 170$. What is the measure of angle BAD?

(A) 75   (B) 80   (C) 85   (D) 90   (E) 95

12. 2011 AMC 10A Problem 24

Two distinct regular tetrahedra have all their vertices among the vertices of the same unit cube. What is the volume of the region formed by the intersection of the tetrahedra?

(A) $\dfrac{1}{12}$ (B) $\dfrac{\sqrt{2}}{12}$ (C) $\dfrac{\sqrt{3}}{12}$ (D) $\dfrac{1}{6}$ (E) $\dfrac{\sqrt{2}}{6}$

6.3   Number Theory

1. 2014 AMC 10A Problem 20 or 2014 AMC 12A Problem 16

The product $(8)(888\ldots 8)$, where the second factor has k digits, is an integer whose digits have a sum of 1000. What is k?

(A) 901   (B) 911   (C) 919   (D) 991   (E) 999

2. 2002 AMC 10A Problem 14 or 2002 AMC 12A Problem 12

Both roots of the quadratic equation $x^2 - 63x + k = 0$ are prime numbers. The number of possible values of k is

(A) 0   (B) 1   (C) 2   (D) 4   (E) more than 4

3. 1990 AHSME Problem 11

How many positive integers less than 50 have an odd number of positive integer divisors?

(A) 3   (B) 5   (C) 7   (D) 9   (E) 11

4. 2002 AMC 12A Problem 20

Suppose that a and b are digits, not both nine and not both zero, and the repeating decimal $0.\overline{ab}$ is expressed as a fraction in lowest terms. How many different denominators are possible?

(A) 3   **(B)** 4   **(C)** 5   **(D)** 8   **(E)** 9

5. 2013 AMC 12B Problem 15

The number 2013 is expressed in the form $2013 = \frac{a_1!a_2!...a_m!}{b_1!b_2!...b_n!}$, where $a_1 \geq a_2 \geq \cdots \geq a_m$ and $b_1 \geq b_2 \geq \cdots \geq b_n$ are positive integers and $a_1 + b_1$ is as small as possible. What is $|a_1 - b_1|$?

(A) 1   **(B)** 2   **(C)** 3   **(D)** 4   **(E)** 5

6. 2012 AMC 12B Problem 11

In the equation below, A and B are consecutive positive integers, and A, B, and $A + B$ represent number bases:

$$132_A + 43_B = 69_{A+B}.$$

What is $A + B$?

(A) 9   **(B)** 11   **(C)** 13   **(D)** 15   **(E)** 17

7. 1987 AHSME Problem 16

A cryptographer devises the following method for encoding positive integers. First, the integer is expressed in base 5. Second, a 1-to-1 correspondence is established between the digits that appear in the expressions in base 5 and the elements of the set $\{V, W, X, Y, Z\}$. Using this correspondence, the cryptographer finds that three consecutive integers in increasing order are coded as VYZ, VYX, VVW, respectively. What is the base-10 expression for the integer coded as XYZ?

(A) 48   **(B)** 71   **(C)** 82   **(D)** 108   **(E)** 113

8. 1992 AHSME Problem 23

Let S be a subset of $\{1, 2, 3, ..., 50\}$ such that no pair of distinct elements in S has a sum divisible by 7. What is the maximum number of elements in S?

(A) 6   **(B)** 7   **(C)** 14   **(D)** 22   **(E)** 23

9. 1982 AHSME Problem 26

Suppose that the base-8 representation of a perfect square is $ab3c$, where $a \neq 0$. What is c?

(A) 0   **(B)** 1   **(C)** 3   **(D)** 4   **(E)** 7

10. 2007 AMC 12B Problem 24

How many pairs of positive integers (a, b) are there such that $\gcd(a, b) = 1$ and

$$\frac{a}{b} + \frac{14b}{9a}$$

is an integer?

(A) 4   **(B)** 6   **(C)** 9   **(D)** 12   **(E)** infinitely many

11. 2010 AMC 12B Problem 25

For every integer $n \geq 2$, let $\text{pow}(n)$ be the largest power of the largest prime that divides n. For example $\text{pow}(144) = \text{pow}(2^4 \cdot 3^2) = 3^2$. What is the largest integer m such that 2010^m divides $\prod_{n=2}^{5300} \text{pow}(n)$?

(A) 74   (B) 75   (C) 76   (D) 77   (E) 78

12. 2015 AMC 12B Problem 23

A rectangular box measures $a \times b \times c$, where a, b, and c are integers and $1 \leq a \leq b \leq c$. The volume and the surface area of the box are numerically equal. How many ordered triples (a, b, c) are possible?

(A) 4   (B) 10   (C) 12   (D) 21   (E) 26

6.4 Counting

1. 2004 AMC 10A Problem 13

At a party, each man danced with exactly three women and each woman danced with exactly two men. Twelve men attended the party. How many women attended the party?

(A) 8   (B) 12   (C) 16   (D) 18   (E) 24

2. 2001 AMC 10 Problem 19

Pat wants to buy four donuts from an ample supply of three types of donuts: glazed, chocolate, and powdered. How many different selections are possible?

(A) 6   (B) 9   (C) 12   (D) 15   (E) 18

3. 1998 AHSME Problem 24

Call a 7-digit telephone number $d_1 d_2 d_3 - d_4 d_5 d_6 d_7$ memorable if the prefix sequence $d_1 d_2 d_3$ is exactly the same as either of the sequences $d_4 d_5 d_6$ or $d_5 d_6 d_7$ (possibly both). Assuming that each d_i can be any of the ten decimal digits $0, 1, 2, \ldots, 9$, the number of different memorable telephone numbers is

(A) $19,810$ (B) $19,910$ (C) $19,990$ (D) $20,000$ (E) $20,100$

4. 2011 AMC 10A Problem 22

Each vertex of convex pentagon $ABCDE$ is to be assigned a color. There are 6 colors to choose from, and the ends of each diagonal must have different colors. How many different colorings are possible?

(**A**) 2520   (**B**) 2880   (**C**) 3120   (**D**) 3250   (**E**) 3750

5. 1994 AHSME Problem 22

Nine chairs in a row are to be occupied by six students and Professors Alpha, Beta and Gamma. These three professors arrive before the six students and decide to choose their chairs so that each professor will be between two students. In how many ways can Professors Alpha, Beta and Gamma choose their chairs?

(**A**) 12   (**B**) 36   (**C**) 60   (**D**) 84   (**E**) 630

6. 2003 AMC 12A Problem 20

How many 15-letter arrangements of 5 A's, 5 B's, and 5 C's have no A's in the first 5 letters, no B's in the next 5 letters, and no C's in the last 5 letters?

(**A**) $\displaystyle\sum_{k=0}^{5} \binom{5}{k}^3$ (**B**) $3^5 \cdot 2^5$ (**C**) 2^{15} (**D**) $\dfrac{15!}{(5!)^3}$ (**E**) 3^{15}

7. 2001 AMC 12 Problem 17

A point P is selected at random from the interior of the pentagon with vertices $A = (0, 2)$, $B = (4, 0)$, $C = (2\pi + 1, 0)$, $D = (2\pi + 1, 4)$, and $E = (0, 4)$. What is the probability that $\angle APB$ is obtuse?

(**A**) $\dfrac{1}{5}$ (**B**) $\dfrac{1}{4}$ (**C**) $\dfrac{5}{16}$ (**D**) $\dfrac{3}{8}$ (**E**) $\dfrac{1}{2}$

8. 2003 AMC 12B Problem 19

Let S be the set of permutations of the sequence $1, 2, 3, 4, 5$ for which the first term is not 1. A permutation is chosen randomly from S. The probability that the second term is 2, in lowest terms, is a/b. What is $a + b$?

(**A**) 5   (**B**) 6   (**C**) 11   (**D**) 16   (**E**) 19

9. 2004 AMC 10A Problem 10

Coin A is flipped three times and coin B is flipped four times. What is the probability that the number of heads obtained from flipping the two fair coins is the same?

(**A**) $\dfrac{29}{128}$ (**B**) $\dfrac{23}{128}$ (**C**) $\dfrac{1}{4}$ (**D**) $\dfrac{35}{128}$ (**E**) $\dfrac{1}{2}$

10. 2001 AMC 12 Problem 11

A box contains exactly five chips, three red and two white. Chips are randomly removed one at a time without replacement until all the red chips are drawn or all the white chips are drawn. What is the probability that the last chip drawn is white?

(**A**) $\dfrac{3}{10}$ (**B**) $\dfrac{2}{5}$ (**C**) $\dfrac{1}{2}$ (**D**) $\dfrac{3}{5}$ (**E**) $\dfrac{7}{10}$

11. 2014 AMC 12B Problem 22

In a small pond there are eleven lily pads in a row labeled 0 through 10. A frog is sitting on pad 1. When the frog is on pad N, $0 < N < 10$, it will jump to pad $N-1$ with probability $\frac{N}{10}$ and to pad $N+1$ with probability $1 - \frac{N}{10}$. Each jump is independent of the previous jumps. If the frog reaches pad 0 it will be eaten by a patiently waiting snake. If the frog reaches pad 10 it will exit the pond, never to return. What is the probability that the frog will escape without being eaten by the snake?

(A) $\frac{32}{79}$ **(B)** $\frac{161}{384}$ **(C)** $\frac{63}{146}$ **(D)** $\frac{7}{16}$ **(E)** $\frac{1}{2}$

12. 2012 AMC 10A Problem 23 or 2012 AMC 12A Problem 19

Adam, Benin, Chiang, Deshawn, Esther, and Fiona have internet accounts. Some, but not all, of them are internet friends with each other, and none of them has an internet friend outside this group. Each of them has the same number of internet friends. In how many different ways can this happen?

(A) 60 **(B)** 170 **(C)** 290 **(D)** 320 **(E)** 660

13. 2013 AMC 12A Problem 20

Let S be the set $\{1, 2, 3, ..., 19\}$. For $a, b \in S$, define $a \succ b$ to mean that either $0 < a - b \le 9$ or $b - a > 9$. How many ordered triples (x, y, z) of elements of S have the property that $x \succ y$, $y \succ z$, and $z \succ x$?

(A) 810 **(B)** 855 **(C)** 900 **(D)** 950 **(E)** 988

6.5 Trigonometry and Logarithms

1. 2010 AMC 12B Problem 13

In $\triangle ABC$, $\cos(2A - B) + \sin(A + B) = 2$ and $AB = 4$. What is BC?

(A) $\sqrt{2}$ **(B)** $\sqrt{3}$ **(C)** 2 **(D)** $2\sqrt{2}$ **(E)** $2\sqrt{3}$

2. 1993 AHSME Problem 23

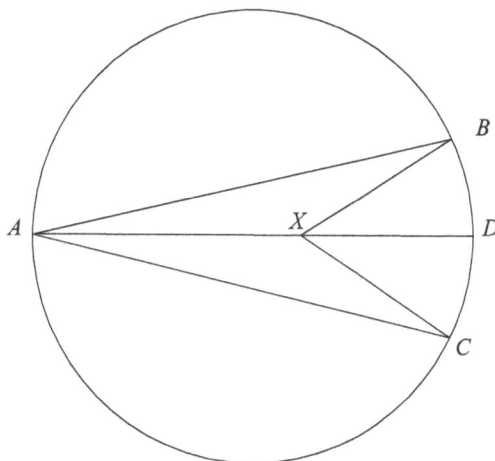

Points A, B, C and D are on a circle of diameter 1, and X is on diameter $\overline{AD}$.

If $BX = CX$ and $3\angle BAC = \angle BXC = 36°$, then $AX =$

(A) $\cos(6°)\cos(12°)\sec(18°)$ **(B)** $\cos(6°)\sin(12°)\csc(18°)$ **(C)** $\cos(6°)\sin(12°)\sec(18°)$

(D) $\sin(6°)\sin(12°)\csc(18°)$ **(E)** $\sin(6°)\sin(12°)\sec(18°)$

3. 2003 AMC 12B Problem 21

An object moves 8 cm in a straight line from A to B, turns at an angle α, measured in radians and chosen at random from the interval $(0, \pi)$, and moves 5 cm in a straight line to C. What is the probability that $AC < 7$?

(A) $\dfrac{1}{6}$ **(B)** $\dfrac{1}{5}$ **(C)** $\dfrac{1}{4}$ **(D)** $\dfrac{1}{3}$ **(E)** $\dfrac{1}{2}$

4. 2004 AMC 12A Problem 21

If $\sum_{n=0}^{\infty} \cos^{2n}\theta = 5$, what is the value of $\cos 2\theta$?

(A) $\dfrac{1}{5}$ **(B)** $\dfrac{2}{5}$ **(C)** $\dfrac{\sqrt{5}}{5}$ **(D)** $\dfrac{3}{5}$ **(E)** $\dfrac{4}{5}$

5. 2006 AMC 12B Problem 24

Let S be the set of all point (x, y) in the coordinate plane such that $0 \le x \le \frac{\pi}{2}$ and $0 \le y \le \frac{\pi}{2}$. What is the area of the subset of S for which

$$\sin^2 x - \sin x \sin y + \sin^2 y \le \frac{3}{4}?$$

(A) $\dfrac{\pi^2}{9}$ **(B)** $\dfrac{\pi^2}{8}$ **(C)** $\dfrac{\pi^2}{6}$ **(D)** $\dfrac{3\pi^2}{16}$ **(E)** $\dfrac{2\pi^2}{9}$

6. 2002 AMC 12B Problem 22

For all integers n greater than 1, define $a_n = \frac{1}{\log_n 2002}$. Let $b = a_2 + a_3 + a_4 + a_5$ and $c = a_{10} + a_{11} + a_{12} + a_{13} + a_{14}$. Then $b - c$ equals

(A) -2 **(B)** -1 **(C)** $\dfrac{1}{2002}$ **(D)** $\dfrac{1}{1001}$ **(E)** $\dfrac{1}{2}$

7. 2014 AMC 12A Problem 18

The domain of the function $f(x) = \log_{\frac{1}{2}}(\log_4(\log_{\frac{1}{4}}(\log_{16}(\log_{\frac{1}{16}} x))))$ is an interval of length $\frac{m}{n}$, where m and n are relatively prime positive integers. What is $m + n$?

(A) 19 **(B)** 31 **(C)** 271 **(D)** 319 **(E)** 511

8. 2011 AMC 12B Problem 17

Let $f(x) = 10^{10x}$, $g(x) = \log_{10}\left(\frac{x}{10}\right)$, $h_1(x) = g(f(x))$, and $h_n(x) = h_1(h_{n-1}(x))$ for integers $n \geq 2$. What is the sum of the digits of $h_{2011}(1)$?

(A) 16081 **(B)** 16089 **(C)** 18089 **(D)** 18098 **(E)** 18099

9. 1997 AHSME Problem 21

For any positive integer n, let

$$f(n) = \begin{cases} \log_8 n, & \text{if } \log_8 n \text{ is rational,} \\ 0, & \text{otherwise.} \end{cases}$$

What is $\sum_{n=1}^{1997} f(n)$?

(A) $\log_8 2047$ **(B)** 6 **(C)** $\dfrac{55}{3}$ **(D)** $\dfrac{58}{3}$ **(E)** 585

10. 2006 AMC 12B Problem 20

Let x be chosen at random from the interval $(0, 1)$. What is the probability that $\lfloor \log_{10} 4x \rfloor - \lfloor \log_{10} x \rfloor = 0$? Here $\lfloor x \rfloor$ denotes the greatest integer that is less than or equal to x.

(A) $\dfrac{1}{8}$ **(B)** $\dfrac{3}{20}$ **(C)** $\dfrac{1}{6}$ **(D)** $\dfrac{1}{5}$ **(E)** $\dfrac{1}{4}$

11. 2014 AMC 12B Problem 25

Find the sum of all the positive solutions of $2\cos 2x \left(\cos 2x - \cos\left(\frac{2014\pi^2}{x}\right)\right) = \cos 4x - 1$.

(A) π **(B)** 810π **(C)** 1008π **(D)** 1080π **(E)** 1800π

6.6 Complex Number

1. 1981 AHSME Problem 24

Suppose that n is a positive integer and that $z + \frac{1}{z} = 2 \cos \theta$, where $0 < \theta < \pi$. What is the value of $z^n + \frac{1}{z^n}$?

(A) $2 \cos \theta$ **(B)** $2^n \cos \theta$ **(C)** $2(\cos \theta)^n$ **(D)** $2 \cos n\theta$ **(E)** $2^n (\cos \theta)^n$

2. 1983 AHSME Problem 17

The diagram shows several numbers in the complex plane. The circle is the unit circle centered at the origin. Which of these numbers might be the reciprocal of F?

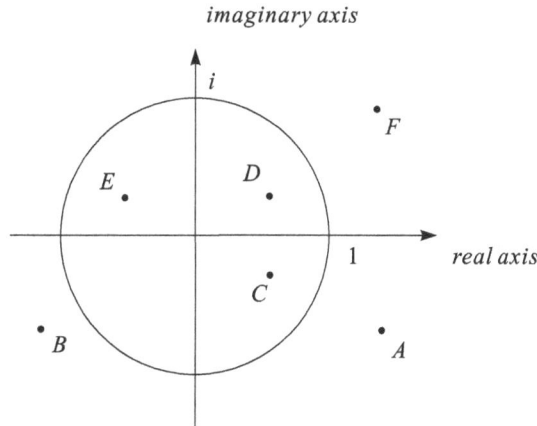

(A) A **(B)** B **(C)** C **(D)** D **(E)** E

3. 1992 AHSME Problem 15

Let $i = \sqrt{-1}$. Define a sequence of complex numbers by

$$z_1 = 0, \quad z_{n+1} = z_n^2 + i \text{ for } n \geq 1.$$

In the complex plane, how far from the origin is z_{111}?

(A) 1 **(B)** $\sqrt{2}$ **(C)** $\sqrt{3}$ **(D)** $\sqrt{110}$ **(E)** $\sqrt{2^{55}}$

4. 1974 AHSME Problem 17

If $i^2 = -1$, then $(1 + i)^{20} - (1 - i)^{20}$ equals

(A) -1024 **(B)** $-1024i$ **(C)** 0 **(D)** 1024 **(E)** $1024i$

5. 2009 AMC 12A Problem 21

Let $p(x) = x^3 + ax^2 + bx + c$, where a, b, and c are complex numbers. Suppose that

$$p(2009 + 9002\pi i) = p(2009) = p(9002) = 0$$

What is the number of nonreal zeros of $x^{12} + ax^8 + bx^4 + c$?

(A) 4 **(B)** 6 **(C)** 8 **(D)** 10 **(E)** 12

6. 1977 AHSME Problem 16

What is the value of $\sum\limits_{n=0}^{40} i^n \cos{(45 + 90n)}°$

(A) $\dfrac{\sqrt{2}}{2}$ (B) $-10i\sqrt{2}$ (C) $\dfrac{21\sqrt{2}}{2}$ (D) $\dfrac{\sqrt{2}}{2}(21 - 20i)$ (E) $\dfrac{\sqrt{2}}{2}(21 + 20i)$

7. 1988 AHSME Problem 21

The complex number z satisfies $z + |z| = 2 + 8i$. What is $|z|^2$? Note: if $z = a + bi$, then $|z| = \sqrt{a^2 + b^2}$.

(A) 68   (B) 100   (C) 169   (D) 208   (E) 289

8. 1985 AHSME Problem 23

If $x = \frac{-1+i\sqrt{3}}{2}$ and $y = \frac{-1-i\sqrt{3}}{2}$, where $i^2 = -1$, then which of the following is not correct?

(A) $x^5 + y^5 = -1$ (B) $x^7 + y^7 = -1$ (C) $x^9 + y^9 = -1$

(D) $x^{11} + y^{11} = -1$ (E) $x^{13} + y^{13} = -1$

9. 2008 AMC 12B Problem 19

A function f is defined by $f(z) = (4 + i)z^2 + \alpha z + \gamma$ for all complex numbers z, where α and γ are complex numbers and $i^2 = -1$. Suppose that $f(1)$ and $f(i)$ are both real. What is the smallest possible value of $|\alpha| + |\gamma|$?

(A) 1   (B) $\sqrt{2}$ (C) 2   (D) $2\sqrt{2}$ (E) 4

10. 2011 AMC 12B Problem 24

Let $P(z) = z^8 + \left(4\sqrt{3} + 6\right)z^4 - \left(4\sqrt{3} + 7\right)$. What is the minimum perimeter among all the 8-sided polygons in the complex plane whose vertices are precisely the zeros of $P(z)$?

(A) $4\sqrt{3} + 4$ (B) $8\sqrt{2}$ (C) $3\sqrt{2} + 3\sqrt{6}$

(D) $4\sqrt{2} + 4\sqrt{3}$ (E) $4\sqrt{3} + 6$

PART II

ESSAYS AND PRACTISE PROBLEMS

CHAPTER 7

COMBINATORICS ESSAY

Using the two basic principles, the multiplication principle and the sum rule, we can derive four formulas that enable us to count the number of samples of a given size (r) taken from a population of a given size (n). Which samples are different from one another depends on the context of the problem, and this discussion is intended to help you decide how to match these formulas with the problems that follow.

Throughout we use both the notation $\binom{n}{r}$ and C_r^n for the number $\dfrac{n!}{(n-r)!r!}$. That is

$$
\begin{aligned}
\binom{n}{r} &= \frac{n!}{(n-r)!r!} \\
&= \frac{n \cdot (n-1) \cdot (n-2) \cdots 2 \cdot 1}{((n-r) \cdot (n-r-1) \cdot (n-r-2) \cdots 2 \cdot 1)(r \cdot (r-1) \cdot (r-2) \cdots 2 \cdot 1)} \\
&= \frac{n \cdot (n-1) \cdot (n-2) \cdots (n-r+1)}{r \cdot (r-1) \cdot (r-2) \cdots 2 \cdot 1}.
\end{aligned}
$$

We are ready to discuss the general idea of counting samples taken from a population of objects. In doing such sampling we are allowed to make a distinction between the order in which the objects became a part of the sample or not. We are also allowed to sample with replacement or not. This leads to four different types of samples.

If we count as different two samples that have the same elements but in a different order, we call these **arrangements**, and if we don't distinguish on this basis, we call the samples **selections**. Let's classify each of the counting problems below using these two conditions.

conditions	order matters arrangements $()$	order does not matter selections $\{\}$
with repetitions	A: Exponations $E_r^n = n^r$ all $9 \cdot 10^4 = 90000$ five-digit numbers	D: Stars and Bars $Y_r^n = C_r^{n+r-1} = \begin{pmatrix} n+r-1 \\ r \end{pmatrix}$ all $\begin{pmatrix} 9+5-1 \\ 5 \end{pmatrix} = 1287$ four-digit nondecreasing numbers
without repetitions	B: Permutations $P_r^n = \dfrac{n!}{(n-r)!}$ all $9 \cdot P_4^9 = 9 \cdot 9!/5! = 27216$ five-digit numbers with five different digits	C: Combinations $C_r^n = \dfrac{n!}{(n-r)!r!} = \begin{pmatrix} n \\ r \end{pmatrix}$ all $\begin{pmatrix} 9 \\ 5 \end{pmatrix} = 126$ five-digit increasing numbers

The four examples below will help you in making decisions about which formulas to use. For convenience, we are discussing all five-digit numbers ($r = 5$). Our population is the set of decimal digits $P = \{0, 1, 2, 3, 4, 5, 6, 7, 8, 9\}$. The examples below correspond to the four quadrants in the table above.

Example 7.1

How many five-digit numbers are there?

Solution Each five-digit number can be viewed as a sample of five objects taken from P where order matters and the sampling is done with replacement. The only restriction is that the number cannot begin with the digit 0. Order matters because, for example, $12345 \neq 54321$ and we are sampling with replacement because we want to count 11222 as a sample. Therefore there are $9 \cdot 10^4 = 90000$ five-digit numbers.

Example 7.2

How many five-digit numbers have five different digits?

Solution You can see that the condition that changes is the replacement condition. Now we're sampling without replacement. Thus we have 9 choices for the first digit, 9 for the second, 8 for the third, and 7 for the fourth, and 6 for the final digit. But $9 \cdot P_4^9 = 9 \cdot 9!/5! = 27216$ as shown in the lower left quadrant above. So, there are 27216 five-digit numbers that have five different digits.

How many five-digit increasing numbers are there?

Solution This is the case where order does not matter and repetitions are not allowed. Note that zero cannot be one of our digits. Since it seems counter-intuitive, many students will ask at this point why order does not matter. The answer is that each five-digit collection can be arranged in exactly one way so that the digits are increasing. For example, the sample $2, 8, 6, 4, 3$ and the sample $3, 4, 2, 8, 6$ are the same because the same increasing number is produced from them. So if we start with a five element set, we can arrange these five digits into a five digit number in exactly $5! = 120$ ways. This means that the number we would have found in Example 7.2 had we omitted the digit 0 in the calculations is 120 times as big as the number we seek. That number P_5^9 is the number of five-digit numbers we can make using just the non-zero digits. Thus we have

$$C_5^9 = \binom{9}{5} = P_5^9 \div 5! = \frac{9!}{5!4!} = 126.$$

So, there are 126 five-digit increasing numbers.

Example 7.4

How many five-digit non-decreasing numbers are there?

Solution This is the case where order does not matter and repetitions are allowed. Notice that each five element collection can be arranged to form a non-decreasing number in exactly one way. This is what some call the 'stars and bars' model. Imagine that we want to code a five digit collection taken from P. Since an increasing number cannot use the digit 0, our population is the set of non-zero digits. Let $*$ denote a selected digit and $|$ means 'go on to the next digit'. Then $**||||*|*||*|$ means two 1's, no 2's, no 3's, no 4's, one 5, one 6, one 8 and no 7's or 9's. So the collection described is $\{1, 1, 5, 6, 8\}$ and the non-decreasing number it codes is 11568. Each coded collection is a string of 13 items each of which is a $*$ or a $|$. There are $\binom{13}{5} = 495$ such strings. So, there are 495 five-digit non-decreasing numbers.

Example 7.5 (This problem does not correspond directly to any of the formulas above.)

Four points A, B, C and D on one line segment are jointed by line segments to each of five points E, F, G, H, and I on a second line segment as shown. What is the maximum number of points **interior** to the angle belonging to two of these twenty segments.

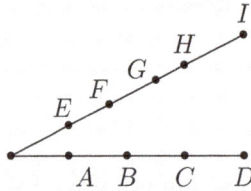

Solution Each point of intersection is determined by a taking a pair from the set $\{A, B, C, D\}$ and a pair from the set $\{E, F, G, H, I\}$.

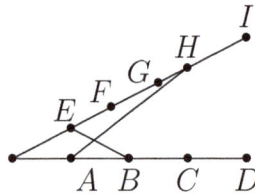

Notice that the point of intersection inside the angle is determined precisely by the two pairs $\{E, H\}$ and $\{A, B\}$ and that each such pair of pairs determines one such point. Thus there are $\binom{4}{2} \cdot \binom{5}{2} = 6 \cdot 10 = 60$ such points.

CHAPTER 8

PLACE VALUES AND FUSING DOTS ESSAY

8.1 Introduction

In the United States mathematics contests often present students with problems related to place value. One reason is that place value notation is the foundation of arithmetic. We use it to understand the arithmetic operations in both the systems of integers and decimals. Understanding arithmetic is essential to understanding algebra. After all, algebra is little more than generalized arithmetic. Yet many elementary school teachers do not appreciate these ideas, so contest preparers feel the need to include such problems in middle and high school contests.

8.2 Representing Integers

We discuss several methods of representing numbers, and several ways to understand these methods of representation. We begin with what is called *decimal representation*, the ordinary method we use to represent integers and fractions. Because the method of representation is an important starting point in learning the arithmetic of integers and decimals, we shall explore alternative methods of representation, that is, representation using bases other than our usual base 10. This is roughly akin to the idea that we do not really understand our own language until we learn a second language. Instead of trying to develop representation in an arbitrary base b, we select a specific base for the sake of clarity. This is base 5 representation. Later we will discuss other representation including those for which the base b is not a positive integer. We also explore the system of enumeration when b is a rational number but not an integer, and then when b is a negative integer. Finally, we'll also see that it is even possible for b to be irrational.

The *place value* interpretation of 4273 is $4000 + 200 + 70 + 3$, which is a *sum* of *multiples* of *powers* of 10. The relevant powers of 10 are $10^3 = 1000, 10^2 = 100, 10^1 = 10$, and $10^0 = 1$. Each one has a *coefficient* or multiplier, $4, 2, 7$, and 3, respectively. Thus $4 \cdot 10^3, 2 \cdot 10^2, 7 \cdot 10^1$, and $3 \cdot 10^0$ are **multiples** of **powers** of 10 and therefore 4273 is a **sum** of **multiples** of **powers** of 10.

Once we learn how to do arithmetic with single place numbers, we can use that knowledge along with the *distribution property* of multiplication over addition, to do arithmetic with decimal numbers in general. This represents a key virtue of place value: it enables arithmetic computation. The fact that the basic arithmetic operations can be efficiently performed by effectively teachable algorithms was the reason that the place value system, which was only introduced into Europe in the late middle ages (around 1200), supplanted the well entrenched system of Roman numerals. Here is an example. Find the product $23 \cdot 41$. First recognize each of these numbers as a place value number, $23 = 20 + 3$ and $41 = 40 + 1$. Then

$$
\begin{aligned}
23 \cdot 41 &= (20 + 3) \cdot (40 + 1) \\
&\stackrel{1}{=} (20 + 3)40 + (20 + 3)1 \\
&\stackrel{2}{=} 20 \cdot 40 + 3 \cdot 40 + 20 \cdot 1 + 3 \cdot 1 \\
&\stackrel{3}{=} 2 \cdot 10 \cdot 4 \cdot 10 + 3 \cdot 4 \cdot 10 + 2 \cdot 10 \cdot 1 + 3 \cdot 1 \\
&\stackrel{4}{=} 8 \cdot 10^2 + 12 \cdot 10 + 2 \cdot 10 + 3 \cdot 10^0 \\
&\stackrel{5}{=} 8 \cdot 10^2 + (10 + 2) \cdot 10 + 2 \cdot 10 + 3 \cdot 10^0 \\
&\stackrel{6}{=} 9 \cdot 10^2 + 1 \cdot 10^2 + 2 \cdot 10 + 3 \cdot 10^0 \\
&\stackrel{7}{=} 9 \cdot 10^2 + 4 \cdot 10 + 3 \cdot 10^0 \\
&\stackrel{8}{=} 943,
\end{aligned}
$$

where, we have used the distribution property of multiplication over addition in 1, 2 and 6; commutativity of multiplication and addition in 3 and 6; and place value notation in 6, 7, and 8. Of course we have also used the digit multiplication table 4 and the digit addition table in 7.

Another objective here is to establish methods of translating between decimal representations and base b representations. In other words, we are given a number expresses as a sum of multiples of powers of 10 and wish to rewrite the number as a **sums of multiples of powers** of b, where b is an integer bigger than 1. For convenience, let us assume for sections 2, 3, and 4 that $b = 5$. The same procedures work no matter what the value of b is, but fixing the value of b here makes discussion much easier. Of course there is also the problem of translating from a base b representation into a decimal representation, and this process is called *interpretation*.

The notation 2113_5 is interpreted as a sum of multiples of powers of 5, just as the decimal number 4273 was a sum of multiples of powers of 10. The *subscript* 5 must be attached unless we are using base 10, because 10 is the default value of the base. Thus $2113_5 = 2 \cdot 5^3 + 1 \cdot 5^2 + 1 \cdot 5^1 + 3 \cdot 5^0 =$

$250 + 25 + 5 + 3 = 283$. Thus we interpreted 2113_5 as 283. The reverse process, that of finding the base 5 representation of an integer expressed in decimal notation is harder and more interesting. There are two methods, (a) *repeated subtraction* and (b) *repeated division*. Each method has some advantages over the other.

To see how to use **repeated subtraction**, first make a list of all the integer powers of 5 that are not bigger than the number we are given. In the case of 283, we need the powers $5^0 = 1, 5^1 = 5, 5^2 = 25$, and $5^3 = 125$. Next repeatedly subtract the largest power of 5 that is less than or equal to the *current number* (which changes during the process). So we have $283 = 125 + 158$. At this point our current number becomes 158 and we repeat the process. Then $283 = 125 + 158 = 125 + 125 + 33$, and our current number is 33. Repeating the process on 33 gives $33 = 25 + 8$ and incorporating that in the above gives $283 = 2 \cdot 125 + 25 + 8 = 2 \cdot 125 + 1 \cdot 25 + 8$. Continuing this with 8 leads to $283 = 2 \cdot 5^3 + 1 \cdot 5^2 + 1 \cdot 5^1 + 3 \cdot 5^0$, which is a sum of multiples of powers of 5, just what we want. Thus $283 = 2113_5$, just as we saw above. Repeated subtraction has two advantages over the repeated division method. First, it is closely related to the definition, hence it leads to a better conceptualization. Second, it can be used in other situations when repeated division cannot, as in the case of Fibonacci representation.

The repeated division method requires that we repeatedly divide the given integer by base 5 and record the remainder at each stage. First we divide 283 by 5 to get $283 \div 5 = 56.6$. We can interpret this as $283 = 5 \cdot 56 + 3$, so the *quotient* is 56 and the *remainder* is 3. Notice that the remainder can never exceed 5 since in such a case the quotient would have been larger. Next divide the quotient by 5 and record the new quotient and the remainder. Thus $56 = 5 \cdot 11 + 1$. Repeat the process with the new quotient $11 = 5 \cdot 2 + 1$ and finally, $2 = 5 \cdot 0 + 2$. Next write the remainders in reverse order, $2, 1, 1$, and 3 to get 2113_5 as the base 5 representation of 283. You'll see why the order must be reversed in the following example.

Example 8.1 Repeated Division

Verify that $283 = 2113_5$.

Solution we can repeatedly replace each quotient with its value obtained during the division process. Thus

$$
\begin{aligned}
283 &= 5 \cdot 56 + 3 = 5(5 \cdot 11 + 1) + 3 \\
&= 5(5(5 \cdot 2 + 1) + 1) + 3 = 5(5 \cdot 5 \cdot 2 + 5 \cdot 1 + 1) + 3 \\
&= 5 \cdot 5 \cdot 5 \cdot 2 + 5 \cdot 5 \cdot 1 + 5 \cdot 1 + 3 \\
&= 2 \cdot 5^3 + 1 \cdot 5^2 + 1 \cdot 5^1 + 3 \cdot 5^0 = 2113_5.
\end{aligned}
$$

The advantage of repeated division is that it is computationally more efficient. Also, the method of justification can be applied in other situations (synthetic division and Euclidean algorithm). When we get to the section on fusing dots, you'll see yet another reason why it makes sense to record the remainders upon division by b.

8.3 Representing Fractions

In the paragraphs above, we saw two methods (algorithms) for writing a given integer in a base different from 10. Before we consider representing fractions, let's review the place value ideas in decimal notation. For example, 5.234 is, as in the first part, a sum of multiples of powers of 10. This time, the powers are (except one) negative exponents:

$$5.234 = 5 \cdot 10^0 + 2 \cdot 10^{-1} + 3 \cdot 10^{-2} + 4 \cdot 10^{-3}.$$

Using this interpretation as a guide, we can interpret 0.124_5 similarly, as a sum of multiples of (negative) powers of 5. Thus

$$
\begin{aligned}
0.124_5 &= 1 \cdot 5^{-1} + 2 \cdot 5^{-2} + 4 \cdot 5^{-3} \\
&= \frac{1}{5} + \frac{2}{25} + \frac{4}{125} = \frac{25 + 10 + 4}{125} \\
&= \frac{39}{125}.
\end{aligned}
$$

As in the discussion of integers, there are two methods for dealing with numbers in the range $0 < x < 1$. They are called (a) *repeated subtraction* and (b) *repeated multiplication*. As before, each has advantages over the other.

> **Example 8.2 Repeated Subtraction**
>
> Show that $\frac{39}{125} = 0.124_5$.

✎ **Solution** To use the method of repeated subtraction on $39/125$, first list the powers of 5 with negative integer exponents:

$$5^{-1} = 1/5, \ 5^{-2} = 1/25, \ 5^{-3} = 1/125, \dots .$$

Find the largest of these powers of 5 and subtract it from the original number. Thus $39/125 - 1/5 = 14/125$. Therefore, $39/125 = 1/5 + 14/125$. Now repeat the process on the number $14/125$. Note that $1/25 = 5/125$. Thus, $14/125 - 1/25 = 8/125$. Therefore, $14/125 = 1/25 + 9/125$. Putting this together with the arithmetic above, we have

$$\frac{39}{125} = \frac{1}{5} + \frac{1}{25} + \frac{1}{25} + \frac{4}{125}.$$

Again dealing with the extra part, $4/125 - 1/125 = 3/125$, etc . At this point we can anticipate the final arithmetic:

$$\begin{aligned}
\frac{39}{125} &= \frac{1}{5} + \frac{1}{25} + \frac{1}{25} + \frac{1}{125} + \frac{1}{125} + \frac{1}{125} + \frac{1}{125} \\
&= 1 \cdot 5^{-1} + 2 \cdot 5^{-2} + 4 \cdot 5^{-3} \\
&= 0.124_5.
\end{aligned}$$

The method of repeated multiplication is much quicker and does not require so much fraction arithmetic.

Example 8.3 Repeated Multiplication

Find the base 5 representation of $39/125$.

✎ **Solution** We repeatedly multiply by 5. Following each multiplication by 5, split the result into its integer part and its fractional part:

$$\frac{39}{125} \cdot 5 = \frac{39 \cdot 5}{25 \cdot 5} = \frac{39}{25} = 1 + \frac{14}{25}.$$

Each integer part is a digit in the representation. Thus $39/125 = 0.1\ldots_5$. Now repeat the process using the new fractional part, $14/25$:

$$\frac{14}{25} \cdot 5 = \frac{14}{5} = 2 + \frac{4}{5}.$$

Thus $39/125 = 0.12\ldots_5$. Repeating the process, $\frac{4}{5} \cdot 5 = 4 + 0$. Since the fractional part is 0, we are done (why?). Thus, $\frac{39}{125} = 0.124_5$.

Of course, not all rational numbers have base 5 representations that terminate (i.e., end in all 0's from some point on). But there is an easy way to tell, and a great notation to use when the representation does not terminate. Consider the problem of finding the *binary* (that is, base 2) representation of $\frac{1}{3}$. Using repeated multiplication, we get $\frac{1}{3} \cdot 2 = 0 + \frac{2}{3}$. Then $\frac{2}{3} \cdot 2 = 1 + \frac{1}{3}$. Thus we see the same fractional part $\frac{1}{3}$ occurs again. The first two digits are 0 and 1, so we have $\frac{1}{3} = 0.01\ldots_2$, but we can see that the block 01 continues to recur. The slick way to write this number $0.01010101\ldots$ is $0.\overline{01}_2$. When the representation repeats in blocks, the number can be regarded as the sum of an infinite geometric series. In this case it is $2^{-2} + 2^{-4} + 2^{-6} + \cdots$. There is a formula for finding the sum of the geometric series $a + ar + ar^2 + ar^3 + \cdots$. It is $\frac{a}{1-r}$, and this holds whenever $|r| < 1$. Thus $2^{-2} + 2^{-4} + 2^{-6} + \cdots = \frac{2^{-2}}{1-2^{-2}} = \frac{1/4}{3/4} = \frac{1}{3}$, just as we knew.

8.4 Fusing Dots

Many thanks to Australian mathematician James Tanton for the idea of exploding dots, the name of which we have changed to Fusing Dots. We're going to explore several machines that enable us to represent positive integers and some other real numbers in some odd ways. Initially, we are given a two-way infinite tape with empty squares, with a heavy line (a vertical bar) at one place on the tape:

⬜...⬜⬜⬜❚⬜⬜ ...

To represent a number n, we put n dots in the square just to the left of the bar, and let the machine go to work. This square, also called a box, is named the *unit box*.

8.4.1 The base 5 machine

The $\boxed{1 \leftrightarrow 5}$, is called the base 5 machine. In this machine, whenever five dots occupy the same square, they are erased (they 'fuse') and they are replaced with one dot in the square to their left. Thus the five dots in ⬜⬜ fuse to become one dot in ⬜⬜. There will also be times when we need to reverse the process in which case one dot in a square is replaced by 5 dots in the square to the right. We'll call this process *explosion*. Thus ⬜⬜ explodes to become ⬜⬜.

How can we use this machine to represent a positive integer, like 27? What happen when we put 27 dots in the unit box? The answer is that we can assign to each box to the left of the bar a value. The integer is the sum of the products of the values times the number of dots in each box with the given value. For example, ⬜⬜⬜❚ $\cdots$, has the value $25 + 0 + 2 = 27$. We agree to write this as 102 instead of putting dots in boxes. Here is another example. Go back to the example we saw above, but add a few dots to the right of the bar. ⬜⬜⬜❚ $\cdots$, has the value $25+0+2+3/5 = 27.6$. Example 8.11 will help you understand base 5 representation.

How can we use this model to add two numbers? Find the values of the numbers represented as 2432 and 2341. Find the sum of 2341 and 2432 using the exploding dot model. Solution. To add two positive integers, m and n, *amalgamate* their representations, and use the $\boxed{1 \leftrightarrow 5}$ idea to build the representation of the sum $m+n$. In the special case, we have 4773 which of course is not a legitimate representation. So, fusing as we go, $4773 \Rightarrow 4823 \Rightarrow 5323 \Rightarrow 10323$.

How can we use this model to subtract two numbers? In particular, work out $2432 - 2341$. Solution. The main idea is the *antidot*, ∘, whose value is the negative of the dot in the same box. That is, dots and antidots collide to produce zero. So, if $m > n$, then $m - n$ is represented by the final set of dots we obtain by representing $-n$ using antidots and then adding m and $-n$. Now, $2432 - 2341$ is represented by ⬜⬜⬜⬜❚ $\cdots$. Can you finish the job?

How can we use this model to multiply two numbers? Solution. What does multiplication by 5 look like here. One way to think about this is that every dot (and antidot) turns into five. And of course each five dots fuse. So the effect is to push all the dots one unit to the left. This means we get a zero added at the right end. So how does multiplication in general work?

How can we use this model to understand fractions? For example, consider what happens when we *explode* a dot repeatedly. Thus ⬜ becomes ⬜. And this becomes ⬜. So, we have $1 = 0.5 = 0.45 = 0.445 = 0.4445 = \ldots = 0.\overline{4}$. In fact, this infinite geometric series converges to $\frac{4/5}{1-1/5} = 1$, as we know.

8.4.2 Binary Representation

The $\boxed{1 \leftrightarrow 2}$ is called base 2 machine. In this machine, whenever two dots occupy the same square, they are fused together to be one dot in the square to their left. Thus ⬜ becomes ⬜. In case we start with 7 dots, we get the following string ⬜ $\mapsto$ ⬜ $\mapsto$ ⬜ $\mapsto$ ⬜ $\mapsto$ ⬜. Instead of constructing a string of squares and dots, we call this representation 111. As an exercise, see what you get for 19 dots. Also, check to see if the order in which the fusions take place affects the final distribution of dots. Since each dot in a square is worth two dots in the square to its right, we can assign values to each square to see what number is represented. For example, the dot configuration ⬜, has the value $8 + 2 + 1 = 11$. Of course it is not a surprise to us that this is just binary representation.

8.4.3 Fractional Bases

The $\boxed{2 \leftrightarrow 3}$, is called the three-halves machine. In this machine, whenever three dots occupy the same square, they fuse together to become two dots in the square to their left. Let's work out the notation for each of the numbers from 1 to 15.

n	1	2	3	4	5	6	7	8	9	10	11	12	13	14	15
$R(n)$	1	2	20	21	22	210	211	212	2100	2101	2102	2120	2121	2122	21010

Work your way up to the representation for 24. Notice that the number of digits in the representation jumps as we move to $3, 6, 9$ and 15. Where is the next jump. Why? Is this machine a base-b representation machine for some number b? If so, then ⬜ would have the value $2b^3 + b^2 + 1$. Compute the value of the representation 2101 without help from the chart above.

Realizing that each pair of dots in a box is worth three in the next box, we can derive the equations $2b = 3, 2b^2 = 3b, 2b^3 = 3b^2$, etc, all of which give us $b = 3/2$.

Find the representation of 123 for this machine. Solution. Again using digits in place of dots, $123 = 82\,0 = 54\,1\,0 = 36\,0\,1\,0 = 24\,0\,0\,1\,0 = 16\,0\,0\,0\,1\,0 = 10\,1\,0\,0\,0\,1\,0 = 61\,1\,0\,0\,0\,1\,0 = 4\,0\,1\,1\,0\,0\,0\,1\,0 = 2\,1\,0\,1\,1\,0\,0\,0\,1\,0$, so the base three halves representation of 123 is 2101100010.

Alternatively, solve the equation $\log 1.5^k = \log 123$ to get an idea of how many digits k to use, then find the value of $210...0$, where there are k digits.

8.4.4 The polynomial machine.

The $\boxed{1 \leftrightarrow x}$, is called the polynomial machine. In this machine, whenever x dots occupy the same square, they fuse, and they are replaced with one dot in the square to their left. This leads to polynomial arithmetic. Let's work out an example of polynomial division using the $\boxed{1 \leftrightarrow x}$ machine.

1. Represent $3x^2 + 8x + 4$ in the $\boxed{1 \leftrightarrow x}$ machine.

2. Represent $x + 2$ in the $\boxed{1 \leftrightarrow x}$ machine.

3. Now find all instances of $\boxed{\cdot \mid :}$ in $\boxed{\begin{array}{c|c|c} \cdot\cdot & {\vdots\vdots\atop\vdots\vdots} & :: \end{array}}$.

4. Next try $\boxed{\begin{array}{c|c|c} \cdot & \circ & 88 \end{array}} \div \boxed{\begin{array}{c|c} \cdot & : \end{array}}$.

8.4.5 Fibonacci Representation and Irrationals bases

The $\boxed{1 \leftrightarrow 1,1}$ is called base ϕ machine and produces Fibonacci representation. For each of the next two problems, we have the same 'fusion scheme': $\boxed{\begin{array}{c|c|c} & \cdot & \end{array}} + \boxed{\begin{array}{c|c|c} & & \cdot \end{array}} = \boxed{\begin{array}{c|c|c} \cdot & & \end{array}}$. In other words, when dots belong to adjacent boxes, they fuse to give a dot in the next box over: $\boxed{\begin{array}{c|c|c} & \cdot & \cdot \end{array}} = \boxed{\begin{array}{c|c|c} \cdot & & \end{array}}$.

Let's call this the $\boxed{1 \leftrightarrow 1,1}$ machine. In the final representations, we are not allowed to have more than one dot in a box. In the first part, we also need a two-way infinite row of boxes. You will see why we need both directions as we start to count. Of course, 1 is represented as usual, $\boxed{\begin{array}{c|c|c} & \cdot & \end{array}}$ The bold vertical segment represents a special location, which for base b, we call a *radix point*.

Instead of using dots in boxes, its more convenient from here on to express integers as digit strings. So 2 is $2 \Rightarrow 1.11 \Rightarrow 10.01$. The box/dot diagram here is $\boxed{\begin{array}{c|c|c} & \cdot\cdot & \end{array}} \Rightarrow \boxed{\begin{array}{c|c|c} & \cdot & \cdot & \cdot \end{array}} \Rightarrow \boxed{\begin{array}{c|c|c} \cdot & & \cdot \end{array}}$. In words, one of the two dots in the first box exploded producing dots in the two previous boxes. So $2 = 10.01$. Then $3 = 2 + 1 = 10 + 0.01 = 11.01 \Rightarrow 100.01, 4 = 3 + 1 = 100.01 + 1 = 101.01$. Now 5 is tricky: $5 = 4 + 1 = 101.01 + 1 = 102.01 \Rightarrow 101.12 \Rightarrow 101.1111 \Rightarrow 110.0111 \Rightarrow 1000.1001$. Then $6 = 1001.1001 \Rightarrow 1010.0001$. A representation is complete only if all digits are either 0 or 1 and no two adjacent digits are 1.

Is this a base system in the usual sense. In particular, is there a real number b for which

$$6 = b^3 + b + b^{-4}?$$

Example 8.4

Find the representations of the next 5 integers, $7, 8, 9, 10$, and 11.

Solution Working from $6 = 1010.0001$, we have $7 = 10000.0001$, $8 = 10001.0001$, $9 = 10002.0001 = 10001.1101 = 10010.0101$, $10 = 10100.0101$, and finally $11 = 10101.0101$. To answer the question above, let $b = \frac{1+\sqrt{5}}{2}$ and compute $b^4 + b^2 + 1 + b^{-2} + b^{-4}$ to get exactly 11. Indeed the machine in this problem is the base golden mean (i.e. ϕ) machine.

Here's the $\boxed{1 \leftrightarrow 1, 1}$ machine with a *black hole*. The only difference between this machine and the base ϕ machine is the black hole. It only takes left infinite strings of boxes with two extra boxes to the right of the radix point, one of which is a black hole: $\begin{array}{|c|c|c|} \hline & & \infty \\ \hline \end{array}$ Here's how this works. Dots in the box marked ∞ at the end disappear. So, $1 = 1, 2 = 2.00 \Rightarrow 1.11 \Rightarrow 10.01 \Rightarrow 10.00 = 10$, $3 = 2 + 1 = 10 + 1 = 11 = 100$, $4 = 12.00 \Rightarrow 11.11 \Rightarrow 100.11 \Rightarrow 101.00$. To find the representation of 5, start with the representation 4, and add one: $5 = 102.00 \Rightarrow 101.11 \Rightarrow 110.01 \Rightarrow 1000.01 \Rightarrow 1000.00 = 1000$.

Example 8.5

Find the representations of the next 5 integers, $7, 8, 9, 10$, and 11.

Solution $7 = 1010; 8 = 10000; 9 = 10001; 10 = 10010$ and $11 = 10100$.

The Fibonacci numbers $F_1 = 1, F_2 = 2, F_3 = 3, F_4 = 5 \ldots$ are defined so that after the first two, every one is the sum of the previous two. In other words, $F_1 = 1, F_2 = 2$ and $F_{n+2} = F_n + F_{n+1}$. Thus the sequence is $1, 2, 3, 5, 8, 13, 21, 34, 55, 89 \ldots$. In the case of Fibonacci representation, we need only two digits, 0 and 1. These represent the absence or presence of the corresponding Fibonacci number. To represent a number in Fibonacci representation, use the method of repeated subtraction.

Example 8.6 Fibonacci Representation

Find the Fibonacci representation of 100.

Solution To find the Fibonacci representation of 100, we find the largest Fibonacci number less than or equal to 100. Then subtract it and repeat the process. Thus $100 = 89 + 11$. Thus $100 = 89 + 11 = 89 + 8 + 3 = 1000010100_f$. Of course, the 1's tell us which Fibonacci numbers are added, and the 0's tell us to leave out the number: 1000010100_f means $1F_{10} + 0F_9 + 0F_8 + 0F_7 + 0F_6 + 1F_5 + 0F_4 + 1F_3 + 0F_2 + 0F_1$.

Notice that the representation 1000010100_f has at least one 0 between each pair of 1's. Try to figure out why this is always the case before reading on. Of course, you have realized by now that this is the representation provided by the machine with the black hole above. You're sure to be curious as to why we can just throw away the dot in the black hole in case the computation puts one there. Try to figure this out on your own. How can we do arithmetic with numbers represented this way?

Example 8.7 Fibonacci Arithmetic

Addition is not very hard. Let's try the addition $87 + 31$. In the notation we (slightly) abuse the notation by using the coefficient 2 at times.

	89	55	34	21	13	8	5	3	2	1
87		1	0	1	0	1	0	1	0	0
+31				1	0	1	0	0	1	0
		1	0	2	0	2	0	1	1	0
		1	0	2	0	2	1	0	0	0
		1	0	2	1	1	0	0	0	0
		1	1	1	0	1	0	0	0	0
118	1	0	0	1	0	1	0	0	0	0

The addition process repeatedly makes use of the fact that the sum of two successive Fibonacci numbers is the next one. In the representation, therefore, you never need to have two successive 1's. Another example might be helpful here. How would you carry out $21 + 21$? That is $1000000_f + 1000000_f = 2000000_f = 1110000_f = 10010000_f = 34 + 8 = 42$. Can you devise an algorithm for multiplication?

8.4.6 Cantor's Representation

This deals with factorial representation, sometimes called Cantor's Representation. It was problem 25 on the 1999 AMC 12, which was then called the American High School Math Exam (AHSME).

Example 8.8

There are unique integers $a_2, a_3, a_4, a_5, a_6, a_7$ such that

$$\frac{5}{7} = \frac{a_2}{2!} + \frac{a_3}{3!} + \frac{a_4}{4!} + \frac{a_5}{5!} + \frac{a_6}{6!} + \frac{a_7}{7!},$$

where $0 \le a_i < i$ for $i = 2, 3, \ldots, 7$. Find a_2, a_3, a_4, a_5, a_6, and a_7.

✎ **Solution** Multiply both sides of the equation by 7! to obtain

$$3600 = 2520a_2 + 840a_3 + 210a_4 + 42a_5 + 7a_6 + a_7.$$

It follows that $3600 - a_7$ is a multiple of 7, which implies that $a_7 = 2$. Thus,

$$\frac{3598}{7} = 514 = 360a_2 + 120a_3 + 30a_4 + 6a_5 + a_6.$$

Reason as above to show that $514 - a_6$ is a multiple of 6, which implies that $a_6 = 4$. Thus, $510/6 = 85 = 60a_2 + 20a_3 + 5a_4 + a_5$. Then it follows that $85 - a_5$ is a multiple of 5, whence $a_5 = 0$. Continue in this fashion to obtain $a_4 = 1, a_3 = 1$, and $a_2 = 1$. And the other values are $a_5 = 0, a_6 = 4$ and $a_7 = 2$.

Another way to get these numbers is to use the greedy algorithm. Note that

$$\frac{5}{7} = \frac{a_2}{2!} + \frac{a_3}{3!} + \frac{a_4}{4!} + \frac{a_5}{5!} + \frac{a_6}{6!} + \frac{a_7}{7!},$$

but the sum of the last five fractions is very small. Therefore we must have $a_2 = 1$, whence

$$\frac{3}{14} = \frac{a_3}{3!} + \frac{a_4}{4!} + \frac{a_5}{5!} + \frac{a_6}{6!} + \frac{a_7}{7!},$$

and we can continue this way to get all the values of the a_is.

8.4.7 Place Value with Negative Bases

The $\boxed{-1 \leftarrow 4}$ $\boxed{1 \leftarrow -4}$ machine enables us to represent numbers and do arithmetic in base negative 4. Here we allow ourselves the digits $0, 1, 2, 3$. Let us first **interpret** a number written in base -4. For example take 113.3_{-4}. We interpret this as a sum of multiples of powers of -4: $1 \cdot (-4)^2 + 1 \cdot (-4)^1 + 3 \cdot (-4)^0 + 3 \cdot (-4)^{-1} = 16 - 4 + 2 - 3/4 = 13.25$. Thus, we write $13.25 = 113.3_{-4}$. The methods for finding the base negative four representation of a positive integer are interesting. Also of interest are methods for finding the base -4 representation of rational numbers r satisfying $0 < r < 1$. We can find the base -4 representation of 13.25 by combining these two methods.

Our machine is denoted by $\boxed{-1 \leftarrow 4}$ machine because whenever four dots accumulate in a box, they fuse, causing an anti-dot to be formed in the next box to the left. Similarly when four antidots accumulate, they fuse to give a dot in the box to the left. This machine can be drawn as follows:

and

> **Example 8.9 Repeated Division**
>
> Show that $477 = 21211_{-4}$.

✎ **Solution** To see why $477 = 21211_{-4}$, we can repeatedly replace each quotient with its value obtained during the division process. Its important to remember that the remainders cannot be negative numbers. Thus

$$
\begin{aligned}
477 &= -4 \cdot -119 + 1 \\
&= -4(-4 \cdot 30 + 1) + 1 \\
&= -4(-4(-4 \cdot -7 + 2) + 1) + 1 \\
&= -4(-4(-4(-4 \cdot 2) + 1 + 1) + 1 \\
&= -4(-4(-4(-4 \cdot 2 + 1) + 2) + 1) + 1 \\
&= 2(-4)^4 + 1(-4)^3 + 2(-4)^2 + 1(-4)^1 + 1(-4)^0 \\
&= 21211_{-4}.
\end{aligned}
$$

Here's how the fusing dot machine above would process the number 477. First, there would be 119 fusions that would produce 119 antidots in the second box and one dot in the right box. Then 29 fusions would take place, producing 29 dots in the third box and 3 antidots in the second box. Then 7 fusions would take place to produce 7 antidots in the fourth box with one dot left in the third box: [dot diagram] The final fusion produces: [dot diagram]. So, can we say that the base -4 representation of 477 is $1 - 31 - 31$? Of course not. We can use only positive digits. So what can we do? Try adding some dot-antidot pairs. [dot diagram]. From here its easy: [dot diagram]. In other words, 21211_{-4}.

The algorithms for finding the base -4 representation of fractions is even more interesting. The following algorithm is related to repeated multiplication. To find the base -4 representation of $7/20$, first note that our number is positive, so it looks like $1.abcd....$ That means the $.abcd...$ has value $7/20 - 1 = -13/20$. Multiply $-13/20$ by -4 to get $52/20 = 13/5 = 3 - 2/5$, so the digit a is 3. Then multiply $-2/5$ by -4 to get $8/5$ which we can write as $2 - 2/5$. Our representation is $1.32cd....$ Now $-2/5 \cdot -4 = 8/5$ again, and we can see that the digit 2 repeats. Thus $7/20 = 1.3\overline{2}_{-4}$. Can you prove that this is correct? Which rational numbers less than 1 require a digit 1 in the unit's position? To see what it is, ask yourself the question, What is the largest rational number representable as $0.x_1 x_2 \ldots$?

In algebra, you learn a method for converting a repeating decimal to a ratio of two integers. We can do that here also. Let $t = 1.3\overline{2}_{-4}$. Then $16t = 132.\overline{2}$ and $16t - t = 15t = 132.2_{-4} - 1.3_{-4} = 132.3 = 21/4$. So $t = 21/4 \div 15 = 7/20$. This algorithm is not perfect, however because the subtraction idea can lead to digits larger than 3. Can you devise another method that avoids this problem?

Here's an idea. To find the base -4 representation of a fraction, we repeatedly multiply by 16, and produce two digits at a time. This way we can avoid the difficulties posed by the negative numbers.

Example 8.10

Find the base -4 representation of $1/3$.

Solution Since multiplication by 16 just moves the radix point two places, we get an idea of the answer by finding the base -4 representation of $16/3$, which begins 132..... Thus we try the number $x = 1.\overline{32}_{-4}$. In the usual way, multiply by $(-4)^2 = 16$ and subtract to get $16x - x = 132.\overline{32} - 1.\overline{32} = 131$, which means that $15x = 5$. So $x = \frac{1}{3}$. Alternatively, write one third as a sum of powers of 4: $1/3 = (1/4) + (1/4)^2 + (1/4)^3 \cdots = 1.3 + .01 + .013 + .0001 + .00013 + .00001 + \cdots = 1.\overline{32}_{-4}$.

Example 8.11

Infinitely many empty boxes (also called cells or squares), each capable of holding four balls are lined up from right to left. At each step we place a ball in the rightmost box that still has room for it and at the same time, we empty all the boxes to the right of it. How many balls are in the boxes after 2010 steps.

Solution This is simply base 5 representation. By repeated division, we find that $2010 = 31020_5$, so there are $3 + 1 + 0 + 2 + 0 = 6$ balls in the boxes at the end.

CHAPTER 9

SYMMETRY ESSAY

9.1 Introduction

This discussion is about symmetry, and its use in plane geometry. But the properties and functions we'll discuss are objects encountered in algebra, so this discussion can be considered a link between the two branches of mathematics. We'll study *binary relations*, which are sets of ordered pairs of real numbers. Technically, a relation from R to R, where R represents the set of real numbers, is a set of ordered pairs (x, y), where x and y are real numbers. For brevity, a relation from R to R is said to be a relation *on* R. The functions (relations that pass the vertical line test) that we study in high school are all relations from R to R, so this essay will enable you to distinguish functions on R from relations that are not functions. Such binary relations may exhibit various types of symmetry. We'll define a few of these below. But first, there are four important properties that any binary relation might or might not satisfy. These four are:

R Reflexivity, for all x, $x \sim x$. Another way to say this is that for all x, the ordered pair (x, x) belongs to the relation.

S Symmetry, for all x, y, if $x \sim y$, then $y \sim x$.

T Transitivity, for all x, y, z, if $x \sim y$ and $y \sim z$, then $x \sim z$.

A Antisymmetry, for all x, y, if $x \sim y$ and $y \sim x$, then $x = y$.

Here, we are using a convenient notation. A relation is a set of ordered pairs, so each ordered pair (x, y) either belongs or does not belong. Instead of writing $(x, y) \in \sim$, where $\sim$ represents the relation in question and $\in$ means "contained in", we write simply $x \sim y$. Relations that satisfy R, S and T are called *equivalence relations*, and those satisfying R, A, and T are called *partial orders*. Both these types of relations are very important and useful throughout mathematics. There are three other symmetry properties that a relation in the plane can exhibit.

S_0 Symmetry with respect to the origin, for all x, y, if $x \sim y$, then $-x \sim -y$. In other words, if (x, y) belongs to the relation, then so does $(-x, -y)$.

S_x Symmetry with respect to the x-axis, for all x, y, if $x \sim y$, then $x \sim -y$.

S_y Symmetry with respect to the y-axis, for all x, y, if $x \sim y$, then $-x \sim y$.

S Symmetry, repeated for emphasis. This is symmetry with respect to the line $y = x$.

9.2 Definitions

Recall the two functions *floor* ($\lfloor \rfloor$) and *fractional part* ($\langle \rangle$), defined by $\lfloor \rfloor$ is the greatest integer that is less than or equal to x, and $\langle x \rangle = x - \lfloor x \rfloor$. For example, $\lfloor \pi \rfloor = 3$ and $\langle \pi \rangle = \pi - 3$, $\lfloor -\pi \rfloor = -4$ and $\langle -\pi \rangle = -\pi - (-4) = 4 - \pi$. Also, notice that for any real number x, $x = \lfloor x \rfloor + \langle x \rangle$. For example, $\lfloor 3.15 \rfloor + \langle 3.15 \rangle = 3 + 0.15 = 3.15$. The *ceiling function*, denoted $\lceil x \rceil$ is defined by $\lceil x \rceil = -\lfloor -x \rfloor$, or the smallest integer that is greater than or equal to x. Finally, the *absolute value* of a number x is defined by $|x| = \sqrt{x^2}$.

Suppose $\mathcal{T}$ is a relation on R. For each real number x, the set $[x]$, called the **cell** of x, is defined by

$$[x] = \{y \mid x \mathcal{T} y\}.$$

Now it may turn out that each x belongs to the cell it names, and it may also turn out that any two cells which overlap are identical. The set of cells is then called a **partition** of R. This happens precisely when $\mathcal{T}$ is an equivalence relation. But this is really just the tip of the iceberg. Take a course in discrete math to get the full picture.

9.3 Examples

Example 9.1

Consider the relation $\leq$. As usual, instead of $(0, 1) \in \leq$ we write the much simpler $0 \leq 1$. Now the three properties (a) $x \leq x$, (b) $x \leq y$ and $y \leq x$ imply that $x = y$, and (c) $x \leq y, y \leq z$ imply $x \leq z$ are actually axioms in most versions of the definition of the real number system. Thus the real numbers are *partially ordered* ($\leq$ satisfies R, A, T), by the relation $\leq$. The graph of the relation $\leq$ in the coordinate plane is the set of points below the line $y = x$, which is a half plane.

Example 9.2

Consider the relation defined by $T = \{(x, y)| \lfloor x \rfloor = \lfloor y \rfloor\}$. The symbol | means "such that". In other words, T contains any ordered pair (x, y) that satisfies the equation $\lfloor x \rfloor = \lfloor y \rfloor$. Which of the properties $R, S, T, A, S_0, S_x, S_y$ does T satisfy?

Solution Since T is defined by $\lfloor x \rfloor = \lfloor y \rfloor$, it satisfies R, S, T. But $(\frac{1}{2}, \frac{2}{3})$ belongs to T and of course $(\frac{2}{3}, \frac{1}{2})$ does as well, but $1/2 \neq 2/3$, so T does not satisfy antisymmetry. With a little work we can show that T satisfies just S_0, but not S_x or S_y.

Example 9.3

The solution set $S = \{(x, y)|(xy - 1)(x^2 + y^2 - 2) = 0\}$ divides the plane into some bounded and some unbounded regions. How many connected regions comprise the complement (the ordered pairs not contained in S) of S?

Solution The answer is 5. The graph of $xy - 1 = 0$ or $xy = 1$ is our favorite equilateral hyperbola, and the graph of $x^2 + y^2 = 2$ is a circle centered at the origin of radius $\sqrt{2}$, and it is tangent to the two branches of the hyperbola. The five points $(\pm 2, \pm 2), (0, 0)$ all belong to different regions determined by S.

Can you work out the properties of this relation: $R, S, T, A, S_0, S_x, S_y$? Not R, S, not T, not A, not S_x and not S_y. To see that S is not transitive, for example, note that $(0, 2)$ and $(2, 0)$ both belong to S but $(0, 0)$ does not.

Example 9.4

Into how many regions does the solution set Q of

$$xy(y - x)(y + x)(x^2 + y^2 - 1) = 0$$

divide the plane?

Solution Note that some of the regions are bounded (surrounded by points of Q), and some are unbounded. The Zero Product Property applies. Thus one or more of the five factors is zero. Hence Q consists of all points satisfying any of the following five equations: $x = 0, y = 0, y = -x, y = x, x^2 + y^2 = 1$ lines and the fifth is a circle. The diagram shows that Q divides the plane into 16 regions.

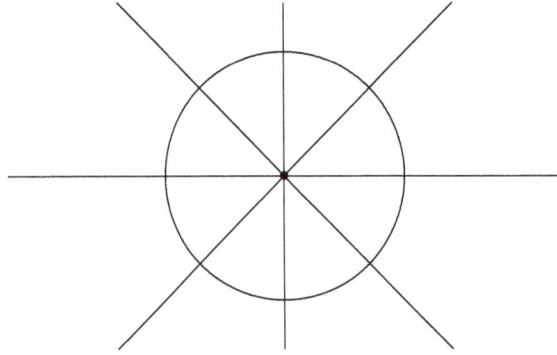

Which of the properties $R, S, T, A, S_0, S_x, S_y$ does $\mathcal{Q}$ satisfy? The answer is all of them except transitivity and antisymmetry.

Example 9.5 This is problem $2(d)$ below

Find the area of region $\mathcal{R}_9$ determined by $\mathcal{R}_9 = \{(x,y)|\ |x+y|+|x|+|y| \leq 6\}$.

Solution First, replace the $\leq$ with $=$. Then note that $|x+y|+|x|+|y| = |-x+(-y)|+|-x|+|-y|$, so our relation is symmetric with respect to the origin. This means that we can draw the graph in just the first and second quadrants. In the first quadrant where both x and y are positive, we can write $x+y+x+y = 6$ and see that this is just a line with slope -1 and y-intercept 3. In the second quadrant, where x is negative and y is positive, we can write $|x+y|-x+y = 6$, which we have to further divide into two parts based on the value of $x+y$. If $x+y \geq 0$, then we have $x+y-x+y = 6$ or $y = 3$ and in case $x+y < 0$, we have $-x-y-x+y = -2x = 6$, so $x = -3$. Putting this all together we get the hexagon below. The next step is to solve the inequality. We can use a test point to determine which of the two regions determined by the hexagon we want. Note that $(0,0)$ satisfies $|x+y|+|x|+|y| \leq 6$, so we simply find the area of the hexagon, which is $9+9+\frac{9}{2}+\frac{9}{2} = 27$.

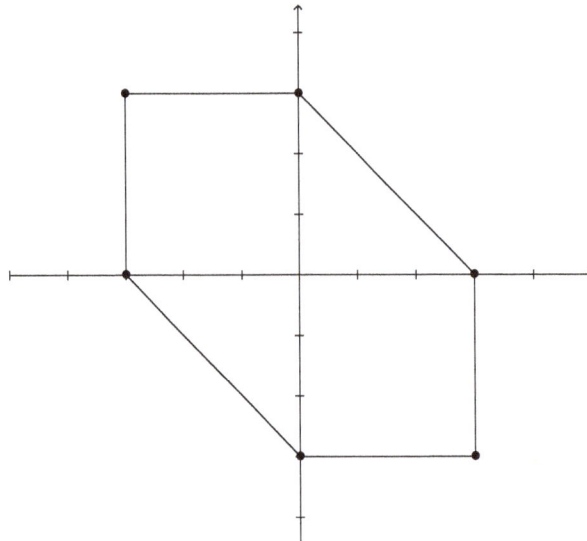

9.4 Problems

1. For each relation on the real numbers R defined below, sketch the graph, and describe the properties listed above that the relation satisfies.

   (a) $\mathcal{R}_1 = \{(x,y)| \frac{xy}{xy} = 1\}$

   Solution Everything except the axes.

   (b) $\mathcal{R}_2 = \{(x,y)| x - y = \lfloor x - y \rfloor\}$

   Solution The union of all lines $y = x + n$ where n is an integer.

   (c) $\mathcal{R}_3 = \{(x,y)| \langle x^2 \rangle + \langle y^2 \rangle = 0\}$

   Solution $Z \times Z$, also known as all integer lattice points.

   (d) $\mathcal{R}_4 = \{(x,y)| \langle x \rangle \cdot \langle y \rangle = 0\}$

   Solution All the lines of the form $x = n$ or $y = n$ where $b \in Z$.

   (e) $\mathcal{R}_5 = \{(x,y)| \lfloor \frac{\lfloor x \rfloor + \lfloor y \rfloor}{2} \rfloor = \frac{1}{2}(\lfloor x \rfloor + \lfloor y \rfloor)\}$

   Solution A checker board.

2. Find the area of the region determined.

   (a) $\mathcal{R}_6 = \{(x,y)| x^2 + y^2 \leq 1\}$

   Solution A circular disc with area π.

   (b) i. $\mathcal{R}_7 = \{(x,y)| \lfloor x^2 \rfloor + \lfloor y^2 \rfloor = 0\}$

   Solution $8(\sqrt{2} - 1)$.

   ii. $\mathcal{R}'_7 = \{(x,y)| \lfloor x^2 \rfloor + \lfloor y^2 \rfloor = 1\}$

   iii. $\mathcal{R}''_7 = \{(x,y)| \lfloor x^2 \rfloor + \lfloor y^2 \rfloor = 2\}$

   (c) $\mathcal{R}_8 = \{(x,y)| \lfloor x \rfloor^2 + \lfloor y \rfloor^2 = 1\}$

   Solution Area 4.

   (d) $\mathcal{R}_9 = \{(x,y)| |x + y| + |x| + |y| \leq 6\}$

   Solution Let P denote the polygonal solution to

$$|x + y| + |x| + |y| = 6.$$

   Examining the equality quadrant by quadrant, we see that P is a convex hexagon with vertices $(\pm 3, 0), (0, \pm 3), (-3, 3), (3, -3)$ and area 27. See Example 5.

   (e) $\mathcal{R}_{10} = \{(x,y)| |x - y| + |x + y| + |x| + |y| \leq 12\}$

3. In case the graph splits the plane into regions, count the regions and decide which of the properties $R, S, T, A, S_0, S_x, S_y$ it has.

(a) $\mathcal{R}_{12} = \{(x, y)| \ xy = 0\}$

Solution Four regions.

(b) $\mathcal{R}_{13} = \{(x, y)| \ (y^2 - x^4)(x^2 - y^4) = 0\}$

Solution 12 regions.

(c) $\mathcal{R}_{14} = \{(x, y)| \ (y^2 - x^4)(x^2 - y^4)(x^2 + y^2 - 1) = 0\}$

Solution 20 regions.

(d) $\mathcal{R}_{15} = \{(x, y)| \ (xy^3 - yx^3)(x^2y^2 + x^6 - y^6 + x^4y^4) = 0\}$

Solution 24 regions.

CHAPTER 10

COUNTING WITH CUBES

10.1 Introduction

A very popular contest topic in the United States is problems about arrangements of cubes. Its popularity derives from the lovely mixture of combinatorics, spacial geometry, and number theory usually required for solving such problems. This essay focuses on four problems, the *visible cubes problem*, the *chameleon cubes problems*, the *drilled cube problem*, and a problem about faces and vertices of a cubical die.

10.2 Visible Cubes Problem

Consider a large cube made from $n \times n \times n$ unit cubes. Look at the large cube from a corner so that you can see three faces. How many unit cubes are in your line of vision? Of course, the number of cubes visible when looking at a vertex is the same as the number of distinct cubes that make up the three faces adjacent to that vertex.

Our objective here is to solve the problem in three different ways. To get a better feel for the problem, let us build a table that shows, for small values of n, the number of cubes visible from one corner.

n	n^3	number visible
1	1	1
2	8	7
3	27	19
4	64	37

How does the table continue? Make some guesses and then try to prove your answer. We can use a technique called the *method of finite differences*. This method fits a polynomial to a sequence of numbers. Let's name the numbers we're looking for. Let $G(n)$ denote the number of cubes visible from a corner of the $n \times n \times n$ cube. Notice that the sequence of first order differences $G(2) - G(1) = 6; G(3) - G(2) = 19 - 7 = 12; G(4) - G(3) = 37 - 19 = 18$ has an interesting property. The differences are all multiples of 6. When we explore such a sequence in which the sequence of successive differences is eventually constant, we can build a polynomial that produces the sequence. Since the second order differences are constant, we propose that $G(n)$ is a quadratic polynomial, $G(n) = an^2 + bn + c$. From the table we can write $G(1) = 1, G(2) = 7$ and $G(3) = 19$. So we can write

$$
\begin{aligned}
1^2 a + 1b + c &= 1 \\
2^2 a + 2b + c &= 7 \\
3^2 a + 3b + c &= 19
\end{aligned}
$$

We can solve this without great difficulty to get $G(n) = 3n^2 - 3n + 1$. But what do these coefficients have to do with the problem? One way to see this is to extend the chart by one more column that shows the cubes that are not visible.

n	n^3	number not visible	number visible
1	1	0	1
2	8	1	7
3	27	8	19
4	64	27	37

Now you can see that we can count the number of visible cubes by counting the invisible ones first. So, in general, $G(n) = n^3 - (n-1)^3 = n^3 - (n^3 - 3n^2 + 3n - 1) = 3n^2 - 3n + 1$. This doesn't completely answer the question however. What are the coefficients telling us? How does the algebra help us reason geometrically? The answer is below.

Our third solution invokes the famous Inclusion/Exclusion Principle. We can first estimate the number of visible cubes by noticing that we can see three faces each with area n^2. But the number $3n^2$ clearly overcounts, because all the cubes on each of the three visible edges are being counted twice. The number $3n^2 - 3n$ works fine except that the corner cube, the one nearest us has been counted 3 times in $3n^2$ and then removed 3 times in $-3n$, so it must be added back into the count. We get $3n^2 - 3n + 1$ as our final count. Finally, we can say what the coefficients mean. The 3 in $3n^2$ is the number of visible faces, the 3 in $3n$ is the number of visible edges, and the 1 is the single corner cube that belongs to all three visible edges.

10.3 Chameleon Cubes Problems

The $3 \times 3 \times 3$ Chameleon Cubes Problem.

You are given 27 unpainted cubes. Can you paint the faces with three colors, red, white, and blue, so that when you're done, you can assemble an all red $3 \times 3 \times 3$ cube, an all white $3 \times 3 \times 3$ cube and an all blue $3 \times 3 \times 3$ cube?

Yes, you can do this in essentially just one way.

Here is one solution. Since the total number of faces we can paint is $27 \cdot 6 = 162$, and since the all-red cube requires $6 \cdot 9 = 54$ red faces as do the other two colors, we must be perfectly efficient in the following sense. Each face we paint red must appear on the outside of the red cube. This implies that there must be exactly one unit cube, the one not visible from the outside, that has no red faces, and similarly exactly one that has no white faces, and one that has no blue faces. These cubes must have exactly three faces of the other two colors. They all look like

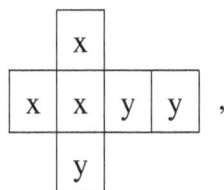

	x		
x	x	y	y
	y		

,

where x is one color and y is another. Why can't they look like

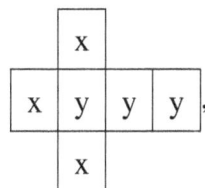

	x		
x	y	y	y
	x		

,

you might ask yourself. The two cubes of this type with red faces account for two of the eight corners. The six other red corner cubes must have faces with the two other colors, so two faces must be one color and one face the third color. Reasoning similarly, there are 6 more cubes with three white and

6 more cubes with three blue faces as shown in the table below. The numbers a and b are 1 and 2 in some order.

n	R	W	B
1	3	3	0
1	3	0	3
1	0	3	3
6	3	a	b
6	a	3	b
6	a	b	3

Notice that exactly 21 cubes have been accounted for, leaving just six more to determine. None of the six can have three faces of the same color. Why? This means they must all have two adjacent faces of each color. Since 12 edge cubes of each color must be available, we can now determine that among the six cubes with three faces of one color, two of another and one of the third color, there are two types, one with two of one color and one with two of the other color. For example, when we look at the six cubes that have three red faces, we'll see three cubes of each pattern

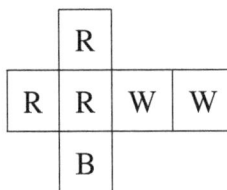

	R		
R	R	W	W
	B		

and

	R		
R	R	B	W
	B		

Thus, the complete table is

n	R	W	B
1	3	3	0
1	3	0	3
1	0	3	3
3	3	1	2
3	3	2	1
3	1	3	2
3	2	3	1
3	1	2	3
3	2	1	3
6	2	2	2

We'll see shortly that there is a much easier and compelling solution. And the new solution generalizes in several ways.

Some readers may realize that we started with the wrong problem. What if you are given 8 unpainted cubes. Can you paint the faces with two colors, red and blue, so that when you're done, you can assemble both an all red cube and an all blue cube? Of course, this is an easy one, right?

These two 3-dimensional problems are just a part of an infinite collection of problems, one for each positive integer n. We call this the space problem or the **3-D Problem:** Color each of the 6 faces of n^3 unit cubes one of the n colors $1, 2, 3, \cdots, n$ in such a way that for each color $i \in \{1, 2, 3, \cdots, n\}$ the n^3 unit cubes can be stacked to form a $n \times n \times n$ big cube in such a way that the big cube is completely colored i.

The 2-D Problem, also called the planar problem: Color each of the 4 sides of n^2 unit squares one of the n colors $1, 2, 3, \cdots, n$ in such a way that for each color $i \in \{1, 2, 3, \cdots, n\}$, the n^2 unit squares can be stacked to form a $n \times n$ big square in such a way that the 4 (outer) sides of this big square are completely colored i.

Maybe even the two-dimensional problem is the wrong place to start.
The 1-D Problem, also called the linear problem: Color each of the 2 end points of n unit intervals one of the n colors $1, 2, 3, \cdots, n$ in such a way that for each color $i \in \{1, 2, 3, \cdots, n\}$, the n unit intervals can be laid side by side to form a big interval of length n in such a way that the two end points of the big interval have the color i.

There are several ways to solve these 3 problems. By far the easiest general solution that we know of is to first find the general solution to the 1-dimensional problem. Then we can use any solution to the 1-dimensional problem to solve the other two.

Solution to Linear Problem. We want to assign the colors $1, 2, 3, \cdots, n$ to the end points of each of the n unit intervals $[0, 1], [1, 2], \ldots, [n - 1, n]$ in such a way that for each color $i \in \{1, 2, 3, \cdots, n\}$, there is an arrangement of intervals so that the endpoints of the union of the intervals both have color i. Start by coloring the endpoints at 0 and n with color 1. Then color the right endpoint of $[0, 1]$ and the left endpoint of $[1, 2]$ with color 2, and so on so that the right endpoint of interval $[k - 1, k]$ and the left endpoint of $[k, k + 1]$ are colored $k + 1$, for $k = 1, 2, \ldots n - 1$. Now it is clear that we can arrage the intervals so that the interval that is their union has each of the i colors as endpoints.

Solution to the Planar Problem. We will now illustrate the solution to Problem 2 for $3^2 = 9$ unit squares. The general solution is exactly the same.

Solution (a)

Solution (b)

$$
\begin{array}{c}
1 \bullet \\
\big\uparrow \\
3 \bullet
\end{array}
$$

$$
\begin{array}{c}
3 \bullet \\
\big\uparrow \\
2 \bullet
\end{array}
$$

$$
\begin{array}{c}
2 \bullet \\
\big\uparrow \\
1 \bullet
\end{array}
$$

We now color the sides of the squares according to the above arrangements (a), (b).

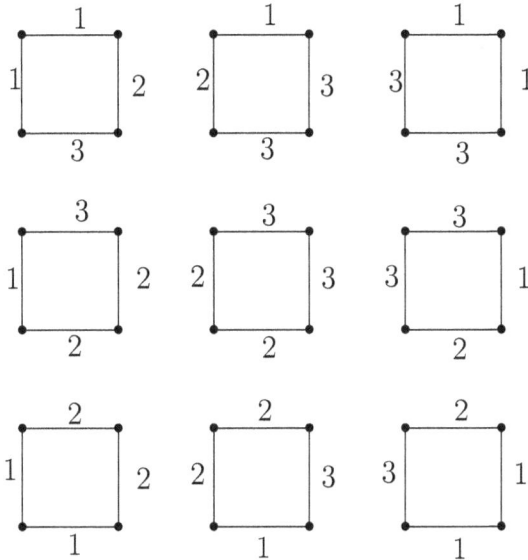

Note that the vertical sides of the columns unit squares are colored according to solution (a). Thus, in column 1, the left side of all squares are colored 1.

Also, in column 1, the right side of all squares are colored 2. Also, in column 2, the left side of all squares are colored 2. Also, in column 2, the right side of all squares are colored 3.

The horizontal sides of the rows of unit squares are colored according to solution (b). Thus, in row 1, the top side of all squares are colored 2. Also, in row 1, the bottom side of all squares are colored 1.

We could achieve this solution in much the same was we achieved the linear solution. Start by coloring the entire outside with color 1, then move the left column of squares to the right side and then move the bottom row to the top. You get a new 3×3 square the sides of which have not been colored. Use color 2 to do the job. Then move the left column to the right side, and the bottom to the top, once again exposing all uncolored sides. Finish the job using color 3.

It is now clear how we can solve the 3D problem.

10.4 The Drilled Cube Problem

The $5 \times 5 \times 5$ cube shown below is built from 125 unit cubes. The dots on the surface show the places where the big cube is drilled through. When all these 'drilled' cubes have been removed, the cube is dipped into a bucket of paint. Then the cubes are cut apart and one is selected at random and rolled. What is the probability that a painted face comes up?

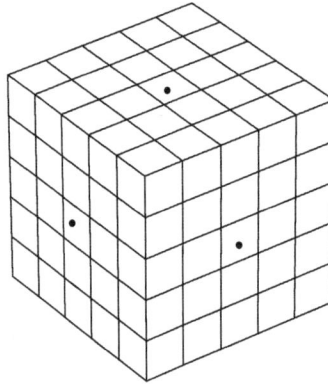

The number of cubes removed by the drilling is $5 + 4 + 4 = 13$. Why not $5 + 5 + 5$? Because the cube in the very center need be counted only once. The surface area of the structure that remains is $6 \cdot 25 - 6 + 8 \cdot 4 = 176$ while the total number of faces is $(125 - 13)6 = 672$, so the probability of a painted face is $176/672 = 11/42$.

10.5   A Die Problem

In this problem you're given a cube. You can assign to each face one of the numbers 1 through 6. You must use all six numbers.

For example, it could be

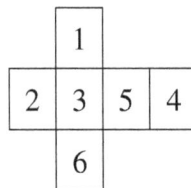

just as the integers 1 through 6 appear on a standard die. Next assign to each vertex the product of the numbers on the faces that vertex belongs to. For example, the vertex in the top right corner of the face with 3 would have an assigned value of $1 \cdot 3 \cdot 5 = 15$. Let T denote the sum of the eight values of the vertices for the die above. Can T be a prime number? What is the largest possible value for T?

Here's a diagram to help you understand what is going on.

Write down the four numbers associated with the top face (where the a is located) $acf + ace + ade + adf$. Next record the four numbers associated with the face at the bottom: $bcf + bce + bde + bdf$. Next note that we can write these two sums in factored form: $a(cf + ce + de + df)$ and $b(cf + ce + de + df)$, so the sum of the eight numbers is $(a + b)[(cf + ce + de + df)]$. But notice that the four numbers inside can be factored also: $(cf + ce + de + df) = (c + d)(e + f)$. This means that we can write our final answer is the following way.

$$T = (a + b)(c + d)(e + f).$$

We prove that the largest possible value of T is 343. First, note that there is some rectangular solid with maximal volume. This follows from the continuity of the volume function, which means essentially that small changes in the edge lengths results in small changes in the volume. Next note that among all rectangles with fixed perimeter P, the one with largest area is the square with side $P/4$. Do this as follows: $(P/4 - x)(P/4 + x) = P^2/16 - x^2 \geq P^2/16$, so we should build the rectangle with sides $P/4$ in order to maximize the area. Next suppose we have a rectangular parallelepiped with sum of edges 21. In case the rectangular parallelepiped with largest volume is not a cube, then one of the faces of such a rectangular parallelepiped is a non-square rectangle. Leaving the third side fixed, we can make the area of the rectangle larger be making it into a square with the same perimeter.

CHAPTER 11

GEOMETRY

One of the fun aspects of geometry problems is the variety of techniques that can be used to solve a single problem. One specific problem can often have several very different looking solutions. For example, consider the following problem.

Example 11.1

Triangle ABC has sides $AB = 45$, $AC = 53$, and $BC = 35$. Point D is on side $\overline{AB}$ so that $AD = 30$. Points E and F trisect segment $\overline{CD}$, as shown. Points G and H are the intersections of side $\overline{BC}$ with lines AE and AF, respectively. Find GH.

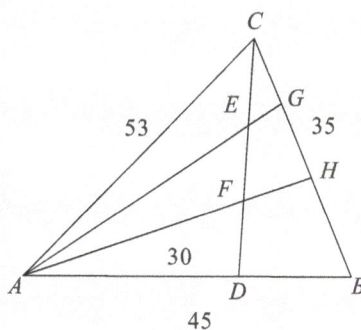

Solution The first solution uses properties of similar triangles. Let points K and L be on segment $\overline{DB}$ so that $\overline{KF} \parallel \overline{LE} \parallel \overline{BC}$, as shown.

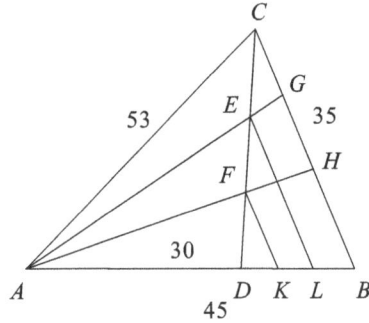

Because their corresponding sides are parallel, the three triangles $\triangle DKF$, $\triangle DLE$, and $\triangle DBC$ are all similar to each other. Because $\frac{DF}{DC} = \frac{1}{3}$ and $\frac{DE}{DC} = \frac{2}{3}$, it follows from $DB = 45 - 30 = 15$ that $DK = 5$ and $DL = 10$. Similarly, $KF = \frac{1}{3}BC = \frac{35}{3}$ and $LE = \frac{2}{3} \cdot BC = \frac{70}{3}$. Again, because their corresponding sides are parallel, the triangle $\triangle AKF$ is similar to $\triangle ABH$, and $\triangle ALE$ is similar to $\triangle ABG$. Thus, $\frac{KF}{AK} = \frac{BH}{AB}$ and $\frac{LE}{AL} = \frac{BG}{AB}$, so

$$BH = \frac{KF}{AK} \cdot AB = \frac{KF}{AD + DK} \cdot AB = \frac{35 \cdot 45}{3 \cdot (30 + 5)} = 15 \qquad \text{and}$$

$$BG = \frac{LE}{AL} \cdot AB = \frac{LE}{AD + DL} \cdot AB = \frac{70 \cdot 45}{3 \cdot (30 + 10)} = \frac{105}{4}.$$

The requested distance is $GH = BG - BH = \frac{105}{4} - 15 = \frac{45}{4} = 11\frac{1}{4}$.

OR

For a completely different looking solution, let us solve this problem using vectors. The strategy is to select two vectors with an obvious connection to $\triangle ABC$ in the problem, and then to represent all the segments in the given diagram as linear combinations of those two vectors. The goal will be to represent a vector pointing from G to H as a multiple of a vector pointing from B to C. Let $\boldsymbol{u}$ be the vector pointing from A to B and $\boldsymbol{v}$ be the vector pointing from A to C, as shown.

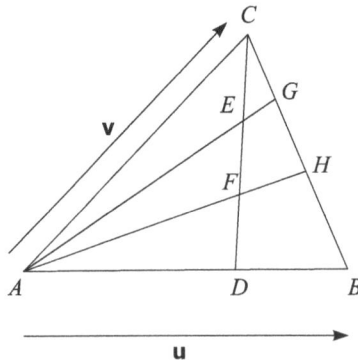

The vector pointing from B to C is $\boldsymbol{v} - \boldsymbol{u}$, the vector pointing from A to D is $\frac{2}{3}\boldsymbol{u}$, and the vector pointing from D to C is $\boldsymbol{v} - \frac{2}{3}\boldsymbol{u}$. Thus, the vector pointing from A to F must be the vector pointing from A to

D plus $\frac{1}{3}$ the vector pointing from D to C which is $\frac{2}{3}\boldsymbol{u}+\frac{1}{3}\left(\boldsymbol{v}-\frac{2}{3}\boldsymbol{u}\right)=\frac{4}{9}\boldsymbol{u}+\frac{1}{3}\boldsymbol{v}$. The vector pointing from A to H can be represented in two different ways: as a multiple of the vector pointing from A to F, and as the vector $\boldsymbol{u}$ plus a multiple of the vector pointing from B to C. That is, there must be real numbers s and t such that the vector pointing from A to H is $t\left(\frac{4}{9}\boldsymbol{u}+\frac{1}{3}\boldsymbol{v}\right)=\boldsymbol{u}+s(\boldsymbol{v}-\boldsymbol{u})$. Because $\boldsymbol{u}$ and $\boldsymbol{v}$ are linearly independent vectors, the coefficients of $\boldsymbol{u}$ on the two sides of this equation must be equal, and the coefficients of $\boldsymbol{v}$ on the two sides of the equation must also be equal. This means that $\frac{4}{9}t=1-s$ and $\frac{1}{3}t=s$. Solving this system of equations shows that $s=\frac{3}{7}$ and $t=\frac{9}{7}$. It follows that $BH=\frac{3}{7}\cdot 35=15$.

A similar construction shows that the vector pointing from A to E is $\frac{2}{3}\boldsymbol{u}+\frac{2}{3}\left(\boldsymbol{v}-\frac{2}{3}\boldsymbol{u}\right)=\frac{2}{9}\boldsymbol{u}+\frac{2}{3}\boldsymbol{v}$. Then there are real numbers s and t so that $t\left(\frac{2}{9}\boldsymbol{u}+\frac{2}{3}\boldsymbol{v}\right)=\boldsymbol{u}+s(\boldsymbol{v}-\boldsymbol{u})$. Solving this systems shows $s=\frac{3}{4}$ and $t=\frac{9}{8}$, and $BG=\frac{3}{4}\cdot 35=\frac{105}{4}$ as in the first solution.

The following advanced AMC 12 problem has a surprising number of very different looking solutions. We present four of them here.

Example 11.2 2014 AMC 12B Problem 21

In the figure, $ABCD$ is a square of side length 1. The rectangles $JKHG$ and $EBCF$ are congruent. What is BE?

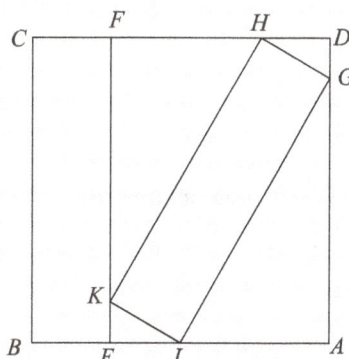

(A) $\frac{1}{2}(\sqrt{6}-2)$ (B) $\frac{1}{4}$ (C) $2-\sqrt{3}$ (D) $\frac{\sqrt{3}}{6}$ (E) $1-\frac{\sqrt{2}}{2}$

Solution (C). The first solution exploits the similarity of all four right triangles in the diagram, $\triangle EJK\sim\triangle AGJ\sim\triangle DHG\sim\triangle FKH$. In fact, $\triangle EJK$ and $\triangle DHG$ are congruent as are $\triangle AGJ$ and $\triangle FKH$. Let $x=BE=JK=GH$, $y=EJ=DH$, and $z=EK=DG$. Then by similar triangles $AJ=\frac{z}{x}$ and $AG=\frac{y}{x}$. Because $AD=AG+DG$, we have $1=\frac{y}{x}+z$, so $z=1-\frac{y}{x}$. Applying the Pythagorean Theorem to $\triangle EJK$ gives $x^2=y^2+z^2=y^2+\left(1-\frac{y}{x}\right)^2$, which can be written as $\frac{y^2}{x^2}\cdot(1+x^2)-2\cdot\frac{y}{x}+(1-x^2)=0$. The quadratic formula then gives

$$\frac{y}{x}=\frac{2\pm\sqrt{(-2)^2-4(1+x^2)(1-x^2)}}{2(1+x^2)}=\frac{1\pm x^2}{1+x^2}.$$

Because $y < x$, conclude that $\frac{y}{x} = \frac{1-x^2}{1+x^2}$. Then $z = 1 - \frac{y}{x} = \frac{2x^2}{1+x^2}$. Because $AB = BE + EJ + AJ$, we have $1 = x + y + \frac{z}{x}$. Thus,

$$1 = x + \frac{1-x^2}{1+x^2} \cdot x + \frac{2x}{1+x^2}.$$

This simplifies to $x^2 - 4x + 1 = 0$, which has solutions $2 \pm \sqrt{3}$. Because $x < 1$, conclude that $BE = 2 - \sqrt{3}$.

OR

The second solution depends on the symmetry in the given diagram. Let L be the point of $\overline{AB}$ so that $\overline{HL} \perp \overline{AB}$. Note that $\triangle BEF$ and $\triangle JLH$ are both right triangles with two equal legs $EF = LH = 1$ and with equal hypotenuses $BF = JH$ because both are diagonals of congruent rectangles $EBCF$ and $JKHG$.

Thus, $\triangle BEF$ is congruent to $\triangle JLH$, and, as a result, $BE = JL$. Let $x = BE$ and $y = EJ$. Because $\triangle EJK$ is congruent to $\triangle DHG$, $EJ = DH = AL = y$. This means that $1 = AB = AL + JL + EJ + BE = 2(x + y)$, so $x + y = \frac{1}{2}$. Because the hypotenuse of right triangle AGJ is $GJ = 1$ and one leg is $AJ = x + y = \frac{1}{2}$, it follows that $\triangle AGJ$ is a $30-60-90°$ triangle with $\angle AJG = 60°$. Thus, $\angle EJK = 30°$, and it follows that $y = EJ = JK \cdot \frac{\sqrt{3}}{2} = \frac{x\sqrt{3}}{2}$. Hence, $\frac{1}{2} = x + y = x + \frac{x\sqrt{3}}{2}$. Solving this equation for x gives $BE = x = 2 - \sqrt{3}$.

OR

The third solution uses a different symmetry from the given diagram. As in the Solution 2, it can be shown that $\triangle BEF$ is congruent to $\triangle JLH$. From this $\angle LJH = \angle EBF$ showing that $\overline{BF} \parallel \overline{JH}$ as well as $BF = JH$. This proves that quadrilateral $BFHJ$ is a parallelogram, and, thus, $FH = BJ$. Because $\triangle AGJ$ is similar to $\triangle FKH$ and their hypotenuses both have length 1, the two triangles are congruent. Therefore, $AJ = FH = BE$, and it follows that each of these must equal $\frac{1}{2}$. Conclude that $\triangle AHJ$ is a $30-60-90°$, and the solution continues as in Solution 2.

OR

The fourth solution is trigonometric. As in Solution 1, it used the fact that $\triangle EJK \sim \triangle AGJ \sim \triangle DHG \sim \triangle FKH$. Again, let $x = BE$, and let $\theta = \angle KJE = \angle JGA = GHD$. It follows from $AD = AG + DG$ that $1 = \cos\theta + x\sin\theta$, so $x = \frac{1-\cos\theta}{\sin\theta}$. It then follows from $AB = BE + EJ + AJ$ that $1 = x + x\cos\theta + \sin\theta = x(1 + \cos\theta) + \sin\theta$. Substituting for x gives

$$1 = \frac{1 - \cos\theta}{\sin\theta}(1 + \cos\theta) + \sin\theta = \frac{1 - \cos^2\theta}{\sin\theta} + \sin\theta = 2\sin\theta.$$

Therefore, $\sin\theta = \frac{1}{2}$, and $\theta = 30°$. Finally, $BE = x = \frac{1-\cos\theta}{\sin\theta} = \frac{1 - \frac{\sqrt{3}}{2}}{\frac{1}{2}} = 2 - \sqrt{3}$.

The last example is a very challenging problem generally handled using trigonometry, but it also has a very clever solution that avoids the use of any discussion of trigonometry. The problem appeared on the American Invitational Mathematics Exam (AIME) in the year 2000.

Example 11.3 2000 AIME I Problem 14

In $\triangle ABC$, it is given that angles B and C are congruent. Points P and Q lie on $\overline{AC}$ and $\overline{AB}$, respectively, so that $AP = PQ = QB = BC$. Angle ACB is r times as large as angle APQ, where r is a positive real number. Find the greatest integer that does not exceed $1000r$.

✎ **Solution** Because it is given that $\angle ABC = \angle ACB$, the triangle is isosceles, so $AB = AC$. Without loss of generality it can be assumed that $BC = 1$. Let $\theta = \angle ACB = \angle ABC$ and $s = AB = AC$, as shown.

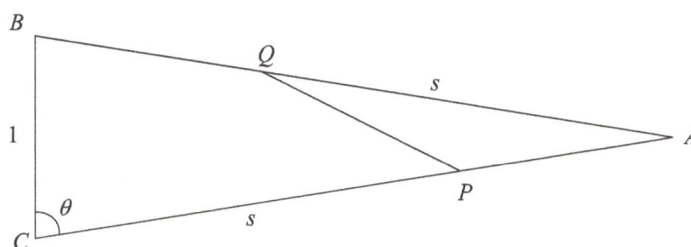

Because $\triangle ABC$ is isosceles, its altitude from A is a median. This implies that $\cos\theta = \frac{1}{2}\frac{BC}{AC} = \frac{1}{2s}$. Note that $\angle BAC = 180° - \angle ABC - \angle ACB = 180° - 2\theta$. The Law of Cosines applied to $\triangle APQ$ says that $PQ^2 = AQ^2 + AP^2 - 2 \cdot AQ \cdot AP\cos(\angle BAC)$, so

$$1 = (s - 1)^2 + 1 - 2(s - 1)\cos(180° - 2\theta).$$

Subtracting 1 from each side of the equation, dividing by $s-1$ (clearly, $s > 1$, so $s-1 \neq 0$), and using the fact that $\cos(180° - 2\theta) = -\cos(2\theta) = -(2\cos^2\theta - 1)$, this simplifies to $0 = s + 4\cos^2\theta - 3$. Now substitute $s = \frac{1}{2\cos\theta}$ and simplify to get $4\cos^3\theta - 3\cos\theta = -\frac{1}{2}$. Here you can apply a useful identity $\cos(3\theta) = \cos(\theta + 2\theta) = \cos\theta\cos(2\theta) - \sin\theta\sin(2\theta) = \cos\theta(2\cos^2\theta - 1) - \sin\theta(2\sin\theta\cos\theta) = $

$4\cos^3\theta - 3\cos\theta$. Thus, for this triangle, $\cos(3\theta) = -\frac{1}{2}$, which implies that 3θ is either $120°$ or $240°$, and θ must be $40°$ or $80°$. Because $AB > BC$, angle $\theta = \angle ABC$ must be greater than $45°$, so you can conclude that $\theta = 80°$. Then $\angle APQ = 180° - 2\angle BAC = 4\theta - 180° = 140°$. The requested answer is $\left\lfloor 1000 \cdot \frac{80}{140} \right\rfloor = 571$.

OR

Some geometry problems can be solved by adding one or more lines to an existing drawing and, thus, pointing out some relationship that was not evident in the original problem. This is the case for the second solution of this AIME problem. In this case, let H be the midpoint of side $\overline{BC}$, and let $x = \angle BAH = \angle CAH$. Also add a point G so that the line segment $\overline{BG}$ contains the point P and $\angle AGB$ is a right angle, as shown.

Because $\triangle APQ$ is isosceles, $\angle AQP = \angle PAQ = 2x$. Then because $\triangle BPQ$ is also isosceles, $\angle PBQ = \angle BPQ$, and since $\angle PBQ + \angle BPQ = \angle AQP$, it follows that $\angle PBQ = x$. Because $\triangle AHB$ and $\triangle BGA$ are both right triangles that share a hypotenuse and have an angle with measure x, the two triangles are congruent. Therefore, $AP = BC = 2BH = 2AG$ which implies that $\cos(\angle GAP) = \frac{1}{2}$, so $\angle GAP = 60°$. From this it follows that $\angle BAG = \angle BAH + \angle CAH + \angle GAP = 60° + 2x$. But $\angle BAG = \angle ABH = 90° - x$. Hence, $60° + 2x = 90° - x$, which implies that $x = 10°$. This gives us $\angle ACB = 90° - x = 80°$, and $\angle APQ = 180° - 2\angle BAC = 180° - 2 \cdot (2 \cdot 10°) = 140°$ as in the first solution.

CHAPTER 12

ALGEBRA

12.1 Polynomials and their Zeros

1. Compute the sum of all the roots of $(2x + 3)(x - 4) + (2x + 3)(x - 6) = 0$.

(A) $7/2$ (B) 4 (C) 7 (D) 13 (E) none of **A, B, C** or **D**

2. Compute the sum of the roots of $x^2 - 5x + 6 = 0$.

(A) 3 (B) $7/2$ (C) 4 (D) $9/2$ (E) 5

3. The product of the zeros of $f(x) = (2x - 24)(6x - 18) - (x - 12)$ is

(A) -72 (B) 5 (C) 6 (D) 37 (E) 432

4. What is the product of the roots of

$$(x - 1)(x - 3) + (x - 4)(x + 5) + (x - 3)(x - 7) = 0?$$

(A) -1260 (B) -420 (C) $4/3$ (D) 10 (E) 36

5. Factor $x^4 + 4y^4$ over the real numbers.

(A) $(x^2 - 2xy + 2y^2)(x^2 + 2xy + 2y^2)$

(B) $(x^2 + 2xy + 2y^2)^2$

(C) $(x^2 + 2xy - 2y^2)(x^2 + 2xy + 2y^2)$

(D) $(x^2 - 2xy - 2y^2)(x^2 + 2xy + 2y^2)$

(E) none of **A, B, C** or **D**

6. Let m be an integer such that $1 \le m \le 1000$. Find the probability of selecting at random an integer m such that the quadratic equation

$$6x^2 - 5mx + m^2 = 0$$

has at least one integer solution.

(**A**) 0.333   (**B**) 0.5   (**C**) 0.667   (**D**) 0.778   (**E**) 0.883

7. Let w be a real number. What is the sum of the (possibly complex) roots of the equation

$$x^2 + 13x + w = 0?$$

(**A**) w (**B**) $-w$ (**C**) $13 + w$ (**D**) $-13 - w$ (**E**) -13

8. Let a and b be the two roots of the equation $x^2 + 3x - 3 = 0$. Evaluate the value $a^2 + b^2$.

(**A**) 4   (**B**) 9   (**C**) 10   (**D**) 12   (**E**) 15

9. Which of the equations below has roots that are the reciprocals of the roots of the equation

$$x^2 - 3x - 2 = 0?$$

(**A**) $2x^2 + 3x - 1 = 0$ (**B**) $2x^2 - 3x - 1 = 0$ (**C**) $2x^2 + 3x + 1 = 0$

(**D**) $2x^2 - 3x + 1 = 0$ (**E**) none of **A**, **B**, **C** or **A**

10. Let r and s be the two solutions to the equation $x^2 - 3x + 1 = 0$. Find $r^3 + s^3$.

(**A**) 12   (**B**) 14   (**C**) 16   (**D**) 18   (**E**) 24

11. Suppose that the equation $x^2 - px + q = 0$ has roots $x = a$ and $x = b$. Which of the following equations has roots $x = a + \dfrac{1}{b}$ and $x = b + \dfrac{1}{a}$?

(**A**) $x^2 - \left(p + \dfrac{p}{q}\right)x + \left(q + \dfrac{1}{q} + 2\right) = 0.$ (**B**) $x^2 - \left(q + \dfrac{p}{q}\right)x + \left(p + \dfrac{1}{q} + 2\right) = 0.$

(**C**) $x^2 - \left(p + \dfrac{q}{p}\right)x + \left(q + \dfrac{1}{p} + 2\right) = 0.$ (**D**) $x^2 - \left(p + \dfrac{q}{p}\right)x + \left(p + \dfrac{1}{p} + 2\right) = 0.$

(**E**) $x^2 - \left(q + \dfrac{q}{p}\right)x + \left(q + \dfrac{1}{p} + 2\right) = 0.$

12. A cubic equation $x^3 - 4x^2 - 11x + a = 0$ has three roots, x_1, x_2, x_3. If $x_1 = x_2 + x_3$, what is a?

(**A**) 24   (**B**) 28   (**C**) 30   (**D**) 32   (**E**) 36

13. Suppose the roots of the equation $(x^2 - 2x + m)(x^2 - 2x + n) = 0$, where m, n are two real numbers, form an arithmetic sequence with the first term being $\dfrac{1}{4}$. Then $|m - n| =$

(A) 1   (B) $\dfrac{3}{4}$ (C) $\dfrac{1}{2}$ (D) $\dfrac{3}{8}$ (E) 2

14. If $P(x) = 2x^3 + kx^2 + x$, find k such that $x - 1$ is a factor of $P(x)$.

(A) -3 (B) $-1/3$ (C) 0   (D) $1/3$ (E) 3

15. The polynomial $p(x) = 2x^4 - x^3 - 7x^2 + ax + b$ is divisible by $x^2 - 2x - 3$ for certain values of a and b. What is the sum of a and b?

(A) -34 (B) -30 (C) -26 (D) -18 (E) 30

16. A non-constant polynomial function $f(x)$ satisfies

$$f(-4) = f(-2) = f(1) = f(3) = 2.$$

What is the smallest possible degree of f?

(A) 1   (B) 3   (C) 4   (D) 5   (E) 6

17. What is the coefficient of x^7 in the polynomial $(x + 3)^{10}$?

(A) 120   (B) 2187   (C) 3240   (D) 3402   (E) 5670

12.2 Exponentials and Radicals

1. For all integers n, $(-1)^{n^4+n+1}$ is equal to

(A) -1 (B) $(-1)^{n+1}$ (C) $(-1)^n$ (D) $(-1)^{n^2}$ (E) $+1$

2. The numbers x and y satisfy $2^x = 15$ and $15^y = 32$. What is the value xy?

(A) 3   (B) 4   (C) 5   (D) 6   (E) none of **A, B, C** or **D**

3. Given that $a = 1/x, b = 9a, c = 1/b, d = 9c, e = 1/d$, and $a, b, c,$ and d are all distinct non-zero numbers, then x must be the same as

(A) a (B) b (C) c (D) d (E) e

4. If N is the cube of a certain positive integer, which of the following is the square of the next positive integer?

(A) $\sqrt{(N+1)}$ (B) $\sqrt[3]{(N+1)}$ (C) $N^2 + 1$ (D) $N^{2/3} + 2N^{1/3} + 1$

(E) $N^{2/3} - 2N^{1/3} + 1$

5. If n is a positive perfect square, which of the following represents the largest perfect square less than n?

(A) $n - 1$ (B) n (C) $n^2 - 2n + 1$ (D) $n^2 + n$ (E) $n - 2\sqrt{n} + 1$

6. The square of $2^{\sqrt{2}}$ equals

(A) 2^2 (B) $4^{\sqrt{2}}$ (C) 4^2 (D) $4^{2\sqrt{2}}$ (E) $4^{\sqrt{2}^2}$

7. If $(2^x - 4^x)^2 + (2^x + 4^x)^2 = 144$, what is x?

(A) $\dfrac{1}{4}$ (B) $\dfrac{1}{2}$ (C) $\dfrac{3}{4}$ (D) $\dfrac{5}{4}$ (E) $\dfrac{3}{2}$

8. If $\sqrt{2 + \sqrt{x}} = 3$, then $x =$

(A) 1 (B) 7 (C) 11 (D) 49 (E) 121

9. Given that $\frac{3}{2} < x < \frac{5}{2}$, find the value of

$$\sqrt{x^2 - 2x + 1} + \sqrt{x^2 - 6x + 9}$$

(A) 1 (B) 2 (C) $2x - 4$ (D) $4 - 2x$ (E) none of **A, B, C** or **D**

10. Evaluate exactly $\sqrt{5 + 2\sqrt{6}} + \sqrt{5 - 2\sqrt{6}}$.

(A) $2\sqrt{2}$ (B) $3\sqrt{2}$ (C) $2\sqrt{3}$ (D) $2\sqrt{5}$ (E) $\frac{3}{2}\sqrt{6}$

11. Solve the equation $\dfrac{\sqrt{x+1} + \sqrt{x-1}}{\sqrt{x+1} - \sqrt{x-1}} = 3$ for x.

(A) $4/3$ (B) $5/3$ (C) $7/5$ (D) $9/5$ (E) none of **A, B, C** or **D**

12. What is the sum of the digits of the integer solution to $\sqrt{14 + \sqrt{27 - \sqrt{x - 1}}} = 4$?

(A) 5 (B) 6 (C) 8 (D) 9 (E) 11

13. When $\sqrt{4 - 2\sqrt{3}}$ is expressed in the form $\sqrt{a} - b$, where a and b are integers, the value of $a + b$ is

(A) 2 (B) 3 (C) 4 (D) 5 (E) 6

14. Suppose

$$8^{\frac{1}{6}} + x^{\frac{1}{3}} = \frac{7}{3 - \sqrt{2}}.$$

Then $x =$

(A) 24 (B) 27 (C) 32 (D) 64 (E) none of **A, B, C** or **D**

15. What is the nth digit after the decimal point in the decimal representation of $(\sqrt{26} - 5)^n$.

(A) 0 (B) 1 (C) 6 (D) 5 (E) Can not be determined

12.3 Equations and Inequalities

1. If a, b and c satisfy $a^2 + b^2 = 208$, $b^2 + c^2 = 164$, and $c^2 + a^2 = 244$, then $a^2 + b^2 - c^2 =$

(A) -36 (B) 20 (C) 108 (D) 120 (E) 180

2. If $z = -x$, what are all the values of y for which

$$(x + y)^2 + (y + z)^2 = 2x^2 \text{ ?}$$

(A) 0 (B) 0.1 (C) $-1, 0, 1$ (D) All positive numbers

(E) There are no values of y for which the equation is true

3. If $a > 0$ and $ab = 3$, $bc = 5$, and $ac = 7$, what is c?

(A) 3 (B) $\sqrt{3}$ (C) $\sqrt{\frac{35}{3}}$ (D) 2 (E) 1

4. Let x and y be two real numbers satisfying $x + y = 6$ and $xy = 7$. Find the value of $x^3 + y^3$.

(A) 55 (B) 62 (C) 78 (D) 90 (E) 216

5. Let x and y be two positive real numbers satisfying

$$x + y + xy = 10 \quad \text{and} \quad x^2 + y^2 = 40.$$

What integer is nearest $x + y$?

(A) 4 (B) 5 (C) 6 (D) 7 (E) 8

6. Given that

$$x(y - a) = 0, \qquad z(y - b) = 0, \quad \text{and} \quad a < b,$$

which of the following must be true?

(A) $xz < 0$ (B) $xz > 0$ (C) $x = 0$ (D) $z = 0$ (E) $xz = 0$

7. The sum $a + b$, the product $a \cdot b$ and the difference of squares $a^2 - b^2$ of two positive numbers a and b is the same nonzero number. What is b?

(A) 1 (B) $\dfrac{1 + \sqrt{5}}{2}$ (C) $\sqrt{3}$ (D) $\dfrac{7 - \sqrt{5}}{2}$ (E) 8

8. Suppose $a < 0$ and $|a| \cdot x \leq a$. Evaluate $|x + 1| - |x - 2|$.

(A) -3 (B) -1 (C) $-2x + 1$ (D) $2x + 3$ (E) 3

9. Given $3 = \sqrt{a} + \dfrac{1}{\sqrt{a}}$, where $a \neq 0$, find $a - \dfrac{1}{a}$.

(A) 5 (B) 6 (C) $3\sqrt{5}$ (D) 7 (E) $5\sqrt{2}$

10. How many points (x, y) satisfy the equation

$$|x^2 - 1| + |y^2 - 4| = 0?$$

(A) 2   **(B)** 4   **(C)** 6   **(D)** 8   **(E)** infinitely many

11. Find the sum of all values of x that satisfy

$$|x + 1| + 3|x - 2| + 5|x - 4| = 20.$$

(A) 2   **(B)** 5   **(C)** 6   **(D)** 9   **(E)** 11

12. How many ordered triples (x, y, z) satisfy the equation

$$(x^2 - 1)^2 + (y^2 - 4)^2 + (z^2 - 9)^2 = 0?$$

(A) 0   **(B)** 2   **(C)** 4   **(D)** 6   **(E)** 8

13. How many of elements of the set $\{-12.3, -9, -5, 1, 2.14, 3.6, 5.2, 7.8, 101\}$ are **not** solutions of the equation $|2x^2 - 9x + 6| = 2x^2 - 9x + 6$?

(A) 2   **(B)** 3   **(C)** 5   **(D)** 6   **(E)** 8

14. The sum of the reciprocals of four different positive integers is 1.85. Which of the following could be the sum of the four integers?

(A) 15   **(B)** 16   **(C)** 17   **(D)** 18   **(E)** 19

15. $\dfrac{10^{12} - 10^{11}}{9} =$

(A) $\frac{1}{9}$ **(B)** $\frac{10}{9}$ **(C)** 10^3 **(D)** $\frac{10^{11}}{9}$ **(E)** 10^{11}

16. Which of the five fractions has the smallest value?

(A) $\dfrac{250,386,765,412}{250,384,765,412}$ **(B)** $\dfrac{250,384,765,412}{250,383,765,412}$ **(C)** $\dfrac{250,385,765,412}{250,384,765,412}$

(D) $\dfrac{250,386,765,412}{250,385,765,412}$ **(E)** $\dfrac{250,387,765,412}{250,386,765,412}$

17. Which of the five fractions is largest?

(A) $\dfrac{25038876541}{25038876543}$ **(B)** $\dfrac{25038876543}{25038876545}$ **(C)** $\dfrac{25038876545}{25038876547}$

(D) $\dfrac{25038876547}{25038876549}$ **(E)** $\dfrac{25038876549}{25038876551}$

18. If $x > 5$, which of the following is smallest?

(A) $\dfrac{5}{x}$ **(B)** $\dfrac{5}{x - 1}$ **(C)** $\dfrac{x}{5}$ **(D)** $\dfrac{5}{x + 1}$ **(E)** $\dfrac{x + 1}{5}$

19. What is the sum of the three positive integers a, b, and c that satisfy

$$a + \cfrac{1}{b + \cfrac{1}{c}} = 7.5?$$

(A) 6   (B) 7   (C) 8   (D) 9   (E) 11

20. If a, b, c, and d are nonzero real numbers, $\dfrac{a}{b} = \dfrac{c}{d}$, and $\dfrac{a}{d} = \dfrac{b}{c}$, then which one of the following must be true?

(A) $a = \pm b$ (B) $a = \pm c$ (C) $a = \pm d$ (D) $b = \pm c$ (E) none of **A, B, C** or **D**

21. The non-zero real numbers a, b, c, d have the properties that $ad - bc \neq 0$ and $\dfrac{ax + b}{cx + d} = 1$ has no solution in x. What is the value of $\dfrac{a^2}{a^2 + c^2}$?

(A) 0   (B) 1/2   (C) 1   (D) 2   (E) an irrational number

22. If x, y, and z satisfy $\dfrac{x}{y - 6} = \dfrac{y}{z - 8} = \dfrac{z}{x - 10} = 3$. What is the value of $x + y + z =$

(A) 24   (B) 30   (C) 32   (D) 36   (E) 40

23. How many ordered triples (x, y, z) satisfy the equation

$$3x^2 + 3y^2 + z^2 - 2xy + 2yz = 0 ?$$

(A) 0   (B) 1   (C) 3   (D) 4   (E) infintely many

24. Let x, y be positive integers with $x > y$. If $1/(x + y) + 1/(x - y) = 1/3$, find $x^2 + y^2$.

(A) 52   (B) 58   (C) 65   (D) 73   (E) 80

25. A recent poll showed that nearly 30% of European school children think that $\dfrac{1}{2} + \dfrac{1}{3} = \dfrac{2}{5}$. This is wrong, of course. Is it possible that $\dfrac{1}{a} + \dfrac{1}{b} = \dfrac{2}{a + b}$ for some real numbers a and b?

(A) Yes, but only if $a + b = 1$ (B) Yes, but only if $a + b = 2$

(C) Yes, but only if $a^2 + b^2 = 1$ (D) Yes, but only if $a^2 + b^2 = 0$

(E) No, it is not possible

26. Let x and y be the positive integer solution to the equation

$$\frac{1}{x + 1} + \frac{1}{y - 1} = 5/6.$$

Find $x + y$.

(A) 2   (B) 3   (C) 4   (D) 5   (E) 6

27. If (x, y, z) satisfy the three equations below, what is $x + y + z$?

$$\left.\begin{array}{rcl} 2x - y + z & = & 7 \\ x + 2y + 3z & = & 1 \\ -x + y + 5z & = & 6 \end{array}\right\}$$

(A) -1 (B) 0 (C) 1 (D) 2 (E) 3

28. If a, b and c are three distinct numbers such that

$$a^2 - bc = 7, \quad b^2 + ac = 7, \quad \text{and} \quad c^2 + ab = 7,$$

then $a^2 + b^2 + c^2 =$

(A) 8 (B) 10 (C) 12 (D) 14 (E) 17

29. Suppose x satisfies $|x^2 - 2x - 3| = |x^2 - 2x + 5|$. Then x belongs to

(A) $[0, 2)$ (B) $[2, 4)$ (C) $[4, 6)$ (D) $[6, 8)$ (E) $[8, \infty)$

30. What is the length of the interval of solutions to the inequality $1 \le 3 - 4x \le 11$?

(A) 1.75 (B) 2.00 (C) 2.25 (D) 2.50 (E) 3.25

31. If $x^2 + 2x + n > 10$ for all real numbers x, then which of the following conditions must be true?

(A) $n > 11$ (B) $n < 11$ (C) $n = 10$ (D) $n = \infty$ (E) $n > -11$

32. If a, b, c and d are four positive numbers such that $\dfrac{a}{b} < \dfrac{c}{d}$, then

(A) $ab < dc$ (B) $a + c < b + d$ (C) $a + d < b + c$ (D) $\dfrac{a + c}{b + d} < \dfrac{c}{d}$

(E) $\dfrac{c - a}{d - b} < \dfrac{c}{d}$

33. Suppose a, b and c are real numbers for which $\dfrac{a}{b} > 1$ and $\dfrac{a}{c} < 0$. Which of the following must be correct?

(A) $a + b - c > 0$ (B) $a > b$ (C) $(a - c)(b - c) > 0$ (D) $a + b + c > 0$ (E) $abc > 0$

34. The inequality $|x + 3| \le 2$ is equivalent to the inequality $a \le \dfrac{6}{x + 7} \le b$. Find the value of $a^2 + b$.

(A) 3 (B) 4 (C) 5 (D) 6 (E) 7

35. The set of all x such that $(|x| - 2)(1 + x) > 0$ is exactly

(A) $x > 2$ (B) $|x| > 2$ (C) $-2 < x < -1$ or $x > 2$ (D) $-1 < x < 2$

(E) $x < -2$ or $x > 2$

36. The number of elements in the intersection of $A = \left\{-1, -\frac{1}{2}, 0, \frac{1}{2}, 1, \frac{3}{2}, 2\right\}$ with the solution set

of $\left(1 + \dfrac{x+2}{x-1}\right)\left(1 + \dfrac{x-3}{2}\right) > 0$ is

(A) 0   (B) 1   (C) 2   (D) 3   (E) 4

37. The numbers x, y, and z satisfy $|x+2| + |y+3| + |z-5| = 1$. Which of the following could be $|x+y+z|$?

(A) 0   (B) 2   (C) 5   (D) 7   (E) 10

38. If $x \neq y$ and $\dfrac{x^3 - y^3}{x - y} = 8$, then $x^2 + xy + y^2 =$

(A) 2   (B) 5   (C) 8   (D) 64

(E) It cannot be determined from the information given.

39. The product of a positive number, its reciprocal, and its square is 7. Which of the following is closest to the sum of the number and its reciprocal?

(A) 2.64   (B) 2.86   (C) 3.02   (D) 3.33   (E) 3.51

40. Suppose $a < 0$ and $|a| \cdot x \leq a$. Evaluate $|x+1| - |x-2|$.

(A) -3 (B) -1 (C) $-2x + 1$ (D) $2x + 3$ (E) 3

41. A rational number $3.abc$ when rounded to the nearest tenth is 3.5. If $a + b + c = 8$ and $a > b > c > 0$, what is the value of $2a + bc$?

(A) 17   (B) 14   (C) 12   (D) 11   (E) 10

42. What is the sum of the three positive integers a, b, and c that satisfy

$$a + \frac{1}{b + \frac{1}{c}} = 5.4?$$

(A) 8   (B) 9   (C) 10   (D) 11   (E) 12

43. Which of the following numbers is closest to the value of the continued fraction $\dfrac{2}{1 + \frac{2}{1+\frac{2}{1+\dots}}}$?

(A) 1.00   (B) 1.02   (C) 1.04   (D) 1.06   (E) 1.08

44. Evaluate the continued fraction

$$2 + \cfrac{1}{2 + \cfrac{1}{2 + \cfrac{1}{2 + \cdots}}}$$

(A) $\sqrt{5}$ (B) $\sqrt{6}$ (C) $1 + \sqrt{2}$ (D) $1 + \sqrt{3}$ (E) 3

45. Consider the number x defined by the periodic continued fraction

$$x = \cfrac{1}{2 + \cfrac{1}{3 + \cfrac{1}{2 + \cfrac{1}{3 + \cdots}}}}.$$

Then $x =$

(A) $\dfrac{1}{3}$ (B) $\dfrac{3}{7}$ (C) $\dfrac{-3 - \sqrt{15}}{2}$ (D) $\dfrac{-3 + \sqrt{15}}{2}$ (E) None of these

46. Suppose a is a real number for which $a^2 + \dfrac{1}{a^2} = 14$. What is the largest possible value of $a^3 + \dfrac{1}{a^3}$?

(A) 48   (B) 52   (C) 56   (D) 60   (E) 64

47. Given that (x, y) satisfies $x^2 + y^2 = 9$, what is the largest possible value of $x^2 + 3y^2 + 4x$?

(A) 22   (B) 24   (C) 36   (D) 27   (E) 29

48. Given the following system of equations

$$\begin{aligned}
\frac{1}{x} + \frac{1}{y} &= \frac{1}{3} \\
\frac{1}{x} + \frac{1}{z} &= \frac{1}{5} \\
\frac{1}{y} + \frac{1}{z} &= \frac{1}{7}
\end{aligned}$$

What is the value of the ratio $\dfrac{z}{y}$?

(A) 17   (B) 23   (C) 29   (D) 31   (E) 36

49. Find the minimum value of

$$1 \circ 2 \circ 3 \circ 4 \circ 5 \circ 6 \circ 7 \circ 8 \circ 9$$

where each "$\circ$" is either a "$+$" or a "$\times$".

(A) 36   (B) 40   (C) 44   (D) 45   (E) 84

50. The fraction $\dfrac{5x - 11}{2x^2 + x - 6}$ was obtained by adding the two fractions $\dfrac{A}{x + 2}$ and $\dfrac{B}{2x - 3}$. Find the value of $A + B$.

(A) -4 (B) -2 (C) 1   (D) 2   (E) 4

51. Suppose x, y, z, and w are real numbers satisfying $x/y = 4/7$, $y/z = 14/3$, and $z/w = 3/11$. When $(x + y + z)/w$ is written in the form m/n where m and n are positive integers with no common divisors bigger than 1, what is $m + n$?

(A) 20   (B) 26   (C) 32   (D) 36   (E) 37

52. The number of elements in the intersection of the solution set of

$$\frac{\sqrt{2} \cdot x - 1}{2x^2 - 4\sqrt{2} \cdot x + 4} > 0$$

with the set $\{-7, -5, -2, -1, 1, \sqrt{2}, \sqrt{107}, \sqrt{108}\}$ is

(A) 0   **(B)** 1   **(C)** 2   **(D)** 3   **(E)** 4

53. Two rational numbers r and s are given. The numbers $r + s, r - s, rs$, and s/r are computed and arranged in order by value to get the list $1/3, 3/4, 4/3, 7/3$. What is the sum of the squares of r and s?

(A) $9/25$ **(B)** $4/9$ **(C)** $9/4$ **(D)** $25/9$ **(E)** 6

12.4 Word Problems

1. Oil is pumped into a non-empty tank at a changing rate. The volume of oil in the tank doubles every minute and the tank is filled in 10 minutes. How many minutes did it take for the tank to be half full?

(A) 2   **(B)** 5   **(C)** 7   **(D)** 8   **(E)** 9

2. A two-inch cube ($2 \times 2 \times 2$) of silver weighs 3 pounds and is worth \$320. How much is a three-inch cube of silver worth?

(A) \$480   **(B)** \$600   **(C)** \$800   **(D)** \$900   **(E)** \$1080

3. A grocer has c pounds of coffee that are divided equally among k sacks. She finds n empty sacks and decides to redistribute the coffee equally among the $k + n$ sacks. When this is done, how many fewer pounds of coffee does each of the original sacks hold?

(A) $\dfrac{c}{k + n}$ **(B)** $\dfrac{c}{k + cn}$ **(C)** $\dfrac{c}{k^2 + kn}$ **(D)** $\dfrac{cn}{k + n}$ **(E)** $\dfrac{cn}{k^2 + kn}$

4. Dick and Nick share their food with Albert. Dick has 5 loaves of bread, and Nick has 3 loaves. They share the bread equally. Albert gives Dick and Nick 8 dollars which they agree to share fairly. How should they divide the \$8 between them?

(A) Dick should get \$3 of Albert's money.

(B) Dick should get \$4 of Albert's money.

(C) Dick should get \$5 of Albert's money.

(D) Dick should get \$6 of Albert's money.

(E) Dick should get \$7 of Albert's money.

5. Cara has 162 coins in her collection of nickels, dimes, and quarters, which has a total value of $22.00. If Cara has twelve fewer nickels than quarters, how many dimes does she have?

(A) 50   **(B)** 60   **(C)** 70   **(D)** 74   **(E)** 78

6. Suppose the value of a new car declines linearly over a ten year period from the original value of $20,000 to the value $2,000. What is the value of the car after six years?

(A) $8,800   **(B)** $9,200   **(C)** $11,000   **(D)** $12,800   **(E)** $13,200

7. I have several quarters. If I divide my quarters into two unequal piles, then the difference of the squares of the number of coins in each pile is 24 times the difference of the number of coins in each pile. How much is my collection of quarters worth?

(A) $9   **(B)** $8   **(C)** $7   **(D)** $6   **(E)** $5

8. (*This question needs a calculator) An amount of $2000 is invested at $r\%$ interest compounded continuously. After four years, the account has grown to $2800. Assuming that it continues to grow at this rate for 16 more years, how much will be in the account?

(A) $8976.47   **(B)** $9874.23   **(C)** $10001.99   **(D)** $10756.48   **(E)** $2004.35

9. The students in Professor Einstein's class decided to reward the fine teacher with a CD player at the end of the course. A total of $529 was collected from the students, with each student contributing the same amount, which was equal to the total number of students in the class. Only ordinary US bills were used and none of these were $2 dollar bills. In addition, each student paid using the same five bills. How many ten-dollar bills were collected?

(A) 10   **(B)** 12   **(C)** 15   **(D)** 23   **(E)** 46

10. Mr Green sells apples for $1.50 each at the local Farmers Market and Ms Blue sells slightly smaller apples for $1 each. One day Ms Blue had to leave early so she asked Mr Green to manage her stall as the two were side-by-side. To make calculations easier, Mr Green mixed the apples together and changed the signs to read "5 apples for $6". At that point they had the same number of apples left. By the end of the day he had sold all the apples, but oddly (to him) when he compared how much each would have made by selling separately and how much he had in the till, he found he was 80 dollars short. He had no clue what the problem was, so he split the money evenly and apologized to Ms Blue for messing things up. Certainly, at least one of them lost money. Did both lose money on the deal, or did one come out ahead, and how much did each lose/gain?

(A) Blue lost $120, Green made $40 extra   **(B)** Blue lost $32 and Green lost $48

(C) Both lost $40   **(D)** Blue made $80 extra and Green lost $160

(E) Blue made $160 extra, Green lost $240

11. In a certain class, two-thirds of the female students and half of the male students speak Spanish. If there are three-fourths as many girls as boys in the class, what fraction of the entire class speaks Spanish?

(A) $\dfrac{5}{6}$ (B) $\dfrac{4}{7}$ (C) $\dfrac{4}{5}$ (D) $\dfrac{1}{3}$ (E) none of **A**, **B**, **C** or **D**

12. Three crazy painters painted the floor in three different colors. One painted 75% of the floor in red on Monday, the second painted 70% of the floor in green on Tuesday, and the third painted 65% of the floor in blue on Wednesday. At least what percent of the floor must be painted in all the three colors?

(A) 10% (B) 25% (C) 30% (D) 35% (E) 65%

13. A school has b boys and g girls, where $g < b$. How many girls must be enrolled so that 60% of the student body is female.

(A) $0.6g$ (B) $0.6g - 0.4b$ (C) $0.6b - 0.4g$ (D) $1.5b - g$ (E) $2b - g$

14. A chemist has a solution consisting of 5 ounces of propanol and 17 ounces of water. She would like to change the solution into a 40% propanol solution by adding z ounces of propanol. Which of the following equations should she solve in order to determine the value of z?

(A) $\dfrac{5}{z+17} = \dfrac{40}{100}$ (B) $\dfrac{z+5}{22} = \dfrac{40}{100}$ (C) $\dfrac{z+5}{17} = \dfrac{40}{100}$

(D) $\dfrac{z+5}{z+17} = \dfrac{40}{100}$ (E) $\dfrac{z+5}{z+22} = \dfrac{40}{100}$

15. Cucumbers contain 99% of water. After being exposed to the sun the amount of water drops to 98%. What percentage of weight did the cucumbers lose?

(A) 0.98% (B) 1% (C) 2% (D) 4% (E) 50%

16. Statistics have shown that in a certain college course, 65% of the students pass the first time they take it. Among those who have to repeat it, 70% pass on the second attempt, and among those who have to take it three times, 50% pass on the third attempt. What percentage of students have to take the course more than three times?

(A) 50% (B) 35% (C) 22.75% (D) 5.25% (E) 1%

17. The quantities x, y, and z are positive and $xy = \dfrac{z}{4}$. If x is increased by 50% and y is decreased by 25%, how must z be changed so that the relation $xy = \dfrac{z}{4}$ remains true?

(A) z must be decreased by 12.5% (B) z must be increased by 12.5%

(C) z must be decreased by 25% (D) z must be increased by 25%

(E) z must be increased by 50%

18. A running track has the shape shown below. The ends are semicircular with diameter 100 yards. Suppose that the lanes are each 1 yard wide and numbered from the inside to the outside. The competitor in the inside lane runs 700 yards counter clockwise. The other runners start ahead of the inside lane runner, and also run 700 yards, with all runners finishing at the same place. Approximately how much of a head start should a runner in the fifth lane receive over a runner in the first lane?

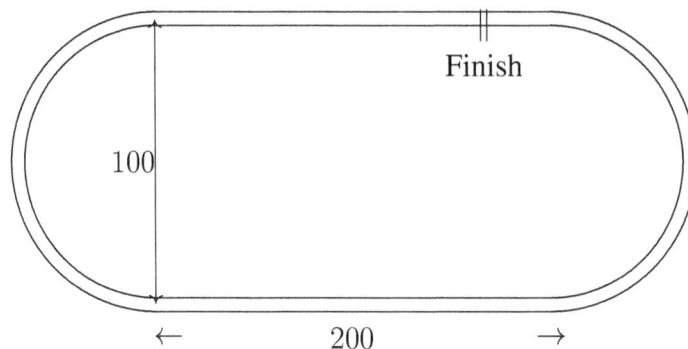

(A) 15 yards   **(B)** 20 yards   **(C)** 25 yards   **(D)** 30 yards   **(E)** 35 yards

19. A taxicab driver charges $3.00 for the first half mile or less and $0.75 for each quarter mile after that. If f represents the fare in dollars, which of the following functions models the fare for a ride of x miles, where x is a positive integer?

(A) $f(x) = 3.00 + 0.75(x - 1)$ **(B)** $f(x) = 3.00 + 0.75(x/4 - 1)$

(C) $f(x) = 4.50 + 3(x - 1)$ **(D)** $f(x) = 3.00 + 3(x - 1)$

(E) $f(x) = 3.00(x + 1)$

20. Elizabeth just completed a 10 mile bike trip. If she had been able to ride 2 miles per hour faster, she would have completed her trip in 20 fewer minutes. Find her speed to the nearest tenth of a mile per hour.

(A) 6.2   **(B)** 6.3   **(C)** 6.5   **(D)** 6.7   **(E)** 6.8

21. Jeremy starts jogging at a constant rate of five miles per hour. Half an hour later, David starts running along the same route at seven miles per hour. For how many minutes must David run to catch Jeremy?

(A) 75 minutes   **(B)** 80 minutes   **(C)** 90 minutes   **(D)** 95 minutes   **(E)** 105 minutes

22. Vic can beat Harold by one tenth of a mile in a two mile race. Harold can beat Charlie by one fifth of a mile in a two mile race. If Vic races Charlie, how far ahead will Vic finish?

(A) 0.15 miles   **(B)** 0.22 miles   **(C)** 0.25 miles   **(D)** 0.29 miles   **(E)** 0.33 miles

23. It takes Mathias and Anders 1188 hours to paint the Gaffney Peach. It takes Anders and Tellis 1540 hours to paint the peach; for Tellis and Hal, it takes 1890 hours; and for Hal and Mathias, it takes 1386 hours. How long would it take all four of them working together to paint the peach?

(A) 364.7 hours   (B) 412.3 hours   (C) 670.7 hours   (D) 729.5 hours   (E) 824.6 hours

24. A cyclist rides his bicycle over a route which is $\frac{1}{3}$ uphill, $\frac{1}{3}$ level, and $\frac{1}{3}$ downhill. If he covers the uphill part of the route at the rate of 16 miles per hour and the level part at the rate of 24 miles per hour, what rate in miles per hour would he have to travel the downhill part of the route in order to average 24 miles per hour for the entire route?

(A) 32   (B) 36   (C) 40   (D) 44   (E) 48

25. Benny eats a box of cereal in 14 days. He eats the same size box of cereal with his younger brother Nathan in 10 days. How many days will it take Nathan to finish the box of cereal alone?

(A) 20   (B) 25   (C) 30   (D) 35   (E) 40

26. Some hikers start on a walk at 9 a.m. and return at 2 p.m. One quarter of the distance walked is uphill, one half is level, and one quarter is downhill. If their speed is 4 miles per hour on level land, 2 miles per hour uphill, and 6 miles per hour downhill, approximately how far did they walk?

(A) 16.4 miles   (B) 17.1 miles   (C) 18.9 miles   (D) 20.0 miles

(E) 21.2 miles

27. Danica was driving in a 500 mile race. After 250 miles, Danica's average speed was 150 miles per hour. Approximately how fast should Danica drive the second half of the race if she wants to attain an overall average of 180 miles per hour?

(A) 210   (B) 215   (C) 220   (D) 225   (E) 230

28. Two armies are advancing towards each other, each one at 1 miles per hour (mph). A messenger leaves the first army when the two armies are 10 miles apart and runs towards the second at 9 mph. Upon reaching the second army, he immediately turns around and runs towards the first army at 9mph. How many miles apart are the two armies when the messenger gets back to first army?

(A) 5.6   (B) 5.8   (C) 6   (D) 6.2   (E) 6.4

29. Jack and Lee walk around a circular track. It takes Jack and Lee respectively 6 and 10 minutes to finish each lap. They start at the same time, at the same point on the track, and walk in the same direction around the track. After how many minutes will they be at the same spot again (not necessarily at the starting point) for the first time after they start walking?

(A) 15   (B) 16   (C) 30   (D) 36   (E) 60

30. Bill and his dog walk home from the shopping center. It takes Bill 36 minutes and his dog walks twice as fast. They start together, but the dog reaches home before Bill and returns to meet Bill. After meeting Bill, the dog walks home, again at double speed, and then turns back to meet Bill again. Bill starts at noon to walk home. How many minutes later does he meet the dog for the second time?

(A) 24   (B) 27   (C) 30   (D) 32   (E) 34

31. On a recent trip from Northburg to Southtown, Jill decided to make a detour so she could pass through Center City. Forty minutes after she left Northburg, she noted that the remaining distance to Center City was twice as much as what she had traveled so far. After traveling another twenty one miles, she calculated that the remaining distance to Southtown was twice as much as what she had left to get to Center City. She arrived in Southtown an hour and a half later. Assuming she traveled at a constant speed, how long was this trip from Northburg to Southtown?

(A) 99 miles   (B) 108 miles   (C) 112 miles   (D) 127 miles   (E) 142 miles

32. A ball is dropped onto a floor from a height of 1 meter. Each time that the ball hits the floor it rebounds to half its previous height. (After falling one meter it rebounds to a height of 1/2 meter. The next time it hits the floor, it rebounds to a height of 1/4 meter, etc.). How far has the ball traveled when it hits the floor for the 40th time?

(A) $T = 2 + (2^{38} - 1)/2^{38}$ (B) $T = 1 + (2^{38} - 1)/2^{39}$ (C) $T = 2$

(D) $T = 3$ (E) $T > 3$

33. An athlete covers three consecutive miles by swimming the first, running the second and cycling the third. He runs twice as fast as he swims and cycles one and a half times as fast as he runs. He takes ten minutes longer than he would do if he cycled the whole three miles. How many minutes does he take?

(A) 16   (B) 22   (C) 30   (D) 46   (E) 70

34. The hour and minute hands of a clock are exactly 50 degrees apart at which of these times?

(A) $2:20$ (B) $5:36$ (C) $7:50$ (D) $9:30$ (E) $11:10$

35. How many times in a 24 hour period do the hour and minute hands of a clock form a right angle?

(A) 48   (B) 44   (C) 34   (D) 24   (E) None of these.

36. One hundred monkeys have 100 apples to divide. Each adult gets three apples while three children share one. How many adult monkeys are there?

(A) 10   (B) 20   (C) 25   (D) 30   (E) 33

37. A pine tree is 14 yards high, and a bird is sitting on its top. The wind blows away a feather of the bird. The feather moves uniformly along a straight line at the speed 4 yards per second; it falls on the ground 4.5 seconds later at a distance D yards from the pine tree's base. Which of the following intervals contains the number D?

(**A**) $(0, 10]$ (**B**) $(10, 11]$ (**C**) $(11, 12]$ (**D**) $(12, 13]$ (**E**) $(13, \inf)$

38. Three cowboys entered a saloon. The first ordered 4 sandwiches, a cup of coffee, and 10 doughnuts for $8.45. The second ordered 3 sandwiches, a cup of coffee, and 7 doughnuts for $6.30. How much did the third cowboy pay for a sandwich, a cup of coffee, and a doughnut?

(**A**) $2.00   (**B**) $2.05   (**C**) $2.10   (**D**) $2.15   (**E**) $2.20

39. A treasure is located at a point along a straight road with towns $A, B, C,$ and D in that order. A map gives the following instructions for locating the treasure:

(a) Start at town A and go $1/2$ of the way to C.

(b) Then go $1/3$ of the way towards D.

(c) Then go $1/4$ of the way towards B, and dig for the treasure.

If $AB = 6$ miles, $BC = 8$ miles, and the treasure is buried midway between A and D, find the distance from C to D.

(**A**) 4   (**B**) 6   (**C**) 8   (**D**) 10   (**E**) None of the above

40. An army is moving along in a convoy that stretches for three miles. Observing radio silence, the general at the rear of the convoy sends a message to the front via a special courier. After delivering the message, the courier returns to the rear. Both the convoy and the courier travel at (different but) constant rates. If the front (and rear) of the convoy travel six miles in the time it takes for the courier to go to the front and return to the back, what is the total distance in miles the courier travels?

(**A**) $5\sqrt{3}$ (**B**) $12 - 3\sqrt{3/4}$ (**C**) $3 + 3\sqrt{5}$ (**D**) $6\sqrt{3}$ (**E**) $6 + 3\sqrt{3}$

41. Surveyors are laying out a rather unusual road through a park. The park is completely flat and forms a disc of radius 20 miles. From the center C, the road is to go exactly two miles north to a point B_0, then make a 90° left turn and go another two miles to a point B_1. At B_1 the road turns left again (not nearly as sharply), this time perpendicular to $\overline{CB_1}$. As before, it goes exactly two miles in this direction to a point B_2. This pattern is followed for the entire road – at B_k, the road makes a left turn that is perpendicular to $\overline{CB_k}$ and goes exactly two miles in this direction to B_{k+1}. Eventually the road reaches a point A (one of the B_js) that is exactly 10 miles from C. Starting from A, how many **more** two-mile segments will be needed before the road gets out of the park?

[In the figure, two consecutive segments are shown starting from an arbitrary point X to the point Y and then from Y to Z.]

(A) fewer than 30   (B) between 30 and 49   (C) between 50 and 69

(D) between 70 and 89   (E) at least 30

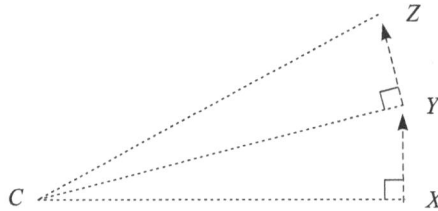

42. Joe lives near a river where he goes swimming every day: he swims 1 mile upstream, 1 mile downstream, and exits the river at the same place as he entered it. Recently Joe went on a vacation to a lake, where he noticed that during his workouts lasting the same time, swimming at the same constant speed, he is able to swim 2.2 miles each day. How much faster is Joe swimming than the speed of the river? Round your answer to two decimal places.

(A) 2.5 times   (B) 3.32 times   (C) 3.5 times   (D) 4.1 times   (E) 11 times

43. John was contracted to work A days. For each of these A days that John actually worked, he received B dollars. For each of these A days that John didn't work, he had to pay a penalty of C dollars. After the A days of contracted work was over, John received a net amount of D dollars for his work. How many of the A days of contracted work did John not work?

(A) $(AB - D)/(B + C)$ (B) $(AB + D)/(B + C)$ (C) $(AB - D)/(B - C)$

(D) $(AB + D)/(B - C)$ (E) $(AC - B)/(D - C)$

44. The Mainter brothers, Abe, Ben and Cal paint houses. They have been in the business so long that each knows exactly how many square feet he (and each of his brothers) paints in one hour and these rates never change. For their latest job, they calculated that if Abe and Ben did the job together, it would take exactly 11 hours. On the other hand, if Abe and Cal did the job together, it would take exactly 9 hours. Finally, Ben and Cal could do it in exactly 9.9 hours (or if you prefer, 9 hours and 54 minutes). They decided all three would paint this particular house. Ben and Cal started the job at 8 AM and Abe joined them at 9:00. Cal left at 1:30 and so Ben and Abe finished the job. After deducting the supply costs (paint etc.), the brothers split the net profit based on what percentage of the total square footage each painted. Who earned the most and who earned the least for this job?

(A) Abe the most, Ben the least   (B) Abe the most, Cal the least

(C) Ben the most, Abe the least   (D) Cal the most, Ben the least

(E) Cal the most, Abe the least

12.5 Sequence and Series

1. In a 10-team baseball league, each team plays each of the other teams 18 times. No game ends in a tie, and, at the end of the season, each team is the same positive number of games ahead of the next best team. What is the greatest number of games that the last place team could have won.

(**A**) 27 (**B**) 36 (**C**) 54 (**D**) 72 (**E**) 90

2. For how many positive integers n do there exist n consecutive integers that sum to -1? (The sum of 1 consecutive integer is just the number itself.)

(**A**) 0 (**B**) 1 (**C**) 2 (**D**) 3 (**E**) 4

3. On January 1st, there is one bean in a bin. On the 2nd, we add two beans, and then on the 3rd, we add six beans. Continuing, we add twelve beans on the 4th and thirty six beans on the 5th. Following this pattern, we continue to add beans that alternate between doubling and tripling the number added on the previous day. What is the total number of beans in the bin after we make the addition on the 24th?

(**A**) $(6^{12}-1)/5$ (**B**) $2(6^{12}-1)/5$ (**C**) $3(6^{12}-1)/5$

(**D**) $2(6^{13}-1)/5$ (**E**) $3(6^{13}-1)/5$

4. Let $a_n = \dfrac{1}{\sqrt{n}+\sqrt{n+1}}$. Find the sum $a_1 + a_2 + a_3 + \ldots + a_{99}$.

(**A**) 6 (**B**) 8 (**C**) 9 (**D**) 12 (**E**) 15

5. Find the sum

$$\frac{1}{1\cdot 2} + \frac{1}{2\cdot 3} + \cdots + \frac{1}{10\cdot 11}.$$

(**A**) 1 (**B**) 2 (**C**) $\frac{10}{11}$ (**D**) $\frac{11}{12}$ (**E**) none of **A**, **B**, **C** or **D**

6. What is the smallest value of the positive integer n for which

$$\frac{1}{1\cdot 2} + \frac{1}{2\cdot 3} + \frac{1}{3\cdot 4} + \cdots + \frac{1}{n\cdot (n+1)}$$

is at least 1?

Hint: Rewrite each $\dfrac{1}{k\cdot (k+1)}$ in the form $\dfrac{1}{a} - \dfrac{1}{b}$.

(**A**) 10 (**B**) 100 (**C**) 1000 (**D**) 2002 (**E**) there is no such value of n

7. After a ship wreck, a surviving mouse finds himself on an uninhabited island with one kilogram of cheese and no other food. On the 1st day, he eats $1/4$ of the cheese, on the 2nd, $1/9$ of the remaining cheese, on the 3rd, $1/16$ of the remainder, and so on. Let M be the total amount of cheese (measured in kilograms) consumed during the first 6 weeks. Which of the following statements is true?

(A) $0.4878 < M \leq 0.4881$ **(B)** $0.4881 < M \leq 0.4884$

(C) $0.4884 < M \leq 0.4887$ **(D)** $0.4887 < M \leq 0.4890$

(E) $0.4890 < M \leq 0.4893$

8. Consider the Fibonacci sequence $1, 1, 2, 3, 5, 8, 13, 21 \ldots$ where each term, after the first two, is the sum of the two previous terms. How many of the first 1000 terms are divisible by 3?

(A) 200 **(B)** 250 **(C)** 299 **(D)** 300 **(E)** 333

9. Find the value of the expression

$$S = 1! \cdot 3 - 2! \cdot 4 + 3! \cdot 5 - 4! \cdot 6 + \ldots - 2006! \cdot 2008 + 2007!$$

(A) -2007 **(B)** -1 **(C)** 0 **(D)** 1 **(E)** 2007

10. It is known that $\sum_{n=1}^{\infty} \frac{1}{n^2} = \frac{\pi^2}{6}$. What is the value of

$$\sum_{n=1}^{\infty} \frac{1}{(2n-1)^2}?$$

(A) $\frac{\pi^2}{36}$ **(B)** $\frac{\pi^2}{12}$ **(C)** $\frac{\pi^2}{8}$ **(D)** $\frac{\pi^2}{7}$ **(E)** $\frac{2\pi^2}{9}$

11. Let $S = 1 + 1/2^2 + 1/3^2 + \cdots + 1/100^2$. Which of the following is true?

(A) $S < 1.40$ **(B)** $1.40 \leq S < 2$ **(C)** $2 \leq S < 4$ **(D)** $4 \leq S < 100$

(E) None of the above

12.6 Defined Operations and Functions

1. If the operation $\oplus$ is defined for all positive x and y by $x \oplus y = (xy)/(x+y)$, which of the following must be true for positive x, y, and z?

   i. $x \oplus x = x/2$

  ii. $x \oplus y = y \oplus x$

 iii. $x \oplus (y \oplus z) = (x \oplus y) \oplus z$

(A) i. only **(B)** i. and ii. only **(C)** i. and iii. only **(D)** ii. and iii. only

(E) all three

2. If f is a function such that $f(3) = 2$, $f(4) = 2$ and $f(n+4) = f(n+3) \cdot f(n+2)$ for all the integers $n \geq 0$, what is the value of $f(6)$?

(A) 4   (B) 5   (C) 6   (D) 8   (E) cannot be determined

3. Suppose that
$$f(n+1) = f(n) + f(n-1)$$
for $n = 2, 3, \ldots$. Given that $f(6) = 23$ and $f(4) = 8$, what is $f(1) + f(3)$?

(A) 6   (B) 7   (C) 8   (D) 12   (E) 13

4. If
$$f(x, y) = (\max(x, y))^{\min(x,y)}$$
and
$$g(x, y) = \max(x, y) - \min(x, y),$$
then
$$f\left(g\left(-1, -\frac{3}{2}\right), g\left(-4, -1.75\right)\right) =$$

(A) -0.5 (B) 0   (C) 0.5   (D) 1   (E) 1.5

5. Consider the function $F : N \to N$ defined by
$$F(n) = \begin{cases} n/3 & \text{if } n \text{ is a multiple of } 3 \\ 2n + 1 & \text{if otherwise} \end{cases}$$

For how many positive integers k is it true that $F(F(F(k))) = k$?

(A) 0   (B) 1   (C) 2   (D) 3   (E) 4

6. Let
$$g(x) = \begin{cases} |x| - 2 & \text{if } x \leq 0 \\ x - 3 & \text{if } 0 < x < 4 \\ 3 - x & \text{if } 4 \leq x \end{cases}$$

Find a number x such that $g(x) = -4$.

(A) -2 (B) -1 (C) 3   (D) 4   (E) 7

7. Suppose that
$$f(n) = 2f(n+1) - f(n-1)$$
for $n = 0, \pm1, \pm2, \pm3, \ldots$ and $f(1) = 4$ and $f(-1) = 2$. Evaluate $f(2)$.

(A) 1   (B) 2   (C) 3   (D) 4   (E) 5

8. Let f be the function whose graph is shown. Which of the following represents the graph of $f(|x|)$?

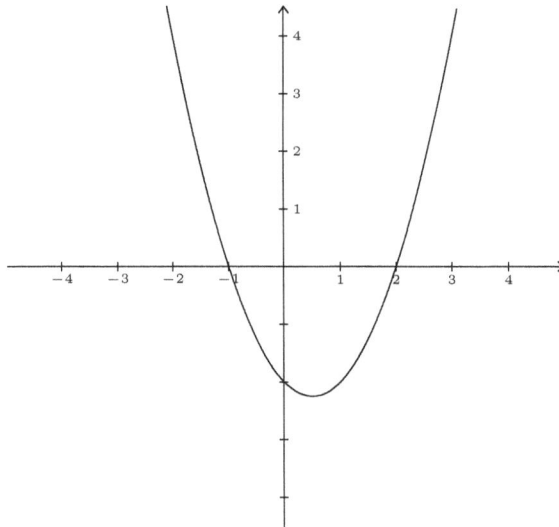

(A) 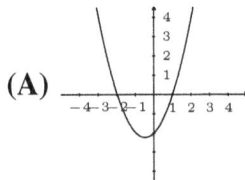   **(B)** 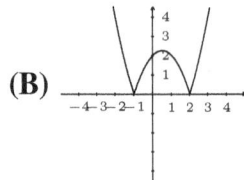   **(C)**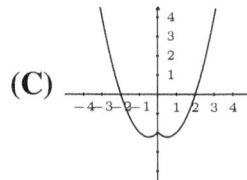

(D) 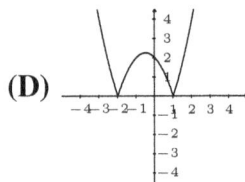   **(E)**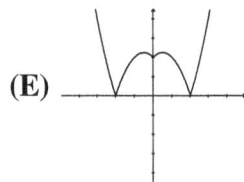

9. How many positive integers are in the range of the function $f(x) = -2x^2 + 8x - 3$?

(A) 0   **(B)** 2   **(C)** 3   **(D)** 5   **(E)** 6

10. Let $f(x) = (x - b)/(x - a)$ for constants a and b. If $f(2) = 0$ and $f(1)$ is undefined. What is $f(1/2)$?

(A) 0   **(B)** 1   **(C)** 2   **(D)** 3   **(E)** 4

11. The graph of the quadratic function $y = ax^2 + bx + c$ is shown below.

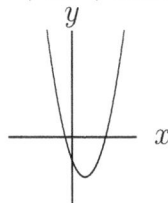

Which of the following is true?

(A) $ac < 0$ and $ab < 0$ **(B)** $ac < 0$ and $ab > 0$ **(C)** $ac > 0$ and $ab < 0$

(D) $ac > 0$ and $ab > 0$ **(E)** At least one of $a, b,$ and c could be zero.

12. Suppose $f(0) = 3$ and $f(n) = f(n-1) + 2$. Let $T = f(f(f(f(5))))$. What is the sum of the digits of T?

(**A**) 6   (**B**) 7   (**C**) 8   (**D**) 9   (**E**) 10

13. Suppose f is a real function satisfying $f(x + f(x)) = 4f(x)$ and $f(1) = 4$. What is $f(21)$?

(**A**) 16   (**B**) 21   (**C**) 64   (**D**) 105   (**E**) none of **A, B, C** or **D**

14. Suppose that $f(x) = ax + b$, where a and b are real numbers. Given that $f(f(f(x))) = 8x + 21$, what is the value of $a + b$?

(**A**) 2   (**B**) 3   (**C**) 4   (**D**) 5   (**E**) 6

15. Let $f(x) = \dfrac{x-1}{x+1}$ and let $f^{(n)}(x)$ denote the n-fold composition of f with itself. That is, $f^{1}(x) = f(x)$ and $f^{(n)}(x) = f\left(f^{(n-1)}(x)\right)$. Which of the following is $f^{(2007)}(x)$?

(**A**) $-\dfrac{1}{x}$ (**B**) $-\dfrac{x+1}{x-1}$ (**C**) $\dfrac{1}{x}$ (**D**) $\dfrac{1-x}{1+x}$ (**E**) $\dfrac{x-1}{x+1}$

16. Let $f(n)$ be the integer closest to $\sqrt[4]{n}$. Then $\displaystyle\sum_{i=1}^{1995} \dfrac{1}{f(i)} =$

(**A**) 375   (**B**) 400   (**C**) 425   (**D**) 450   (**E**) 500

CHAPTER 13

GEOMETRY

13.1 Triangle Geometry

1. For what positive value of x is there a right triangle with sides $x + 1, 4x$, and $4x + 1$?

(A) 4   (B) 6   (C) 8   (D) 10   (E) 12

2. Let ABC be an equilateral triangle with an inscribed circle of radius 1, find the length of AB.

(A) $\sqrt{2}$ (B) $2\sqrt{2}$ (C) $\sqrt{3}$ (D) $2\sqrt{3}$ (E) $3\sqrt{3}$

3. In the right triangle ABC shown, D is on $\overline{AC}$, $\angle A = 30°$, $\angle BDC = 60°$, and $AD = 4$. Find BC

(A) 3   (B) $2\sqrt{3}$ (C) $\sqrt{14}$ (D) 4   (E) $3\sqrt{2}$

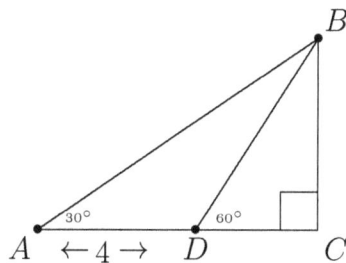

4. In right triangle ABC, the point D on $\overline{AB}$ is 4 units from A, $\angle CDB = 60°$ and $\angle CAB = 30°$. What is the altitude h?

(A) 3   (B) $2\sqrt{3}$ (C) $\sqrt{14}$ (D) 4   (E) $3\sqrt{2}$

C

h

$30°$ 4 $60°$

A D B

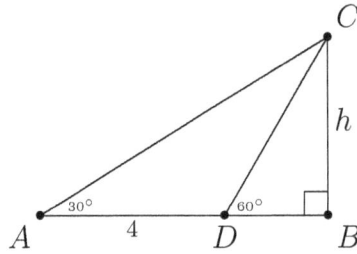

5. In the right triangle ABC the segment CD bisects angle C, $AC = 15$, and $BC = 9$. Find the length of $\overline{CD}$.

(A) $9\sqrt{5}/2$ **(B)** 11 **(C)** $9\sqrt{6}/2$ **(D)** 12 **(E)** $15\sqrt{3}/2$

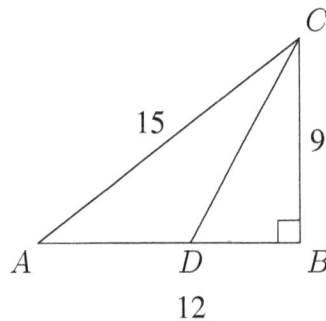

C

15 9

A D B

12

6. A triangle is inscribed in a semi-circle of radius r as shown in the figure:

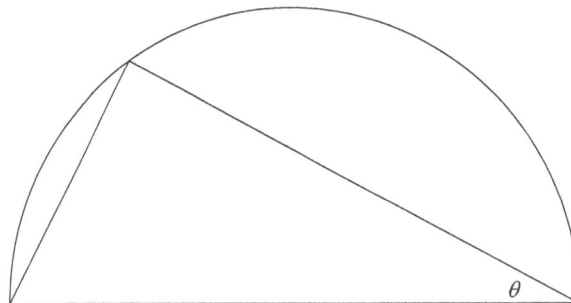

θ

The area of the triangle is

(A) $r^2 \sin 2\theta$ **(B)** $\pi r^2 - \sin\theta$ **(C)** $r \sin\theta \cos\theta$ **(D)** $\pi r^2/4$ **(E)** $\pi r^2/2$

7. There are three triangles of different sizes: small, medium and large. The small one is inscribed in the medium one such that its vertices are at the midpoints of the three edges of the medium one. The medium triangle is inscribed in the large triangle in the same way as shown in the figure. If the small triangle has area 1, what is the sum of the areas of the three triangles?

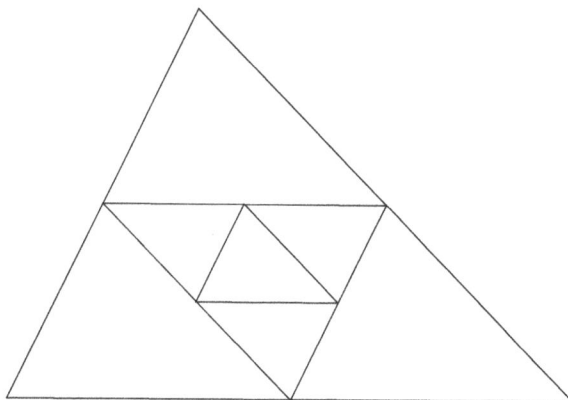

(A) 14   (B) 16   (C) 19   (D) 21   (E) 25

8. What is the area of the largest rectangular region that can be inscribed in a right triangle with legs of length 3 and 4?

   (A) 2   (B) 2.5   (C) 3   (D) 3.5   (E) 4

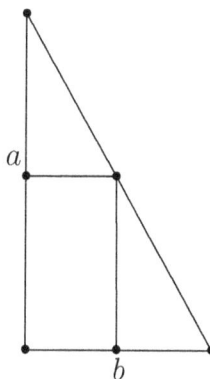

9. The hypotenuse of an isosceles right triangle has a length of b units. What is the area of the triangle.

   (A) b (B) b^2 (C) $b^2/4$ (D) $2b^2$ (E) $\sqrt{2b^2}$

10. The area of $\triangle ABC$ is 144 square units. Point U is on $\overline{AB}$ such that the ratio AU to UB is 5 to 7. Point V is on $\overline{BC}$ such that the ratio BV to VC is 2 to 1. What is the area of $\triangle UVB$?

    (A) 50   (B) 54   (C) 55   (D) 56   (E) 60

11. Let A be the area of a triangle with sides $5, 5$, and 8, and let B denote the area of a triangle with sides $5, 5$, and 6. Which of the following is true.

    (A) $A < B < 12$ (B) $B < A < 12$ (C) $A = B$ (D) $12 < A < B$ (E) $12 < B < A$

12. An isosceles right triangular region of area 25 is cut from a corner of a rectangular region with sides of length $5\sqrt{2}$ and $5(1 + \sqrt{2})$. What is the perimeter of the resulting trapezoid?

    (A) 25   (B) 35   (C) $20 + 10\sqrt{2}$ (D) $10 + 20\sqrt{2}$ (E) $15 + 15\sqrt{2}$

13. One leg of a right triangle is two meters longer than twice the length of the other leg. The hypotenuse is four meters less than the sum of the lengths of both legs. What is the perimeter of the triangle, in meters?

(A) 20   (B) 25   (C) 30   (D) 35   (E) 50

14. The sides a, b, and c of a triangle satisfy $\sqrt{a} + \sqrt{b} = \sqrt{c}$. Which of the following best describes the triangle?

(A) acute   (B) scalene   (C) isosceles   (D) non-existent   (E) equilateral

15. The sides of a right triangle are $a, 2a + 2d$ and $2a + 3d$, with a and $d > 0$. The ratio of a to d is:

(A) $5 : 1$ (B) $27 : 2$ (C) $4 : 1$ (D) $1 : 5$ (E) $2 : 3$

16. Consider triangle ABC with $AB = x, BC = x + 1$, and $CA = x + 2$. Which of the following statements must be true?

I. $x \geq 1$

II. $x \leq 2\sqrt{3}$

III. Angle $C < 60$ degrees.

(A) I only   (B) II only   (C) III only   (D) I and II   (E) I and III

17. In triangle ABC, $AB = x$, $BC = x + 1$, and $AC = x + 2$. Which of the following **must** be true?

i. $x \geq 1$

ii. $x \leq 5\sqrt{2}$

iii. $\angle C \leq 60°$

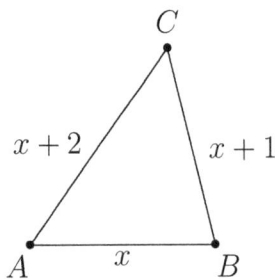

(A) i only   (B) ii only   (C) iii only   (D) i and ii only   (E) i and iii only

18. Four congruent triangular corners are cut off an 11×13 rectangle. The resulting octagon has eight edges of equal length. What is this length?

(A) 3   (B) 4   (C) 5   (D) 6   (E) 7

19. Two sides of an isosceles triangle have length 5 and the third has length 6. What is the radius of the inscribed circle?

(A) 1   (B) 1.25   (C) 1.5   (D) 1.75   (E) 2

13.2 Circle Geometry

1. The area of an annular region bounded by two concentric circles is 5π square centimeters. The difference between the radii of the circles is one centimeter. What is the radius of the smaller circle, in centimeters?

(**A**) 1   (**B**) 2   (**C**) 3   (**D**) 4   (**E**) 6

2. Three mutually tangent circles have centers A, B, C and radii a, b, and c respectively. The lengths of segments AB, BC, CA are 12, 17, and 23 respectively. Find the lengths of the radii.

(**A**) $a = 9, b = 7, c = 13$ (**B**) $a = 5, b = 6, c = 12$ (**C**) $a = 9, b = 6, c = 12$

(**D**) $a = 8, b = 4, c = 12$ (**E**) $a = 9, b = 3, c = 14$

3. Five circles with equal radii are situated in the plane so that each is tangent to two others and externally tangent to a unit circle. Find the radius of each of the five circles, rounded to two decimal places.

(**A**) 1.39   (**B**) 1.41   (**C**) 1.43   (**D**) 1.45   (**E**) 1.47

4. What is the area of the region common to two unit circles whose centers are $\sqrt{2}$ apart?

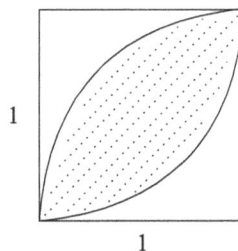

(A) $\dfrac{1}{2}\left(1-\dfrac{\pi}{4}\right)$ **(B)** $\dfrac{\pi}{2}$ **(C)** $1-\dfrac{\pi}{4}$ **(D)** $\dfrac{\pi}{2}-1$ **(E)** $\dfrac{1}{2}$

5. Two circles, A and B, overlap each other as shown. The area of the common part is $2/5$ of the area of Circle A, and $5/8$ of the area of Circle B. What is the ratio of the radius of Circle A to that of Circle B?

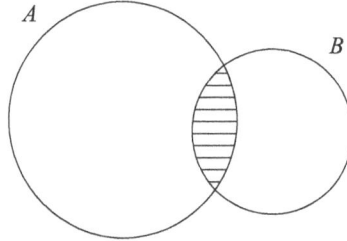

(A) $2:1$ **(B)** $3:2$ **(C)** $4:3$ **(D)** $5:4$ **(E)** $6:5$

6. A 12×12 square is divided into n^2 congruent squares by equally spaced lines parallel to its sides. Circles are inscribed in each of the squares. Find the sum of the areas of the circles.

(A) 6π **(B)** 12π **(C)** 24π **(D)** 36π **(E)** the answer depends on n

7. Let A, B be the centers of two circles with radii 1 and assume that $AB = 1$. Find the area of the region enclosed by arcs CAD and CBD.

(A) 1 **(B)** $\dfrac{2\pi}{3}$ **(C)** $\dfrac{2\pi}{3}-\dfrac{\sqrt{3}}{2}$ **(D)** $\dfrac{\sqrt{3}}{2}$ **(E)** $\sqrt{3}$

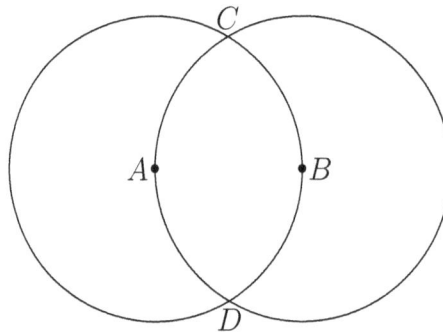

8. A right triangle ABC is given. Semicircles are constructed with the sides of the triangle as diameters, as shown below. Suppose the area of the largest semicircle is 36 and the area of the smallest one is 16. What is the area of the other one?

(A) 20 **(B)** 24 **(C)** 25 **(D)** 26 **(E)** 30

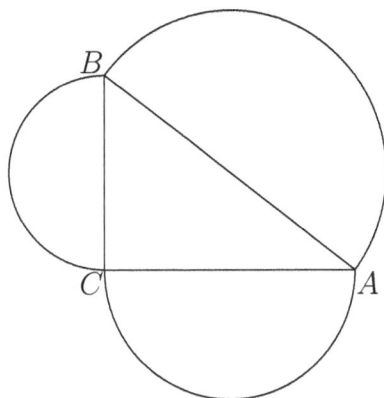

9. In the figure below, the two triangles are right triangles with sides of lengths x, y, p, and q, as shown. Given that $x^2 + y^2 + p^2 + q^2 = 72$, find the circumference of the circle.

(A) 8π (B) 9π (C) 12π (D) 24π (E) 36π

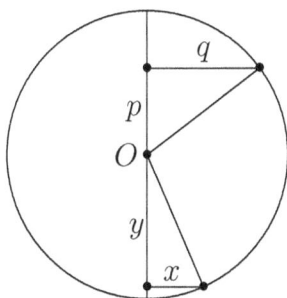

10. Two circles with centers O and O' have radii of 9 inches and 12 inches, respectively. The centers are 28 inches apart. How far from the center of circle O is the intersection of the line joining the centers with the common internal tangent PP'?

(A) 9 in.   (B) 10 in.   (C) 11 in.   (D) 12 in.   (E) 13 in.

11. A chain with two links is 13 cm long. A chain made from three links of the same type is 18 cm long. How long is a chain made from 25 such links?

(A) 120   (B) 128   (C) 136   (D) 144   (E) 150

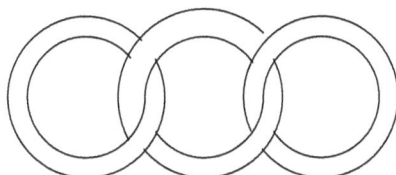

12. Three chords in a circle have lengths a, b, and c, where $c = a + b$. If the chord of length a subtends an arc of 30° and the chord of length b subtends an arc of 90°, then the number of degrees in the smaller arc subtended by the chord of length c is

(A) 120   (B) 130   (C) 140   (D) 150   (E) 160

13. Two parallel lines are one unit apart. A circle of radius 2 touches one of the lines and cuts the other line. The area of the circular cap between the two parallel lines can be written in the form $a\pi/3 - b\sqrt{3}$. Find the sum $a + b$ of the two integers a and b.

(A) 3 **(B)** 4 **(C)** 5 **(D)** 6 **(E)** 7

14. The area of the $a \times b$ rectangle shown in the picture below is $\dfrac{6}{5\pi}$ of the area of the circle. Assuming $b > a$, what is the value of $\dfrac{b}{a}$?

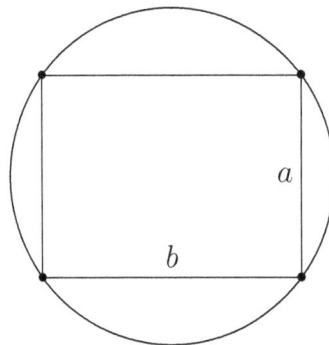

(A) 2 **(B)** 2.5 **(C)** 3 **(D)** 3.25 **(E)** 3.5

15. A robot arm in a plane has two sections, $\overline{RS}$ and $\overline{ST}$. The arm is fixed at a pivot at point R and can turn all the way around there. The arm is hinged at S and the two pieces can make any angle there. If section $\overline{RS}$ has length a and section $\overline{ST}$ has length b with $b < a$, what is the area of the region in the plane that is composed of all the points that end T of the second section of the arm can touch? [Two possible positions are shown in the figure below.]

(A) $(a^2 - b^2)\pi$ **(B)** $(a^2 + b^2)\pi$ **(C)** $ab\pi$ **(D)** $2ab\pi$ **(E)** $4ab\pi$

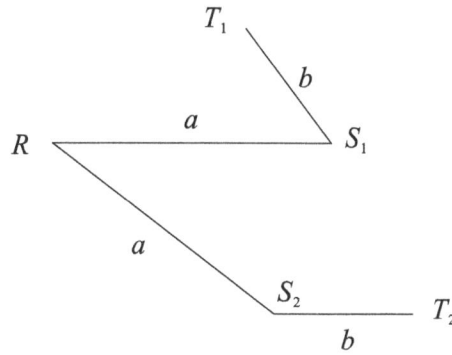

16. A circle of radius r is 'rolled' horizontally until the point A at the top becomes the bottom A' as shown. The distance $A'F$ is r. The point C is the intersection of the vertical line through A' and the circle with diameter FG. What is the area of the square $A'CDE$?

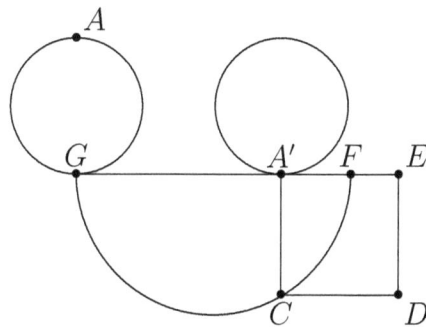

(A) $3r^2$ **(B)** πr^2 **(C)** $4\pi r^2/3$ **(D)** $2\pi r^2$ **(E)** $4r^2$

17. Find the radius of the circle inscribed in a triangle whose sides are 8, 15, and 17.

(A) 2.5   **(B)** 2.7   **(C)** 2.9   **(D)** 3.0   **(E)** 3.2

18. In the figure below, the circle is inscribed in an isosceles triangle ABC, with segment $\overline{AP}$ passing through the center O of the circle, $AC = AB = 12$ and $BP = 4$. Find the radius r of the circle.

(A) $2\sqrt{2}$ **(B)** $3\sqrt{2}$ **(C)** $4\sqrt{2}$ **(D)** $6\sqrt{2}$ **(E)** $12 + \sqrt{2}$

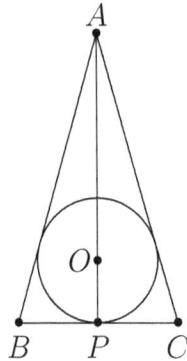

19. Seven circles of radius 10 are arranged as in the figure. Note that the six outer circles all pass through the center of the inner circle, the inner circle passes through the center of each outer circle, and each outer circle passes through the center of the two outer circles it is adjacent to. The area of the shaded region is A. Which of the following is true about A?

(A) $40 \leq A < 60$ **(B)** $60 \leq A < 80$ **(C)** $80 \leq A < 100$

(D) $100 \leq A < 120$ **(E)** $120 \leq A < 140$

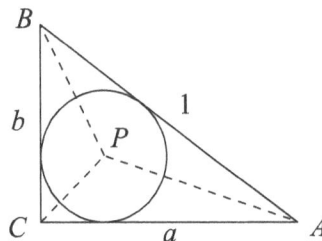

20. A circle is inscribed in a right triangle whose hypotenuse is 1. What is the largest possible radius of such a circle?

(A) $\left(\sqrt{2} - 1\right)/2$ **(B)** $\sqrt{2} - 1$ **(C)** $1/2$ **(D)** $\left(\sqrt{2} + 1\right)/4$ **(E)** $\sqrt{2}/2$

13.3 Polygons

1. Three adjacent squares rest on a line. Line L passes through a corner of each square as shown in the following figure. The lengths of the sides of the two smaller squares are 4 cm and 6 cm. Find the length of one side of the largest square.

   (A) 8   **(B)** 9   **(C)** 10   **(D)** 12   **(E)** 14

2. We want to divide the L shaped region shown below into two pieces with equal areas by means of a line from P to Q. The point P is always in the upper left hand corner of the region and the point Q must lie along the bottom edge as shown. When this is done, which of the following numbers is closest to the distance from A to Q?

   (A) 1.2   **(B)** 1.3   **(C)** 1.4   **(D)** 1.5   **(E)** 1.6

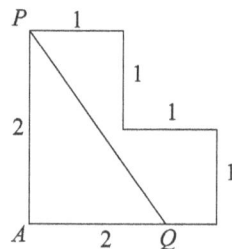

3. Consider a quadrilateral $ABCD$ with $AB = 4$, $BC = 10\sqrt{3}$, and $\angle DAB = 150°$, $\angle ABC = 90°$, and $\angle BCD = 30°$. Find DC.

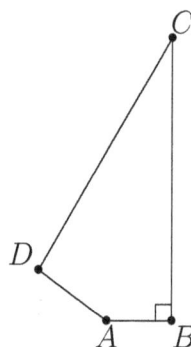

   (A) 16   **(B)** 17   **(C)** 18   **(D)** 19   **(E)** 20

4. An equilateral triangle and a regular hexagon have the same perimeter. What is the ratio of the area of the triangle to the area of the hexagon?

(A) $1/2$ (B) $2/3$ (C) $3/4$ (D) $\sqrt{2}/2$ (E) $\sqrt{3}/3$

5. In the diagram, $ABCD$ is a square and P is a point on the circle with diameter CD, $CP = 7$, and $PD = 11$. What is the area of the square?

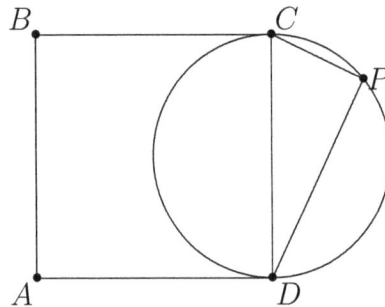

(A) 144   (B) 169   (C) 170   (D) 180   (E) 225

6. Quadrilateral $ABCD$ with the sides $AB = 20, BC = 7, CD = 24$ and $DA = 15$ has right angles at A and C. What is the area of $ABCD$?

(A) 154   (B) 186   (C) 200   (D) 234   (E) 286

7. Let $ABCD$ be a convex quadrilateral with the area s and let $P, Q, R,$ and S be the midpoints of sides $AB, BC, CD,$ and DA respectively. The sum of the areas of the triangles PBQ and RDS equals

(A) $3s/4$ (B) $2s/3$ (C) $s/2$ (D) $s/4$ (E) the ratio in question cannot be determined

8. Two squares, each with side length 12 inches, are placed so that the corner of one lies at the center of the other (see the diagram below). Suppose the length of BJ is 3. What is the area of the quadrilateral EJCK?

(A) 25 in^2 (B) 30 in^2 (C) 36 in^2 (D) 40 in^2 (E) 49 in^2

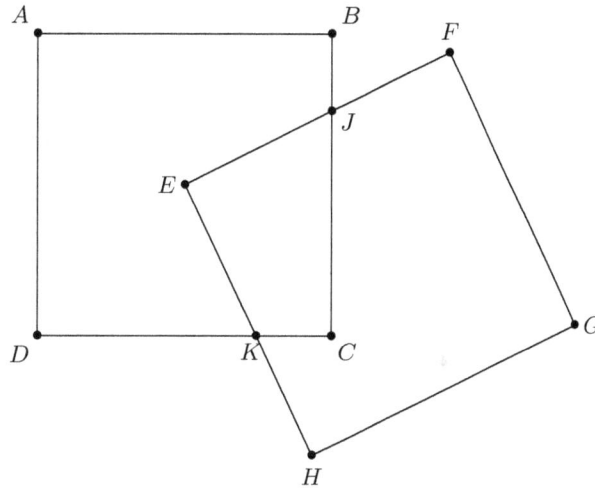

9. Consider a square with side length s. Let a denote the area of the inscribed circle (which touches all four edges) and let A denote the area of the circumscribed circle (which goes through all four corners). Which of the following holds?

(A) $a = \dfrac{1}{2}A$ **(B)** $a = \dfrac{1}{3}A$ **(C)** $a = \dfrac{2}{3}A$ **(D)** $a = \dfrac{2}{\pi}A$ **(E)** $a = \dfrac{3}{4}A$

10. A square and an equilateral triangle have the same perimeter. The area of the triangle is 1. What is the area of the square? Express your answer as a decimal to the nearest hundredth.

(A) 1.29   **(B)** 1.30   **(C)** 1.31   **(D)** 1.32   **(E)** 1.33

11. Two adjacent sides of the unit square and two sides of an equilateral triangle bisect each other. Find the area of the equilateral triangle.

(A) $\sqrt{3}/2$ **(B)** 1   **(C)** 0.8   **(D)** $\sqrt{2}$ **(E)** $\sqrt{2}/2$

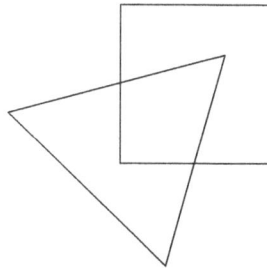

12. The line segment AB divided the rectangle in the picture into two parts. The proportion of the areas of these parts is $2 : 9$.

In what proportion are the lengths x and y?

(A) $2:9$ (B) $2:7$ (C) $4:7$ (D) $4:9$ (E) $5:9$

13. Consider the trapezoid $ABCD$ (see the diagram). Suppose $\overline{MN}$ is parallel to $\overline{DC}$, $AB = a$, $DC = b$, and $MN = x$. If the area of $ABNM$ is half the area of $ABCD$, express x as a function of a and b.

(A) $x = \dfrac{a+b}{2}$ (B) $x = \dfrac{b-a}{2}$ (C) $x = \dfrac{3b-a}{2}$ (D) $x = \sqrt{a^2 + b^2}$

(E) $x = \sqrt{\dfrac{a^2 + b^2}{2}}$

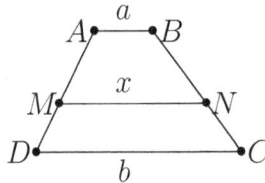

14. Three squares are arranged as in the diagram. The largest($\square ABCD$) has area 36 and the smallest ($\square DEFG$) has area 8. If ($\triangle CDE$) has area 6, what is the area of the other square ($\square CEHI$)?

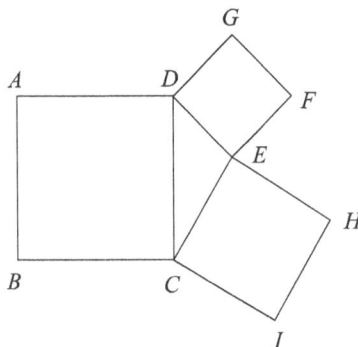

(A) 18   (B) 20   (C) 22   (D) 24   (E) 28

15. The area of a circle circumscribed about a regular hexagon is 200π. What is the area of the hexagon?

(A) $60\sqrt{3}$ (B) 600 (C) 1200 (D) $300\sqrt{3}$ (E) $600\sqrt{3}$

16. The four angles of a quadrilateral form an arithmetic sequence. The largest is 15 degrees less than twice the smallest. What is the degree measure of the largest angle?

(**A**) $95°$ (**B**) $100°$ (**C**) $105°$ (**D**) $115°$ (**E**) $125°$

17. $ABCD$ is an isosceles trapezoid with $\overline{AB}$ parallel to $\overline{DC}$, $AC = DC$, and $AD = BC$. If the height h of the trapezoid is equal to AB, find the ratio $AB : DC$.

(**A**) $2:3$ (**B**) $3:5$ (**C**) $4:5$ (**D**) $5:7$ (**E**) $5:9$

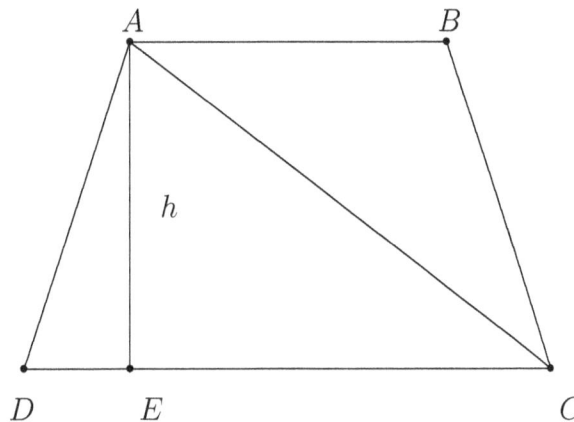

18. We subdivide the rectangle $ABCD$ with two pairs of parallel lines into 9 parts, as shown in the figure. The areas of five of these parts, measured in square centimeters is provided. What is the area of the rectangle $ABCD$, measured in square centimeters?

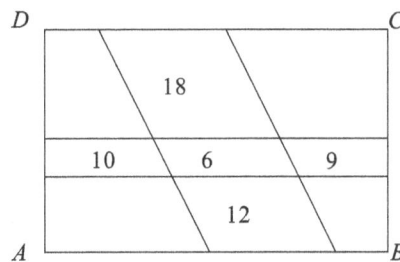

(**A**) 150 (**B**) 190 (**C**) 180 (**D**) 155 (**E**) 200

19. A farmer has 200 yards of fencing material. What is the largest rectangular area he can enclose if he wants to use a 4 yard wide gate that does not need to be covered by the fencing material?

(**A**) $2,500$ square yards

(**B**) $2,601$ square yards

(**C**) $2,704$ square yards

(**D**) $2,809$ square yards

(**E**) $2,916$ square yards

20. The diagonals of the convex four-gon $ABCD$ divide the four-gon into four triangles, whose areas are a_1, a_2, a_3 and a_4 respectively, as shown in the figure.

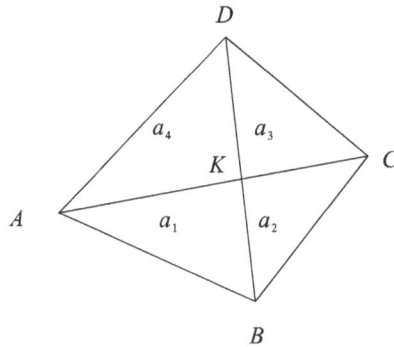

Which of the following identities must hold for these areas?

(A) $a_1 + a_2 = a_3 + a_4$ **(B)** $a_1 + a_3 = a_2 + a_4$ **(C)** $a_1 \cdot a_2 = a_3 \cdot a_4$

(D) $a_1 \cdot a_3 = a_2 \cdot a_4$ **(E)** $a_1 - a_3 = a_2 - a_4$

21. A rectangular table $PQRS$ is 5 units long and 3 units wide. A ball is rolled from point P at an angle of $45°$ toward the point E, and bounces off SR at an angle of $45°$ as shown below. The ball continues to bounce off the sides at $45°$ until it reaches a corner. How many times does the ball bounce?

(A) 4   **(B)** 5   **(C)** 6   **(D)** 7   **(E)** 9

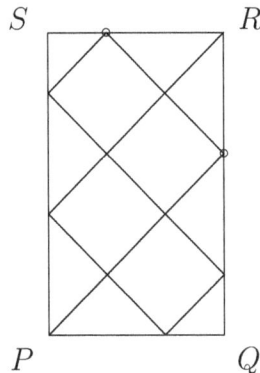

22. Consider a convex 4-gon $ABCD$. Reflect A about B to get E, B about C to get F, C about D to get G and D about A to get H. What percentage is the area of $ABCD$ of the resulting 4-gon $EFGH$?

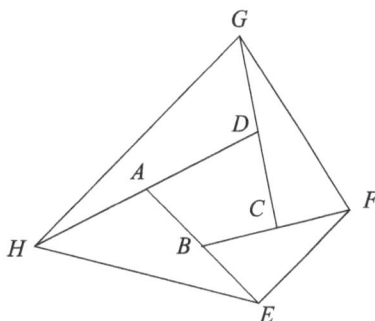

(A) 10% (B) 20% (C) 25% (D) 30% (E) 33.5%

23. For a tetrahedron $ABCD$, a plane P is called a *middle plane* if all four distances from the vertices $A, B, C,$ and D to the plane P are the same. How many middle planes are there for a given tetrahedron?

(A) 1 (B) 3 (C) 4 (D) 6 (E) 7

24. Five regular polygons, a triangle, a square, a pentagon, a hexagon, and a dodecagon (a 12-sided polygon), all have the same perimeter. Which one has the greatest area?

(A) the triangle   (B) the square   (C) the pentagon   (D) the hexagon

(E) the dodecagon

25. In the following figure, what is the sum $m(\angle A) + m(\angle B) + m(\angle C) + m(\angle D) + m(\angle E)$ of the measures of the angles A, B, C, D, and E?

(A) $180°$ (B) $360°$ (C) $540°$ (D) $720°$ (E) $900°$

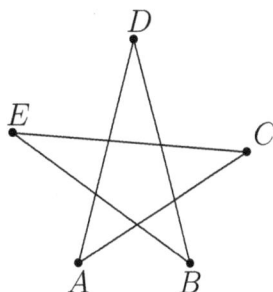

26. If a convex n-gon has no adjacent obtuse angles then n is at most

(A) 4 (B) 5 (C) 6 (D) 7 (E) 8

13.4   Three-Dimensional Geometry

1. A room is shaped like a cube with sides of length 4 meters. Suppose that A and B denote two corners of the room that are farthest from each other. A caterpillar is crawling from A to B along the walls. What is the shortest possible length of the trip?

(**A**) $4\sqrt{2}$ meters   (**B**) $4\sqrt{3}$ meters   (**C**) 8 meters   (**D**) $4\sqrt{5}$ meters   (**E**) $4\sqrt{6}$ meters

2. A plane passes through the center of a cube and is perpendicular to one of the cube's diagonals. How many edges of the cube does the plane intersect? (An edge is a line connecting adjacent corners.)

   (**A**) 3   (**B**) 4   (**C**) 5   (**D**) 6   (**E**) 8

3. How many lattice points (ordered triples of integers) (x, y, z) are exactly 13 units from the origin $(0, 0, 0)$?

   (**A**) 11   (**B**) 30   (**C**) 54   (**D**) 72   (**E**) 78

4. Let Γ be a plane containing three points $A(1, 0, 0), B(0, 2, 0), C(0, 0, 1)$. Find the distance from the origin $(0, 0, 0)$ to the plane Γ.

   (**A**) $1/3$ (**B**) $2/3$ (**C**) 1   (**D**) $\pi/3$ (**E**) $2\pi/3$

13.5   Analytical Geometry

1. What is the area of the region of the plane determined by the inequality $3 \le |x| + |y| \le 4$?

   (**A**) 7   (**B**) 9   (**C**) 14   (**D**) 16   (**E**) 32

2. The line that passes through the points $(2, 5)$ and $(7, -2)$ also passes through the point $(17, y)$ for some y. What is y?

   (**A**) -16 (**B**) -15 (**C**) -14 (**D**) -5 (**E**) 5

3. The slope of the line through the two points that satisfy both $y = 8 - x^2$ and $y = x^2$ is

   (**A**) 2   (**B**) 4   (**C**) 0   (**D**) -2 (**E**) -4

4. Which of the following lines has a slope that is less than the sum of its x- and y- intercepts?

   (**A**) $y = 2x + 1$ (**B**) $y = 3x/2 - 1$ (**C**) $y = -4x - 1$

   (**D**) $y = 4x + 16/3$ (**E**) $y = 3x$

5. For what value of k are the lines $2x + 3y = 4k$ and $x - 2ky = 7$ perpendicular?

   (**A**) $-3/4$ (**B**) $1/6$ (**C**) $1/3$ (**D**) $1/2$ (**E**) $2/3$

6. Find the slope of the line connecting the two points that satisfy

$$2x + 3y = 6 \text{ and } x^2 + y^2 = 36.$$

   (**A**) -1 (**B**) $-2/3$ (**C**) $-1/2$ (**D**) $-1/3$ (**E**) 0

7. The point $A = (2,3)$ is reflected about the x-axis to a point B. Then B is reflected about the line $y = x$ to a point C. What is the area of the triangle ABC?

(A) 12   (B) 14   (C) 15   (D) 16   (E) 24

8. The slope of the line tangent to the graph of $y = x^2$ at the point $(2,4)$ is 4. What is the y-intercept of the line?

(A) -12 (B) -4 (C) 0   (D) 4   (E) 12

9. A point (x,y) is selected at random from the rectangular region shown. What is the probability that $x < y$?

(A) $1/5$ (B) $1/4$ (C) $1/3$ (D) $1/2$ (E) $2/3$

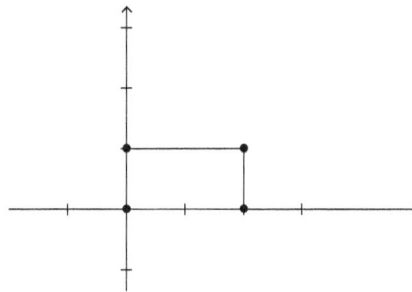

10. In how many points can a line intersect the graph of the function $f(x) = x^2 \sin(x)$?

I. no points

II. one point

III. infinitely many points

(A) II only   (B) III only   (C) I and II only   (D) I and III only

(E) II and III only

11. Determine the sum of the y-coordinates of the three intersection points of $y = x^2 - x - 5$ and $y = -5/x$.

(A) -5 (B) -3 (C) 0   (D) 3   (E) 5

12. A line l_1 has a slope of -2 and passes through the point $(r, -3)$. A second line, l_2, is perpendicular to l_1, intersects l_1 at (a,b), and passes through the point $(6,r)$. The value of a is

(A) r (B) $\frac{2}{5}r$ (C) 1   (D) $2r - 3$ (E) $\frac{5}{2}r$

13. Three points $A = (0,1)$, $B = (2,a)$, and $C = (3,7)$ are on a straight line. What is the value of a ?

(A) 4   (B) 5   (C) 6   (D) 7   (E) 8

14. The vertices of a triangle T are $(0,0)$, $(0,y)$, and $(x,0)$, where x and y are positive. The area of T is 30 and the perimeter is also 30. What is $x + y$?

(**A**) 12   (**B**) 13   (**C**) 15   (**D**) 17   (**E**) 18

15. A circle C contains the points $(0,6)$, $(0,10)$, and $(8,0)$. What is the x-coordinate of the center?

(**A**) 6.75   (**B**) 7.25   (**C**) 7.50   (**D**) 7.75   (**E**) 8.25

16. The radius of the circle given by $x^2 - 6x + y^2 + 4y = 12$ is

(**A**) 5   (**B**) 6   (**C**) 7   (**D**) 8   (**E**) 36

17. The radius of the circle given by $x^2 - 6x + y^2 + 4y = 36$ is

(**A**) 5   (**B**) 6   (**C**) 7   (**D**) 8   (**E**) 36

18. A circle with radius $\sqrt{2}$ is centered at $(0,0)$. The area of the smaller region cut from the circle by the chord from $(-1,1)$ to $(1,1)$ is

(**A**) π (**B**) $\sqrt{2} - 1$ (**C**) $\dfrac{\pi}{2} - 1$ (**D**) $\sqrt{2}\left(1 - \dfrac{\pi}{4}\right)$ (**E**) $\dfrac{\sqrt{2}}{\pi}$

19. Triangle T has vertices $(0,30)$, $(4,0)$, and $(30,0)$. Circle C with radius r circumscribes T. Which of the following is the closest to r?

(**A**) 20   (**B**) 21   (**C**) 22   (**D**) 23   (**E**) 24

20. Consider the circle $x^2 - 14x + y^2 - 4y = -49$. Let L_1 and L_2 be lines through the origin O that are tangent to the circle at points A and B. Which of the following is closest to the measure of the angle AOB?

(**A**) 35.1°   (**B**) 34.8°   (**C**) 33.6°   (**D**) 32.8°   (**E**) 31.9°

21. Let (h,k) denote the center and let r denote the radius of the circle given by

$$x^2 + 2x + y^2 - 4y = 4.$$

What is the sum $h + k + r$?

(**A**) 2   (**B**) 4   (**C**) 6   (**D**) 8   (**E**) 10

22. For what positive value of c does the line

$$y = -x + c$$

intersect the circle $x^2 + y^2 = 1$ in *exactly* one point?

(**A**) $\ln 4$ (**B**) $4^{1/3}$ (**C**) $\dfrac{3}{2}$ (**D**) $\sqrt{2}$ (**E**) $\sin^{-1}(1)$

23. Which of the following is an equation of the line tangent to the circle $x^2 + y^2 = 2$ at the point $(1, 1)$?

(A) $x = 1$ (B) $y = 1$ (C) $x + y = 2$ (D) $x - y = 0$ (E) $x + y = 0$

24. Let C be a circle that intersects each of the circles

$$(x + 2)^2 + y^2 = 2^2, \quad (x - 4)^2 + (y - 2)^2 = 2^2 \quad \text{and} \quad (x - 4)^2 + (y + 2)^2 = 2^2$$

in exactly one point and does not contain any of these circles inside it. If the radius r of C is of the form $r = p/q$ where p and q are relatively prime integers, what is $p + q$?

(A) 5   (B) 7   (C) 9   (D) 11   (E) 13

25. The graphs of

$$x^2 + y^2 = 24x + 10y - 120$$

and

$$x^2 + y^2 = k^2$$

intersect when k satisfies $0 \leq a \leq k \leq b$, and for no other positive values of k. Find $b - a$.

(A) 10   (B) 14   (C) 26   (D) 34   (E) 144

26. A triangle is formed by the coordinate axes and the tangent line at the point with coordinates (x_0, y_0) lying on the curve $\mathscr{C} = \{(x, y) : xy = k\}$, as shown in the figure.

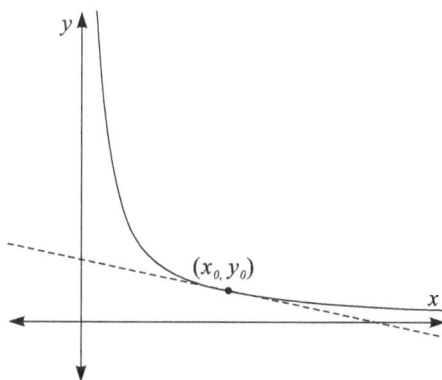

The numbers x_0, y_0, and the slope m of the tangent line satisfy $y_0 = -mx_0$. The area of the triangle is

(A) $2k$ (B) $2k^2$ (C) $x_0 + 2k$ (D) $k(x_0 + y_0)$ (E) $k/(x_0 + y_0)$

27. The graph of $|x| + |y| = 4$ encloses a region in the plane. What is the area of the region?

(A) 4   (B) 8   (C) 16   (D) 32   (E) 64

28. The set of points satisfying the three inequalities

$$y \geq 0, \quad y \leq x, \text{ and } \quad y \leq 6 - x/2$$

is a triangular region with an area of

(A) 12   (B) 18   (C) 24   (D) 36   (E) 48

29. What is the area of the triangular region in the first quadrant bounded on the left by the y-axis, bounded above by the line $7x + 4y = 168$ and bounded below by the line $5x + 3y = 121$?

(A) 16   (B) 50/3   (C) 17   (D) 52/3   (E) 53/3

30. Find the area of the polygon $ABCDEF$ whose vertices are $A = (-1, 0), B = (0, 2), C = (1, 1), D = (2, 2), E = (2, 0)$, and $F = (0, -1)$.

(A) 4   (B) 4.5   (C) 5   (D) 5.5   (E) 6

31. What is the area of the pentagonal region $ABCDE$, where $A = (0, 0), B = (12, 0), C = (12, 2), D = (6, 7)$, and $E = (0, 5)$?

(A) 60   (B) 62   (C) 63   (D) 65   (E) 66

32. The area bounded by the graph of the function $y = |x|$ and by the line $y = c$ is 5. What is c?

(A) $\sqrt{5}$ (B) $2\sqrt{5}$ (C) 5   (D) 10   (E) 25

33. A quadrilateral $ABCD$ has vertices with coordinates $A(0, 0), B(6, 0), C(5, 4), D(3, 6)$. What is its area?

(A) 18   (B) 19   (C) 20   (D) 21   (E) 22

34. What is the area (in square units) of the region of the first quadrant defined by $18 \leq x + y \leq 20$?

(A) 36   (B) 38   (C) 40   (D) 42   (E) 44

35. The midpoints of the sides of a triangle are $(1, 1), (4, 3)$, and $(3, 5)$. Find the area of the triangle.

(A) 14   (B) 16   (C) 18   (D) 20   (E) 22

36. A sphere of radius 2 is centered at $(4, 4, 7)$. What is the distance from the origin $(0, 0, 0)$ to the point on the sphere farthest from the origin?

(A) 8   (B) 9   (C) 10   (D) 11   (E) 12

37. Let A be the point $(7, 4)$ and D be the point $(5, 3)$. What is the length of the shortest path $ABCD$, where B is a point $(x, 2)$ and C is a point $(x, 0)$? This path consists of three connected segments, with the middle one vertical.

(A) $2 + \sqrt{29}$ (B) $\sqrt{31}$ (C) $2 + \sqrt{31}$ (D) $2 + \sqrt{33}$ (E) $\sqrt{41}$

38. Let P denote the point on the circle $x^2 + 2x + y^2 - 4y = 20$ that is closest to $(7, 8)$. What is the slope of the line passing through both P and $(7, 8)$?

(A) $\dfrac{2}{3}$ (B) $\dfrac{3}{4}$ (C) $\dfrac{4}{5}$ (D) 1   (E) 2

39. Inside the unit circle $D = \{(x, y) \mid x^2 + y^2 = 1\}$ there are three smaller circles of equal radius a, tangent to each other and to D. If $a = p\sqrt{3} - q$, find the sum of the integers $p + q$.

(A) 4   (B) 5   (C) 7   (D) 10   (E) 12

40. The origin and the points where the line ℓ intersects the x-axis and the y-axis are the vertices of a right triangle T whose area is 5. Also the line ℓ is perpendicular to the line given by the equation $5x - y = 15$. What is the length of the hypotenuse of T?

(A) $\sqrt{20}$ (B) $\sqrt{26}$ (C) $\sqrt{29}$ (D) $\sqrt{45}$ (E) $\sqrt{52}$

41. Into how many regions does the solution set S of

$$xy(y - x)(y + x)(x^2 + y^2 - 1) = 0$$

divide the plane? Note that some of the regions are bounded (surrounded by points of S), and some are unbounded.

(A) 4   (B) 8   (C) 16   (D) 18   (E) 22

42. A particle P moves from the point $A = (0, 4)$ to the point $B = (10, -4)$. The particle P can travel the upper half place $\{(x, y) \mid y \geq 0\}$ at the speed of 1 and travel the lower half plane $\{(x, y) \mid y \leq 0\}$ at the speed of 2. Find the point $C = (c, 0)$ on the x-axis which would minimize the sum of squares of the travel times in the upper and lower half plane.

(A) 1   (B) 2   (C) 3   (D) 4   (E) 5

43. For which of the following values of a does the line $y = a(x - 3)$ and the circle $(x - 3)^2 + y^2 = 25$ have two points of intersection, one in the 1st quadrant and one in the 4th quadrant?

(A) -1 (B) 0   (C) 1   (D) 2   (E) none of **A, B, C** or **D**

44. The cells of an infinite chess board are labeled with two integers each (the number of the column and that of the row). A child rolls a very small ball that starts from the center of the cell $(8, 11)$. There are tiny bugs at the centers of some cells. The ball rolls along a straight line; it hits a bug if and only if it rolls exactly through the center of a cell containing a bug. This does not affect the subsequent motion of the ball. It is known that the ball has hit a bug at the center of the cell $(20, 24)$. The other bugs are at the centers of the cells listed below. Which of them will be hit?

(A) $(68, 75)$ (B) $(69, 76)$ (C) $(67, 75)$ (D) $(69, 77)$ (E) $(68, 76)$

45. The circle shown is a unit circle centered at the origin. The segment BC is a diameter and C is the point $(1, 0)$. The angle α has measure 30 degrees. What is the x-coordinate of the point A?

(A) $\dfrac{1}{4}$ (B) $\dfrac{1}{3}$ (C) $\dfrac{1}{2}$ (D) $\dfrac{2}{3}$ (E) $\dfrac{\sqrt{3}}{2}$

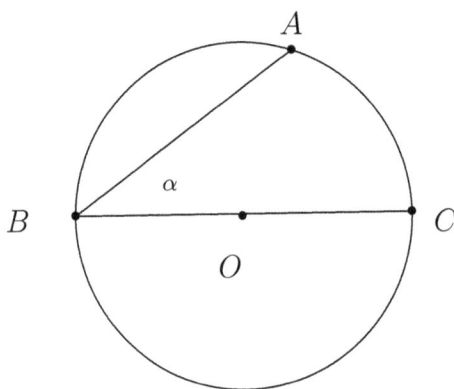

46. What is the length of the shortest path $APQB$ in the plane, where $A = (2, 3)$, $B = (5, 1)$, P lies on the y-axis, and Q lies on the x-axis.

(A) 7   (B) 8   (C) $\sqrt{65}$ (D) $5\sqrt{3}$ (E) 9

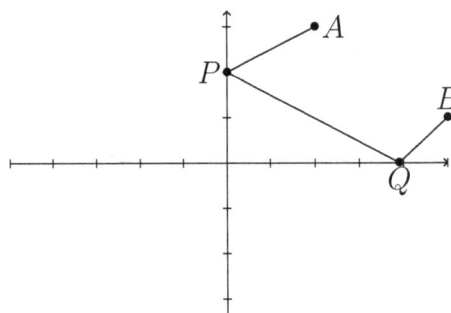

CHAPTER 14

COUNTING

14.1 Counting

1. The points A, B, C and D are the vertices of a unit square. How many squares (including $ABCD$ itself) in the same plane have two or more of these points as vertices?

(A) 13   **(B)** 12   **(C)** 9   **(D)** 5   **(E)** 4

2. Imagine that Rubik's cube consists of 27 equal cubes. Find the minimal amount of hits of an axe that is needed in order to divide Rubik's cube into 27 equal cubes?

(A) 6   **(B)** 8   **(C)** 10   **(D)** 12   **(E)** 16

3. A triangular pyramid is placed inside a sphere so that it does not intersect or touch the sphere. Each of the faces of the pyramid is extended so that they become planes. In how many pieces will the extensions cut the sphere?

(A) 9   **(B)** 10   **(C)** 12   **(D)** 14   **(E)** 15

4. In a box there are red and blue balls. If you select a handful of them with eyes closed, you have to grab at least 5 of them to make sure at least one of them is red and you have to grab at least 10 of them to make sure both colors appear among the balls selected. How many balls are there in the box?

(A) 10   **(B)** 11   **(C)** 12   **(D)** 13   **(E)** 14

5. Let S denote the set $\{(-2, -2), (2, -2), (-2, 2), (2, 2)\}$. How many circles of radius 3 in the plane have exactly two points of S on them?

(A) 6   (B) 8   (C) 10   (D) 12   (E) 16

6. The diagram shows six congruent circles with collinear centers in the plane. Each circle touches each of its nearest neighbors at exactly one point. How many paths of length 3π along the circular arcs are there from $A = (0,0)$ to $B = (6,0)$?

   (A) 16   (B) 32   (C) 64   (D) 128   (E) 256

7. How many squares of all sizes have sides determined by the grid lines below?

   (A) 48   (B) 72   (C) 75   (D) 89   (E) 91

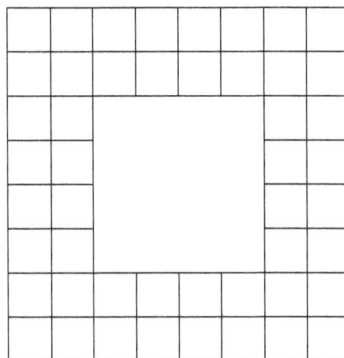

8. How many two-digit integers are there where the tens digit is greater than the units digit?

   (A) 35   (B) 36   (C) 45   (D) 55   (E) 85

9. How many non-empty sets T of natural numbers satisfy the property: $s \in T$ implies $\frac{8}{s} \in T$?

   (A) 0   (B) 1   (C) 2   (D) 3   (E) 4

10. The integers $3, 4, 5, 6, 12$, and 13 are arranged, without repetition, in a horizontal row so that the sum of any two numbers in adjoining positions is a perfect square (the square of an integer). Find the sum of the first and last.

    (A) 9   (B) 10   (C) 11   (D) 12   (E) 17

11. In the language BadSpeak, the alphabet contains only the letters a, b, and c. How many 4-letter words in BadSpeak contain the letter "c" ?

    (A) 81   (B) 108   (C) 108   (D) 65   (E) 1

12. Let $T = \{0, 1, 2, 3, 5, 7, 11\}$. How many different numbers can be obtained as the sum of three different members of T?

(A) 19   **(B)** 20   **(C)** 21   **(D)** 22   **(E)** 23

13. An archer misses the target on his first shot and hits the target on the next three shots. What is the least number of consecutive hits he must achieve following the first four shots in order to hit the target on more than nine tenths of his shots?

(A) 6   **(B)** 7   **(C)** 9   **(D)** 10   **(E)** 11

14. A two-digit number is written at random. What is the probability that the sum of its digits is 5?

(A) $\dfrac{5}{89}$ **(B)** $\dfrac{1}{18}$ **(C)** $\dfrac{6}{89}$ **(D)** $\dfrac{1}{15}$ **(E)** $\dfrac{4}{89}$

15. How many different sums can you get by adding three different numbers from the set $\{3, 6, 9, \ldots, 21, 24\}$?

(A) 15   **(B)** 16   **(C)** 18   **(D)** 20   **(E)** 22

16. It takes 852 digits to number the pages of a book consecutively. How many pages are there in the book?

(A) 184   **(B)** 235   **(C)** 320   **(D)** 368   **(E)** 425

17. The language of a pre-historic tribe is very simple and it contains only the 4 letters A, M, N, and O. Every word in this language has at most 4 letters. What is the largest number of words that can be created if every word must contain at least one vowel (A or O)?

(A) 256   **(B)** 310   **(C)** 340   **(D)** 400   **(E)** 420

18. What is the fewest crickets that must hop to new locations so that each row and each column has three crickets? Crickets can jump from any square to any other square.

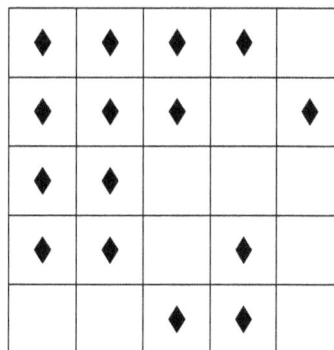

(A) 0   **(B)** 1   **(C)** 2   **(D)** 3   **(E)** 4

19. The figure below is built with eight line segments, each with either A or B as an endpoint. Triangles of various sizes are formed with parts of these segments as boundary.

How many triangular regions are there?

(A) 19   (B) 32   (C) 36   (D) 38   (E) 42

20. The stairs leading up to to the entrance of Joe's house have 5 steps. Playful Joe is able to go up 1 step or 2 steps at a time. How many ways are there for him to reach the top of the stairs? In other words, how many ways are there to write 5 as a sum of 1's and 2's if the order of the summands matters?

    (A) 5   (B) 6   (C) 7   (D) 8   (E) 10

21. Two red, two white, and two blue faces, all unit squares, are available for building a cube. How many distinguishable cubes can be built?

    (A) 5   (B) 6   (C) 7   (D) 8   (E) 9

22. How many odd three-digit numbers have three digits different?

    (A) 60   (B) 288   (C) 300   (D) 320   (E) 405

23. An eight-bit binary word is a sequence of eight digits each of which is either 0 or 1. The number of different eight-bit binary words is

    (A) 32   (B) 64   (C) 128   (D) 256   (E) 512

24. The number of integers from 1 to 10000 (inclusive) which are divisible neither by 13 nor by 51 is

    (A) 9030   (B) 9050   (C) 9070   (D) 9090   (E) 9110

25. Twelve lattice points are arranged along the edges of a 3×3 square as shown. How many triangles have all three of their vertices among these points? One such triangle is shown.

    (A) 48   (B) 64   (C) 204   (D) 220   (E) 256

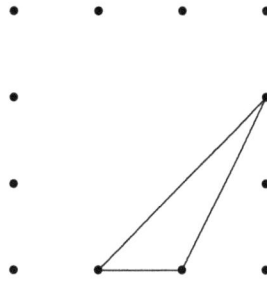

26. Seven women and five men attend a party. At this party each man shakes hands with each other person once. Each woman shakes hands only with men. How many handshakes took place at the party?

    (A) 31   **(B)** 35   **(C)** 45   **(D)** 56   **(E)** 66

27. How many two-element subsets $\{a, b\}$ of $\{1, 2, 3, \ldots, 16\}$ satisfy ab is a perfect square?

    (A) 4   **(B)** 5   **(C)** 6   **(D)** 7   **(E)** 8

28. How many positive integers can be represented as a product of two distinct members of the set $\{1, 2, 3, 4, 5, 6\}$?

    (A) 9   **(B)** 10   **(C)** 11   **(D)** 12   **(E)** 13

29. Let V denote the set of vertices of a cube. There are $\binom{8}{3} = 56$ triangles all of whose vertices belong to V. How many of these are right triangles?

    (A) 24   **(B)** 28   **(C)** 32   **(D)** 48   **(E)** 56

30. Three fair (six sided) dice are colored green, red and yellow respectively. Each die is rolled once. Let g, r and y be the resulting values of the green, red and yellow dice respectively. What is the probability that $g \leq r \leq y$?

    (A) $\dfrac{5}{27}$ **(B)** $\dfrac{6}{27}$ **(C)** $\dfrac{7}{27}$ **(D)** $\dfrac{8}{27}$ **(E)** $\dfrac{5}{18}$

31. Three points are selected simultaneously and randomly from the 3 by 3 grid of lattice points shown. What is the probability that they are collinear? Express your answer as a common fraction.

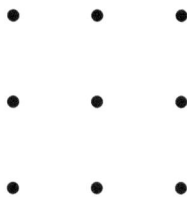

    (A) 1/42   **(B)** 1/21   **(C)** 2/21   **(D)** 1/7   **(E)** 1/6

32. Eight people Amy, Bee, Cindy, Dennis, Eli, Fay, Gil, and Hilary attend a dinner party. They need to be seated around a circular table, but Cindy and Gil, the hosts, choose not to be seated next to one another. How many different arrangements are there which seat Gil in the seat nearest the kitchen (so he can serve the dinner)?

(A) 3600 **(B)** 4320 **(C)** 4800 **(D)** 38880 **(E)** 43200

33. Football teams score $1, 2, 3,$ or 6 points at a time. They can score 1 point (point-after-touchdown) only immediately after scoring 6 points (a touchdown). A scoring sequence is a sequence of numbers $1, 2, 3, 6$, where all the 1's are immediately preceded by 6. Both $2, 6, 1, 3, 2$ and $3, 3, 3, 3, 2$ are scoring sequences with *value* 14. How many scoring sequences have value 10?

(A) 12 **(B)** 14 **(C)** 15 **(D)** 16 **(E)** 18

34. Seven women and seven men attend a party. At this party, each man shakes hands with each other person once. Each woman shakes hands only with men. How many handshakes took place at the party?

(A) 49 **(B)** 70 **(C)** 91 **(D)** 133 **(E)** 182

35. A triangular grid of 11 points is given. How many triangles have all three vertices among the 11 points?

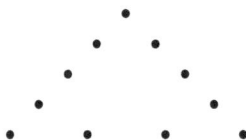

(A) 140 **(B)** 141 **(C)** 142 **(D)** 150 **(E)** 165

36. How many six-digit multiples of 5 can be formed from the digits $1, 2, 3, 4, 5,$ and 6 using each of the digits exactly once?

(A) 21 **(B)** 32 **(C)** 36 **(D)** 64 **(E)** 120

37. In his last will, a farmer asked that his horses be distributed among his four sons. The oldest was to get one third of the herd, the second oldest, one fourth of the herd, and each of the two youngest ones was to get one fifth of the herd. When the sons read the will, they were puzzled because none of them were going to get an integer number of horses. At that moment, they discovered that a baby horse had just been born. Each son would receive an integer number of horses, but the baby horse would be left over. How many horses did the farmer have originally?

(A) 29 **(B)** 59 **(C)** 89 **(D)** 119 **(E)** 239

38. A drawer contains exactly six socks–two are green, two are red, two are blue. If two socks are selected at random without replacement, what is the probabilty that they match?

(A) $\dfrac{1}{6}$ (B) $\dfrac{1}{5}$ (C) $\dfrac{1}{4}$ (D) $\dfrac{1}{3}$ (E) $\dfrac{1}{2}$

39. You bought a big cake for a party and expect 10 or 11 people to come. What is the minimal number of pieces (perhaps of different sizes) you need to divide the cake evenly if exactly 10 guests attend and also evenly among 11 guests?

(A) 11 (B) 20 (C) 30 (D) 55 (E) 110

40. How many four-digit numbers between 6000 and 7000 are there for which the thousands digit equals the sum of the other three digits?

(A) 20 (B) 22 (C) 24 (D) 26 (E) 28

41. An octahedral net is a collection of adjoining triangles that can be folded into a regular octahedron. When the net below is folded to form an octahedron, what is the sum of the numbers on the faces adjacent to one marked with a 3?

(A) 13 (B) 15 (C) 17 (D) 18 (E) 19

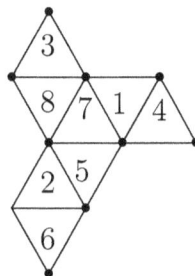

42. The 4×168 rectangular grid of squares shown below contains a shaded square. Let N denote the number of rectangular subregions that contain the shaded square. What is the sum of the digits of N?

(A) 6 (B) 16 (C) 17 (D) 26 (E) 27

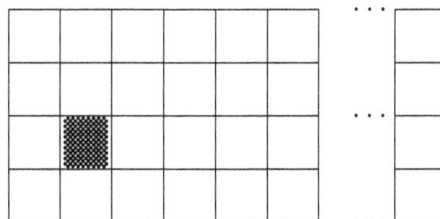

43. How many four-digit positive integers are there with exactly two even and two odd, pairwise different digits?

(A) 2008   (B) 2100   (C) 2160   (D) 2240   (E) 2640

44. A doodlebug is an insect which crawls among the lattice points (points with integer coordinates) of the plane. Each move of a doodlebug is 5 units horizontally or vertically followed by 3 units in a perpendicular direction. For example, from $(0,0)$ a doodlebug can move to any of the eight locations $(\pm5, \pm3), (\pm3, \pm5)$. What is the fewest number of moves required to get from $(0,0)$ to $(8,0)$?

(A) 6   (B) 7   (C) 8   (D) 10   (E) No such sequence of moves exists

45. A machine was programmed to transmit a certain sequence of five digits, all zeros and ones, five times. One time it did it correctly; one time it did so with one mistake; one time it did so with two mistakes; one time it did so with three mistakes; one time it did so with four mistakes. The five transmissions are listed below. Which is the correct sequence?

(A) 00001   (B) 00100   (C) 01100   (D) 10010   (E) 10011

46. The National Assembly of a country has 10 committees. Every member of the National Assembly works in exactly two committees and any pair of committees has exactly one member in common. How many members does the National Assembly have?

(A) 25   (B) 30   (C) 45   (D) 50   (E) 60

47. We roll three (six-sided) dice at once. What is the probability that at least two of the dice will show the same number?

(A) 1/6   (B) 1/3   (C) 1/2   (D) 4/9   (E) 2/3

48. We draw all diagonals of a convex 12-gon. Each intersection of diagonals is contained in exactly two diagonals. How many intersections of diagonals are there inside the 12-gon?

(A) 216   (B) 480   (C) 495   (D) 500   (E) 512

49. A convex polyhedron P has 47 faces, 35 of which are triangles, 5 of which are quadrilaterals, and 7 of which are pentagons. How many vertices does P have? Hint: Recall that Euler's theorem provides a relationship among the number f of faces, the number e of edges, and the number v of vertices of a polyhedron:

$$e + 2 = f + v.$$

(A) 32   (B) 34   (C) 35   (D) 38   (E) 40

50. At a picnic there were c children, f adult females, and m adult males, where $2 \leq c < f < m$. Every person shook hands with every other person. The sum of the number of handshakes between children, the number of handshakes between adult females and the number of handshakes between adult males is 57. How many handshakes were there altogether?

(A) 153   **(B)** 171   **(C)** 190   **(D)** 210   **(E)** 231

51. In the English alphabet of capital letters, there are 15 "stick" letters which contain no curved lines, and 11 "round" letters which contain at least some curved segment. How many different 3-letter sequences can be made of two different stick letters and one curved letter?

Stick: A E F H I K L M N T V W X Y Z
Round: B C D G J O P Q R S U

(A) 2310   **(B)** 4620   **(C)** 6930   **(D)** 13860   **(E)** none of these

52. Consider a five-digit integer of the form $a11bc$ (with $a \geq 1$). It is known that it is divisible by 45. How many such numbers are there?

(A) 21   **(B)** 20   **(C)** 19   **(D)** 18   **(E)** 10

53. How many of the 1024 integers in the set $1024, 1025, 1026, \ldots, 2047$ have more 1's than 0's in their binary representation?

(A) 252   **(B)** 512   **(C)** 638   **(D)** 768   **(E)** 772

54. Consider the domain in the figure below, which has been separated into five regions. Four different colors (red, yellow, blue and green) are available to color the regions. The only restrictions are that each region must be entirely one color and no adjacent regions are allowed to be the same color, so at least three of the four colors must be used. How many different coloring schemes are possible?

(A) 60   **(B)** 72   **(C)** 96   **(D)** 100   **(E)** 120

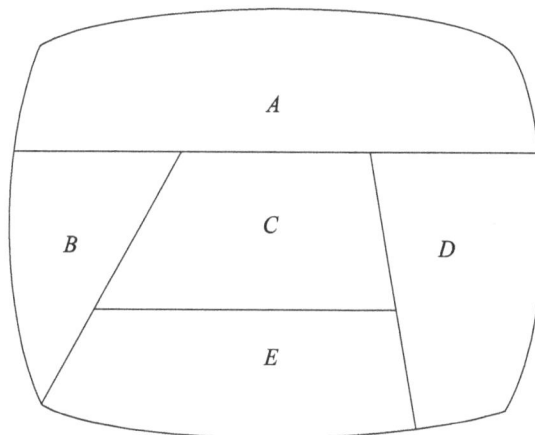

55. Of the members of three athletic teams at Harding High, 21 are on the basketball team, 26 are on the baseball team, and 29 are on the football team. A total of 14 play both baseball and basketball;

15 play both baseball and football; and 12 play football and basketball. There are eight who play all three sports. How many students play on at least one of the teams?

(A) 43 **(B)** 49 **(C)** 51 **(D)** 76 **(E)** 84

56. In an interstellar store, a customer is buying construction materials for his granddaughter who is going to build a 6-dimensional cube. The customer only needs the edges for the cube. A big sign says edges are on sale for 2 ISD (interstellar dollar) each. The grandfather (using a cheat sheet provided by the granddaughter) asks for the correct number of edges, but is surprised at the total price. The clerk explains that even though he is buying only edges, he is required by law to also pay for the vertices necessary to build the cube. If each vertex costs 1 ISD, what did the grandfather pay?

(A) 288 ISD **(B)** 448 ISD **(C)** 576 ISD **(D)** 768 ISD **(E)** 832 ISD

57. Conard High School has 50 students who play on the baseball, football, and tennis teams. Some students play more than one sport. If 15 play tennis, 25 play baseball, 30 play football, 8 play tennis and baseball, 5 play tennis and football, and 10 play baseball and football, determine how many students play all three sports.

(A) 2 **(B)** 3 **(C)** 4 **(D)** 5 **(E)** 6

58. An $a \times b \times c$ block of unit cubes has faces with surface areas 48, 60, and 80. Suppose the six faces are painted, after which the block is cut into unit cubes. Which of the following is the probability that a randomly selected cube is painted on exactly two faces?

(A) $1/2$ **(B)** $3/20$ **(C)** $2/21$ **(D)** $7/60$ **(E)** $1/8$

59. An unlimited supply of struts of lengths $3, 4, 5, 6,$ and 7 are available from which to build (non-degenerate) triangles. How many noncongruent triangles can be built? For example, we want to count both a $3, 3, 3$ triangle and a $4, 4, 4$ triangle.

(A) 21 **(B)** 25 **(C)** 28 **(D)** 30 **(E)** 32

60. Let $L(n)$ denote the smallest number of vertical and horizontal line segments needed to construct exactly n non-overlapping unit squares in the plane. Thus, $L(1) = 4, L(2) = 5, L(3) = 6, L(4) = 6$, and $L(100) = 22$. What is $L(2004)$?

(A) 92 **(B)** 93 **(C)** 94 **(D)** 95 **(E)** 96

61. Color the surface of a cube of dimension $5\times5\times5$ red, and then cut the cube into unit cubes. Remove all the unit cubes with no red faces. Use the remaining cubes to build a cuboid (a rectangular brick), keeping the outer surface of the cuboid red. What is the maximum possible volume of the cuboid?

(A) 70 **(B)** 80 **(C)** 92 **(D)** 96 **(E)** 98

62. A person is given a square-shaped pizza and a knife. He cuts the pizza with 40 straight cuts into as many pieces as possible. Denote the resulting number of pieces by N. Which of the following statements is true?

(A) $831 \le N \le 840$ (B) $821 \le N \le 830$ (C) $811 \le N \le 820$

(D) $801 \le N \le 810$ (E) $791 \le N \le 800$

63. Consider an infinite binary tree that begins at level 0 and has an additional horizontal edge connecting the two nodes at level 1. At level n, the nodes are labeled from left to right as $X_{n,1}$, $X_{n,2}$, ..., $X_{n,2^n}$. How many different paths of exactly ten moves are there that start at $X_{0,1}$ and never go from an upper vertex to a lower one (but can pass back and forth on the connection between $X_{1,1}$ and $X_{1,2}$)? [The first five levels (levels 0 through 4) are shown in the figure below. The path $X_{0,1} \to X_{1,1} \to X_{1,2} \to X_{1,1} \to X_{1,2} \to X_{2,4}$ is an example of a path with five moves.]

(A) 1024   (B) 1600   (C) 2046   (D) 2048   (E) 4096

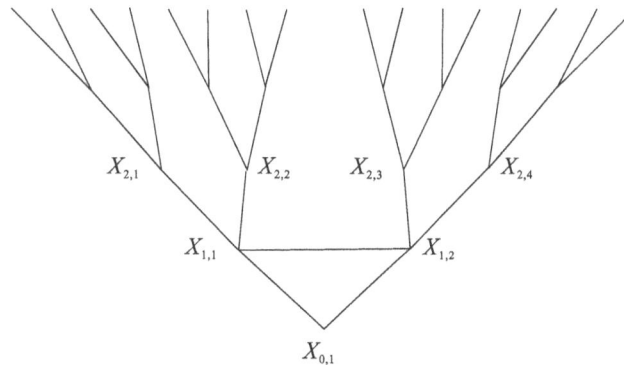

64. The faces of a cube are colored red and blue, one at a time, with equal probability. What is the probability that the resulting cube has at least one vertex P such that all three faces containing P are colored red?

(A) $1/4$ (B) $5/16$ (C) $27/64$ (D) $1/2$ (E) None of the above

65. A space diagonal of a dodecahedron (ie, a regular 12-sided polyhedron as shown below) is a segment connecting two vertices that do not lie on the same face. How many space diagonals does a dodecahedron have?

(A) 64   (B) 90   (C) 100   (D) 120   (E) 150

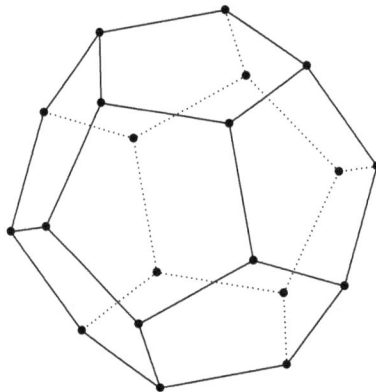

66. A *snickel* is a bug which crawls among the lattice points (points with only integer coordinates) of the plane. Each move of a snickel is eight units horizontally or vertically followed by three units in a perpendicular direction. For example, from $(0,0)$ the snickel could move to any of the eight locations $(\pm 8, \pm 3), (\pm 3, \pm 8)$. What is the least number of moves required to get from $(0,0)$ to $(19,0)$?

(A) 7   **(B)** 9   **(C)** 11   **(D)** 13   **(E)** no such sequence of moves exists

14.2 Probability and Statistics

1. Two cards are selected randomly and simultaneously from a set of four cards numbered $2, 3, 4$, and 6. What is the probability that both cards selected are prime numbered cards? Express your answer as a fraction.

(A) $1/6$ **(B)** $1/4$ **(C)** $1/3$ **(D)** $1/2$ **(E)** $2/3$

2. You have 10 coins, all of different weights and you can weigh them only in pairs in a two-pan balance. What is the minimal numbers of weighings needed to find the heaviest coin?

(A) 5   **(B)** 9   **(C)** 10   **(D)** 12   **(E)** 45

3. From a group of three female students and two male students, a three student committee is selected. If the selection is random, what is the probability that exactly 2 females and 1 male are selected?

(A) 0.3 **(B)** 0.4 **(C)** 0.5 **(D)** 0.6 **(E)** 0.7

4. Fifteen numbers are picked from the 21-element set

$$\{1, 2, 3, \ldots 20, 21\}.$$

What is the probability that at least three of those numbers are consecutive?

(A) 0   **(B)** 0.5   **(C)** 0.9   **(D)** 0.99   **(E)** 1

5. A $4 \times 4 \times 4$ cube is made from 32 white unit cubes and 32 black unit cubes. What is the largest possible percent of black surface area?

(A) 50%   **(B)** 60%   **(C)** 64%   **(D)** $66\frac{2}{3}\%$ **(E)** 75%

6. 99 fair coins are tossed simultaneously. Let P be the probability that the number of heads is odd. Which of the following statements is true?

(A) $0 \leq P < 0.2$ **(B)** $0.2 \leq P < 0.4$ **(C)** $0.4 \leq P < 0.6$

(D) $0.6 \leq P < 0.8$ **(E)** $0.8 \leq P \leq 1$

7. English and American spellings are 'rigour' and 'rigor', respectively. A randomly chosen man staying at a Parisian hotel writes this word, and a letter taken at random from his spelling is found to be a vowel. If 40 percent of the English-speaking men at the hotel are English and 60 percent are Americans, what is the probability that the writer is an Englishman?

(A) 1/4   **(B)** 1/3   **(C)** 2/5   **(D)** 5/11   **(E)** 1/2

8. What is the probability of obtaining an ace on both the first and second draws from an ordinary deck of 52 playing cards when the first card is not replaced before the second is drawn? There are four aces in the deck.

(A) 1/221   **(B)** 4/221   **(C)** 1/13   **(D)** 1/17   **(E)** 30/221

9. Two people toss fair coin.One threw it 10 times while the other − 11 times.What is the probability that the second coin's tail appeared more times than the first?

(A) $\dfrac{511}{1024}$ **(B)** $\dfrac{513}{1024}$ **(C)** $\dfrac{1025}{2028}$ **(D)** $\dfrac{1027}{2028}$ **(E)** $\dfrac{1}{2}$

10. Maggie has 2 quarters, 3 nickels, and 3 pennies. If she selects 3 coins at random, what is the probability the total value is exactly 35 cents?

(A) 3/56   **(B)** 2/28   **(C)** 5/56   **(D)** 3/28   **(E)** 7/56

11. When a missile is fired from a ship, the probability it is intercepted is $1/3$. The probability that the missile hits the target, given that it is not intercepted, is $3/4$. If three missiles are fired independently from the ship, what is the probability that all three hit their targets?

(A) 1/12   **(B)** 1/8   **(C)** 9/64   **(D)** 3/8   **(E)** 3/4

12. A $4 \times 4 \times 4$ wooden cube is painted on all 6 faces and then cut into 64 unit cubes. One unit cube is randomly selected and rolled. What is the probability that exactly one of the five visible faces is painted?

(A) 5/16   **(B)** 7/16   **(C)** 15/31   **(D)** 31/64   **(E)** 1/2

13. Three digits a, b, and c are selected from the set $\{0, 1, 2, 3, 4, 5, 6, 7, 8, 9\}$, one at a time, with repetition allowed. What is the probability that $a < b < c$?

(A) $3/25$ (B) $4/25$ (C) $1/5$ (D) $6/25$ (E) $7/25$

14. A standard deck of 52 cards contains 13 hearts. Twenty six cards have already been dealt, eight of which are hearts. If you are dealt 13 of the remaining cards, what is the probability that you will get exactly 2 of the remaining 5 hearts? (Round your answer.)

(A) 22% (B) 26% (C) 30% (D) 34% (E) 38%

15. Three fair dice are rolled. What is the probability that the product of the three outcomes is a prime number? Recall that 1 is not considered to be prime.

(A) 0 (B) $1/72$ (C) $1/36$ (D) $1/24$ (E) $1/8$

16. Two different unit squares are randomly selected from the 16 squares in the 4×4 grid shown in the following figure. What is the probability that they have at least one point in common?

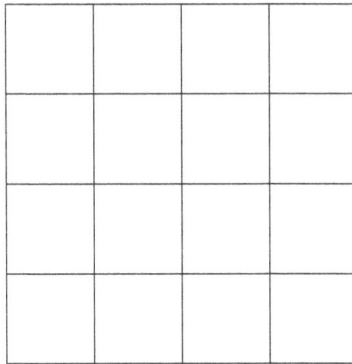

(A) $\frac{1}{4}$ (B) $\frac{1}{3}$ (C) $\frac{7}{20}$ (D) $\frac{11}{30}$ (E) $\frac{3}{8}$

17. A $4 \times 4 \times 4$ wooden cube is painted on five of its faces and is then cut into 64 unit cubes. One unit cube is randomly selected and rolled. What is the probability that exactly two of the five visible faces are painted?

(A) $15/64$ (B) $17/64$ (C) $29/128$ (D) $23/96$ (E) $71/256$

18. If Sam and Peter are among 6 men who are seated at random in a row, the probability that exactly 2 men are seated between them is

(A) $1/10$ (B) $1/8$ (C) $1/5$ (D) $1/4$ (E) $4/15$

19. What is the probability of obtaining an ace on both the first and second draws from a deck of cards when the first card is not replaced before the second is drawn?

(A) $1/13$ (B) $1/17$ (C) $1/221$ (D) $30/221$ (E) $4/221$

20. John and Bill toss a biased coin that has a 60% chance of coming up heads and a 40% chance of coming up tails. They flip the coin until either two heads or two tails in a row are observed. Bill is a winner if two heads in a row are observed first. Which of the following numbers is closest to the probability that Bill will win?

(A) $61/95$ (B) $63/95$ (C) $64/95$ (D) $67/95$ (E) $69/95$

21. A box of coins contains two with heads on both sides, one standard coin with heads on one side and tails on the other, and one coin with tails on both sides. A coin is randomly selected and flipped twice. What is the probability that the second flip results in heads given that the first flip results in heads?

(A) 0.6 (B) 0.7 (C) 0.8 (D) 0.9 (E) 0.95

22. In the Carolina Cash 5 game, you have to select five different numbers from 1 to 39. You win some prize if two or more of your five chosen numbers match the five numbers drawn (order does not matter). What is the probability that you win some prize in a single drawing? Round your answer to the nearest percent.

(A) (7%) (B) (8%) (C) (9%) (D) (11%) (E) (12%)

23. Five test scores have a mean (average score) of 91, a median (middle score) of 92 and a mode (most frequent score) of 95. The sum of the two lowest test scores is

(A) 172 (B) 173 (C) 174 (D) 178 (E) 179

24. The mean test score in a math class with 27 students was 72. A student who scored 85 was moved to another class. What was the mean score of the remaining 26 students?

(A) 69.5 (B) 70 (C) 70.5 (D) 71 (E) 71.5

25. The following table gives the distribution of families in the town of Colville in 1991 by the number of children. If there were 5000 families, how many families had no children.

Number of Children	0	1	2	3	4 or more
Percent of Families	n%	19%	18%	10%	9%

(A) 2000 (B) 2200 (C) 2350 (D) 2500 (E) 2800

26. In a math class of size 50, the average score on the final exam is 68. The best ten exams are all 100. The average of the other 40 is

(A) 50 (B) 55 (C) 60 (D) 65 (E) 70

27. For the final exam in Professor Ahlin's class, the average (= arithmetic mean) score of the group of failing students was 62 and the average score among the passing students was 92. The overall average for the 20 students in the class was 80. How many students passed the final?

(**A**) 9   (**B**) 10   (**C**) 11   (**D**) 12   (**E**) 13

28. Four numbers are written in a row. The average of the first two numbers is 5. The average of the middle two numbers is 4 and the average of the last two numbers is 10. What is the average of the first and last numbers?

(**A**) 9   (**B**) 10   (**C**) 10.5   (**D**) 11   (**E**) 11.5

CHAPTER 15

NUMBER THEORY

15.1 Prime factors and Factorization

1. Find the number of odd divisors of 7!.

(A) 4   (B) 6   (C) 10   (D) 12   (E) 24

2. What is the largest prime divisor of $2^{16} - 16$?

(A) 7   (B) 13   (C) 17   (D) 19   (E) 23

3. Let x denote the smallest positive integer satisfying $12x = 25y^2$ for some positive integer y. What is $x + y$?

(A) 75   (B) 79   (C) 81   (D) 83   (E) 88

4. The number $N = 700,245$ can be expressed as the product of three two-digit integers, $x, y,$ and z. What is $x + y + z$?

(A) 210   (B) 267   (C) 269   (D) 271   (E) 272

5. What is the largest power of 2 that divides the number $K = 75! - 71!$?

(A) 2^{62} (B) 2^{63} (C) 2^{65} (D) 2^{66} (E) 2^{67}

6. The product of three consecutive non-zero integers is 33 times the sum of the three integers. What is the sum of the digits of this product?

(A) 5   (B) 6   (C) 12   (D) 16   (E) 18

7. Suppose a and b are positive numbers different from 1 satisfying $ab = a^b$ and $a/b = a^{2b}$. Then the value of $8a + 3b$ is

(A) 26   (B) 27   (C) 28   (D) 29   (E) 30

8. How many positive two-digit integers have an odd number of positive divisors?

(A) 3   (B) 4   (C) 5   (D) 6   (E) 7

9. Three integers a, b, and c have a product of 27,846 and the property that the same number N results from each of the following operations:

- a is divided by 6.

- 4 is added to b.

- 4 is subtracted from c.

What is $a + b + c$?

(A) 102   (B) 136   (C) 152   (D) 160   (E) 177

10. The number $((N - 2)(N - 4)(N - 6)(N - 8) - 1)/2$ is an integer if N is

(A) 1 only   (B) 2 only   (C) 9 only   (D) any odd integer

(E) any even integer

11. For how many integers n is the value of $\frac{n}{50-n}$ the square of an integer?

(A) 1   (B) 2   (C) 3   (D) 4   (E) 5

12. How many positive integer divisors does $N = 250 \cdot 88$ have?

(A) 24   (B) 28   (C) 30   (D) 32   (E) 40

13. Which of the following could be the exact value of n^4, where n is a positive integer?

(A) 1.6×10^{20} (B) 1.6×10^{21} (C) 1.6×10^{22} (D) 1.6×10^{23}

(E) 1.6×10^{24}

14. Let

$$r = 11 \cdot \sqrt{10!}, s = 10 \cdot \sqrt{11!}, t = \sqrt{12!}.$$

Rank the numbers r, s, t from smallest to largest.

(A) $r < t < s$ (B) $r < s < t$ (C) $t < r < s$ (D) $t < s < r$

(E) $s < t < r$

15. What is the largest factor of 11! that is one bigger than a multiple of 6?

(A) 55   (B) 77   (C) 385   (D) 463   (E) 1925

16. Four positive integers a, b, c and d satisfy $abcd = 10!$. What is the smallest possible sum $a + b + c + d$?

(**A**) 170   (**B**) 175   (**C**) 178   (**D**) 183   (**E**) 185

17. Three integers a, b and c have a product of 9! and satisfy $a \leq b \leq c$. What is the smallest possible value of $c - a$?

(**A**) 0   (**B**) 1   (**C**) 2   (**D**) 42   (**E**) 51

18. The product of five consecutive positive integers divided by the sum of the five integers is a multiple of 100. What is the least possible sum of the five integers?

(**A**) 605   (**B**) 615   (**C**) 620   (**D**) 625   (**E**) 645

19. Let x and y be two integers that satisfy all of the following properties:

(a) $5 < x < y$,

(b) x is a power of a prime and y is a power of a prime, and

(c) the quantities $xy + 3$ and $xy - 3$ are both primes.

Among all the solutions, let (x, y) be the one with the smallest product. Which of the following statements is true? Note the list of primes on the last page.

(**A**) $x + y$ is a perfect square   (**B**) the number xy is prime   (**C**) $y = x + 3$

(**D**) $y = x + 1$ (**E**) $x + y = 17$

20. On a die, 1 and 6, 2 and 5, 3 and 4 appear on opposite faces. When 2 dice are thrown, multiply the numbers appearing on the top and bottom faces of the dice as follows:

(a) number on top face of 1st die $\times$ number on top face of 2nd die

(b) number on top face of 1st die $\times$ number on bottom face of 2nd die

(c) number on bottom face of 1st die $\times$ number on top face of 2nd die

(d) number on bottom face of 1st die number $\times$ on bottom face of 2nd die.

What can be said about the sum S of these 4 products?

(**A**) The value of S depends on luck and its expected value is 48

(**B**) The value of S depends on luck and its expected value is 49

(**C**) The value of S depends on luck and its expected value is 50

(**D**) The value of S is 49

(**E**) The value of S is 50

21. How many pairs of positive integers (a, b) with $a + b \leq 100$ satisfy $\dfrac{a + b^{-1}}{a^{-1} + b} = 13$?

(A) 0   (B) 1   (C) 3   (D) 5   (E) 7

22. The number 240,240 can be expressed as a product of k consecutive integers. A possible value of k is

(A) 4   (B) 5   (C) 6   (D) 7   (E) 8

23. Two consecutive positive integers n and $n + 1$, both with exactly four divisors, have the same sum of divisors. What is the number of divisors of their product?

(A) 6   (B) 12   (C) 16   (D) 20   (E) 24

15.2   Place value and Digits

1. Find the smallest five-digit integer such that the product of its digits is 2520. The sum of its digits is

(A) 25   (B) 26   (C) 27   (D) 29   (E) 30

2. Let N be the smallest four digit number such that the three digit number obtained by removing the leftmost digit is one ninth of the original number. What is the sum of the digits of N?

(A) 6   (B) 7   (C) 8   (D) 9   (E) 10

3. Let M and N denote the two integers that are respectively twice and three times the sum of their digits. What is $M + N$?

(A) 27   (B) 36   (C) 45   (D) 54   (E) 60

4. If a, b, c and d represent distinct nonzero base ten digits for which

$$\text{aaaa}_{\text{ten}} + \text{bbb}_{\text{ten}} + \text{cc}_{\text{ten}} + \text{d}_{\text{ten}} = 1995,$$

then $\text{a} \cdot \text{b} \cdot \text{c} \cdot \text{d} =$

(A) 336   (B) 432   (C) 486   (D) 504   (E) 567

5. Suppose you visit Mars and meet some aliens who teach you their system of counting. You notice that they use a true place value system that is similar to ours, but the Martians use base 6 instead of base 10. Table 15.1 shows how to translate their characters into our digits.

#	&	<	@	/	*
0	1	2	3	4	5

Table 15.1   Martian characters

Convert the number 54 (base 10) to its Martian equivalent.

(**A**) &@   (**B**) &@#   (**C**) @&   (**D**) #@&   (**E**) ∗/

6. Let N be the largest 7-digit number that can be constructed using each of the digits $1, 2, 3, 4, 5, 6,$ and 7 such that the sum of each two consecutive digits is a prime number. What is the reminder when N is divided by 7?

(**A**) 0   (**B**) 1   (**C**) 2   (**D**) 3   (**E**) 4

7. For each positive integer N, define $S(N)$ as the sum of the digits of N and $P(N)$ as the product of the digits. For example, $S(1234) = 10$ and $P(1234) = 24$. How many four digit numbers N satisfy $S(N) = P(N)$?

(**A**) 0   (**B**) 6   (**C**) 12   (**D**) 24   (**E**) None of the above

8. What is the sum of all two-digit numbers whose tens digit and units digit differ by exactly one?

(**A**) 878   (**B**) 890   (**C**) 900   (**D**) 990   (**E**) 991

9. Let x and b be positive integers. Suppose that x is represented as 324 in base b, and x is represented as 155 in base $b + 2$. What is b?

(**A**) 5   (**B**) 6   (**C**) 7   (**D**) 8   (**E**) 9

10. There are six ways to insert two multiplication signs in the string 33223 keeping the digits in the same order. For example, $33 \cdot 22 \cdot 3 = 2178$. If one puts these numbers in order from least to largest, in which place does 2007 occur?

(**A**) 1st (**B**) 2nd (**C**) 3rd (**D**) 4th (**E**) 5th

11. Let N denote the two-digit number whose cube root is the square root of the sum of its digits. How many positive divisors does N have?

(**A**) 2   (**B**) 3   (**C**) 4   (**D**) 5   (**E**) 6

12. Let $D(a, b, c)$ denote the number of multiples of a that are less than c and greater than b. For example, $D(2, 3, 8) = 2$ because there are two multiples of 2 between 3 and 8. What is $D(9^3, 9^4, 9^6)$?

(**A**) 71   (**B**) 719   (**C**) 720   (**D**) 7200   (**E**) 72000

13. Let N denote the smallest four-digit number with all different digits that is divisible by each of its digits. What is the sum of the digits of N?

(**A**) 9   (**B**) 10   (**C**) 11   (**D**) 12   (**E**) 13

14. What is the sum of the digits of the decimal representation of $\dfrac{10^{27} + 2}{3}$?

(**A**) 80   (**B**) 82   (**C**) 84   (**D**) 86   (**E**) 87

15. The six-digit number $5ABB7A$ is a multiple of 33 for digits A and B. Which of the following could be $A + B$?

(A) 8   (B) 9   (C) 10   (D) 11   (E) 14

16. Cifarelli Builders has just completed developing a section of homes in Southeast Charlotte. The homes are numbered consecutively starting with the address 1. The contractor in charge of ordering the single-digit brass numerals that will be used on each house for its address has determined that 999 numerals need to be ordered. How many homes are in the development?

(A) 200   (B) 369   (C) 379   (D) 381   (E) 999

17. Define the sequence $a_1, a_2, a_3, \ldots$ by

$$a_i = \left\lfloor 10^i \times \frac{1}{13} \right\rfloor - 10 \times \left\lfloor 10^{i-1} \times \frac{1}{13} \right\rfloor, \text{ for } i = 1, 2, \ldots.$$

The largest value of any a_i is

(A) 6   (B) 7   (C) 8   (D) 9   (E) 10

18. An integer between 100,000 and 199,999 becomes three times as big when we move the 1 from the leftmost position to the rightmost position. Find the sum of the digits of the number

(A) 22   (B) 24   (C) 27   (D) 28   (E) 29

19. Suppose a and b are digits satisfying $1 < a < b < 8$. The sum $1111 + 111a + 111b + \cdots$ of the smallest eight four-digit numbers that use only the digits $\{1, a, b, 8\}$ is 8994. What is $a + b$?

(A) 6   (B) 7   (C) 8   (D) 9   (E) 10

20. For how many n in $\{1, 2, 3, \ldots, 100\}$ is the tens digit of n^2 odd?

(A) 16   (B) 17   (C) 18   (D) 19   (E) 20

21. Infinitely many empty boxes, each capable of holding six dots are lined up from right to left. Each minute a new dot appears in the rightmost box. Whenever six dots appear in the same box, they **fuse** together to form one dot in the next box to the left. How many dots are there after 2012 minutes? For example, after seven minutes we have just two dots, one in the rightmost box and one in the next box over.

(A) 8   (B) 9   (C) 10   (D) 11   (E) 12

22. Use each of the digits $2, 3, 4, 6, 7, 8$ exactly once to construct two three-digit numbers M and N so that $M - N$ is positive and is as small as possible. Compute $M - N$.

(A) 33   (B) 35   (C) 39   (D) 41   (E) 47

23. Exactly one four digit number N satisfies $9 \cdot N = \overline{N}$ where $\overline{N}$ is obtained from N by reversing the digits. What is the sum of the digits of N?

(**A**) 15   (**B**) 16   (**C**) 17   (**D**) 18   (**E**) 20

24. Let $S(n) = n$ in case n is a single digit integer. If $n \geq 10$ is an integer, $S(n)$ is the sum of the digits of n. Let N denote the smallest positive integer such that $N + S(N) + S(S(N)) = 99$. What is $S(N)$?

(**A**) 9   (**B**) 10   (**C**) 12   (**D**) 15   (**E**) 18

25. Let $N = \underline{abcdef}$ be a six-digit number such that $\underline{defabc}$ is six times the value of $\underline{abcdef}$. What is the sum of the digits of N?

(**A**) 27   (**B**) 29   (**C**) 31   (**D**) 33   (**E**) 35

26. Charlie's current age is a prime number less than 100. The product of the digits of Charlie's age is the same number as it was seven years ago. In how many years will the product of the digits be the same again?

(**A**) 7   (**B**) 8   (**C**) 9   (**D**) 11   (**E**) 13

27. After each mile on a highway from Charlotte (CLT) to the City of Mathematics Students (CMS) there is a two sided distance marker. One side of the marker shows the distance from CLT while the opposite side shows the distance to CMS. A curious student noticed that the sum of the digits from the two sides of each marker stays a constant 13. Find the distance d between the cities.

(**A**) $10 \leq d < 20$ (**B**) $20 \leq d < 30$ (**C**) $30 \leq d < 40$

(**D**) $40 \leq d < 50$ (**E**) $50 \leq d < 60$

28. Consider the 2700 digit number

$$N = 100101102\ldots999$$

obtained by listing all the three digit numbers in order. What is the remainder when N is divided by 11?

(**A**) 2   (**B**) 4   (**C**) 6   (**D**) 8   (**E**) 10

29. How many three digit numbers can be written as a sum of a three digit number and its (one-, two-, or three-digit) reversal?

(**A**) 75   (**B**) 80   (**C**) 85   (**D**) 90   (**E**) 95

30. Every Monday, Harvey includes a puzzle on his blog. The puzzle for today goes like this: "My two sisters, all of my children and my younger brother and I were born between Jan. 1, 1901 and Dec. 31, 1999, each of us in a different year. Oddly, we all satisfy a very peculiar property. Each of us turned yx in some year $19xy$ where $0 \le y < x \le 9$ (a different year for each of us of course). And odder still, if someone born in $19ab$ satisfies this peculiar property, then one of us was born in that year. My sisters, my brother and I take care of the first four such years $19ab$, so in how many of the years $19xy$ (with $0 \le y < x \le 9$) has one of my children turned yx?"

(A) 5   **(B)** 6   **(C)** 10   **(D)** 12   **(E)** 15

31. The numbers $1, 2, 4, 8, 16, 32$ are arranged in a multiplication table, with three along the top and the other three down the left hand column. The multiplication table is completed and the sum of the nine entries is tabulated. What is the largest possible sum obtainable.

$\times$	a	b	c
d			
e			
f			

(A) 902   **(B)** 940   **(C)** 950   **(D)** 980   **(E)** 986

32. Let $a, b, c,$ and d denote four digits, not all of which are the same, and suppose that $a \le b \le c \le d$. Let n denote any four digit integer that can be built using these digits. Define $K(n) = \underline{dcba} - \underline{abcd}$. The function K is called the *Kaprekar* function. For example $K(1243) = 4321 - 1234 = 3087$. A four-digit integer M is called a Kaprekar number if there is a four-digit integer N such that $K(N) = M$. Which of the following is not a Kaprekar number?

(A) 2936   **(B)** 7263   **(C)** 5265   **(D)** 3996   **(E)** 6264

15.3   Modular arithmetic and Divisibility

1. If $2^{100} = 5m + k$, where k and m are integers and $0 \le k \le 4$, then k is

(A) 0   **(B)** 1   **(C)** 2   **(D)** 3   **(E)** 4

2. What is the units digit of 3^{2013}?

(A) 1   **(B)** 3   **(C)** 5   **(D)** 7   **(E)** 9

3. In order that 9986860883748524N5070273447265625 equal 1995^{10}, the letter N should be replaced by the digit

(A) 1   **(B)** 2   **(C)** 4   **(D)** 7   **(E)** 8

4. The sum of the odd positive integers from 1 to n is 9,409. What is n?

(**A**) 93   (**B**) 97   (**C**) 103   (**D**) 167   (**E**) 193

5. There exist positive integers x, y, and z satisfying $28x + 30y + 31z = 365$. Compute the value of $z - 2x$ for some such triplet.

   (**A**) 5   (**B**) 6   (**C**) 7   (**D**) 8   (**E**) 9

6. Which of the following numbers is the sum of the squares of three consecutive odd numbers?

   (**A**) 1281   (**B**) 1441   (**C**) 1595   (**D**) 1693   (**E**) 1757

7. Suppose a and b are positive integers neither of which is a multiple of 3. Then the remainder when $a^2 + b^2$ is divided by 3

   (**A**) must be 0   (**B**) must be 1   (**C**) must be 2   (**D**) may be 1 or 2 but not 0

   (**E**) may be $0, 1$ or 2

8. Determine the sum of all the natural numbers less than 45 that are not divisible by 3.

   (**A**) 600   (**B**) 625   (**C**) 650   (**D**) 675   (**E**) 700

9. Which one of the following five numbers can be expressed as the sum of the squares of six odd integers (repetitions allowed).

   (**A**) 1996   (**B**) 1997   (**C**) 1998   (**D**) 1999   (**E**) 2000

10. There are some elevens in a collection of numbers and the rest of the numbers are twelves. There are three more elevens than twelves. Which of the following could be the sum of the numbers in the collection?

    (**A**) 230   (**B**) 232   (**C**) 234   (**D**) 235   (**E**) 240

11. Let N denote a six-digit integer whose 6 digits are $1, 2, 3, 4, 5, 6$ in random order. What is the probability that N is divisible by 6?

    (**A**) 1/6   (**B**) 1/3   (**C**) 2/5   (**D**) 1/2   (**E**) 3/5

12. Let n be the product of four consecutive positive integers. How many of the numbers 4, 8, 10, 12, and 15 must be divisors of n?

    (**A**) 1   (**B**) 2   (**C**) 3   (**D**) 4   (**E**) 5

13. How many pairs of positive integers satisfy the equation $3x + 6y = 95$?

    (**A**) none   (**B**) one   (**C**) two   (**D**) three   (**E**) four

14. The number 839 can be written as $19q + r$ where q and r are positive integers. What is the largest possible value of $q - r$?

    (**A**) 37   (**B**) 39   (**C**) 41   (**D**) 45   (**E**) 47

15. If we divide 344 by d the remainder is 3, and if we divide 715 by d the remainder is 2. Which of the following is true about d?

(A) $10 \leq d \leq 19$ (B) $20 \leq d \leq 29$ (C) $30 \leq d \leq 39$ (D) $40 \leq d \leq 49$

(E) $50 \leq d \leq 59$

16. A number N is divisible by $90,\ 98$ and 882 but it is NOT divisible by $50, 270, 686$ and 1764. It is known that N is a factor of 9261000. What is N?

(A) 4410 (B) 8820 (C) 22050 (D) 44100 (E) 88200

17. How many positive integers less than one million have all digits equal and are divisible by 9?

(A) 5 (B) 6 (C) 8 (D) 10 (E) 18

18. Consider the sequence $a_1 = 1, a_2 = 13, ...$, where each term a_n is obtained from the previous term a_{n-1} by appending the nth odd number. So $a_3 = 135, a_4 = 1357$, etc. Find the number m so that a_m is the 30th multiple of 9 in the sequence.

(A) 66 (B) 77 (C) 81 (D) 90 (E) 99

19. What is the sum of all integers n, for which

$$Q(n) = \frac{4n^2 - 4n - 24}{n^3 - 3n^2 - 4n + 12}$$

is also an integer?

(A) 10 (B) 11 (C) 12 (D) 13 (E) infinity

20. You are in a large room with 50 ceiling lights (numbered from 1 to 50) that are changed from on to off or off to on by pulling a cord hanging from each light. Initially, all the lights are off. You begin by pulling the cord on every light (now they are all on). Then you pull the cord on light 2, 4, ..., 50. After you finish that, you pull the cord on light 3, 6, ..., 48. You repeat this with every fourth light, every fifth light, etc. until you pull the cord for every 50th light (only number 50, of course). How many lights are on at the end?

(A) 1 (B) 2 (C) 3 (D) 5 (E) 7

21. Consider the sum $1^2 + 2^2 + \cdots + 2012^2$. What is its last digit?

(A) 6 (B) 5 (C) 3 (D) 1 (E) 0

22. Let N denote the 180-digit number obtained by listing the 90 two-digit numbers from 10 to 99 in order. Thus $N = 10111213\ldots99$. What is the remainder when N is divided by 99?

(A) 0 (B) 10 (C) 45 (D) 54 (E) 90

15.4   Diophantine equations (and Euclidean algorithm)

1. How many integers x with $1 \le x \le 100$ satisfy the equation $x^2 + x^3 = y^2$ for some integer y?

(A) 6   (B) 7   (C) 8   (D) 9   (E) 10

2. Suppose a and b are positive integers for which

$$(2a + b)^2 - (a + 2b)^2 = 9.$$

What is ab?

(A) 2   (B) 6   (C) 9   (D) 12   (E) 24

3. Suppose a, b, c are integers satisfying $1 \le a < b < c$ and $a^2 + b^2 + c^2 = 14(a + b + c)$. What is the sum of $a + b + c$?

(A) 38   (B) 39   (C) 40   (D) 41   (E) 42

4. How many ordered pairs of positive integers (x, y) satisfy the equation $\dfrac{1}{x} - \dfrac{2}{y} = \dfrac{1}{6}$?

(A) 1   (B) 2   (C) 3   (D) 4   (E) 5

5. The sum of the cubes of ten consecutive integers is 405. What is the sum of the ten integers?

(A) 10   (B) 15   (C) 20   (D) 23   (E) 24

6. For how many real numbers x is it true that

$$\left\lfloor \frac{x}{2} \right\rfloor + \left\lfloor \frac{2x}{3} \right\rfloor = x?$$

(A) 2   (B) 3   (C) 4   (D) 5   (E) 6

7. How many pairs (x, y) of positive integers satisfy $2x + 7y = 1000$?

(A) 70   (B) 71   (C) 72   (D) 73   (E) 74

8. How many integers n satisfy $|n^3 - 222| < 888$?

(A) 11   (B) 17   (C) 18   (D) 19   (E) 20

9. It is possible that the difference of two cubes is a perfect square. For example, $28^2 = a^3 - b^3$ for certain positive integers, a and b. In this example, what is $a + b$?

(A) 12   (B) 14   (C) 16   (D) 18   (E) 20

10. If x and y are positive integers for which $2(x - y)^2 + 4y^2 = 54$, then x could be

(A) 2   (B) 5   (C) 6   (D) 8   (E) 10

11. Let (m, n) be an ordered pair of integers such that to $5m^2 + 2n^2 = 2002$. Which of the following digits could be the units digit of n?

(A) 2   (B) 3   (C) 4   (D) 7   (E) 9

12. Suppose x and y are integers satisfying both $x^2 + y = 62$ and $y^2 + x = 176$. What is $x + y$?

(A) 20   (B) 21   (C) 22   (D) 23   (E) 24

13. How many ordered pairs (a, b) of integers satisfy $a^2 = b^3 + 1$?

(A) 1   (B) 2   (C) 3   (D) 4   (E) at least 5

14. Margaret and Cyprian both have some nickels, dimes and quarters, at least one of each type and a different number of each type. Margaret has the same number of quarters as Cyprian has dimes, and she has the same number of dimes as Cyprian has nickels. She also has the same number of nickels as Cyprian has quarters. The value of their coins is the same. What is the smallest possible total value of Margaret's coins in cents?

(A) 85   (B) 95   (C) 105   (D) 125   (E) 135

15. The product of four distinct positive integers, $a, b, c,$ and d is 8!. The numbers also satisfy

$$ab + a + b + 1 = 323$$
$$bc + b + c + 1 = 399.$$

What is d?

(A) 7   (B) 14   (C) 21   (D) 28   (E) 35

16. Let p denote the smallest prime number greater than 200 for which there are positive integers a and b satisfying $a^2 + b^2 = p$. What is $a + b$?

(A) 16   (B) 17   (C) 18   (D) 19   (E) 20

17. Suppose $a, b, c,$ and d are positive integers satisfying

$$ab + cd = 38$$
$$ac + bd = 34$$
$$ad + bc = 43$$

What is $a + b + c + d$?

(A) 15   (B) 16   (C) 17   (D) 18   (E) 20

18. How many pairs of positive integers x and y satisfy the equation $xy + 8x + y = 83$?

(A) 1   (B) 2   (C) 3   (D) 4   (E) more that 4

19. Find the real numbers m, n such that $m+n = 3$, and $m^3+n^3 = 117$. What is the value of m^2+n^2?

(**A**) 29   (**B**) 3   (**C**) 9   (**D**) 17   (**E**) 45

20. Find an ordered pair (n, m) of positive integers satisfying

$$\frac{1}{n} - \frac{1}{m} + \frac{1}{mn} = \frac{2}{5}.$$

What is mn?

(**A**) 5   (**B**) 10   (**C**) 15   (**D**) 20   (**E**) 45

21. Each letter in the long division below stands for a single digit of a decimal number. The letter M is not zero and no leading digit is zero. Different letters may be used for the same digit. What is the value of B?

```
              J  K  8  L  M
      _____
A  B )  C  D  E  F  G  H  I
        N  P  Q
      _____
              R  S
              T  U
              _____
                 V  W  X
                 Y  O  Z
                 _____
                       1
```

(**A**) 1   (**B**) 2   (**C**) 3   (**D**) 4   (**E**) 5

CHAPTER 16

TRIGONOMETRY

1. If $\sin x + \cos x = 1.2$, then what is the value of $\sin 2x$?

(A) 0.22   (B) 0.88   (C) -0.2 (D) 0.44   (E) -0.88

2. Suppose $\sin \theta + \cos \theta = 0.8$. What is the value of $\sin(2\theta)$?

(A) -0.36 (B) -0.16 (C) 0   (D) 0.16   (E) 0.36

3. Consider the equation $\log(2) + \log(\sin(\theta)) + \log(\cos(\theta)) = 0$ where θ is in radians. Which one of the following formulas describes all solutions to this equation where the "k" represents all integers?

(A) $\theta = 2k\pi + (\pi/4)$ (B) $\theta = k\pi + (\pi/4)$ (C) $\theta = k\pi + (\pi/2)$

(D) $\theta = 2k\pi \pm (\pi/4)$ (E) $\theta = k\pi \pm (\pi/4)$

4. Consider the function

$$f(x) = (\sin(x) - \cos(x) - 1)(\sin(x) + \cos(x) - 1)$$

where $0 \leq x \leq 2\pi$ is measured in radians. What is the minimum value of $f(x)$?

(A) 0   (B) $-1/4$ (C) $-\sqrt{3}/4$ (D) $-1/2$ (E) $-\sqrt{2}/2$

5. What is the area of a triangle with sides of length 13, 16, and $\sqrt{41}$?

(A) 36   (B) 38   (C) 40   (D) 42   (E) 44

6. In triangle ABC, $AB = 9, BC = 10$, and $AC = 11$. Among the following, which number is closest to $\cos(\angle ABC)$?

(A) 2/9   (B) 1/4   (C) 2/7   (D) 3/10   (E) 1/3

7. *(Need calculator) A family is traveling due west on a straight road that passes a famous landmark, L in the figure below. At a given time, the bearing on the landmark is $62°$ west of north. After the family has travelled 5 miles farther, the landmark is $38°$ west of north. What is the closest the family can come to the landmark if they remain on the road?

(A) 4.25   (B) 4.35   (C) 4.55   (D) 4.76   (E) 4.85

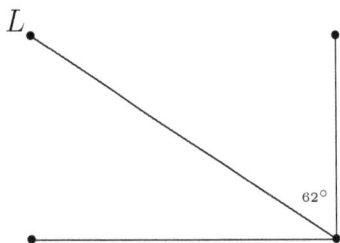

8. You are walking across a very large field with no obstructions when you see your friend, Sandy, Y yards to your north. Sandy is walking at 1 yard per second toward the north-east. If you walk at a steady speed of 2 yards per second and Sandy does not alter her speed or direction, what is the shortest distance (measured in yards) you must walk before you catch up with her?

(A) $2Y\left(\frac{\sqrt{3}}{2}\right)$ (B) $2Y\left(\frac{\sqrt{3}+\sqrt{8}}{6}\right)$ (C) $2Y\left(\frac{1+\sqrt{3}}{3}\right)$ (D) $2Y\left(\frac{\sqrt{2}+\sqrt{14}}{6}\right)$

(E) $2Y\left(\frac{\sqrt{2}+\sqrt{15}}{6}\right)$

9. What is the area of a triangle whose sides are $5, 6$, and $\sqrt{13}$?

(A) $5\sqrt{2}$ (B) 8   (C) 9   (D) $6\sqrt{2}$ (E) 10

10. In triangle $ABC, AB = 20, BC = 5$, and $\angle ABC = 60°$. The triangle is reflected in the plane about the bisector of angle ABC to produce a new triangle $A'BC'$ as shown. What is the area of the region enclosed by the union of the two triangles?

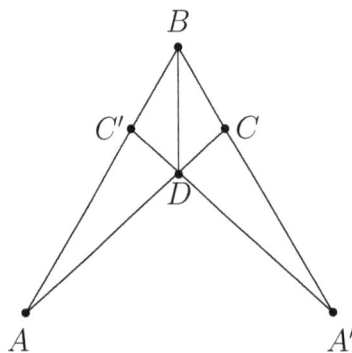

(A) $20\sqrt{3}$ (B) $24\sqrt{3}$ (C) $40\sqrt{2}$ (D) $40\sqrt{3}$ (E) $50\sqrt{2}$

CHAPTER 17

LOGARITHMS

1. If $2^{10x-1} = 1$, what is $\log x$?

(A) -1 (B) 0 (C) 1 (D) 2 (E) 3

2. What is the base x for which $\log_x 729 = 4$?

(A) $3\sqrt{3}$ (B) $7/\sqrt{2}$ (C) 5 (D) $2\sqrt{5}$ (E) $\pi\sqrt{3}$

3. If $\log_2 x + \log_2 5 = \log_2 x^2 - \log_2 14$, then $x =$

(A) 0 (B) 70 (C) both 0 and 70 (D) $\log_2 70$ (E) 2^{70}

4. The product of the two roots of the equation $\log x + \log(x + 2) = 3$ is equal to

(A) $-\log 2$ (B) -10^3 (C) $\log 2$ (D) 10^3

(E) The equation has only one root

5. Find all solutions of the equation $\log_2 x + \log_2(x + 2) = 3$.

(A) There is one solution x, and $x \geq 3$.

(B) There is one solution x, and $1 < x < 3$.

(C) There are two positive solutions.

(D) There is one positive solution and one negative solution.

(E) There is are no solutions.

6. What is the positive zero of the function

$$f(x) = \log \sqrt{5x + 5} + \frac{1}{2}\log(2x + 1) - \log 15?$$

(A) 2   **(B)** $\log 15$ **(C)** 3   **(D)** 4   **(E)** f has no positive zero

7. It is known that $\log_{10} 3 = .4771$, correct to four places. How many digits are there in the decimal representation of 3^{100}?

(A) 46   **(B)** 47   **(C)** 48   **(D)** 49   **(E)** 50

8. Let a and b be two positive integers such that b is a multiple of a. If $\log_{10}(b/a)^{b/2} + \log_{10}\left(\sqrt{a/b}\right)^{9a} = 1$, then $b^2 - a^2 =$

(A) 357   **(B)** 396   **(C)** 1600   **(D)** 5967   **(E)** 8436

CHAPTER 18

MISCELLANEOUS

1. Each person in a group of five makes a statement: Amy says "At least one of us is lying." Ben says "At least two of us are lying." Carrie says "At least four of us are lying." Donna says "All of us are lying." Eddie says "None of us is lying." Based on these statements, how many are telling the truth?

   (A) 1   (B) 2   (C) 3   (D) 4   (E) 5

2. During recess, one of five pupils wrote something nasty on the chalkboard. When questioned by the class teacher, the following ensued:

   A : It was 'B' or 'C'.

   B : Neither 'E' nor I did it.

   C : You are both lying.

   D : No, either A or B is telling the truth.

   E : No, 'D', that is not true.

   The class teacher knows that three of them never lie while the other two cannot be trusted. Who was the culprit?

   (A) A (B) B (C) C (D) D (E) E

3. The odd numbers from 1 to 17 can be used to build a 3×3 magic square (the rows and columns have the same sum). If the $1, 5,$ and 13 are as shown, what is x?

   (A) 7   (B) 9   (C) 11   (D) 15   (E) 17

4. Consider the 4×4 Kenken® puzzle below. The solution uses the numbers 1 to 4 exactly once in each row and each column. The sum of the digits in each *cage* is the number given in the upper left corner of one of the squares. What digit goes in the square with the '?' ?

(A) 1   **(B)** 2   **(C)** 3   **(D)** 4   **(E)** The given puzzle has no solution.

5. Once a strange notebook was found. It contained exactly the following 100 statements:

"This notebook has exactly one false statement."

"This notebook has exactly two false statements."

"This notebook has exactly three false statements."

$\vdots$

"This notebook has exactly 100 false statements." How many true statements are there in the notebook?

(A) 0   **(B)** 1   **(C)** 50   **(D)** 99   **(E)** 100

6. Which of the following statements below could be used to disprove " If p is a prime number, then p is three less than a multiple of four."

(A) Some even numbers are not prime.

(B) Not all odd numbers are prime.

(C) Seven is prime.

(D) Nine is not prime.

(E) Five is prime.

7. Five points lie on a line. When the 10 distances between each pair of them are computed and listed from smallest to largest we obtain $2, 4, 5, 7, 8, k, 13, 15, 17, 19$. What is k?

(A) 9   **(B)** 10   **(C)** 11   **(D)** 12   **(E)** 13

8. Two women and three girls wish to cross a river. Their small rowboat will carry the weight of only one woman or two girls. What is the minimum number of times the boat must cross the river in order to get all five females to the opposite side? At least one person must be in the boat each time it crosses the river.

    (A) 9   **(B)** 10   **(C)** 11   **(D)** 13   **(E)** 15

9. The natural numbers are arranged in groups as follows: {1}, {2,3}, {4,5,6}, {7,8,9,10}, etc. Note that there are always k numbers in the kth group. What is the sum of the numbers in the 10th group?

    (A) 260   **(B)** 369   **(C)** 452   **(D)** 505   **(E)** 638

10. Let $z = x + iy$ be a complex number, where x and y are real numbers. Let A and B be the sets defined by

$$A\{z \mid |z| \le 2\} \text{ and } B = \{z \mid (1-i)z + (1+i)\bar{z} \ge 4\}.$$

Recall that $\bar{z} = x - iy$ is the conjugate of z and that $|z| = \sqrt{x^2 + y^2}$. Find the area of the region $A \cap B$.

    (A) $\pi/4$ **(B)** $\pi - 2$ **(C)** $(\pi - 2)/4$ **(D)** 4π **(E)** $\pi - 4$

11. Each of the cards shown below has a number on one side and a letter on the other. How many of the cards must be turned over to prove the correctness of the statement:

Every card with a vowel on one side has a prime number on the other side.

    $\boxed{A}$ $\boxed{B}$ $\boxed{E}$ $\boxed{4}$ $\boxed{5}$ $\boxed{6}$ $\boxed{8}$

    (A) 2   **(B)** 3   **(C)** 4   **(D)** 5   **(E)** 6

12. The nine numbers $N, N + 3, N + 6, \cdots, N + 24$, where N is a positive integer, can be used to complete a three by three magic square. What is the sum of the entries of a row of such a magic square? A magic square is a square array of numbers such that the sum of the numbers in each row, each column, and the two diagonals is the same.

    (A) $3N$ **(B)** $3N + 6$ **(C)** $3N + 12$ **(D)** $3N + 24$ **(E)** $3N + 36$

13. Find the largest value for $(ac - bd)^2$ where (a, b) is a point on the circle $x^2 + y^2 = 4$ and (c, d) is a point on the circle $x^2 + y^2 = 9$?

    (A) 26   **(B)** 30   **(C)** 35   **(D)** 36   **(E)** 40

14. A standard deck of playing cards with 26 red and 26 black cards is split into two non-empty piles. In pile A there are four times as many black cards as red cards. In pile B, the number of red cards is an integer multiple of the number of black cards. How many red cards are in Pile B?

(A) 16   **(B)** 18   **(C)** 20   **(D)** 22   **(E)** 24

15. Construct a rectangle by putting together nine squares with sides equal to 1, 4, 7, 8, 9, 10, 14, 15 and 18. What is the sum of the areas of the squares on the 4 corners of the resulting rectangle?

(A) 626   **(B)** 746   **(C)** 778   **(D)** 810   **(E)** 826

16. *(Need calculator) Let $D(n)$ denote the leftmost digit of the decimal representation of n. Thus, $D(5^4) = D(625) = 6$. What is $D(6^{2004})$?

(A) 1   **(B)** 2   **(C)** 3   **(D)** 8   **(E)** 9

17. The 8×10 grid below has numbers in half the squares. These numbers indicate the number of mines among the squares that share an edge with the given one. Squares containing numbers do not contain mines. Each square that does not have a number either has a single mine or nothing at all. How many mines are there?

(A) 19   **(B)** 20   **(C)** 21   **(D)** 22   **(E)** 23

	1		1		2		2		1
1		2		3		2		3	
	3		2		3		2		2
1		3		2		1		3	
	3		2		1		1		2
2		4		1		1		2	
	3		3		2		3		1
1		2		2		3		2	

PART III

PAST CONTEST PROBLEMS

CHAPTER 19

AMC 8 PROBLEMS

The followings are problems from 2016 AMC 8.

1. The longest professional tennis match ever played lasted a total of 11 hours and 5 minutes. How many minutes was this?

(**A**) 605   (**B**) 655   (**C**) 665   (**D**) 1005   (**E**) 1105

2. In rectangle $ABCD$, $AB = 6$ and $AD = 8$. Point M is the midpoint of $\overline{AD}$. What is the area of $\triangle AMC$?

(**A**) 12   (**B**) 15   (**C**) 18   (**D**) 20   (**E**) 24

3. Four students take an exam. Three of their scores are $70, 80$, and 90. If the average of their four scores is 70, then what is the remaining score?

(**A**) 40   (**B**) 50   (**C**) 55   (**D**) 60   (**E**) 70

4. When Cheenu was a boy he could run 15 miles in 3 hours and 30 minutes. As an old man he can now walk 10 miles in 4 hours. How many minutes longer does it take for him to walk a mile now compared to when he was a boy?

(**A**) 6   (**B**) 10   (**C**) 15   (**D**) 18   (**E**) 30

5. The number N is a two-digit number. When N is divided by 9, the remainder is 1. When N is divided by 10, the remainder is 3. What is the remainder when N is divided by 11?

(**A**) 0   (**B**) 2   (**C**) 4   (**D**) 5   (**E**) 7

6. The following bar graph represents the length (in letters) of the names of 19 people. What is the median length of these names?

(A) 3   (B) 4   (C) 5   (D) 6   (E) 7

frequency

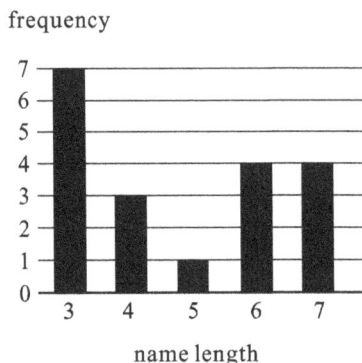

name length

7. Which of the following numbers is not a perfect square?

(A) 1^{2016} (B) 2^{2017} (C) 3^{2018} (D) 4^{2019} (E) 5^{2020}

8. Find the value of the expression

$$100 - 98 + 96 - 94 + 92 - 90 + \cdots + 8 - 6 + 4 - 2.$$

(A) 20   (B) 40   (C) 50   (D) 80   (E) 100

9. What is the sum of the distinct prime integer divisors of 2016?

(A) 9   (B) 12   (C) 16   (D) 49   (E) 63

10. Suppose that $a * b$ means $3a - b$. What is the value of x if

$$2 * (5 * x) = 1?$$

(A) $\frac{1}{10}$ (B) 2   (C) $\frac{10}{3}$ (D) 10   (E) 14

11. Determine how many two-digit numbers satisfy the following property: when the number is added to the number obtained by reversing its digits, the sum is 132.

(A) 5   (B) 7   (C) 9   (D) 11   (E) 12

12. Jefferson Middle School has the same number of boys and girls. Three-fourths of the girls and two-thirds of the boys went on a field trip. What fraction of the students were girls?

(A) $\frac{1}{2}$ (B) $\frac{9}{17}$ (C) $\frac{7}{13}$ (D) $\frac{2}{3}$ (E) $\frac{14}{15}$

13. Two different numbers are randomly selected from the set $-2, -1, 0, 3, 4, 5$ and multiplied together. What is the probability that the product is 0?

(A) $\dfrac{1}{6}$ (B) $\dfrac{1}{5}$ (C) $\dfrac{1}{4}$ (D) $\dfrac{1}{3}$ (E) $\dfrac{1}{2}$

14. Karl's car uses a gallon of gas every 35 miles, and his gas tank holds 14 gallons when it is full. One day, Karl started with a full tank of gas, drove 350 miles, bought 8 gallons of gas, and continued driving to his destination. When he arrived, his gas tank was half full. How many miles did Karl drive that day?

    (A) 525   (B) 560   (C) 595   (D) 665   (E) 735

15. What is the largest power of 2 that is a divisor of $13^4 - 11^4$?

    (A) 8   (B) 16   (C) 32   (D) 64   (E) 128

16. Annie and Bonnie are running laps around a 400-meter oval track. They started together, but Annie has pulled ahead because she is 25% faster than Bonnie. How many laps will Annie have run when she first passes Bonnie?

    (A) $1\frac{1}{4}$ (B) $3\frac{1}{3}$ (C) 4   (D) 5   (E) 25

17. An ATM password at Fred's Bank is composed of four digits from 0 to 9, with repeated digits allowable. If no password may begin with the sequence $9, 1, 1$, then how many passwords are possible?

    (A) 30   (B) 7290   (C) 9000   (D) 9990   (E) 9999

18. In an All-Area track meet, 216 sprinters enter a $100-$meter dash competition. The track has 6 lanes, so only 6 sprinters can compete at a time. At the end of each race, the five non-winners are eliminated, and the winner will compete again in a later race. How many races are needed to determine the champion sprinter?

    (A) 36   (B) 42   (C) 43   (D) 60   (E) 72

19. The sum of 25 consecutive even integers is $10,000$. What is the largest of these 25 consecutive integers?

    (A) 360   (B) 388   (C) 412   (D) 416   (E) 424

20. The least common multiple of a and b is 12, and the least common multiple of b and c is 15. What is the least possible value of the least common multiple of a and c?

    (A) 20   (B) 30   (C) 60   (D) 120   (E) 180

21. A box contains 3 red chips and 2 green chips. Chips are drawn randomly, one at a time without replacement, until all 3 of the reds are drawn or until both green chips are drawn. What is the probability that the 3 reds are drawn?

    (A) $\frac{3}{10}$ (B) $\frac{2}{5}$ (C) $\frac{1}{2}$ (D) $\frac{3}{5}$ (E) $\frac{7}{10}$

22. Rectangle $DEFA$ below is a 3×4 rectangle with $DC = CB = BA$. The area of the "bat wings" is

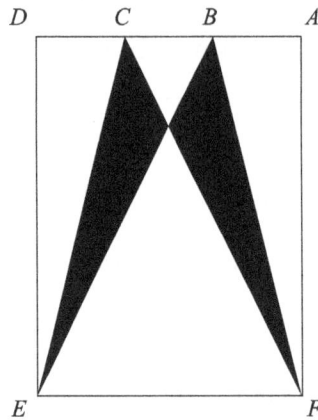

(A) 2   (B) $2\frac{1}{2}$ (C) 3   (D) $3\frac{1}{2}$ (E) 5

23. Two congruent circles centered at points A and B each pass through the other circle's center. The line containing both A and B is extended to intersect the circles at points C and D. The circles intersect at two points, one of which is E. What is the degree measure of $\angle CED$?

(A) 90   (B) 105   (C) 120   (D) 135   (E) 150

24. The digits 1, 2, 3, 4, and 5 are each used once to write a five-digit number $PQRST$. The three-digit number PQR is divisible by 4, the three-digit number QRS is divisible by 5, and the three-digit number RST is divisible by 3. What is P?

(A) 1   (B) 2   (C) 3   (D) 4   (E) 5

25. A semicircle is inscribed in an isosceles triangle with base 16 and height 15 so that the diameter of the semicircle is contained in the base of the triangle as shown. What is the radius of the semicircle?

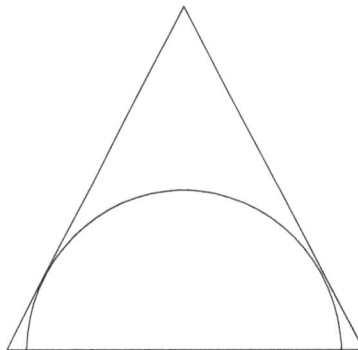

(A) $4\sqrt{3}$ (B) $\dfrac{120}{17}$ (C) 10   (D) $\dfrac{17\sqrt{2}}{2}$ (E) $\dfrac{17\sqrt{3}}{2}$

CHAPTER 20

AMC 10A/10B PROBLEMS

20.1 2016 AMC 10A

1. What is the value of $\dfrac{11! - 10!}{9!}$?

   (A) 99   (B) 100   (C) 110   (D) 121   (E) 132

2. For what value of x does $10^x \cdot 100^{2x} = 1000^5$?

   (A) 1   (B) 2   (C) 3   (D) 4   (E) 5

3. For every dollar Ben spent on bagels, David spent 25 cents less. Ben paid $12.50 more than David. How much did they spend in the bagel store together?

   (A) $37.50   (B) $50.00   (C) $87.50   (D) $90.00   (E) $92.50

4. The remainder can be defined for all real numbers x and y with $y \neq 0$ by

$$\operatorname{rem}(x, y) = x - y \left\lfloor \frac{x}{y} \right\rfloor$$

   where $\left\lfloor \frac{x}{y} \right\rfloor$ denotes the greatest integer less than or equal to $\frac{x}{y}$. What is the value of $\operatorname{rem}(\frac{3}{8}, -\frac{2}{5})$?

   (A) $-\dfrac{3}{8}$ (B) $-\dfrac{1}{40}$ (C) 0   (D) $\dfrac{3}{8}$ (E) $\dfrac{31}{40}$

5. A rectangular box has integer side lengths in the ratio $1 : 3 : 4$. Which of the following could be the volume of the box?

   (A) 48   (B) 56   (C) 64   (D) 96   (E) 144

6. Ximena lists the whole numbers 1 through 30 once. Emilio copies Ximena's numbers, replacing each occurrence of the digit 2 by the digit 1. Ximena adds her numbers and Emilio adds his numbers. How much larger is Ximena's sum than Emilio's?

(A) 13   (B) 26   (C) 102   (D) 103   (E) 110

7. The mean, median, and mode of the 7 data values $60, 100, x, 40, 50, 200, 90$ are all equal to x. What is the value of x?

(A) 50   (B) 60   (C) 75   (D) 90   (E) 100

8. Trickster Rabbit agrees with Foolish Fox to double Fox's money every time Fox crosses the bridge by Rabbit's house, as long as Fox pays 40 coins in toll to Rabbit after each crossing. The payment is made after the doubling, Fox is excited about his good fortune until he discovers that all his money is gone after crossing the bridge three times. How many coins did Fox have at the beginning?

(A) 20   (B) 30   (C) 35   (D) 40   (E) 45

9. A triangular array of 2016 coins has 1 coin in the first row, 2 coins in the second row, 3 coins in the third row, and so on up to N coins in the Nth row. What is the sum of the digits of N?

(A) 6   (B) 7   (C) 8   (D) 9   (E) 10

10. A rug is made with three different colors as shown. The areas of the three differently colored regions form an arithmetic progression. The inner rectangle is one foot wide, and each of the two shaded regions is 1 foot wide on all four sides. What is the length in feet of the inner rectangle?

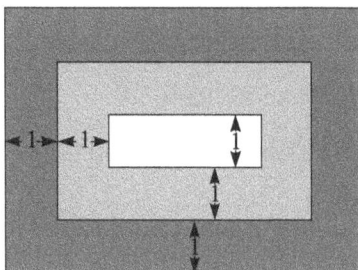

(A) 1   (B) 2   (C) 4   (D) 6   (E) 8

11. What is the area of the shaded region of the given 8×5 rectangle?

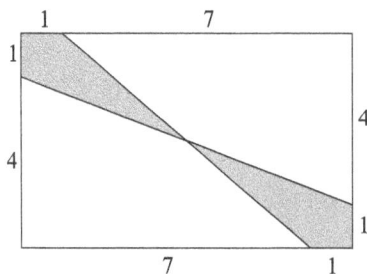

(A) $4\frac{3}{5}$ (B) 5   (C) $5\frac{1}{4}$ (D) $6\frac{1}{2}$ (E) 8

12. Three distinct integers are selected at random between 1 and 2016, inclusive. Which of the following is a correct statement about the probability p that the product of the three integers is odd?

(A) $p < \frac{1}{8}$ (B) $p = \frac{1}{8}$ (C) $\frac{1}{8} p < \frac{1}{3}$ (D) $p = \frac{1}{3}$ (E) $p > \frac{1}{3}$

13. Five friends sat in a movie theater in a row containing 5 seats, numbered 1 to 5 from left to right. (The directions "left" and "right" are from the point of view of the people as they sit in the seats.) During the movie Ada went to the lobby to get some popcorn. When she returned, she found that Bea had moved two seats to the right, Ceci had moved one seat to the left, and Dee and Edie had switched seats, leaving an end seat for Ada. In which seat had Ada been sitting before she got up?

(A) 1   (B) 2   (C) 3   (D) 4   (E) 5

14. How many ways are there to write 2016 as the sum of twos and threes, ignoring order? (For example, $1008 \cdot 2 + 0 \cdot 3$ and $402 \cdot 2 + 404 \cdot 3$ are two such ways.)

(A) 236   (B) 336   (C) 337   (D) 403   (E) 672

15. Seven cookies of radius 1 inch are cut from a circle of cookie dough, as shown. Neighboring cookies are tangent, and all except the center cookie are tangent to the edge of the dough. The leftover scrap is reshaped to form another cookie of the same thickness. What is the radius in inches of the scrap cookie?

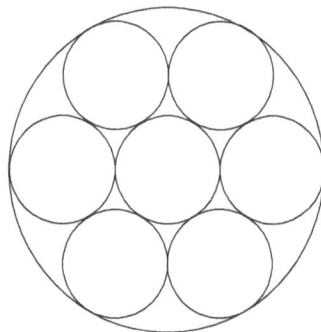

(A) $\sqrt{2}$ (B) 1.5   (C) $\sqrt{\pi}$ (D) $\sqrt{2\pi}$ (E) π

16. A triangle with vertices $A(0,2)$, $B(-3,2)$, and $C(-3,0)$ is reflected about the x-axis, then the image $\triangle A'B'C'$ is rotated counterclockwise about the origin by $90°$ to produce $\triangle A''B''C''$. Which of the following transformations will return $\triangle A''B''C''$ to $\triangle ABC$?

(A) counterclockwise rotation about the origin by $90°$

(B) clockwise rotation about the origin by $90°$

(C) reflection about the $x -$ axis

(D) reflection about the line $y = x$

(E) reflection about the $y - $ axis

17. Let N be a positive multiple of 5. One red ball and N green balls are arranged in a line in random order. Let $P(N)$ be the probability that at least $\frac{3}{5}$ of the green balls are on the same side of the red ball. Observe that $P(5) = 1$ and that $P(N)$ approaches $\frac{4}{5}$ as N grows large. What is the sum of the digits of the least value of N such that $P(N) < \frac{321}{400}$?

    (A) 12   **(B)** 14   **(C)** 16   **(D)** 18   **(E)** 20

18. Each vertex of a cube is to be labeled with an integer 1 through 8, with each integer being used once, in such a way that the sum of the four numbers on the vertices of a face is the same for each face. Arrangements that can be obtained from each other through rotations of the cube are considered to be the same. How many different arrangements are possible?

    (A) 1   **(B)** 3   **(C)** 6   **(D)** 12   **(E)** 24

19. In rectangle $ABCD$, $AB = 6$ and $BC = 3$. Point E between B and C, and point F between E and C are such that $BE = EF = FC$. Segments $\overline{AE}$ and $\overline{AF}$ intersect $\overline{BD}$ at P and Q, respectively. The ratio $BP : PQ : QD$ can be written as $r : s : t$ where the greatest common factor of $r, s,$ and t is 1. What is $r + s + t$?

    (A) 7   **(B)** 9   **(C)** 12   **(D)** 15   **(E)** 20

20. For some particular value of N, when $(a+b+c+d+1)^N$ is expanded and like terms are combined, the resulting expression contains exactly 1001 terms that include all four variables $a, b, c,$ and d, each to some positive power. What is N?

    (A) 9   **(B)** 14   **(C)** 16   **(D)** 17   **(E)** 19

21. Circles with centers P, Q and R, having radii $1, 2$ and 3, respectively, lie on the same side of line l and are tangent to l at P', Q' and R', respectively, with Q' between P' and R'. The circle with center Q is externally tangent to each of the other two circles. What is the area of triangle PQR?

    (A) 0   **(B)** $\sqrt{\frac{2}{3}}$ **(C)** 1   **(D)** $\sqrt{6} - \sqrt{2}$ **(E)** $\sqrt{\frac{3}{2}}$

22. For some positive integer n, the number $110n^3$ has 110 positive integer divisors, including 1 and the number $110n^3$. How many positive integer divisors does the number $81n^4$ have?

    (A)110   **(B)**191   **(C)** 261   **(D)** 325   **(E)** 425

23. A binary operation $\diamondsuit$ has the properties that $a \diamondsuit (b \diamondsuit c) = (a \diamondsuit b) \cdot c$ and that $a \diamondsuit a = 1$ for all nonzero real numbers $a, b,$ and c. (Here $\cdot$ represents multiplication). The solution to the equation $2016 \diamondsuit (6 \diamondsuit x) = 100$ can be written as $\frac{p}{q}$, where p and q are relatively prime positive integers. What is $p + q$?

(A) 109   **(B)** 201   **(C)** 301   **(D)** 3049   **(E)** 33, 601

24. A quadrilateral is inscribed in a circle of radius $200\sqrt{2}$. Three of the sides of this quadrilateral have length 200. What is the length of the fourth side?

(A) 200   **(B)** $200\sqrt{2}$ **(C)** $200\sqrt{3}$ **(D)** $300\sqrt{2}$ **(E)** 500

25. How many ordered triples (x, y, z) of positive integers satisfy $\text{lcm}(x, y) = 72, \text{lcm}(x, z) = 600$ and $\text{lcm}(y, z) = 900$?

(A) 15   **(B)** 16   **(C)** 24   **(D)** 27   **(E)** 64

20.2   2016 AMC 10B

1. What is the value of $\frac{2a^{-1}+\frac{a^{-1}}{2}}{a}$ when $a = \frac{1}{2}$?

(A) 1   **(B)** 2   **(C)** $\dfrac{5}{2}$ **(D)** 10   **(E)** 20

2. If $n \heartsuit m = n^3 m^2$, what is $\frac{2 \heartsuit 4}{4 \heartsuit 2}$?

(A) $\dfrac{1}{4}$ **(B)** $\dfrac{1}{2}$ **(C)** 1   **(D)** 2   **(E)** 4

3. Let $x = -2016$. What is the value of $\left| \left| |x| - x \right| - |x| \right| - x$?

(A) -2016 **(B)** 0   **(C)** 2016   **(D)** 4032   **(E)** 6048

4. Zoey read 15 books, one at a time. The first book took her 1 day to read, the second book took her 2 days to read, the third book took her 3 days to read, and so on, with each book taking her 1 more day to read than the previous book. Zoey finished the first book on a Monday, and the second on a Wednesday. On what day the week did she finish her 15th book?

(A) Sunday   **(B)** Monday   **(C)** Wednesday   **(D)** Friday   **(E)** Saturday

5. The mean age of Amanda's 4 cousins is 8, and their median age is 5. What is the sum of the ages of Amanda's youngest and oldest cousins?

(A) 13   **(B)** 16   **(C)** 19   **(D)** 22   **(E)** 25

6. Laura added two three-digit positive integers. All six digits in these numbers are different. Laura's sum is a three-digit number S. What is the smallest possible value for the sum of the digits of S?

(A) 1   **(B)** 4   **(C)** 5   **(D)** 15   **(E)** 20

7. The ratio of the measures of two acute angles is $5 : 4$, and the complement of one of these two angles is twice as large as the complement of the other. What is the sum of the degree measures of the two angles?

(A) 75   **(B)** 90   **(C)** 135   **(D)** 150   **(E)** 270

8. What is the tens digit of $2015^{2016} - 2017$?

   (A) 0   **(B)** 1   **(C)** 3   **(D)** 5   **(E)** 8

9. All three vertices of $\triangle ABC$ lie on the parabola defined by $y = x^2$, with A at the origin and $\overline{BC}$ parallel to the x-axis. The area of the triangle is 64. What is the length of BC?

   (A) 4   **(B)** 6   **(C)** 8   **(D)** 10   **(E)** 16

10. A thin piece of wood of uniform density in the shape of an equilateral triangle with side length 3 inches weighs 12 ounces. A second piece of the same type of wood, with the same thickness, also in the shape of an equilateral triangle, has side length of 5 inches. Which of the following is closest to the weight, in ounces, of the second piece?

    (A) 14.0   **(B)** 16.0   **(C)** 20.0   **(D)** 33.3   **(E)** 55.6

11. Carl decided to fence in his rectangular garden. He bought 20 fence posts, placed one on each of the four corners, and spaced out the rest evenly along the edges of the garden, leaving exactly 4 yards between neighboring posts. The longer side of his garden, including the corners, has twice as many posts as the shorter side, including the corners. What is the area, in square yards, of Carl's garden?

    (A)256   **(B)**336   **(C)**384   **(D)**448   **(E)**512

12. Two different numbers are selected at random from $\{1, 2, 3, 4, 5\}$ and multiplied together. What is the probability that the product is even?

    (A) 0.2   **(B)** 0.4   **(C)** 0.5   **(D)** 0.7   **(E)** 0.8

13. At Megapolis Hospital one year, multiple-birth statistics were as follows: Sets of twins, triplets, and quadruplets accounted for 1000 of the babies born. There were four times as many sets of triplets as sets of quadruplets, and there was three times as many sets of twins as sets of triplets. How many of these 1000 babies were in sets of quadruplets?

    (A) 25   **(B)** 40   **(C)** 64   **(D)** 100   **(E)** 160

14. How many squares whose sides are parallel to the axes and whose vertices have coordinates that are integers lie entirely within the region bounded by the line $y = \pi x$, the line $y = -0.1$ and the line $x = 5.1$?

    (A) 30   **(B)** 41   **(C)** 45   **(D)** 50   **(E)** 57

15. All the numbers $1, 2, 3, 4, 5, 6, 7, 8, 9$ are written in a 3×3 array of squares, one number in each square, in such a way that if two numbers are consecutive then they occupy squares that share an edge. The numbers in the four corners add up to 18. What is the number in the center?

(A) 5   (B) 6   (C) 7   (D) 8   (E) 9

16. The sum of an infinite geometric series is a positive number S, and the second term in the series is 1. What is the smallest possible value of S?

(A) $\dfrac{1+\sqrt{5}}{2}$ (B) 2   (C) $\sqrt{5}$ (D) 3   (E) 4

17. All the numbers $2,3,4,5,6,7$ are assigned to the six faces of a cube, one number to each face. For each of the eight vertices of the cube, a product of three numbers is computed, where the three numbers are the numbers assigned to the three faces that include that vertex. What is the greatest possible value of the sum of these eight products?

(A) 312   (B) 343   (C) 625   (D) 729   (E) 1680

18. In how many ways can 345 be written as the sum of an increasing sequence of two or more consecutive positive integers?

(A) 1   (B) 3   (C) 5   (D) 6   (E) 7

19. Rectangle $ABCD$ has $AB = 5$ and $BC = 4$. Point E lies on $\overline{AB}$ so that $EB = 1$, point G lies on $\overline{BC}$ so that $CG = 1$. and point F lies on $\overline{CD}$ so that $DF = 2$. Segments $\overline{AG}$ and $\overline{AC}$ intersect $\overline{EF}$ at Q and P, respectively. What is the value of $\frac{PQ}{EF}$?

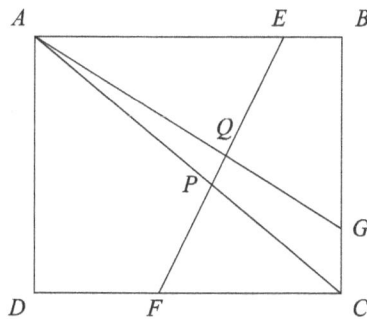

(A) $\dfrac{\sqrt{13}}{16}$ (B) $\dfrac{\sqrt{2}}{13}$ (C) $\dfrac{9}{82}$ (D) $\dfrac{10}{91}$ (E) $\dfrac{\sqrt{1}}{9}$

20. A dilation of the plane—that is, a size transformation with a positive scale factor—sends the circle of radius 2 centered at $A(2,2)$ to the circle of radius 3 centered at $A.(5,6)$. What distance does the origin $O(0,0)$, move under this transformation?

(A) 0   (B) 3   (C) $\sqrt{13}$ (D) 4   (E) 5

21. What is the area of the region enclosed by the graph of the equation $x^2 + y^2 = |x| + |y|$?

(A) $\pi + \sqrt{2}$ (B) $\pi + 2$ (C) $\pi + 2\sqrt{2}$ (D) $2\pi + \sqrt{2}$ (E) $2\pi + 2\sqrt{2}$

22. A set of teams held a round-robin tournament in which every team played every other team exactly once. Every team won 10 games and lost 10 games; there were no ties. How many sets of three teams $\{A, B, C\}$ were there in which A beat B, B beat C, and C beat A?

(A) 385   **(B)** 665   **(C)** 945   **(D)** 1140   **(E)** 1330

23. In regular hexagon $ABCDEF$, points W, X, Y, and Z are chosen on sides $\overline{BC}$, $\overline{CD}$, $\overline{EF}$, and $\overline{FA}$ respectively, so lines AB, ZW, YX, and ED are parallel and equally spaced. What is the ratio of the area of hexagon $WCXYFZ$ to the area of hexagon $ABCDEF$?

(A) $\dfrac{1}{3}$ **(B)** $\dfrac{10}{27}$ **(C)** $\dfrac{11}{27}$ **(D)** $\dfrac{4}{9}$ **(E)** $\dfrac{13}{27}$

24. How many four-digit integers $abcd$, with $a \neq 0$, have the property that the three two-digit integers $ab < bc < cd$ form an increasing arithmetic sequence? One such number is 4692, where $a = 4$, $b = 6$, $c = 9$, and $d = 2$.

(A) 9   **(B)** 15   **(C)** 16   **(D)** 17   **(E)** 20

25. Let $f(x) = \sum_{k=2}^{10}(\lfloor kx \rfloor - k\lfloor x \rfloor)$, where $\lfloor r \rfloor$ denotes the greatest integer less than or equal to r. How many distinct values does $f(x)$ assume for $x \geq 0$?

(A) 32   **(B)** 36   **(C)** 45   **(D)** 46   **(E)** infinitely many

CHAPTER 21

AMC 12A/12B PROBLEMS

21.1 2016 AMC 12A

1. What is the value of $\dfrac{11! - 10!}{9!}$?

   (A) 99   (B) 100   (C) 110   (D) 121   (E) 132

2. For what value of x does $10^x \cdot 100^{2x} = 1000^5$?

   (A) 1   (B) 2   (C) 3   (D) 4   (E) 5

3. The remainder can be defined for all real numbers x and y with $y \neq 0$ by

   $$\mathrm{rem}(x, y) = x - y \left\lfloor \frac{x}{y} \right\rfloor$$

   where $\left\lfloor \frac{x}{y} \right\rfloor$ denotes the greatest integer less than or equal to $\frac{x}{y}$. What is the value of $\mathrm{rem}(\frac{3}{8}, -\frac{2}{5})$?

   (A) $-\dfrac{3}{8}$ (B) $-\dfrac{1}{40}$ (C) 0   (D) $\dfrac{3}{8}$ (E) $\dfrac{30}{40}$

4. The mean, median, and mode of the 7 data values $60, 100, x, 40, 50, 200, 90$ are all equal to x. What is the value of x?

   (A) 50   (B) 60   (C) 75   (D) 90   (E) 100

5. Goldbach's conjecture states that every even integer greater than 2 can be written as the sum of two prime numbers (for example, $2016 = 13 + 2003$). So far, no one has been able to prove that the conjecture is true, and no one has found a counterexample to show that the conjecture is false. What would a counterexample consist of?

(**A**) an odd integer greater that 2 that can be written as the sum of two prime numbers

(**B**) an odd integer greater that 2 that cannot be written as the sum of two prime numbers

(**C**) an even integer greater that 2 that can be written as the sum of two numbers that are not prime

(**D**) an even integer greater that 2 that can be written as the sum of two prime numbers

(**E**) an even integer greater that 2 that cannot be written as the sum of two prime numbers

6. A triangular array of 2016 coins has 1 coin in the first row, 2 coins in the second row, 3 coins in the third row, and so on up to N coins in the Nth row. What is the sum of the digits of N?

(**A**) 6 (**B**) 7 (**C**) 8 (**D**) 9 (**E**) 10

7. Which of these describes the graph of $x^2(x + y + 1) = y^2(x + y + 1)$?

(**A**) two parallel lines

(**B**) two intersecting lines

(**C**) three lines that all pass through a common point

(**D**) three lines that do not all pass through a comment point

(**E**) a line and a parabola

8. What is the area of the shaded region of the given 8×5 rectangle?

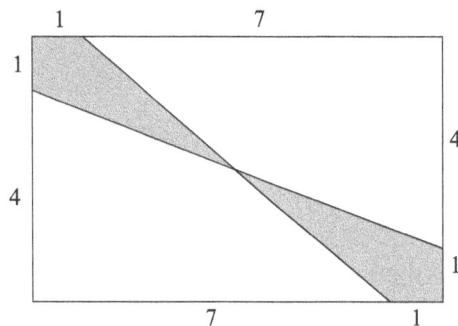

(**A**) $4\frac{3}{5}$ (**B**) 5 (**C**) $5\frac{1}{4}$ (**D**) $6\frac{1}{2}$ (**E**) 8

9. The five small shaded squares inside this unit square are congruent and have disjoint interiors. The midpoint of each side of the middle square coincides with one of the vertices of the other four small squares as shown. The common side length is $\frac{a-\sqrt{2}}{b}$, where a and b are positive integers. What is $a + b$?

(A) 7   (B) 8   (C) 9   (D) 10   (E) 11

10. Five friends sat in a movie theater in a row containing 5 seats, numbered 1 to 5 from left to right. (The directions "left" and "right" are from the point of view of the people as they sit in the seats.) During the movie Ada went to the lobby to get some popcorn. When she returned, she found that Bea had moved two seats to the right, Ceci had moved one seat to the left, and Dee and Edie had switched seats, leaving an end seat for Ada. In which seat had Ada been sitting before she got up?

(A) 1   (B) 2   (C) 3   (D) 4   (E) 5

11. Each of the 100 students in a certain summer camp can either sing, dance, or act. Some students have more than one talent, but no student has all three talents. There are 42 students who cannot sing, 65 students who cannot dance, and 29 students who cannot act. How many students have two of these talents?

(A) 16   (B) 25   (C) 36   (D) 49   (E) 64

12. In $\triangle ABC$, $AB = 6$, $BC = 7$, and $CA = 8$. Point D lies on $\overline{BC}$, and $\overline{AD}$ bisects $\angle BAC$. Point E lies on $\overline{AC}$, and $\overline{BE}$ bisects $\angle ABC$. The bisectors intersect at F. What is the ratio $AF : FD$?

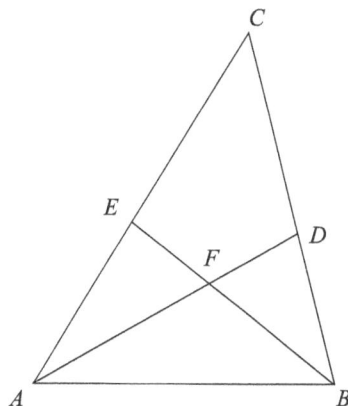

(A) $3:2$ (B) $5:3$ (C) $2:1$ (D) $7:3$ (E) $5:2$

13. Let N be a positive multiple of 5. One red ball and N green balls are arranged in a line in random order. Let $P(N)$ be the probability that at least $\frac{3}{5}$ of the green balls are on the same side of the red ball. Observe that $P(5) = 1$ and that $P(N)$ approaches $\frac{4}{5}$ as N grows large. What is the sum of the digits of the least value of N such that $P(N) < \frac{321}{400}$?

(A) 12   (B) 14   (C) 16   (D) 18   (E) 20

14. Each vertex of a cube is to be labeled with an integer 1 through 8, with each integer being used once, in such a way that the sum of the four numbers on the vertices of a face is the same for each face. Arrangements that can be obtained from each other through rotations of the cube are considered to be the same. How many different arrangements are possible?

(A) 1   (B) 3   (C) 6   (D) 12   (E) 24

15. Circles with centers P, Q and R, having radii $1, 2$ and 3, respectively, lie on the same side of line l and are tangent to l at P', Q' and R', respectively, with Q' between P' and R'. The circle with center Q is externally tangent to each of the other two circles. What is the area of triangle PQR?

(A) 0   (B) $\sqrt{\frac{2}{3}}$ (C) 1   (D) $\sqrt{6} - \sqrt{2}$ (E) $\sqrt{\frac{3}{2}}$

16. The graphs of $y = \log_3 x$, $y = \log_x 3$, $y = \log_{\frac{1}{3}} x$, and $y = \log_x \frac{1}{3}$ are plotted on the same set of axes. How many points in the plane with positive x-coordinates lie on two or more of the graphs?

(A) 2   (B) 3   (C) 4   (D) 5   (E) 6

17. Let $ABCD$ be a square. Let E, F, G and H be the centers, respectively, of equilateral triangles with bases $\overline{AB}, \overline{BC}, \overline{CD}$, and $\overline{DA}$, each exterior to the square. What is the ratio of the area of square $EFGH$ to the area of square $ABCD$?

(A) 1   (B) $\dfrac{2 + \sqrt{3}}{3}$ (C) $\sqrt{2}$ (D) $\dfrac{\sqrt{2} + \sqrt{3}}{2}$ (E) $\sqrt{3}$

18. For some positive integer n, the number $110n^3$ has 110 positive integer divisors, including 1 and the number $110n^3$. How many positive integer divisors does the number $81n^4$ have?

(A) 110   (B) 191   (C) 261   (D) 325   (E) 425

19. Jerry starts at 0 on the real number line. He tosses a fair coin 8 times. When he gets heads, he moves 1 unit in the positive direction; when he gets tails, he moves 1 unit in the negative direction. The probability that he reaches 4 at some time during this process $\frac{a}{b}$, where a and b are relatively prime positive integers. What is $a + b$? (For example, he succeeds if his sequence of tosses is $HTHHHHHH$.)

(A) 69   (B) 151   (C) 257   (D) 293   (E) 313

20. A binary operation $\Diamond$ has the properties that $a \Diamond (b \Diamond c) = (a \Diamond b) \cdot c$ and that $a \Diamond a = 1$ for all nonzero real numbers $a, b,$ and c. (Here $\cdot$ represents multiplication). The solution to the equation $2016 \Diamond (6 \Diamond x) = 100$ can be written as $\frac{p}{q}$, where p and q are relatively prime positive integers. What is $p + q$?

(**A**) 109 (**B**) 201 (**C**) 301 (**D**) 3049 (**E**) $33,601$

21. A quadrilateral is inscribed in a circle of radius $200\sqrt{2}$. Three of the sides of this quadrilateral have length 200. What is the length of the fourth side?

(**A**) 200 (**B**) $200\sqrt{2}$ (**C**) $200\sqrt{3}$ (**D**) $300\sqrt{2}$ (**E**) 500

22. How many ordered triples (x, y, z) of positive integers satisfy $\operatorname{lcm}(x, y) = 72, \operatorname{lcm}(x, z) = 600$ and $\operatorname{lcm}(y, z) = 900$?

(**A**) 15 (**B**) 16 (**C**) 24 (**D**) 27 (**E**) 64

23. Three numbers in the interval $[0, 1]$ are chosen independently and at random. What is the probability that the chosen numbers are the side lengths of a triangle with positive area?

(**A**) $\dfrac{1}{6}$ (**B**) $\dfrac{1}{3}$ (**C**) $\dfrac{1}{2}$ (**D**) $\dfrac{2}{3}$ (**E**) $\dfrac{5}{6}$

24. There is a smallest positive real number a such that there exists a positive real number b such that all the roots of the polynomial $x^3 - ax^2 + bx - a$ are real. In fact, for this value of a the value of b is unique. What is this value of b?

(**A**) 8 (**B**) 9 (**C**) 10 (**D**) 11 (**E**) 12

25. Let k be a positive integer. Bernardo and Silvia take turns writing and erasing numbers on a blackboard as follows: Bernardo starts by writing the smallest perfect square with k+1 digits. Every time Bernardo writes a number, Silvia erases the last k digits of it. Bernardo then writes the next perfect square, Silvia erases the last k digits of it, and this process continues until the last two numbers that remain on the board differ by at least 2. Let f(k) be the smallest positive integer not written on the board. For example, if k = 1, then the numbers that Bernardo writes are $16, 25, 36, 49, 64$, and the numbers showing on the board after Silvia erases are 1, 2, 3, 4, and 6, and thus f(1) = 5. What is the sum of the digits of

$$f(2) + f(4) + f(6) + \cdots + f(2016)?$$

(**A**) 7986 (**B**) 8002 (**C**) 8030 (**D**) 8048 (**E**) 8064

21.2   2016 AMC 12B

1. What is the value of $\frac{2a^{-1}+\frac{a^{-1}}{2}}{a}$ when $a = \frac{1}{2}$?

   (A) 1   (B) 2   (C) $\frac{5}{2}$ (D) 10   (E) 20

2. The harmonic mean of two numbers can be calculated as twice their product divided by their sum. The harmonic mean of 1 and 2016 is closest to which integer?

   (A) 2   (B) 45   (C) 504   (D) 1008   (E) 2015

3. Let $x = -2016$. What is the value of $\left| \Big| |x| - x \Big| - |x| \right| - x$?

   (A) -2016 (B) 0   (C) 2016   (D) 4032   (E) 6048

4. The ratio of the measures of two acute angles is $5 : 4$, and the complement of one of these two angles is twice as large as the complement of the other. What is the sum of the degree measures of the two angles?

   (A) 75   (B) 90   (C) 135   (D) 150   (E) 270

5. The War of 1812 started with a declaration of war on Thursday, June 18, 1812. The peace treaty to end the war was signed 919 days later, on December 24, 1814. On what day of the week was the treaty signed?

   (A) Friday   (B) Saturday   (C) Sunday   (D) Monday   (E) Tuesday

6. All three vertices of $\triangle ABC$ lie on the parabola defined by $y = x^2$, with A at the origin and $\overline{BC}$ parallel to the x-axis. The area of the triangle is 64. What is the length of BC?

   (A) 4   (B) 6   (C) 8   (D) 10   (E) 16

7. Josh writes the numbers $1, 2, 3, \ldots, 99, 100$. He marks out 1, skips the next number (2), marks out 3, and continues skipping and marking out the next number to the end of the list. Then he goes back to the start of his list, marks out the first remaining number (2), skips the next number (4), marks out 6, skips 8, marks out 10, and so on to the end. Josh continues in this manner until only one number remains. What is that number?

   (A) 13   (B) 32   (C) 56   (D) 64   (E) 96

8. A thin piece of wood of uniform density in the shape of an equilateral triangle with side length 3 inches weighs 12 ounces. A second piece of the same type of wood, with the same thickness, also in the shape of an equilateral triangle, has side length of 5 inches. Which of the following is closest to the weight, in ounces, of the second piece?

   (A) 14.0   (B) 16.0   (C) 20.0   (D) 33.3   (E) 55.6

9. Carl decided to fence in his rectangular garden. He bought 20 fence posts, placed one on each of the four corners, and spaced out the rest evenly along the edges of the garden, leaving exactly 4 yards between neighboring posts. The longer side of his garden, including the corners, has twice as many posts as the shorter side, including the corners. What is the area, in square yards, of Carl's garden?

(A) 256   (B) 336   (C) 384   (D) 448   (E) 512

10. A quadrilateral has vertices $P(a, b)$, $Q(b, a)$, $R(-a, -b)$, and $S(-b, -a)$, where a and b are integers with $a > b > 0$. The area of $PQRS$ is 16. What is $a + b$?

(A) 4   (B) 5   (C) 6   (D) 12   (E) 13

11. How many squares whose sides are parallel to the axes and whose vertices have coordinates that are integers lie entirely within the region bounded by the line $y = \pi x$, the line $y = -0.1$ and the line $x = 5.1$?

(A) 30   (B) 41   (C) 45   (D) 50   (E) 57

12. All the numbers $1, 2, 3, 4, 5, 6, 7, 8, 9$ are written in a 3×3 array of squares, one number in each square, in such a way that if two numbers of consecutive then they occupy squares that share an edge. The numbers in the four corners add up to 18. What is the number in the center?

(A) 5   (B) 6   (C) 7   (D)8   (E) 9

13. Alice and Bob live 10 miles apart. One day Alice looks due north from her house and sees an airplane. At the same time Bob looks due west from his house and sees the same airplane. The angle of elevation of the airplane is 30° from Alice's position and 60° from Bob's position. Which of the following is closest to the airplane's altitude, in miles?

(A) 3.5   (B) 4   (C) 4.5   (D) 5   (E) 5.5

14. The sum of an infinite geometric series is a positive number S, and the second term in the series is 1. What is the smallest possible value of S?

(A) $\dfrac{1 + \sqrt{5}}{2}$ (B) 2   (C) $\sqrt{5}$ (D) 3   (E) 4

15. All the numbers $2, 3, 4, 5, 6, 7$ are assigned to the six faces of a cube, one number to each face. For each of the eight vertices of the cube, a product of three numbers is computed, where the three numbers are the numbers assigned to the three faces that include that vertex. What is the greatest possible value of the sum of these eight products?

(A) 312   (B) 343   (C) 625   (D) 729   (E) 1680

16. In how many ways can 345 be written as the sum of an increasing sequence of two or more consecutive positive integers?

(A) 1 **(B)** 3 **(C)** 5 **(D)** 6 **(E)** 7

17. In $\triangle ABC$ shown in the figure, $AB = 7$, $BC = 8$, $CA = 9$, and $\overline{AH}$ is an altitude. Points D and E lie on sides $\overline{AC}$ and $\overline{AB}$, respectively, so that $\overline{BD}$ and $\overline{CE}$ are angle bisectors, intersecting $\overline{AH}$ at Q and P, respectively. What is PQ?

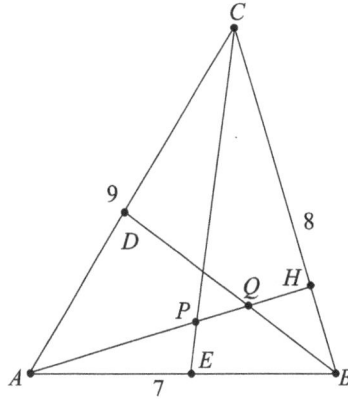

(A) 1 **(B)** $\frac{5}{8}\sqrt{3}$ **(C)** $\frac{4}{5}\sqrt{2}$ **(D)** $\frac{8}{15}\sqrt{5}$ **(E)** $\frac{6}{5}$

18. What is the area of the region enclosed by the graph of the equation $x^2 + y^2 = |x| + |y|$?

(A) $\pi + \sqrt{2}$ **(B)** $\pi + 2$ **(C)** $\pi + 2\sqrt{2}$ **(D)** $2\pi + \sqrt{2}$ **(E)** $2\pi + 2\sqrt{2}$

19. Tom, Dick, and Harry are playing a game. Starting at the same time, each of them flips a fair coin repeatedly until he gets his first head, at which point he stops. What is the probability that all three flip their coins the same number of times?

(A) $\frac{1}{8}$ **(B)** $\frac{1}{7}$ **(C)** $\frac{1}{6}$ **(D)** $\frac{1}{4}$ **(E)** $\frac{1}{3}$

20. A set of teams held a round-robin tournament in which every team played every other team exactly once. Every team won 10 games and lost 10 games; there were no ties. How many sets of three teams $\{A, B, C\}$ were there in which A beat B, B beat C, and C beat A?

(A) 385 **(B)** 665 **(C)** 945 **(D)** 1140 **(E)** 1330

21. Let $ABCD$ be a unit square. Let Q_1 be the midpoint of $\overline{CD}$. For $i = 1, 2, \ldots$, let P_i be the intersection of $\overline{AQ_i}$ and $\overline{BD}$, and let Q_{i+1} be the foot of the perpendicular from P_i to $\overline{CD}$. What is

$$\sum_{i=1}^{\infty} \text{Area of } \triangle DQ_iP_i\,?$$

(A) $\frac{1}{6}$ **(B)** $\frac{1}{4}$ **(C)** $\frac{1}{3}$ **(D)** $\frac{1}{2}$ **(E)** 1

22. For a certain positive integer n less than 1000, the decimal equivalent of $\frac{1}{n}$ is $0.\overline{abcdef}$, a repeating decimal of period of 6, and the decimal equivalent of $\frac{1}{n+6}$ is $0.\overline{wxyz}$, a repeating decimal of period 4. In which interval does n lie?

(A) $[1, 200]$ **(B)** $[201, 400]$ **(C)** $[401, 600]$ **(D)** $[601, 800]$ **(E)** $[801, 999]$

23. What is the volume of the region in three-dimensional space defined by the inequalities $|x| + |y| + |z| \le 1$ and $|x| + |y| + |z - 1| \le 1$

    (A) $\dfrac{1}{6}$ **(B)** $\dfrac{1}{4}$ **(C)** $\dfrac{1}{3}$ **(D)** $\dfrac{1}{2}$ **(E)** 1

24. There are exactly $77,000$ ordered quadruplets (a, b, c, d) such that $\gcd(a, b, c, d) = 77$ and $\text{lcm}(a, b, c, d) = n$. What is the smallest possible value for n?

    (A) $13,860$ **(B)** $20,790$ **(C)** $21,560$ **(D)** $27,720$ **(E)** $41,580$

25. The sequence (a_n) is defined recursively by $a_0 = 1$, $a_1 = \sqrt[19]{2}$, and $a_n = a_{n-1} a_{n-2}^2$ for $n \ge 2$. What is the smallest positive integer k such that the product $a_1 a_2 \cdots a_k$ is an integer?

    (A) 17 **(B)** 18 **(C)** 19 **(D)** 20 **(E)** 21

CHAPTER 22

AIME I/II PROBLEMS

22.1 2016 AIME I Problems

1. For $-1 < r < 1$, let $S(r)$ denote the sum of the geometric series

$$12 + 12r + 12r^2 + 12r^3 + \cdots .$$

Let a between -1 and 1 satisfy $S(a)S(-a) = 2016$. Find $S(a) + S(-a)$.

2. Two dice appear to be standard dice with their faces numbered from 1 to 6, but each die is weighted so that the probability of rolling the number k is directly proportional to k. The probability of rolling a 7 with this pair of dice is $\frac{m}{n}$, where m and n are relatively prime positive integers. Find $m + n$.

3. A *regular icosahedron* is a 20-faced solid where each face is an equilateral triangle and five triangles meet at every vertex. The regular icosahedron shown below has one vertex at the top, one vertex at the bottom, an upper pentagon of five vertices all adjacent to the top vertex and all in the same horizontal plane, and a lower pentagon of five vertices all adjacent to the bottom vertex and all in another horizontal plane. Find the number of paths from the top vertex to the bottom vertex such that each part of a path goes downward or horizontally along an edge of the icosahedron, and no vertex is repeated.

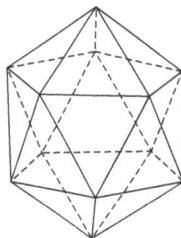

4. A right prism with height h has bases that are regular hexagons with sides of length 12. A vertex A of the prism and its three adjacent vertices are the vertices of a triangular pyramid. The dihedral angle (the angle between the two planes) formed by the face of the pyramid that lies in a base of the prism and the face of the pyramid that does not contain A measures $60°$. Find h^2.

5. Anh read a book. On the first day she read n pages in t minutes, where n and t are positive integers. On the second day Anh read $n+1$ pages in $t+1$ minutes. Each day thereafter Anh read one more page than she read on the previous day, and it took her one more minute than on the previous day until she completely read the 374 page book. It took her a total of 319 minutes to read the book. Find $n+t$.

6. In $\triangle ABC$ let I be the center of the inscribed circle, and let the bisector of $\angle ACB$ intersect $\overline{AB}$ at L. The line through C and L intersects the circumscribed circle of $\triangle ABC$ at the two points C and D. If $LI = 2$ and $LD = 3$, then $IC = \frac{p}{q}$, where p and q are relatively prime positive integers. Find $p+q$.

7. For integers a and b consider the complex number

$$\frac{\sqrt{ab+2016}}{ab+100} - \left(\frac{\sqrt{|a+b|}}{ab+100}\right)i.$$

Find the number of ordered pairs of integers (a, b) such that this complex number is a real number.

8. For a permutation $p = (a_1, a_2, \ldots, a_9)$ of the digits $1, 2, \ldots, 9$, let $s(p)$ denote the sum of the three 3-digit numbers $a_1a_2a_3$, $a_4a_5a_6$, and $a_7a_8a_9$. Let m be the minimum value of $s(p)$ subject to the condition that the units digit of $s(p)$ is 0. Let n denote the number of permutations p with $s(p) = m$. Find $|m - n|$.

9. Triangle ABC has $AB = 40$, $AC = 31$, and $\sin A = \frac{1}{5}$. This triangle is inscribed in rectangle $AQRS$ with B on $\overline{QR}$ and C on $\overline{RS}$. Find the maximum possible area of $AQRS$.

10. A strictly increasing sequence of positive integers $a_1, a_2, a_3, \ldots$ has the property that for every positive integer k, the subsequence $a_{2k-1}, a_{2k}, a_{2k+1}$ is geometric and the subsequence $a_{2k}, a_{2k+1}, a_{2k+2}$ is arithmetic. Suppose that $a_{13} = 2016$. Find a_1.

11. Let $P(x)$ be a nonzero polynomial such that $(x-1)P(x+1) = (x+2)P(x)$ for every real x, and $\left(P(2)\right)^2 = P(3)$. Then $P(\frac{7}{2}) = \frac{m}{n}$, where m and n are relatively prime positive integers. Find $m+n$.

12. Find the least positive integer m such that $m^2 - m + 11$ is a product of at least four not necessarily distinct primes.

13. Freddy the frog is jumping around the coordinate plane searching for a river, which lies on the horizontal line $y = 24$. A fence is located at the horizontal line $y = 0$. On each jump Freddy randomly chooses a direction parallel to one of the coordinate axes and moves one unit in that direction. When he is at a point where $y = 0$, with equal likelihoods he chooses one of three directions where he either jumps parallel to the fence or jumps away from the fence, but he never chooses the direction that would have him cross over the fence to where $y < 0$. Freddy starts his search at the point $(0, 21)$ and will stop once he reaches a point on the river. Find the expected number of jumps it will take Freddy to reach the river.

14. Centered at each lattice point in the coordinate plane are a circle radius $\frac{1}{10}$ and a square with sides of length $\frac{1}{5}$ whose sides are parallel to the coordinate axes. The line segment from $(0,0)$ to $(1001, 429)$ intersects m of the squares and n of the circles. Find $m + n$.

15. Circles ω_1 and ω_2 intersect at points X and Y. Line ℓ is tangent to ω_1 and ω_2 at A and B, respectively, with line AB closer to point X than to Y. Circle ω passes through A and B intersecting ω_1 again at $D \neq A$ and intersecting ω_2 again at $C \neq B$. The three points C, Y, and D are collinear, $XC = 67$, $XY = 47$, and $XD = 37$. Find AB^2.

22.2 2016 AIME II Problems

1. Initially Alex, Betty, and Charlie had a total of 444 peanuts. Charlie had the most peanuts, and Alex had the least. The three numbers of peanuts that each person had form a geometric progression. Alex eats 5 of his peanuts, Betty eats 9 of her peanuts, and Charlie eats 25 of his peanuts. Now the three numbers of peanuts that each person has form an arithmetic progression. Find the number of peanuts Alex had initially.

2. There is a 40% chance of rain on Saturday and a 30% chance of rain on Sunday. However, it is twice as likely to rain on Sunday if it rains on Saturday than if it does not rain on Saturday. The probability that it rains at least one day this weekend is $\frac{a}{b}$, where a and b are relatively prime positive integers. Find $a + b$.

3. Let x, y, and z be real numbers satisfying the system

$$\log_2(xyz - 3 + \log_5 x) = 5$$
$$\log_3(xyz - 3 + \log_5 y) = 4$$
$$\log_4(xyz - 3 + \log_5 z) = 4.$$

Find the value of $|\log_5 x| + |\log_5 y| + |\log_5 z|$.

4. An $a \times b \times c$ rectangular box is built from $a \cdot b \cdot c$ unit cubes. Each unit cube is colored red, green, or yellow. Each of the a layers of size $1 \times b \times c$ parallel to the $(b \times c)$-faces of the box contains exactly 9 red cubes, exactly 12 green cubes, and some yellow cubes. Each of the b layers of size $a \times 1 \times c$ parallel to the $(a \times c)$-faces of the box contains exactly 20 green cubes, exactly 25 yellow cubes, and some red cubes. Find the smallest possible volume of the box.

5. Triangle ABC_0 has a right angle at C_0. Its side lengths are pairwise relatively prime positive integers, and its perimeter is p. Let C_1 be the foot of the altitude to $\overline{AB}$, and for $n \geq 2$, let C_n be the foot of the altitude to $\overline{C_{n-2}B}$ in $\triangle C_{n-2}C_{n-1}B$. The sum $\sum_{n=1}^{\infty} C_{n-1}C_n = 6p$. Find p.

6. For polynomial $P(x) = 1 - \frac{1}{3}x + \frac{1}{6}x^2$, define

$$Q(x) = P(x)P(x^3)P(x^5)P(x^7)P(x^9) = \sum_{i=0}^{50} a_i x^i.$$

Then $\sum_{i=0}^{50} |a_i| = \frac{m}{n}$, where m and n are relatively prime positive integers. Find $m + n$.

7. Squares $ABCD$ and $EFGH$ have a common center and $\overline{AB} \parallel \overline{EF}$. The area of $ABCD$ is 2016, and the area of $EFGH$ is a smaller positive integer. Square $IJKL$ is constructed so that each of its vertices lies on a side of $ABCD$ and each vertex of $EFGH$ lies on a side of $IJKL$. Find the difference between the largest and smallest possible integer values for the area of $IJKL$.

8. Find the number of sets $\{a, b, c\}$ of three distinct positive integers with the property that the product of a, b, and c is equal to the product of 11, 21, 31, 41, 51, and 61.

9. The sequences of positive integers $1, a_2, a_3, \ldots$ and $1, b_2, b_3, \ldots$ are an increasing arithmetic sequence and an increasing geometric sequence, respectively. Let $c_n = a_n + b_n$. There is an integer k such that $c_{k-1} = 100$ and $c_{k+1} = 1000$. Find c_k.

10. Triangle ABC is inscribed in circle ω. Points P and Q are on side $\overline{AB}$ with $AP < AQ$. Rays CP and CQ meet ω again at S and T (other than C), respectively. If $AP = 4$, $PQ = 3$, $QB = 6$, $BT = 5$, and $AS = 7$, then $ST = \frac{m}{n}$, where m and n are relatively prime positive integers. Find $m + n$.

11. For positive integers N and k, define N to be *k-nice* if there exists a positive integer a such that a^k has exactly N positive divisors. Find the number of positive integers less than 1000 that are neither 7-nice nor 8-nice.

12. The figure below shows a ring made of six small sections which you are to paint on a wall. You have four paint colors available and will paint each of the six sections a solid color. Find the number of ways you can choose to paint the sections if no two adjacent sections can be painted with the same color.

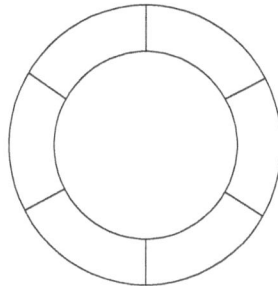

13. Beatrix is going to place six rooks on a 6×6 chessboard where both the rows and columns are labeled 1 to 6; the rooks are placed so that no two rooks are in the same row or the same column. The *value* of a square is the sum of its row number and column number. The *score* of an arrangement of rooks is the least value of any occupied square. The average score over all valid configurations is $\frac{p}{q}$, where p and q are relatively prime positive integers. Find $p + q$.

14. Equilateral $\triangle ABC$ has side length 600. Points P and Q lie outside the plane of $\triangle ABC$ and are on the opposite sides of the plane. Furthermore, $PA = PB = PC$, and $QA = QB = QC$, and the planes of $\triangle PAB$ and $\triangle QAB$ form a $120°$ dihedral angle (the angle between the two planes). There is a point O whose distance from each of A, B, C, P, and Q is d. Find d.

15. For $1 \le i \le 215$ let $a_i = \frac{1}{2^i}$ and $a_{216} = \frac{1}{2^{215}}$. Let $x_1, x_2, \ldots, x_{216}$ be positive real numbers such that

$$\sum_{i=1}^{216} x_i = 1 \quad \text{and} \quad \sum_{1 \le i < j \le 216} x_i x_j = \frac{107}{215} + \sum_{i=1}^{216} \frac{a_i x_i^2}{2(1 - a_i)}.$$

The maximum possible value of $x_2 = \frac{m}{n}$, where m and n are relatively prime positive integers. Find $m + n$.

CHAPTER 23

MATHCOUNTS PROBLEMS

1. What is the tenth digit to the right of the point in the decimal representation of $1/98$?

2. Fill in the blanks with digits so that each sentence becomes true.

The number of times the digit 0 appears in the puzzle is ___.

The number of times the digit 1 appears in the puzzle is ___.

The number of times the digit 2 appears in the puzzle is ___.

The number of times the digit 3 appears in the puzzle is ___.

The number of times the digit 4 appears in the puzzle is ___.

3. From the Master's Round 2008 where students have 15 minutes to work on their solutions before presenting it to an audience. Recall that a base -4 representation of a number is a string of digits $\{0, 1, 2, 3\}$ representing a sum of multiples of powers of -4. For example,

$$
\begin{aligned}
321.21_{-4} &= 3(-4)^2 + 2(-4)^1 + 1(-4)^0 + 2(-4){-1} + 1(-4)^{-2} \\
&= 3 \cdot 16 - 2 \cdot 4 + 1 \cdot 1 - 2 \cdot \frac{1}{4} + 1 \cdot \frac{1}{16} \\
&= 48 - 8 + 1 - \frac{1}{2} + \frac{1}{16} \\
&= 40 + \frac{9}{16} = 649/16.
\end{aligned}
$$

(a) In the space provided, construct the addition and multiplications tables for the base -4 digits.

+	0	1	2	3
0				
1				
2				
3				

×	0	1	2	3
0				
1				
2				
3				

(b) List, in ascending order, the representations of the integers from 1 to 14.

(c) Note that 12_{-4} represents $1 \cdot (-4)^1 + 2 \cdot (-4)^0 = -2$. List, in descending order, the representations of the first 14 negative integers.

(d) How can one determine whether a number in the system is positive or negative?

(e) Note that the sum of digits table, above, indicates that every carry from addition involves two carry digits. Use this fact to explain why the sum of two 'positives' is 'positive' and the sum of two 'negatives' is 'negative'.

(f) Use the notion that the product of two positive integers may be regarded as successive addition and use your explanation from (e) to argue that the product of two 'positives' is 'positive' and a 'positive' times a 'negative' is 'negative'.

Definition. If two symbols represent the integers that have a sum of zero, then the two integers are called *additive inverses* of each other. Note that from questions (a) and (b) we have that 1_{-4} and 13_{-4} represent additive inverses since $1 + 13 = 0$.

(g) Find a quick method to determine the additive inverse of a given integer. Then use this method to work the subtraction problem $1132003_{-4} - 1202313_{-4}$.

(h) Interpret 123.32_{-4} as we did above for 321.21_{-4}.

(i) Find the base -4 representation of 99.

(j) Find the base -4 representation of 17.5.

(k) Find the base -4 representation of $1/2, 1/4$ and $1/16$.

(l) Find the base -4 representation of $1/3$.

(m) Devise a method to determine if a given integer is a multiple of 5 based on its base -4 representation. Find a digit d that makes $23231123d_{-4}$ a multiple of 5.

(n) Carry out the arithmetic $2312.12_{-4} + 13202.31_{-4}$.

(o) Write the symbol that represents the additive inverse of the number 1202313_{-4}.

(p) Using your answer from the previous question and the definition of subtraction, rewrite and then solve the addition problem defined by $1132003_{-4} - 1202313_{-4}$.

(q) Carry out the arithmetic $112.3_{-4} \times 33.2_{-4}$.

4. Read the following ten problems. What do they have in common? Try them. Then you'll see how they are related.

(a) What is the value of $\dfrac{7!}{(7-3)!3!}$?

(b) How many 3-element subsets does the set $\{A, B, C, D, E, F, G\}$ have?

(c) How many solutions are there to

$$x + y + u + v = 4$$

where x, y, u, and v are nonnegative integers. For example, $(2, 1, 0, 1)$ is such a solution.

(d) How many solutions does

$$x + y + u + v = 8$$

have subject to the condition that each of the variables is a positive integer?

(e) How many ways can a 3-person committee be selected from a 7-member club?

(f) Let $P_1, P_2, P_3, P_4, P_5, P_6, P_7$ be seven points distributed around a circle. How many triangles have all three vertices in the set.

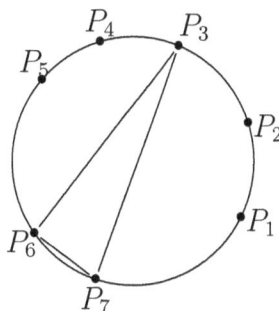

(g) How many paths of length 7 are there from A to B in the grid below?

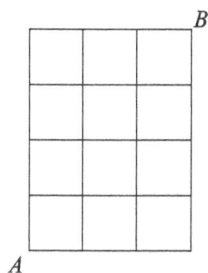

(h) Seven points are distributed around a circle. All pairs of them are joined by a secant line. What is the largest possible number of points of intersection *inside* the circle?

(i) What is the coefficient of x^3 in the expanded form of $(x + 1)^7$?

(j) What is the third entry of the seventh row of Pascal's triangle?

CHAPTER 24

ARML CONTEST PROBLEMS

Instructions: The power question is worth 50 points; each part's point value is given in brackets next to the part. To receive full credit, the presentation must be legible, orderly, clear, and concise. If a problem says "list" or "compute," you need not justify your answer. If a problem says "determine," "find," or "show," then you must show your work or explain your reasoning to receive full credit, although such explanations do not have to be lengthy. If a problem says "justify" or "prove," then you must prove your answer rigorously. Even if not proved, earlier numbered items may be used in solutions to later numbered items, but not vice versa.

24.1 Power Question 2009: Sign on the Label

An n-**label** is a permutation of the numbers 1 through n. For example, $J = 35214$ is a 5-label and $K = 132$ is a 3-label. For a fixed positive integer p, where $p \leq n$, consider consecutive blocks of p numbers in an n-label. For example, when $p = 3$ and $L = 263415$, the blocks are $263, 634, 341$, and 415. We can associate to each of these blocks a p-label that corresponds to the relative order of the numbers in that block. For $L = 263415$, we get the following:

$$\underline{263}415 \to 132; \quad 2\underline{634}15 \to 312; \quad 26\underline{341}5 \to 231; \quad 263\underline{415} \to 213.$$

Moving from left to right in the n-label, there are $n - p + 1$ such blocks, which means we obtain an $(n - p + 1)$-tuple of p-labels. For $L = 263415$, we get the 4-tuple $(132, 312, 231, 213)$. We will call this $(n - p + 1)$-tuple the **p-signature** of L (or **signature**, if p is clear from the context) and denote it by $S_p[L]$; the p-labels in the signature are called **windows**. For $L = 263415$, the windows are $132, 312, 231$, and 213, and we write

$$S_3[263415] = (132, 312, 231, 213).$$

More generally, we will call *any* $(n - p + 1)$-tuple of p-labels a p-signature, even if we do not know of an n-label to which it corresponds (and even if no such label exists). A signature that occurs for exactly one n-label is called **unique**, and a signature that doesn't occur for any n-labels is called **impossible**. A **possible** signature is one that occurs for at least one n-label.

In this power question, you will be asked to analyze some of the properties of labels and signatures.

The Problems 2009

1. (a) Compute the 3-signature for 52341.   [1]

   (b) Find another 5-label with the same 3-signature as in part (a).   [2]

   (c) Compute two other 6-labels with the same 4-signature as 462135.   [2]

2. (a) Explain why the label 1234 has a unique 3-signature.   [1]

   (b) List three other 4-labels with unique 3-signatures.   [1]

   (c) Explain why the 3-signature $(123, 321)$ is impossible.   [1]

   (d) List three other impossible 3-signatures that have exactly two windows.   [1]

We can associate a **shape** to a given 2-signature: a diagram of up and down steps that indicates the relative order of adjacent numbers. For example, the following shape corresponds to the 2-signature $(12, 12, 12, 21, 12, 21)$:

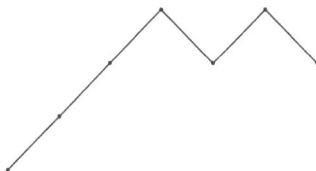

A 7-label with this 2-signature corresponds to placing the numbers 1 through 7 at the nodes above so that numbers increase with each up step and decrease with each down step. The 7-label 2347165 is shown below:

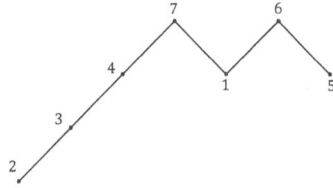

3. Consider the shape below:

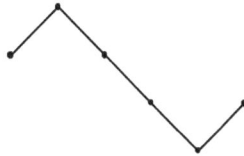

(a) Find the 2-signature that corresponds to this shape. [2]

(b) Compute two different 6-labels with the 2-signature you found in part (a). [3]

4. (a) List all 5-labels with 2-signature $(12, 12, 21, 21)$. [2]

(b) Find a formula for the number of $(2n + 1)$-labels with the 2-signature [3]

$$(\underbrace{12, 12, \ldots, 12}_{n}, \underbrace{21, 21, \ldots, 21}_{n}).$$

5. (a) Compute the number of 5-labels with 2-signature $(12, 21, 12, 21)$. [2]

(b) Determine the number of 9-labels with 2-signature

$$(12, 21, 12, 21, 12, 21, 12, 21).$$

Justify your answer. [3]

6. (a) Determine whether the following signatures are possible or impossible:

   (i) $(123, 132, 213)$, [1]

   (ii) $(321, 312, 213)$. [1]

(b) Notice that a $(p + 1)$-label has only two windows in its p-signature. For a given window ω_1, compute the number of windows ω_2 such that $S_p[L] = (\omega_1, \omega_2)$ for some $(p + 1)$-label L. [1]

(c) Justify your answer from part (b). [2]

7. (a) For a general n, determine the number of distinct possible p-signatures. [4]

(b) If a randomly chosen p-signature is 575 times more likely of being impossible than possible, determine p and n. [1]

8. (a) Show that $(312, 231, 312, 132)$ is not a unique 3-signature.   [1]

  (b) Show that $(231, 213, 123, 132)$ is a unique 3-signature.   [2]

  (c) Find two 5-labels with unique 2-signatures.   [1]

  (d) Find a 6-label with a unique 4-signature but which has the 3-signature from part (a).   [2]

9. (a) For a general $n \geq 2$, compute all n-labels that have unique 2-signatures.   [1]

  (b) Determine whether or not $S_5[495138627]$ is unique.   [1]

  (c) Determine the smallest p for which the 20-label

$$L = 3, 11, 8, 4, 17, 7, 15, 19, 6, 2, 14, 1, 10, 16, 5, 12, 20, 9, 13, 18$$

    has a unique p-signature.   [3]

10. Show that for each $k \geq 2$, the number of unique 2^{k-1}-signatures on the set of 2^k-labels is at least $2^{2^k - 3}$.   [5]

24.2 Power Question 2010: Power of Circular Subdivisions

A king strapped for cash is forced to sell off his kingdom $U = \{(x, y) : x^2 + y^2 \leq 1\}$. He sells the two circular plots C and C' centered at $\left(\pm\frac{1}{2}, 0\right)$ with radius $\frac{1}{2}$. The retained parts of the kingdom form two regions, each bordered by three arcs of circles; in what follows, we will call such regions *curvilinear triangles,* or *c-triangles* (c$\triangle$) for short.

This sad day marks day 0 of a new fiscal era. Unfortunately, these drastic measures are not enough, and so each day thereafter, court geometers mark off the largest possible circle contained in each c-triangle in the remaining property. This circle is tangent to all three arcs of the c-triangle, and will be referred to as the *incircle* of the c-triangle. At the end of the day, all incircles demarcated that day are sold off, and the following day, the remaining c-triangles are partitioned in the same manner.

The Problems 2010

1. Without using Descartes' Circle Formula (see below):

  a. Show that the circles marked off and sold on day 1 are centered at $\left(0, \pm\frac{2}{3}\right)$ with radius $\frac{1}{3}$.   [2]

  b. Find the combined area of the six remaining curvilinear territories.   [2]

On day 2, the plots bounded by the incircles of the six remaining curvilinear territories are sold.

2a. Determine the number of curvilinear territories remaining at the *end* of day 3.   [2]

2b. Let X_n be the number of plots sold on day n. Find a formula for X_n in terms of n.   [2]

2c. Determine the total number of plots sold up to and including day n.   [2]

Some notation: when discussing mutually tangent circles (or arcs), it is convenient to refer to the curvature of a circle rather than its radius. We define *curvature* as follows. Suppose that circle A of radius r_a is externally tangent to circle B of radius r_b. Then the curvatures of the circles are simply the reciprocals of their radii, $\frac{1}{r_a}$ and $\frac{1}{r_b}$. If circle A is internally tangent to circle B, however, as in the right diagram below, the curvature of circle A is still $\frac{1}{r_a}$, while the curvature of circle B is $-\frac{1}{r_b}$, the opposite of the reciprocal of its radius.

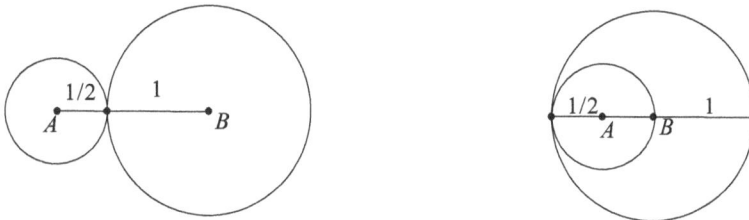

Circle A has curvature 2; circle B has curvature 1.   Circle A has curvature 2; circle B has curvature -1.

Using these conventions allows us to express a beautiful theorem of Descartes: when four circles A, B, C, D are pairwise tangent, with respective curvatures a, b, c, d, then

$$(a + b + c + d)^2 = 2\left(a^2 + b^2 + c^2 + d^2\right),$$

where (as before) a is taken to be negative if B, C, D are internally tangent to A, and correspondingly for b, c, or d. This Power Question does not involve the proof of Descartes' Circle Formula, but the formula may be used for all problems below.

3a. Two unit circles and a circle of <u>radius $\frac{2}{3}$</u> are mutually externally tangent. Compute all possible values of r such that a circle of radius r is tangent to all three circles.   [2]

3b. Given three mutually tangent circles with curvatures $a, b, c > 0$, suppose that $(a, b, c, 0)$ does not satisfy Descartes' Circle Formula. Show that there are two distinct values of r such that there is a circle of radius r tangent to the given circles.   [3]

3c. Algebraically, it is possible for a quadruple $(a, b, c, 0)$ to satisfy Descartes' Circle Formula, as occurs when $a = b = 1$ and $c = 4$. Find a geometric interpretation for this situation.   [2]

4. Let $\phi = \dfrac{1 + \sqrt{5}}{2}$, and let $\rho = \phi + \sqrt{\phi}$.

   a. Prove that $\rho^4 = 2\rho^3 + 2\rho^2 + 2\rho - 1$.   [2]

b. Show that four pairwise externally tangent circles with nonequal radii in geometric progression must have common ratio ρ. [2]

As shown in problem 3, given A, B, C, D as above with $s = a+b+c+d$, there is a second circle A' with curvature a' also tangent to $B, C,$ and D. We can describe A and A' as *conjugate circles*.

5. Use Descartes' Circle Formula to show that $a' = 2s-3a$ and therefore $s' = a'+b+c+d = 3s-4a$. [4]

In the context of this problem, a *circle configuration* is a quadruple of real numbers (a,b,c,d) representing curvatures of mutually tangent circles A, B, C, D. In other words, a circle configuration is a quadruple (a,b,c,d) of real numbers satisfying Descartes' Circle Formula.

The result in problem 5 allows us to compute the curvatures of the six plots removed on day 2. In this case, $(a,b,c,d) = (-1,2,2,3)$, and $s = 6$. For example, one such plot is tangent to both of the circles C and C' centered at $\left(\pm\frac{1}{2},0\right)$, and to one of the circles of radius $\frac{1}{3}$ removed on day 1; it is conjugate to the unit circle (the boundary of the original kingdom U). If this new plot's curvature is a, we can write $(-1,2,2,3) \vdash (a,2,2,3)$, and we say that the first circle configuration *yields* the second.

6a. Use the result of problem 5 to compute the curvatures of all circles removed on day 2, and the corresponding values of s'. [2]

6b. Show that by area, 12% of the kingdom is sold on day 2. [1]

6c. Find the areas of the circles removed on day 3. [2]

6d. Show that the plots sold on day 3 have mean curvature of 23. [1]

7. Prove that the curvature of each circular plot is an integer. [3]

Descartes' Circle Formula can be extended by interpreting the coordinates of points on the plane as complex numbers in the usual way: the point (x,y) represents the complex number $x + yi$. On the complex plane, let z_A, z_B, z_C, z_D be the centers of circles A, B, C, D respectively; as before, a, b, c, d are the curvatures of their respective circles. Then Descartes' Extended Circle Formula states

$$(a \cdot z_A + b \cdot z_B + c \cdot z_C + d \cdot z_D)^2 = 2\left(a^2 z_A^2 + b^2 z_B^2 + c^2 z_C^2 + d^2 z_D^2\right).$$

8a. Suppose that A' is a circle conjugate to A with center $z_{A'}$ and curvature a', and $\hat{s} = a \cdot z_A + b \cdot z_B + c \cdot z_C + d \cdot z_D$. Use Descartes' Extended Circle Formula to show that $a' \cdot z_{A'} = 2\hat{s} - 3a \cdot z_A$ and therefore $a' \cdot z_{A'} + b \cdot z_B + c \cdot z_C + d \cdot z_D = 3\hat{s} - 4a \cdot z_A$. [2]

8b. Prove that the center of each circular plot has coordinates $\left(\dfrac{u}{c}, \dfrac{v}{c}\right)$ where u and v are integers, and c is the curvature of the plot. [2]

Given a c-triangle T, let a, b, and c be the curvatures of the three arcs bounding T, with $a \leq b \leq c$, and let d be the curvature of the incircle of T. Define the *circle configuration associated with T* to be $\mathcal{C}(T) = (a, b, c, d)$. Define the c-triangle T to be *proper* if $c \leq d$. For example, circles of curvatures -1, 2, and 3 determine two c-triangles. The incircle of one has curvature 6, so it is proper; the incircle of the other has curvature 2, so it is not proper.

Let P and Q be two c-triangles, with associated configurations $\mathcal{C}(P) = (a, b, c, d)$ and $\mathcal{C}(Q) = (w, x, y, z)$. We say that P *dominates* Q if $a \leq w$, $b \leq x$, $c \leq y$, and $d \leq z$. (The term "dominates" refers to the fact that the radii of the arcs defining Q cannot be larger than the radii of the arcs defining P.)

Removing the incircle from T gives three c-triangles, $T^{(1)}$, $T^{(2)}$, $T^{(3)}$, each bounded by the incircle of T and two of the arcs that bound T. These triangles have associated configurations

$$\mathcal{C}(T^{(1)}) = (b, c, d, a'),$$
$$\mathcal{C}(T^{(2)}) = (a, c, d, b'),$$
$$\mathcal{C}(T^{(3)}) = (a, b, d, c'),$$

where a', b', and c' are determined by the formula in problem 5.

9. Let P and Q be two proper c-triangles such that P dominates Q. Let $\mathcal{C}(P) = (a, b, c, d)$ and $\mathcal{C}(Q) = (w, x, y, z)$.

  a. Show that $P^{(3)}$ dominates $P^{(2)}$ and that $P^{(2)}$ dominates $P^{(1)}$. [2]

  b. Prove that $P^{(1)}$ dominates $Q^{(1)}$. [2]

  c. Prove that $P^{(3)}$ dominates $Q^{(3)}$. [2]

10a. Prove that the largest plot sold by the king on day n has curvature $n^2 + 2$. [3]

10b. If $\rho = \phi + \sqrt{\phi}$, as in problem 4, prove that the curvature of the smallest plot sold by the king on day n does not exceed $2\rho^n$. [3]

24.3 Power Question 2011: Power of Triangles

The arrangement of numbers known as Pascal's Triangle has fascinated mathematicians for centuries. In fact, about 700 years before Pascal, the Indian mathematician Halayudha wrote about it in his commentaries to a then-1000-year-old treatise on verse structure by the Indian poet and mathematician Pingala, who called it the *Meruprastāra*, or "Mountain of Gems". In this Power Question, we'll explore some properties of Pingala's/Pascal's Triangle ("PT") and its variants.

Unless otherwise specified, the **only** definition, notation, and formulas you may use for PT are the definition, notation, and formulas given below.

PT consists of an infinite number of rows, numbered from 0 onwards. The nth row contains $n+1$ numbers, identified as $Pa(n, k)$, where $0 \leq k \leq n$. For all n, define $Pa(n, 0) = Pa(n, n) = 1$. Then for $n > 1$ and $1 \leq k \leq n - 1$, define $Pa(n, k) = Pa(n - 1, k - 1) + Pa(n - 1, k)$. It is convenient to define $Pa(n, k) = 0$ when $k < 0$ or $k > n$. We write the nonzero values of PT in the familiar pyramid shown below.

$$1$$
$$Pa(0, 0)$$

$$1 \qquad\qquad 1$$
$$Pa(1, 0) \qquad Pa(1, 1)$$

$$1 \qquad\qquad 2 \qquad\qquad 1$$
$$Pa(2, 0) \qquad Pa(2, 1) \qquad Pa(2, 2)$$

$$1 \qquad\qquad 3 \qquad\qquad 3 \qquad\qquad 1$$
$$Pa(3, 0) \qquad Pa(3, 1) \qquad Pa(3, 2) \qquad Pa(3, 3)$$

As is well known, $Pa(n, k)$ gives the number of ways of choosing a committee of k people from a set of n people, so a simple formula for $Pa(n, k)$ is $Pa(n, k) = \frac{n!}{k!(n-k)!}$. You may use this formula or the recursive definition above throughout this Power Question.

The Problems 2011

1a. For $n = 1, 2, 3, 4$, and $k = 4$, find $Pa(n, n) + Pa(n + 1, n) + \cdots + Pa(n + k, n)$.   [2]

1b. If $Pa(n, n) + Pa(n + 1, n) + \cdots + Pa(n + k, n) = Pa(m, j)$, find and justify formulas for m and j in terms of n and k.   [2]

Consider the *parity* of each entry: define

$$PaP(n, k) = \begin{cases} 1 & \text{if } Pa(n, k) \text{ is odd,} \\ 0 & \text{if } Pa(n, k) \text{ is even.} \end{cases}$$

2a. Prove that $PaP(n, 0) = PaP(n, n) = 1$ for all nonnegative integers n.   [1]

2b. Compute rows $n = 0$ to $n = 8$ of PaP.   [2]

You may have learned that the array of parities of Pascal's Triangle forms the famous fractal known as Sierpinski's Triangle. The first 128 rows are shown below, with a dot marking each entry where $\text{Pa}P(n, k) = 1$.

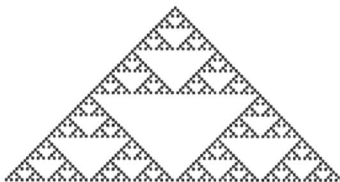

The next problem helps explain this surprising connection.

3a. If $n = 2^j$ for some nonnegative integer j, and $0 < k < n$, show that $\text{Pa}P(n, k) = 0$.   [2]

3b. Let $j \geq 0$, and suppose $n \geq 2^j$. Prove that $\text{Pa}(n, k)$ has the same parity as the sum $\text{Pa}(n - 2^j, k - 2^j) + \text{Pa}(n - 2^j, k)$, i.e., either both $\text{Pa}(n, k)$ and the given sum are even, or both are odd.   [2]

3c. If j is an integer such that $2^j \leq n < 2^{j+1}$, and $k < 2^j$, prove that

$$\text{Pa}P(n, k) = \text{Pa}P(n - 2^j, k).$$ [2]

Clark's Triangle: If the left side of PT is replaced with consecutive multiples of 6, starting with 0, but the right entries (except the first) and the generating rule are left unchanged, the result is called *Clark's Triangle*. If the kth entry of the nth row is denoted by $\text{Cl}(n, k)$, then the formal rule is:

$$\begin{cases} \text{Cl}(n, 0) = 6n & \text{for all } n, \\ \text{Cl}(n, n) = 1 & \text{for } n \geq 1, \\ \text{Cl}(n, k) = \text{Cl}(n - 1, k - 1) + \text{Cl}(n - 1, k) & \text{for } n \geq 1 \text{ and } 1 \leq k \leq n - 1. \end{cases}$$

The first four rows of Clark's Triangle are given below.

$$0$$
$$\text{Cl}(0,0)$$

$$6 \qquad\qquad 1$$
$$\text{Cl}(1,0) \qquad \text{Cl}(1,1)$$

$$12 \qquad\qquad 7 \qquad\qquad 1$$
$$\text{Cl}(2,0) \qquad \text{Cl}(2,1) \qquad \text{Cl}(2,2)$$

$$18 \qquad\qquad 19 \qquad\qquad 8 \qquad\qquad 1$$
$$\text{Cl}(3,0) \qquad \text{Cl}(3,1) \qquad \text{Cl}(3,2) \qquad \text{Cl}(3,3)$$

4a. Compute the next three rows of Clark's Triangle. [2]

4b. If $\text{Cl}(n,1) = an^2 + bn + c$, determine the values of a, b, and c. [2]

4c. Prove the formula you found in 4b. [2]

5a. Compute $\text{Cl}(11,2)$. [1]

5b. Find and justify a formula for $\text{Cl}(n,2)$ in terms of n. [2]

5c. Compute $\text{Cl}(11,3)$. [1]

5d. Find and justify a formula for $\text{Cl}(n,3)$ in terms of n. [2]

6. Find and prove a closed formula (that is, a formula with a fixed number of terms and no "...") for $\text{Cl}(n,k)$ in terms of n, k, and the Pa function. [3]

Leibniz's Harmonic Triangle: Consider the triangle formed by the rule

$$\begin{cases} \text{Le}(n,0) = \frac{1}{n+1} & \text{for all } n, \\ \text{Le}(n,n) = \frac{1}{n+1} & \text{for all } n, \\ \text{Le}(n,k) = \text{Le}(n+1,k) + \text{Le}(n+1,k+1) & \text{for all } n \text{ and } 0 \le k \le n. \end{cases}$$

This triangle, discovered first by Leibniz, consists of reciprocals of integers as shown below.

$$1$$
$$\mathrm{Le}(0,0)$$

$$\frac{1}{2} \qquad\qquad \frac{1}{2}$$
$$\mathrm{Le}(1,0) \qquad\qquad \mathrm{Le}(1,1)$$

$$\frac{1}{3} \qquad\qquad \frac{1}{6} \qquad\qquad \frac{1}{3}$$
$$\mathrm{Le}(2,0) \qquad\qquad \mathrm{Le}(2,1) \qquad\qquad \mathrm{Le}(2,2)$$

$$\frac{1}{4} \qquad \frac{1}{12} \qquad \frac{1}{12} \qquad \frac{1}{4}$$
$$\mathrm{Le}(3,0) \qquad \mathrm{Le}(3,1) \qquad \mathrm{Le}(3,2) \qquad \mathrm{Le}(3,3)$$

For this contest, you may assume that $\mathrm{Le}(n,k) > 0$ whenever $0 \le k \le n$, and that $\mathrm{Le}(n,k)$ is undefined if $k < 0$ or $k > n$.

7a. Compute the entries in the next two rows of Leibniz's Triangle. [1]

7b. Compute $\mathrm{Le}(17,1)$. [1]

7c. Compute $\mathrm{Le}(17,2)$. [1]

8a. Find and justify a formula for $\mathrm{Le}(n,1)$ in terms of n. [2]

8b. Compute $\displaystyle\sum_{n=1}^{2011} \mathrm{Le}(n,1)$. [2]

8c. Find and justify a formula for $\mathrm{Le}(n,2)$ in terms of n. [2]

9a. If $\displaystyle\sum_{i=1}^{\infty} \mathrm{Le}(i,1) = \mathrm{Le}(n,k)$, determine the values of n and k. [2]

9b. If $\displaystyle\sum_{i=m}^{\infty} \mathrm{Le}(i,m) = \mathrm{Le}(n,k)$, compute expressions for n and k in terms of m. [2]

9c. Justify your result in 9b. [2]

10. Find three distinct sets of positive integers $\{a,b,c,d\}$ with $a < b < c < d$ such that
$$\frac{1}{3} = \frac{1}{a} + \frac{1}{b} + \frac{1}{c} + \frac{1}{d}.$$ [3]

11. Find and prove a closed formula (that is, a formula with a fixed number of terms and no "…") for $\mathrm{Le}(n,k)$ in terms of n, k, and the Pa function. [4]

24.4 Power Question 2012: Cops and Robbers

Every day, each Cop searches one hideout: the Cops know the location of all hideouts and which hideouts are adjacent to which. Cops are thorough searchers, so if the Robber is present in the hideout searched, he is found and arrested. If the Robber is not present in the hideout searched, his location is *not* revealed. That is, the Cops only know that the Robber was not caught at any of the hideouts searched; they get no specific information (other than what they can derive by logic) about what hideout he was in. Cops are not constrained by edges on the map: a Cop may search any hideout on any day, regardless of whether it is adjacent to the hideout searched the previous day. A Cop may search the same hideout on consecutive days, and multiple Cops may search different hideouts on the same day. In the map above, a Cop could search A on day 1 and day 2, and then search C on day 3.

The focus of this Power Question is to determine, given a hideout map and a fixed number of Cops, whether the Cops can be sure of catching the Robber within some time limit.

Map Notation: The following notation may be useful when writing your solutions. For a map M, let $h(M)$ be the number of hideouts and $e(M)$ be the number of edges in M. The *safety* of a hideout H is the number of hideouts adjacent to H, and is denoted by $s(H)$.

The Problems 2012

1a. Consider the hideout map M below.

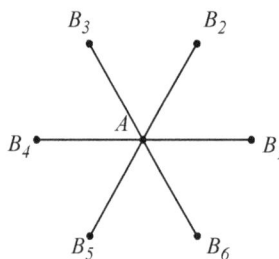

Show that one Cop can always catch the Robber. [1]

The *Cop number* of a map M, denoted $C(M)$, is the minimum number of Cops required to guarantee that the Robber is caught. In 1a, you have shown that $C(M) = 1$.

1b. The map shown below is $\mathcal{C}_6$, the cyclic graph with six hideouts. Show that three Cops are sufficient to catch the Robber on $\mathcal{C}_6$, so that $C(\mathcal{C}_6) \leq 3$. [2]

1c. Show that for all maps M, $C(M) < h(M)$. [2]

2a. Find $C(M)$ for the map below. [2]

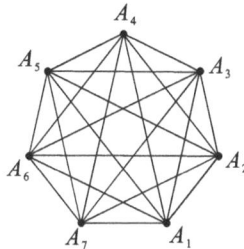

2b. Show that $C(M) \leq 3$ for the map below. [2]

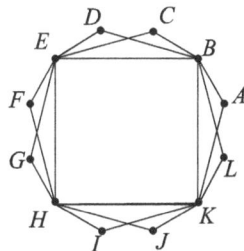

2c. Generalize the result of 2a: if $\mathcal{K}_n$ is a map with n hideouts in which every hideout is adjacent to every other hideout (also called the *complete map* on n hideouts), determine $C(\mathcal{K}_n)$. [3]

The police want to catch the Robber with a minimum number of Cops, but time is of the essence. For a map M and a fixed number of Cops $c \geq C(M)$, define the *capture time*, denoted $D(M, c)$, to be the minimum number of days required to *guarantee* a capture using c Cops. For example, in the graph from 1b, if three Cops are deployed, they might catch the Robber in the first day, but if they don't, there is a strategy that will guarantee they will capture the Robber within two days. Therefore the capture time is $D(\mathcal{C}_6, 3) = 2$.

3. A *path on n hideouts* is a map with n hideouts, connected in one long string. (More formally, a map is a path if and only if two hideouts are adjacent to exactly one hideout each and all other hideouts are adjacent to exactly two hideouts each.) It is denoted by $\mathcal{P}_n$. The maps $\mathcal{P}_3$ through $\mathcal{P}_6$ are shown below.

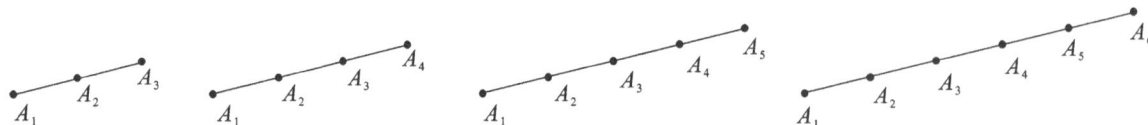

Show that $D(\mathcal{P}_n, 1) \leq 2n$ for $n \geq 3$.   [4]

4. A *cycle on n hideouts* is a map with n hideouts, connected in one loop. (More formally, a map is a cycle if every hideout is adjacent to exactly two hideouts.) It is denoted by $\mathcal{C}_n$. The maps $\mathcal{C}_3$ through $\mathcal{C}_6$ are shown below.

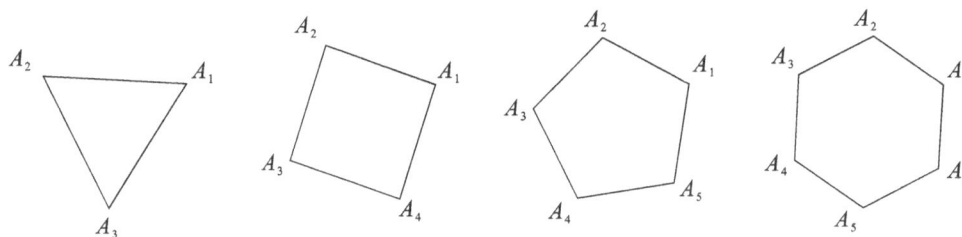

   a. Determine $C(\mathcal{C}_n)$ for $n \geq 3$. (Note that this is *determine* and not *compute*.)   [2]

   b. Show that $D(\mathcal{C}_n, C(\mathcal{C}_n)) < \frac{3n}{2}$ for $n \geq 3$.   [2]

Definition: The *workday number* of M, denoted $W(M)$, is the minimum number of *Cop workdays* needed to guarantee the Robber's capture. For example, a strategy that guarantees capture within three days using 17 Cops on the first day, 11 Cops on the second day, and only 6 Cops on the third day would require a total of $17 + 11 + 6 = 34$ Cop workdays.

5. Determine $W(M)$ for each of the maps in problem 2a and 2b.   [4]

6. Let M be a map with $n \geq 3$ hideouts. Prove that $2 \leq W(M) \leq n$, and that these bounds cannot be improved. In other words, prove that for each $n \geq 3$, there exist maps M_1 and M_2 such that $W(M_1) = 2$ and $W(M_2) = n$.   [6]

Definition: A map is *bipartite* if it can be partitioned into two sets of hideouts, $\mathcal{A}$ and $\mathcal{B}$, such that $\mathcal{A} \cap \mathcal{B} = \emptyset$, and each hideout in $\mathcal{A}$ is adjacent only to hideouts in $\mathcal{B}$, and each hideout in $\mathcal{B}$ is adjacent only to hideouts in $\mathcal{A}$.

7a. Prove that if M is bipartite, then $C(M) \leq n/2$.   [3]

7b. Prove that $C(M) \leq n/2$ for *any* map M with the property that, for all hideouts H_1 and H_2, either all paths from H_1 to H_2 contain an odd number of edges, or all paths from H_1 to H_2 contain an even number of edges.   [3]

8. A map M is called k-*perfect* if its hideout set H can be partitioned into equal-sized subsets $\mathcal{A}_1, \mathcal{A}_2, \ldots, \mathcal{A}_k$ such that for any j, the hideouts of $\mathcal{A}_j$ are only adjacent to hideouts in $\mathcal{A}_{j+1}$ or

$\mathcal{A}_{j-1}$. (The indices are taken modulo k: the hideouts of $\mathcal{A}_1$ may be adjacent to the hideouts of $\mathcal{A}_k$.) Show that if M is k-perfect, then $C(M) \leq \frac{2n}{k}$. [4]

9. Find an example of a map M with 2012 hideouts such that $C(M) = 17$ and $W(M) = 34$, or prove that no such map exists. [5]

10. Find an example of a map M with 4 or more hideouts such that $W(M) = 3$, or prove that no such map exists. [5]

24.5 Power Question 2013: Power of (Urban) Planning

In ARMLopolis, every house number is a positive integer, and City Hall's address is 0. However, due to the curved nature of the cowpaths that eventually became the streets of ARMLopolis, the distance $d(n)$ between house n and City Hall is not simply the value of n. Instead, if $n = 3^k n'$, where $k \geq 0$ is an integer and n' is an integer not divisible by 3, then $d(n) = 3^{-k}$. For example, $d(18) = 1/9$ and $d(17) = 1$. Notice that even though no houses have negative numbers, $d(n)$ is well-defined for negative values of n. For example, $d(-33) = 1/3$ because $-33 = 3^1 \cdot -11$. By definition, $d(0) = 0$. Following the dictum "location, location, location," this Power Question will refer to "houses" and "house numbers" interchangeably.

Curiously, the arrangement of the houses is such that the distance from house n to house m, written $d(m, n)$, is simply $d(m - n)$. For example, $d(3, 4) = d(-1) = 1$ because $-1 = 3^0 \cdot -1$. In particular, if $m = n$, then $d(m, n) = 0$.

The Problems 2013

1a. Compute $d(6)$, $d(16)$, and $d(72)$. [1 pt]

1b. Of the houses with positive numbers less than 100, find, with proof, the house or houses which is (are) closest to City Hall. [2 pts]

1c. Find four houses at a distance of $1/9$ from house number 17. [1 pt]

1d. Find an infinite sequence of houses $\{h_n\}$ such that
$$d(17, h_1) > d(17, h_2) > \cdots > d(17, h_n) > d(17, h_{n+1}) > \cdots.$$
[2 pts]

The *neighborhood* of a house n, written $\mathcal{N}(n)$, is the set of all houses that are the same distance from City Hall as n. In symbols, $\mathcal{N}(n) = \{m \mid d(m) = d(n)\}$. Geometrically, it may be helpful to think of $\mathcal{N}(n)$ as a circle centered at City Hall with radius $d(n)$.

2a. Find four houses in $\mathcal{N}(6)$. [1 pt]

2b. Suppose that n is a house with $d(n) = 1$. Find four houses in $\mathcal{N}(n)$. [1 pt]

2c. Suppose that n is a house with $d(n) = 1/27$. Determine the ten smallest positive integers m (in the standard ordering of the integers) such that $m \in \mathcal{N}(n)$. [2 pts]

3a. Notice that $d(16) = d(17) = 1$ and that $d(16, 17) = 1$. Is it true that, for all m, n such that $d(m) = d(n) = 1$ and $m \neq n$, $d(m, n) = 1$? Either prove your answer or find a counterexample. [1 pt]

3b. Suppose that $d(m) = d(n) = 1/3^k$ and that $m \neq n$. Determine the possible values of $d(m, n)$. [2 pts]

3c. Suppose that $d(17, m) = 1/81$. Determine the possible values of $d(16, m)$. [2 pts]

A visitor to ARMLopolis is surprised by the ARMLopolitan distance formula, and starts to wonder about ARMLopolitan geometry. An *ARMLopolitan triangle* is a triangle, all of whose vertices are ARMLopolitan houses.

4a. Show that $d(17, 51) \leq d(17, 34) + d(34, 51)$ and that $d(17, 95) \leq d(17, 68) + d(68, 95)$. [1 pt]

4b. If $a = 17$ and $b = 68$, determine all values of c such that $d(a, c) = d(a, b)$. [2 pts]

5a. Prove that, for all a and b, $d(a, b) \leq \max\{d(a), d(b)\}$. [2 pts]

5b. Prove that, for all a, b, and c, $d(a, c) \leq \max\{d(a, b), d(b, c)\}$. [2 pts]

5c. Prove that, for all a, b, and c, $d(a, c) \leq d(a, b) + d(b, c)$. [1 pt]

5d. After thinking about it some more, the visitor announces that all ARMLopolitan triangles have a special property. What is it? Justify your answer. [2 pts]

Unfortunately for new development, ARMLopolis is full: every nonnegative integer corresponds to (exactly one) house (or City Hall, in the case of 0). However, eighteen families arrive and are looking to move in. After much debate, the connotations of using negative house numbers are deemed unacceptable, and the city decides on an alternative plan. On July 17, Shewad Movers arrive and relocate every family from house n to house $n + 18$, for all positive n (so that City Hall does not move). For example, the family in house number 17 moves to house number 35.

6a. Find at least one house whose distance from City Hall changes as a result of the move. [1 pt]

6b. Prove that the distance between houses with consecutive numbers does not change after the move. [2 pts]

7a. The residents of $\mathcal{N}(1)$ value their suburban location and protest the move, claiming it will make them closer to City Hall. Will it? Justify your answer. [2 pts]

7b. Some residents of $\mathcal{N}(9)$ claim that their tightly-knit community will be disrupted by the move: they will be scattered between different neighborhoods after the change. Will they? Justify your answer. [2 pts]

7c. Other residents of $\mathcal{N}(9)$ claim that their community will be disrupted by newcomers: they say that after the move, their new neighborhood will also contain residents previously from several different old neighborhoods (not just the new arrivals to ARMLopolis). Will it? Justify your answer. [2 pts]

7d. Determine all values of n such that $\mathcal{N}(n)$ will either entirely relocate (i.e., all residents r of $\mathcal{N}(n)$ are at a different distance from City Hall than they were before), lose residents (as in 7b), or gain residents besides the new ARMLopolitans (as in 7c). [2 pts]

8a. One day, Paul (house 23) and Sally (house 32), longtime residents of $\mathcal{N}(1)$, are discussing the "2 side" of the neighborhood, that is, the set of houses $\{n \mid n = 3k + 2, \ k \in \mathbb{Z}\}$. Paul says "I feel like I'm at the center of the '2 side': when I looked out my window, I realized that the '2 side' consists of exactly those houses whose distance from me is at most $1/3$." Prove that Paul is correct. [2 pts]

8b. Sally replies "It's not all about you: I have the same experience!" Justify Sally's claim; in other words, prove that $\{n \mid n = 3k + 2, \ k \in \mathbb{Z}\}$ consists exactly of those houses whose distance from Sally's house is at most $1/3$. [2 pts]

8c. Paul's and Sally's observations can be generalized. For any x, let $\mathcal{D}_r(x) = \{y \mid d(x,y) \le r\}$, that is, $\mathcal{D}_r(x)$ is the *disk* of radius r. Prove that if $d(x,z) = r$, then $\mathcal{D}_r(x) = \mathcal{D}_r(z)$. [2 pts]

Paul's and Sally's experiences may seem incredible to a newcomer to ARMLopolis. Given a circle, any point on the circle is the center of the circle (so really *a* center of the circle). But ARMLopolitan geometry is even stranger than that!

9a. Ross takes a walk starting at his house, which is number 34. He first visits house n_1, such that $d(n_1, 34) = 1/3$. He then goes to another house, n_2, such that $d(n_1, n_2) = 1/3$. Continuing in that way, he visits houses $n_3, n_4, \ldots$, and each time, $d(n_i, n_{i+1}) = 1/3$. At the end of the day, what is his maximum possible distance from his original house? Justify your answer. [2 pts]

9b. Generalize your answer to 9a: given a value of n_0, determine the rational values of r for which the sequence $n_1, n_2, \ldots$ with $d(n_i, n_{i+1}) = r$ is entirely contained in a circle of finite radius centered at n_0. [2 pts]

It turns out that the eighteen newcomers are merely the first wave of a veritable deluge of enthusiastic arrivals: infinitely many, in fact. ARMLopolis finally decides on a drastic expansion plan: now house numbers will be *rational* numbers. To define $d(p/q)$, with p and q integers such that $pq \neq 0$, write $p/q = 3^k p'/q'$, where neither p' nor q' is divisible by 3 and k is an integer (not necessarily positive); then $d(p/q) = 3^{-k}$.

10a. Compute $d(3/5), d(5/8)$, and $d(7/18)$. [1 pt]

10b. Determine all pairs of relatively prime integers p and q such that $p/q \in \mathcal{N}(4/3)$. [1 pt]

10c. A longtime resident of IMOpia moves to ARMLopolis and hopes to keep his same address. When asked, he says "My old address was e, that is, the sum $\frac{1}{0!} + \frac{1}{1!} + \frac{1}{2!} + \ldots$. I'd really like to keep that address, because the addresses of my friends here are all partial sums of this series: Alice's house is $\frac{1}{0!}$, Bob's house is $\frac{1}{0!} + \frac{1}{1!}$, Carol's house is $\frac{1}{0!} + \frac{1}{1!} + \frac{1}{2!}$, and so on. Just let me know what I have to do in order to be near my friends!" After some head-scratching, the ARMLopolitan planning council announces that this request cannot be satisfied: there is no number, rational or otherwise, that corresponds to the (infinite) sum, or that is arbitrarily close to the houses in this sequence. Prove that the council is correct (and not just bureaucratic). [2 pts]

24.6 Power Question 2014: Power of Potlucks

In each town in ARMLandia, the residents have formed groups, which meet each week to share math problems and enjoy each others' company over a potluck-style dinner. Each town resident belongs to exactly one group. Every week, each resident is required to make one dish and to bring it to his/her group.

It so happens that each resident knows how to make precisely two dishes. Moreover, no two residents of a town know how to make the same pair of dishes. Shown below are two example towns. In the left column are the names of the town's residents. Adjacent to each name is the list of dishes that the corresponding resident knows how to make.

ARMLton

Resident	Dishes
Paul	pie, turkey
Arnold	pie, salad
Kelly	salad, broth

ARMLville

Resident	Dishes
Sally	steak, calzones
Ross	calzones, pancakes
David	steak, pancakes

The *population* of a town T, denoted $\mathrm{pop}(T)$, is the number of residents of T. Formally, the town itself is simply the set of its residents, denoted by $\{r_1, \ldots, r_{\mathrm{pop}(T)}\}$ unless otherwise specified.

The set of dishes that the residents of T collectively know how to make is denoted dish(T). For example, in the town of ARMLton described above, pop(ARMLton) $= 3$, and dish(ARMLton) $=$ {pie, turkey, salad, broth}.

A town T is called *full* if for every pair of dishes in dish(T), there is exactly one resident in T who knows how to make those two dishes. In the examples above, ARMLville is a full town, but ARMLton is not, because (for example) nobody in ARMLton knows how to make both turkey and salad.

Denote by $\mathcal{F}_d$ a full town in which collectively the residents know how to make d dishes. That is, $|\text{dish}(\mathcal{F}_d)| = d$.

The Problems 2014

1a. Compute pop($\mathcal{F}_{17}$). [1 pt]

1b. Let $n = \text{pop}(\mathcal{F}_d)$. In terms of n, compute d. [1 pt]

1c. Let T be a full town and let $D \in \text{dish}(T)$. Let T' be the town consisting of all residents of T who do not know how to make D. Prove that T' is full. [2 pts]

In order to avoid the embarrassing situation where two people bring the same dish to a group dinner, if two people know how to make a common dish, they are forbidden from participating in the same group meeting. Formally, a *group assignment* on T is a function $f : T \to \{1, 2, \ldots, k\}$, satisfying the condition that if $f(r_i) = f(r_j)$ for $i \neq j$, then r_i and r_j do not know any of the same recipes. The *group number* of a town T, denoted gr(T), is the least positive integer k for which there exists a group assignment on T.

For example, consider once again the town of ARMLton. A valid group assignment would be $f(\text{Paul}) = f(\text{Kelly}) = 1$ and $f(\text{Arnold}) = 2$. The function which gives the value 1 to each resident of ARMLton is **not** a group assignment, because Paul and Arnold must be assigned to different groups.

2a. Show that gr(ARMLton) $= 2$. [1 pt]

2b. Show that gr(ARMLville) $= 3$. [1 pt]

3a. Show that gr($\mathcal{F}_4$) $= 3$. [1 pt]

3b. Show that gr($\mathcal{F}_5$) $= 5$. [2 pts]

3c. Show that gr($\mathcal{F}_6$) $= 5$. [2 pts]

4. Prove that the sequence gr($\mathcal{F}_2$), gr($\mathcal{F}_3$), gr($\mathcal{F}_4$), $\ldots$ is a non-decreasing sequence. [2 pts]

For a dish D, a resident is called a D-*chef* if he or she knows how to make the dish D. Define $\text{chef}_T(D)$ to be the set of residents in T who are D-chefs. For example, in ARMLville, David is a steak-chef and a pancakes-chef. Further, $\text{chef}_{\text{ARMLville}}(\text{steak}) = \{\text{Sally, David}\}$.

5. Prove that

$$\sum_{D \in \text{dish}(T)} |\text{chef}_T(D)| = 2\text{pop}(T). \qquad \text{[2 pts]}$$

6. Show that for any town T and any $D \in \text{dish}(T)$, $\text{gr}(T) \geq |\text{chef}_T(D)|$.   [2 pts]

If $\text{gr}(T) = |\text{chef}_T(D)|$ for some $D \in \text{dish}(T)$, then T is called *homogeneous*. If $\text{gr}(T) > |\text{chef}_T(D)|$ for each dish $D \in \text{dish}(T)$, then T is called *heterogeneous*. For example, ARMLton is homogeneous, because $\text{gr}(\text{ARMLton}) = 2$ and exactly two chefs make pie, but ARMLville is heterogeneous, because even though each dish is only cooked by two chefs, $\text{gr}(\text{ARMLville}) = 3$.

7. For $n = 5, 6$, and 7, find a heterogeneous town T of population 5 for which $|\text{dish}(T)| = n$.   [3 pts]

A *resident cycle* is a sequence of distinct residents $r_1, \ldots, r_n$ such that for each $1 \leq i \leq n-1$, the residents r_i and r_{i+1} know how to make a common dish, residents r_n and r_1 know how to make a common dish, and no other pair of residents r_i and r_j, $1 \leq i, j \leq n$ know how to make a common dish. Two resident cycles are *indistinguishable* if they contain the same residents (in any order), and distinguishable otherwise. For example, if r_1, r_2, r_3, r_4 is a resident cycle, then r_2, r_1, r_4, r_3 and r_3, r_2, r_1, r_4 are indistinguishable resident cycles.

8a. Compute the number of distinguishable resident cycles of length 6 in $\mathcal{F}_8$.   [1 pt]

8b. In terms of k and d, find the number of distinguishable resident cycles of length k in $\mathcal{F}_d$. [1 pt]

9. Let T be a town with at least two residents that has a single resident cycle that contains every resident. Prove that T is homogeneous if and only if $\text{pop}(T)$ is even.   [3 pts]

10. Let T be a town such that, for each $D \in \text{dish}(T)$, $|\text{chef}_T(D)| = 2$.

    a. Prove that there are finitely many resident cycles $C_1, C_2, \ldots, C_j$ in T so that each resident belongs to exactly one of the C_i.   [3 pts]

    b. Prove that if $\text{pop}(T)$ is odd, then T is heterogeneous.   [3 pts]

11. Let T be a town such that, for each $D \in \text{dish}(T)$, $|\text{chef}_T(D)| = 3$.

    a. Either find such a town T for which $|\text{dish}(T)|$ is odd, or show that no such town exists.   [2 pts]

b. Prove that if T contains a resident cycle such that for every dish $D \in \text{dish}(T)$, there exists a chef in the cycle that can prepare D, then $\text{gr}(T) = 3$. [3 pts]

12. Let k be a positive integer, and let T be a town in which $|\text{chef}_T(D)| = k$ for every dish $D \in \text{dish}(T)$. Suppose further that $|\text{dish}(T)|$ is odd.

   a. Show that k is even. [1 pt]
   b. Prove the following: for every group in T, there is some dish $D \in \text{dish}(T)$ such that no one in the group is a D-chef. [3 pts]
   c. Prove that $\text{gr}(T) > k$. [3 pts]

13a. For each odd positive integer $d \geq 3$, prove that $\text{gr}(\mathcal{F}_d) = d$. [3 pts]

13b. For each even positive integer d, prove that $\text{gr}(\mathcal{F}_d) = d - 1$. [4 pts]

CHAPTER 25

UNC CHARLOTTE SUPER COMPETITION

25.1 2017 Comprehensive

1. Suppose f, g and h are polynomials of degrees $7, 8$ and 9 respectively. Let d be the degree of the quotient $(f \circ g) \cdot h \div (f + g + h)$, where $\circ$ means composition. Which of the following statements is true?

   (A) $d \le 40$ **(B)** $40 < d \le 50$ **(C)** $50 < d \le 60$ **(D)** $60 < d \le 70$ **(E)** $d > 70$

2. A ball of radius 1 is rolling on the floor $Q: -3 \le x \le 3, -3 \le y \le 3$ of a square room, getting reflected elastically off its walls. Otherwise, it moves straight. Initially, it touches the Eastern wall of the room at the point $(3, 0)$, then moves in the North-West direction and hits the Northern wall at the point $(0, 3)$, then gets reflected and moves South-West, etc. Let L be the distance the ball travels before returning to the initial position. Which of the following statements is correct?

   (A) $11 < L \le 12$ **(B)** $12 < L \le 14$ **(C)** $14 < L \le 15$ **(D)** $15 < L \le 16$

   (E) $16 < L \le 18$

3. There are exactly 2 positive values of r for which the system of equations

$$\begin{cases} x^2 + y^2 & = 9, \\ (x - 8)^2 + (y - 6)^2 & = r^2 \end{cases}$$

   has a unique solution (x, y). Let us denote these values of r by r_1 and r_2, and put $s = r_1 + r_2$. Which of the following statements is true?

(A) $s \leq 10$ (B) $10 < s \leq 17$ (C) $17 < s \leq 21$ (D) $21 < s \leq 26$ (E) $s > 26$

4. What is the area of the region determined by $|x - 1| + |y - 1| \leq 2$?

   (A) 2   (B) 4   (C) $4\sqrt{2}$ (D) 8   (E) $8\sqrt{2}$

5. Six points are distributed around a circle. In how many ways can you build two disjoint (i.e, non-intersecting) triangles using the six points as vertices?

   (A) 1   (B) 2   (C) 3   (D) 4   (E) more than 4

6. The top of a rectangular box has area 40 square inches, the front has area 48 square inches, and the side has area 30 square inches. How high is the box?

   (A) 2   (B) 3   (C) 4   (D) 5   (E) 6

7. The lower two vertices of a square lie on the x-axis and the upper two vertices of the square lie on the parabola $y = 15 - x^2$. What is the area of the square?

   (A) 9   (B) $10\sqrt{2}$ (C) 16   (D) 25   (E) 36

8. A math class has between 15 and 40 students. Exactly 25% of the class knows how to play poker. On a certain Wednesday, 3 students were absent (because they were participating in a math contest). On that day, exactly 20% of the students attending the class knew how to play poker. How many students attending the class on that day knew how to play poker?

   (A) 3   (B) 4   (C) 5   (D) 6   (E) 7

9. Pansies have five petals and lilacs have four petals. A bouquet has twenty flowers with a total of 92 petals. How many pansies does the bouquet have?

   (A) $3 \leq P \leq 7$ (B) $8 \leq P \leq 10$ (C) $11 \leq P \leq 14$ (D) $15 \leq P \leq 17$ (E) $P \geq 18$

10. A dictionary contains pages numbered 1 through 852. How many times does the number 8 appear as a digit in this numbering?

    (A) 85   (B) 146   (C) 165   (D) 217   (E) 218

11. Two married couples have purchased theater tickets and are seated in a row consisting of just four seats. If they take their seats in a completely random order, what is the probability that Jim and Paula (husband and wife) sit in the two seats on the far left?

    (A) $\dfrac{1}{12}$ (B) $\dfrac{1}{24}$ (C) $\dfrac{1}{4}$ (D) $\dfrac{1}{2}$ (E) $\dfrac{1}{6}$

12. Inside a large circle of radius 10 are four small circles and one medium size circle as in the diagram below. The four small circles all have the same radius and each is tangent to the large circle and to two other small circles. The center of the medium circle is the same as the center of the large

circle and the medium circle passes through the center of each small circle. What is the radius of the medium circle?

(A) $10\left(2-\sqrt{2}\right)$

(B) $5\sqrt{2}$

(C) $3\sqrt{5}$

(D) $10\left(\sqrt{3}-1\right)$

(E) $2\left(2+\sqrt{2}\right)$

13. Suppose N is the size of a certain caterpillar population in the UNC Charlotte botanical gardens, M is the maximum possible population size (the carrying capacity), and $x = N/M$ is the relative size of the population. Scientists have observed that if the population one spring has relative size x, then the following spring the population has relative size $4x(1-x)$. There is at most one value x_0 $(0 < x_0 < 1)$ with the property that if the relative size one spring is x_0, then the relative size for all following springs will be x_0 as well. Which of the following statements is true?

(A) $0 < x_0 \le 1/2$ (B) $1/2 < x_0 \le 3/5$ (C) $3/5 < x_0 \le 7/8$ (D) $7/8 < x_0 < 1$

(E) no such x_0 exists

14. Suppose a, b, and c are positive integers with $a < b < c$ such that $1/a + 1/b + 1/c = 1$. What is $a + b + c$?

(A) 7   (B) 8   (C) 9   (D) 11   (E) no such integers exist

15. One of the roots of the quadratic equation $x^2 - 9x + a = 0$ is twice the other root. Which of the following statements is true?

(A) $a \le 5$ (B) $15 < a \le 10$ (C) $10 < a \le 15$ (D) $15 < a \le 20$ (E) $a > 20$

16. In the quadratic equation $x^2 - 7x + a = 0$ the sum of the squares of the roots equals 39. Find a.

(A) 8   (B) 7   (C) 6   (D) 5   (E) 4

17. The sides of a right triangle form an arithmetic sequence, while their sum equals 48. Find the area of the triangle.

(A) 24   (B) 96   (C) 48   (D) 54   (E) 84

18. The ratio of the legs in a right triangle equals $3/2$, while the length of the hypotenuse is $\sqrt{52}$. Find the area of the triangle.

(A) 12   (B) 13   (C) 26   (D) 30   (E) 169

19. Fresh cucumbers are 90% water. It is known that in a week after they are picked the amount of water reduces to 80%. How much will 20 pounds of such cucumbers weigh after a week?

(A) 8   (B) 10   (C) 12   (D) 15   (E) 18

20. For how many pairs of digits (a, b) does $\sqrt{0.aaaaa\ldots} = 0.bbbbb\ldots$?

(A) 1   (B) 2   (C) 3   (D) 4   (E) 5

25.2   2017 Level 3

1. How many points (x, y) in the plane satisfy both $x^2 + y^2 = 25$ and $x^2 - 10x + y^2 - 24y = -105$?

(A) none   (B) 1   (C) 2   (D) 3   (E) more than 3

2. Let $f(x) = (2x + 3)^3$ and $g(x) = x^3 + x^2 - x - 1$. Denote the sum of the coefficients of the polynomial $h(x) = f(g(x))$ by s. Which of the following statements is true?

(A) $s \leq 0$ (B) $1 \leq s \leq 6$ (C) $7 \leq s \leq 20$ (D) $21 \leq s \leq 36$ (E) $s > 36$

3. The graph of the function $f(x) = ||2x| - 10|$ on the interval $[-10, 10]$ looks like

(A) M   (B) W   (C) V   (D) Λ (E) none of these

4. There is a unique positive number r such that the two equations $y + 2x = 0$ and $(x-3)^2 + (y-6)^2 = r^2$ have exactly one simultaneous solution. Which of the following statements is true?

(A) $0 < r < 1$ (B) $1 \leq r < 3$ (C) $3 \leq r < 5$ (D) $5 \leq r < 6$ (E) $r \geq 6$

5. The vertices of a triangle are the centers of the circles $C_1 = \{(x, y) \mid x^2 + y^2 = 1\}$, $C_2 = \{(x, y) \mid (x - 4)^2 + y^2 = 1\}$ and $C_3 = \{(x, y) \mid x^2 - 14x + y^2 - 16y = 0\}$. Let S be the area of the triangle. Which of the following statements is true?

(A) $S \leq 6$ (B) $6 < S \leq 9$ (C) $9 < S \leq 12$ (D) $12 < S \leq 15$ (E) $S > 15$

6. How many real solutions does the following system have?

$$\begin{cases} x + y = 2, \\ xy - z^2 = 1. \end{cases}$$

(A) 0   (B) 1   (C) 2   (D) 3   (E) 4

7. Let $a > 1$. How many positive solutions has the equation

$$\sqrt{a - \sqrt{a + x}} = x \, ?$$

(A) 1   (B) 2   (C) 0   (D) 3   (E) 4

8. The top of a rectangular box has area 40 square inches, the front has area 48 square inches, and the side has area 30 square inches. How high is the box?

   (A) 3   (B) 4   (C) 5   (D) 6   (E) 8

9. The lower two vertices of a square lie on the x-axis, while the upper two vertices of the square lie on the parabola $y = 15 - x^2$. What is the area of the square?

   (A) 9   (B) $10\sqrt{2}$ (C) 16   (D) 25   (E) 36

10. Pansies have 5 petals while lilacs have 4 petals. A bouquet has 20 flowers with a total of 92 petals. Let P be the number of pansies in the bouquet. Which of the following statements does P satisfy?

   (A) $3 \le P \le 7$ (B) $8 \le P \le 10$ (C) $11 \le P \le 14$ (D) $15 \le P \le 17$ (E) $P \ge 18$

11. A three-digit number abc is *palindromic* if $a = c$. What is the number of distinct three-digit palindromic numbers?

   (A) 72   (B) 84   (C) 88   (D) 90   (E) 100

12. The double of a positive number is the triple of its cube. The number is:

   (A) $\sqrt{2/3}$ (B) 1   (C) $\sqrt{3/2}$ (D) $\sqrt[3]{2}/\sqrt{3}$ (E) $\sqrt[3]{3}/\sqrt{2}$

13. Suppose a, b and c are positive integers with $a < b < c$ such that $1/a + 1/b + 1/c = 1$. What is $a + b + c$?

   (A) 6   (B) 8   (C) 9   (D) 11   (E) no such integers exist

14. A quadratic equation $x^2 - 9x + a = 0$ has two distinct roots, one of them being twice the other. Which of the following statements is true?

   (A) $a \le 5$ (B) $5 < a \le 10$ (C) $10 < a \le 15$ (D) $15 < a \le 20$ (E) $a > 20$

15. In the quadratic equation $x^2 - 7x + a = 0$ the sum of the squares of the roots equals 39. Find a.

   (A) 8   (B) 7   (C) 6   (D) 5   (E) 4

16. Evaluate $S = \cot 1° \cot 2° \cot 3° \ldots \cot 89°$.

   (A) $\dfrac{\pi}{2}$ (B) $\dfrac{2}{\pi}$ (C) 1   (D) $\dfrac{\sqrt{2}}{2}$ (E) 2

17. The sides of a right triangle form an arithmetic sequence, while their sum equals 48. Find the area of the triangle.

   (A) 24   (B) 96   (C) 48   (D) 54   (E) 84

18. The ratio of the legs in a right triangle equals $3/2$, while the length of the hypotenuse is $\sqrt{52}$. Find the area of the triangle.

    (A) 12   **(B)** 13   **(C)** 26   **(D)** 30   **(E)** 169

19. The Chebyshev polynomial of the first kind of order n is defined by $T_n(\cos \alpha) = \cos n\alpha$, so that $T_0(\cos \alpha) = 1$ and hence $T_0(x) = 1$; $T_1(\cos \alpha) = \cos \alpha$, hence $T_1(x) = x$; $T_2(\cos \alpha) = \cos 2\alpha = 2\cos^2 \alpha - 1$ so that $T_2(x) = 2x^2 - 1$, etc. What is the value of $T_{10}(\sin \alpha)$?

    (A) $\cos 10\alpha$ **(B)** $\sin 10\alpha$ **(C)** $-\sin 10\alpha$ **(D)** $-\cos 10\alpha$ **(E)** $\dfrac{\sin 10\alpha}{\cos \alpha}$

20. Find the maximum value of the expression $f(x,y) = x\sqrt{1-y^2} + y\sqrt{1-x^2}$ over the square $Q: -1 \le x \le 1, \ -1 \le y \le 1$.

    (A) 1   **(B)** $\sqrt{\dfrac{3}{2}}$ **(C)** $\dfrac{3}{2}$ **(D)** $\dfrac{2}{\sqrt{3}}$ **(E)** $\dfrac{4}{3}$

CHAPTER 26

STATE MATH CONTEST OF NORTH CAROLINA PROBLEMS

26.1 2016 PART I: 20 MULTIPLE CHOICE PROBLEMS

1. The product

$$\left(1+\frac{1}{2}\right)\left(1-\frac{1}{3}\right)\left(1+\frac{1}{4}\right)\left(1-\frac{1}{5}\right)\cdots\left(1-\frac{1}{n-1}\right)\left(1+\frac{1}{n}\right)$$

is equal to

(A) 1 (B) $1+\dfrac{1}{n}$ (C) -1 (D) $\dfrac{1}{n}$

(E) None of the answers (A) through (D) is correct.

2. Find the area of the region above the x-axis and below the graph of $x^2 + y^2 = 1 - 2y$.

(A) $\dfrac{\pi}{2}$ (B) $2-\dfrac{\pi}{2}$ (C) $\dfrac{\pi}{2}-1$ (D) $\dfrac{3\pi}{2}+1$

(E) None of the answers (A) through (D) is correct.

3. Let x be a real number greater than 1 such that $x - \dfrac{1}{x} = \sqrt{x} + \dfrac{1}{\sqrt{x}}$. Determine the value of $x + \dfrac{1}{x}$.

(A) $\sqrt{6}$ (B) 4 (C) 3 (D) 5

(E) None of the answers (A) through (D) is correct.

4. Points A and B are on the parabola $y = 2x^2 + 4x - 2$. The origin is the midpoint of the line segment joining A and B. Find the length of this line segment.

(A) $2\sqrt{17}$ (B) 8 (C) $\sqrt{70}$ (D) 9

(E) None of the answers (A) through (D) is correct.

5. Find the sum of all 3-digit positive integers that are 34 times the sum of their digits.

(A) 102   (B) 306   (C) 510   (D) 612

(E) None of the answers (A) through (D) is correct.

6. Let $f : \mathbb{R} \to \mathbb{R}$ be a function which satisfies $f(x + 31) = f(31 - x)$ for all real numbers x. If f has exactly three real roots, then their sum is:

(A) 0   (B) 31   (C) 62   (D) 93

(E) None of the answers (A) through (D) is correct.

7. Let a, b, and c be three consecutive members of a geometric progression (in the given order). Assume $a > 1, b > 1, c > 1$ and the common ration of the geometric progression is greater than 1. Then $\dfrac{\log_b 3 \left(\log_{a^2} c - \log_c \sqrt{a} \right)}{\log_a 9 - 2 \log_c 3}$ equals to

(A) $\dfrac{1}{2}$ (B) $\dfrac{1}{3}$ (C) $\log_b \left(c\sqrt{a} \right)$ (D) 3

(E) None of the answers (A) through (D) is correct.

8. Determine the number of integer numbers a for which the equation $x^3 - 13x + a = 0$ has three integer roots.

(A) 0   (B) 1   (C) 2   (D) 4

(E) None of the answers (A) through (D) is correct.

9. Adam and David are playing a game on a circular board with n spaces. Both players place their chip at the same starting space. First Adam moves his chip forward five spaces from the starting space, then David moves his chip forward seven, then Adam five, then David seven, and so on. The first player to finish his turn on the starting space wins the game. If n is a random two-digit number, what is the probability that Adam wins?

(A) $\dfrac{77}{90}$ (B) $\dfrac{75}{90}$ (C) $\dfrac{32}{90}$ (D) $\dfrac{31}{90}$ (E) $\dfrac{13}{90}$

10. Determine the value of $\sin^3 18° + \sin^2 18°$.

(A) $\dfrac{1}{4}$ (B) $\dfrac{1}{8}$ (C) $\dfrac{3\sqrt{3}}{2}$ (D) $\dfrac{1}{2}$

(E) None of the answers (A) through (D) is correct.

11. Let I denote the center of the inscribed circle in the triangle ABC. If one of the triangles AIB, BIC, or CIA is similar to triangle ABC, find the largest angle (in radians) of $\triangle ABC$.

(A) $\dfrac{5\pi}{7}$ (B) $\dfrac{4\pi}{7}$ (C) $\dfrac{3\pi}{5}$ (D) $\dfrac{4\pi}{5}$

(E) None of the answers (A) through (D) is correct.

12. The value of $\arctan \dfrac{1}{2} + \arctan \dfrac{1}{4} + \arctan \dfrac{1}{13}$ is

(A) $\dfrac{\pi}{6}$ (B) $\dfrac{\pi}{4}$ (C) $\dfrac{\pi}{3}$ (D) $\dfrac{\pi}{5}$

(E) None of the answers (A) through (D) is correct.

13. Let A, B, and C be points on a circle of radius 3. In the triangle ABC, $\angle ACB = 30°$ and $AC = 2$. Find the length of the segment $\overline{BC}$.

(A) $\sqrt{2} + \sqrt{3}$ (B) $\sqrt{2} + 2\sqrt{3}$ (C) $2\sqrt{2} + \sqrt{3}$ (D) $2\sqrt{2}$

(E) None of the answers (A) through (D) is correct.

14. Let x and y be nonzero real numbers such that

$$x^2 + y\cos^2 \alpha = x \sin \alpha \cos \alpha \text{ and } x \cos 2\alpha + y \sin 2\alpha = 0.$$

Then the relationship between x and y is:

(A) $x^2 + 4y^2 = 0$ (B) $x^2 + 4y = 0$ (C) $4x^2 + 4y = 1$ (D) $x^2 - 4y = 1$

(E) None of the answers (A) through (D) is correct.

15. Find the number of the integer solutions (x, y) of the equation $x^2 y^3 = 6^{12}$.

(A) 9   (B) 6   (C) 18   (D) 12

(E) None of the answers (A) through (D) is correct.

16. In a triangle ABC let M be the midpoint of the segment $\overline{AB}$ and N be the midpoint of the segment $\overline{AC}$. Let T be the intersection of $\overline{BN}$ and $\overline{CM}$. Let P be the midpoint the segment $\overline{CT}$ and let Q be the intersection of the lines BP and AC. Determine the value of $\frac{CQ+AN}{NQ}$.

(A) $\dfrac{4}{3}$ (B) $\dfrac{5}{3}$ (C) 2   (D) $\dfrac{7}{3}$

(E) None of the answers (A) through (D) is correct.

17. Find the minimum value of the real function

$$f(x, y) = x^2 + 2xy + 3y^2 + 2x + 6y + 4.$$

(A) -1 (B) 0   (C) 1   (D) 4   (E) 2

18. Let a, b, and c be real numbers in the interval $(0, 1)$ such that $ab + bc + ca = 1$. Find the largest value of $a + b + c + abc$.

(A) does not have the largest value   (B) 2   (C) $\dfrac{26\sqrt{3}}{9}$ (D) $\dfrac{28\sqrt{3}}{9}$

(E) None of the answers (A) through (D) is correct.

19. Determine the number of all ordered pairs of prime numbers (p, q) such that $p, q < 100$ and $p + 6$, $p + 10$, $q + 4$, $q + 10$, and $p + q + 1$ are all prime numbers.

   (A) 2   **(B)** 3   **(C)** 4   **(D)** 5

   (E) None of the answers (A) through (D) is correct.

20. Let a, b, and c be real numbers from the interval $\left(0, \dfrac{\pi}{2}\right)$ such that $\cos a = a$, $\sin(\cos b) = b$, and $\cos(\sin c) = c$. Order the numbers a, b, and c from the smallest to the largest.

   (A) $b < a < c$ **(B)** $b < c < a$ **(C)** $c < b < a$ **(D)** $c < a < b$

   (E) None of the answers (A) through (D) is correct.

26.2   2016 PART II: 10 INTEGER ANSWER PROBLEMS

1. Three cards each have one of the digits from 1 through 9 written on them; the three digits written on the cards are distinct. When the three cards are arranged in some order, they make a three digit number. The largest number that can be made plus the second largest number that can be made is 1233. What is the largest number that can be made?

2. A closed right circular cylinder has an integer radius and an integer hight. The numerical value of its volume is four times the numerical value of its total surface area (including the top and the bottom). If the numerical value of the smallest possible volume of the cylinder is written as $K\pi$, where K is a positive integer, find the value of K.

3. Find the smallest integer number n such that $n + 2002$ and $n - 2002$ are perfect squares.

4. Find the smallest positive integer n so that there are exactly 25 integers i satisfying $2 \leq \dfrac{n}{i} \leq 5$.

5. A lattice point is a point (x, y) in the coordinate plane with each of x and y an integer. Determine the number of lattice points inside the region $|x| + |y| \leq 100$.

6. Let $\lfloor x \rfloor$ denote the greatest integer less than or equal to x. Let $\displaystyle\sum_{i=1}^{n} \lfloor \sqrt{i} \rfloor = 217$. Find the value of n.

7. Find the greatest integer which divides $(a - b)(b - c)(c - d)(d - a)(a - c)(b - d)$ for any integers a, b, c, and d.

8. Let $A_0, A_1, \ldots, A_{100}$ be distinct points on one side of an angle and let $B_0, B_1, \ldots, B_{100}$ be distinct points on the other side of the same angle such that $A_0 A_1 = A_1 A_2 = \cdots = A_{99} A_{100}$ and $B_0 B_1 = B_1 B_2 = \cdots = B_{99} B_{100}$. Find the area of the quadrilateral $A_{99} A_{100} B_{100} B_{99}$ if the areas of the quadrilaterals $A_0 A_1 B_1 B_0$ and $A_1 A_2 B_2 B_1$ are equal to 5 and 7 respectively.

9. Let n be an integer with the property that if m is randomly chosen integer from the set $\{1, 2, 3, \ldots, 1000\}$, the probability that m is a divisor of n is $\frac{1}{100}$. If $n \leq 1000$, determine the maximum possible value of n.

10. Determine the exact number of digits in the decimal expression of 2^{100}.

The following problem, will be used only as part of a tie-breaking procedure. Do not work on it until you have completed the rest of the test.

TIE BREAKER PROBLEM

If m and n are any two integer numbers, let $m \circ n$ denote an integer number determined by m and n. Assume that the operation $\circ$ satisfies the following three rules:

(i) $m \circ (n + s) = (m \circ n) - s$ for all integer numbers m, n, s;

(ii) $(n + s) \circ m = (n \circ m) + 2s$ for all integer numbers m, n, s;

(iii) $1 \circ 1 = 1$

Calculate $47 \circ 20$.

26.3 2017 PART I: 20 MULTIPLE CHOICE PROBLEMS

1. A survey of 100 recent college graduates was made to determine their mean salary. The mean salary found was $45,000. It turns out that one of the alumnus incorrectly answered the survey. He said he earns $35,000 when in fact he earns $53,000. What is the actual mean salary of the 100 graduates?

(**A**) $45,150$ (**B**) $45,165$ (**C**) $45,180$ (**D**) $45,200$

(**E**) None of the answers (A) through (D) is correct.

2. The areas of three faces of a rectangular parallelepiped are 18, 40, and 80. Find its volume.

(**A**) 220 (**B**) 228 (**C**) 230 (**D**) 240

(**E**) None of the answers (A) through (D) is correct.

3. Find the sum of the digits of $10^{2017} - 2017$.

(**A**) $18,135$ (**B**) $18,144$ (**C**) $18,149$ (**D**) $18,153$

(**E**) None of the answers (A) through (D) is correct.

4. A box contains three red, six blue, and four yellow balls. If two balls are selected at random, what is the probability that they are both yellow, given that they are the same color?

(A) $\dfrac{1}{4}$ (B) $\dfrac{1}{3}$ (C) $\dfrac{2}{3}$ (D) $\dfrac{3}{4}$ (E) None of the answers (A) through (D) is correct.

5. Calculate $(\log_3 5 + \log_9 25 + \log_{27} 125 + \cdots + \log_{3^n} 5^n)\log_{25} \sqrt[2n]{27}$.

   (A) $\dfrac{3}{5}$ (B) $(\log_3 5)^n$ (C) $\dfrac{9}{5}$ (D) $\dfrac{3}{4}$ (E) None of the answers (A) through (D) is correct.

6. The nonzero integers x, y, and z, in the given order, are three consecutive terms of a geometric progression, while the numbers x, $2y$, and $3z$, in the given order, are three consecutive terms of an arithmetic progression. Find the sum of all possible ratios of the geometric progression.

   (A) $\dfrac{1}{3}$ (B) $\dfrac{1}{2}$ (C) 1 (D) $\dfrac{4}{3}$ (E) None of the answers (A) through (D) is correct.

7. Find the sum of all real numbers a for which the equation $2x^2 + ax + 5x + 7 = 0$ has only one solution.

   (A) -10 (B) 0 (C) 10 (D) 31 (E) None of the answers (A) through (D) is correct.

8. The lengths of the heights in a triangle are 12, 15, and 20. Find the area of the triangle.

   (A) 120 (B) 150 (C) 180 (D) 240

   (E) None of the answers (A) through (D) is correct.

9. How many six-digit numbers can be formed using the digits 1, 2, and 3 that do not contain two consecutive 1's?

   (A) 224 (B) 256 (C) 416 (D) 448

   (E) None of the answers (A) through (D) is correct.

10. Find the product of the real solutions of the equation $1 + x^2 - x^4 = x^5 - x^3 - x$.

   (A) -1 (B) $\dfrac{7}{2}$ (C) $\dfrac{1+\sqrt{5}}{2}$ (D) $-\dfrac{1+\sqrt{5}}{2}$

   (E) None of the answers (A) through (D) is correct.

11. Compute the product

$$\left(1 - \frac{4}{1}\right)\left(1 - \frac{4}{3^2}\right)\left(1 - \frac{4}{5^2}\right)\left(1 - \frac{4}{7^2}\right)\cdots\left(1 - \frac{4}{99^2}\right).$$

   (A) $-\dfrac{101}{\cdot 99}$ (B) $-\dfrac{101}{33}$ (C) $-\dfrac{99}{97}$ (D) $-\dfrac{101}{97}$

   (E) None of the answers (A) through (D) is correct.

12. Let ABC be a right triangle and the lengths of its legs are the roots of the equation $ax^2 + bx + c = 0$. Find the area of the circumscribed circle of the triangle ABC.

   (A) $\pi\dfrac{b^2 - 2ac}{4a^2}$ (B) $\pi\dfrac{b^2 - 4ac}{4a^2}$ (C) $\pi\dfrac{b^2 - 2ac}{2a^2}$ (D) $\pi\dfrac{b^2 + 2ac}{4a^2}$

(E) None of the answers (A) through (D) is correct.

13. Let a be a positive real number. Find the sum

$$\log_2 a \log_4 a + \log_4 a \log_8 a + \log_8 a \log_{16} a + \cdots + \log_{2^{n-1}} a \log_{2^n} a$$

(A) $\log_2^2 a \left(1 - \dfrac{1}{n+1}\right)$ **(B)** $\log_2^2 a \left(1 - \dfrac{1}{n}\right)$ **(C)** $\dfrac{\log_2^2 a}{n}$ **(D)** $\dfrac{\log_2^2 a}{n}$

(E) None of the answers (A) through (D) is correct.

14. Let a, b, and c be positive integer numbers such that $\dfrac{a\sqrt{3}+b}{b\sqrt{3}+c}$ is a rational number. Then $\dfrac{a^2+b^2+c^2}{a+b+c}$ is equal to

(A) $a + b - c$ **(B)** $a - b + c$ **(C)** $-a + b + c$ **(D)** $a - b - c$

(E) None of the answers (A) through (D) is correct.

15. Let $ABCD$ be a trapezoid such that $\overline{AB} \parallel \overline{CD}$, $BC = CD = 7$, $AD = 8$, and $BD \perp AD$. Find the length of AB.

(A) 11   **(B)** 12   **(C)** 13   **(D)** 14

(E) None of the answers (A) through (D) is correct.

16. How many triples of positive integers (x, y, z) satisfy the equations $x^2 + y - z = 100$ and $x + y^2 - z = 124$?

(A) 0   **(B)** 1   **(C)** 2   **(D)** 3   **(E)** 4

17. A regular pentagon is inscribed in a circle of radius 1. Let a and d be the lengths of the side and the diagonal of the pentagon. Find $a^2 + d^2$.

(A) $2\sqrt{3}$ **(B)** $3\sqrt{2}$ **(C)** 4   **(D)** 5   **(E)** None of the answers (A) through (D) is correct.

18. Evaluate the product

$$(1 - \cot 1°)(1 - \cot 2°)(1 - \cot 3°) \cdots (1 - \cot 44°).$$

(A) $\left(\dfrac{\sqrt{2}}{2}\right)^{44}$ **(B)** $\left(\dfrac{\sqrt{3}}{2}\right)^{44}$ **(C)** 3^{22} **(D)** $(\sqrt{2})^{22}$ **(E)** 2^{22}

19. Let $\{a_n\}$ be a finite sequence of real numbers (n is a positive integer number) given with $a_{n+1} = \dfrac{n+1}{n}a_n + 1$ for $1 \le n \le 2016$, and $a_{2017} = 2017$. Find the sum $a_1 + a_2 + \cdots + a_{2016}$.

(A) $1008 \cdot 2017$ **(B)** $504 \cdot 2017$ **(C)** $1009 \cdot 2017$ **(D)** $\dfrac{2017 \cdot 2018}{4}$

(E) None of the answers (A) through (D) is correct.

20. Let x and y be real numbers such that $3x^2 + 2y^2 \le 6$. Find the greatest value of $2x + y$.

(A) $\dfrac{3\sqrt{6}}{\sqrt{5}}$ **(B)** $2\sqrt{3}$ **(C)** $\sqrt{11}$ **(D)** $\sqrt{13}$

(E) None of the answers (A) through (D) is correct.

26.4 2017 PART II: 10 INTEGER ANSWER PROBLEMS

1. Determine the sum of all positive three-digit integer numbers that give a reminder 2 when divided by 7, a reminder 4 when divided by 9, and reminder 7 when divided by 12.

2. Let $f : \mathbb{R} \to \mathbb{R}$ be a function such that $f(1) = 1$ and $f(x + y) = 3^y f(x) + 2^x f(y)$ for all real numbers x and y. Find $f(4)$.

3. Let x, y, and z be positive integer numbers such that $x < y < z$ and $3^x + 3^y + 3^z = 21897$. Find $x + y + z$.

4. Let $z_1 = \sqrt{a - 5} + ai$ and $z_2 = 2\cos\alpha + 3i\sin\alpha$ be two complex numbers where a is a real number such that $a \geq 5$ and $i^2 = -1$. Find the minimum value of $\mid z_1 - z_2 \mid$.

5. The number 2^{29} has 9 distinct digits. Which digit is missing?

6. Let D be a point on the side $\overline{BC}$ of the isosceles triangle ABC ($AB = BC$) such that $DC = 4BD$. Let E be a point on the side $\overline{AC}$ such that BE is a height in $\triangle ABC$. Let F be the point of intersection of $\overline{AD}$ and $\overline{BE}$. Find $\frac{EF}{BF}$.

7. Find the value of m for which the function $f(x) = \mid x^2 - 6x \mid - m$ has exactly three x-intercepts.

8. Let a and b be positive real numbers such that

$$\log(1 + a^2) - \log a - 2\log 2 = 1 - \log(100 + b^2) + \log b.$$

Find $a + b$.

9. Let a, b, and c be distinct positive integer numbers greater than 1. Let $\frac{m}{n}$ be the maximum of $\left(1 + \frac{1}{a}\right)\left(2 + \frac{1}{b}\right)\left(3 + \frac{1}{c}\right)$, where m and n are relatively prime positive integers. Find $m + n$.

10. Find the sum of the squares of the real roots of the equation $x^{256} - 256^{32} = 0$.

TIE BREAKER PROBLEM

Find the sum of all positive integers n such that $n = d_1^2 + d_2^2 + d_3^2 + d_4^2$ where $d_1 < d_2 < d_3 < d_4$ are the four smallest divisors of n.

CHAPTER 27

BAMO PROBLEMS

The followings are problems from 2014 BAMO.

A. The four bottom corners of a cube are colored red, green, blue, and purple. How many ways are there to color the top four corners of the cube so that every face has four different colored corners? Prove that your answer is correct.

B. There are n holes in a circle. The holes are numbered $1, 2, 3$ and so on to n. In the beginning, there is a peg in every hole except for hole 1. A peg can jump in either direction over one adjacent peg to an empty hole immediately on the other side. After a peg moves, the peg it jumped over is removed. The puzzle will be solved if all pegs disappear except for one. For example, if $n = 4$ the puzzle can be solved in two jumps: peg 3 jumps peg 4 to hole 1, then peg 2 jumps the peg in 1 to hole 4. (See illustration below, in which black circles indicate pegs and white circles are holes.)

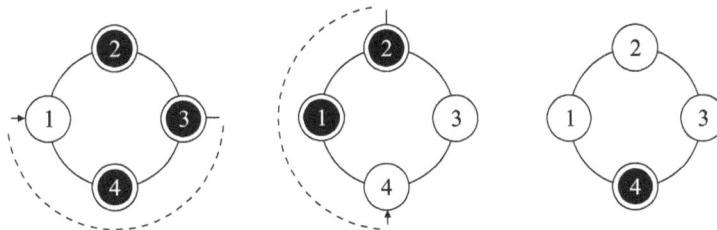

(a) Can the puzzle be solved when $n = 5$?

(b) Can the puzzle be solved when $n = 2014$?

In each part (a) and (b) either describe a sequence of moves to solve the puzzle or explain why it is impossible to solve the puzzle.

Reprinted by permission of Bay Area Mathematical Olympiad.

C. and 1 Amy and Bob play a game. They alternate turns, with Amy going first. At the start of the game, there are 20 cookies on a red plate and 14 on a blue plate. A legal move consists of eating two cookies taken from one plate, or moving one cookie from the red plate to the blue plate (but never from the blue plate to the red plate). The last player to make a legal move wins; in other words, if it is your turn and you cannot make a legal move, you lose, and the other player has won.

Which player can guarantee that they win no matter what strategy their opponent chooses? Prove that your answer is correct.

D. and 2 Let ABC be a scalene triangle with the longest side AC. (A *scalene* triangle has sides of different lengths.) Let P and Q be the points on the side AC such that $AP = AB$ and $CQ = CB$. Thus we have a new triangle BPQ inside triangle ABC. Let k_1 be the circle *circumscribed* around the triangle BPQ (that is, the circle passing through the vertices B, P, and Q of the triangle BPQ); and let k_2 be the circle *inscribed* in triangle ABC (that is, the circle inside triangle ABC that is tangent to the three sides AB, BC, and CA). Prove that the two circles k_1 and k_2 are *concentric*, that is, they have the same center.

3. Suppose that for two real numbers x and y the following equality is true:

$$(x + \sqrt{1 + x^2})(y + \sqrt{1 + y^2}) = 1.$$

Find (with proof) the value of $x + y$.

4. Let $F_1, F_2, F_3, \ldots$ be the *Fibonacci sequence*, the sequence of positive integers satisfying

$$F_1 = F_2 = 1 \quad \text{and} \quad F_{n+2} = F_{n+1} + F_n \text{ for all } n \geq 1.$$

Does there exist an $n \geq 1$ for which F_n is divisible by 2014?

5. A chess tournament took place between $2n + 1$ players. Every player played every other player once, with no draws. In addition, each player had a numerical rating before the tournament began, with no two players having equal ratings.

It turns out there were exactly k games in which the lower-rated player beat the higher-rated player. Prove that there is some player who won no less than $n - \sqrt{2k}$ and no more than $n + \sqrt{2k}$ games.

CHAPTER 28

PURPLE COMET MATH MEET

1. Let x satisfy $(6x + 7) + (8x + 9) = (10 + 11x) + (12 + 13x)$. There are relatively prime positive integers so that $x = -\frac{m}{n}$. Find $m + n$.

2. The prime factorization of $12 = 2 \cdot 2 \cdot 3$ has three prime factors. Find the number of prime factors in the factorization of $12! = 12 \cdot 11 \cdot 10 \cdot 9 \cdot 8 \cdot 7 \cdot 6 \cdot 5 \cdot 4 \cdot 3 \cdot 2 \cdot 1$.

3. The grid below contains six rows with six points in each row. Points that are adjacent either horizontally or vertically are a distance two apart. Find the area of the irregularly shaped ten sided figure shown.

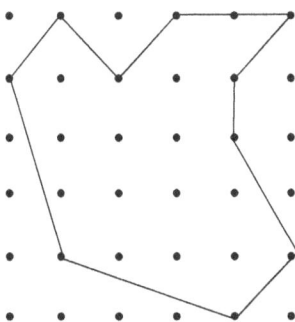

4. Sally's salary in 2006 was $37,500. For 2007 she got a salary increase of x percent. For 2008 she got another salary increase of x percent. For 2009 she got a salary decrease of $2x$ percent. Her 2009 salary is $34,825. Suppose instead, Sally had gotten a $2x$ percent salary decrease for 2007, an x percent salary increase for 2008, and an x percent salary increase for 2009. What would her 2009 salary be then?

5. If a and b are positive integers such that $a \cdot b = 2400$, find the least possible value of $a + b$.

6. Evaluate the sum $1 + 2 - 3 + 4 + 5 - 6 + 7 + 8 - 9 \cdots + 208 + 209 - 210$.

7. Find the sum of the digits in the decimal representation of the number $5^{2010} \cdot 16^{502}$.

8. The diagram below shows some small squares each with area 3 enclosed inside a larger square. Squares that touch each other do so with the corner of one square coinciding with the midpoint of a side of the other square. Find integer n such that the area of the shaded region inside the larger square but outside the smaller squares is $\sqrt{n}$.

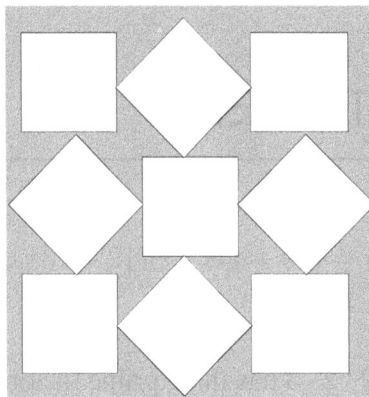

9. Find positive integer n so that $\frac{80 - 6\sqrt{n}}{n}$ is the reciprocal of $\frac{80 + 6\sqrt{n}}{n}$.

10. The set S contains nine numbers. The mean of the numbers in S is 202. The mean of the five smallest of the numbers in S is 100. The mean of the five largest numbers in S is 300. What is the median of the numbers in S?

11. A jar contains one white marble, two blue marbles, three red marbles, and four green marbles. If you select two of these marbles without replacement, the probability that both marbles will be the same color is $\frac{m}{n}$ where m and n are relatively prime positive integers. Find $m + n$.

12. A good approximation of π is 3.14. Find the least positive integer d such that if the area of a circle with diameter d is calculated using the approximation 3.14, the error will exceed 1.

13. Let S be the set of all 10-term arithmetic progressions that include the numbers 4 and 10. For example, $(-2, 1, 4, 7, 10, 13, 16, 19, 22, 25)$ and $(10, 8\frac{1}{2}, 7, 5\frac{1}{2}, 4, 2\frac{1}{2}, 1, -\frac{1}{2}, -2, -3\frac{1}{2})$ are both members of S. Find the sum of all values of a_{10} for each $(a_1, a_2, a_3, a_4, a_5, a_6, a_7, a_8, a_9, a_{10}) \in S$, that is,

$$\sum_{(a_1, a_2, a_3, \ldots, a_{10}) \in S} a_{10}.$$

14. There are positive integers b and c such that the polynomial $2x^2 + bx + c$ has two real roots which differ by 30. Find the least possible value of $b + c$.

15. Find the smallest possible sum $a+b+c+d+e$ where $a, b, c, d,$ and e are positive integers satisfying the conditions

- each of the pairs of integers (a, b), (b, c), (c, d), and (d, e) are **not** relatively prime
- all other pairs of the five integers **are** relatively prime.

16. The triangle ABC has sides lengths $AB = 39$, $BC = 57$, and $CA = 70$ as shown. Median $\overline{AD}$ is divided into three congruent segments by points E and F. Lines BE and BF intersect side $\overline{AC}$ at points G and H, respectively. Find the distance from G to H.

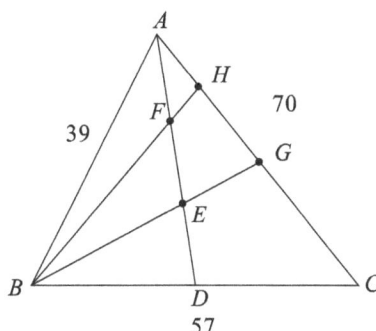

17. Alan, Barb, Cory, and Doug are on the golf team; Doug, Emma, Fran, and Greg are on the swim team; and Greg, Hope, Inga, and Alan are on the tennis team. These nine people sit in a circle in random order. The probability that no two people from the same team sit next to each other is $\frac{m}{n}$ where m and n are relatively prime positive integers. Find $m + n$.

18. When $4\cos\theta - 3\sin\theta = \frac{13}{3}$, it follows that $7\cos 2\theta - 24\sin 2\theta = \frac{m}{n}$ where m and n are relatively prime positive integers. Find $m + n$.

19. The centers of the three circles A, B, and C are collinear with the center of circle B lying between the centers of circles A and C. Circles A and C are both externally tangent to circle B, and the three circles share a common tangent line. Given that circle A has radius 12 and circle B has radius 42, find the radius of circle C.

20. How many of the rearrangements of the digits 123456 have the property that for each digit, no more than two digits smaller than that digit appear to the right of that digit? For example, the rearrangement 315426 has this property because digits 1 and 2 are the only digits smaller than 3 which follow 3, digits 2 and 4 are the only digits smaller than 5 which follow 5, and digit 2 is the only digit smaller than 4 which follows 4.

21. Let a be the sum of the numbers:

99×0.9

999×0.9

9999×0.9

$\vdots$

$999\cdots9 \times 0.9$

where the final number in the list is 0.9 times a number written as a string of 101 digits all equal to 9. Find the sum of the digits in the number a.

22. Ten distinct points are placed in a circle. All ten of the points are paired so that the line segments connecting the pairs do not intersect. In how many different ways can this pairing be done?

23. A disk with radius 10 and a disk with radius 8 are drawn so that the distance between their centers is 3. Two congruent small circles lie in the intersection of the two disks so that they are tangent to each other and to each of the larger circles as shown. The radii of the smaller circles are both $\frac{m}{n}$ where m and n are relatively prime positive integers. Find $m + n$.

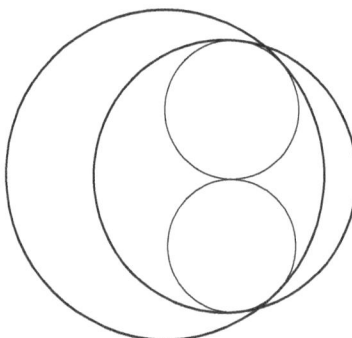

24. Find the number of ordered pairs of integers (m, n) that satisfy $20m - 10n = mn$.

25. Let x_1, x_2, and x_3 be the three roots of the polynomial $x^3 + 3x + 1$. There are relatively prime positive integers m and n so that

$$\frac{m}{n} = \frac{x_1^2}{(5x_2 + 1)(5x_3 + 1)} + \frac{x_2^2}{(5x_1 + 1)(5x_3 + 1)} + \frac{x_3^2}{(5x_1 + 1)(5x_2 + 1)}.$$

Find $m + n$.

26. In the coordinate plane a parabola passes through the points $(7,6)$, $(7,12)$, $(18,19)$, and $(18,48)$. The axis of symmetry of the parabola is a line with slope $\frac{r}{s}$ where r and s are relatively prime positive integers. Find $r + s$.

27. Let a and b be real numbers satisfying $2(\sin a + \cos a)\sin b = 3 - \cos b$. Find $3\tan^2 a + 4\tan^2 b$.

28. There are relatively prime positive integers p and q such that

$$\frac{p}{q} = \sum_{n=3}^{\infty} \frac{1}{n^5 - 5n^3 + 4n}.$$

Find $p + q$.

29. Square $ABCD$ is shown in the diagram below. Points E, F, and G are on sides $\overline{AB}$, $\overline{BC}$ and $\overline{DA}$, respectively, so that lengths BE, BF, and DG are equal. Points H and I are the midpoints of segments $\overline{EF}$ and $\overline{CG}$, respectively. Segment $\overline{GJ}$ is the perpendicular bisector of segment $\overline{HI}$. The ratio of the areas of pentagon $AEHJG$ and quadrilateral $CIHF$ can be written as $\frac{m}{n}$ where m and n are relatively prime positive integers. Find $m + n$.

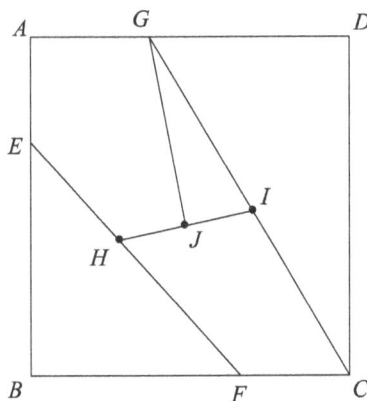

30. Let x and y be real numbers satisfying

$$\left(x^2 + x - 1\right)\left(x^2 - x + 1\right) = 2\left(y^3 - 2\sqrt{5} - 1\right)$$

and

$$\left(y^2 + y - 1\right)\left(y^2 - y + 1\right) = 2\left(x^3 + 2\sqrt{5} - 1\right)$$

Find $8x^2 + 4y^3$.

CHAPTER 29

WISCONSIN MATH TALENT SEARCH

1. Let us define a process which replaces each triple of real numbers $t = (a, b, c)$ by a new triple $t' = (a', b', c')$ where $a' = a + b$, $b' = b + c$ and $c' = c + a$. Suppose we start with a triple and apply this process again and again. Show that if we ever return to the original triple, then we will return after just six steps.

2. Find all 3-digit numbers m which are equal to the arithmetic mean of the six numbers one obtains by rearranging the digits of m in all possible ways.

3. Are there functions $f(x)$ and $g(y)$ defined on all the real numbers so that $f(x)g(y) = x + y + 1$ for every choice of x and y?

4. You have 5 green and 7 red balls, and two empty boxes. You place all the balls in the two boxes so that each box contains at least one ball. Your friend then chooses one of the two boxes randomly, and picks a ball randomly from the chosen box. If the chosen ball is green, then you win a prize. How should you arrange the balls between the two boxes initially to maximize the probability of winning? What is this probability?

5. Consider the sequence $a_0, a_1, a_2, \ldots$ where $a_0 = 1$, and $a_{k+1} = a_k + \frac{2}{a_k}$ for $k \geq 0$. (Thus $a_1 = 1 + 2/1 = 3$ and $a_2 = 3 + 2/3 = 11/3$. Prove that $a_{2015} > 89$.

Reprinted by permission of the Wisconsin Mathematics, Engineering and Science Talent Search.

CHAPTER 30

USA MATHEMATICAL TALENT SEARCH(2015-2016)

30.1 USAMTS-Round 1 Problems

1. Fill in the spaces of the grid to the right with positive integers so that in each 2×2 square with top left number a, top right number b, bottom left number c, and bottom right number d, either $a + d = b + c$ or $ad = bc$.

   You do not need to prove that your answer is the only one possible; you merely need to find an answer that satisfies the constraints above. (Note: In any other USAMTS problem, you need to provide a full proof. Only in this problem is an answer without justification acceptable.)

3	9			
	11		7	2
10				16
15				
20	36			32

2. Suppose a, b, and c are distinct positive real numbers such that

$$abc = 1000,$$

$$bc(1 - a) + a(b + c) = 110.$$

If $a < 1$, show that $10 < c < 100$.

3. Let P be a convex n-gon in the plane with vertices labeled $V_1, \ldots, V_n$ in counterclockwise order. A point Q not outside P is called a *balancing point* of P if, when the triangles $QV_1V_2, QV_2V_3, \ldots, QV_{n-1}V_n, QV_nV_1$ are alternately colored blue and green, the total areas of the blue and green regions are the same. Suppose P has exactly one balancing point. Show that the balancing point must be a vertex of P.

4. Several players try out for the USAMTS basketball team, and they all have integer heights and weights when measured in centimeters and pounds, respectively. In addition, they all weigh less in pounds than they are tall in centimeters. All of the players weigh at least 190 pounds and are at most 197 centimeters tall, and there is exactly one player with every possible height-weight combination.

    The USAMTS wants to field a competitive team, so there are some strict requirements.

    (i) If person P is on the team, then anyone who is at least as tall and at most as heavy as P must also be on the team.

    (ii) If person P is on the team, then no one whose weight is the same as P's height can also be on the team.

    Assuming the USAMTS team can have any number of members (including zero), how many distinct basketball teams can be constructed?

5. Find all positive integers n that have distinct positive divisors $d_1, d_2, \ldots, d_k$, where $k > 1$, that are in arithmetic progression and

$$n = d_1 + d_2 + \cdots + d_k.$$

    Note that $d_1, \ldots, d_k$ do not need to be all the divisors of n.

30.2 USAMTS-Round 2 Problems

1. In the grid to the right, the shortest path through unit squares between the pair of 2's has length 2. Fill in some of the unit squares in the grid so that

    (i) exactly half of the squares in each row and column contain a number,

    (ii) each of the numbers 1 through 12 appears exactly twice, and

    (iii) for $n = 1, 2, \ldots, 12$, the shortest path between the pair of n's has length exactly n.

You do not need to prove that your answer is the only one possible; you merely need to find an answer that satisfies the constraints above. (Note: In any other USAMTS problem, you need to provide a full proof. Only in this problem is an answer without justification acceptable.)

2. A net for a polyhedron is cut along an edge to give two **pieces**. For example, we may cut a cube net along the red edge to form two pieces as shown.

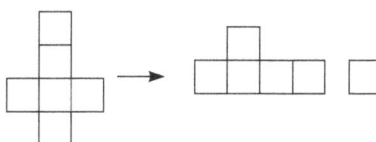

Are there two distinct polyhedra for which this process may result in the same two pairs of pieces?

3. For all positive integers n, show that

$$\frac{1}{n} \sum_{k=1}^{n} \frac{k \cdot k! \cdot \binom{n}{k}}{n^k} = 1.$$

4. Find all polynomials $P(x)$ with integer coefficients such that, for all integers a and b, $P(a+b) - P(b)$ is a multiple of $P(a)$.

5. Let $n > 1$ be an even positive integer. A $2n \times 2n$ grid of unit squares is given, and it is partitioned into n^2 contiguous 2×2 blocks of unit squares. A subset S of the unit squares satisfies the following properties:

  (i) For any pair of squares A, B in S, there is a sequence of squares in S that starts with A, ends with B, and has any two consecutive elements sharing a side; and

  (ii) In each of the 2×2 blocks of squares, at least one of the four squares is in S.

An example for $n = 2$ is shown below, with the squares of S shaded and the four 2×2 blocks of squares outlined in bold.

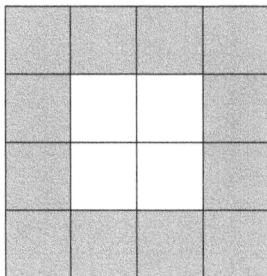

In terms of n, what is the minimum possible number of elements in S?

CHAPTER 31

HARVARD-MIT MATHEMATICS TOURNAMENT

31.1 Harvard-MIT Mathematics Tournament 2015

31.1.1 HMMT 2015 - Individual

1. Find the number of triples (a, b, c) of positive integers such that $a + ab + abc = 11$.

2. Let a and b be real numbers randomly (and independently) chosen from the range $[0, 1]$. Find the probability that a, b and 1 form the side lengths of an obtuse triangle.

3. Neo has an infinite supply of red pills and blue pills. When he takes a red pill, his weight will double, and when he takes a blue pill, he will lose one pound. If Neo originally weighs one pound, what is the minimum number of pills he must take to make his weight 2015 pounds?

4. Chords AB and CD of a circle are perpendicular and intersect at a point P. If $AP = 6, BP = 12$, and $CD = 22$, find the area of the circle.

5. Let S be a subset of the set $\{1, 2, 3, \ldots, 2015\}$ such that for any two elements $a, b \in S$, the difference $a - b$ does not divide the sum $a + b$. Find the maximum possible size of S.

6. Consider all functions $f : \mathbb{Z} \to \mathbb{Z}$ satisfying

$$f(f(x) + 2x + 20) = 15.$$

Call an integer n *good* if $f(n)$ can take any integer value. In other words, if we fix n, for any integer m, there exists a function f such that $f(n) = m$. Find the sum of all good integers x.

7. Let $\triangle ABC$ be a right triangle with right angle C. Let I be the incenter of ABC, and let M lie on AC and N on BC, respectively, such that M, I, N are collinear and $\overline{MN}$ is parallel to AB. If $AB = 36$ and the perimeter of CMN is 48, find the area of ABC.

8. Let $ABCD$ be a quadrilateral with an inscribed circle ω that has center I. If $IA = 5, IB = 7, IC = 4, ID = 9$, find the value of $\frac{AB}{CD}$.

9. Rosencrantz plays $n \leq 2015$ games of question, and ends up with a win rate $\left(\text{i.e. } \frac{\text{\# of games won}}{\text{\# of games played}}\right)$ of k. Guildenstern has also played several games, and has a win rate less than k. He realizes that if, after playing some more games, his win rate becomes higher than k, then there must have been some point in time when Rosencrantz and Guildenstern had the exact same win-rate. Find the product of all possible values of k.

10. Let N be the number of functions f from $\{1, 2, \ldots, 101\} \to \{1, 2, \ldots, 101\}$ such that $f^{101}(1) = 2$. Find the remainder when N is divided by 103.

31.1.2 HMMT 2015 - Guts round

1. Farmer Yang has a 2015×2015 square grid of corn plants. One day, the plant in the very center of the grid becomes diseased. Every day, every plant adjacent to a diseased plant becomes diseased. After how many days will all of Yang's corn plants be diseased?

2. The three sides of a right triangle form a geometric sequence. Determine the ratio of the length of the hypotenuse to the length of the shorter leg.

3. A parallelogram has 2 sides of length 20 and 15. Given that its area is a positive integer, find the minimum possible area of the parallelogram.

4. Eric is taking a biology class. His problem sets are worth 100 points in total, his three midterms are worth 100 points each, and his final is worth 300 points. If he gets a perfect score on his problem sets and scores $60\%, 70\%$, and 80% on his midterms respectively, what is the minimum possible percentage he can get on his final to ensure a passing grade? (Eric passes if and only if his overall percentage is at least 70%).

5. James writes down three integers. Alex picks some two of those integers, takes the average of them, and adds the result to the third integer. If the possible final results Alex could get are 42, 13, and 37, what are the three integers James originally chose?

6. Let AB be a segment of length 2 with midpoint M. Consider the circle with center O and radius r that is externally tangent to the circles with diameters AM and BM and internally tangent to the circle with diameter AB. Determine the value of r.

7. Let n be the smallest positive integer with exactly 2015 positive factors. What is the sum of the (not necessarily distinct) prime factors of n? For example, the sum of the prime factors of 72 is

$$2 + 2 + 2 + 3 + 3 = 14.$$

8. For how many pairs of nonzero integers (c, d) with $-2015 \leq c, d \leq 2015$ do the equations $cx = d$ and $dx = c$ both have an integer solution?

9. Find the smallest positive integer n such that there exists a complex number z, with positive real and imaginary part, satisfying $z^n = (\overline{z})^n$.

10. Call a string of letters S an *almost palindrome* if S and the reverse of S differ in exactly two places. Find the number of ways to order the letters in

$$HMMTTHEMETEAM$$

to get an almost palindrome.

11. Find all integers n, not necessarily positive, for which there exist positive integers a, b, c satisfying $a^n + b^n = c^n$.

12. Let a and b be positive real numbers. Determine the minimum possible value of

$$\sqrt{a^2 + b^2} + \sqrt{(a-1)^2 + b^2} + \sqrt{a^2 + (b-1)^2} + \sqrt{(a-1)^2 + (b-1)^2}$$

13. Consider a 4×4 grid of squares, each of which are originally colored red. Every minute, Piet can jump on one of the squares, changing the color of it and any adjacent squares (two squares are adjacent if they share a side) to blue. What is the minimum number of minutes it will take Piet to change the entire grid to blue?

14. Let ABC be an acute triangle with orthocenter H. Let D, E be the feet of the A, B-altitudes respectively. Given that $AH = 20$ and $HD = 15$ and $BE = 56$, find the length of BH.

15. Find the smallest positive integer b such that 1111_b (1111 in base b) is a perfect square. If no such b exists, write "No solution".

16. For how many triples (x, y, z) of integers between -10 and 10 inclusive do there exist reals a, b, c that satisfy

$$ab = x$$

$$ac = y$$

$$bc = z?$$

17. Unit squares $ABCD$ and $EFGH$ have centers O_1 and O_2 respectively, and are originally situated such that B and E are at the same position and C and H are at the same position. The squares then rotate clockwise about their centers at the rate of one revolution per hour. After 5 minutes, what is the area of the intersection of the two squares?

18. A function f satisfies, for all nonnegative integers x and y:

- $f(0, x) = f(x, 0) = x$
- If $x \geq y \geq 0$, $f(x, y) = f(x - y, y) + 1$
- If $y \geq x \geq 0$, $f(x, y) = f(x, y - x) + 1$

Find the maximum value of f over $0 \leq x, y \leq 100$.

19. Each cell of a 2×5 grid of unit squares is to be colored white or black. Compute the number of such colorings for which no 2×2 square is a single color.

20. Let n be a three-digit integer with nonzero digits, not all of which are the same. Define $f(n)$ to be the greatest common divisor of the six integers formed by any permutation of ns digits. For example, $f(123) = 3$, because $\gcd(123, 132, 213, 231, 312, 321) = 3$. Let the maximum possible value of $f(n)$ be k. Find the sum of all n for which $f(n) = k$.

21. Consider a 2×2 grid of squares. Each of the squares will be colored with one of 10 colors, and two colorings are considered equivalent if one can be rotated to form the other. How many distinct colorings are there?

22. Find all the roots of the polynomial $x^5 - 5x^4 + 11x^3 - 13x^2 + 9x - 3$.

23. Compute the smallest positive integer n for which

$$0 < \sqrt[4]{n} - \lfloor \sqrt[4]{n} \rfloor < \frac{1}{2015}.$$

24. Three ants begin on three different vertices of a tetrahedron. Every second, they choose one of the three edges connecting to the vertex they are on with equal probability and travel to the other vertex on that edge. They all stop when any two ants reach the same vertex at the same time. What is the probability that all three ants are at the same vertex when they stop?

25. Let ABC be a triangle that satisfies $AB = 13, BC = 14, AC = 15$. Given a point P in the plane, let P_A, P_B, P_C be the reflections of A, B, C across P. Call P *good* if the circumcircle of $P_A P_B P_C$ intersects the circumcircle of ABC at exactly 1 point. The locus of good points P encloses a region $\mathcal{S}$. Find the area of $\mathcal{S}$.

26. Let $f : \mathbb{R}^+ \to \mathbb{R}$ be a *continuous* function satisfying $f(xy) = f(x) + f(y) + 1$ for all positive reals x, y. If $f(2) = 0$, compute $f(2015)$.

27. Let $ABCD$ be a quadrilateral with

$$A = (3, 4), B = (9, -40), C = (-5, -12), D = (-7, 24).$$

Let P be a point in the plane (not necessarily inside the quadrilateral). Find the minimum possible value of

$$AP + BP + CP + DP.$$

28. Find the shortest distance between the lines

$$\frac{x+2}{2} = \frac{y-1}{3} = \frac{z}{1}$$

and

$$\frac{x-3}{-1} = \frac{y}{1} = \frac{z+1}{2}.$$

29. Find the largest real number k such that there exists a sequence of positive reals $\{a_i\}$ for which $\sum_{n=1}^{\infty} a_n$ converges but

$$\sum_{n=1}^{\infty} \frac{\sqrt{a_n}}{n^k}$$

does not.

30. Find the largest integer n such that the following holds: there exists a set of n points in the plane such that, for any choice of three of them, some two are unit distance apart.

31. Two random points are chosen on a segment and the segment is divided at each of these two points. Of the three segments obtained, find the probability that the largest segment is more than three times longer than the smallest segment.

32. Find the sum of all positive integers $n \leq 2015$ that can be expressed in the form

$$\left\lceil \frac{x}{2} \right\rceil + y + xy,$$

where x and y are positive integers.

33. How many ways are there to place four points in the plane such that the set of pairwise distances between the points consists of exactly 2 elements? (Two configurations are the same if one can be obtained from the other via rotation and scaling.)

34. Let n be the **second** smallest integer that can be written as the sum of two positive cubes in two different ways. Compute n. If your guess is a, you will receive $\max \left(25 - 5 \cdot \max \left(\frac{a}{n}, \frac{n}{a} \right), 0 \right)$ points, rounded up.

35. Let n be the smallest positive integer such that any positive integer can be expressed as the sum of n integer 2015th powers. Find n. If your answer is a, your score will be $\max \left(20 - \frac{1}{5} |\log_{10} \frac{a}{n}|, 0 \right)$, rounded up.

36. Consider the following seven false conjectures with absurdly high counterexamples. Pick any subset of them, and list their labels in order of their smallest counterexample (the smallest n for which the conjecture is false) from smallest to largest. For example, if you believe that the below list is already ordered by counterexample size, you should write "PECRSGA".

- P. (**Polya's conjecture**) For any integer n, at least half of the natural numbers below n have an odd number of prime factors.

- E. (**Euler's conjecture**) There is no perfect cube n that can be written as the sum of three positive cubes.

- C. (**Cyclotomic**) The polynomial with minimal degree whose roots are the primitive nth roots of unity has all coefficients equal to -1, 0, or 1.

- R. (**Prime race**) For any integer n, there are more primes below n equal to 2 (mod 3) than there are equal to 1 (mod 3).

- S. (**Seventeen conjecture**) For any integer n, $n^{17} + 9$ and $(n + 1)^{17} + 9$ are relatively prime.

- G. (**Goldbach's (other) conjecture**) Any odd composite integer n can be written as the sum of a prime and twice a square.

- A. (**Average square**) Let $a_1 = 1$ and
$$a_{k+1} = \frac{1 + a_1^2 + a_2^2 + \ldots + a_k^2}{k}.$$
Then a_n is an integer for any n.

If your answer is a list of $4 \leq n \leq 7$ labels in the correct order, your score will be $(n - 2)(n - 3)$. Otherwise, it will be 0.

31.1.3   HMMT 2015 - Team round

1. Triangle ABC is isosceles, and $\angle ABC = x°$. If the sum of the possible measures of $\angle BAC$ is $240°$, find x.

2. Bassanio has three red coins, four yellow coins, and five blue coins. At any point, he may give Shylock any two coins of different colors in exchange for one coin of the other color; for example, he may give Shylock one red coin and one blue coin, and receive one yellow coin in return. Bassanio wishes to end with coins that are all the same color, and he wishes to do this while having as many coins as possible. How many coins will he end up with, and what color will they be?

3. Let $\lfloor x \rfloor$ denote the largest integer less than or equal to x, and let $\{x\}$ denote the fractional part of x. For example, $\lfloor \pi \rfloor = 3$, and $\{\pi\} = 0.14159\ldots$, while $\lfloor 100 \rfloor = 100$ and $\{100\} = 0$. If n is the largest solution to the equation $\dfrac{\lfloor n \rfloor}{n} = \dfrac{2015}{2016}$, compute $\{n\}$.

4. Call a set of positive integers *good* if there is a partition of it into two sets S and T, such that there do not exist three elements $a, b, c \in S$ such that $a^b = c$ and such that there do not exist three elements $a, b, c \in T$ such that $a^b = c$ (a and b need not be distinct). Find the smallest positive integer n such that the set

$$\{2, 3, 4, \ldots, n\}$$

is *not* good.

5. Kelvin the Frog is trying to hop across a river. The river has 10 lilypads on it, and he must hop on them in a specific order (the order is unknown to Kelvin). If Kelvin hops to the wrong lilypad at any point, he will be thrown back to the wrong side of the river and will have to start over. Assuming Kelvin is infinitely intelligent, what is the minimum number of hops he will need to guarantee reaching the other side?

6. Marcus and four of his relatives are at a party. Each pair of the five people are either *friends* or *enemies*. For any two enemies, there is no person that they are both friends with. In how many ways is this possible?

7. Let $ABCD$ be a convex quadrilateral whose diagonals AC and BD meet at P. Let the area of triangle APB be 24 and let the area of triangle CPD be 25. What is the minimum possible area of quadrilateral $ABCD$?

8. Find **any** quadruple of positive integers (a, b, c, d) satisfying

$$a^3 + b^4 + c^5 = d^{11}$$

and

$$abc < 10^5.$$

9. A graph consists of 6 vertices. For each pair of vertices, a coin is flipped, and an edge connecting the two vertices is drawn if and only if the coin shows heads. Such a graph is *good* if, starting from any vertex V connected to at least one other vertex, it is possible to draw a path starting and ending at V that traverses each edge exactly once. What is the probability that the graph is good?

10. A number n is *bad* if there exists some integer c for which $x^x \equiv c \pmod{n}$ has no integer solutions for x. Find the number of bad integers between 2 and 42 inclusive.

31.1.4   HMMT 2015 - Theme round

1. Consider a 1×1 grid of squares. Let A, B, C, D be the vertices of this square, and let E be the midpoint of segment CD. Furthermore, let F be the point on segment BC satisfying $BF = 2CF$, and let P be the intersection of lines AF and BE. Find $\dfrac{AP}{PF}$.

2. Consider a 2×2 grid of squares. David writes a positive integer in each of the squares. Next to each row, he writes the product of the numbers in the row, and next to each column, he writes the product of the numbers in each column. If the sum of the eight numbers he writes down is 2015, what is the minimum possible sum of the four numbers he writes in the grid?

3. Consider a 3×3 grid of squares. A circle is inscribed in the lower left corner, the middle square of the top row, and the rightmost square of the middle row, and a circle O with radius r is drawn such that O is externally tangent to each of the three inscribed circles. If the side length of each square is 1, compute r.

4. Consider a 4×4 grid of squares. Aziraphale and Crowley play a game on this grid, alternating turns, with Aziraphale going first. On Aziraphale's turn, he may color any uncolored square red, and on Crowley's turn, he may color any uncolored square blue. The game ends when all the squares are colored, and Aziraphale's score is the area of the largest closed region that is entirely red. If Aziraphale wishes to maximize his score, Crowley wishes to minimize it, and both players play optimally, what will Aziraphale's score be?

5. Consider a 5×5 grid of squares. Vladimir colors some of these squares red, such that the centers of any four red squares do **not** form an axis-parallel rectangle (i.e. a rectangle whose sides are parallel to those of the squares). What is the maximum number of squares he could have colored red?

6. Consider a 6×6 grid of squares. Edmond chooses four of these squares uniformly at random. What is the probability that the centers of these four squares form a square?

7. Consider a 7×7 grid of squares. Let

$$f : \{1, 2, 3, 4, 5, 6, 7\} \to \{1, 2, 3, 4, 5, 6, 7\}$$

be a function; in other words,

$$f(1), f(2), \ldots, f(7)$$

are each (not necessarily distinct) integers from 1 to 7. In the top row of the grid, the numbers from 1 to 7 are written in order; in every other square, $f(x)$ is written where x is the number above the square. How many functions have the property that the bottom row is identical to the top row, and no other row is identical to the top row?

8. Consider an 8×8 grid of squares. A rook is placed in the lower left corner, and every minute it moves to a square in the same row or column with equal probability (the rook must move; i.e. it cannot stay in the same square). What is the expected number of minutes until the rook reaches the upper right corner?

9. Consider a 9×9 grid of squares. Haruki fills each square in this grid with an integer between 1 and 9, inclusive. The grid is called a *super-sudoku* if each of the following three conditions hold:

  - Each column in the grid contains each of the numbers $1, 2, 3, 4, 5, 6, 7, 8, 9$ exactly once.

  - Each row in the grid contains each of the numbers $1, 2, 3, 4, 5, 6, 7, 8, 9$ exactly once.

  - Each 3×3 subsquare in the grid contains each of the numbers $1, 2, 3, 4, 5, 6, 7, 8, 9$ exactly once.

  How many possible super-sudoku grids are there?

10. Consider a 10×10 grid of squares. One day, Daniel drops a burrito in the top left square, where a wingless pigeon happens to be looking for food. Every minute, if the pigeon and the burrito are in the same square, the pigeon will eat 10% of the burrito's original size and accidentally throw it into a random square (possibly the one it is already in). Otherwise, the pigeon will move to an adjacent square, decreasing the distance between it and the burrito. What is the expected number of minutes before the pigeon has eaten the entire burrito?

31.2 HMMT 2016

31.2.1 HMMT 2016 - Algebra

1. Let z be a complex number such that $|z| = 1$ and $|z - 1.45| = 1.05$. Compute the real part of z.

2. For which integers $n \in \{1, 2, \ldots, 15\}$ is $n^n + 1$ a prime number?

3. Let A denote the set of all integers n such that $1 \le n \le 10000$, and moreover the sum of the decimal digits of n is 2. Find the sum of the squares of the elements of A.

4. Determine the remainder when

$$\sum_{i=0}^{2015} \left\lfloor \frac{2^i}{25} \right\rfloor$$

  is divided by 100, where $\lfloor x \rfloor$ denotes the largest integer not greater than x.

5. An infinite sequence of real numbers $a_1, a_2, \ldots$ satisfies the recurrence

$$a_{n+3} = a_{n+2} - 2a_{n+1} + a_n$$

  for every positive integer n. Given that $a_1 = a_3 = 1$ and $a_{98} = a_{99}$, compute $a_1 + a_2 + \cdots + a_{100}$.

6. Call a positive integer $N \ge 2$ "special" if for every k such that $2 \le k \le N$, N can be expressed as a sum of k positive integers that are relatively prime to N (although not necessarily relatively prime to each other). How many special integers are there less than 100?

7. Determine the smallest positive integer $n \geq 3$ for which

$$A \equiv 2^{10n} \pmod{2^{170}}$$

where A denotes the result when the numbers

$$2^{10}, 2^{20}, \ldots, 2^{10n}$$

are written in decimal notation and concatenated (for example, if $n = 2$ we have $A = 10241048576$).

8. Define $\phi^!(n)$ as the product of all positive integers less than or equal to n and relatively prime to n. Compute the number of integers $2 \leq n \leq 50$ such that n divides $\phi^!(n) + 1$.

9. For any positive integer n, S_n be the set of all permutations of $\{1, 2, 3, \ldots, n\}$. For each permutation $\pi \in S_n$, let $f(\pi)$ be the number of ordered pairs (j, k) for which $\pi(j) > \pi(k)$ and $1 \leq j < k \leq n$. Further define $g(\pi)$ to be the number of positive integers $k \leq n$ such that

$$\pi(k) \equiv k \pm 1 \pmod{n}.$$

Compute

$$\sum_{\pi \in S_{999}} (-1)^{f(\pi) + g(\pi)}.$$

10. Let a, b and c be real numbers such that

$$a^2 + ab + b^2 = 9$$
$$b^2 + bc + c^2 = 52$$
$$c^2 + ca + a^2 = 49.$$

Compute the value of

$$\frac{49b^2 + 39bc + 9c^2}{a^2}.$$

31.2.2   HMMT 2016 - Combinatorics

1. For positive integers n, let S_n be the set of integers x such that n distinct lines, no three concurrent, can divide a plane into x regions (for example, $S_2 = \{3, 4\}$, because the plane is divided into 3 regions if the two lines are parallel, and 4 regions otherwise). What is the minimum i such that S_i contains at least 4 elements?

2. Starting with an empty string, we create a string by repeatedly appending one of the letters H, M, T with probabilities $\frac{1}{4}, \frac{1}{2}, \frac{1}{4}$, respectively, until the letter M appears twice consecutively. What is the expected value of the length of the resulting string?

3. Find the number of ordered pairs of integers (a, b) such that a, b are divisors of 720 but ab is not.

4. Let R be the rectangle in the Cartesian plane with vertices at $(0,0)$, $(2,0)$, $(2,1)$, and $(0,1)$. R can be divided into two unit squares, as shown; the resulting figure has seven edges.

How many subsets of these seven edges form a connected figure?

5. Let a, b, c, d, e, f be integers selected from the set $\{1, 2, \ldots, 100\}$, uniformly and at random with replacement. Set
$$M = a + 2b + 4c + 8d + 16e + 32f.$$
What is the expected value of the remainder when M is divided by 64?

6. Define the sequence $a_1, a_2 \ldots$ as follows: $a_1 = 1$ and for every $n \geq 2$,
$$a_n = \begin{cases} n - 2 & \text{if } a_{n-1} = 0 \\ a_{n-1} - 1 & \text{if } a_{n-1} \neq 0 \end{cases}$$

A non-negative integer d is said to be *jet-lagged* if there are non-negative integers r, s and a positive integer n such that $d = r + s$ and that $a_{n+r} = a_n + s$. How many integers in $\{1, 2, \ldots, 2016\}$ are jet-lagged?

7. Kelvin the Frog has a pair of standard fair 8-sided dice (each labelled from 1 to 8). Alex the sketchy Kat also has a pair of fair 8-sided dice, but whose faces are labelled differently (the integers on each Alex's dice need not be distinct). To Alex's dismay, when both Kelvin and Alex roll their dice, the probability that they get any given sum is equal!

Suppose that Alex's two dice have a and b total dots on them, respectively. Assuming that $a \neq b$, find all possible values of $\min\{a, b\}$.

8. Let X be the collection of all functions
$$f : \{0, 1, \ldots, 2016\} \to \{0, 1, \ldots, 2016\}.$$
Compute the number of functions $f \in X$ such that
$$\max_{g \in X} \left(\min_{0 \leq i \leq 2016} \left(\max(f(i), g(i)) \right) - \max_{0 \leq i \leq 2016} \left(\min(f(i), g(i)) \right) \right) = 2015.$$

9. Let $V = \{1, \ldots, 8\}$. How many permutations $\sigma : V \to V$ are automorphisms of some tree?

 (A *graph* consists of a some set of vertices and some edges between pairs of distinct vertices. It is *connected* if every two vertices in it are connected by some path of one or more edges. A *tree* G on V is a connected graph with vertex set V and exactly $|V| - 1$ edges, and an *automorphism* of G is a permutation $\sigma : V \to V$ such that vertices $i, j \in V$ are connected by an edge if and only if $\sigma(i)$ and $\sigma(j)$ are.)

10. Kristoff is planning to transport a number of indivisible ice blocks with positive integer weights from the north mountain to Arendelle. He knows that when he reaches Arendelle, Princess Anna and Queen Elsa will name an ordered pair (p, q) of nonnegative integers satisfying $p + q \leq 2016$. Kristoff must then give Princess Anna *exactly* p kilograms of ice. Afterward, he must give Queen Elsa *exactly* q kilograms of ice.

 What is the minimum number of blocks of ice Kristoff must carry to guarantee that he can always meet Anna and Elsa's demands, regardless of which p and q are chosen?

31.2.3 HMMT 2016 - Geometry

1. Dodecagon $QWARTZSPHINX$ has all side lengths equal to 2, is not self-intersecting (in particular, the twelve vertices are all distinct), and moreover each interior angle is either $90°$ or $270°$. What are all possible values of the area of $\triangle SIX$?

2. Let ABC be a triangle with $AB = 13$, $BC = 14$, $CA = 15$. Let H be the orthocenter of ABC. Find the distance between the circumcenters of triangles AHB and AHC.

3. In the below picture, T is an equilateral triangle with a side length of 5 and ω is a circle with a radius of 2. The triangle and the circle have the same center. Let X be the area of the shaded region, and let Y be the area of the starred region. What is $X - Y$?

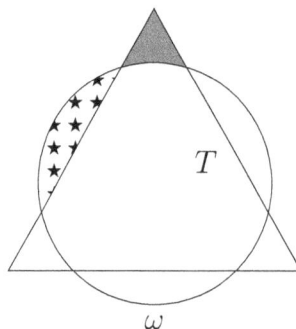

4. Let ABC be a triangle with $AB = 3$, $AC = 8$, $BC = 7$ and let M and N be the midpoints of $\overline{AB}$ and $\overline{AC}$, respectively. Point T is selected on side BC so that $AT = TC$. The circumcircles of triangles BAT, MAN intersect at D. Compute DC.

5. Nine pairwise noncongruent circles are drawn in the plane such that any two circles intersect twice. For each pair of circles, we draw the line through these two points, for a total of $\binom{9}{2} = 36$ lines. Assume that all 36 lines drawn are distinct. What is the maximum possible number of points which lie on at least two of the drawn lines?

6. Let ABC be a triangle with incenter I, incircle γ and circumcircle Γ. Let M, N, P be the midpoints of sides $\overline{BC}$, $\overline{CA}$, $\overline{AB}$ and let E, F be the tangency points of γ with $\overline{CA}$ and $\overline{AB}$, respectively. Let U, V be the intersections of line EF with line MN and line MP, respectively, and let X be the midpoint of arc $\overarc{BAC}$ of Γ. Given that $AB = 5$, $AC = 8$, and $\angle A = 60°$, compute the area of triangle XUV.

7. Let
$$S = \{(x, y) | x, y \in \mathbb{Z}, 0 \le x, y, \le 2016\}.$$
Given points $A = (x_1, y_1), B = (x_2, y_2)$ in S, define
$$d_{2017}(A, B) = (x_1 - x_2)^2 + (y_1 - y_2)^2 \pmod{2017}.$$
The points $A = (5, 5), B = (2, 6), C = (7, 11)$ all lie in S. There is also a point $O \in S$ that satisfies
$$d_{2017}(O, A) = d_{2017}(O, B) = d_{2017}(O, C).$$
Find $d_{2017}(O, A)$.

8. For $i = 0, 1, \ldots, 5$ let l_i be the ray on the Cartesian plane starting at the origin, an angle $\theta = i\frac{\pi}{3}$ counterclockwise from the positive x-axis. For each i, point P_i is chosen uniformly at random from the intersection of l_i with the unit disk. Consider the convex hull of the points P_i, which will (with probability 1) be a convex polygon with n vertices for some n. What is the expected value of n?

9. In cyclic quadrilateral $ABCD$ with $AB = AD = 49$ and $AC = 73$, let I and J denote the incenters of triangles ABD and CBD. If diagonal $\overline{BD}$ bisects $\overline{IJ}$, find the length of IJ.

10. The incircle of a triangle ABC is tangent to BC at D. Let H and Γ denote the orthocenter and circumcircle of $\triangle ABC$. The B-mixtilinear incircle, centered at O_B, is tangent to lines BA and BC and internally tangent to Γ. The C-mixtilinear incircle, centered at O_C, is defined similarly. Suppose that $\overline{DH} \perp \overline{O_B O_C}$, $AB = \sqrt{3}$ and $AC = 2$. Find BC.

31.2.4   HMMT 2016 - Guts

1. Let x and y be complex numbers such that $x + y = \sqrt{20}$ and $x^2 + y^2 = 15$. Compute $|x - y|$.

2. Sherry is waiting for a train. Every minute, there is a 75% chance that a train will arrive. However, she is engrossed in her game of sudoku, so even if a train arrives she has a 75% chance of not noticing it (and hence missing the train). What is the probability that Sherry catches the train in the next five minutes?

3. Let $PROBLEMZ$ be a regular octagon inscribed in a circle of unit radius. Diagonals MR, OZ meet at I. Compute LI.

4. Consider a three-person game involving the following three types of fair six-sided dice.

   - Dice of type A have faces labelled $2, 2, 4, 4, 9, 9$.
   - Dice of type B have faces labelled $1, 1, 6, 6, 8, 8$.
   - Dice of type C have faces labelled $3, 3, 5, 5, 7, 7$.

   All three players simultaneously choose a die (more than one person can choose the same type of die, and the players don't know one another's choices) and roll it. Then the score of a player P is the number of players whose roll is less than P's roll (and hence is either 0, 1, or 2). Assuming all three players play optimally, what is the expected score of a particular player?

5. Patrick and Anderson are having a snowball fight. Patrick throws a snowball at Anderson which is shaped like a sphere with a radius of 10 centimeters. Anderson catches the snowball and uses the snow from the snowball to construct snowballs with radii of 4 centimeters. Given that the total volume of the snowballs that Anderson constructs cannot exceed the volume of the snowball that Patrick threw, how many snowballs can Anderson construct?

6. Consider a $2 \times n$ grid of points and a path consisting of $2n - 1$ straight line segments connecting all these $2n$ points, starting from the bottom left corner and ending at the upper right corner. Such a path is called *efficient* if each point is only passed through once and no two line segments intersect. How many efficient paths are there when $n = 2016$?

7. A contest has six problems worth seven points each. On any given problem, a contestant can score either 0, 1, or 7 points. How many possible total scores can a contestant achieve over all six problems?

8. For each positive integer n and non-negative integer k, define $W(n, k)$ recursively by

$$
W(n, k) = \begin{cases} n^n & k = 0 \\ W(W(n, k - 1), k - 1) & k > 0. \end{cases}
$$

Find the last three digits in the decimal representation of $W(555, 2)$.

9. Victor has a drawer with two red socks, two green socks, two blue socks, two magenta socks, two lavender socks, two neon socks, two mauve socks, two wisteria socks, and 2000 copper socks, for a total of 2016 socks. He repeatedly draws two socks at a time from the drawer at random, and stops if the socks are of the same color. However, Victor is red-green colorblind, so he also stops if he sees a red and green sock.

   What is the probability that Victor stops with two socks of the same color? Assume Victor returns both socks to the drawer at each step.

10. Let ABC be a triangle with $AB = 13$, $BC = 14$, $CA = 15$. Let O be the circumcenter of ABC. Find the distance between the circumcenters of triangles AOB and AOC.

11. Define $\phi^!(n)$ as the product of all positive integers less than or equal to n and relatively prime to n. Compute the remainder when

$$\sum_{\substack{2 \le n \le 50 \\ \gcd(n,50)=1}} \phi^!(n)$$

   is divided by 50.

12. Let R be the rectangle in the Cartesian plane with vertices at $(0,0), (2,0), (2,1)$, and $(0,1)$. R can be divided into two unit squares, as shown; the resulting figure has seven edges.

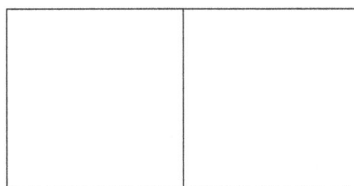

   Compute the number of ways to choose one or more of the seven edges such that the resulting figure is traceable without lifting a pencil. (Rotations and reflections are considered distinct.)

13. A right triangle has side lengths a, b, and $\sqrt{2016}$ in some order, where a and b are positive integers. Determine the smallest possible perimeter of the triangle.

14. Let ABC be a triangle such that $AB = 13$, $BC = 14$, $CA = 15$ and let E, F be the feet of the altitudes from B and C, respectively. Let the circumcircle of triangle AEF be ω. We draw three lines, tangent to the circumcircle of triangle AEF at A, E, and F. Compute the area of the triangle these three lines determine.

15. Compute $\tan\left(\dfrac{\pi}{7}\right) \tan\left(\dfrac{2\pi}{7}\right) \tan\left(\dfrac{3\pi}{7}\right)$.

16. Determine the number of integers $2 \le n \le 2016$ such that $n^n - 1$ is divisible by 2, 3, 5, 7.

17. Compute the sum of all integers $1 \le a \le 10$ with the following property: there exist integers p and q such that p, q, $p^2 + a$ and $q^2 + a$ are all distinct prime numbers.

18. Alice and Bob play a game on a circle with 8 marked points. Alice places an apple beneath one of the points, then picks five of the other seven points and reveals that none of them are hiding the apple. Bob then drops a bomb on any of the points, and destroys the apple if he drops the bomb either on the point containing the apple or on an adjacent point. Bob wins if he destroys the apple, and Alice wins if he fails. If both players play optimally, what is the probability that Bob destroys the apple?

19. Let

$$A = \lim_{n \to \infty} \sum_{i=0}^{2016} (-1)^i \cdot \frac{\binom{n}{i}\binom{n}{i+2}}{\binom{n}{i+1}^2}$$

Find the largest integer less than or equal to $\frac{1}{A}$.

The following decimal approximation might be useful: $0.6931 < \ln(2) < 0.6932$, where $\ln$ denotes the natural logarithm function.

20. Let ABC be a triangle with $AB = 13$, $AC = 14$, and $BC = 15$. Let G be the point on AC such that the reflection of BG over the angle bisector of $\angle B$ passes through the midpoint of AC. Let Y be the midpoint of GC and X be a point on segment AG such that $\frac{AX}{XG} = 3$. Construct F and H on AB and BC, respectively, such that $FX \parallel BG \parallel HY$. If AH and CF concur at Z and W is on AC such that $WZ \parallel BG$, find WZ.

21. Tim starts with a number n, then repeatedly flips a fair coin. If it lands heads he subtracts 1 from his number and if it lands tails he subtracts 2. Let E_n be the expected number of flips Tim does before his number is zero or negative. Find the pair (a, b) such that

$$\lim_{n \to \infty} (E_n - an - b) = 0.$$

22. On the Cartesian plane $\mathbb{R}^2$, a circle is said to be *nice* if its center is at the origin $(0, 0)$ and it passes through at least one lattice point (i.e. a point with integer coordinates). Define the points $A = (20, 15)$ and $B = (20, 16)$. How many nice circles intersect the open segment AB?

For reference, the numbers 601, 607, 613, 617, 619, 631, 641, 643, 647, 653, 659, 661, 673, 677, 683, 691 are the only prime numbers between 600 and 700.

23. Let $t = 2016$ and $p = \ln 2$. Evaluate in closed form the sum

$$\sum_{k=1}^{\infty} \left(1 - \sum_{n=0}^{k-1} \frac{e^{-t}t^n}{n!}\right)(1 - p)^{k-1} p.$$

Hint. You may use the fact that $e^x = \sum_{m=0}^{\infty} \frac{x^m}{m!}$ for all real x.

24. Let $\Delta A_1 B_1 C$ be a triangle with $\angle A_1 B_1 C = 90°$ and $\dfrac{CA_1}{CB_1} = \sqrt{5} + 2$. For any $i \geq 2$, define A_i to be the point on the line $A_1 C$ such that $A_i B_{i-1} \perp A_1 C$ and define B_i to be the point on the line $B_1 C$ such that $A_i B_i \perp B_1 C$. Let Γ_1 be the incircle of $\Delta A_1 B_1 C$ and for $i \geq 2$, Γ_i be the circle tangent to $\Gamma_{i-1}, A_1 C, B_1 C$ which is smaller than Γ_{i-1}.

    How many integers k are there such that the line $A_1 B_{2016}$ intersects Γ_k?

25. A particular coin can land on heads (H), on tails (T), or in the middle (M), each with probability $\dfrac{1}{3}$. Find the expected number of flips necessary to observe the contiguous sequence HMMTH-MMT...HMMT, where the sequence HMMT is repeated 2016 times.

26. For positive integers a, b, $a \uparrow\uparrow b$ is defined as follows: $a \uparrow\uparrow 1 = a$, and $a \uparrow\uparrow b = a^{a\uparrow\uparrow(b-1)}$ if $b > 1$.

    Find the smallest positive integer n for which there exists a positive integer a such that $a \uparrow\uparrow 6 \not\equiv a \uparrow\uparrow 7 \bmod n$.

27. Find the smallest possible area of an ellipse passing through $(2, 0)$, $(0, 3)$, $(0, 7)$, and $(6, 0)$.

28. Among citizens of Cambridge there exist 8 different types of blood antigens. In a crowded lecture hall are 256 students, each of whom has a blood type corresponding to a distinct subset of the antigens; the remaining of the antigens are foreign to them.

    Quito the Mosquito flies around the lecture hall, picks a subset of the students uniformly at random, and bites the chosen students in a random order. After biting a student, Quito stores a bit of any antigens that student had. A student bitten while Quito had k blood antigen foreign to him/her will suffer for k hours. What is the expected total suffering of all 256 students, in hours?

29. Katherine has a piece of string that is 2016 millimeters long. She cuts the string at a location chosen uniformly at random, and takes the left half. She continues this process until the remaining string is less than one millimeter long. What is the expected number of cuts that she makes?

30. Determine the number of triples $0 \leq k, m, n \leq 100$ of integers such that

$$2^m n - 2^n m = 2^k.$$

31. For a positive integer n, denote by $\tau(n)$ the number of positive integer divisors of n, and denote by $\phi(n)$ the number of positive integers that are less than or equal to n and relatively prime to n. Call a positive integer n *good* if $\varphi(n) + 4\tau(n) = n$. For example, the number 44 is good because

$$\varphi(44) + 4\tau(44) = 44.$$

    Find the sum of all good positive integers n.

32. How many equilateral hexagons of side length $\sqrt{13}$ have one vertex at $(0,0)$ and the other five vertices at lattice points?

(A lattice point is a point whose Cartesian coordinates are both integers. A hexagon may be concave but not self-intersecting.)

33. (**Lucas Numbers**) The Lucas numbers are defined by $L_0 = 2$, $L_1 = 1$, and $L_{n+2} = L_{n+1} + L_n$ for every $n \geq 0$. There are N integers $1 \leq n \leq 2016$ such that L_n contains the digit 1. Estimate N.

An estimate of E earns $\lfloor 20 - 2|N - E| \rfloor$ or 0 points, whichever is greater.

34. (**Caos**) A cao [sic] has 6 legs, 3 on each side. A walking pattern for the cao is defined as an ordered sequence of raising and lowering each of the legs exactly once (altogether 12 actions), starting and ending with all legs on the ground. The pattern is safe if at any point, he has at least 3 legs on the ground and not all three legs are on the same side. Estimate N, the number of safe patterns.

An estimate of $E > 0$ earns $\lfloor 20 \min(N/E, E/N)^4 \rfloor$ points.

35. (**Maximal Determinant**) In a 17×17 matrix M, all entries are ± 1. The maximum possible value of $|\det M|$ is N. Estimate N.

An estimate of $E > 0$ earns $\lfloor 20 \min(N/E, E/N)^2 \rfloor$ points.

36. (**Self-Isogonal Cubics**) Let ABC be a triangle with $AB = 2$, $AC = 3$, $BC = 4$. The *isogonal conjugate* of a point P, denoted P^*, is the point obtained by intersecting the reflection of lines PA, PB, PC across the angle bisectors of $\angle A$, $\angle B$, and $\angle C$, respectively.

Given a point Q, let $\mathfrak{K}(Q)$ denote the unique cubic plane curve which passes through all points P such that line PP^* contains Q. Consider:

(a) the M'Cay cubic $\mathfrak{K}(O)$, where O is the circumcenter of $\triangle ABC$,

(b) the Thomson cubic $\mathfrak{K}(G)$, where G is the centroid of $\triangle ABC$,

(c) the Napoleon-Feuerbach cubic $\mathfrak{K}(N)$, where N is the nine-point center of $\triangle ABC$,

(d) the Darboux cubic $\mathfrak{K}(L)$, where L is the de Longchamps point (the reflection of the orthocenter across point O),

(e) the Neuberg cubic $\mathfrak{K}(X_{30})$, where X_{30} is the point at infinity along line OG,

(f) the nine-point circle of $\triangle ABC$,

(g) the incircle of $\triangle ABC$, and

(h) the circumcircle of $\triangle ABC$.

Estimate N, the number of points lying on at least two of these eight curves. An estimate of E earns $\lfloor 20 \cdot 2^{-|N-E|/6} \rfloor$ points.

31.2.5 HMMT 2016 - Team

1. Let a and b be integers (not necessarily positive). Prove that $a^3 + 5b^3 \neq 2016$.

2. For positive integers n, let c_n be the smallest positive integer for which $n^{c_n} - 1$ is divisible by 210, if such a positive integer exists, and $c_n = 0$ otherwise. What is $c_1 + c_2 + \cdots + c_{210}$?

3. Let ABC be an acute triangle with incenter I and circumcenter O. Assume that $\angle OIA = 90°$. Given that $AI = 97$ and $BC = 144$, compute the area of $\triangle ABC$.

4. Let $n > 1$ be an odd integer. On an $n \times n$ chessboard the center square and four corners are deleted. We wish to group the remaining $n^2 - 5$ squares into $\frac{1}{2}(n^2 - 5)$ pairs, such that the two squares in each pair intersect at exactly one point (i.e. they are diagonally adjacent, sharing a single corner).

   For which odd integers $n > 1$ is this possible?

5. Find all prime numbers p such that $y^2 = x^3 + 4x$ has exactly p solutions in integers modulo p.

   In other words, determine all prime numbers p with the following property: there exist exactly p ordered pairs of integers (x, y) such that $x, y \in \{0, 1, \dots, p - 1\}$ and

   $$p \text{ divides } y^2 - x^3 - 4x.$$

6. A nonempty set S is called *well-filled* if for every $m \in S$, there are fewer than $\frac{1}{2}m$ elements of S which are less than m. Determine the number of well-filled subsets of $\{1, 2, \dots, 42\}$.

7. Let $q(x) = q^1(x) = 2x^2 + 2x - 1$, and let $q^n(x) = q(q^{n-1}(x))$ for $n > 1$. How many negative real roots does $q^{2016}(x)$ have?

8. Compute
   $$\int_0^\pi \frac{2\sin\theta + 3\cos\theta - 3}{13\cos\theta - 5}\,d\theta.$$

9. Fix positive integers $r > s$, and let F be an infinite family of sets, each of size r, no two of which share fewer than s elements. Prove that there exists a set of size $r - 1$ that shares at least s elements with each set in F.

10. Let ABC be a triangle with incenter I whose incircle is tangent to $\overline{BC}, \overline{CA}, \overline{AB}$ at D, E, F. Point P lies on $\overline{EF}$ such that $\overline{DP} \perp \overline{EF}$. Ray BP meets $\overline{AC}$ at Y and ray CP meets $\overline{AB}$ at Z. Point Q is selected on the circumcircle of $\triangle AYZ$ so that $\overline{AQ} \perp \overline{BC}$.

    Prove that P, I, Q are collinear.

31.2.6   HMMT 2016 - Invitational Competition

1. Theseus starts at the point $(0,0)$ in the plane. If Theseus is standing at the point (x, y) in the plane, he can step one unit to the north to point $(x, y + 1)$, one unit to the west to point $(x - 1, y)$, one unit to the south to point $(x, y - 1)$, or one unit to the east to point $(x + 1, y)$. After a sequence of more than two such moves, starting with a step one unit to the south (to point $(0, -1)$), Theseus finds himself back at the point $(0, 0)$. He never visited any point other than $(0, 0)$ more than once, and never visited the point $(0, 0)$ except at the start and end of this sequence of moves.

   Let X be the number of times that Theseus took a step one unit to the north, and then a step one unit to the west immediately afterward. Let Y be the number of times that Theseus took a step one unit to the west, and then a step one unit to the north immediately afterward. Prove that $|X - Y| = 1$.

2. Let ABC be an acute triangle with circumcenter O, orthocenter H, and circumcircle Ω. Let M be the midpoint of AH and N the midpoint of BH. Assume the points M, N, O, H are distinct and lie on a circle ω. Prove that the circles ω and Ω are internally tangent to each other.

3. Denote by $\mathbb{N}$ the positive integers. Let $f : \mathbb{N} \to \mathbb{N}$ be a function such that, for any $w, x, y, z \in \mathbb{N}$,

$$f(f(f(z)))f(wxf(yf(z))) = z^2 f(xf(y))f(w).$$

   Show that $f(n!) \geq n!$ for every positive integer n.

4. Let P be an odd-degree integer-coefficient polynomial. Suppose that $xP(x) = yP(y)$ for infinitely many pairs x, y of integers with $x \neq y$. Prove that the equation $P(x) = 0$ has an integer root.

5. Let $S = \{a_1, \ldots, a_n\}$ be a finite set of positive integers of size $n \geq 1$, and let T be the set of all positive integers that can be expressed as sums of perfect powers (including 1) of distinct numbers in S, meaning

$$T = \left\{ \sum_{i=1}^{n} a_i^{e_i} \mid e_1, e_2, \ldots, e_n \geq 0 \right\}.$$

   Show that there is a positive integer N (only depending on n) such that T contains no arithmetic progression of length N.

PART IV

SOLUTION TO PART I EXERCISES

CHAPTER 32

SOLUTIONS TO CHAPTER 6 EXERCISES

32.1 Algebra

1. 2013 AMC 10A Problem 21

Solution (D). For $1 \leq k \leq 11$, the number of coins remaining in the chest before the kth pirate takes a share is $\frac{12}{12-k}$ times the number remaining afterward. Thus if there are n coins left for the 12th pirate to take. the number of coins originally in the chest is

$$\frac{12^{11} \cdot n}{11!} = \frac{2^{22} \cdot 3^{11} \cdot n}{2^8 \cdot 3^{11} \cdot 5^2 \cdot 7 \cdot 11} = \frac{2^{14} \cdot 3^7 \cdot n}{5^2 \cdot 7 \cdot 11}.$$

The smallest value of n for which this is a positive integer is $5^2 \cdot 7 \cdot 11 = 1925$. In this case there are

$$2^{14} \cdot 3^7 \cdot \frac{11!}{(12-k)! \cdot 12^{k-1}}$$

coins left for the kth pirate to take, and note that this amount is an integer for each k. Hence the 12th pirate receives 1925 coins.

2. 1974 AHSME Problem 4

Solution (D). By the remainder theorem $x^{51} + 51$ divided by $x + 1$ leaves a remainder of $(-1)^{51} + 51 = 50$. This can also be seen quite easily by long division.

3. 2001 AMC 12 Problem 13

Solution (E). The equation of the first parabola can be written in the form

$$y = a(x - h)^2 + k = ax^2 - 2axh + ah^2 + k,$$

and the equation for the second(having the same shape and vertex, but opening in the opposite direction) can be written in the form

$$y = -a(x - h)^2 + k = -ax^2 + 2axh - ah^2 + k.$$

Hence

$$\begin{aligned}
&a + b + c + d + e + f \\
= \ &a + (-2ah) + (ah^2 + k) + (-a) + (2ah) + (-ah^2 + k) \\
= \ &2k.
\end{aligned}$$

OR

The reflection of a point (x, y) about the line $y = k$ is $(x, 2k - y)$. Thus the equation of the reflected parabola is

$$2k - y = ax^2 + bx + c,$$

or equivalently,

$$y = 2k - (ax^2 + bx + c).$$

Hence $a + b + c + d + e + f = 2k$.

Note : The case $y = x^2 + 2$ demonstrates that none of the other answer choices is true in every case.

4. 2013 AMC 10B Problem 19

Solution (D). Let the common difference in the arithmetic sequence be d, so that $a = b + d$, $c = b - d$. Because the quadratic has exactly one root, $b^2 - 4ac = 0$. Substitution gives $b^2 = 4(b + d)(b - d)$, and therefore $3b^2 = 4d^2$. Because $b \geq 0$ and $d \geq 0$, it follows that $\sqrt{3}b = 2d$. Thus the real root is

$$\frac{-b \pm \sqrt{b^2 - 4ac}}{2a} = \frac{-b}{2a} = \frac{-b}{2(b+d)} = \frac{-b}{2(b + \frac{\sqrt{3}}{2}b)} = -2 + \sqrt{3}.$$

Note that the quadratic equation $x^2 + (4 - 2\sqrt{3})x + 7 - 4\sqrt{3}$ satisfies the given conditions.

5. 2008 AMC 12A Problem 19

Solution (C). Each term in the expansion has the form x^{a+b+c}, where $0 \leq a \leq 27$, $0 \leq b \leq 14$, and $0 \leq c \leq 14$. There are $(14 + 1)^2 = 225$ possible combinations of values for b and c, and for every combination except $(b, c) = (0, 0)$, there is a unique a with $a + b + c = 28$. Thus the coefficient of x^{28} is 224.

OR

Let $P(x) = (1 + x + x^2 + \cdots + x^{14})^2 = 1 + r_1 x + r_2 x^2 + \cdots + r_{28} x^{28}$ and $Q(x) = 1 + x + x^2 + \cdots + x^{27}$. The coefficient of x^{28} in the product $P(x)Q(x)$ is $r_1 + r_2 + \cdots + r_{28} = P(1) - 1 = 15^2 - 1 = 224$.

6. 1983 AHSME Problem 25

Solution (B). $12 = 60/5 = 60/60^b = 60^{1-b}$. So

$$12^{[(1-a-b)/2(1-b)]} = \left[60^{1-b} \right]^{(1-a-b)/2(1-b)}$$
$$= 60^{(1-a-b)/2} = \sqrt{\frac{60}{60^a 60^b}}$$
$$= \sqrt{\frac{60}{3 \cdot 5}} = 2.$$

7. 2013 AMC 10B Problem 14

Solution (E). The equation $x \clubsuit y = y \clubsuit x$ is equivalent to $x^2 y - xy^2 = y^2 x - yx^2$. This equation is equivalent to gives $2xy(x - y) = 0$. This equation will hold exactly if $x = 0, y = 0$, or $x = y$. The solution set consists of three lines: the x-axis, the y-axis, and the line $x = y$.

8. 2006 AMC 12A Problem 18

Solution (E). The conditions on f imply that both

$$x = f(x) + f\left(\frac{1}{x}\right) \quad \text{and} \quad \frac{1}{x} = f\left(\frac{1}{x}\right) + f\left(\frac{1}{1/x}\right) = f\left(\frac{1}{x}\right) + f(x).$$

Thus if x is in the domain of f, then $x = 1/x$, so $x = \pm 1$. The conditions are satisfied if and only if $f(1) = 1/2$ and $f(-1) = -1/2$.

9. 2009 AMC 12A Problem 17

Solution (C). The sum of the first series is

$$\frac{a}{1 - r_1} = r_1,$$

from which $r_1^2 - r_1 + a = 0$, and $r_1 = \frac{1}{2}(1 \pm \sqrt{1 - 4a})$. Similarly, $r_2 = \frac{1}{2}(1 \pm \sqrt{1 - 4a})$. Because r_1 and r_2 must be different, $r_1 + r_2 = 1$. Such series exist as long as $0 < a < \frac{1}{4}$.

10. 1992 AHSME Problem 18

Solution (D). If $a_1 = a$ and $a_2 = b$ then

$$(a_3, a_4, a_5, a_6, a_7, a_8)$$
$$= (a + b, a + 2b, 2a + 3b, 3a + 5b, 5a + 8b, 8a + 13b).$$

Therefore $5a + 8b = a_7 = 120$. Since $5a = 8(15 - b)$ and 8 is relatively prime to 5, a must be a multiple of 8. Similarly, b must be a multiple of 5. Let $a = 8j$ and $b = 5k$ to obtain $40j + 40k = 120$ which has two solutions in positive integers, $(j, k) = (1, 2)$ and $(2, 1)$. When $(j, k) = (2, 1)$, $(a, b) = (16, 5)$, which is impossible since the sequence is increasing, so $(j, k) = (1, 2)$ and $(a, b) = (8, 10)$. Consequently, $a_8 = 8a + 13b = 194$.

Note. This sequence begins with the eight terms

$$8, 10, 18, 28, 46, 74, 120, 194.$$

11. 2010 AMC 10B Problem 25

Solution (B). Because $1, 3, 5$, and 7 are roots of the polynomial $P(x) - a$, it follows that

$$P(x) - a = (x - 1)(x - 3)(x - 5)(x - 7)Q(x),$$

where $Q(x)$ is a polynomial with integer coefficients. The previous identity must hold for $x = 2, 4, 6$ and 8, thus

$$-2a = -15Q(2) = 9Q(4) = -15Q(6) = 105Q(8).$$

Therefore $315 = \text{lcm}(15, 9, 105)$ divides a, that is a is an integer multiple of 315. Let $a = 315A$. Because $Q(2) = Q(6) = 42A$, it follows that $Q(x) - 42A = (x - 2)(x - 6)R(x)$ where $R(x)$

is a polynomial with integer coefficients. Because $Q(4) = -70A$ and $Q(8) = -6A$ it follows that $-112A = -4R(4)$ and $-48A = 12R(8)$, that is $R(4) = 28A$ and $R(8) = -4A$. Thus $R(x) = 28A + (x-4)(-6A + (x-8)T(x))$ where $T(x)$ is a polynomial with integer coefficients. Moreover, for any polynomial $T(x)$ and any integer A, the polynomial $P(x)$ constructed this way satisfies the required conditions. The required minimum is obtained when $A = 1$ and so $a = 315$.

12. 2010 AMC 12B Problem 23

Solution (A). Because both $P(Q(x))$ and $Q(P(x))$ have four distinct real zeros, both $P(x)$ and $Q(x)$ must have two distinct real zeros, so there are real numbers h_1, k_1, h_2 and k_2 such that $P(x) = (x-h_1)^2 - k_1^2$ and $Q(x) = (x-h_2)^2 - k_2^2$. The zeros of $P(Q(x))$ occur when $Q(x) = h_1 \pm k_1$. The solution of each equation are equidistant from h_2, so $h_2 = -19$. It follows that

$$Q(-15) - Q(-17) = (16 - k_2^2) - (4 - k_2^2) = 12,$$

and also

$$Q(-15) - Q(-17) = 2k_1,$$

so $k_1 = 6$. Similarly $h_1 = -54$, so

$$2k_2 = P(-49) - P(-51) = (25 - k_1^2) - (9 - k_1^2) = 16,$$

and $k_2 = 8$. Thus the sum of the minimum values of $P(x)$ and $Q(x)$ is $-k_1^2 - k_2^2 = -100$.

13. 2004 AMC 10B Problem 21

Solution (A). The smallest number that appears in both sequences is 16. The two sequences have common differences 3 and 7, whose least common multiple is 21, so a number appears in both sequences if and only if it is in the form

$$16 + 21k, \quad \text{where } k \text{ is a nonnegative integer.}$$

Such a number is in the first 2004 terms of both sequences if and only if

$$16 + 21k \leq 1 + 2003(3) = 6010.$$

Thus $0 \leq k \leq 285$, so there are 286 duplicate numbers. Therefore the number of distinct number is $4008 - 286 = 3722$.

32.2 Geometry

32.2.1 Triangle and Geometry

1. 1997 AHSME Problem 5

Solution (C). Let x and y denote the width and height of one of the five rectangles, with $x < y$. Then $5x + 4y = 176$ and $3x = 2y$. Solve simultaneously to get $x = 16$ and $y = 24$. The perimeter in equation is $2 \cdot 16 + 2 \cdot 24 = 80$.

2. 2007 AMC 12B Problem 14

Solution (D). Let the side length of $\triangle ABC$ is s. Then area of $\triangle APB$, $\triangle BPC$, and $\triangle CPA$ are, respectively, $s/2, s$, and $3s/2$. the area of $\triangle ABC$ is the sum of these, which is $3s$. The area of $\triangle ABC$ may also be expressed as $(\sqrt{3}/4)s^2$, so $3s = (\sqrt{3}/4)s^2$. The unique positive solution for s is $4\sqrt{3}$.

3. 2015 AMC 10A Problem 19

Solution (D). Because the area is 12.5, it follows that $AC = BC = 5$. Label D and E so that D is closer to A than to B. Let F be the foot of the perpendicular to $\overline{AC}$ passing through D. Let $h = FD$. Then $AF = h$ because $\triangle ADF$ is an isosceles right triangle, and $CF = h\sqrt{3}$ because $\triangle CDF$ is a $30 - 60 - 90°$ triangle. So $h + h\sqrt{3} = AC = 5$ and

$$h = \frac{5}{1 + \sqrt{3}} = \frac{5\sqrt{3} - 5}{2}.$$

Thus the area of $\triangle CDE$ is

$$\frac{25}{2} - 2 \cdot \frac{1}{2} \cdot 5 \cdot \frac{5\sqrt{3} - 5}{2} = \frac{50 - 25\sqrt{3}}{2}.$$

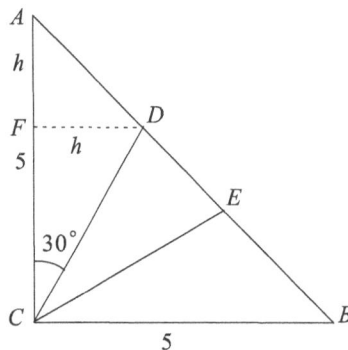

4. 1983 AHSME Problem 19

Solution (A). Let $AD = y$. Since AD bisects $\angle BAC$, we have $\frac{DB}{CD} = \frac{AB}{BC} = 2$; so we may set $CD = x$, $DB = 2x$ as in the figure. Applying the Law of Cosines to $\triangle CAD$ and $\triangle DAB$, we have

$$x^2 = 3^2 + y^2 - 3y,$$

$$(2x)^2 = 6^2 + y^2 - 6y.$$

Subtracting 4 times the first equation from the second yields $0 = -3y^2 + 6y = -3y(y-2)$. Since $y \neq 0$, $y = 2$.

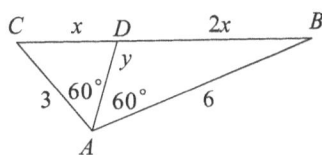

OR

Extend CA to E so that $BE \parallel DA$ as in the new figure. Then $\triangle ABE$ is equilateral: $\angle BEA = \angle DAC$ by corresponding angles, $\angle ABE = \angle BAD$ by alternate interior angles, and $\angle EAB = 180° - 120°$. Since $\triangle BEC \sim \triangle DAC$, we have $\frac{DA}{BE} = \frac{CA}{CE}$, or $\frac{DA}{6} = \frac{3}{9}$. So $DA = 2$.

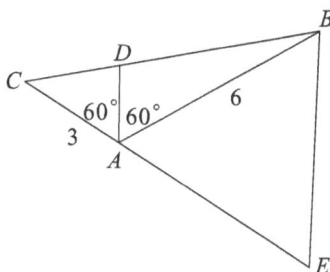

5. 1990 AHSME Problem 14

Solution (A). Angles BAC, BCD and CBD all intercept the same circular arc, minor arc BC of measure $2x$ since $\angle A = x$, therefore $\angle BCD = \angle CBD = x$, and considering the sum of the angles in $\triangle BCD$, $\angle D = \pi - 2x$. Since $\angle ABC = \angle ACB$, considering the sum of the angles in $\triangle ABC$ we have $\angle ABC = (\pi - x)/2$. The given condition, $\angle ABC = 2\angle D$, now becomes

$$\frac{\pi - x}{2} = 2(\pi - 2x), \text{ so } x = \frac{3\pi}{7}.$$

OR

Let O be the center of the circle. Since $\triangle OAB$ is isosceles, $\angle ABO = \angle BAO = x/2$. Since $\overline{BD}$ is tangent to the circle, $\angle OBD = \pi/2$. Thus

$$\angle ABD = \angle ABO + \angle OBD = \frac{x}{2} + \frac{\pi}{2}.$$

Since $\triangle ABC$ is isosceles,

$$\angle ABC = \frac{\pi - x}{2},$$

so

$$\angle D = \frac{1}{2}\angle ABC = \frac{\pi - x}{4}.$$

Sum the angles of quadrilateral $ABDC$ to obtain

$$
\begin{aligned}
2\pi &= \angle CAB + \angle ABD + \angle D + \angle DCA \\
&= x + \left(\frac{x}{2} + \frac{\pi}{2}\right) + \frac{\pi - x}{4} + \left(\frac{x}{2} + \frac{\pi}{2}\right)
\end{aligned}
$$

which we solve to obtain $x = 3\pi/7$.

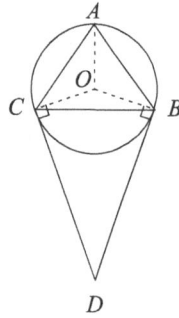

OR

Let O be the center of the circle. Then $\angle COB = 2x$ and, from the sum of the angles of the quadrilateral $COBD$, we obtain $2x + \angle D = \pi$. The conditions of the problem yield $x + 4\angle D = \pi$ to be the sum of the angles of $\triangle ABC$. Solve these two equations in x and $\angle D$ simultaneously to find $x = 3\pi/7$.

Query. What is x if $\triangle ABC$ is an obtuse isosceles triangle?

6. 1997 AHSME Problem 9

Solution (C). In right triangle BAE, $BE = \sqrt{2^2 + 1^2} = \sqrt{5}$. Since $\triangle CFB \sim \triangle BAE$, it follows that

$$
[CFB] = (CB/BE)^2 \cdot [BAE] = \left(\frac{2}{\sqrt{5}}\right)^2 \cdot \left(\frac{1}{2}\right)(2 \cdot 1) = \frac{4}{5}.
$$

Then

$$
[CDEF] = [ABCD] - [BAE] - [CFB] = 4 - 1 - \frac{4}{5} = \frac{11}{5}.
$$

OR

Draw the figure in the plane as shown with B at the origin. An equation of the line BE is $y = 2x$, and, since the lines are perpendicular, an equation of the line CF is $y = -(x-2)/2$. Solve these two equations simultaneously to get $F = (2/5, 4/5)$ and

$$
[CDEF] = [DEF] + [CDF] = \frac{1}{2}(1)\left(2 - \frac{4}{5}\right) + \frac{1}{2}(2)\left(2 - \frac{2}{5}\right) = \frac{11}{5}
$$

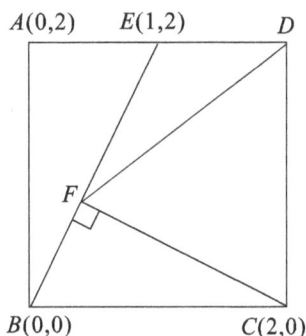

$A(0,2)$ $E(1,2)$ D

F

$B(0,0)$ $C(2,0)$

7. 2008 AMC 10A Problem 16 or 2008 AMC 12A Problem 13

Solution (B). Let r and R be the radii of the smaller and larger circles, respectively. Let E be the center of the smaller circle, Let $\overline{OC}$ be the radius of the larger circle that contains E, and let D be the point of tangency of the smaller circle to $\overline{OA}$. Then $OE = R - r$, and because $\triangle EDO$ is a 30-60-90° triangle, $OE = 2DE = 2r$. Thus $2r = R - r$, so $\frac{r}{R} = \frac{1}{3}$. The ratio of the areas is $\left(\frac{1}{3}\right)^2 = \frac{1}{9}$.

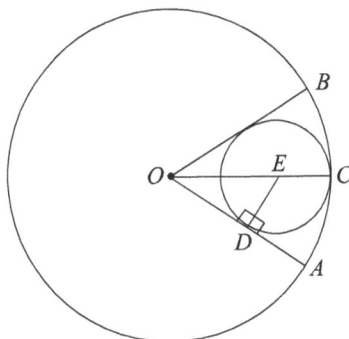

8. 2003 AMC 12A Problem 15

Solution (C). First note that the area of the region determined by the triangle topped by the semicircle of diameter 1 is

$$\frac{1}{2} \cdot \frac{\sqrt{3}}{2} + \frac{1}{2}\pi\left(\frac{1}{2}\right)^2 = \frac{\sqrt{3}}{4} + \frac{\pi}{8}.$$

The area of the lune results from subtracting from this the area of the sector of the larger semicircle,

$$\frac{1}{6}\pi(1)^2 = \frac{1}{6}\pi.$$

So the area of the lune is

$$\frac{\sqrt{3}}{4} + \frac{\pi}{8} - \frac{\pi}{6} = \frac{\sqrt{3}}{4} - \frac{\pi}{24}.$$

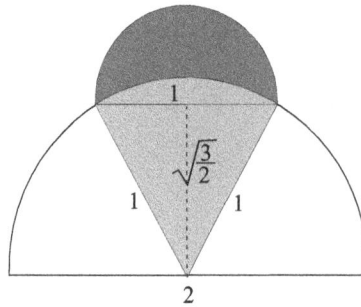

Note that the answer does not depend on the position of the lune on the semicircle.

9. 1997 AHSME Problem 26

Solution (A). Construct a circle with center P and radius PA. Then C lies on the circle, since the angle ACB is half angle APB. Extend $\overline{BP}$ through P to get a diameter $\overline{BE}$. Since $A, B, C,$ and E are concyclic,

$$
\begin{aligned}
AD \cdot CD &= ED \cdot BD \\
&= (PE + PD)(PB - PD) \\
&= (3 + 2)(3 - 2) \\
&= 5.
\end{aligned}
$$

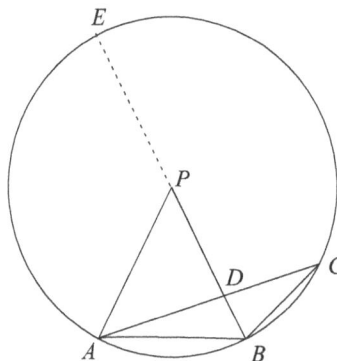

OR

Let E denote the point where $\overline{AC}$ intersects the angle bisector of angle APB. Note that $\triangle PED \sim \triangle CBD$. Hence $DE/2 = 1/DC$ so $DE \cdot DC = 2$. Apply the *Angle Bisector Theorem* to $\triangle APD$ to obtain

$$
\frac{EA}{DE} = \frac{PA}{PD} = \frac{3}{2}.
$$

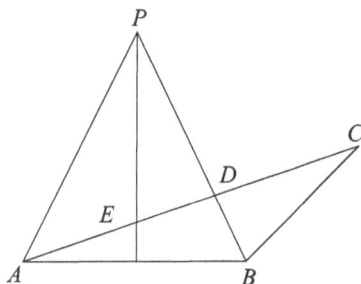

Thus $DA \cdot DC = (DE + EA) \cdot DC = (DE + 1.5DE) \cdot DC = 2.5DE \cdot DC = 5$.

10. 2015 AMC 12A Problem 11

Solution (D). If the smaller circle is in the interior of the larger circle, there are no common tangent lines. If the smaller circle is internally tangent to the larger circle, there is exactly one common tangent line. If the circles intersect at two points, there are exactly two common tangent lines. If the circles are externally tangent, there are exactly three tangent lines. Finally, if the circles do not intersect, there are exactly four tangent lines. Therefore, k can be any of the numbers 0, 1, 2, 3, or 4, which gives 5 possibilities.

11. 2008 AMC 12A Problem 24

Solution (D). Let $C = (0, 0)$, $B = (2, 2\sqrt{3})$, and $A = (x, 0)$ with $x > 0$. Then $D = (1, \sqrt{3})$. Let P be on the positive x-axis to the right of A. Then $\angle BAD = \angle PAD - \angle PAB$. Provided $\angle PAD$ and $\angle PAB$ are not right angles, it follows that

$$
\begin{aligned}
\tan(\angle BAD) &= \tan(\angle PAD - \angle PAB) = \frac{\tan(\angle PAD) - \tan(\angle PAB)}{1 + \tan(\angle PAD)\tan(\angle PAB)} \\
&= \frac{m_{AD} - m_{AB}}{1 + m_{AD}m_{AB}} \frac{\frac{\sqrt{3}}{1-x} - \frac{2\sqrt{3}}{2-x}}{1 + \frac{\sqrt{3}}{1-x} \cdot \frac{2\sqrt{3}}{2-x}} = \frac{\sqrt{3}x}{x^2 - 3x + 8} \\
&= \frac{\sqrt{3}}{\left(\sqrt{x} - \frac{2\sqrt{2}}{\sqrt{x}}\right)^2 + (4\sqrt{2} - 3)} \le \frac{\sqrt{3}}{4\sqrt{2} - 3},
\end{aligned}
$$

with equality when $x = 2\sqrt{2}$. If $\angle PAD = 90°$, then

$$
\tan(\angle BAD) = \cot(\angle PAB) = \frac{1}{2\sqrt{3}} < \frac{\sqrt{3}}{4\sqrt{2} - 3}.
$$

If $\angle PAB = 90°$, then

$$
\tan(\angle BAD) = -\cot(\angle PAD) = \frac{1}{\sqrt{3}} < \frac{\sqrt{3}}{4\sqrt{2} - 3}.
$$

Therefore the largest possible value of $\tan(\angle BAD)$ is $\sqrt{3}/(4\sqrt{2} - 3)$.

OR

Because the circle with diameter $\overline{BD}$ does not intersect the line AC, it follows that $\angle BAD < 90°$. Thus the value of $\tan(\angle BAD)$ is greatest when $\angle BAD$ is greatest. This occurs when A is placed to minimize the size of the circle passing through A, B, and D, so the maximum is attained when that circle is tangent to $\overline{AC}$ at A. For this location of A, the Power of a Point Theorem implies that

$$AC^2 = CB \cdot CD = 4 \cdot 2 = 8, \quad \text{and} \quad AC = \sqrt{8} = 2\sqrt{2}.$$

Because $\frac{CA}{CB} = \frac{CD}{CA}$, it follows that $\triangle CAD$ is similar to $\triangle CBA$. Thus $AB = \sqrt{2}AD$. The Law of Cosines, applied to $\triangle ADC$, gives

$$AD^2 = CD^2 + CA^2 - 2CD \cdot CA \cdot \cos 60° = 12 - 4\sqrt{2}.$$

Let O be the center of the circle passing through A, B, and D. The Extended Law of Sines, applied to $\triangle ABD$ and $\triangle ADC$, gives

$$
\begin{aligned}
2OB &= \frac{AB}{\sin(\angle BDA)} = \frac{AB}{\sin(\angle ADC)} \\
&= \frac{AB \cdot AD}{AC \cdot \sin 60°} = \frac{2AB \cdot AD}{\sqrt{3}AC} \\
&= \frac{2\sqrt{2}AD^2}{2\sqrt{2}\sqrt{3}} = \frac{AD^2}{\sqrt{3}}.
\end{aligned}
$$

Let M be the midpoint of BD. Because $\angle BAD = \frac{1}{2}\angle BOD = \angle BOM$, it follows that

$$
\begin{aligned}
\tan(\angle BAD) &= \tan(\angle BOM) = \frac{MB}{OM} \\
&= \frac{1}{\sqrt{OB^2 - 1}} = \frac{1}{\sqrt{\frac{AD^4}{12} - 1}} \\
&= \frac{\sqrt{3}}{\sqrt{(6 - 2\sqrt{2})^2 - 3}} = \frac{\sqrt{3}}{4\sqrt{2} - 3}.
\end{aligned}
$$

12. 2000 AMC 12 Problem 24

Solution (D). Construct the circle with center A and radius AB. Let F be the point of tangency of the two circles. Draw $\overline{AF}$, and Let E be the point of intersection of $\overline{AF}$ and the given circle. By the *Power of a Point Theorem*, $AD^2 = AF \cdot AE$(see Note below). Let r be the radius of the smaller circle. Since $\overline{AF}$ and $\overline{AB}$ are radii of the larger circle, $AF = AB$ and $AE = AF - EF = AB - 2r$. Because $AD = AB/2$, substitution into the first equation yields

$$(AB/2)^2 = AB(AB - 2r),$$

or, equivalently, $\frac{r}{AB} = \frac{3}{8}$. Points A, B, and C are equidistant from each other, so $\widehat{BC} = 60°$ and thus the circumference of the larger circle is $6\cdot$ (length of $\widehat{BC}$)$= 6 \cdot 12$. Let c be the circumference of the smaller circle. Since the circumferences of the two circles are in the same ratio as their radii, $\frac{c}{72} = \frac{r}{AB} = \frac{3}{8}$. Therefore

$$c = \frac{3}{8}(72) = 27.$$

Note. From any exterior point P, a secant PAB and a tangent PT are drawn. Consider triangles PAT and PTB. They have a common angle P. Since angle ATP and PBT intercept the same arc $\overset{\frown}{AT}$, they are congruent. Therefore triangles PAT and PTB are similar, and it follows that $PA/PT = PT/PB$ and $PA \cdot PB = PT^2$. The number PT^2 is called *the power of the point P* with respect to the circle. Intersecting secants, tangents, and chords, paired in any manner create various cases of this theorem, which is sometimes called *Crossed Chords*.

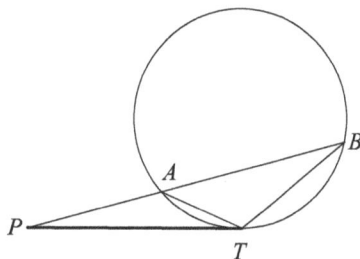

OR

The construction given in the problem is the classic way to construct an equilateral triangle, $\triangle ABC$, with side length AB. The arc length BC is one-sixth the circumference of the circle with radii AB, so

$$12 = \frac{1}{6}(2\pi \cdot AB) \quad AB = \frac{36}{\pi}.$$

Let O be the center of the circle, r be the radius, and D be the midpoint of AB. The symmetry of the region implies that $\overline{OD}$ is a perpendicular bisector of AB. Construct $\overline{AE}$, the line segment passing through O and intersecting the arc $\overline{BC}$ at E. Then $AE = AB$ and $r = OE = OD$, so in the right $\triangle ADO$ we have

$$
\begin{aligned}
\frac{36}{\pi} = AE &= OE + AO = OE + \sqrt{AD^2 + DO^2} \\
&= r + \sqrt{\left(\frac{18}{\pi}\right)^2 + r^2}.
\end{aligned}
$$

Hence

$$
\begin{aligned}
0 &= \left(\frac{36}{\pi} - r\right)^2 - \left(\left(\frac{18}{\pi}\right)^2 + r^2\right) \\
&= 3\left(\frac{18}{\pi}\right)^2 - \frac{72}{\pi}r.
\end{aligned}
$$

and

$$
r = \frac{\pi}{72} \cdot 3\left(\frac{18}{\pi}\right)^2 = \frac{27}{2\pi}.
$$

The circumference of the circle is $2\pi r = 2\pi \left(\frac{27}{2\pi}\right) = 27$.

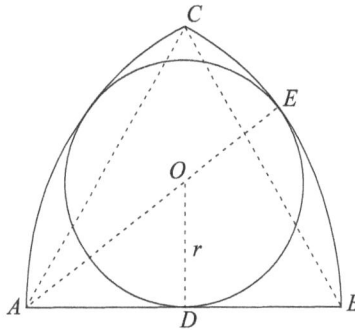

32.2.2 Polygons and Solid Geometry

1. 1993 AHSME Problem 14

Solution (B). Draw $\overline{CE}$. Since $EA = BC$ and $\angle A = \angle B$, it follows that $ABCE$ is an isosceles trapeziod. Let F be the foot of the perpendicular from A to $\overline{CE}$, and G be the foot of the perpendicular from B to $\overline{CE}$. Then $EF = CG$. Since $\angle GBC = 30°$, we have

$$
CG = \frac{1}{2}(BC) = 1 \quad \text{and} \quad BG = \frac{\sqrt{3}}{2}(BC) = \sqrt{3}.
$$

Now

$$
CE = CG + GF + FE = 1 + 2 + 1 = 4,
$$

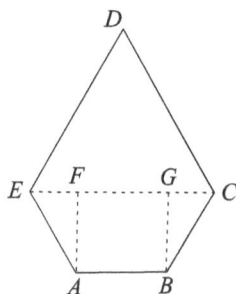

so CDE is an equilateral triangle. Thus,

$$[ABCE] = \frac{1}{2}(BG)(AB + CE) = \frac{1}{2}\sqrt{3}(2+4) = 3\sqrt{3},$$

and

$$[CDE] = \frac{\sqrt{3}}{4}(CE)^2 = \frac{\sqrt{3}}{4}(16) = 4\sqrt{3}.$$

Therefore,

$$[ABCDE] = [ABCE] + [CDE] = 7\sqrt{3}.$$

OR

Draw $\overline{HI}$ where H is the midpoint of $\overline{ED}$ and I is the midpoint of $\overline{CD}$.Then $ABCIHE$ is a regular hexagon and $\triangle HDI$ is congruent to any of the six equilateral triangles of side 2 that make up $ABCIHE$. Thus, the area of $ABCDE$ is the sum of the areas of 7 equilateral triangles of side 2, so it is $7\left(2^2\frac{\sqrt{3}}{4}\right) = 7\sqrt{3}$.

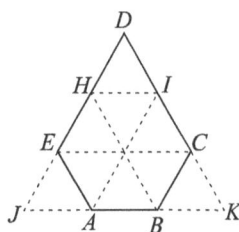

Note. A glance at the diagram for the previous solution shows that the area of $ABCDE$ is tiled by 7 of the 9 congruent equilateral triangles that tile equilateral triangle DJK. Since $JK = 3\cdot 2 = 6$, the area of $ABCDE$ can by computed as

$$\frac{7}{9}\left(6^2\frac{\sqrt{3}}{4}\right) = 7\sqrt{3}.$$

OR

Extend $\overline{EA}$ and $\overline{CB}$ to meet at P. Since $\angle ABP = \angle BAP = 60°$, $\triangle ABP$ is equilateral as is $\triangle ECP$, and since $EC = CP = CB + BP = 4 = CD = DE$, $\triangle ECD$ is also equilateral. Now,

$$[ABCDE] = ([ECD] + [ECP]) - [ABP]$$
$$= 2(\frac{\sqrt{3}}{4}4^2) - \frac{\sqrt{3}}{4}2^2 = 7\sqrt{3}.$$

OR

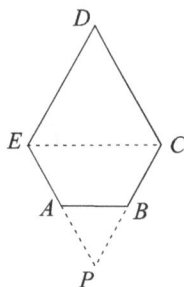

Construct P, and note that $\triangle PAB$ and $\triangle PEC$ are both equilateral, as above. Thus, $\angle AEC = 60°$. Because $\triangle ABC$ is isosceles, $\angle BAC = 30°$, so $\angle CAE = 90°$. Therefore, $\triangle CEA$ is a $30° - 60° - 90°$ triangle, so $CE = 4, AC = 2\sqrt{3}$ and the altitude from B to $\overline{AC}$ has length 1. Finally, $\triangle CDE$ is equilateral since $CE = 4 = CD = DE$.

Hence,

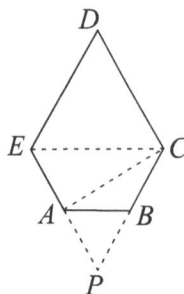

$$[ABCDE] = ([ABC] + [CEA]) + [CDE]$$
$$= \frac{1}{2} \cdot 2\sqrt{3} \cdot 1 + \frac{1}{2} \cdot 2 \cdot 2\sqrt{3} + \frac{\sqrt{3}}{4} \cdot 4^2 = 7\sqrt{3}.$$

2. 1995 AHSME Problem 17

Solution (E). Let O be the center of the circle. Since the sum of the interior in any $n-$gon is $(n - 2)180°$, the sum of the angles in $ABCDO$ is $540°$. Since $\angle ABC = \angle BCD = 108°$ and $\angle OAB = \angle ODC = 90°$,it follows that the measure of $\angle AOD$,and thus the measure of minor arc AD, equal $144°$.

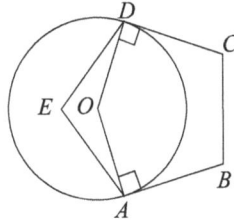

OR

Draw $\overline{AD}$.Since $\triangle AED$ is isosceles with $\angle AED = 108°$,it follows that $\angle EDA = \angle EAD = 36°$.Consequently,$\angle ADC = 108° - 36° = 72°$.Since $\angle ADC$ is a tangent-chord angle for the arc in question,the measure of the arc is $2(72°) = 144°$.

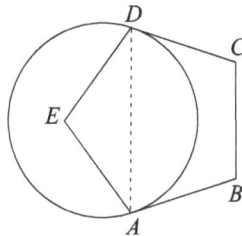

OR

Let O be the center of the circle,and extend $\overline{DC}$ and $\overline{AB}$ to meet at F.Since $\angle DCB = 108°$ and $\triangle BCF$ is isosceles,it follows that $\angle AFD = [180° - 2(180° - 108°)] = 36°$. Since$\angle ODF = \angle OAF = 90°$, in quadrilateral $OAFD$ we have angles AOD and AFD supplementary, so the measures of the angle AOD and the minor arc AD are $180° - 36° = 144°$.

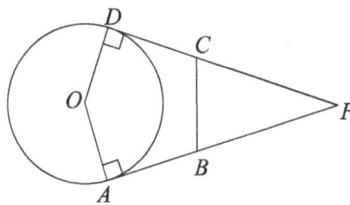

Note : A circle can be drawn tangent to two intersecting lines at given points on those lines if and only if those points are equidistant from the point of intersection of the lines.

3. 1992 AHSME Problem 20

Solution (D). Partition the n-pointed regular star into the regular n-gon $B_1B_2\cdots B_n$ and n isosceles triangles congruent to $\triangle B_1A_2B_2$. The sum of the star's interior angles is the sum of the interior angles of the regular n-gon plus the sum of the interior angles of the n triangles, which is

$$(n - 2)180° + n180° = (2n - 2)180°.$$

Since the interior angles of the star consist of n angles congruent to A_1 and n angles congruent to $360° - B_1$.

$$
\begin{aligned}
(2n-2)180° &= n\angle A_1 + n(360° - \angle B_1) \\
n(\angle B_1 - \angle A_1) &= 2 \cdot 180°.
\end{aligned}
$$

Since $\angle B_1 - \angle A_1 = 10°$, $n = 36$.

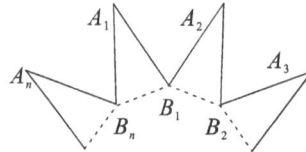

Note. In general, the sum of the interior angles of *any* N-sided simple closed polygon, convex or not, is $(N-2)180°$.

OR

Extend $\overline{A_1 B_1}$ and $\overline{A_2 B_2}$ to their intersection C_1, $\overline{A_2 B_2}$ and $\overline{A_3 B_3}$ to their intersection C_2,..., $\overline{A_n B_n}$ and $\overline{A_1 B_1}$ to their intersection C_n. Note that the n triangles $A_1 C_1 A_2$, $A_2 C_2 A_3$, ..., $A_n C_n A_1$ can be translated (moved without rotation) so all n points, C_i, coincide, and the angles with these vertices form a $360°$ angle. Since each $\angle C_i = 10°$, $n \cdot 10° = 360°$, so $n = 36$.

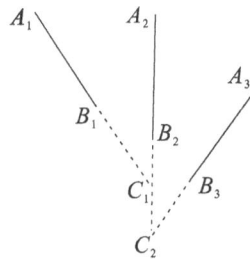

OR

Let the measure of the acute angle at each A_i be $A°$ and the measure of the acute angle at each B_i be $B°$. Let O be the center of the star. The $2n$ triangles $A_i B_i O$ and $A_{i+1} B_i O$ (where $A_{n+1} = A_1$) are all congruent triangles whose angles measure

$$
\frac{A°}{2}, \quad 180° - \frac{B°}{2} \quad \text{and} \quad \frac{360°}{2n}.
$$

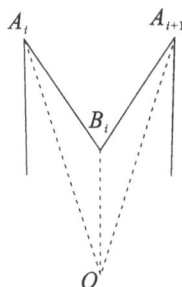

Therefore

$$\frac{A^\circ}{2} + \left(180^\circ - \frac{B^\circ}{2}\right) + \frac{360^\circ}{2n} = 180^\circ$$

or

$$\frac{360^\circ}{2n} = \frac{B^\circ - A^\circ}{2} = \frac{10^\circ}{2},$$

so $n = 36$.

OR

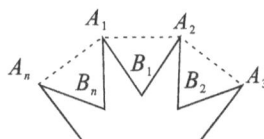

Draw the segments $\overline{A_1 A_2}, \overline{A_2 A_3}, ..., \overline{A_n A_1}$ to form a regular n-gon. In $\triangle A_1 A_2 B_1$, $\angle B_1 A_1 A_2 = \angle B_1 A_2 A_1$, so

$$2\angle B_1 A_1 A_2 + \angle A_1 B_1 A_2 = 180^\circ.$$

The measure of an interior angle of regular polygon $A_1 A_2 \cdots A_n$ is $\frac{n-2}{n} 180^\circ$, so

$$\begin{aligned} \angle A_1 A_2 B_1 + \angle B_1 A_2 B_2 + \angle B_2 A_2 A_3 &= \angle A_1 A_2 A_3, \\ 2\angle B_1 A_1 A_2 + \angle B_1 A_2 B_2 &= \frac{n-2}{n} \cdot 180^\circ. \end{aligned}$$

Subtract the second equation to the first one:

$$\angle A_1 B_1 A_2 - \angle B_1 A_2 B_2 = 180^\circ - \left(\frac{n-2}{n} 180^\circ\right) = \frac{2}{n} \cdot 180^\circ.$$

But $\angle A_1 B_1 A_2 - \angle B_1 A_2 B_2 = 10^\circ$, so $n = 36$.

4. 2014 AMC 10A Problem 23

Solution (C). Without loss of generality, assume that the rectangle has dimensions 3 by $\sqrt{3}$. Then the fold has length 2, and the overlapping areas are equilateral triangles each with area $\frac{\sqrt{3}}{4} \cdot 2^2$. The new shape has area $3\sqrt{3} - \frac{\sqrt{3}}{4} \cdot 2^2 = 2\sqrt{3}$, and the desired ratio is $2\sqrt{3} : 3\sqrt{3} = 2 : 3$.

5. 2015 AMC 10B Problem 22

Solution (D). Triangles AGB and CHJ are isosceles and congruent, so $AG = HC = HJ = 1$. Triangles AFG and BGH are congruent, so $FG = GH$. Triangles AGF, AHJ and ACD are similar, so $\frac{a}{b} = \frac{a+b}{c} = \frac{2a+b}{d}$.

Since $a = c = 1$ the first equation becomes $\frac{1}{b} = \frac{1+b}{1}$ and $b^2 + b - 1 = 0$ or $b = \frac{-1+\sqrt{5}}{2}$. Substituting this in the second equation gives $d = \frac{1+\sqrt{5}}{2}$, so $b + c + d = 1 + \sqrt{5}$.

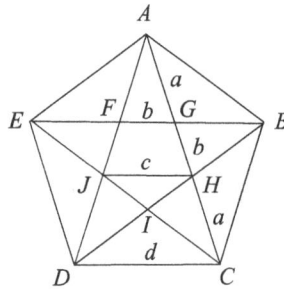

6. 2002 AMC 12B Problem 24

Solution (E). We have

$$\text{Area}(ABCD) \le \frac{1}{2}AC \cdot BD,$$

with equality if and only if $AC \perp BD$. Since

$$
\begin{aligned}
2002 = \text{Area}(ABCD) \ &\le\ \frac{1}{2}AC \cdot BD \\
&\le\ \frac{1}{2}(AP + PC)(BP + PD) = \frac{52 \cdot 77}{2} = 2002,
\end{aligned}
$$

it follows that the diagonals AC and BD are perpendicular and intersect at P. Thus $AB = \sqrt{24^2 + 32^2} = 40$, $BC = \sqrt{28^2 + 32^2} = 4\sqrt{113}$, $CD = \sqrt{28^2 + 45^2} = 53$, and $DA = \sqrt{24^2 + 45^2} = 51$. The perimeter of $ABCD$ is therefore

$$144 + 4\sqrt{113} = 4(36 + \sqrt{113}).$$

7. 2006 AMC 10B Problem 23

Solution (D). Partition are quadrilateral into two triangles and let the areas of the triangles be R and S as shown. Then the required area is $T = R + S$.

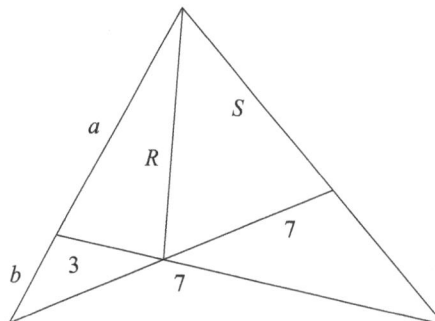

Let a and b, respectively, be the bases of the triangles with areas R and 3, as indicated. If two triangles have the same altitude, then the ratio of their areas is the same as the ratio of their bases. Thus

$$\frac{a}{b} = \frac{R}{3} = \frac{R+S+7}{3+7}, \quad \text{so} \quad \frac{R}{3} = \frac{T+7}{10}.$$

Similarly,

$$\frac{S}{7} = \frac{S+R+3}{7+7}, \quad \text{so} \quad \frac{S}{7} = \frac{T+3}{14}.$$

Thus

$$T = R + S = 3\left(\frac{T+7}{10}\right) + 7\left(\frac{T+3}{14}\right).$$

From this we obtain

$$10T = 3(T+7) + 5(T+3) = 8T + 36,$$

and it follows that $T = 18$.

8. 1994 AHSME Problem 11

Solution (D). The sum of the surface areas of three cubes is $6 + 24 + 54 = 84$. We minimize the surface area by attaching each cube to the other two along an entire face of the smaller cube. The figure shows how this is possible. Each attachment subtracts twice the area of a face of the smaller cube from the total. The remaining surface area is

$$84 - 2(1) - 2(1) - 2(4) = 72.$$

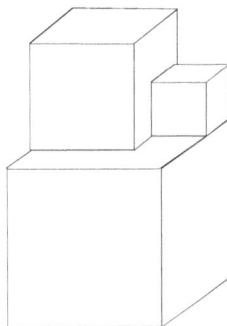

9. 1996 AHSME Problem 28

Solution (C). Let h be the required distance. Find the volume of pyramid $ABCD$ as a third of the area of a triangular base times the altitude to that base in two different ways, and equate these volumes. Use the altitude $\overline{AD}$ to $\triangle BCD$ to find that the volume is 8. Next, note that h is the length of the altitude of the pyramid form D to $\triangle ABC$. since the sides of $\triangle ABC$ are $5,5$, and $4\sqrt{2}$, by the Pythagorean Theorem the altitude to the side of length $4\sqrt{2}$ is $a = \sqrt{17}$. Thus, the area of $\triangle ABC$ is $2\sqrt{34}$, and the volume of the pyramid is $2\sqrt{34}h/3$. Equating the volumes yields

$$2\sqrt{34}h/3 = 8. \quad \text{and thus} \quad h = 12/\sqrt{34} \approx 2.1.$$

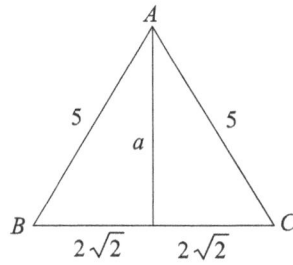

OR

Imagine the parallelepiped embedded in a coordinate system as shown in the diagram. The equation for the plane (in intercept form) is $\frac{x}{4} + \frac{y}{4} + \frac{z}{3} = 1$. Thus, it can be expressed as $3x + 3y + 4z - 12 = 0$. The formula for the distence d form a point (a, b, c) to the plane $Rx + Sy + Tz + U = 0$ is given by

$$d = \frac{|Ra + Sb + Tc + U|}{\sqrt{R^2 + S^2 + T^2}}.$$

which in this case is

$$\frac{|-12|}{\sqrt{3^2 + 3^2 + 4^2}} = \frac{12}{\sqrt{34}} \approx 2.1.$$

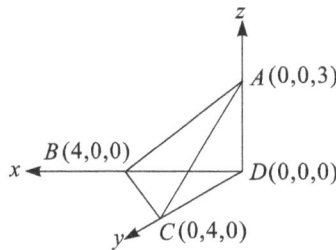

10. 2004 AMC 10A Problem 25 or 2004 AMC 12A Problem 22

Solution (B). Let A, B, C and E be the centers of the three small spheres and the large spheres, respectively.

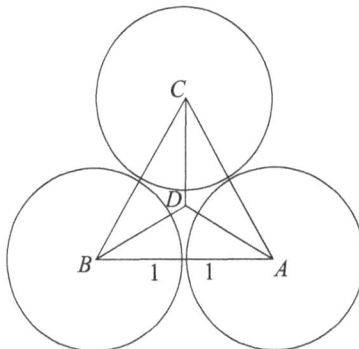

Then $\triangle ABC$ is equilateral with side length 2. Let D be the intersection of the medians of $\triangle ABC$. Then E is directly above D. Because $AE = 3$ and $AD = 2\sqrt{3}/3$, it follows that

$$DE = \sqrt{3^2 - (\frac{2\sqrt{3}}{3})^2} = \frac{\sqrt{69}}{3}.$$

Because D is 1 unit above the plane and the top of the larger sphere is 2 units above E, the distance from the plane to the top of the larger sphere is

$$3 + \frac{\sqrt{69}}{3}.$$

11. **2008 AMC 10B Problem 24**

Solution (C). let M be on the same side of the line BC as A, such that $\triangle BMC$ is equilateral. Then $\triangle ABM$ and $\triangle MCD$ are isosceles with $\angle ABM = 10°$ and $\angle MCD = 110°$. Hence $\angle AMB = 85°$ and $\angle CMD = 35°$. Therefore

$$\angle AMD = 360° - \angle AMB - \angle BMC - \angle CMD = 360° - 85° - 60° - 35° = 180°$$

It follows that M lies on $\overline{AD}$ and $\angle BAD = \angle BAM = 85°$

OR

Let $\triangle ABO$ be equilateral as shown.

Then

$$\angle OBC = \angle ABC - \angle ABO = 70° - 60° = 10°.$$

Because $\angle BCD = 170°$ and $OB = BC = CD$, the quadrilateral $BCDO$ is a parallelogram. Thus $OD = BC = AO$ and $\triangle AOD$ is isosceles. Let $\alpha = \angle ODA = \angle OAD$. The sum of the interior angles of $ABCD$ is $360°$, so we have

$$360 = (\alpha + 60) + 70 + 170 + (\alpha + 10) \quad \text{and} \quad \alpha = 25.$$

Thus $\angle DAB = 60° + \alpha = 85°$

12. 2011 AMC 10A Problem 24

Solution (D). Let the tetrahedra be T_1 and T_2, and let R be their intersection. Let squares $ABCD$ and $EFGH$, respectively, be the top and bottom faces of the unit cube, with E directly under A and F directly under B. Without loss of generality, T_1 has vertices $A, C, F,$ and H, and T_2 has vertices $B, D, E,$ and G. One face of T_1 is $\triangle ACH$, which intersects edges of T_2 at the midpoints $J, K,$ and L of $\overline{AC}, \overline{CH},$ and $\overline{HA}$, respectively. Let S be the tetrahedron with vertices $J, K, L,$ and D. Then S is similar to T_2 and is contained in T_2, but not in R. The other three faces of T_1 each cut off from T_2 a tetrahedron congruent to S. Therefore the volume of R is equal to the volume of T_2 minus four times the volume of S.

A regular tetrahedron of edge length s has base area $\frac{\sqrt{3}}{4}s^2$ and altitude $\frac{\sqrt{6}}{3}s$, so its volume is $\frac{1}{3}\left(\frac{\sqrt{3}}{4}s^2\right)\left(\frac{\sqrt{6}}{3}s\right) = \frac{\sqrt{2}}{12}s^3$. Because the edges of tetrahedron T_2 are face diagonals of the cube, T_2 has edge length $\sqrt{2}$. Because J and K are centers of adjacent faces of the cube, tetrahedron S has edge length $\frac{\sqrt{2}}{2}$. Thus the volume of R is

$$\frac{\sqrt{2}}{12}\left((\sqrt{2})^3 - 4\left(\frac{\sqrt{2}}{2}\right)^3\right) = \frac{1}{6}.$$

OR

Let T_1 and T_2 be labeled as in the previous solution. The cube is partitioned by T_1 and T_2 into 8 tetrahedra congruent to $DJKL$ (one for every vertex of the cube), 12 tetrahedra congruent to $AJLD$ (one for every edge of the cube), and the solid $T_1 \cap T_2$. Because the based AJL and JLK are equilateral triangles with the same area, and the altitudes to vertex D of the tetrahedra $AJLD$ and $DJKL$ are the same, it follows that the volumes of $AJLD$ and $DJKL$ are equal. Moreover,

$$\text{Volume}(AJLD) = \frac{1}{3}\text{Area}(ALD) \cdot h_J,$$

where $h_J = \frac{1}{2}$ is the distance from J to the face ALD, and $\text{Area}(ALD) = \frac{1}{4}$. Therefore $\text{Volume}(AJLD) = \frac{1}{3}\cdot\frac{1}{4}\cdot\frac{1}{2} = \frac{1}{24}$, and thus the volume of $T_1 \cap T_2$ is equal to $1 - (8+12)\cdot\frac{1}{24} = \frac{1}{6}$.

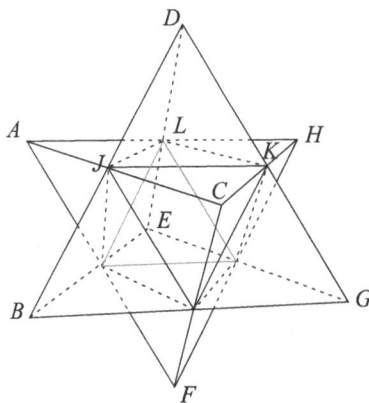

32.3 Number Theory

1. 2014 AMC 10A Problem 20 or 2014 AMC 12A Problem16

Solution (D). By direct multiplication, $8 \cdot 888\ldots8 = 7111\ldots104$, where the product has 2 fewer ones than the number of digits in $888\ldots8$. Because $7 + 4 = 11$, the product must have $1000 - 11 = 989$ ones, so $k - 2 = 989$ and $k = 991$.

2. 2002 AMC 10A Problem 14 or 2002 AMC 12A Problem 12

Solution (B). Let p and q be two primes that are roots of $x^2 - 63x + k = 0$. Then

$$x^2 - 63x + k = (x - p)(x - q) = x^2 - (p+q)x + p \cdot q,$$

so $p + q = 63$ and $p \cdot q = k$. Since 63 is odd, one of the primes must be 2 and the other 61. Thus there is exactly one possible value for k, namely $k = p \cdot q = 2 \cdot 61 = 122$.

3. 1990 AHSME Problem 11

Solution (C). Let N be a positive integer and d a divisor of N. Then N/d is also a divisor of N. Thus the divisors of N occur in pairs d, N/d and these two divisors will be distinct unless N is a perfect square and $d = \sqrt{N}$. It follows that N has an odd number of divisors if and only if N is a perfect square. There are 7 perfect squares, $1^2, 2^2, 3^2, 4^2, 5^2, 6^2, 7^2$, among the numbers $1, 2, 3, \ldots, 50$.

Note. If $N > 1$ is an integer then

$$N = p_1^{r_1} \cdot p_2^{r_2} \cdot \ldots \cdot p_k^{r_k}$$

where p_i is the ith prime. The divisors of N are those

$$d = p_1^{s_1} \cdot p_2^{s_2} \cdot \ldots \cdot p_k^{s_k}$$

with $0 \le s_i \le r_i$ for all i. Thus, N has

$$(r_1 + 1) \cdot (r_2 + 1) \cdot \ldots \cdot (r_k + 1)$$

divisors, a product which will be an odd number only when each r_i is even. Each r_i is even if and only if N is a perfect square.

4. 2002 AMC 12A Problem 20

Solution (C). Since $0.\overline{ab} = ab/99$, the denominator must be a factor of $99 = 3^2 \cdot 11$. The factor of 99 are 1,3,9,11,33, and 99. Since a and b are not both nine, the reduced denominator cannot be 1. To see that each of the other denominators is possible, note that

$$0.\overline{01} = \frac{1}{99}, 0.\overline{03} = \frac{1}{33}, 0.\overline{09} = \frac{1}{11}, 0.\overline{11} = \frac{1}{9}, 0.\overline{33} = \frac{1}{3}$$

5. 2013 AMC 12B Problem 15

Solution (B). The prime factorization of 2013 is $3 \cdot 11 \cdot 61$. There must be a factor of 61 in the numerator, so $a_1 \geq 61$. Since $a_1!$ will have a factor of 59 and 2013 does not, there must be a factor of 59 in the denominator, and $b_1 \geq 59$. Thus $a_1 + b_1 \geq 120$, and this minimum value can be achieved only if $a_1 = 61$ and $b_1 = 59$. Furthermore, this minimum value is attainable because

$$2013 = \frac{(61!)(11!)(3!)}{(59!)(10!)(5!)}.$$

Thus $|a_1 - b_1| = a_1 - b_1 = 61 - 59 = 2$.

6. 2012 AMC 12B Problem 11

Solution (C). First assume $B = A - 1$. By the definition of number bases,

$$A^2 + 3A + 2 + 4(A - 1) + 3 = 6(A + A - 1) + 9.$$

Simplifying yields $A^2 - 5A - 2 = 0$, which has no integer solutions.

Next assume $B = A + 1$. In this case

$$A^2 + 3A + 2 + 4(A + 1) + 3 = 6(A + A + 1) + 9,$$

which simplifies to $A^2 - 5A - 6 = (A - 6)(A + 1) = 0$. The only positive solution is $A = 6$. Letting $A = 6$ and $B = 7$ in the original equation produces $132_6 + 43_7 = 69_{13}$, or $56 + 31 = 87$, which is true. The required sum is $A + B = 13$.

7. 1987 AHSME Problem 16

Solution (D). The fact that VVW follows VYX establishes $W \leftrightarrow 0$ and $X \leftrightarrow 4$. Since VYX follows VYZ, X immediately follows Z, and this establishes that $Z \leftrightarrow 3$. That VVW follows VYX also establishes that V follows Y, so $Y \leftrightarrow 1$ and $V \leftrightarrow 2$. Hence $XYZ \leftrightarrow 413_5 = 108_{10}$.

8. 1992 AHSME Problem 23

Solution (E). Partition $F = \{1, 2, 3, ..., 50\}$ into seven subsects, $F_0, F_1, ..., F_6$, so that all the elements of F_i leave a remainder of i when divided by 7:

$$
\begin{aligned}
F_0 &= \{7, 14, 21, 28, 35, 42, 49\}, \\
F_1 &= \{1, 8, 15, 22, 29, 36, 43, 50\}, \\
F_2 &= \{2, 9, 16, 23, 30, 37, 44\}, \\
F_3 &= \{3, 10, 17, 24, 31, 38, 45\}, \\
F_4 &= \{4, 11, 18, 25, 32, 39, 46\}, \\
F_5 &= \{5, 12, 19, 26, 33, 40, 47\} \\
F_6 &= \{6, 13, 20, 27, 34, 41, 48\}.
\end{aligned}
$$

Note that S can contain at most one member of F_0, but that if S contains some member of any of the other subsets, then it can contain all of the members of that subset. Also, S cannot contain members of both F_1 and F_6, or both F_2 and F_5, or both F_3 and F_4. Since F_1 contains 8 members and each of other subsets contains 7 members, the largest subset, S, can be constructed by selecting one member of F_0, all the members of F_1, all the members of either F_2 or F_5, and all of the members of either F_3 or F_4. Thus the largest subset, S, contains $1 + 8 + 7 + 7 = 23$ elements.

9. 1982 AHSME Problem 26

Soluton (B). If $n^2 = (ab3c)_8$, let $n = (de)_8$. Then $n^2 = (8d + e)^2 = 64d^2 + 8(2de) + e^2$. Thus, the 3 in $ab3c$ is the first digit (in base 8) of the sum of the eights digit of e^2 (in base 8) and the units digit of $(2de)$ (in base 8). The latter is even, so the former is odd. The entire table of base 8 representations of squares of base 8 digits appears below.

e	1	2	3	4	5	6	7
e^2	1	4	11	20	31	44	61

The eights digit of e^2 is odd only if e is 3 or 5; in either case c, which is the units digit of e^2, is 1. (In fact, there are three choices for n: $(33)_8$, $(73)_8$ and $(45)_8$. The squares are $(1331)_8$, $(6631)_8$ and $(2531)_8$, respectively.)

OR

We are given

$$n^2 = (ab3c)_8 = 8^3 a + 8^2 b + 8 \cdot 3 + c.$$

If n is even, n^2 is divisible by 4, and its remainder upon division by 8 is 0 or 4. If n is odd, say $n = 2k+1$, then $n^2 = 4(k^2 + k) + 1$, and since $k^2 + k = k(k+1)$ is always even, n^2 has remainder 1 upon division by 8. Thus in all cases, the only possible values of c are 0, 1, or 4. If $c = 0$, then $n^2 = 8(8K + 3)$, an impossibility since 8 is not a square. If $c = 4$, then $n^2 = 4(8L + 7)$ another impossibility since no odd squares have the form $8L + 7$. Thus $c = 1$.

10. 2007 AMC 12B Problem 24

Solution (A). Let $u = a/b$. Then the problem is equivalent to finding all positive rational numbers u such that

$$u + \frac{14}{9u} = k$$

for some integer k. This equation is equivalent to $9u^2 - 9uk + 14 = 0$. whose solutions are

$$u = \frac{9k \pm \sqrt{81k^2 - 504}}{18} = \frac{k}{2} \pm \frac{1}{6}\sqrt{9k^2 - 56}.$$

Hence u is rational if and only if $\sqrt{9k^2 - 56}$ is rational, which is true if and only if $9k^2 - 56$ is a perfect square. Suppose that $9k^2 - 56 = s^2$ for some positive integer s. Then $(3k-s)(3k+s) = 56$. The only factors of 56 are $1, 2, 4, 7, 8, 14, 28, 56$, so $(3k - s, 3k + s)$ is one of the ordered pairs $(1, 56), (2, 28), (4, 14), (7, 8)$. The cases $(1, 56)$ and $(7, 8)$ yield no integer solutions. The cases $(2, 28)$ and $(4, 14)$ yield $k = 5$ and $k = 3$, respectively, if $k = 5$, then $u = 1/3$ or $u = 14/3$. If $k = 3$, then $u = 2/3$ or $u = 7/3$. Therefore there are four pairs (a, b) that satisfy the given conditions, namely $(1, 3), (2, 3), (7, 3)$, and $(14, 3)$.

OR

Rewrite the equation
$$\frac{a}{b} + \frac{14b}{9a} = k$$
in two different forms. First, multiply both sides by b and subtract a to obtain
$$\frac{14b^2}{9a} = bk - a.$$

Because a, b and k are integers, $14b^2$ must be a multiple of a, and because a and b have no common factors greater than 1, if follows that 14 is divisible by a. Next, multiply both sides of the original equation by $9a$ and subtract $14b$ to obtain
$$\frac{9a^2}{b} = 9ak - 14b.$$

This shows that $9a^2$ is a multiple of b, so 9 must be divisible by b. Thus if (a, b) is a solution, then $b = 1, 3$, or 9, and $a = 1, 2, 7$, or 14. This gives a total of twelve possible solutions (a, b), each of which can be checked quickly. The only such pairs for which
$$\frac{a}{b} + \frac{14b}{9a}$$
is an integer are when (a, b) is $(1, 3), (2, 3), (7, 3)$, or $(14, 3)$.

11. **2010 AMC 12B Problem 25**

   Solution (D). Observe that $2010 = 2 \cdot 3 \cdot 5 \cdot 67$. Let $P = \prod_{n=2}^{5300} \text{pow}(n) = 2^a \cdot 3^b \cdot 5^c \cdot 67^d \cdot Q$ where Q is relatively prime with 2, 3, 5, and 67. The largest power of 2010 that divides P is equal to 2010^m where $m = \min(a, b, c, d)$.

   By definition $\text{pow}(n) = 2^k$ if and only if $n = 2^k$. Because
   $$2^{12} = 4096 < 5300 < 8912 = 2^{13},$$
   it follows that
   $$a = 1 + 2 + \cdots + 12 = \frac{12 \cdot 13}{2} = 78.$$

Similarly, $\text{pow}(n) = 67$ if and only if $n = 67N$ and the largest prime dividing N is smaller than 67. Because $5300 = 79 \cdot 67 + 7$ and $71, 73$, and 79 are the only primes p in the range $67 < p \le 79$; it follows that for $n \le 5300$, $\text{pow}(n) = 67$ if and only if

$$n \in \{67k : 1 \le k \le 79\} \setminus \{67^2, 67 \cdot 71, 67 \cdot 73, 67 \cdot 79\}.$$

Because $67^2 < 5300 < 2 \cdot 67^2$, the only $n \le 5300$ for which $\text{pow}(n) = 67^k$ with $k \ge 2$, is $n = 67^2$. Therefore

$$d = 79 - 4 + 2 = 77.$$

If $n = 2^j \cdot 3^k$ for $j \ge 0$ and $k \ge 1$ then $\text{pow}(n) = 3^k$. Moreover, if $0 \le j \le 2$ and $1 \le k \le 6$, or if $0 \le j \le 1$ and $k = 7$; then $n = 2^j \cdot 3^k \le 2 \cdot 3^7 = 4374 < 5300$.

Thus

$$b \ge 3(1 + 2 + \cdots 6) + 7 + 7 = 3 \cdot 21 + 14 = 77.$$

If $n = 2^i \cdot 3^j \cdot 5^k$ for $i, j \ge 0$ and $k \ge 1$, then $\text{pow}(n) = 5^k$. Moreover, if $2^i \cdot 3^j \in \{1, 2, 3, 2^2, 2 \cdot 3, 2^3, 3^2, 2^2 \cdot 3\}$ and $1 \le k \le 3$, or if $2^i \cdot 3^j \in \{1, 2, 3, 2^2, 2 \cdot 3, 2^3\}$ and $k = 4$, or if $2^i \cdot 3^j = 1$ and $k = 5$; then $n = 2^i \cdot 3^j \cdot 5^k \le 8 \cdot 5^4 = 5000 < 5300$.

Thus

$$c \ge 8(1 + 2 + 3) + 6 \cdot 4 + 5 = 77.$$

Therefore $m = d = 77$.

12. 2015 AMC 12B Problem 23

Solution (B). Because the volume and surface area are numerically equal, $abc = 2(ab + ac + bc)$. Rewriting the equation as $ab(c - 6) + ac(b - 6) + bc(a - 6) = 0$ shows that $a \le 6$. The original equation can also be written as $(a - 2)bc - 2ab - 2ac = 0$. Note that if $a = 2$, this becomes $b + c = 0$, and there are no solutions. Otherwise, multiplying both sides by $a - 2$ and adding $4a^2$ to both sides gives $[(a - 2)b - 2a][(a - 2)c - 2a] = 4a^2$. Consider the possible values of a.

$a = 1$: $(b + 2)(c + 2) = 4$ There are no solutions in positive integers.

$a = 3$: $(b - 6)(c - 6) = 36$ The 5 solutions for (b, c) are $(7, 42), (8, 24), (9, 18), (10, 15)$, and $(12, 12)$.

$a = 4$: $(2b - 8)(2c - 8) = 64 \to (b - 4)(c - 4) = 16$ The 3 solutions for (b, c) are $(5, 20), (6, 12)$, and $(8, 8)$.

$a = 5$: $(3b - 10)(3c - 10) = 100$ Each factor must be congruent to 2 modulo 3, so the possible pairs of factors are $(2, 50)$ and $(5, 20)$. The solutions for (b, c) are $(4, 20)$ and $(5, 10)$, but only $(5, 10)$ has $a \le b$.

$a = 6$: $(4b - 12)(4c - 12) = 144 \to (b - 3)(c - 3) = 9$ The solutions for (b, c) are $(4, 12)$ and $(6, 6)$, but only $(6, 6)$ has $a \le b$.

Thus in all there are 10 ordered triples (a, b, c): $(3, 7, 42)$, $(3, 8, 24)$, $(3, 9, 18)$, $(3, 10, 15)$, $(3, 12, 12)$, $(4, 5, 20)$, $(4, 6, 12)$, $(4, 8, 8)$, $(5, 5, 10)$, and $(6, 6, 6)$.

OR

Because the volume and surface area are numerically equal, $abc = 2(ab + ac + bc)$. Rewriting the equation as

$$\frac{1}{2} = \frac{1}{a} + \frac{1}{b} + \frac{1}{c}$$

shows that

$$\frac{1}{6} \le \frac{1}{a} < \frac{1}{2},$$

so $3 \le a \le 6$. Also,

$$\frac{a-2}{2a} = \frac{1}{2} - \frac{1}{a} = \frac{1}{b} + \frac{1}{c} \le \frac{2}{b},$$

so $a \le b \le \frac{2a}{a-2}$. Solving the original equation for c gives $c = \frac{2ab}{(a-2)b-2a}$. The total number of integer ordered triples (a, b, c) is 10 as shown in the following table.

a	$a \le b \le \frac{4a}{a-2}$	$c = \frac{2ab}{(a-2)b-2a}$	(a, b, c)
3	$3 \le b \le 12$	$c = \frac{6b}{b-6}$	$(3, 7, 42), (3, 8, 24), (3, 9, 18)$
			$(3, 10, 15), (3, 12, 12)$
4	$4 \le b \le 8$	$c = \frac{4b}{b-4}$	$(4, 5, 20), (4, 6, 12), (4, 8, 8)$
5	$5 \le b \le 6$	$c = \frac{10b}{3b-10}$	$(5, 5, 10)$
6	$6 \le b \le 6$	$c = \frac{3b}{b-3}$	$(6, 6, 6)$

32.4 Counting

1. 2004 AMC 10A Problem 13

Solution (D). Because each man danced with exactly three women, there were $12 \cdot 3 = 36$ pairs of men and women who danced together. Each women had two partners, so $36/2 = 18$ women attended.

2. 2001 AMC 10 Problem 19

Solution (D). The number of possible selections is the number of solution of the equation

$$g + c + p = 4$$

where $g.c$ and p represent, respectively, the number of glazed, chocolate, and powdered donuts. The 15 possible solutions to this equation are

$$(4, 0, 0), (0, 4, 0), (0, 0, 4), (3, 0, 1), (3, 1, 0)(1, 3, 0), (0, 3, 1),$$

$$(1, 0, 3), (0, 1, 3), (2, 2, 0)(2, 0, 2), (0, 2, 2), (2, 1, 1), (1, 2, 1), \text{ and } (1, 1, 2).$$

OR

Code each selection as a sequence of four $*$'s and two $|$'s, where each $*$ represents a donut and each $|$ denotes a "separator" between types of donuts. For example, $**|*|*$ represents two glazed donuts, one chocolate donut, and one powered donut. From the six slots that can be occupied by a $|$ or a $*$, we must choose two places for the $|$'s to determine a selection. Thus there are

$$\frac{6!}{2!4!} = 15$$

selections.

Note : There are various notations used for choosing n objects for a collection of m, where $m \geq n$. The most common is $\binom{m}{n}$, but C_n^m, and mCn are also used. In any case, the number of ways to choose n items from m items is $\frac{m!}{n!(m-n)!}$.

3. 1998 AHSME Problem 24

Solution (C). There are 10000 ways to write the last four digits $d_4d_5d_6d_7$, and among these there are 10000-10=9990 for which not all the digits are the same. For each of these, there are exactly two ways to adjoin the three digits $d_1d_2d_3$ to obtain a memorable number. There are ten memorable numbers for which the last four digits are the same, for a total of $2 \cdot 9990 + 10 = 19990$.

OR

Let A denote the set of telephone numbers for which $d_1d_2d_3$ and $d_4d_5d_6$ are identical and B the set for which $d_1d_2d_3$ is the same as $d_5d_6d_7$. A number $d_1d_2d_3 - d_4d_5d_6d_7$ belongs to $A \cap B$ if and only if $d_1 = d_4 = d_5 = d_2 = d_6 = d_3 = d_7$. Hence, $n(A \cap B) = 10$. Thus, by the *Inclusion − Exclusion Principle*,

$$n(A \cup B) = n(A) + n(B) - n(A \cap B) = 10^3 \cdot 1 \cdot 10 + 10^3 \cdot 10 \cdot 1 - 10 = 19990.$$

4. 2011 AMC 10A Problem 22

Solution (C). If five distinct colors are used, then there are $\binom{6}{5} = 6$ different color choices possible. They may be arranged in $5! = 120$ ways on the pentagon, resulting in $120 \cdot 6 = 720$ colorings.

If four distinct colors are used, then there is one duplicated color, so there are $\binom{6}{4}\binom{4}{1} = 60$ different color choices possible. Then duplicated color must appear on neighboring vertices. There are 5 neighbor choices and $3! = 6$ ways to color the remaining three vertices, resulting in a total of $60 \cdot 5 \cdot 6 = 1800$ colorings.

If three distinct colors are used, then there must be two duplicated colors, so there are $\binom{6}{3}\binom{3}{2} = 60$ different color choices possible. The non-duplicated color may appear in 5 locations. As before, a duplicated color must appear on neighboring vertices, so there are 2 ways left to color the remaining vertices. In this case there are $60 \cdot 5 \cdot 2 = 600$ colorings possible.

There are no colorings with two or fewer colors. The total number of colorings is $720 + 1800 + 600 = 3120$.

5. 1994 AHSME Problem 22

Solution (C). The two end chairs must be occupied by students, so the professors have seven middle chairs from which to choose, with no two adjacent. If these chairs are numbered from 2 to 8, the three chairs can be

$$(2,4,6), (2,4,7), (2,4,8), (2,5,7), (2,5,8)$$

$$(2,6,8), (3,5,7), (3,5,8), (3,6,8), (4,6,8)$$

Within each triple, the professors can arrange themselves in 3! ways, so the total number is $6*10 = 60$.

OR

Imaging the six students standing in a row before they are seated. There are 5 spaces between them, each of which may be occupied by at most one of the 3 professors. Therefore, there are $P(5,3) = 5 \times 4 \times 3 = 60$ ways the three professors can select their places.

6. 2003 AMC 12A Problem 20

Solution (A). Since the first group of five letters contains no A's, it must contain k B's and $(5-k)$ C's for some integer k with $0 \leq k \leq 5$. Since the third group of five letters contains no C's, the remaining k C's must be in the second group, along with $(5-k)$ A's.

Similarly, the third group of five letters must contain k A's and $(5-k)$ B's. Thus each arrangement that satisfies the conditions is determined uniquely by the location of the k B's in the first group, the k C's in the second group, and the k A's in the third group.

For each k, the letters can be arranged in $\binom{5}{k}^3$ ways, so the total number of arrangements is

$$\sum_{k=0}^{5} \binom{5}{k}^3.$$

7. 2001 AMC 12 Problem 17

Solution (C). Since $\angle APB = 90°$ if and only if P lies on the semicircle with center $(2,1)$ and radius $\sqrt{5}$, the angle is obtuse if and only if the point P lies inside this semicircle. The semicircle lies entirely inside the pentagon, since the distance, 3, from $(2, 1)$ to $\overline{DE}$ is greater than the radius of the circle. Thus the probability that the angle is obtuse is the ratio of the area of the semicircle to the area of the pentagon.

Let $O = (0, 0)$. Then the area of the pentagon is

$$[ABCDE] = [OCDE] - [OAB] = 4 \cdot (2\pi + 1) - \frac{1}{2}(2 \cdot 4) = 8\pi,$$

and the area of the semicircle is

$$\frac{1}{2}\pi \left(\sqrt{5}\right)^2 = \frac{5}{2}\pi.$$

The probability is

$$\frac{\frac{5}{2}\pi}{8\pi} = \frac{5}{16}.$$

8. 2003 AMC 12B Problem 19

Solution (E). Since the first term is not 1, the probability that it is 2 is $1/4$. If the first term is 2, then the second term cannot be 2. If the first term is not 2, there are four equally likely values, including 2, for the second term. Thus the probability that the second term is 2 is

$$\frac{1}{4} \cdot 0 + \frac{3}{4} \cdot \frac{1}{4} = \frac{3}{16},$$

so $a + b = 3 + 16 = 19$.

OR

The set S contains $(4)(4!) = 96$ permutations, since there are 4 choices for the first term, and for each of these choices there are $4!$ arrangements of the remaining terms. The number of permutations in S whose second term is 2 is $(3)(3!) = 18$. Since there are 3 choices for the first term and for each of these choices there are $3!$ arrangements of the last 3 terms. Thus the requested probability is $18/96 = 3/16$, and $a + b = 19$.

9. 2004 AMC 10A Problem 10

Solution (D). The result will occur when both A and B have either $0, 1, 2,$ or 3 heads, and these possibilities are shown in the table.

Heads	0	1	2	3
A	$\frac{1}{8}$	$\frac{3}{8}$	$\frac{3}{8}$	$\frac{1}{8}$
B	$\frac{1}{16}$	$\frac{4}{16}$	$\frac{6}{16}$	$\frac{4}{16}$

The probability of both coins having the same number of heads is

$$\frac{1}{8} \cdot \frac{1}{16} + \frac{3}{8} \cdot \frac{4}{16} + \frac{3}{8} \cdot \frac{6}{16} + \frac{1}{8} \cdot \frac{4}{16} = \frac{35}{128}.$$

10. 2001 AMC 12 Problem 11

Solution (D). Thinking of continuing the drawing until all five chips are removed from the box. There are ten possible orderings of the colors: **RRRWW, RRWRW, RWRRW, WRRRW, RRWWR, RWRWR, WRRWR, RWWRR, WRWRR,** and **WWRRR**. The six ordering that end in R represent drawings that would have ended when the second white chip was drawn.

OR

Imagine drawing until only one chip remains. If the remaining chip is red, then that draw would have ended when the second white chip was removed. The last chip will be red with probability $3/5$.

11. 2014 AMC 12B Problem 22

Solution (C). First note that once the frog is on pad 5, it has probability $\frac{1}{2}$ of eventually being eaten by the snake, and a probability $\frac{1}{2}$ of eventually exiting the pond without being eaten. It is therefore necessary only to determine the probability that the frog on pad 1 will reach pad 5 before being eaten.

Consider the frog's jumps in pairs. The frog on pad 1 will advance to pad 3 with probability $\frac{9}{10} \cdot \frac{8}{10} = \frac{72}{100}$, will be back at pad 1 with probability $\frac{9}{10} \cdot \frac{2}{10} = \frac{18}{100}$, and will retreat to pad 0 and be eaten with probability $\frac{1}{10}$. Because the frog will eventually make it to pad 3 or make it to pad 0, the probability that it ultimately makes it to pad 3 is $\frac{72}{100} \div \left(\frac{72}{100} + \frac{10}{100}\right) = \frac{36}{41}$, and the probability that it ultimately makes it to pad 0 is $\frac{10}{100} \div \left(\frac{72}{100} + \frac{10}{100}\right) = \frac{5}{41}$. Similarly, in a pair of jumps the frog will advance from pad 3 to pad 5 with probability $\frac{7}{10} \cdot \frac{6}{10} = \frac{42}{100}$, will be back at pad 3 with probability $\frac{7}{10} \cdot \frac{4}{10} + \frac{3}{10} \cdot \frac{8}{10} = \frac{52}{100}$, and will retreat to pad 1 with probability $\frac{3}{10} \cdot \frac{2}{10} = \frac{6}{100}$. Because the frog will ultimately make it to pad 5 or pad 1 from pad 3, the probability that it ultimately makes it to pad 5 is $\frac{42}{100} \div \left(\frac{42}{100} + \frac{6}{100}\right) = \frac{7}{8}$, and the probability that it ultimately makes it to pad 1 is $\frac{6}{100} \div \left(\frac{42}{100} + \frac{6}{100}\right) = \frac{1}{8}$.

The sequences of pairs of moves by which the frog will advance to pad 5 without being eaten are

$$1 \to 3 \to 5, 1 \to 3 \to 1 \to 3 \to 5, 1 \to 3 \to 1 \to 3 \to 1 \to 3 \to 5,$$

and so on. The sum of the respective probabilities of reaching pad 5 is then

$$
\begin{aligned}
&\frac{36}{41} \cdot \frac{7}{8} + \frac{36}{41} \cdot \frac{1}{8} \cdot \frac{36}{41} \cdot \frac{7}{8} + \frac{36}{41} \cdot \frac{1}{8} \cdot \frac{36}{41} \cdot \frac{1}{8} \cdot \frac{36}{41} \cdot \frac{7}{8} \\
&= \frac{63}{82} \cdot \left(1 + \frac{9}{82} + \left(\frac{9}{82} \right)^2 + \cdots \right) \\
&= \frac{63}{82} \div \left(1 - \frac{9}{82} \right) \\
&= \frac{63}{73}.
\end{aligned}
$$

Therefore the requested probability is $\frac{1}{2} \cdot \frac{63}{73} = \frac{63}{146}$.

OR

For $1 \le j \le 5$, let p_j be the probability that the frog eventually reaches pad 10 starting at pad j. By symmetry $p_5 = \frac{1}{2}$. For the frog to reach pad 10 starting from pad 4, the frog goes either to pad 3 with probability $\frac{2}{5}$ or to pad 5 with probability $\frac{3}{5}$, and then continues on a successful sequence from either of these pads. Thus $p_4 = \frac{2}{5}p_3 + \frac{3}{5}p_5 = \frac{2}{5}p_3 + \frac{3}{10}$. Similarly, to reach pad 10 starting from pad 3, the frog goes either to pad 2 with probability $\frac{3}{10}$ or to pad 4 with probability $\frac{7}{10}$. Thus $p_3 = \frac{3}{10}p_2 + \frac{7}{10}p_4$, and substituting from the previous equation for p_4 gives $p_3 = \frac{5}{12}p_2 + \frac{7}{24}$. In the same way, $p_2 = \frac{1}{5}p_1 + \frac{4}{5}p_3$ and after substituting for p_3 gives $p_2 = \frac{3}{10}p_1 + \frac{7}{20}$. Lastly, for the frog to escape starting from pad 1, it is necessary for it to get to pad 2 with probability $\frac{9}{10}$, and then escape starting from pad 2. Thus $p_1 = \frac{9}{10}p_2 = \frac{9}{10}\left(\frac{3}{10}p_1 + \frac{7}{20} \right)$, and solving the equation gives $p_1 = \frac{63}{146}$.

Note : This type of random process is called a Markov process.

12. **2012 AMC 10A Problem 23 or 2012 AMC 12A Problem 19**

Solution (B). This situation can be modeled with a graph having these six people as vertices, in which two vertices are joined by an edge if and only if the corresponding people are internet friends. Let n be the number of friends each person has; then $1 \le n \le 4$. If $n = 1$, then the graph consists of three edges sharing no endpoints. There are 5 choices for Adam's friend and then 3 ways to partition the remaining 4 people into 2 pairs of friends, for a total of $5 \cdot 3 = 15$ possibilities. The case $n = 4$ is complementary, with non-friendship playing the role of friendship, so there are 15 possibilities in that case as well.

For $n = 2$, the graph must consist of cycles, and the only two choices are two triangles (3-cycles) and a hexagon (6-cycle). In the former case, there are $\binom{5}{2} = 10$ ways to choose two friends for Adam and that choice uniquely determines the triangles. In the latter case, every permutation of the six vertices determines a hexagon, but each hexagon is counted $6 \cdot 2 = 12$ times, because the hexagon can start at any vertex and be traversed in either direction. This gives $\frac{6!}{12} = 60$ hexagons, for a total of $10 + 60 = 70$ possibilities. The complementary case $n = 3$ provides 70 more. The total is therefore $15 + 15 + 70 + 70 = 170$.

13. 2013 AMC 12A Problem 20

Solution (B). Consider the elements of S as integers modulo 19. Assume $a \succ b$. If $a > b$, then $a - b \leq 9$. If $a < b$, then $b - a > 9$; that is $b - a \geq 10$ and so $a + 19 - b \leq 9$. Thus $a \succ b$ if and only if $0 < (a - b)(\mathrm{mod}\ 19) \leq 9$.

Suppose that (x, y, z) is a triple in $S \times S \times S$ such that $x \succ y$, $y \succ z$, and $z \succ x$. There are 19 possibilities for the first entry x. Once x is chosen, y can equal $x + i$ for any i, $1 \leq i \leq 9$. Then z is at most $x + 9 + i$ and at least $x + 10$, so once y is chosen, there are i possibilities for the third element z.

The number of required triples is equal to $19(1 + 2 + \cdots + 9) = 19 \cdot \frac{1}{2} \cdot 9 \cdot 10 = 19 \cdot 45 = 855$.

32.5 Trigonometry and Logarithms

1. 2010 AMC 12B Problem 13

Solution (C). The maximum value for $\cos x$ and $\sin x$ is 1; hence $\cos(2A - B) = 1$ and $\sin(A + B) = 1$. Therefore $2A - B = 0°$ and $A + B = 90°$, and solving gives $A = 30°$ and $B = 60°$. Hence $\triangle ABC$ is a $30 - 60 - 90°$ right triangle and $BC = 2$.

2. 1993 AHSME Problem 23

Solution (B). The center of the circle is not X since $2\angle BAC \neq \angle BXC$. Thus $\overline{AD}$ bisects $\angle BXC$ and $\angle BAC$. Since $\angle ABD$ is inscribed in a semicircle, $\angle ABD = 90°$, and thus

$$AB = AD \cos \angle BAD = 1 \cdot \cos\left(\frac{1}{2}\angle BAC\right) = \cos 6°.$$

Also,

$$\angle AXB = 180° - \angle DXB = 180° - \frac{36°}{2} = 162°.$$

Since the sum of the angles in $\triangle AXB$ is $180°$,

$$\angle ABX = 180° - (162° + 6°) = 12°.$$

By the Law of Sines,

$$\frac{AB}{\sin \angle AXB} = \frac{AX}{\sin \angle ABX} \quad so \quad \frac{\cos 6°}{\sin 162°} = \frac{AX}{\sin 12°}.$$

Since $\sin 162° = \sin 18°$, we have

$$AX = \frac{\cos 6° \cdot \sin 12°}{\sin 18°} = \cos 6° \sin 12° \csc 18°.$$

Note: If X were the center of the circle, then $\overline{BX}$ and $\overline{CX}$ could be any radii. Therefore, to establish $\overline{AD}$ as the bisector of $\angle BAC$ and $\angle BXC$, it is necessary first to establish that X is not the center of the circle.

3. 2003 AMC 12B Problem 21

Solution (D). Let $\beta = \pi - \alpha$. Apply the Law of Cosines to $\triangle ABC$ to obtain

$$(AC)^2 = 8^2 + 5^2 - 2(8)(5)\cos\beta = 89 - 80\cos\beta.$$

Thus $AC < 7$ if and only if

$$89 - 80\cos\beta < 49, \quad \text{that is, if and only if,} \quad cos\beta > \frac{1}{2}.$$

Therefore we must have $0 < \beta < \frac{\pi}{3}$, and the requested probability is $\frac{\pi/3}{\pi} = \frac{1}{3}$.

4. 2004 AMC 12A Problem 21

Solution (D). The given series is geometric with an initial term of 1 and a common ratio of $\cos^2\theta$, so its sum is

$$5 = \sum_{n=0}^{\infty} \cos^{2n}\theta = \frac{1}{1-\cos^2\theta} = \frac{1}{\sin^2\theta}.$$

Therefore $\sin^2\theta = \frac{1}{5}$ and

$$\cos 2\theta = 1 - 2\sin^2\theta = 1 - \frac{2}{5} = \frac{3}{5}.$$

5. 2006 AMC 12B Problem 24

Solution (C). For a fixed value of y, the value of $\sin x$ for which $\sin^2 x - \sin x \sin y + \sin^2 y = \frac{3}{4}$ can be determined by the quadratic formula. Namely,

$$\sin x = \frac{\sin y \pm \sqrt{\sin^2 y - 4(\sin^2 y - \frac{3}{4})}}{2} = \frac{1}{2}\sin y \pm \frac{\sqrt{3}}{2}\cos y.$$

Because $\cos(\frac{\pi}{3}) = \frac{1}{2}$ and $\sin(\frac{\pi}{3}) = \frac{\sqrt{3}}{2}$, this implies that

$$\sin x = \cos\left(\frac{\pi}{3}\right)\sin y \pm \sin\left(\frac{\pi}{3}\right)\cos y = \sin\left(y \pm \left(\frac{\pi}{3}\right)\right).$$

Within S, $\sin x = \sin(y - \frac{\pi}{3})$ implies $x = y - \frac{\pi}{3}$. However, the case $\sin x = \sin(y + \frac{\pi}{3})$ implies $x = y + \frac{\pi}{3}$ when $y \leq \frac{\pi}{6}$, and $x = -y + \frac{2\pi}{3}$ when $y \geq \frac{\pi}{6}$. Those three lines divides the region S into four subregions, within each of which the truth value of the inequality is constant. Testing the points $(0,0)$, $\left(\frac{\pi}{2}, 0\right)$, $\left(0, \frac{\pi}{2}\right)$ and $\left(\frac{\pi}{2}, \frac{\pi}{2}\right)$ shows that the inequality is true only in the shaded subregion. The area of this subregion is

$$\left(\frac{\pi}{2}\right)^2 - \frac{1}{2} \cdot \left(\frac{\pi}{3}\right)^2 - 2 \cdot \frac{1}{2} \cdot \left(\frac{\pi}{6}\right)^2 = \frac{\pi^2}{6}.$$

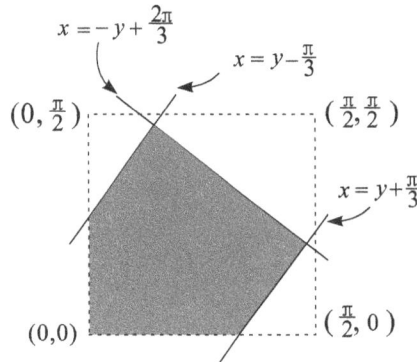

6. 2002 AMC 12B Problem 22

Solution (B). We have $a_n = \frac{1}{\log_n 2002} = \log_{2002} n$, so

$$
\begin{aligned}
b - c &= (\log_{2002} 2 + \log_{2002} 3 + \log_{2002} 4 + \log_{2002} 5) \\
&\quad - (\log_{2002} 10 + \log_{2002} 11 + \log_{2002} 12 + \log_{2002} 13 + \log_{2002} 14) \\
&= \log_{2002} \frac{2 \cdot 3 \cdot 4 \cdot 5}{10 \cdot 11 \cdot 12 \cdot 13 \cdot 14} = \log_{2002} \frac{1}{11 \cdot 13 \cdot 14} \\
&= \log_{2002} \frac{1}{2002} = -1.
\end{aligned}
$$

7. 2014 AMC 12A Problem 18

Solution (C). For every $a > 0$, $a \neq 1$, the domain of $\log_a x$ is the set $\{x : x > 0\}$. Moreover, for $0 < a < 1$, $\log_a x$ is a decreasing function on its domain, and for $a > 1$, $\log_a x$ is an increasing function on its domain. Thus the function $f(x)$ is defined if and only if $\log_4 \left(\log_{\frac{1}{4}} \left(\log_{16} \left(\log_{\frac{1}{16}}(x) \right) \right) \right) > 0$, and this inequality is equivalent to each of the following:

$$\log_{\frac{1}{4}} \left(\log_{16} \left(\log_{\frac{1}{16}}(x) \right) \right) > 1, \quad 0 < \log_{16} \left(\log_{\frac{1}{16}}(x) \right) < \frac{1}{4},$$

$$1 < \log_{\frac{1}{16}} x < 2, \quad \text{and} \quad \frac{1}{256} < x < \frac{1}{16}.$$

Thus $\frac{m}{n} = \frac{1}{16} - \frac{1}{256} = \frac{15}{256}$ and $m + n = 271$.

8. 2011 AMC 12B Problem 17

Solution (B). Note that

$$h_1(x) = \log_{10} \left(\frac{10^{10x}}{10} \right) = \log_{10} \left(10^{10x-1} \right) = 10x - 1.$$

Therefore $h_2(x) = 10^2 x - (1+10)$, $h_3(x) = 10^3 x - (1+10+10^2)$, and in general,

$$h_n(x) = 10^n x - \sum_{k=0}^{n-1} 10^k.$$

Hence $h_n(1)$ is an n-digit integer whose units digit is 9 and whose other digits are all 8's. The sum of the digits of $h_{2011}(1)$ is $8 \cdot 2010 + 9 = 16,089$.

9. 1997 AHSME Problem 21

Solution (C). Since $\log_8 n = \frac{1}{3}(\log_2 n)$, it follows that $\log_8 n$ is rational if and only if $\log_2 n$ is rational. The nonzero numbers in the sum will therefore be all numbers of the form $\log_8 n$, where n is an integral power of 2. The highest power of 2 that does nor exceed 1997 is 2^{10}, so the sum is:

$$\log_8 1 + \log_8 2 + \log_8 2^2 + \log_8 2^3 + \cdots + \log_8 2^{10}$$
$$= 0 + \frac{1}{3} + \frac{2}{3} + \frac{3}{3} + \cdots + \frac{10}{3} = \frac{55}{3}.$$

Challenge. Prove that $\log_2 3$ is irrational. Prove that, for every integer n, $\log_2 n$ is rational if and only if n is an integral power of 2.

10. 2006 AMC 12B Problem 20

Solution (C). The given condition is equivalent to $\lfloor \log_{10} x \rfloor = \lfloor \log_{10} 4x \rfloor$. Thus the condition holds if and only if

$$n \le \log_{10} x < \log_{10} 4x < n+1$$

for some negative integer n. Equivalently,

$$10^n \le x < 4x < 10^{n+1}.$$

This inequality is true if and only if

$$10^n \le x < \frac{10^{n+1}}{4}.$$

Hence in each interval $[10^n, 10^{n+1})$, the given condition holds with probability

$$\frac{(10^{n+1}/4) - 10^n}{10^{n+1} - 10^n} = \frac{10^n(10/4 - 1)}{10^n(10 - 1)} = \frac{1}{6}.$$

Because each number in $(0,1)$ belongs to a unique interval $[10^n, 10^{n+1})$ and the probability is the same on each interval, the required probability is also $1/6$.

11. 2014 AMC 12B Problem 25

Solution (D). If $x = \frac{1}{2}\pi y$, then the given equation is equivalent to

$$2\cos(xy)\left[\cos(xy) - \cos\left(\frac{4028\pi}{y}\right)\right] = \cos(2\pi y) - 1.$$

Dividing both sides by 2 and using the identity $\frac{1}{2}(1 - \cos(2\pi y)) = \sin^2(\pi y)$ yields

$$\cos^2(\pi y) - \cos(\pi y)\cos\left(\frac{4028\pi}{y}\right) = \frac{1}{2}\left(\cos(2\pi y) - 1\right) = -\sin^2(\pi y).$$

This is equivalent to

$$1 = \cos(\pi y)\cos\left(\frac{4028\pi}{y}\right).$$

Thus either $\cos(\pi y) = \cos(\frac{4028\pi}{y}) = 1$ or $\cos(\pi y) = \cos(\frac{4028\pi}{y}) = -1$. It follows that y and $\frac{4028}{y}$ are both integers having the same parity. Therefore y cannot be odd or a multiple of 4. Finally, let $y = 2a$ with a a positive odd divisor of $4028 = 2^2 \cdot 19 \cdot 53$, that is $a \in \{1, 19, 53, 19 \cdot 53\}$. Then $\cos(\pi y) = \cos(2a\pi) = 1$ and $\cos(\frac{4028\pi}{y}) = \cos(\frac{2014\pi}{a}) = 1$. Therefore the sum of all solutions x is $\pi(1 + 19 + 53 + 19 \cdot 53) = \pi(19 + 1)(53 + 1) = 1080\pi.$

32.6 Complex Number

1. 1981 AHSME Problem 24

Solution (D). Write $x + \frac{1}{x} = 2\cos\theta$ as $x^2 - 2x\cos\theta + 1 = 0$. Then

$$x = \cos\theta \pm \sqrt{\cos^2\theta - 1} = \cos\theta \pm i\sin\theta \left(= e^{\pm i\theta}\right).$$

By De Moivre's theorem

$$x^n = \cos n\theta \pm i\sin n\theta \left(= e^{\pm in\theta}\right),$$

$$\frac{1}{x^n} = \frac{1}{\cos n\theta \pm i\sin n\theta} = \cos n\theta \mp i\sin n\theta \left(= e^{\mp in\theta}\right).$$

Thus

$$x^n + \frac{1}{x^n} = 2\cos n\theta.$$

Note 1: Squaring both sides of the given equation, we obtain

$$x^2 + 2 + \frac{1}{x^2} = 4\cos^2\theta.$$

Hence

$$x^2 + \frac{1}{x^2} = 4\cos^2\theta - 2 = 2(2\cos^2\theta - 1) = 2\cos 2\theta.$$

This eliminates all choices except (D).

Note 2: Several contestants wrote to the Contests Committee to say that this problem had no solution because, for $0 < \theta < \pi$ as given, $2\cos\theta < 2$, yet $x + \frac{1}{x} \geq 2$ for all x. This claim is true for *real* x, but nowhere in the problem did it say that x must be real. When such a restriction is intended, it is always stated.

2. 1983 AHSME Problem 17

Solution (C). Write F as $a + bi$, where we see from the diagram that $a, b > 0$ and $a^2 + b^2 > 1$. Since

$$\frac{1}{a + bi} = \frac{a - bi}{a^2 + b^2} = \frac{a}{a^2 + b^2} - \frac{b}{a^2 + b^2}i,$$

we see that the reciprocal of F is in quadrant IV, since the real part on the right is positive and the coefficient of the imaginary part is negative. Also, the magnitude of the reciprocal is

$$\frac{1}{a^2 + b^2}\sqrt{a^2 + (-b)^2} = \frac{1}{\sqrt{a^2 + b^2}} < 1.$$

Thus the only possibility is point C.

OR

For any complex number $z \neq 0$, the argument (standard reference angle) of $1/z$ is the negative of the argument of z, and the modulus (magnitude) of $1/z$ is the reciprocal of the modulus of z. (Quick proof: if $z = re^{i\theta}$, then $\frac{1}{z} = \frac{1}{re^{i\theta}} = \frac{1}{r}e^{i(-\theta)}$.) Applying this to the point F, its reciprocal must be in quadrant IV, inside the unit circle. The only possibility is point C.

3. **1992 AHSME Problem 15**

   Solution (B). We compute

   $$z_1 = 0, z_2 = i, z_3 = i - 1, z_4 = -i, z_5 = i - 1.$$

   Since $z_5 = z_3$, it follows that

   $$z_{111} = z_{109} = z_{107} = \cdots = z_5 = z_3 = i - 1,$$

   which is $\sqrt{2}$ units form the origin.

   Note. The *Mandelbrot set* is defined to be the set of complex numbers c for which all the terms of the sequence defined by $z_1 = 0$, $z_{n+1} = z_n^2 + c$ for $n \geq 1$, stay close to the origin. Thus $c = i$ is in the Mandelbrot set.

4. **1974 AHSME Problem 17**

   Solution (C). Since $i^2 = -1$, $(1 + i)^2 = 2i$ and $(1 - i)^2 = -2i$. Writing

   $$(1 + i)^{20} - (1 - i)^{20} = \left((1 + i)^2\right)^{10} - \left((1 - i)^2\right)^{10},$$

   we have

   $$(1 + i)^{20} - (1 - i)^{20} = (2i)^{10} - (-2i)^{10} = 0.$$

OR

Observe that $i(1-i) = i - i^2 = 1 + i$, and $i^4 = (-1)^2 = 1$. Therefore

$$(1+i)^4 = i^4(1-i)^4 = (1-i)^4,$$

and

$$(1+i)^{4n} - (1-i)^{4n} = 0 \text{ for } n = 1, 2, 3, \ldots$$

OR

$$(1+i)^{20} = \sqrt{2}^{20} \left[\cos 900° + i \sin 900°\right] = -1024,$$
$$(1-i)^{20} = \sqrt{2}^{20} \left[\cos(-900)° + i \sin(-900)°\right] = -1024,$$

and the difference is 0.

5. 2009 AMC 12A Problem 21

Solution (C). Because $x^{12} + ax^8 + bx^4 + c = p(x^4)$, the value of this polynomial is 0 if and only if

$$x^4 = 2009 + 9002\pi i \quad \text{or} \quad x^4 = 2009 \quad \text{or} \quad x^4 = 9002.$$

The first of these three equations has four distinct nonreal solutions, and the second and third each have two distinct nonreal solutions. Thus $p(x^4) = x^{12} + ax^8 + bx^4 + c$ has 8 distinct nonreal zeros.

6. 1977 AHSME Problem 16

Solution (D). Since

$$
\begin{aligned}
i^{n+2} \cos(45 + 90(n+2))° &= -i^n \left(-\cos(45 + 90n)°\right) \\
&= i^n \left(\cos(45 + 90n)°\right),
\end{aligned}
$$

every other term has the same value. The first is $\sqrt{2}/2$, and there are 21 terms with this value ($n = 0, 2, 4, \ldots, 40$). The second term is $i \cos 135° = -i\sqrt{2}/2$, and there are 20 terms with this value ($n = 1, 3, \ldots, 39$). Thus the sum is

$$\frac{\sqrt{2}}{2}(21 - 20i).$$

7. 1988 AHSME Problem 21

Solution (E). Set $z = a + bi$. We seek $|z|^2 = a^2 + b^2$. So

$$
\begin{aligned}
z + |z| &= 2 + 8i, \\
a + bi + \sqrt{a^2 + b^2} &= 2 + 8i, \\
a + \sqrt{a^2 + b^2} = 2, &\qquad b = 8, \\
a + \sqrt{a^2 + 64} &= 2, \\
a^2 + 64 = (2-a)^2 &= a^2 - 4a + 4, \\
4a = -60, &\qquad a = -15.
\end{aligned}
$$

Thus $a^2 + b^2 = 225 + 64 = 289$.

OR

In the complex plane, z and $|z|$ are vectors of equal length, and $z+|z| = 2+8i$ is their vector sum. So 0, z, $2+8i$, and $|z|$ form the vertices of a rhombus, as in the figure. The diagonals of a rhombus bisect each other and are perpendicular. The diagonal form 0 to $2+8i$ has slope 4 and midpoint $1+4i$. Thus the diagonal form z to $|z|$ passes through $1+4i$ with slope $-1/4$. Therefore this diagonal intersects the real axis at $x = 17$. Since $|z|$ is on the real axis, we conclude that $|z| = 17$ and $|z|^2 = 289$.

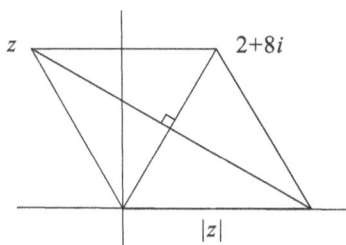

8. 1985 AHSME Problem 23

Solution (C). Note that $x^3 = y^3 = 1$, because x and y are the complex roots of $0 = x^3 - 1 = (x-1)(x^2+x+1)$. Alternatively, plot x and y in the complex plane and observe that they have modulus 1 and arguments $2\pi/3$ and $4\pi/3$. Thus $x^9 + y^9 = 1 + 1 \neq -1$. (To show that the other equations are correct is easy if one also notes that $y = x^2$ and $x + y = -1$. For instance, $x^{11} + y^{11} = x^2 + x^{22} = y + x = -1$.)

9. 2008 AMC 12B Problem 19

Solution (B). Let $\alpha = a + bi$ and $\gamma = c + di$ where a, b, c and d are real numbers. Then $f(1) = (4 + a + c) + (1 + b + d)i$, and $f(i) = (-4 - b + c) + (-1 + a + d)i$. Because both $f(1)$ and $f(i)$ are real, it follows that $a = 1 - d$ and $b = -1 - d$ Thus

$$\begin{aligned} |\alpha| + |\gamma| &= \sqrt{a^2 + b^2} + \sqrt{c^2 + d^2} \\ &= \sqrt{(1-d)^2 + (-1-d)^2} + \sqrt{c^2 + d^2} \\ &= \sqrt{2 + 2d^2} + \sqrt{c^2 + d^2} \end{aligned}$$

The minimum value of $|\alpha| + |\gamma|$ is consequently $\sqrt{2}$, which is achieved when $c = d = 0$. In this case we also have $a = 1$ and $b = -1$.

10. 2011 AMC 12B Problem 24

Solution (B). Factoring or using the quadratic formula with z^4 as the variable yields $P(z) = (z^4 - 1)(z^4 + (4\sqrt{3} + 7))$. Moreover, $4\sqrt{3} + 7 = (\sqrt{3} + 2)^2$ and $2(\sqrt{3} + 2) = 2\sqrt{3} + 4 =$

$(\sqrt{3}+1)^2$; thus $4\sqrt{3}+7 = \left(\frac{1}{2}(\sqrt{6}+\sqrt{2})\right)^4$. If $w = \frac{1}{2}(\sqrt{3}+1)$, then the eight zeros of $P(z)$ are $1, -1, i, -i, w(1+i), w(-1+i), w(-1-i)$, and $w(1-i)$.

The distances from 1 to the other zeros are

$$|1-(-1)| = 2, \quad |1 \pm i| = \sqrt{2},$$

$$|1 - w(1 \pm i)| = \sqrt{(1-w)^2 + w^2} = \sqrt{2},$$

and

$$|1 - w(-1 \pm i)| = \sqrt{(1+w)^2 + w^2} = \sqrt{2\sqrt{3}+4} = \sqrt{3}+1.$$

Similarly, the distances from $w(1+i)$ to the other zeros are

$$|w(1+i) - w(1-i)| = |w(1+i) - w(-1+i)| = 2w = \sqrt{3}+1,$$

$$|w(1+i) - w(-1-i)| = 2\sqrt{2}w = \sqrt{6}+\sqrt{2},$$

and by symmetry,

$$|w(1+i) - 1| = |w(1+i) - i| = \sqrt{2},$$

and

$$|w(1+i) + 1| = |w(1+i) + i| = \sqrt{3}+1.$$

Because the set of zeros is 4-fold symmetric with respect to the origin, it follows that every line segment joining two of the zeros has length at least $\sqrt{2}$. This shows that any polygon with vertices at the zeros has perimeter at least $8\sqrt{2}$. Finally, note that the polygon with consecutive vertices $1, w(1+i), i, w(-1+i), -1, w(-1-i), -i$, and $w(1-i)$ has perimeter $8\sqrt{2}$.

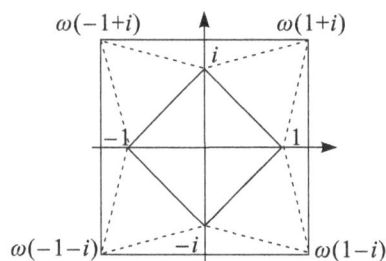

PART V

SOLUTION TO PART II PRACTICE PROBLEMS

CHAPTER 33

ALGEBRA

33.1 Polynomials and their Zeros

1. **Solution** (A). Factor to get $(2x + 3)(2x - 10) = 0$, so the two roots are $-3/2$ and 5.

2. **Solution** (E). The sum of the roots of the quadratic $ax^2 + bx + c = 0$ is $-b/a$ which, for this example, is 5. Alternatively, factor the quadratic into $(x - 3)(x - 2) = 0$ and find the roots.

3. **Solution** (D). The roots are 12 and $37/12$ because

$$f(x) = 12 \cdot (x - 12) \cdot (x - 3) - (x - 12) = (x - 12)(12x - 37)$$

so the product is $12 \cdot (37/12) = 37$.

4. **Solution** (C). Group the terms together and factor $(x - 3)$ from the first two to get

$$
\begin{aligned}
& (x - 1)(x - 3) + (x - 4)(x + 5) + (x - 3)(x - 7) \\
= \ & (x - 3)[(x - 1) + (x - 7)] + (x - 4)(x + 5) \\
= \ & (x - 3)[(2x - 8)] + (x - 4)(x + 5) \\
= \ & 2(x - 3)[(x - 4)] + (x - 4)(x + 5) \\
= \ & (x - 4)[2(x - 3) + (x + 5)] \\
= \ & (x - 4)(3x - 1) = 0,
\end{aligned}
$$

so the product of the two roots is $4 \cdot 1/3 = 4/3$.

OR

Combine the terms to get

$$x^2 - 4x + 3 + x^2 + x - 20 + x^2 - 10x + 21 = 3x^2 - 13x + 4 = 0,$$

which is equivalent to

$$x^2 - 13x/3 + 4/3 = 0,$$

the product of whose roots is the constant term, $4/3$.

5. **Solution** (A). Add and subtract $4x^2y^2$ to get

$$
\begin{aligned}
x^4 + 4y^4 &= x^4 + 4x^2y^2 + 4y^4 - 4x^2y^2 \\
&= (x^2 + 2y^2)^2 - (2xy)^2 \\
&= (x^2 - 2xy + 2y^2)(x^2 + 2xy + 2y^2).
\end{aligned}
$$

6. **Solution** (C). The quadratic equation has two solutions $x = \dfrac{m}{2}$, $x = \dfrac{m}{3}$. There are 500 multiples of 2, 333 multiples of 3, and 166 multiples of 6 between 1 and 1000. Therefore, the probability is

$$p = \frac{500 + 333 - 166}{1000} = 0.667.$$

7. **Solution** (E). Let the roots be c, d. Then

$$x^2 + 13x + w = (x - c)(x - d) = x^2 - (c + d)x - cd.$$

Hence the sum of the roots is $c + d = -13$.

8. **Solution** (E). Since $a + b = -3$ and $ab = -3$ so we have

$$a^2 + b^2 = (a + b)^2 - 2ab = (-3)^2 - 2 \cdot (-3) = 15.$$

9. **Solution** (A). Let x_1, x_2 be the roots. Then $x_1 + x_2 = 3$ and $x_1 \cdot x_2 = -2$. So

$$\frac{1}{x_1} \cdot \frac{1}{x_2} = -\frac{1}{2}$$

and

$$\frac{1}{x_1} + \frac{1}{x_2} = \frac{x_1 + x_2}{x_1 \cdot x_2} = \frac{3}{-2}.$$

Therefore the quadratic equation with roots x_1 and x_2 must be

$$x^2 + \frac{3}{2}x - \frac{1}{2} = 0,$$

or equivalently,

$$2x^2 + 3x - 1 = 0.$$

10. Solution (D). The easiest way to solve this is to use the calculator to find r and s and then to compute $r^3 + s^3$. What follows is a more elegant solution. The sum of the roots of $ax^2 + bx + c = 0$ is $-b/a$ and the product is c/a. Therefore $r + s = 3$ and $rs = 1$, and it follows that

$$r^3 + s^3 = (r+s)(r^2 - rs + s^2) = (r+s)[(r+s)^2 - 3rs] = 18.$$

Alternatively, $27 = (r+s)^3 = r^3 + s^3 + 3rs(r+s) = r^3 + s^3 + 3 \cdot 1 \cdot 3.$

11. Solution (A). Note that $q = ab$ and $p = a + b$. If the new equation is $x^2 - Px + Q = 0$, we require that

$$P = a + b + \frac{1}{a} + \frac{1}{b} = a + b + \frac{a+b}{ab} = p + \frac{p}{q}$$

and

$$Q = \left(a + \frac{1}{b}\right)\left(b + \frac{1}{a}\right) = ab + 2 + \frac{1}{ab} = q + 2 + \frac{1}{q}.$$

12. Solution (C). Because the coefficient of x^2 is -4, we have $x_1 + x_2 + x_3 = 4$. We also know that $x_1 = x_2 + x_3$, so $2x_1 = 4$ and $x_1 = 2$. Since $x = 2$ is a root, we have $2^3 - 4(2^2) - 11(2) + a = 0$ so $a = 30$.

13. Solution (C). Suppose the 4 roots are given by

$$\frac{1}{4}, \frac{1}{4} + d, \frac{1}{4} + 2d, \frac{1}{4} + 3d,$$

where d is a real number. Further suppose the roots of

$$x^2 - 2x + m = 0$$

are r_1, r_2 and the roots of

$$x^2 - 2x + n = 0$$

are r_3, r_4, then from

$$r_1 + r_2 + r_3 + r_4 = 2 + 2,$$

we have $d = \frac{1}{2}$. Since

$$r_1 + r_2 = 2, \quad r_3 + r_4 = 2$$

and

$$r_1, r_2, r_3, r_4 \in \left\{\frac{1}{4}, \frac{3}{4}, \frac{5}{4}, \frac{7}{4}\right\},$$

there are only two possibilities: either

$$r_1, r_2 \in \left\{\frac{3}{4}, \frac{5}{4}\right\}, r_3, r_4 \in \left\{\frac{1}{4}, \frac{7}{4}\right\}$$

or

$$r_1, r_2 \in \left\{\frac{1}{4}, \frac{7}{4}\right\}, r_3, r_4 \in \left\{\frac{3}{4}, \frac{5}{4}\right\}.$$

In both two cases, we have $|m - n| = |r_1 \cdot r_2 - r_3 \cdot r_4| = \frac{1}{2}$.

14. **Solution** (A). The value of k that works is the one for which $P(1) = 0$ since $x - r$ is a factor of P exactly when $P(r) = 0$. So we have $P(1) = 2 + k + 1 = 0$, and $k = -3$.

15. **Solution** (A). Because $x^2 - 2x - 3 = (x - 3)(x + 1)$, $p(x)$ has zeros of 3 and -1, $p(3) = 72 + 3a + b = 0$ and $p(-1) = -4 - a + b = 0$ which we can solve simultaneously to get $a = -19$ and $b = -15$.

Alternatively, simply start by dividing $2x^4 - x^3 - 7x^2 + ax + b$ by $x^2 - 2x - 3$ by long division. The "last step" is to subtract $5x^2 - 10x - 15$ from $5x^2 + (9 + a)x + b$. The difference must be 0. So $b = -15$ and $9 + a = -10$. Thus $a = -19$ and $a + b + -34$. No factoring is done.

Yet another method is to factor $x^2 - 2x - 3$ and see that it has two zeros, $x = 3$ and $x = -1$. Use synthetic division (starting with either zero–here starting with $x = -1$) to see that

$$b - a - 4 = 0, \tag{1}$$

and

$$2x^4 - x^3 - 7x^2 + ax + b = (x + 1)(2x^3 - 3x^2 - 4x + a + 4). \tag{2}$$

Now use synthetic division with $x = 3$ on the quotient $2x^3 - 3x^2 - 4x + a + 4$ to see that $a + 19 = 0$. Thus $a = -19$. Put this into (1) to see that $b = -15$, etc.

16. **Solution** (C). The polynomial $h(x) = f(x) - 2$ has four distinct zeros, and is not constant. Therefore its degree is at least 4, and so is f's degree.

17. **Solution** (C). Using the Binomial Theorem or by constructing the 10th row of Pascal's triangle, the expansion of $(x + 3)^{10}$ can be found to be

$$x^{10} + 10(3)x^9 + 45(3^2)x^8 + 210(3^3)x^7 \cdots .$$

So the coefficient of x^7 is $120 \cdot 3^3 = 3240$.

33.2 Exponentials and Radicals

1. **Solution** (A). Since n and n^4 are both even or both odd, their sum is even. Hence

$$(-1)^{n^4+n+1} = (-1)^{n^4+n}(-1)^1 = (1)(-1) = -1.$$

2. **Solution** (C). Note that $(2^x)^y = 15^y = 32$, so $2^{xy} = 2^5$ and $xy = 5$.

3. **Solution** (D). Note that $b = 9(1/x) = 9/x$. Thus, $c = x/9$ and $x = 9c = d$.

4. Solution (D). Let x denote the integer in question. Then $N^{1/3} = x$, so

$$(x+1)^2 = (N^{1/3}+1)^2 = N^{2/3} + 2 \cdot N^{1/3} + 1.$$

5. Solution (E). Since n is a perfect square, $\sqrt{n}$ is an integer, and the largest perfect square less than n has a square root of $\sqrt{n} - 1$. Hence that number is $(\sqrt{n} - 1)^2 = n - 2\sqrt{n} + 1$.

6. Solution (B). $(2^{\sqrt{2}})^2 = 2^{2\sqrt{2}} = (2^2)^{\sqrt{2}} = 4^{\sqrt{2}}$.

7. Solution (E). To simplify the calculation, let $t = 2^x$. Then

$$144 = (t - t^2)^2 + (t + t^2)^2 = 2(t^2 + t^4)$$

Thus

$$t^4 + t^2 - 72 = (t^2 + 9)(t^2 - 8) = 0$$

and hence $t^2 = 2^{2x} = 8 = 2^3$. Therefore $x = 3/2$.

8. Solution (D). $\sqrt{2 + \sqrt{x}} = 3 \Rightarrow 2 + \sqrt{x} = 9 \Rightarrow \sqrt{x} = 7 \Rightarrow x = 49$.

9. Solution (B). Under the condition we have

$$\sqrt{x^2 - 2x + 1} + \sqrt{x^2 - 6x + 9} = x - 1 + 3 - x = 2.$$

Note that $\sqrt{x^2}$ is not always equal to x.

10. Solution (C). Denoting $x = \sqrt{5 + 2\sqrt{6}} + \sqrt{5 - 2\sqrt{6}}$, we have

$$x^2 = 5 + 2\sqrt{6} + 5 - 2\sqrt{6} + 2\sqrt{5 + 2\sqrt{6}} \cdot \sqrt{5 - 2\sqrt{6}} = 10 + 2\sqrt{5^2 - 4 \cdot 6} = 12.$$

Hence $x = \sqrt{12} = 2\sqrt{3}$.

11. Solution (B). Cross multiplying and simplifying, we obtain $2\sqrt{x-1} = \sqrt{x+1}$. Squaring both sides yields $4(x-1) = x+1$ or $3x = 5$. Thus, $x = 5/3$.

12. Solution (C). Square both sides to get $14 + \sqrt{27 - \sqrt{x-1}} = 16$, then massage it, square again, and solve $27 - \sqrt{x-1} = 4$ to get $x - 1 = 529$. Thus $x = 530$ and the sum of the digits is $5 + 3 + 0 = 8$.

13. Solution (C). Note that

$$
\begin{aligned}
\sqrt{4 - 2\sqrt{3}} &= \sqrt{(\sqrt{3})^2 - 2\sqrt{3} + 1^2} \\
&= \sqrt{(\sqrt{3} - 1)^2} \\
&= \sqrt{3} - 1,
\end{aligned}
$$

so $a = 3$ and $b = 1$.

Alternatively, solve

$$\sqrt{4 - 2\sqrt{3}} = \sqrt{a} - b$$

by squaring both sides to get

$$4 - 2\sqrt{3} = a + b^2 - 2b\sqrt{a}.$$

Then it is easy to guess that $a = 3$ and $b = 1$, and check these values.

14. Solution (B). Note that $8^{\frac{1}{6}} = (2^3)^{\frac{1}{6}} = 2^{\frac{1}{2}}$. Rationalizing the denominator of the right side gives

$$\frac{7}{3 - \sqrt{2}} = \frac{7(3 + \sqrt{2})}{(3 - \sqrt{2})(3 + \sqrt{2})} = \frac{21 + 7\sqrt{2}}{7} = 3 + \sqrt{2}.$$

Thus the equation reduces to $x^{\frac{1}{3}} = 3$ or $x = 27$.

15. Solution (A). We have

$$\sqrt{26} - 5 = \frac{26 - 25}{\sqrt{26} + 5} = \frac{1}{\sqrt{26} + 5} < \frac{1}{5 + 5} = 10^{-1}.$$

That is, we have

$$0 < \sqrt{26} - 5 < 10^{-1}.$$

Thus

$$0 < (\sqrt{26} - 5)^n < 10^{-n}.$$

Therefore, in the decimal expansion of $(\sqrt{26} - 5)^n$ the n-th digit after the decimal point must be a zero.

33.3   Equations and Inequalities

1. Solution (C). Subtract the second equation from the first to get $a^2 - c^2 = 44$. Combine this with the third equation to get $a^2 = 144$; then it follows that

$$c^2 = 244 - 144 = 100, \text{ and } b^2 = 164 - 100 = 64.$$

2. Solution (A). Replace z with $-x$ in

$$(x + y)^2 + (y + z)^2 = 2x^2$$

to get

$$(x + y)^2 + (y - x)^2 = 2x^2$$

or

$$2x^2 + 2y^2 = 2x^2$$

which implies that $2y^2 = 0$. Hence $y = 0$.

3. **Solution** (C). We have

$$a^2 b^2 c^2 = 3 \cdot 5 \cdot 7 = 105.$$

Since $a > 0$, so are b and c. Hence $abc = \sqrt{105}$. Thus

$$c = \frac{abc}{ab} = \frac{\sqrt{105}}{3} = \sqrt{\frac{35}{3}}.$$

4. **Solution** (D). Expand $(x + y)^3$ to yield

$$
\begin{aligned}
216 = 6^3 &= (x + y)^3 \\
&= x^3 + 3x^2 y + 3xy^2 + y^3 \\
&= x^3 + y^3 + 3xy(x + y) \\
&= x^3 + y^3 + 3 \cdot 7 \cdot 6 \\
&= x^3 + y^3 + 126.
\end{aligned}
$$

So $x^3 + y^3 = 216 - 126 = 90$.

5. **Solution** (D). Square $x + y$ to get

$$(x + y)^2 = x^2 + y^2 + 2xy = 40 + 2(10 - x - y) = 60 - 2(x + y).$$

Let $t = x + y$, and note that $t^2 + 2t - 60 = 0$. Use the quadratic formula to find that $t = -1 + \sqrt{61} \approx 6.81$ or $t = -1 - \sqrt{61} \approx -8.81$.

6. **Solution** (E). From the first condition, either $x = 0$ or $y = a$. Case 1: If $x = 0$, then $xz = 0$. Case 2: If $x \neq 0$, then $y = a$ and since $a < b$, $y - b \neq 0$. Thus, from the second equation, $z = 0$ and so $xz = 0$. In either case $xz = 0$.

7. **Solution** (B). Since

$$a + b = a^2 - b^2 = (a + b) \cdot (a - b),$$

dividing both sides by $a + b \neq 0$ yields $1 = a - b$, so $a = b + 1$. Substituting this into $a + b = a \cdot b$ we get $2b + 1 = (b + 1) \cdot b$. The only positive solution of this quadratic equation is

$$b = \frac{1 + \sqrt{5}}{2}.$$

8. **Solution** (A). Divide both sides of the second inequality by $|a|$ to obtain $x \leq -1$. In this case

$$|x + 1| = -(x + 1) \quad \text{and} \quad |x - 2| = 2 - x,$$

so

$$|x + 1| - |x - 2| = -(x + 1) - (2 - x) = -3.$$

9. Solution (C). Using the identity $(a-b)^2 = (a+b)^2 - 4ab$, we get

$$\left(\sqrt{a} - \frac{1}{\sqrt{a}}\right)^2 = \left(\sqrt{a} + \frac{1}{\sqrt{a}}\right)^2 - 4 = 3^2 - 4 = 5,$$

hence $\sqrt{a} - \dfrac{1}{\sqrt{a}} = \sqrt{5}$. Thus

$$a - \frac{1}{a} = \left(\sqrt{a} + \frac{1}{\sqrt{a}}\right)\left(\sqrt{a} - \frac{1}{\sqrt{a}}\right) = 3\sqrt{5}.$$

10. Solution (B). The sum of two nonnegative numbers is zero if and only if they are both zero. The only values for which $|x^2 - 1| = 0$ are $x = \pm 1$ and those for which $|y^2 - 4| = 0$ are $y = \pm 2$. Hence there are four ordered pair solutions:

$$(1, 2), (1, -2), (-1, 2), (-1, -2).$$

11. Solution (C). Consider the four cases,

$$x < -1, -1 < x < 2, 2 < x < 4, \text{ and } 4 < x.$$

Each of these gives rise to a linear equation in x. Just two of these have solutions in the appropriate intervals, $x = 1$ and $x = 5$. Their sum is 6.

12. Solution (E). Each term must be zero. Thus $x = \pm 1, y = \pm 2$, and $z = \pm 3$. There are 8 ways to put these values into a triplet.

13. Solution (B). The equation is equivalent to the inequality $2x^2 - 9x + 6 \geq 0$, whose solution is given by $x \leq (9 - \sqrt{33})/4 \approx 0.8138\ldots$ or $x > (9 + \sqrt{33})/4 \approx 3.6861\ldots$

14. Solution (C). Because

$$\frac{1}{2} + \frac{1}{3} + \frac{1}{4} + \frac{1}{5} < 1.85,$$

one of the integers must be 1. Let a, b and c denote the others. Since

$$\frac{1}{3} + \frac{1}{4} + \frac{1}{5} = \frac{47}{60} < 0.85,$$

one of the numbers, say a, must be 2. Then

$$\frac{1}{b} + \frac{1}{c} = 0.35.$$

Since

$$0.35 - 1/3 = 5/300 = 1/60,$$

one set of integers that works is $\{1, 2, 3, 60\}$. But

$$1 + 2 + 3 + 60 = 66$$

is not one of the options. If $b = 4$, then

$$c = (.35 - .25)^{-1} = 10$$

also works, and

$$1 + 2 + 4 + 10 = 17$$

is one of the options. Note that these are the only solutions since

$$\frac{1}{5} + \frac{1}{6} \neq 0.35$$

and all other possible sums are less than 0.35.

15. Solution (E).

$$\frac{10^{12} - 10^{11}}{9} = \frac{10^{11}(10 - 1)}{9} = 10^{11}.$$

16. Solution (E). Option E is the smallest value:

$$A = \frac{250,386,765,412}{250,384,765,412} = 1 + \frac{2,000,000}{250,384,765,412},$$
$$B = \frac{250,384,765,412}{250,383,765,412} = 1 + \frac{1,000,000}{250,383,765,412},$$
$$C = \frac{250,385,765,412}{250,384,765,412} = 1 + \frac{1,000,000}{250,384,765,412},$$
$$D = \frac{250,386,765,412}{250,385,765,412} = 1 + \frac{1,000,000}{250,385,765,412},$$
$$E = \frac{250,387,765,412}{250,386,765,412} = 1 + \frac{1,000,000}{250,386,765,412}.$$

17. Solution (E). The numbers are all of the form $\dfrac{n}{n+2}$ for n equal to five consecutive odd numbers. Since $\dfrac{n}{n+2}$ gets larger as n gets larger, the fraction (E) is the largest.

18. Solution (D). The fractions $\dfrac{x}{5}$ and $\dfrac{x+1}{5}$ are both larger than 1. The other three are less than 1 and have the same numerator, so the smallest is the one with the largest denominator.

19. Solution (D). First note that $a = 7$ and that $\dfrac{1}{b + \dfrac{1}{c}} = 0.5$. So $b + \dfrac{1}{c} = 2$ and it follows that $b = c = 1$. Thus $a + b + c = 9$.

20. Solution (A). From the second equation, we have $\dfrac{a}{b} = \dfrac{d}{c}$. But the first equation asserts that $\dfrac{a}{b} = \dfrac{c}{d}$. Thus $\dfrac{c}{d} = \dfrac{d}{c}$, and so $\dfrac{a}{b} = \pm 1$. To see that none of the other choices can be correct, let $a = 1$, $b = -1$, $c = 2$, and $d = -2$.

21. **Solution** (B). The equation is solvable if $ax + b = cx + d$ which is equivalent to $(a - c)x = d - b$. This has a solution unless $a = c$. Therefore $a = c$ and it follows that $\dfrac{a^2}{a^2 + c^2} = \dfrac{1}{2}$.

22. **Solution** (D). Add together the three equations,

$$x = 3(y - 6), \ y = 3(z - 8), \text{ and } z = 3(x - 10)$$

to get $x + y + z = 3(x + y + z - 24)$, from which it follows that $x + y + z = 36$. This does not prove that such x, y, z exist. However, you can check to see that $x = 180/13$, $y = 138/13$ and $z = 150/13$ does work.

23. **Solution** (B). We can express the left side as a sum of squares:

$$
\begin{aligned}
& 3x^2 + 3y^2 + z^2 - 2xy + 2yz \\
= \ & x^2 - 2xy + y^2 + 2x^2 + y^2 + y^2 + 2yz + z^2 \\
= \ & (x - y)^2 + 2x^2 + y^2 + (y + z)^2.
\end{aligned}
$$

A sum of squares can be zero only if each one is zero. Thus $x = 0, y = 0, z = 0$ is the only solution.

24. **Solution** (E). Write $u = x + y$ and $v = x - y$, then u and v are positive integers and $v < u$. Now $1/u + 1/v = 1/3$ so $3 < v < 6$, so $v = 5$ or $v = 4$. If $v = 5$ then $u = 7.5$ which is not an integer. If $v = 4$ then $u = 12$, so we will have $x = 8, y = 4$. Therefore, $x^2 + y^2 = 64 + 16 = 80$.

25. **Solution** (E). The given equation is equivalent to

$$\frac{a + b}{ab} = \frac{2}{a + b},$$

which is true only if $a^2 + 2ab + b^2 = 2ab$, but this can happen only when $a^2 + b^2 = 0$, which implies that $a = b = 0$, in which case the equation is nonsense.

26. **Solution** (D). Since x and y are natural numbers, $y \geq 2$. Let $u = x + 1$ and $v = y - 1$, u and v are natural numbers with $u \geq 2$ and $v \geq 2$. One of $1/u$ and $1/v$ is at least half of $5/6$, so either $u \leq 12/5 = 2.4$ or $v \leq 12/5 = 2.4$. Consider the two cases: case one, $u = 2$ so $x = 1$ then $v = 3$ then $y = 4$; case two, $v = 2$ so $y = 3$ then $u = 3$ then $x = 2$. In either case, $x + y = 5$.

27. **Solution** (B). Subtracting twice the second equation from the first equation, we obtain $-5y - 5z = 5$ so that $y = -1 - z$. Adding the second and third equations yields $3y + 8z = 7$. We now have $3(-1 - z) + 8z = 7$. Therefore $z = 2, y = -3$, and $x = 1$. The sum is zero.

28. **Solution** (D). Subtract the second given equation from the third to get

$$c^2 - b^2 - a(c - b) = 0. \tag{1}$$

Dividing equation (1) by $(c - b)$ we have

$$b + c - a = 0. \tag{2}$$

Add the three given equations to obtain

$$a^2 + b^2 + c^2 - bc + a(c + b) = 21. \tag{3}$$

Using equation (2) we can substitute a for $c + b$ in equation (3) to obtain

$$a^2 + b^2 + c^2 - bc + a^2 = 21,$$

and then use the fact that $a^2 - bc = 7$ to obtain $a^2 + b^2 + c^2 = 14$. For completeness, note that the system *has* a solution, $a = 3, b = 2$, and $c = 1$.

29. Solution (A). Since

$$x^2 - 2x + 5 = (x - 1)^2 + 4$$

we may omit the absolute value sign on the right hand side. Now

$$x^2 - 2x - 3 = (x + 1)(x - 3)$$

is negative for $-1 < x < 3$, elsewhere it is not negative. Hence the equation is of the form $x^2 - 2x - 3 = x^2 - 2x + 5$, i.e, $-3 = 5$ when $x \leq -1$ or $x \geq 3$. In this case we have no solution. If $-1 < x < 3$ then the equation $-x^2 + 2x + 3 = x^2 - 2x + 5$ is equivalent to $2(x - 1)^2 = 0$. Hence $x = 1$, and this is a solution since $-1 < 1 < 3$.

Alternatively, since

$$|x^2 - 2x - 3| = |x^2 - 2x + 5|,$$

it must be the case that either

$$(x^2 - 2x - 3) = (x^2 - 2x + 5) \quad \text{or} \quad (x^2 - 2x - 3) = -(x^2 - 2x + 5).$$

The first has no solutions, and the second is equivalent to $2x^2 - 4x + 2 = 0$ which has just one solution, $x = 1$.

30. Solution (D). Subtract 3 from all parts to get $-2 \leq -4x \leq 8$, then divide all by -4 to get $1/2 \geq x \geq -2$, so the length of the interval is $1/2 - (-2) = 2.50$.

Alternatively, since the length is given by the difference, the shift of 3 and the negative sign on the 4 can be ignored. Simply calculate $(11 - 1)/4 = 2.5$.

31. Solution (A). Complete the square to find that

$$f(x) = x^2 + 2x + n = (x + 1)^2 - 1 + n > 10$$

if and only if $n > 11$.

Alternatively, the minimum value of $x^2 + 2x + n$ occurs at the vertex of the parabola, whose x-coordinate is given by

$$-b/2a = -2/2 = -1.$$

Thus

$$f(-1) = (-1)^2 + 2(-1) + n > 10$$

if and only if $n > 11$.

32. **Solution** (D). Since $\dfrac{a}{b} < \dfrac{c}{d}$, we have $ad < bc$. Add cd to both sides: $ad + cd < bc + dc$ so $d(a + c) < (b + d)c$. Dividing both sides by $d(b + d)$ yields the result. To see that the other four options fail, let $a = 5, b = c = 3$, and $d = 1$. This choice eliminates options A, B, and C while $a = 1, b = c = 2$, and $d = 3$ shows that E is not always true.

33. **Solution** (C). First, note that a and b agree in signs and that c has a different sign. Thus, a, b and $-c$ have the same sign. Therefore, $a - c$ and $b - c$ have the same sign. Hence their product must be positive. None of the other inequalities need be true.

34. **Solution** (B). $|x + 3| \leq 2$ is equivalent to $-2 \leq x + 3 \leq 2$. Adding 4 to each part of this inequality gives $2 \leq x + 7 \leq 6$. Therefore, $\dfrac{1}{6} \leq \dfrac{1}{x + 7} \leq \dfrac{1}{2}$ or $1 \leq \dfrac{6}{x + 7} \leq 3$. Thus, $a = 1$ and $b = 3$.

35. **Solution** (C). The inequality $(|x| - 2)(1 + x) > 0$ is satisfied if both factors are positive or if both are negative. Both are positive if $x > 2$ and both are negative if $-2 < x < -1$.

36. **Solution** (E). The left hand side simplifies to $\dfrac{2x + 1}{x - 1} \cdot \dfrac{x - 1}{2} > 0$ which is meaningless for $x = 1$ and otherwise it is $\dfrac{2x + 1}{2} > 0$. Thus the solution of the inequality is $x > -\dfrac{1}{2}$ but $x \neq 1$.

37. **Solution** (A). By the triangle inequality, note that

$$|x + y + z| = |x + 2 + y + 3 + z - 5| \leq |x + 2| + |y + 3| + |z - 5| = 1.$$

Hence A is the only possible value. On the other hand, $x = -2.5, y = -3$ and $z = 5.5$ satisfies the equation, so $|x + y + z|$ can be zero.

38. **Solution** (C). The expression $\dfrac{x^3 - y^3}{x - y}$ is the same as $x^2 + xy + y^2$ in case $x \neq y$, so $x^2 + xy + y^2 = 8$.

39. **Solution** (C). Let x denote the number. Then it follows from the given information that $x \cdot 1/x \cdot x^2 = 7$. This means that $x = \pm\sqrt{7}$. Since it is given that x is positive,

$$x + \dfrac{1}{x} \approx 2.645 + 1/2.645 \approx 3.02.$$

40. **Solution** (A). Divide both sides of the second inequality by $|a|$ to obtain $x \leq -1$. In this case $|x + 1| = -(x + 1)$ and $|x - 2| = 2 - x$, so

$$|x + 1| - |x - 2| = -(x + 1) - (2 - x) = -3.$$

41. Solution (C). Because $3.abc$ rounds to 3.5, $a = 4$ or $a = 5$. Since $b < a$, we see that $a = 5$. The conditions $a + b + c = 8$ and $a > b > c > 0$ force $b = 2$ and $c = 1$. Thus, $2a + bc = 12$.

42. Solution (B). First note that $a = 5$ and that $\dfrac{1}{b + \dfrac{1}{c}} = 0.4$. So $b + \dfrac{1}{c} = 2.5$ and it follows that $b = 2$ and $c = 2$. Thus $a + b + c = 9$.

43. Solution (A). Let x be the value of the continued fraction. Then $x = \dfrac{2}{1 + x}$ so that $x = 1$.

44. Solution (C). Let
$$x = 2 + \cfrac{1}{2 + \cfrac{1}{2 + \cfrac{1}{2 + \cdots}}}.$$

Then
$$2 + \frac{1}{x} = x$$

from which it follows that
$$x^2 - 2x - 1 = 0.$$

Because $x > 0$, we have
$$x = \frac{2 + \sqrt{4 - 4(-1)}}{2} = 1 + \sqrt{2}.$$

45. Solution (D). Note that x is embedded in its defining fraction, and that $x = \dfrac{1}{2 + \dfrac{1}{3 + x}}$. This is equivalent to the quadratic $2x^2 + 6x - 3 = 0$ which has two solutions,
$$\frac{-3 - \sqrt{15}}{2} \quad \text{and} \quad \frac{-3 + \sqrt{15}}{2},$$
the first of which is extraneous.

46. Solution (B). Compute $\left(a + \dfrac{1}{a}\right)^2$ to get
$$a^2 + 2 + \frac{1}{a^2} = 14 + 2,$$

so $a + \dfrac{1}{a}$ is either 4 or -4. Then
$$\left(a + \frac{1}{a}\right)^3 = a^3 + 3a + \frac{3}{a} + \frac{1}{a^3} = a^3 + \frac{1}{a^3} + 12 = 4^3 = 64,$$

so $a^3 + \dfrac{1}{a^3} = 52$.

47. Solution (E). Replace y^2 with $9 - x^2$ to get
$$x^2 - 3x^2 + 27 + 4x = 29 - 2x^2 + 4x - 2 = 29 - 2(x - 1)^2,$$

which is at most 29.

48. Solution (C). Add the three equations together to get

$$(*) \quad 2\left(\frac{1}{x}+\frac{1}{y}+\frac{1}{z}\right)=\frac{1}{3}+\frac{1}{5}+\frac{1}{7}=\frac{35+21+15}{105}=\frac{71}{105}.$$

Then subtract

$$2\left(\frac{1}{x}+\frac{1}{y}\right)=\frac{2}{3}$$

from both sides to get

$$\frac{2}{z}=\frac{71}{105}-\frac{70}{105}=\frac{1}{105},$$

so $z=210$. Subtracting

$$2\left(\frac{1}{x}+\frac{1}{z}\right)=\frac{2}{5}$$

from both sides of $(*)$ yields

$$\frac{2}{y}=\frac{71}{105}-\frac{42}{105}=\frac{29}{105},$$

so $y=\dfrac{210}{29}$. It follows that

$$\frac{z}{y}=\frac{210}{\frac{210}{29}}=29.$$

OR

Alternatively, change variables to $u=1/x$, $v=1/y$ and $w=1/z$. Then (using the new forms) subtract the second equation from the first to get $v-w=2/15$. Add this to the third to get $v=29/210$. Substitute into the third to get $w=1/210$. Thus $z/y=v/w=29$.

49. Solution (C). For integers larger than 1, addition produces smaller values than multiplication. Make the first operator $\times$ and the others $+$'s to get the minimum value

$$1\times 2+3+4+5+6+7+8+9=44.$$

50. Solution (D). Note that

$$\frac{5x-11}{2x^2+x-6}=\frac{A}{x+2}+\frac{B}{2x-3}=\frac{A(2x-3)}{(x+2)(2x-3)}+\frac{B(x+2)}{(x+2)(2x-3)}.$$

Also, note that

$$5x-11=(2A+B)x-(3A-2B),\quad 2A+B=5\quad\text{and}\quad 3A-2B=11.$$

Solving this system of equations we obtain $A=3$ and $B=-1$, so $A+B=2$.

51. Solution (D). Calculate x/w, y/w and compute the sum. For example,

$$x/w = 4/7 \cdot 14/3 \cdot 3/11 = 8/11$$

and

$$y/w = 14/3 \cdot 3/11 = 14/11,$$

so

$$(x+y+z)/w = (8+14+3)/11.$$

52. Solution (D). The denominator is $2(x - \sqrt{2})^2$, which vanishes at $x = \sqrt{2}$, and it is positive for all other values of x. Hence the inequality is equivalent to $\sqrt{2} \cdot x - 1 > 0$ but $x \neq \sqrt{2}$, that is, $x > 1/\sqrt{2}$ but $x \neq \sqrt{2}$.

53. Solution (D). Because both rs and $r + s$ are positive, both r and s are positive. Since $r - s > 0$, it follows that $s < r$ and thus $s/r < 1$. Thus either $s/r = 1/3$ or $s/r = 3/4$. If $s/r = 1/3$ then $3s = r$ and $r + s = 4s$ and $r - s = 2s$. In case $s/r = 3/4$, then

$$4s = 3r$$

and

$$r + s = r + 3r/4 = 7r/4$$

and

$$r - s = r - 3r/4 = r/4$$

so

$$r + s = 7(r - s).$$

It follows that $r + s = 7/3$ and $r - s = 1/3$. From this it follows that $r = 4/3$, $s = 1$ and $r^2 + s^2 = 25/9$.

An alternate solution starts the same way, with rs and $r + s$ positive, so both r and s are positive. Thus $r > s$ and $s/r < 1$. But then $r + s > r - s$, so $r + s$ must be greater than 1 since only two of the numbers are less than 1. Consider

$$(r + s)2 = r^2 + 2rs + s^2.$$

If $r + s = 4/3$, then s/r and $r - s$ are the two numbers less than one leaving $rs = 7/3$. But then the left side is $(r + s)^2 = 16/9$ and the right side is

$$r^2 + 2rs + s^2 = 14/3 + r^2 + s^2 > 14/3 > 16/9,$$

a contradiction. So we must have $r + s = 7/3$ and

$$49/9 = (r + s)^2 = r^2 + 2rs + s^2 \leq r^2 + s^2 + 8/3.$$

Thus

$$25/9 \leq r^2 + s^2 < 49/9 < 6$$

and therefore the only choice is $25/9$.

33.4   Word Problems

1. **Solution** (E). At the 10 minute mark, the tank has twice as much oil in it as it did at the 9 minute mark. So the tank became half-full at the 9 minute mark.

2. **Solution** (E). The value of a $3 \times 3 \times 3$ cube is

$$(27/8)(320) = 27 \cdot 40 = 1080$$

dollars.

3. **Solution** (E). Originally each sack holds $\dfrac{c}{k}$ pounds of coffee. With $k + n$ sacks, each sack holds $\dfrac{c}{k + n}$ pounds, so the difference is

$$\frac{c}{k} - \frac{c}{k+n} = \frac{c(k+n) - ck}{k(k+n)} = \frac{cn}{k^2 + kn}.$$

4. **Solution** (E). Albert pays $8 for his 8/3 loaves, so loaves must be worth $3 each. Nick eats all but $1/3$ of a loaf of his bread while Dick gives up $7/3$ loaves. Thus Dick should get $7 of Albert's money.

5. **Solution** (C). Solve simultaneously the three equations

$$q - 12 = n, \quad n + d + q = 162 \quad \text{and} \quad 5n + 10d + 25q = 2200$$

to get $d = 70$.

6. **Solution** (B). The value of the car decreases by $1800 each year, so the value after 6 years is $20,000 - 6 \cdot 1800 = \$9200$.

7. **Solution** (D). Let x be the number of coins in the larger pile and y be the number of coins in the smaller pile. Then $x^2 - y^2 = 24(x - y)$. Dividing both sides by $x - y$ (since $x - y$ is nonzero) yields $x + y = 24$. So I have $6 worth of quarters.

8. **Solution** (D). Use the formula $A = Pe^{rt}$ where r is the annual rate of interest, t is the time in years, P is the principle, and A is the amount in the account at time t. Then $2800 = 2000e^{4r}$, which implies that $r = \log 1.4/4$. So the amount in the account after $4 + 16 = 20$ years is

$$A = 2000e^{20r} = 2000 \cdot \left(e^{\ln 1.4/4}\right)^{20} = 2000(1.4)^5 = 10756.48.$$

OR

Since 16 is an integer multiple of 4, all that matters is that we start with \$2000 and have \$2800 after four years–the type of compounding doesn't matter. Based on compounding on a four year cycle, the (decimal) rate is $800/2000 = .4$. So we simply use 5 "compoundings" at this rate to see that the amount in the account will be

$$A = 2000 \cdot (1.4)^5 = 10,756.48.$$

9. **Solution** (E). Let x equal the amount each student contributed. If each student gave an amount equal to the total number of students, then x^2 represents the total value of the collection, \$529. Hence, $x = \$23$. Since each student used the same five bills, the only combination of bills that each student paid must be 2 ten-dollar bills and 3 one-dollar bills. Therefore there are $23 \cdot 2 = 46$ ten dollar bills collected.

10. **Solution** (E). Let x denote the number of apples each has for sale. Then Ms Blue should receive x dollars and Mr Green $1.5x$ dollars. By mixing and selling $2x$ apples at 5 for \$6, Mr Green is effectively charging \$1.20 per apple. At this rate he is losing \$0.30 an apple and Ms Blue is receiving an extra \$0.20 per apple. As he is \$80 short, $80 = 0.3x - 0.2x = 0.1x$ and we have $x = 800$. At the end of the day, Ms Blue made an extra \$160 and Mr Green lost \$240.

11. **Solution** (B). Suppose there are m male students in the class. Then there are $3m/4$ females. The number of Spanish speaking students is therefore $\frac{2}{3} \cdot \frac{3m}{4} + \frac{1}{2}m$. Thus, the fraction of the class who can speak Spanish is
$$(\frac{1}{2}m + \frac{1}{2}m)/(m + \frac{3}{4}m) = \frac{4}{7}.$$

OR

Suppose there are 42 students in the class. Then 18 are female, and 12 of these speak Spanish. There are 24 males and 12 of these speak Spanish, so the fraction of the class who can speak Spanish is $24/42 = 4/7$.

12. **Solution** (A). 25% of the floor is NOT painted in red. 30% of the floor is NOT painted in green. 35% of the floor is NOT painted in blue. $25\% + 30\% + 35\% = 90\%$. That means at least 10% are not painted in the three colors. One can show that such a coloring exists.

13. **Solution** (D). Let x denote the required number of girls. Solve the equation $\frac{g+x}{b+g+x} = .60$ for x to get $x = 1.5b - g$.

14. **Solution** (E). The new volume must be $22 + z$ and the amount of propanol is $5 + z$, so the fraction $(z+5)/(z+22)$ must be 0.4.

15. Solution (E). The non-watery part of fresh cucumbers makes up 1% of their weight. In "dry" cucumbers the same non-watery amount represents 2% of the weight. That means the weight halves.

16. Solution (D). The percent of students who have to take the course twice is 35%. The percent of students who have to take the course 3 times is $.3 \times .35 = .105$. The percent of students who have to take the course more than 3 times is $.3 \times .35 \times .5 = 0.0525$.

17. Solution (B). The quantity xy becomes $\frac{3}{2}x \cdot \frac{3}{4}y$ which represents a $\frac{1}{8} = 12.5\%$ increase in $\frac{z}{4}$, hence also for z.

18. Solution (C). In order to run 700 yards, the runners must traverse both semicircular ends. The runner in the first lane has an inner radius of 50 yards, while the runner in the fifth lane has an inner radius of 54 yards. The difference in distance is

$$2\pi(54 - 50) = 8\pi \approx 25 \text{ yards.}$$

19. Solution (C). The first mile costs $\$4.50$, and every mile thereafter costs $\$3.00$, so item C. is the right choice.

20. Solution (E). Let r and t denote the rate in mph and time in hours, respectively. Then $rt = 10$ and $(r + 2)(t - 1/3) = 10$. Solve these simultaneously to get the quadratic $r^2 + 2r - 60 = 0$. Solve this using the quadratic formula to get the positive root $r = -1 + \sqrt{61} \approx 6.81$.

21. Solution (A). David runs at 7 mph for t hours while Jeremy runs for $t + 1/2$ hours at 5 mph. They run the same distance, so $5 \cdot (t + 1/2) = 7t$, which yields $t = 5/4$ hours, or 75 minutes.

22. Solution (D). Vic is 20/19 times as fast as Harold. Harold is 20/18 times as fast as Charlie, so Vic is $(20/19)(20/18) \approx 1.1696$ times as fast as Charlie. When Vic runs 2 miles, Charlie will have run $2/1.1696 \approx 1.71$ miles. Vic will finish 0.29 miles ahead of Charlie.

23. Solution (D). There is more information given than is needed. Let m, a, t, h denote the rates, in jobs per hour, of Mathias, Anders, Tellis and Hal respectively. Then

$$(m + a) \cdot 1188 = 1$$

and

$$(t + h) \cdot 1890 = 1.$$

From this it follows that

$$1 = (m + a + t + h)x = \left(\frac{1}{1188} + \frac{1}{1890}\right) \cdot 1$$

and from this we get $x = 729.473\ldots \approx 729.5$ hours.

24. Solution (E). Let $3d$ be the number of miles in the entire route. Then the time t, in hours, for the cyclist to cover the entire route is

$$t = \frac{d}{16} + \frac{d}{24} + \frac{d}{r}$$

where r is the rate in miles per hour over the downhill portion of the route. Hence, the average rate in miles per hour over the entire route is

$$\frac{3d}{t} = 3\frac{(48r)}{5r + 48}.$$

Equating this result to 24 gives $r = 48$.

25. Solution (D). Within $10 \times 14 = 140$ days Benny will eat 10 boxes of cereal alone, while together with Nathan they will eat $140 \div 10 = 14$ boxes for the same time period. That means the share of his brother is $14 - 10 = 4$ boxes for 140 days. Therefore, Nathan eats one box of cereal for $140 : 4 = 35$ days.

26. Solution (B). Suppose that they walk $4x$ miles, so that they walk x miles uphill, x miles downhill, and $2x$ level miles. This takes a total time of $\frac{x}{2} + \frac{x}{6} + \frac{2x}{4}$ hours. Setting this equal to 5 hours, we find $x = 30/7$ so that the total distance is $4x = 120/7 \approx 17.1$ miles.

27. Solution (D). Danica has driven for $250/150 = 5/3$ hours so far. If she drives the remaining 250 miles in x hours we need $\frac{500}{5/3 + x} = 180$. This gives $x = 10/9$ so her average speed must be $250/x = 225$ mph.

28. Solution (E). The messenger reaches the second army in one hour (the messenger and the second army advance at each other at combined speed of 10 mph). At that time, the two armies are 8 miles apart. It takes 0.8 hours for the messenger to get back to the first army. At that time the armies are $8 - 0.8 - 0.8 = 6.4$ miles apart.

29. Solution (A). Experimenting with the numbers in turn, dividing 6 into 15 and 10 into 15, we get answers one whole number apart, so they are together again at 15 minutes.

30. Solution (D). The dog is at home for the first time after 18 minutes. Then he runs back to Bill at double speed. It takes $\frac{18}{3} = 6$ minutes for the dog to run back towards Bill. Hence Bill and the dog meet the first time after $18 + 6 = 24$ minutes. Since $24 = \frac{2}{3}36$, at that moment, Bill has still a third of his way to walk, and the dog begins its second leg. The second leg of the dog, forward to home and backward to Bill takes $\frac{1}{3}$ of the time of the first leg, hence $\frac{24}{3} = 8$ minutes. Thus Bill and the dog meet the second time after $24 + 8 = 32$ minutes.

31. Solution (A). Let x denote the number of miles Jill had traveled in the first 40 minutes. Since this was one third of the way to Center City, it took her a total of 2 hours to get to Center City. Next let y be the number of miles she had left to get to Center City after she had gone the additional 21 miles. We have $2x = 21 + y$, and the total trip is

$$3x + y = x + 2y + 21.$$

She can travel $1.5x = 4y/3$ miles in 1 hour. Thus $9x = 8y$. Solve to get $x = 24$ and $y = 27$. The total distance is 99 miles.

32. **Solution** (A). When it hits the floor the first time, it has traveled 1 meter. Then it rebounds to a height of 1/2 meter and falls back to the floor to hit the floor for the second time. This adds 1 more meter distance so it now has traveled 2 meters. After the second hit, the ball rebounds to a height of (1/4) meter and the total distance traveled when the ball hits the floor the third time is

$$1 + 2(1/2) + 2(1/4) = 1 + 1 + (1/2).$$

The total distance traveled when the ball hits the floor for the 40th time is

$$1 + 1 + \ldots + (1/2)^{38} = 1 + \frac{1 - (1/2)^{39}}{1 - 1/2} = 3 - (1/2)^{38}.$$

33. **Solution** (B). Let S denote the time in hours required to swim a mile. Then the time required to run a mile is $\frac{S}{2}$, and the time needed to cycle a mile is $\frac{S}{3}$. It follows that $S + \frac{1}{2}S + \frac{1}{3}S - \frac{1}{6} = 3\left(\frac{1}{3}S\right)$, so $S = \frac{1}{5}$, and $S + \frac{1}{2}S + \frac{1}{3}S = \frac{11}{6}\frac{1}{5} = \frac{11}{30}$ hours, which is 22 minutes.

Alternatively, let s be the swimming speed, r the running speed and c the cycling speed in miles per minute. Then his total time for the three miles is

$$t = (1/s) + (1/r) + (1/c) = 10 + (3/c).$$

Also, $r = 2s$ and $c = 1.5r$. So $c = 3s$, $1/s = 3/c$ and $1/r = 3/2c$. Subbing in yields $5/2c = 10$ and $1/c = 4$. So the total time is $10 + 3(4) = 22$ minutes.

34. **Solution** (A). Every hour, the hour hand covers $360/12 = 30$ degrees. At 2:20, the hour hand is $2 \cdot 30 + (1/3) \cdot 30 = 70$ degrees from the vertical. The minute hand is 120 degrees from the vertical, so there are 50 degrees between the two hands at 2:20.

35. **Solution** (B). The minute hand moves 360 degrees in 60 minutes, or 6 degrees per minute. The hour hand moves 360 degrees in 12 hours or 0.5 degrees per minute. Since the minute hand travels 5.5 degrees per minute faster than the hour hand, it gains 90 degrees every $90/5.5$ minutes. In 24 hours (1440 minutes) 90 degrees is gained $1440/(90/5.5) = 88$ times. For half of these times, the angular difference between the hands is 180 or zero degrees. The hands are at a right angle the remaining 44 times.

36. Solution (C). Assume there are x adults and y children, then $x + y = 100$ and $3x + (1/3)y = 100$. Solving simultaneously leads to $x = 25$ and $y = 75$.

37. Solution (C). The right triangle formed by the trunk of the tree, the path of the feather and its projection on the ground, has hypotenuse 18 and legs 14 and D, so that

$$D^2 + 14^2 = 18^2,$$

or

$$D^2 = 324 - 196 = 128;$$

since $11^2 = 121 < 128 < 144 = 12^2$, we have $11 < D < 12$.

38. Solution (A). From the payment of the first cowboy we find the price of 8 sandwiches, 2 cups of coffee, and 20 doughnuts = $16.90. From the payment of the second cowboy we calculate the price of 9 sandwiches, 3 cups of coffee, and 21 doughnuts = $18.90. The difference of the sums

$$18.90 - 16.90 = 2.00$$

is exactly the price of one sandwich, a cup of coffee, and a doughnut.

39. Solution (B). Let the distance from C to D be x. Using the information given, we can represent the location of the treasure in two ways, which gives the following equation:

$$\frac{x + 14}{2} = 7 + (1/3)(x + 7) - \frac{1}{4}\left[\frac{1}{3}(x + 7) + 1\right].$$

Solve this to get $x = 6$.

40. Solution (C). Let q be the distance traveled by the courier to get to the front of the convoy and let p be the distance traveled by the convoy during the same time. We have a relation: $q = p + 3$. To reach the rear of convoy when the rear has advanced 6 miles from the starting point, the courier needs to travel $q - 6$ miles back (after reaching the front). As a fraction of the forward journey, we have $(q - 6)/q$. Thus the convoy will move $p(q - 6)/q$ while the courier is returning to the rear. We have

$$6 = p + p(q - 6)/q = q - 3 + (q - 3)(q - 6)/q.$$

Multiply both sides by q, and rearrange to get

$$q^2 - 9q + 9 = 0.$$

Thus $q = (9 + 3\sqrt{5})/2$. The total distance the courier travels: $2q - 6 = 3 + 3\sqrt{5}$.

41. Solution (D). If the square of the length $\overline{CB_k}$ is q, then the square of the length of CB_{k+1} is $a + 4$. To have B_{k+1} outside the circle of radius 20 with center C, we need $q + 4$ to be greater than 400. The sequence of distances from $A = B_j$ on runs as follows:

$$\overline{CB_j} = \sqrt{100}, \quad \overline{CB_{j+1}} = \sqrt{104}, \quad \overline{CB_{j+2}} = \sqrt{108}.$$

In general, $\overline{CB_{j+i}} = \sqrt{100 + 4i}$. To have B_{j+i} outside we need $100 + 4i > 400$, so $i > 75$ (if $i = 75$, B_{j+i} is on the park boundary).

42. **Solution** (B). Denote Joe's speed by v_1, the river's speed by v_2. Swimming 1 mile upstream and 1 mile downstream takes

$$1/(v_1 - v_2) + 1/(v_1 + v_2)$$

units of time, swimming 2.2 miles in a lake takes $2.2/v_1$ units of time. The equation

$$\frac{1}{v_1 - v_2} + \frac{1}{v_1 + v_2} = \frac{2.2}{v_1}$$

can be re-written as

$$\frac{1}{1 - v_2/v_1} + \frac{1}{1 + v_2/v_1} = 2.2,$$

which implies that $(v_2/v_1)^2 = 1/11$ and thus $v_1/v_2 = \sqrt{11} \approx 3.32$.

Alternatively, let J be Joe's speed, R be the speed of the river and $Q = J/R$ the number we are looking for. Then (using the same reasoning as in the other solution) have

$$\frac{1}{QR - R} + \frac{1}{QR + R} = \frac{2.2}{QR}.$$

Note that the R cancels, so simply get a common denominator for the left-hand side, cross multiply and use a little algebra to get $Q^2 = 11$.

43. **Solution** (A). Let x denote the number of days John did not work. Then he worked $A - x$ days and so earned

$$B(A - x) - Cx = D \quad \text{dollars.}$$

Solving this for x, we get

$$-(B + C)x = D - AB$$

and so

$$x = (AB - D)/(B + C).$$

Taken from Horatio Nelson Robinson's 1859 book 'A Theoretical and Practical Treatise on Algebra', with thanks to Dave Renfro.

44. **Solution** (A). Let a be the rate at which Abe paints, b the rate Ben paints and c the rate Cal paints. Also let H be the total number of square feet of painting to do in the house. Then

$$11a + 11b = H, 9a + 9c = H \text{ and } 9.9b + 9.9c = H.$$

An equivalent system is

$$a + b = H/11, a + c = H/9 \text{ and } b + c = H/9.9.$$

Solving this systems yields

$$a = H(1/9 + 1/11 - 1/9.9) = 5H/99,$$

$$b = H(1/11 + 1/9.9 - 1.9) = 4H/99$$

and

$$c = H(1/9 + 1/9.9 - 1/11) = 6H/99.$$

Cal paints for 5.5 hours, so he paints exactly $1/3$ of the house. Abe paints in 4 hours what Ben paints in 5. So by $1:30$, Abe has painted more of the house than Ben. Since Abe paints faster than Ben and Cal painted exactly one third of the house, Abe painted the most and Ben the least.

33.5   Sequence and Series

1. Solution (D). The number of games played is $18(1 + 2 + \cdots + 9) = 810$. If n is the number of wins of the last-place team, and d is the common difference of wins between successive teams, then

$$n + (n + d) + (n + 2d) + \cdots + (n + 9d) = 10n + 45d = 810,$$

so $2n + 9d = 162$. Now n is the maximum when d is a minimum (but not zero, because there are no ties). The smallest integral value of d for which n is integral is $d = 2$. Thus $n = 72$.

2. Solution (C). Suppose there is a sequence of n consecutive integers beginning with u having a sum of -1. Regrouping and factoring we obtain,

$$\begin{aligned} u + (u+1) + (u+2) + \cdots + (u+n-1) &= (n) \cdot u + 1 + 2 \cdots + n - 1 \\ &= (n)\left(u + \frac{n-1}{2}\right) \\ &= -1 \end{aligned}$$

Since $n \geq 1$, there are only two possible ways to get -1 as such a product: $n = 1$, in which case we get $u = -1$ and $n = 2$, in which case we get $u = -1$. Thus, there are exactly two ways to write -1 as the sum of one of more consecutive integers: $-1 = -1$ and $-1 + 0 = -1$.

3. Solution (C). We have

$$1, 1+2, 1+2+6, 1+2+6+12, 1+2+6+12+36, 1+2+6+12+36+72\ldots$$

Condense the sequence by looking at only the even days, combining the additions as follows:

$$3 + 3 \cdot 6 + 3 \cdot 36 + 3 \cdot 216.$$

On the 24th, we have

$$3 + 3 \cdot 6 + 3 \cdot 36 + 3 \cdot 216 + \cdots + 3 \cdot 6^{11} = 3(6^{12} - 1)/(6 - 1).$$

4. Solution (C). It is easy to see that

$$a_n = \sqrt{n+1} - \sqrt{n}.$$

Consequently,

$$\sum_{n=1}^{99} a_n = \sqrt{100} - 1 = 9.$$

5. Solution (C). Each summand sum can be split into the difference of two unit fractions:

$$\frac{1}{1 \cdot 2} + \frac{1}{2 \cdot 3} + \cdots + \frac{1}{10 \cdot 11}$$

$$= \left(1 - \frac{1}{2}\right) + \left(\frac{1}{2} - \frac{1}{3}\right) + \left(\frac{1}{3} - \frac{1}{4}\right) + \cdots + \left(\frac{1}{10} - \frac{1}{11}\right)$$

$$= 1 - \frac{1}{11} = \frac{10}{11}.$$

6. Solution (E).

Since $\dfrac{1}{k} - \dfrac{1}{k+1} = \dfrac{1}{k \cdot (k+1)}$, we get

$$\frac{1}{1 \cdot 2} + \frac{1}{2 \cdot 3} + \frac{1}{3 \cdot 4} + \cdots + \frac{1}{n \cdot (n+1)}$$

$$= \left(1 - \frac{1}{2}\right) + \left(\frac{1}{2} - \frac{1}{3}\right) + \cdots + \left(\frac{1}{n} - \frac{1}{n+1}\right)$$

$$= 1 - \frac{1}{n+1},$$

which is always less than 1.

7. Solution (B). The initial amount of cheese is 1kg; since on the kth day the mouse eats $\dfrac{1}{(k+1)^2}$ of the remaining cheese, the amount remaining after n days is

$$\left(1 - \frac{1}{2^2}\right)\left(1 - \frac{1}{3^2}\right)\left(1 - \frac{1}{4^2}\right) \cdots \left(1 - \frac{1}{(n-1)^2}\right)\left(1 - \frac{1}{n^2}\right)\left(1 - \frac{1}{(n+1)^2}\right)$$

$$= \frac{1 \cdot 3}{2 \cdot 2} \cdot \frac{2 \cdot 4}{3 \cdot 3} \cdot \frac{3 \cdot 5}{4 \cdot 4} \cdots \frac{(n-2) \cdot n}{(n-1) \cdot (n-1)} \cdot \frac{(n-1) \cdot (n+1)}{n \cdot n} \cdot \frac{n \cdot (n+2)}{(n+1) \cdot (n+1)}$$

$$= \frac{1}{2} \cdot \frac{n+2}{n+1},$$

so that the consumed amount is

$$1 - \frac{1}{2} \cdot \frac{n+2}{n+1} = \frac{1}{2} \cdot \frac{n}{n+1}.$$

For $n = 42$, this is

$$\frac{1}{2} \cdot \frac{42}{43} \approx 0.48837.$$

8. **Solution** (B). Consider the sequence in terms of its remainders when divided by 3, one gets a repeating cycle of length 8: $1, 1, 2, 0, 2, 2, 1, 0, \ldots$. That is, every fourth number in the sequence is divisible by 3. Thus the answer is $1000/4 = 250$.

9. **Solution** (D). Let us simplify S from the right end:

$$-2006! \cdot 2008 + 2007! = 2006!(-2008 + 2007) = -2006!.$$

Thus,

$$S = 1! \cdot 3 - 2! \cdot 4 + 3! \cdot 5 - 4! \cdot 6 + \ldots + 2005! \cdot 2007 - 2006!.$$

Next,

$$2005! \cdot 2007 - 2006! = 2005!(2007 - 2006) = 2005!$$

Therefore,

$$S = 1! \cdot 3 - 2! \cdot 4 + 3! \cdot 5 - 4! \cdot 6 + \ldots - 2004! \cdot 2006 + 2005!$$

Continuing simplification of S we finally obtain that $S = 1! = 1$.

Alternatively, for each

$$n, \quad n! \cdot (n+2)$$

can be rewritten as

$$n! \cdot (n+2) = n! + n!(n+1) = n! + (n+1)!.$$

Do this from the beginning to get

$$
\begin{aligned}
S &= (1! + 2!) - (2! + 3!) + (3! + 4!) - \cdots \\
&\quad + (2005! + 2006!) - (2006! + 2007!) + 2007! \\
&= 1! = 1.
\end{aligned}
$$

10. **Solution** (C). Notice that

$$\sum_{n=1}^{\infty} \frac{1}{n^2} = \sum_{n=1}^{\infty} \frac{1}{(2n-1)^2} + \sum_{n=1}^{\infty} \frac{1}{(2n)^2}.$$

Therefore,

$$
\begin{aligned}
\sum_{n=1}^{\infty} \frac{1}{(2n-1)^2} &= \sum_{n=1}^{\infty} \frac{1}{n^2} - \frac{1}{4} \sum_{n=1}^{\infty} \frac{1}{n^2} \\
&= \frac{3}{4} \sum_{n=1}^{\infty} \frac{1}{n^2} \\
&= \frac{3}{4} \left(\frac{\pi^2}{6} \right) \\
&= \frac{\pi^2}{8}.
\end{aligned}
$$

11. Solution (B). Since

$$1 + 1/2^2 + 1/3^2 + 1/4^2 > 1.4,$$

we can rule out option A. On the other hand,

$$
\begin{aligned}
S &= \sum_{n=1}^{100} \frac{1}{n^2} \leq \sum_{n=1}^{100} \frac{1}{n^2 - (.5)^2} \\
&= \sum_{n=1}^{100} \left(\frac{1}{n - .5} - \frac{1}{n + .5} \right) \\
&= \frac{1}{1 - \frac{1}{2}} - \frac{1}{100 + \frac{1}{2}} \\
&< \frac{1}{1 - \frac{1}{2}} = 2.
\end{aligned}
$$

Alternatively, the sum $1 + 1/4 + 1/9 + 1/16$ is between 1.42 and 1.43. So the sum is larger than 1.4. For the rest use some rather crude estimates based on the noticing that

$$1/n^2 + 1/(n+1)^2 + \cdots + 1/(2n-1)^2$$

is less than $n \cdot (1/n^2) = 1/n$ and greater than $n \cdot (1/(2n)^2) = 1/4n$. So

$$0.05 = 1/20 < 1/5^2 + 1/6^2 + \cdots 1/9^2 < 1/5 = 0.2,$$

$$1/40 = 0.25 < 1/10^2 + 1/11^2 + \cdots + 1/19^2 < 1/10 = 0.1,$$

$$1/80 = 0.0125 < 1/20^2 + \cdots + 1/39^2 < 1/20 = 0.05,$$

$$1/160 = 0.00625 < 1/40^2 + \cdots + 1/79^2 < 1/40 = 0.025$$

and

$$21 \cdot (1/100^2) = 0.0021 < 1/80^2 + \cdots + 1/100^2 < 21 \cdot (1/80^2) = 0.00328125.$$

Thus the entire sum is between

$$1.42 + 0.05 + 0.025 + 0.0125 + 0.00625 + 0.0021 = 1.50335$$

and

$$1.43 + 0.2 + 0.1 + 0.05 + 0.025 + 0.00328125 = 1.80828125 < 2.$$

33.6 Defined Operations and Functions

1. Solution (E). The operation can be written

$$x \oplus y = \left(\frac{1}{x} + \frac{1}{y} \right)^{-1}.$$

It is not hard to see that it is commutative. A straightforward calculation shows that it is associative as well. Also,

$$x \oplus x = \left(\frac{1}{x} + \frac{1}{x}\right)^{-1} = x/2.$$

2. **Solution** (D). $f(6) = f(2+4) = f(5) \cdot f(4) = (f(4) \cdot f(3)) \cdot f(4) = 2^2 \cdot 2 = 8.$

3. **Solution** (E). Substituting $n = 5$ into the given equation, we find that

$$f(5) = 23 - 8 = 15.$$

Then substituting $n = 4$, we obtain $f(3) = 15 - 8 = 7$. Then $n = 3$ yields

$$f(2) = 8 - 7 = 1.$$

Finally, $n = 2$ gives

$$f(1) = 7 - 1 = 6.$$

Thus, $f(1) + f(3) = 13$.

4. **Solution** (E). First, evaluate the two occurrences of g, and then take f of the results.

$$g\left(-1, -\frac{3}{2}\right) = \max\left(-1, -\frac{3}{2}\right) - \min\left(-1, -\frac{3}{2}\right)$$
$$= -1 - \left(-\frac{3}{2}\right) = \frac{1}{2}$$

and

$$g(-4, -1.75) = \max(-4, -1.75) - \min(-4, -1.75)$$
$$= -1.75 - (-4)$$
$$= 2.25 = \frac{9}{4}.$$

Then,

$$f\left(\frac{1}{2}, \frac{9}{4}\right) = \max\left(\frac{1}{2}, \frac{9}{4}\right)^{\min\left(\frac{1}{2}, \frac{9}{4}\right)}$$
$$= \left(\frac{9}{4}\right)^{\frac{1}{2}} = \frac{3}{2}.$$

Notice that $g(x, y) = |x - y|$ is just the distance between the numbers x and y.

5. **Solution** (C). The two-fold composition $F \circ F(k)$ of F with itself evaluated at k results in one of the four values $\frac{k}{9}, 2\left(\frac{k}{3}\right) + 1, \frac{2k+1}{3}$ or $2(2k+1) + 1$. Only the second and third of these give values of k for which $F(F(k)) = k$. These values are 1 and 3.

6. Solution (E). Set each piece of g equal to -4 : $|x| - 2 = -4$ has no solutions; $x - 3 = -4$ only for $x = -1$ which is not in the domain of that piece; and $3 - x = -4$ only when $x = 7$, and 7 does belong to the domain of that piece.

Alternatively, you can plug into g the five options.

7. Solution (E). We have

$$f(0) = 2f(1) - f(-1) = 8 - 2 = 6.$$

Substituting $f(0) = 6$ into $f(1) = 2f(2) - f(0)$ gives $4 = 2f(2) - 6$. Thus $f(2) = 5$.

8. Solution (C). The functions shown are A, $f(-x)$; B, $|f(x)|$; C, the answer key; D, $|f(-x)|$; and E, $|f(|x|)|$.

9. Solution (D). Since

$$f(x) = -2(x-2)^2 + 5,$$

the maximum value of f is 5, the function is continuous and not bounded from below.

10. Solution (D). Note that

$$f(1) = \frac{1-b}{1-a},$$

so $f(1)$ is undefined for $a = 1$. Because $f(2) = 0$, it follows that $2 - b = 0$, and $b = 2$. Finally,

$$f(1/2) = (1/2 - 2) \div (1/2 - 1) = 3.$$

11. Solution (A). The y-intercept c is negative, and the constant a is positive since the parabola opens upward. The x-coordinate of the vertex which is visibly positive, is given by the formula $x = -b/2a$. Since $a > 0$, the number $-b/2a$ is positive only if $b < 0$. Thus, $ab < 0$ and $ac < 0$.

12. Solution (C). In fact,

$$f(n) = 2n + 3,$$

and

$$T = f(f(f(f(5)))) = f(f(f(13))) = f(f(29)) = f(61) = 125.$$

13. Solution (C). Note that

$$f(1 + f(1)) = 4f(1) = 16,$$

so $f(5) = 16$. Next,

$$f(21) = f(5 + f(5)) = 4f(5) = 64.$$

14. Solution (D).

$$f(f(f(x))) = a^3 x + a^2 b + ab + b = 8x + 21$$

so $a = 2$ and $7b = 21$. Therefore $a + b = 5$.

15. Solution (B). Computing

$$f^{(2)}(x), f^{(3)}(x), f^{(4)}(x) \text{ and } f^{(5)}(x),$$

we see that

$$f^{(2)}(x) = -\frac{1}{x}, f^{(3)}(x) = -\frac{x+1}{x-1}, f^{(4)}(x) = x$$

and

$$f^{(5)}(x) = f(x).$$

Since the remainder r when 2007 is divided by 4 is 3 $(2007 = 4 \cdot 501 + 3)$, it follows

$$f^{(2007)}(x) = f^{(3)}(x) = -\frac{x+1}{x-1}.$$

16. Solution (B). The largest integer with fourth root closest to k is $\left\lfloor \left(k+\frac{1}{2}\right)^4 \right\rfloor$. Why? Then

$$
\begin{aligned}
\left\lfloor \left(k+\frac{1}{2}\right)^4 \right\rfloor &= \left\lfloor k^4 + 2k^3 + \frac{3}{2}k^2 + \frac{1}{2}k + \frac{1}{16} \right\rfloor \\
&= \left\lfloor k^4 + 2k^3 + \frac{1}{2}\left(3k^2 + k\right) + \frac{1}{16} \right\rfloor \\
&= k^4 + 2k^3 + \frac{1}{2}\left(3k^2 + k\right) \quad \text{since } 3k^2 + k \text{ is even.}
\end{aligned}
$$

Therefore, the number of integers with fourth root closest to k is

$$
\begin{aligned}
&\left\lfloor \left(k+\frac{1}{2}\right)^4 \right\rfloor - \left\lfloor \left((k-1)+\frac{1}{2}\right)^4 \right\rfloor \\
&= \left\lfloor \left(k+\frac{1}{2}\right)^4 \right\rfloor - \left\lfloor \left(k-\frac{1}{2}\right)^4 \right\rfloor \\
&= \left\{ k^4 + 2k^3 + \frac{1}{2}\left(3k^2 + k\right) \right\} - \left\{ k^4 - 2k^3 + \frac{1}{2}\left(3k^2 - k\right) \right\} \\
&= 4k^3 + k.
\end{aligned}
$$

That is, $f(n) = k$ for $4k^3 + k$ (consecutive) values of n. Since

$$f(1995) = 7, \sum_{k=1}^{6}(4k^3 + k) = 1785$$

and

$$f(1786) = f(1787) = \ldots = f(1995) = 7,$$

it follows that

$$\sum_{i=1}^{1995}\frac{1}{f(i)} = \sum_{i=1}^{1785}\frac{1}{f(i)} + \frac{210}{7} = \sum_{i=1}^{6}\frac{4k^3 + k}{k} + 30 = \sum_{i=1}^{6}(4k^2 + 1) + 30 = 400.$$

CHAPTER 34

GEOMETRY

34.1 Triangle Geometry

1. Solution (B). The number x must satisfy $(x+1)^2 + (4x)^2 = (4x+1)^2$ since $4x+1$ is certainly the largest of the three numbers. Thus $x^2 + 2x + 1 + 16x^2 - 16x^2 - 8x - 1 = x^2 - 6x = 0$ from which it follows that $x = 6$.

2. Solution (D). Let O be the center of the circle and M be the intersection point of the circle with AB. Then $\angle OAM = 30°$ with $OM = 1$ hence $OA = 2$. Thus $OM = \sqrt{3}$ and $AB = 2OM = 2\sqrt{3}$.

3. Solution (B). Refer to the diagram. Since $\angle ABD = \angle DAB$, it follows that $BD = 4$. Since $\triangle BCD$ is a $30, 60, 90$ triangle, $DC = BD/2 = 2$. Then $BC^2 = 4^2 - 2^2 = 12$, so $BC = 2\sqrt{3}$.

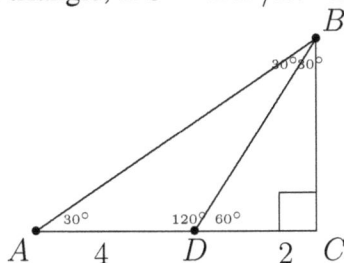

4. Solution (B). Let $x = DC, y = DB, \alpha = \angle ACD$, and $\beta = \angle DCB$. It is easy to see that $\alpha = \beta = 30°$, so $x = 4$ and $y = \frac{1}{2}x = 2$. Then

$$h^2 = x^2 - y^2 = 4^2 - 2^2 = 12,$$

so $h = 2\sqrt{3}$.

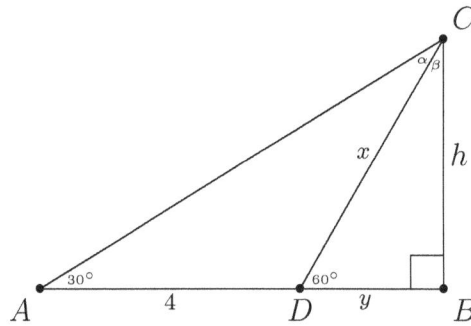

5. Solution (A). Use the Pythagorean Theorem to find that $AB = 12$. Then use the angle bisector theorem that says in a triangle, an angle bisector divides the opposite sides into two parts proportional to the adjacent sides. Thus $AD/15 = DB/9$. It follows from this that $DB = 9/2$. Use the Pythagorean Theorem again to see that $CD = 9\sqrt{5}/2$.

Alternate proof using trigonometry: Let z be the length of segment $\overline{CD}$ and let θ be the measure of angle $\angle DCB$. Then $\cos(\theta) = 9/z$ and

$$\cos(2\theta) = 9/15 = 2\cos^2(\theta) - 1 = (162/z^2) - 1.$$

Solve for z to get $z = 9\sqrt{5}/2$.

6. Solution (A). The triangle is right with hypotenuse a diameter of the circle, hence of length $2r$. The sides a and b of the triangle have lengths $2r\cos\theta$ and $2r\sin\theta$, hence the area of the triangle is

$$A = \frac{1}{2}ab = \frac{1}{2}(2r\cos\theta)(2r\sin\theta) = r^2\sin 2\theta.$$

7. Solution (D). The larger triangle has area 4 times that of the inscribed one. So the medium triangle has area 4 and the large triangle has area 16 and the total area is $1 + 4 + 16 = 21$.

8. Solution (C). Situate the triangle so the right angle is at the origin. Then the hypotenuse is part of the line $y = -\frac{4}{3}x + 4$ and the area of the rectangle with upper right corner (x, y) is given by $A = x(-4x/3 + 4)$, a parabola which opens downward. The x-coordinate of the vertex of this parabola is the midpoint of the two zeros, $x = 0$ and $x = 3$. Therefore the maximal area is given by $A = \frac{3}{2} \cdot 2 = 3$.

9. Solution (C). Since the hypotenuse is b, the Pythagorean Theorem gives sides of length $b/\sqrt{2}$. Since we have a right triangle, the area is half the product of the lengths of the two legs, or $\frac{1}{2} \cdot b/\sqrt{2} \cdot b/\sqrt{2} = b^2/4$.

10. Solution (D). The area of $\triangle ABV$ is $\frac{2}{3} \cdot 144 = 96$ (because the altitude AF to the base $\overline{BC}$ is the same for both triangles ABC and ABV), so the area of $\triangle UVB$ is

$$\frac{7}{12} \cdot 96 = 7 \cdot 8 = 56.$$

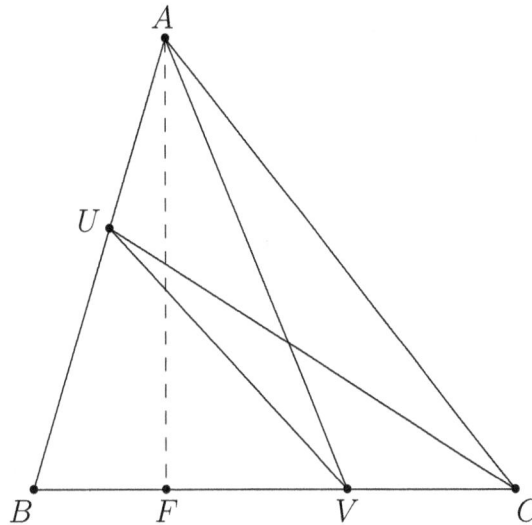

11. Solution (C). Each triangle has area 12. To see this, construct for each triangle the altitude to the even length side and use the Pythagorean Theorem.

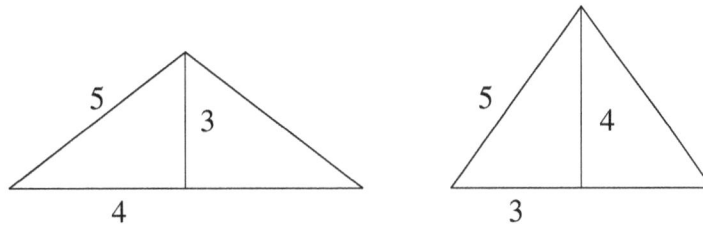

12. Solution (C). Notice from the diagram that all the sides of the trapezoid are easily determined. Adding them gives $20 + 10\sqrt{2}$.

13. Solution (C). The three sides have lengths $x, 2x + 2$, and $3x - 2$, and these numbers satisfy the Pythagorean identity. Thus,

$$x^2 + (2x + 2)^2 = x^2 + 4x^2 + 8x + 4 = 9x^2 - 12x + 4,$$

which is equivalent to $4x^2 - 20x = 0$, so the solutions are $x = 0$ and $x = 5$. Therefore the perimeter is $5 + 12 + 13 = 30$.

14. Solution (D). Square sides to obtain $a+b+2\sqrt{ab} = c$. This means that $a+b < c$ which contradicts to the triangle inequality. Therefore, there is no such triangle existing.

15. Solution (A). For the right triangle, the longest side is the hypotenuse, therefore

$$a^2 + (2a + 2d)^2 = (2a + 3d)^2,$$

$$a^2 + (4a^2 + 8ad + 4d^2) = 4a^2 + 12ad + 9d^2,$$

$$a^2 - 4ad - 5d^2 = 0,$$

$$(a - 5d)(a + d) = 0,$$

$a = 5d$ or $a = -d$ (dropped since both a and $d > 0$.) $a : d = 5d : d = 5 : 1$.

16. Solution (E). The triangle inequality shows that $x + 2 \le x + (x + 1)$, so that I is true. Note that for all x with $x \ge 1$, there is a unique triangle with sides of the given length, so II is false. Finally, note that because of the lengths of the sides opposite angles A, B, and C, we may conclude that

$$\angle C < \angle A < \angle B.$$

Since

$$\angle A + \angle B + \angle C = 180°,$$

we see that $3\angle C < 180°$, hence $\angle C < 60°$ and III is true.

17. Solution (E). Because $AB + BC \ge AC$, it follows that $2x + 1 \ge x + 2$, so i. is true. For any $x \ge 1$, there is a unique triangle with the given lengths, so x could be larger than $5\sqrt{2}$. Hence ii. need not be true. Since

$$\frac{\sin C}{x} = \frac{\sin A}{x + 1} = \frac{\sin B}{x + 2},$$

angle C is opposite the smallest side, so it is smaller than angles A and B. Because the sum of the three angles is $180°$, it follows that $3C \le 180°$, hence iii. is true.

18. Solution (C). Let x and y be the sides of the triangles on the sides of the rectangle of lengths 13 and 11. Because all sides of the octagon are equal

$$\sqrt{x^2 + y^2} = 13 - 2x$$
$$13 - 2x = 11 - 2y$$

The second equation yields $y = x - 1$. Plugging into the first equation and squaring yields

$$x^2 + (x - 1)^2 = (13 - 2x)^2$$
$$2x^2 - 2x + 1 = 169 - 52x + 4x^2$$
$$168 - 50x + 2x^2 = 0$$
$$x^2 - 25x + 84 = 0$$

This quadratic equation has the two solution $x = 21$ or $x = 4$. Only the last solution can occur since $13 - 2x > 0$. One gets $y = x - 1 = 3$ and finally $\sqrt{x^2 + y^2} = 5$ for the side of the octagon.

OR

We can solve directly for the length of the sides of the octagon: Let z be the length of the side of the octagon and let x and y be as above. Then have $2x = 13 - z$ and $2y = 11 - z$. So

$$4z^2 = 4x^2 + 4y^2 = (13 - z)^2 + (11 - z)^2 = 169 - 26z + z^2 + 121 - 22z + z^2.$$

Hence, $0 = 2z^2 + 48z - 290 = (2z - 10)(z + 29)$. Thus $z = 5$.

19. **Solution** (C). The altitude associated to the side of length 6 cuts the isosceles triangle into two congruent right triangles. The hypotenuse in these right triangles is 5, and one of the legs has length 3. By the Pythagorean Theorem, the height of the triangle is 4. Hence the area is $A = (6 \times 4)/2 = 12$ square units, while the semiperimeter is $s = (5+5+6)/2 = 8$ units. The radius ρ of the inscribed circle satisfies $A = s \cdot \rho^*$ (*If P is the incenter of $\triangle ABC$, the triangles APB, BPC, CPA have areas given by $\frac{1}{2}\rho \cdot AB, \frac{1}{2}\rho \cdot BC, \frac{1}{2}\rho \cdot CA$ so the area of $\triangle ABC$ is $\frac{1}{2}\rho(AB + BC + CA) = \rho \cdot s$, and so we get $\rho = A/s = 12/8 = 3/2$.)

OR

Let $\overline{AB}$ and $\overline{AC}$ be the sides of length 5 in the triangle $\triangle ABC$ and let P be the center of the inscribed circle. Let R be the point where the circle is tangent to side $\overline{AB}$ and let D be the foot of the perpendicular from A to the side $\overline{BC}$. By the Pythagorean Theorem, the length of $\overline{AD}$ is 4. Thus the length of segment $\overline{AP}$ is $4 - r$. Finally, let θ be the measure of angle $\angle BAD = \angle RAP$. Then $\sin(\theta) = 3/5 = r/(4 - r)$, the first value based on $\triangle ADB$ and the second based on $\triangle APR$. Solving yields $r = 1.5$.

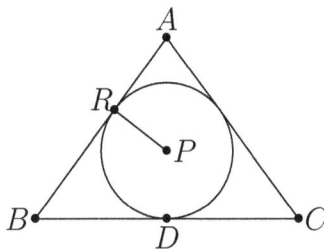

34.2   Circle Geometry

1. **Solution** (B). If r is the radius of the smaller circle, the area **of the annular region** is $(r + 1)^2\pi - r^2\pi = (2r + 1)\pi$. The solution of the equation $2r + 1 = 5$ is $r = 2$.

Alternatively, let s be the radius of the larger circle Then

$$5\pi = s^2\pi - r^2\pi = (s-r)(s+r)\pi.$$

Since $s - r = 1$, $s + r = 5$. So $s = 3$ and $r = 2$.

2. **Solution** (E). We have $a + b = 12$, $b + c = 17$, and $c + a = 23$. Adding the equations gives $2(a + b + c) = 52$ or $a + b + c = 26$. Subtracting each of the original equations from this last equation gives $a = 9, b = 3$, and $c = 14$.

3. **Solution** (C). The centers of the circles are the vertices of a regular pentagon. Thus, the angle between adjacent circles is $360°/5 = 72°$, and in the right triangle OCD we have $\angle COD = 36°$. Hence $\sin 36° = \dfrac{R}{R+1}$ from which it follows that $R \approx 1.43$.

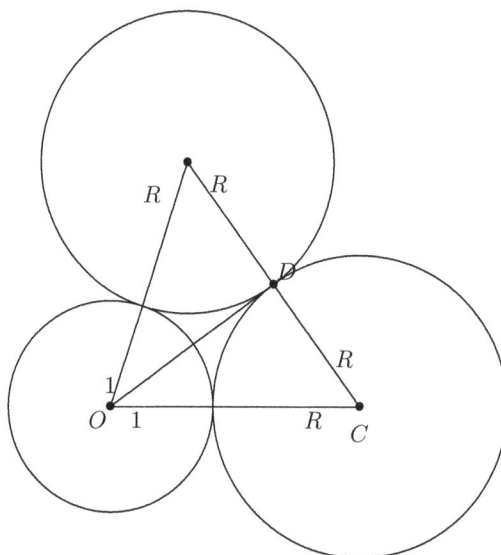

4. **Solution** (D). Half of the unshaded area is $1 - \dfrac{\pi}{4}$ so the shaded area is

$$1 - 2\left(1 - \frac{\pi}{4}\right) = \frac{\pi}{2} - 1.$$

5. **Solution** (D). Let the radii of Circle A and B be r and R. Then we have

$$\frac{2}{5}\pi r^2 = \frac{5}{8}\pi R^2.$$

Thus, $16r^2 = 25R^2$ and $4r = 5R$ The ratio is $5 : 4$.

6. **Solution** (D). The diameter of each of the n^2 circles is $12/n$, and each radius is $6/n$. Therefore, the area of each circle is $\pi(6/n)^2 = 36\pi/n^2$ and the total area of the n^2 circles is $(36\pi/n^2) \times n^2 = 36\pi$.

7. **Solution** (C). Note that $\angle CAD$ is $\dfrac{2\pi}{3}$. This implies that the area is twice the difference between area of the area of the sector CAD and the triangle CAD. In other words, it is

$$2\left(\pi \cdot 1^2/3 - \frac{1}{2}\sqrt{3} \cdot \frac{1}{2}\right)$$

which is $\dfrac{2\pi}{3} - \dfrac{\sqrt{3}}{2}$.

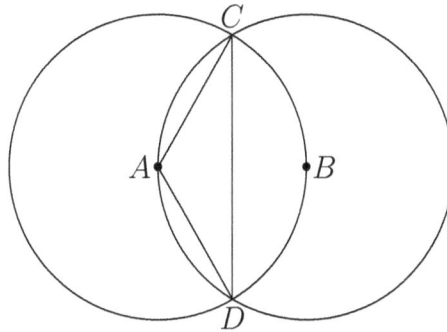

8. **Solution** (A). Let a, b and c denote the lengths of the three sides, $a = BC, b = AC$, and $c = AB$. Now

$$a^2 + b^2 = c^2$$

since the triangle is right. The areas of the three semicircles are

$$(1/2)\pi(a/2)^2, (1/2)\pi(b/2)^2, (1/2)\pi(c/2)^2.$$

Therefore, we have

$$(1/2)\pi(a/2)^2 + (1/2)\pi(b/2)^2 = \pi(a^2 + b^2)/8 = (\pi/8)c^2,$$

so the sum of the areas of the two smaller semicircles is the area of the largest one. Thus the area of the middle one is

$$36 - 16 = 20.$$

9. **Solution** (C). Note that

$$x^2 + y^2 + p^2 + q^2 = 2r^2 = 72,$$

so $r^2 = 36$, and $r = 6$. Hence the circumference is 12π.

10. **Solution** (D). Let A denote the point of intersection of OO' and PP' and let $x = OA$. Using similar triangles we can write $\dfrac{x}{28 - x} = \dfrac{9}{12}$ from which it follows that $x = 12$.

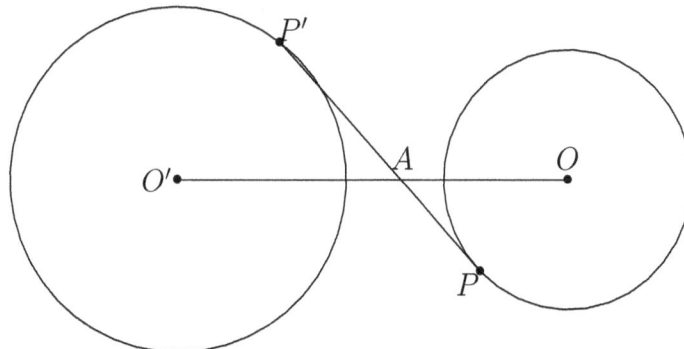

11. **Solution** (B). Look at the diagram, and note the length of a chain with n links is $2R + 2(n-1)r$ where R is the outer radius and r is the inner radius. Thus, $2R + 2r = 13$ and $2R + 4r = 18$, which yields $2r = 5/2$ and $2R = 8$. Thus the length of a 25 link chain is $2R + 24r = 8 + 120 = 128$. Alternatively, note that each link adds 5 cm. to the length of the chain, for a total of $8 + 5 \cdot 24 = 128$.

12. **Solution** (D).

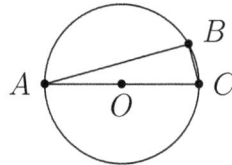

Let O be the center of the circle as shown with diameter AC of length d and chord AB of length c. Let $\overset{\frown}{AB} = \theta$. Then $\angle ACB = \dfrac{\theta}{2}$. Since $\triangle ABC$ is inscribed in a semicircle, $\angle ABC$ is a right angle and $c = d \sin \dfrac{\theta}{2}$. Then for chords of lengths a and b in the same circle, with a subtending an arc of $30°$ and b subtending an arc of $90°$, $a = d \sin 15°$ and $b = d \sin 45°$. Since

$$c = a + b, \sin \frac{\theta}{2} = \sin 15° + \sin 45°.$$

For any angle u,

$$\sin\left(\frac{\pi}{3} + u\right) - \sin\left(\frac{\pi}{3} - u\right) = \sin u.$$

Letting $u = 15°$ and noticing that $\dfrac{\pi}{3} = 60°$, we get

$$\sin 75° = \sin 15° + \sin 45°.$$

Hence $\dfrac{\theta}{2} = 75°$ and the number of degrees in the smaller arc subtended by a chord of length c is 150.

13. **Solution** (C). The cap is a segment minus the triangle OAB as shown below. The central angle of the segment is $120°$. The circular segment covers one third of the circle and hence has area $4\pi/3$. Hence $a = 4$. The triangle is isosceles with height 1 and two congruent sides of length 2. Its third side has length $2\sqrt{3}$ and hence its area is $\sqrt{3}$. Hence $b = 1$ and $a + b = 5$.

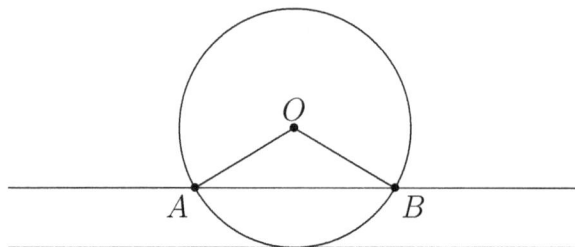

14. Solution (C). By the Pythagorean Theorem, the diameter of the circle is $\sqrt{a^2 + b^2}$, and the radius is $\dfrac{\sqrt{a^2 + b^2}}{2}$. The area of the circle is $\dfrac{a^2 + b^2}{4} \cdot \pi$, so we need to solve the equation

$$a \cdot b = \frac{a^2 + b^2}{4} \cdot \pi \cdot \frac{6}{5\pi}, \quad \text{that is,} \quad a \cdot b = 0.3(a^2 + b^2).$$

Dividing both sides by a^2 and rearranging yields

$$0 = 0.3 \cdot \left(\frac{b}{a}\right)^2 - \frac{b}{a} + 0.3,$$

a quadratic equation in $\dfrac{b}{a}$. According to the quadratic formula

$$\frac{b}{a} = \frac{1 \pm \sqrt{1 - 0.36}}{0.6} = \frac{1 \pm \sqrt{0.64}}{0.6} = \frac{1 \pm 0.8}{0.6} = 3 \text{ or } \frac{1}{3}.$$

Since $b > a$, the only solution is $\dfrac{b}{a} = 3$.

15. Solution (E). The region that T touchs is the annulus formed by the circle with center at R of radius $a + b$ and the one with center R and radius $a - b$. So the total area is

$$\pi(a + b)^2 - \pi(a - b)^2 = 4ab\pi.$$

16. Solution (B). Let x denote $A'C$. Because triangle $A'CF$ is similar to triangle $A'GC$, x is the geometric mean of $A'G$ and $A'F$; that is, $x^2 = \pi r \cdot r = \pi r^2$. Therefore the area of the square is πr^2.

Alternatively, first consider a more general problem where the distance between A' and F is s with s not necessarily equal to r. Then the distance between G and A' is πs since the circle has made half a revolution. Let x be the distance from A' to C, y the distance from G to C and z the distance from F to C. Since C, G and F are on a circle with G and F and endpoints of a diameter, $\angle GCF$ is a right angle. Since $\angle GA'C$ is also a right angle,

$$x^2 + \pi^2 s^2 = y^2, \quad x^2 + r^2 = z^2$$

and

$$y^2 + z^2 = (\pi s + r)^2 = \pi^2 s^2 + 2\pi sr + r^2.$$

Adding the first two together yields $2x^2 + \pi^2 s^2 + r^2 = \pi^2 s^2 + 2\pi sr + r^2$. Thus the area of square in the general setting is πsr. Since $s = r$ in this problem, the area of the square is πr^2.

17. Solution (D). Let r denote the radius. Since the triangle is a right triangle, its area A is easily computed in two ways. First

$$A = (1/2)bh = (1/2)15 \cdot 8 = 60.$$

Also, we can break the triangle into three triangles by connecting the center of the circle with each of the vertices. Then we get A by adding the areas of these triangles together:

$$\frac{15r}{2} + \frac{8r}{2} + \frac{17r}{2} = \frac{40r}{2}.$$

It follows that $40r = 2 \cdot 60$ and $r = 3$.

Alternatively, the distance from a point (s, t) to a line given by the equation

$$Ax + By + C = 0$$

is

$$|As + Bt + C|/\sqrt{A^2 + B^2}.$$

Position the triangle in the plane so that the vertices are $(0, 0)$, $(0, 8)$ and $(15, 0)$. Then the coordinates of the center of the circle are (r, r) where r is the radius. An equation for the line through $(0, 8)$ and $(15, 0)$ is

$$8x + 15y - 120 = 0.$$

By the formula, the distance from (r, r) to this line is

$$|8r + 15r - 120|/\sqrt{8^2 + 15^2} = |120 - 23r|/17.$$

Of course this distance is also equal to r. In this case, $120 - 23r$ is positive. So $120 - 23r = 17r$, and $r = 3$. (The solution of $23r - 120 = 17r$ is $r = 20$ which is too big to be the radius.)

18. **Solution** (A). See the diagram below. Since the circle is inscribed within the triangle and $\overline{AP}$ passes through the circle's center, it follows that $\overline{AP}$ is perpendicular to $\overline{BC}$. By the Pythagorean Theorem, we have $AP^2 + 4^2 = 12^2$ and $AP = 8\sqrt{2}$. Let R be the point of tangency of the circle with $\overline{AC}$. Consider triangles OPC and ORC. They are right triangles that have the same hypotenuse. Since $OP = OR$ is the radius of the circle, they are congruent. Hence, $PC = RC = 4$ and $AR = 12 - 4 = 8$. And in right triangle AOR, we have

$$(8\sqrt{2} - r)^2 = 8^2 + r^2.$$

Solve the equation

$$128 - 2 \cdot 8\sqrt{2}r + r^2 = 64 + r^2$$

for the radius r to get $r = \dfrac{4}{\sqrt{2}} = 2\sqrt{2}$.

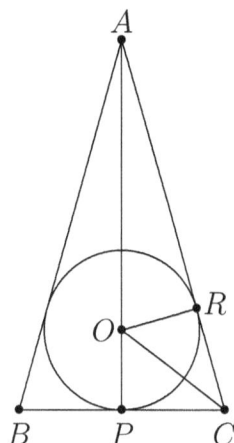

19. **Solution** (D). We may assume the inner circle has center at $(0,0)$ and for the outer circles, the centers are at angles $0, \pi, \pm\pi/3$ and $\pm 2\pi/3$, 10 units from $(0,0)$. The colored portion is made up of twelve identical pieces: start with the wedge of a circle where the angle is $\pi/3$, connect the secant line and then remove the equilateral triangle (in this case with side length 10) formed. The remaining area is $100(\pi/6 - \sqrt{3}/4)$. So the total area is

$$12 \cdot 100(\pi/6 - \sqrt{3}/4) = 100(2\pi - 3\sqrt{3}).$$

To estimate the area, note that $17^2 = 289$, and $17.5^2 = 306.25 (18^2 = 324)$, so $1.7 < \sqrt{3} < 1.75$. Thus the area is between 100 and 120.

20. **Solution** (A). Let r be the radius and P be the center of the inscribed circle. Then the triangle can be subdivided into three triangles using P and the three pairs $\{A, B\}$, $\{A, C\}$ and $\{B, C\}$. The sum of the areas of these three triangles is

$$r/2 + ra/2 + rb/2 = ab/2.$$

Solving for r we have $r = ab/(a + b + 1)$. If you suspect the largest value for the fraction on the right is when $a = b = \sqrt{2}/2$, you are correct. The form in (A) is obtained by rationalizing the denominator. To prove that $a = b$ gives the maximum radius takes a bit more work.

Suppose

$$F(x) = g(x)/h(x)$$

is a variable fraction where both numerator and denominator are positive. If $h(x)$ takes its smallest value when $x = t$ and $g(x)$ takes its largest value when $x = t$, then $F(t)$ is the largest value of $F(t)$. Unfortunately, in the present form, the numerator ab is as large as possible at the same time the denominator $a + b + 1$ is also as large as possible (noting that $b = \sqrt{1 - a^2}$). To fix that, multiply top and bottom by $1/ab$ to get

$$r = 1/((1/b) + (1/a) + (1/ab)).$$

Now all we have to do is show the smallest value for the denominator is when $a = b$. We have $a = \cos(\alpha)$ and $b = \sin(\alpha)$ where α is the angle at A. Thus $ab = \sin(2\alpha)/2$. Certainly, the largest value of ab occurs when $\alpha = 45°$ and so when $a = b = \sqrt{2}/2$. So the smallest value of $1/ab$ is 2. As for $(1/b) + (1/a)$, it is as small as possible at the same time as its square is as small as possible. Squaring yields

$$(1/b^2) + (2/ab) + (1/a^2)$$

and

$$(1/b^2) + (1/a^2) = (a^2 + b^2)/a^2b^2 = (1/ab)^2.$$

So $2/ab$ and the sum $(1/b^2) + (1/a^2)$ are also as small as possible when $a = b$.

34.3 Polygons

1. **Solution** (B). The slope of the line in question is $1/2$, so the largest square is 9×9.

2. **Solution** (D). Place Q 1.5 units to the right of A. Then remove the rectangle $BCDQ$ and place it in the position $FGBH$ as shown in Figure. The line PQ cuts the area in half.

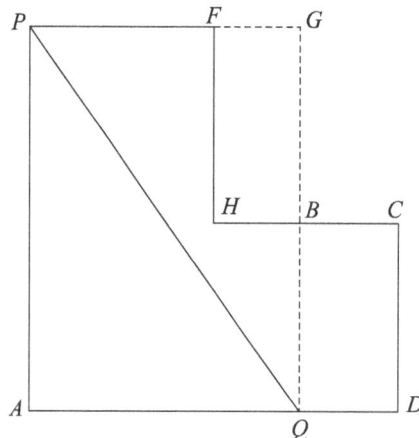

3. **Solution** (B). Refer to the diagram below.

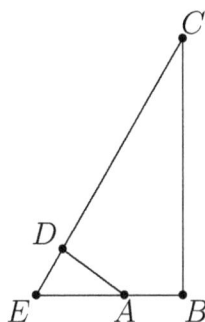

$$EC = BC/\cos 30 = 20, \ EB = EC \sin 30 = 10.$$

Therefore $EA = 6$. Because $\triangle EAD$ is a right triangle with $\angle EAD = 30°$ and hypotenuse $EA = 6$, $ED = 3$. Consequently, $DC = 20 - 3 = 17$.

4. Solution (B). Each of the small triangles has the same area.

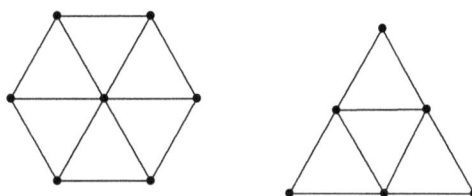

5. Solution (C). Angle CPD is a right angle. Therefore

$$CD = \sqrt{7^2 + 11^2} = \sqrt{49 + 121} = \sqrt{170},$$

and $CD^2 = 170$.

6. Solution (D). The area of the quadrilateral is the sum of the areas of the right triangles ABD and BCD. This sum is

$$\frac{1}{2}(20)(15) + \frac{1}{2}(7)(24) = 234.$$

7. Solution (D). The area of PBQ is a quarter of the area of ABC and the area of RDS is a quarter of ACD.

8. **Solution** (C). Triangle EBJ is congruent to triangle ECK (see below), hence the area of the quadrilateral EJCK is the same as the area of the triangle EBC which is $\frac{1}{2}(6)(12) = 36$ square inches.

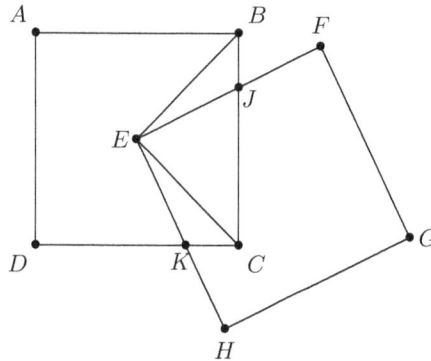

9. **Solution** (A). Note that $a = \frac{1}{2}A$, since the radius of the inscribed circle is $r = \frac{1}{2}s$, while by the Pythagorean theorem, the radius of the circumscribed circle is

$$R = \sqrt{(\frac{1}{2}s)^2 + (\frac{1}{2}s)^2} = s/\sqrt{2},$$

so

$$a = \pi r^2 = \pi \frac{1}{4}s^2, A = \pi R^2 = \pi \frac{1}{4} \cdot 2s^2 = 2a.$$

10. **Solution** (B). Suppose the triangle has perimeter $12a$. Then each side of the square has side $3a$. The area of the triangle is given by $1 = \dfrac{(4a)^2\sqrt{3}}{4}$ from which it follows that $a^2 = 1/(4\sqrt{3})$. Therefore the area of the square is

$$9a^2 = \frac{9}{4\sqrt{3}} = \frac{3\sqrt{3}}{4} \approx 1.299 \approx 1.30.$$

11. **Solution** (A). Let M and N be the two points where the two sides of the unit square and two sides of an equilateral triangle bisect each other. Their distance from each other is $|MN| = \frac{\sqrt{2}}{2}$. Hence the side of the equilateral triangle is $a = \sqrt{2}$. Its area is $\frac{\sqrt{3}}{4}a^2 = \frac{\sqrt{3}}{2}$.

12. **Solution** (C). Introducing w for the width of the rectangle and using the notation in the figure, we get $2t = x \cdot w/2$ and $9t = (x + y + y) \cdot w/2$. Dividing the second equation by the first yields $9/2 = (x + 2y)/x$. Thus $7/2 = 2y/x$ and $y/x = 7/4$.

13. **Solution** (E). Let m and n be lengths of the altitudes as shown in the diagram below. Since the area of $ABMN$ is half the area of $ABCD$, we have

$$(m + n)(\frac{a + b}{2}) = 2m(\frac{x + a}{2}),$$

which is equivalent to

$$(1 + \frac{n}{m})(b + a) = 2(x + a) \qquad \aleph$$

Furthermore, equating the areas of $ABNM$ and $MNCD$ we have

$$m(\frac{a + x}{2}) = n(\frac{b + x}{2}),$$

so that

$$\frac{n}{m} = \frac{a + x}{b + x}.$$

Substituting into $\aleph$, we have

$$(1 + \frac{a + x}{b + x})(b + a) = 2(x + a).$$

After multiplying both sides by $b + x$ and simplifying, we see that $a^2 + b^2 = 2x^2$.

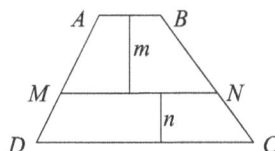

Alternatively, we can use the similarity of the triangles of the figure below.

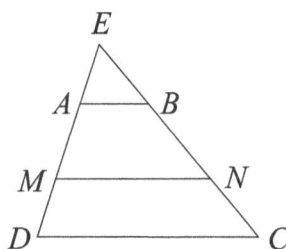

14. **Solution** (B). The area of the middle sized square is the square of the length of the segment $\overline{CE}$. Also $\overline{CD}$ has length 6 and $\overline{DE}$ has length $\sqrt{8}$. Let J be the foot of the perpendicular from E to $\overline{CD}$. And let t be the length of $\overline{EJ}$, z be the length of $\overline{CE}$ and x be the length of $\overline{DJ}$. We have the following relations: $6t = 2 \cdot 6$, $8 = x^2 + t^2$, $z^2 = (6 - x)^2 + t^2 = 36 - 12x = x^2 + t^2 = 36 - 12x + 8 = 44 - 12x$. Then $t = 2$ so $x^2 = 8 - 4 = 4$. Finally $z^2 = 44 - 24 = 20$.

15. **Solution** (D). A circle with an area of $200\pi = \pi r^2$ has a radius of $10\sqrt{2}$. A regular hexagon is made up of six equilateral triangles each with side equal to the radius of the circle. Since a hexagon has six sides the sum of the angles is $(4 - 2)180 = 720$, so that each interior angle has measure $120°$. The side of the triangle bisects this angle resulting in a $30 - 60 - 90$ triangle. Using this information, the area of each triangle is computed to be $50\sqrt{3}$ so the area of the hexagon is $6 \cdot 50\sqrt{3} = 300\sqrt{3}$.

16. Solution (D). Let the angles be $a, a+d, a+2d$, and $a+3d$. Then $4a+6d = 360$ and $a+3d+15 = 2a$. Thus $a = 3d + 15$, so we can replace a with $3d + 15$ in the first equation. Solve this to get $3d = 50$ and $a = 50 + 15 = 65$, so the largest angle is $65 + 50 = 115°$.

17. Solution (B). Let E denote the intersection between the altitude h and DC and let $x = DE$. Since $AB = h$ we have $DC = 2x + h$. Noting that $AC = DC$ and applying the Pythagorean theorem to triangle ACE, we have

$$h^2 + (h + x)^2 = (h + 2x)^2.$$

This is equivalent to $(h - 3x)(h + x) = 0$. Therefore $h = 3x$ and $AB : DC = h : (h + 2x) = 3 : 5$.

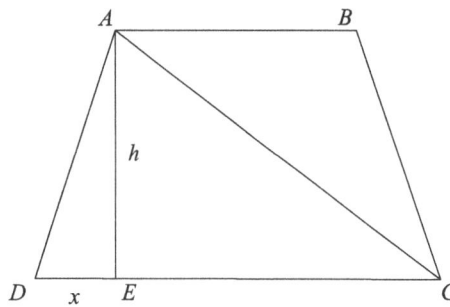

18. Solution (A). Let a be the length of AB and let b be the length of BC. Let x be the length of the side of the parallelogram of area 12 that is parallel to AB. The heights of the parallelograms are then $12/x, 6/x$ and $18/x$. So we have

$$b = 12/x + 6/x + 18/x.$$

The area of the central horizontal stripe is $a \cdot 6/x = 10 + 6 + 9$, so $a = 25x/6$. The area of the rectangle $ABCD$ is $ab = 25x/6 \cdot 36/x = 150$.

19. Solution (B). We may add the length of the gate to the length of the fencing material and then we have 204 yards of fencing material. If the width of the enclosed area is x yards then the length is $(204 - 2x)/2 = 102 - x$ yards. The enclosed area is $x(102 - x) = 102x - x^2 = 51^2 - (x - 51)^2$ square yards. The largest area of $51^2 = 2601$ is achieved when the length and the width are both 51 yards.

20. Solution (D). Let d_A, respectively d_C denote the distance of A, respectively C from the line BD. Then $\frac{a_1}{a_2} = \frac{BK \cdot d_A/2}{BK \cdot d_C/2} = \frac{d_A}{d_C}$ and $\frac{a_4}{a_3} = \frac{DK \cdot d_A/2}{DK \cdot d_C/2} = \frac{d_A}{d_C}$. Thus $a_1/a_2 = a_4/a_3$ and so $a_1 \cdot a_3 = a_2 \cdot a_4$.

21. Solution (C). Think of the 3×5 box as part of a grid of such boxes in the plane. The line from $(0, 0)$ to $(15, 15)$ hits the boundaries of these boxes 6 times, and each one of these corresponds to a bounce. You can see that the ball will hit the pocket R after six bounces.

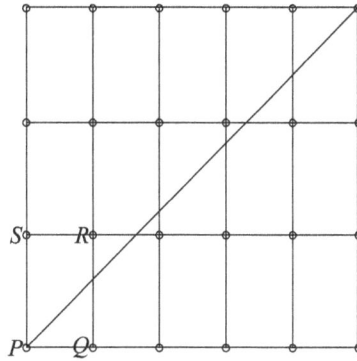

Alternatively, the first bounce is 2 units below R (3 units above Q), so the second is 1 unit to the (viewer's) right of S, the next is one unit below S, the fourth occurs $1+3$ units below R or 1 unit above Q, the fifth is on side $\overline{PQ}$ at the point 1 unit to the left of Q and 2 to the right of P. The sixth and final bounce will be 2 units above P, or 3 below S. From there the ball hits corner R.

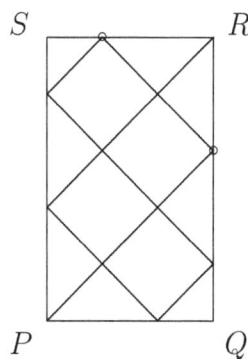

22. **Solution** (B). For example, the area of the triangle BEF is twice the area of the triangle ABC since AB and BE have the same length, but the distance of F from the line BE is twice the distance of C from the line AB. Similarly the area of the triangle DGH is twice the area of ACD, and so on. Hence the combined area of the triangles BEF and DGH is twice the area of $ABCD$. Similarly the combined area of the triangles CFG and AHE is also twice the area of $ABCD$.

23. **Solution** (E). A plane subdivides three-space into two half-spaces. There are just two cases to consider:
    1) Three vertices of tetrahedron lie in one half-space and the fourth one lies in the other half-space. There are 4 middle planes of this type, one for each of the four faces. The middle planes go through middle points of edges with common vertex.
    2) Two vertices of tetrahedron line in one half-space and two other vertices lie in the other half-space. There exist 3 such middle planes because the tetrahedron has 6 edges. Each middle plane goes through the middle lines of two faces which are parallel to the common edge of these faces.

24. Solution (E). Suppose we have an n sided regular polygon with sides of length s. By connecting each vertex to the center of the polygon, we can form n congruent triangles. The central angle for each triangle is $360/n$ degrees. We can use trigonometry to determine that the area of each triangle is

$$\frac{s^2}{4\tan(180/n)}.$$

Setting the perimeter $p = ns$ and adding the areas of the n triangles, we see that the area of the n sided regular polygon is

$$\frac{p^2}{4n\tan(180/n)}.$$

This increases as n increases, so $n = 12$ gives the largest area.

25. Solution (A). Consider angles 1-5 in the center pentagon.

Because the sum of the angles of a triangle is $180°$, it follows that

$$m(\angle A) + m(\angle D) + m(\angle 3) = 180°,$$

$$m(\angle A) + m(\angle C) + m(\angle 1) = 180°,$$

$$m(\angle E) + m(\angle C) + m(\angle 4) = 180°,$$

$$m(\angle E) + m(\angle B) + m(\angle 2) = 180°,$$

and

$$m(\angle D) + m(\angle B) + m(\angle 5) = 180°.$$

Adding these equations yields

$$2\left(m(\angle A) + m(\angle B) + m(\angle C) + m(\angle D) + m(\angle E)\right)$$
$$+m(\angle 1) + m(\angle 2) + m(\angle 3) + m(\angle 4) + m(\angle 5)$$
$$= 900°.$$

But since the measures of the angles of the central pentagon sum to $540°$, we have

$$2(m(\angle A) + m(\angle B) + m(\angle C) + m(\angle D) + m(\angle E)) + 540 = 900$$

and

$$2(m(\angle A) + m(\angle B) + m(\angle C) + m(\angle D) + m(\angle E)) = 360°.$$

Thus

$$m(\angle A) + m(\angle B) + m(\angle C) + m(\angle D) + m(\angle E) = 180°.$$

26. **Solution** (C). Let $\alpha_1, \alpha_2, \ldots, \alpha_n$ be the angles of the n-gon. We distinguish two cases, depending on the parity of n. Assume first n is even. By the stated condition $\alpha_1 + \alpha_2 < 270°$, $\alpha_3 + \alpha_4 < 270°$, $\ldots, \alpha_{n-1} + \alpha_n < 270°$. Thus the sum of the angles is less than $n/2 \cdot 270° = n \cdot 135°$. On the other hand, the sum of the angles of a convex n-gon is $(n-2) \cdot 180°$. We obtain the inequality

$$(n-2) \cdot 180 < n \cdot 135,$$

the solution of which is $n < 8$. Since n is even, this implies $n \le 6$.

Assume n is odd. We may assume that α_n is not an obtuse angle, thus $\alpha_n \le 90°$. Similarly to the previous case, we have $\alpha_1 + \alpha_2 < 270°$, $\alpha_3 + \alpha_4 < 270°$, $\ldots, \alpha_{n-2} + \alpha_{n-1} < 270°$. Thus the sum of the angles is less than $(n-1)/2 \cdot 270° + 90° = n \cdot 135° - 45°$. (The inequality is strict, since $n \ge 3$.) We obtain the inequality

$$(n-2) \cdot 180 < n \cdot 135 - 45,$$

the solution of which is $n < 7$. Since n is odd, this gives $n \le 5$.

Comparing the two cases we see that n can not exceed 6. There is a hexagon with the required properties, for example, we may choose $\alpha_1 = \alpha_3 = \alpha_5 = 90°$ and $\alpha_2 = \alpha_4 = \alpha_6 = 150°$.

34.4 Three-Dimensional Geometry

1. **Solution** (D). A shortest path between A and B is contained in two adjacent faces of the cube and goes through the middle of their common side. The caterpillar travels x units in each of two faces where $x^2 = 4^2 + 2^2 = 20$. Therefore, the distance traveled is $2x = 4\sqrt{5}$.

2. **Solution** (D). Suppose that each edge of the cube has length one and that we place the cube so that one corner is at the origin of a three dimensional coordinate system (x, y, z) so that $0 \le x, y, z \le 1$. Let the diagonal connect the vertices $(0, 0, 0)$ and $(1, 1, 1)$. Then the equation of the plane is $x + y + z = \dfrac{3}{2}$. This plane intersects edges connecting vertices whose coordinates sum to 3. (For example, it intersects the edge connecting $(1, 0, 0)$ to $(1, 1, 0)$ (because at a point halfway between these vertices, we have $x + y + z = \frac{3}{2}$). There are six such edges.

OR

The angle a diagonal of the cube makes with the corresponding diagonal of the square at the bottom is less than 45°. Thus the perpendicular plane through the midpoint of the diagonal will intersect the diagonal of the bottom inside the bottom square and the diagonal of the top inside the top square. Thus the plane must cut through two adjacent edges on the top and two adjacent edges on the bottom. It will also intersect both vertical edges that contain neither endpoint of the diagonal.

3. **Solution** (E). Because the distance from a point (x, y, z) to the origin is just $(x^2 + y^2 + z^2)^{\frac{1}{2}}$, we need to solve the equation

$$x^2 + y^2 + z^2 = 169.$$

The number of triples for which two of the coordinates are zero is 6:

$$(0, 0, \pm13), (0, \pm13, 0), (\pm13, 0, 0).$$

The number with exactly one zero is 24:

$$(0, \pm5, \pm12), (\pm5, 0, \pm12), (\pm5, \pm12, 0), (\pm12, \pm5, 0), (\pm12, 0, \pm5,), (0, \pm12, \pm5).$$

Finally there are 48 triples with all nonzero coordinates since there are six ways to permute the values ±3, ±4, and ±12. Hence there are $6 + 24 + 48 = 78$ lattice points in three-space that are 13 units from the origin.

4. **Solution** (B). Note that the volume of the tetrahedron $OABC$ is $= \frac{1}{3}(\frac{1}{2} \times 2) = 1/3$. Since $AB = \sqrt{5} = BC$ and $AC = \sqrt{2}$, the area of the triangle CAB is $= (1/2) \times \sqrt{2}\sqrt{(\sqrt{5})^2 - (\sqrt{2}/2)^2} = 3/2$. If the distance from O to Γ is h, then the volume can be written as $(1/3)(h)(3/2)$. Thus, the volume $(1/3)(h)(3/2)$ should be $1/3$. Hence $h = 2/3$.

Alternatively, an equation of the plane containing A, B, and C is $2x + y + 2z = 2$. Therefore the square of the distance from the origin to the point (x, y, z) on the plane is

$$d^2 = x^2 + y^2 + z^2 = x^2 + (2 - 2x - 2z)^2 + z^2.$$

By symmetry, the point on the plane that is closest to the origin must satisfy $x = z$. Then $d^2 = 18x^2 - 16x + 4$. The x-coordinate of the vertex of this parabola is

$$x = \frac{16}{36} = \frac{4}{9}.$$

It will minimize d^2 and we get

$$d^2 = 18(\frac{4}{9})^2 - 16(\frac{4}{9}) + 4 = \frac{4}{9},$$

so that $d = \dfrac{2}{3}$.

34.5 Analytical Geometry

1. **Solution** (C). The region is an 'annular square' whose outside bounding square has area 32 and whose inside bounding square has area 18. Hence, the area of the region is $32 - 18 = 14$.

2. **Solution** (A). The slope of the line is $-7/5$ and an equation for it is $y - 5 = -\frac{7}{5}(x - 2)$. Let $x = 17$ to find that $y = -16$.

3. **Solution** (C). The slope is 0 because the two parabolas are intersecting even functions. In fact they are reflections (through the line $y = 4$) of each other. Alternatively,

$$8 - x^2 = x^2 \Rightarrow x = \pm 2, y = 4.$$

4. **Solution** (C). Check these to see that only $y = -4x - 1$ satisfies the condition.

5. **Solution** (C). The slope of the first line is $-2/3$ so the slope of the second, which is $1/2k$, must be $3/2$. Hence, $k = 1/3$.

6. **Solution** (B). The two points both belong to the line $2x + 3y = 6$ whose slope is $-2/3$.

7. **Solution** (C). The coordinates of B are $(2, -3)$, and those of C are $(-3, 2)$, so the triangle can be viewed as one with base $AB = 3 - (-3) = 6$ and altitude of $2 - (-3) = 5$, so the area is $\frac{1}{2}(6 \cdot 5) = 15$.

8. **Solution** (B). The equation is given by $y - 4 = 4(x - 2)$ which is equivalent to $y = 4x - 4$, so the y-intercept is -4.

9. **Solution** (B). Exactly $1/4$ of the rectangular region lies above the line $y = x$.

10. **Solution** (E). The x-axis intersects the graph infinitely many times and vertical lines intersect just once. No line in the plane is disjoint from the graph of f.

11. **Solution** (A). Setting both equations equal to each other you get $x = 1, x = \pm\sqrt{5}$. It follows that the corresponding y-coordinates are -5 and $\pm\sqrt{5}$ so the sum of the three y-coordinates is -5.

12. **Solution** (B). An equation for l_1 is

$$y + 3 = -2(x - r),$$

so

$$y = -2x + (2r - 3).$$

An equation for l_2 is

$$y - r = \frac{1}{2}(x - 6),$$

so

$$y = \frac{1}{2}x + (r - 3).$$

Hence we have

$$-2x + 2r - 3 = \frac{1}{2}x + r - 3,$$

so

$$x = \frac{2}{5}r.$$

13. **Solution** (B). The slope of $\overline{AB}$ is the same as that of $\overline{AC}$. That is

$$\frac{a - 1}{2 - 0} = \frac{7 - 1}{3 - 0},$$

and it follows that $a - 1 = 4$, so $a = 5$. Alternatively, use A and C to get slope and then use A to get slope-intercept equation $y = 2x + 1$. Then just plug in $x = 2$. Also, there is the graphical approach: draw the line and notice that the line goes through $(2, 5)$.

14. **Solution** (D). The two unknowns x and y satisfy $xy = 60$ and $x + y + \sqrt{x^2 + y^2} = 30$. Trying a few pythagorean triples, we see that $5, 12, 13$ works just right. So the sum

$$x + y = 5 + 12 = 17.$$

Alternatively, square both sides of $x + y - 30 = \sqrt{x^2 + y^2}$ and eliminate to get $120 - 60(x + y) + 900 = 0$ from which it follows that $x + y = 17$.

15. **Solution** (D). The center (h, k) of C must lie on the line $y = 8$, because the center is the same distance from $(0, 6)$ as it is from $(0, 10)$. Thus the circle satisfies

$$(x - h)^2 + (y - 8)^2 = r^2,$$

for some number h. It must also lie on the line that perpendicularly bisects the segment from $(0, 6)$ to $(8, 0)$, an equation for which is $y - 3 = \frac{4}{3}(x - 4)$. Since $y = 8$, it follows that $h = 7.75$. Alternatively, the distance from $(h, 8)$ to $(0, 6)$ must be the same as the distance from $(h, 8)$ to $(8, 0)$, which we can solve for h.

16. **Solution** (A). Complete the squares by adding 9 and 4 to both sides to get

$$x^2 - 6x + 9 + y^2 + 4y + 4 = (x - 3)^2 + (y + 2)^2 = 12 + 9 + 4 = 25 = 5^2.$$

So the radius is 5.

17. **Solution** (C). Complete the squares by adding 9 and 4 to both sides to get

$$x^2 - 6x + 9 + y^2 + 4y + 4 = 36 + 9 + 4 = 49 = 7^2.$$

So the radius is 7.

18. **Solution** (C). The area of the circle is 2π and that of the inscribed square 4, so the shaded area is

$$\frac{2\pi - 4}{4} = \frac{\pi}{2} - 1.$$

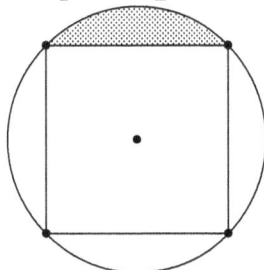

19. **Solution** (B). The lines $y = x$ and $x = 17$ intersect at the center of C, so the radius is $\sqrt{(13^2 + 17^2)} \approx 21.40$ which is closest to 21.

20. **Solution** (E). One of the tangent lines is the x-axis since the radius of the circle is 2 and the center is at $(7, 2)$. Therefore the angle is given by $\angle AOB = 2\tan^{-1}\frac{2}{7} \approx 31.9°$.

21. **Solution** (B). Complete the square to write the equation in the form $(x + 1)^2 + (y - 2)^2 = 3^2$, at which point it is easy to see that $h = -1, k = 2$, and $r = 3$, so the sum is $-1 + 2 + 3 = 4$.

22. **Solution** (D). Since $x^2 + (-x + c)^2 = 1$, we must have $x^2 + x^2 - 2xc + c^2 = 1$, so $2x^2 - 2xc + c^2 - 1 = 0$ for exactly one value of c. Thus the discriminant of the quadratic must be zero. Hence, $4c^2 - 4(2(c^2 - 1)) = 0$ and $c = \pm\sqrt{2}$.

23. **Solution** (C). The center of our circle is $(0, 0)$ and the line connecting the center of the circle with $(1, 1)$ has slope 1. Thus the tangent line must have slope -1 and pass through $(1, 1)$. The point-slope form of the equation of the tangent line is $y - 1 = -(x - 1)$, which may be rearranged into $x + y = 2$.

24. **Solution** (B). The center of C is of the form $(r, 0)$ by symmetry. Finding its distances to the center $(4, 2)$ of $(x - 4)^2 + (y - 2)^2 = 2^2$ and to the center $(-2, 0)$ of $(x + 2)^2 + y^2 = 2^2$ leads to

$$(r + 2)^2 = (r - 4)^2 + (0 - 2)^2.$$

Solving for r, we get $r = \frac{4}{3}$.

25. **Solution** (B). The first circle $(x - 12)^2 + (y - 5)^2 = 49$ is centered at $(12, 5)$ and has radius 7, while the second is centered at $(0, 0)$ and has radius k. The two circle intersect when $6 \leq k \leq 20$, so $b - a = 14$.

Alternatively, let C be the circle given by the equation $x^2 + y^2 = 24x + 10y - 120$. An equivalent equation for C is $(x - 12)^2 + (y - 5)^2 = 49$. So the radius of C is 7 and the center is $(12, 7)$.

Let L be the line through the origin and the point $(12, 5)$, the center of C. Since the radius of C is less than 12, the line L intersects C at two points in the first quadrant, call them P and Q and assume the x-coordinate of P is smaller than the x-coordinate of Q. The point P must be on the circle with center at $(0, 0)$ and radius a and the point Q must be on the circle with center at $(0, 0)$ and radius b. Since P and Q are on the circle C, on a line through the center of C and on the same line through the center of the other two circles, the distance between P and Q is $14 = b - a$.

26. Solution (A). The point-slope form of the equation of the line is

$$y - \frac{k}{x_0} = \left(-\frac{y_0}{x_0}\right)(x - x_0). \quad (1)$$

The triangle is a right triangle with height the value of y when we set $x = 0$ in (1), $H = 2k/x_0$, and base the value of x when we set $y = 0$ in (1), $B = 2x_0$. Thus the area is

$$A = \frac{1}{2}HB = 2k.$$

27. Solution (D). The enclosed region is a square with vertices at $(4, 0)$, $(0, 4)$, $(-4, 0)$, and $(0, -4)$, which has area $(4\sqrt{2})^2 = 32$.

28. Solution (C). The vertices of the triangle can be obtained by solving the *equations* simultaneously in pairs: $(0, 0)$, $(12, 0)$, and $(4, 4)$. The triangle has base 12 and altitude 4, hence area 24.

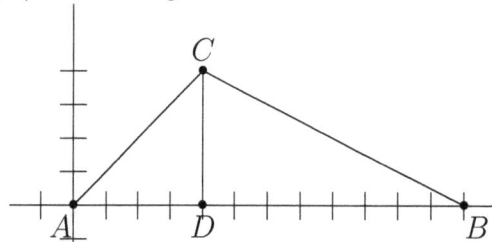

29. Solution (B). The triangle has a base along the y-axis of $42 - 121/3 = 5/3$ and an altitude of 20 (the lines intersect at $(20, 7)$). So the area is $\frac{1}{2} \cdot \frac{5}{3} \cdot 20 = 50/3$.

30. Solution (D). Looking at the diagram, the area of $\triangle ABE$ is 3, the area of $\triangle AEF$ is 1.5, and the area of $\triangle CDE$ is 1. The total area is therefore 5.5.

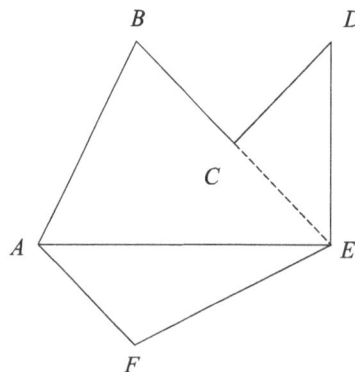

31. Solution (C). Partition the region into two rectangles and two right triangles as shown, with areas $24, 18, 6$, and 15 respectively, for a total area of 63.

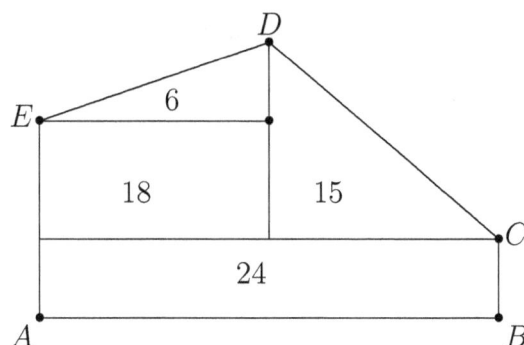

32. Solution (A). The area bounded by the two graphs is a triangle with base $2c$ and height c. Thus its area is $c^2 = 5$ hence $c = \sqrt{5}$.

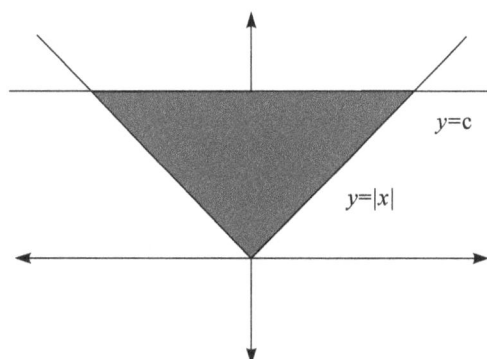

33. Solution (D). The point $E = (2, 4)$ lies on AD. The line segment from $(2, 4)$ to $(5, 4)$ divides the quadrilateral into a trapezoid with area $\left(\dfrac{6+3}{2}\right) 4$ and a triangle with area $\dfrac{1}{2}(3)(2)$. The total area is 21.

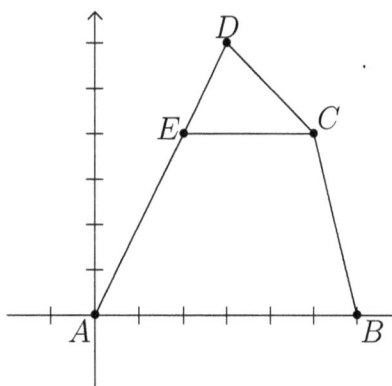

34. Solution (B). In the first quadrant, the inequalities $x + y \leq 20$ and $x + y \leq 18$ define similar right triangles with areas $20^2/2$ and $18^2/2$. The area of the region in question is therefore $(20^2 - 18^2) \div 2 = 38$ square units.

Alternatively, we can also find the area as the area of a trapezoid. The parallel sides have lengths $20\sqrt{2}$ and $18\sqrt{2}$ and the height of the trapezoid is $\sqrt{2}$. Thus the area is $(38\sqrt{2}/2) \cdot \sqrt{2} = 38$.

35. Solution (B). The triangle in question has area four times the one obtained by connecting the midpoints with line segments. (Draw the picture.) That triangle has area 4. You can find the area of any triangle with integer vertices by augmenting triangular regions to build a rectangle, and then subtracting areas. The rectangle here has vertices $(1, 1), (4, 1), (4, 5)$, and $(1, 5)$, so the area of the midpoint triangle is $12 - 3 - 4 - 1 = 4$. Thus the large triangle has area 16.

36. Solution (D). The distance is 2 more than the distance to the center of the sphere, which is $2 + \sqrt{4^2 + 4^2 + 7^2} = 2 + \sqrt{81} = 11$.

37. Solution (A). Distance $= 2 + \sqrt{29}$. Reflect $(5, 3)$ about the x-axis to get $(5, -3)$. Now the shortest distance from $(7, 4)$ to $(5, -3)$ is the same as for the original problem. Imagine shrinking the vertical segment to zero by lowering the point $(7, 4)$ down to $(7, 2)$. The line between $(7, 2)$ and $(5, -3)$ goes through the x-axis at $x = \dfrac{31}{5}$, and the length of this segment is $\sqrt{29}$. Thus the shortest path has length $2 + \sqrt{29}$.

38. Solution (B). Completing the squares, the circle can be written $(x + 1)^2 + (y - 2)^2 = 25$. The line through $(7, 8)$ and the center of the circle $(-1, 2)$ includes the point P, so the slope is $\frac{8-2}{7+1} = \dfrac{3}{4}$.

39. Solution (B). The centers of the three smaller circles form an equilateral triangle with side $2a$ and height $h = \sqrt{3} \cdot a$. The center O of this triangle satisfies $OA = OB = OC = 2h/3 = 2a\sqrt{3}/3$. The radius of the unit circle satisfies $1 = OA + a = \frac{2\sqrt{3}+3}{3}a$. Hence

$$a = \frac{3}{2\sqrt{3}+3} = \frac{3(2\sqrt{3}-3)}{(2\sqrt{3}-3)(2\sqrt{3}+3)} = \frac{3(2\sqrt{3}-3)}{12-9} = 2\sqrt{3} - 3.$$

We see that $p = 2$ and $q = 3$.

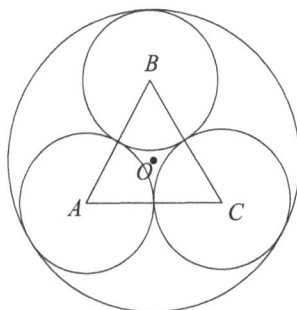

40. Solution (E). Let $(f, 0)$ be the x–intercept of ℓ and let $(0, g)$ be the y–intercept of ℓ. Then the area of T is $|fg/2|$ and the length of the hypotenuse is $\sqrt{f^2 + g^2}$. There is an equation for ℓ of the form $5x - y = C$ for some number C with $f = C = 5g$. So $5 = 5g^2/2$, and from this, we get $g = \pm\sqrt{2}$ and $f = \pm 5\sqrt{2}$. Thus the length of the hypotenuse is $\sqrt{52}$.

41. Solution (C). The Zero Product Property applies. Thus one or more of the five factors is zero. Hence S consists of all points satisfying any of the following five equations: $x = 0, y = 0, y = -x, y = x, x^2 + y^2 = 1$. The first four are lines and the fifth is a circle. The diagram shows that S divides the plane into 16 regions.

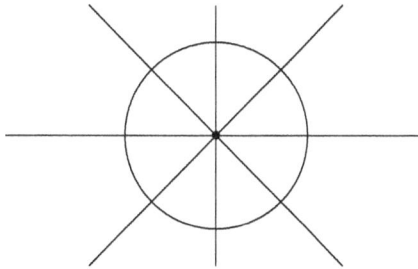

42. Solution (B). The sum of squares of travel times is

$$\left(\frac{\sqrt{c^2 + 16}}{1}\right)^2 + \left(\frac{\sqrt{(10 - c)^2 + 16}}{2}\right)^2 = \frac{5}{4}c^2 - 5c + 45 = \frac{5}{4}((c - 2)^2 + 32)$$

Therefore the sum of squares of travel times is minimized when $c = 2$.

43. Solution (D). The line goes through the center of the circle, $(3, 0)$ so it must intersect the circle twice. For $a = -1$ the intersection includes a point in the second quadrant. For $a = 2$, the line is

$$y = 2x - 6,$$

so

$$(x - 3)^2 + (2x - 6)^2 = 25,$$

which becomes

$$(x - 3)^2 = 5,$$

which has two positive solutions. For $a = 0$ the intersection includes only points on the x-axis. For $a = 1$, the intersection includes a point of the third quadrant.

OR

The intersection of the circle with the y-axis is a pair of points $(0, 4)$ and $(0, -4)$. The slope of the line through the center and $(0, 4)$ is $-4/3$ and the slope of the line thought the center and the point $(0, -4)$ is $4/3$. The lines with slopes $1, 0$ and -1 cross the y-axis between $(0, 4)$ and $(0, -4)$, so cannot intersect the circle in both the first and fourth quadrants. On the other hand, a line through the center whose slope has absolute values greater than $4/3$ intersects the circle in both of these quadrants. Thus $a = 2$ works.

44. **Solution** (E). The bug at the center of the cell (a, b) is hit if and only if the centers of the cells $(8,11)$, $(20,24)$ and (a, b) are on the same line, that is, $\frac{a-8}{20-8} = \frac{b-11}{24-11}$. This is true only for the last pair $(a = 68, b = 76)$.

45. **Solution** (C). Recall a fact from geometry: the angle β (see below) is twice the measure of the given angle. Then $x = \cos 60 = 1/2$. Otherwise, suppose that the coordinates of A are (x, y). Then

$$y = (1 + x) \tan \alpha = \frac{1 + x}{\sqrt{3}}$$

and

$$x^2 + y^2 = 1.$$

Substituting y from the first equation into the second, we have

$$x^2 + \left(\frac{1 + x}{\sqrt{3}}\right)^2 = 1.$$

After squaring and simplifying, we find

$$2x^2 + x - 1 = (2x - 1)(x + 1) = 0.$$

The only positive root is $\frac{1}{2}$. Alternatively, note that angle ACO is $60°$. It follows that $\triangle ACO$ is equilateral. Therefore, the projection of point A onto the x-axis is the midpoint of OC, namely, $1/2$. Therefore the x-coordinate of A is $1/2$.

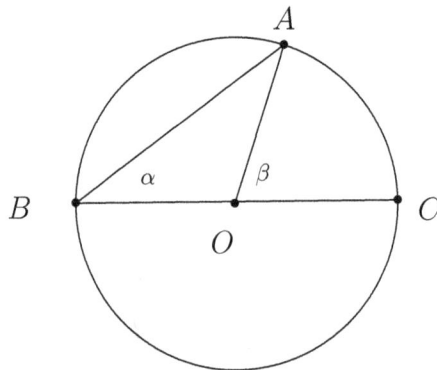

Alternate Solution: Let (r, s) be the coordinates of A. Then $r^2 + s^2 = 1$. Since O is the center of the circle , the segments $\overline{OB}, \overline{OA}$ and $\overline{OC}$ all have length 1. Also $\angle ABC$ is a right angle and the measure of $\angle OCA$ is $60°$. Thus $\triangle AOC$ is equilateral. Then $r^2 + s^2 = 1$ and $(r - 1)^2 + s^2 = 1$. Squaring out $(r - 1)^2$ and subtracting the second equation from the first yields $2r - 1 = 0$. Thus $r = 1/2$.

46. Solution (C). The solution uses the *reflection principle*. Let $A' = (-2, 3)$ and $B' = (5, -1)$ represent the reflections of A and B across the y-axis and x-axis, respectively. The sum $A'P + PQ + QB'$ is the same as the sum $AP + PQ + QB$ for any points P on the y-axis and Q on the x-axis. But the shortest path from $(-2, 3)$ to $(5, -1)$ is a straight line that hits both the y and x axes. The distance between $(-2, 3)$ and $(5, -1)$ is $\sqrt{65}$, so this is the length of the shortest path.

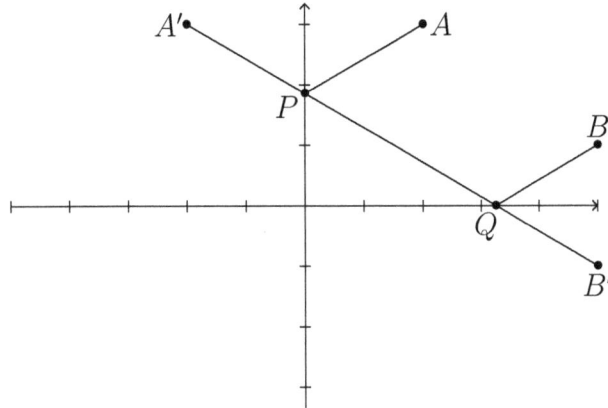

CHAPTER 35

COUNTING

35.1 Counting

1. **Solution** (A). There are squares of three sizes. There are 4 of area $1/2$, 5 of area 1 and 4 of area 2.

2. **Solution** (A). The small cube in the middle has 6 faces. On the other hand it is easy to see how to strike 6 times to divide the cube into 27 pieces.

3. **Solution** (D). A piece of the sphere is situated next to the each vertex, each face and each edge of the pyramid.

4. **Solution** (D). If 5 balls are needed to make sure at least one of them is red, that means that there are 4 blue balls in the box. If you need 10 to make sure both colors appear then there are 9 balls of the more used color –which must be red. Therefore there are $9 + 4 = 13$ balls in the box.

5. **Solution** (D). Each pair of points of S belongs to two circles of radius 3. There are no three points on the same circle of radius 3. There are six pairs of points, so there are 12 circles.

6. **Solution** (C). The shortest path from A to B uses exactly six semi-circles. Selecting either the upper or the lower semicircle from each circle leads to $2^6 = 64$ paths from A to B.

7. **Solution** (E). Count them by size. There are 48 unit squares, 24 with area 4, one with area 16, four with area 25, nine with area 36, four with area 49, and one with area 64, for a total of $48 + 24 + 1 + 4 + 9 + 4 + 1 = 91$.

8. **Solution** (C). If the units digits is 0 then there are 9 possible tens digit which is greater than 0. Now discuss the cases when the units digit is 1 through 9. We have total $9 + 8 + \cdots + 2 + 1 = 45$.

9. **Solution** (D). Both s and $\frac{8}{s}$ should be factors of 8 and T should contain the pair $\{s, \frac{8}{s}\}$. Thus, the possible sets are $T = \{1, 2, 4, 8\}, T = \{1, 8\}$, and $T = \{2, 4\}$.

10. **Solution** (C). Build a graph with the six vertices labeled $3, 4, 5, 6, 12$, and 13. Join two when their sum is a perfect square. Note that 5 and 6 are related to only 4 and 3 respectively, so they must be at the ends. The sequence 6 3 13 12 4 5 works.

11. **Solution** (D). There are $3^4 = 81$ 4-letter words in BadSpeak. Of these, $2^4 = 16$ do not contain the letter "c." Hence $81 - 16 = 665$ words do contain the letter "c."

12. **Solution** (B). The smallest possible sum is $3 = 0 + 1 + 2$ and the largest is $23 = 5 + 7 + 11$. There are twenty one integers in this range, but it is impossible to have a sum of 22. Except for 22, all other sums are possible for a total of twenty.

13. **Solution** (B). If the archer hits the next seven consecutive shots then he will have hit 10 out of 11 shots and $\frac{10}{11} > \frac{9}{10}$. But six consecutive hits will not be enough.

14. **Solution** (B). There are 90 two-digit numbers from 10 to 99. Among them there are five whose digits' sum equals $5 : 50, 41, 32, 21, 14$. Thus, the probability is $\dfrac{5}{90} = \dfrac{1}{18}$.

15. **Solution** (B). The smallest possible sum is

$$3 + 6 + 9 = 18$$

and the largest is

$$18 + 21 + 24 = 63$$

and the set of all possible sums is

$$\{18, 21, 24, \ldots, 63\},$$

that is, the set of multiples of 3 in the range 18 up to 63. There are 16 numbers in this set.

16. **Solution** (C). Pages 1 through 9 use 9 digits and 10 through 99 use

$$90 \times 2 = 180$$

digits, for a total of 189 digits for pages 1 through 99. That leaves 663 digits remaining to make the required total of 852 digits. These are obtained by going 221 pages beyond page 99, through page 320.

17. **Solution** (B). There are 4^n ways of making all n-letter words; 2^n of these contain no vowels. Adding the number of 1-letter, 2-letter, 3-letter, and 4-letter words with vowels, we obtain

$$4 - 2 + 4^2 - 2^2 + 4^3 - 2^3 + 4^4 - 2^4 = 310.$$

18. Solution (C). Notice that two rows have four crickets, so at least two crickets must move. The pair of crickets at $(1,5)$ and $(2,4)$ on the main diagonal can be moved to $(5,3)$ and $(5,1)$ as shown.

19. Solution (E). Each triangle contains A or B or both. There are $6 \cdot 4 = 24$ that contains A and $10 \cdot 3 = 30$ that contain B. But we are double counting $4 \cdot 3 = 12$ triangles so there are

$$24 + 30 - 12 = 42$$

triangular regions.

20. Solution (D). One way is to list all possibilities: $1+2+2 = 5, 1+1+1+2 = 5, 2+1+2 = 5,$ $1+1+2+1 = 5, 2+2+1 = 5, 1+2+1+1 = 5, 1+1+1+1+1 = 5, 2+1+1+1 = 5.$ Alternatively, we can denote by F_n the number of ways Joe can reach the top of the stairs if there are n steps. The first leap is 1 step or 2 steps, and there are F_{n-1}, respectively F_{n-2} ways to complete the procedure. This observation proves the recursion formula $F_n = F_{n-1} + F_{n-2}$, where $F_1 = 1$ and $F_2 = 2$. From here we see that F_n is the n-th Fibonacci number, and so $F_5 = 8$.

21. Solution (B). First, suppose that there is exactly one pair of opposite faces of the same color. In this case, there is just one cube with the blue faces opposite, one with the white faces opposite, and one with the red faces opposite. There are no cubes with exactly two pairs of opposite faces of the same color. There is just one cube where all three pairs of opposite faces are the same color. Finally, there are two cubes where no pair of opposite faces are the same color. We get a total of six distinguishable cubes.

22. Solution (D). Since the number is odd, there are 5 ways to choose the units digit. Next, pick the hundreds digit in any of 8 ways, (you can't use the units digit or 0), and finally, pick the tens digit in any of 8 ways. Thus there are $5 \cdot 8 \cdot 8$ ways to choose the number.

OR

Let the three digit number be $100a + 10b + c$ with digit a being one of $1, 2, 3, 4, 5, 6, 7, 8, 9$; digit b being one of $0, 1, 2, 3, 4, 5, 6, 7, 8, 9$; digit c being one of $1, 3, 5, 7, 9$. One distinguishes four cases

a, b both even: There are $4 \cdot 4 \cdot 5 = 80$ possibilities. a, b both odd: There are $5 \cdot 4 \cdot 3 = 60$ possibilities. a odd and b even: There are $5 \cdot 5 \cdot 4 = 100$ possibilities. a even and b odd: There are $4 \cdot 5 \cdot 4 = 80$ possibilities. Taken together, it adds up to $80 + 60 + 100 + 80 = 320$ possibilities.

23. **Solution** (D). Since there are two choices for each bit, the number of different words is $2^8 = 256$.

24. **Solution** (B). By the principle of inclusion-exclusion, the number is

$$10000 - \left\lfloor \frac{10000}{13} \right\rfloor - \left\lfloor \frac{10000}{51} \right\rfloor + \left\lfloor \frac{10000}{13 \times 51} \right\rfloor =$$

$$10000 - 769 - 196 + 15 = 9050.$$

25. **Solution** (C). Every set of three points except those which are collinear can be the vertices of a triangle. There are $\binom{12}{3} = 220$ three element subsets, and $4 \cdot \binom{4}{3} = 16$ which are collinear, so there are

$$220 - 16 = 204$$

triangles.

26. **Solution** (C). There are $5 \cdot 7 = 35$ handshakes between men and women, and $\binom{5}{2} = 10$ handshakes exchanged among the men, for a total of 45 handshakes.

27. **Solution** (E). Any two-element subset of $\{1, 4, 9, 16\}$ satisfies the condition. There are six of these. There are just two other sets, $\{2, 8\}$ and $\{3, 12\}$.

28. **Solution** (E). Count them one at a time. They are $2, 3, 4, 5, 6, 8, 10, 12, 15, 18, 20, 24$, and 30. Alternatively, there are $\binom{6}{2} = 15$ pairs, but the numbers 6 and 12 are represented twice, so there are $15 - 2 = 13$ products.

29. **Solution** (D). There are 24 right triangles with edge lengths $1, 1, \sqrt{2}$; 24 right triangles with edge lengths $1, \sqrt{2}, \sqrt{3}$; and 8 equilateral triangles with edge lengths $\sqrt{2}, \sqrt{2}, \sqrt{2}$.

30. **Solution** (C). There are $\binom{8}{3} = 56$ successful outcomes, so the probability is $\frac{56}{6^3} = \frac{7}{27}$.

31. **Solution** (C). There are $\binom{9}{3} = 84$ ways to choose the three points, and exactly 8 of these result in collinear points.

32. **Solution** (A). Once Gil's seat is determined, there are 6 ways to choose the person to his right and 5 ways to choose the person to his left. After that there are 5! ways to arrange the others, so the total number of arrangements is

$$6 \cdot 5 \cdot 5! = 3600.$$

33. **Solution** (A). There is either one touchdown or none. With 1 touchdown the scoring sequences with value 10 are $613, 361, 622, 262, 226$. With no touchdowns there is 3322 and its permutations, of which there are 6 total, and 22222 So the total number of sequences with value 10 is

$$5 + 6 + 1 = 12.$$

34. **Solution** (B). Seven men can shake hands with each other in $7 \cdot 6/2 = 21$ ways. The men can shake hands with the women in $7 \cdot 7 = 49$ ways. Adding these, we get 70 handshakes.

35. **Solution** (B). There are $\binom{11}{3} = 165$ three elements subsets of the grid, but

$$\binom{5}{3} + \binom{5}{3} + \binom{4}{3} = 10 + 10 + 4 = 24$$

of these are colinear sets. Therefore there are $165 - 24 = 141$ sets of vertices.

36. **Solution** (E). Only the numbers ending in 5 are divisible by 5. There are $5! = 120$ ways to permute the other five digits.

37. **Solution** (B). Suppose the farmer originally had N horses. Splitting $N + 1$ horses as described in the will and recalling that one horse is not distributed leads to

$$\frac{N+1}{3} + \frac{N+1}{4} + \frac{2(N+1)}{5} = N$$

Adding the fractions yields

$$\frac{59(N+1)}{60} = N$$

Hence $N = 59$.

38. **Solution** (B). Once the first sock has been picked, the probability that the second matches it is one-fifth.

OR

$P(\text{matched pair}) = \#\text{possible matched pairs}/\#\text{possible pairs} = 3/((6 \cdot 5)/2)$.

39. **Solution** (B). In order to accommodate 11 guests, no single slice of cake can be larger that $1/11$ of the cake. Therefore, if ten guests arrive, each guest must get at least two slices of cake. Consequently, we must divide the cake into at least 20 slices. Let us show that this number suffices. Divide the cake into 11 equal pieces. Then divide one of these into 10 equal pieces. (Each of these is $1/110$ of the cake.) We now have a total of 20 pieces. If 10 guests are coming everybody gets one big and one small piece, so that

$$1/11 + 1/110 = 1/10.$$

If 11 guests arrive, then 10 of them get $1/11$ each and the last one receives the $1/11$ that was divided into 10 pieces.

40. Solution (E). The question is equivalent to 'how many three digit numbers $\underline{abc}$ (0 allowed as the hundreds digit) satisfy $a + b + c = 6$. When $a = 6$, there is only one way to do this: $b = c = 0$. When $a = 5$, there are two ways: either $b = 1$ and $c = 0$ or $c = 1$ and $b = 0$. When $a = 4$, there are three ways:

$$(b, c) = (0, 2), (1, 1), \text{ or } (2, 0).$$

Continuing in this way, we find that there are

$$1 + 2 + 3 + \ldots + 7 = 28$$

ways to build the number.

OR

There are three ways to choose three different digits that sum to 6: $1+2+3$, $0+2+4$ and $0+1+5$. Since the order counts, this gives 18 ways. There are three ways to choose two digits the same and a different third digit: $1 + 1 + 4$, $0 + 3 + 3$ and $0 + 0 + 6$. Since two of the digits are the same, there are 9 ways of ordering these. Finally, there is one sum, $2 + 2 + 2$, where all three digits are the same. So the total number of ways is

$$18 + 9 + 1 = 28.$$

41. Solution (B). We can build the net in a different way showing all the faces except the one with the 6, which is at the back in the figure below.

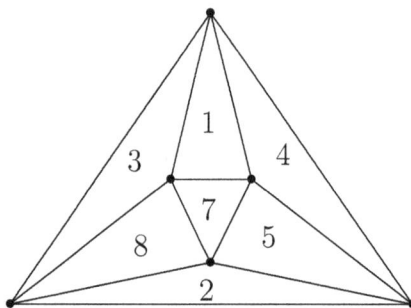

Note first that the face with the 7 is adjacent to those with numbers 1, 5, and 8. Filling in those four values in the center triangles, we can then see that the face with the 4 must be adjacent to those

COUNTING **439**

with the 1 and the 5 and similarly with the faces marked 3 and 2. This shows that the face marked 3 is adjacent to those marked with 1, 8 and 6 so the total value is 15.

42. **Solution** (A). Each rectangular region is determined by two pairs of horizontal and vertical lines. In order that the shaded square is inside the region, the lower bounding line must be one of the bottom two, the upper boundary line must be one of the top three, the left side must be one of the two left-most vertical lines and the right boundary line must be one of the 167 vertical lines to the right on the shaded square. Thus the number of regions is $N = 2 \cdot 3 \cdot 2 \cdot 167 = 2004$ and the sum of the digits is 6.

Alternatively, count based on the height of the rectangle and position of bottom left square from the grid that is part of the rectangle. A rectangle of length one must have its bottom left square in the second column, and one of length 168 must have its bottom left square in the first column. All other lengths may have bottom left square in either of the first two columns. So for rectangles of height 1, there is one of length 1, one of length 168, and two each of all 166 other lengths. The same count works for rectangles of height 4 – one of length 1, one of length 168 and two each of all 166 other lengths. For rectangles of height 2, there are two of length 1, two of length 168, and four each of all 166 other lengths. The same count works for rectangles of height 3 – two of length 1, two of length 168, and four each of all 166 other lengths. Thus there are six rectangles of length 1, six of length 168, and twelve of each other length. Therefore the total number of rectangles is $2004 = 6 + 6 + 12 \cdot 166$. The sum of the digits is 6.

43. **Solution** (C). Let us denote the position of the even digits by e, and the position of the odd digits by o. There are 6 possible patterns:

$$eeoo, \ eoeo, \ eooe, \ oeeo, \ oeoe \ \text{and} \ ooee.$$

If the first digit is even, we may choose it 4 ways (we can not begin with 0), and the other even digit may be any of the hitherto unused 4 digits. Thus, in this case, we may choose the even digits $4 \cdot 4 = 16$ ways. If the first digit is not even, the even digits may be selected $5 \cdot 4 = 20$ ways. In either case, the odd digits may be selected $5 \cdot 4 = 20$ ways. Thus the number of positive integers satisfying all our criteria is

$$3 \cdot 16 \cdot 20 + 3 \cdot 20 \cdot 20 = 2160.$$

44. **Solution** (C). Note that 8 is not a multiple of 5 or a multiple of 3, so the horizontal portion of the moves must contain at least one three and one five. This must then be true of the vertical portion of the moves as well. Since the y-coordinate of (8,0) is zero, we need a multiple of fives and a multiple of threes to add to zero. The least number of moves that allow this requires 5 three's and 3 fives. This works since we can choose the vertical portion to be

$$5 + 5 + 5 - 3 - 3 - 3 - 3 - 3 = 0$$

and the horizontal portion to be

$$3 + 3 - 3 + 5 + 5 - 5 + 5 - 5 = 8.$$

45. **Solution** (B). Let us call " distance " between transmissions the number of positions in which they differ. We need to find which of the five transmissions has distances to the other four transmissions, $1, 2, 3$, and 4 in some order. Now, the distances from (a) to the other transmissions are $2, 3, 2, 3$; from (b):$2, 1, 3, 4$; from (c):$3, 1, 4, 5$; from (d):$3, 3, 4, 1$;from (e):$2, 4, 5, 1$. Thus the answer is B.

46. **Solution** (C). Let us represent each member and each committee with a vertex. Connect each vertex representing a member to each vertex representing a committee, of which (s)he is a member. We obtain a graph, which may also be created as follows. Take 10 vertices, representing the 10 committees, connect each pair of these committees by an edge. Subdivide each edge connecting two committees by inserting the midpoint of the edge and labeling it with the only member of the National Assembly that belongs to both committees. There are $10 \cdot 9/2 = 45$ pairs of committees, so we must have 45 members in the National Assembly.

47. **Solution** (D). There are $6^3 = 216$ possible outcomes, each occurring with probability $1/216$. For six of these, all three numbers will be same. The number of times exactly two are the same is $3 \cdot 6 \cdot 5 = 90$: choose which die is different and that value ($3 \cdot 6 = 18$ choices for this), then choose the number that is the same for the other two die (5 choices). So the probability at least two are the same is $96/216$. Or count how many where all three are different, $6 \cdot 5 \cdot 4 = 120$, subtract this number from 216 and then divide by 216.

48. **Solution** (C). If the intersection of the two diagonals is inside the 12-gon, then the endpoints of these diagonals form a convex 4-gon, whose diagonals are the selected two diagonals. The converse is also true. There are $\binom{12}{4} = 495$ ways to select 4 vertices out of 12.

49. **Solution** (C). The number of edges is $\frac{1}{2}(35 \cdot 3 + 5 \cdot 4 + 7 \cdot 5) = 80$. By Euler's formula, $e + 2 = f + v = 82$, it follows that there are $82 - 47 = 35$ vertices. This polyhedron is called a *gyroelongated pentagonal cupolarotunda* (J47).

50. **Solution** (B). The number of handshakes among children is $T_{c-1} = C_2^c = \binom{c}{2} = \frac{c!}{(c-2)!2!}$, and similarly for the women and the men. The only way to write 57 as a sum of three *different* triangular numbers is $6 + 15 + 36$, so there must have been 4 children, 6 women, and 9 men, and $\binom{19}{2} = 171$ handshakes.

51. **Solution** (C). 6930. Without regard to order, there are $C(15, 2) = 105$ choices for the two stick letters and 11 choices for the round letter. Allowing for different orders, there are $3! = 6$ ways to

order the chosen letters, so the total number with order considered is

$$6 \times 105 \times 11 = 6930.$$

52. **Solution** (B). c can be either 0 or 5 and $a + b + c + 2$ is divisible by 9. Two cases:

a) $c = 0$ and $a + b + 2$ is divisible by 9, so $a + b = 16$ (3 cases: $(7, 9)$, $(8, 8)$, $(9, 7)$) or $a + b = 7$ (7 cases: $(1, 6)$, $(2, 5)$, ..., $(7, 0)$).

b) $c = 5$ and $a + b + 7$ is divisible by 9, so $a + b = 11$ (8 cases: $(2, 9)$, $(3, 8)$, ..., $(9, 2)$) or $a + b = 2$ (2 cases: $(2, 0)$ and $(1, 1)$). Thus the answer is $3 + 7 + 8 + 2 = 20$.

53. **Solution** (C). Each number in the set has an 11-digit binary representation that starts with 1. Therefore it takes five more 1's in order to have more 1's than 0's. Hence we can build such a number by selecting five or more places to put the 1's. Thus, the number is

$$\binom{10}{5} + \binom{10}{6} + \binom{10}{7} + \binom{10}{8} + \binom{10}{9} + \binom{10}{10}$$
$$= \frac{1}{2}\left(2^{10} - \binom{10}{5}\right) + \binom{10}{5} = 386 + 252 = 638.$$

54. **Solution** (B): To start, choose colors for regions A, B and C, all three will have to be different. Since regions A and E do not touch, they can be the same color. In this case, region D can either be the same color as region B or it could be colored with the fourth color. So there are $4 \cdot 3 \cdot 2 \cdot 2$ ways do the coloring this way. If E and A are different colors, then all four colors have been used and now region D must be the same color as region B. For this scheme there are $4 \cdot 3 \cdot 2$ ways to color. Thus the total is $48 + 24 = 72$.

55. **Solution** (A). Construct a Venn diagram as shown below. Start with the fact that 8 students play all three sports.

Since 12 play both football and basketball, we have:

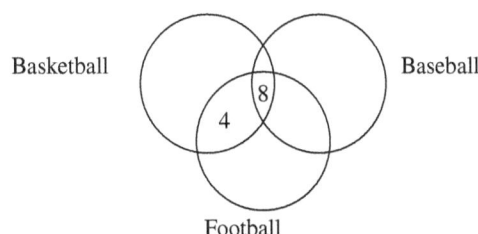

Proceeding in this manner until all the given information is used yields the following totals:

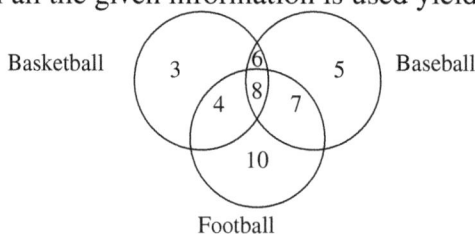

Hence the total number of members is 43.

56. Solution (B). Just as the points $(0,0,0)$, $(1,0,0)$, $(0,1,0)$, $(0,0,1)$, $(1,1,0)$, $(1,0,1)$, $(0,1,1)$ and $(1,1,1)$ are the vertices of a three-dimensional cube, the points of the form $(a_1, a_2, a_3, a_4, a_5, a_6)$ where each $a_i \in \{0,1\}$ are the vertices of a six-dimensional cube (in six-dimensional space). There are $2^6 = 64$ vertices. An edge connects a pair of vertices if and only if the vertices have exactly one coordinate that is different. So each vertex is connected by an edge to six other vertices. Thus the total number of edges is $6 \cdot 2^6 / 2 = 192$. The total price is $2 \cdot 192 + 64$.

57. Solution (B). Let TBF represent the number of students that play all three sports, let B be the number of students that play baseball, let F be the number of students that play football, and let T be the number of students that play tennis. Let TB, BF, and TF represent the number of students who play the appropriate combinations of the individual sports. Then a Venn diagram shows that the total number of students, 50 equals $T + B + F - TB - TF - BF + TBF$. Therefore $50 = 15 + 25 + 30 - 8 - 5 - 10 + TBF$ which implies that $TBF = 3$. Alternatively, label each region of the Venn diagram as shown.

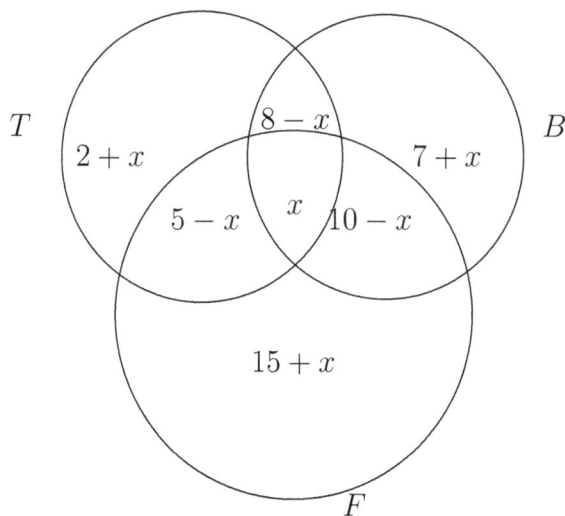

Then $4x + 24 - 3x + 23 = x + 47 = 50$, so $x = 3$.

58. Solution (B). First we compute a, b, and c from $ab = 48$, $bc = 60$ and $ac = 80$. Dividing the product of the second two equations by the first gives $c^2 = 100$, that is, $c = 10$. Using this information we get $b = 6$ from the second equation and a = 8 from the first equation. There are $4(4 + 6 + 8) = 72$ unit cubes with paint on exactly 2 faces. The total number of all cubes is $6 \cdot 8 \cdot 10 = 480$. The probability is $72/480 = 3/20$.

59. Solution (E). If we think of each triangle we can build as a three digit number where the digits are nondecreasing as we move from left to right. They are $333, 334, 335, 344, 345, 346, \ldots$. There are 35 of these, and they fall into three types: 10 of the type (a, b, c) with $a < b < c$, 20 of the type (a, a, b) or (a, b, b) with $a < b$ and 5 of the type (a, a, a). The only triplets of numbers that don't work are those for which the sum of two smaller sides is at most the third side. Every such triplet built from the digits $3, 4, 5, 6$ and 7 corresponds to a triangle in this way except $336, 337$, and 347, because these three fail to satisfy the triangle inequality $a + b > c$. There are $35 - 3 = 32$ of these numbers.

60. Solution (A). We can construct $45^2 = 2025$ unit squares with $45 + 1 + 45 + 1 = 92$ segments, half horizontal and half vertical. Note that $2004 = 45^2 - 25 + 4$, so there is hope to accomplish the construction with 92 lines. Let 84 of the lines have length 45; 4 have length 41; 3 have length 40, and 1 with length 44. On the other hand, the maximum number of unit squares that can be constructed using 91 segments is $44 \cdot 45 = 1980$, so 91 lines is not sufficient.

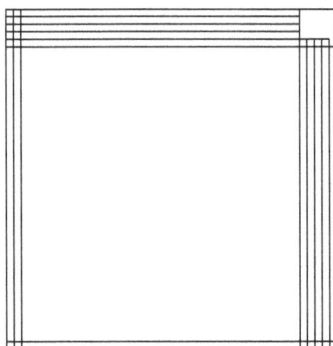

OR

An "additive" approach. For a perfect square m^2, the minimum number of lines needed is $2 \cdot (m + 1)$. The minimum is obtained by making an $m \times m$ grid using $m + 1$ vertical line segments and $m + 1$ horizontal line segments. To make $m^2 + 1$ through $m^2 + m$ unit squares, simply add one more vertical segment and extend the appropriate number of consecutive horizontal ones to get the required number. For more than $m^2 + m$ and up to $(m + 1)^2$, add one horizontal segment to the $m + 1 \times m$ grid and extend the appropriate number of consecutive vertical segments. Since $\sqrt{2004} \approx 44.77$, the largest perfect square that is less than or equal to 2004 is 44. The difference $2004 - 44^2 = 68$ is greater than 44, so the minimum number of lines for 2004 is $92 = 2 \cdot (45 + 1)$, the same minimum for 45^2.

61. **Solution** (D). We have

$$5^3 - 3^3 = 98$$

unit cubes with some red faces. Among these there are 8 corner cubes with three painted faces, and of course, these must be used as the corners of the brick. There are $12 \cdot 3 = 36$ unit cubes with two adjacent red faces, and there are $6 \cdot 9 = 54$ unit cubes with one red face. The maximum volume is therefore no more than $8 + 36 + 54 = 98$. The only brick we could hope to build with volume 98 is a $7 \times 7 \times 2$, but this would require 40 unit cubes with two adjacent red faces. Of course, building a brick with volume 97 is hopeless. How about 96? Could it be a $6 \times 4 \times 4$ brick? How many 'edge cubes' are needed? Answer: $2 \cdot 2 \cdot 4 + 4 \cdot 4 = 32$. Thus we have 4 two-faced cubes left that we can use with the 54 one red face cubes. This is enough to fill out the rest of the brick.

Alternatively, count the total blocks available, and how many of each type. Let $a \leq b \leq c$ be the dimensions of the cuboid. Based on the type of blocks available, $a \geq 2$. Also $a < 5$, since $5^3 = 125$. If $a = 2$, the cuboid would have $4(b - 2) + 4(c - 2)$ edge blocks (that are not corners). As there are 36 edge blocks available, $b + c \leq 13$. The largest volume is with $b = 6$ and $c = 7$, for a volume of 84. If $a = 3$, the cuboid would have $4(b - 2) + 4(c - 2) + 4$ edge blocks. From this $b + c \leq 12$, $b = 6 = c$ uses too many blocks as does $b = 5, c = 7$, but $b = 4, c = 8$ gives a volume of 96 (while $b = 5, c = 6$ has volume 90). If $a = 4$, similar analysis gives $b + c \leq 11$. The pairs $b = 4, c = 7$, $b = 5, c = 6$ and $b = 5 = c$ give volumes that are larger than 98, but $b = 4, c = 6$ also yields a cuboid of volume 96.

62. **Solution** (B). We first assume that the pizza occupies the entire plane. Denote the number of pieces after k cuts by N_k. We have $N_0 = 1$ and $N_1 = 2$. Furthermore, $N_2 = 4$, unless the 2nd cut is parallel to the 1st, in which case $N_2 < 4$. In general, $N_k = N_{k-1} + k$ if the kth cut intersects the first $k - 1$ cuts at $k - 1$ distinct points; otherwise, the difference $N_k - N_{k-1}$ will be smaller than k. Therefore, the largest possible value of N_k (which is attained if no cuts are parallel and no more than 2 cuts pass through a single point) is $1 + 1 + 2 + 3 + \ldots + k = 1 + \frac{k(k+1)}{2}$. For $k = 40$, this is 821.

This result is true if the pizza occupies the entire plane. For a bounded square-shaped pizza, the number of pieces cannot be more than 821. (Indeed, any 40 cuts of the square can be continued to the plane, thus generating a partition of the plane into no more than 821 pieces, as we know; each of them, being intersected with the square, produces no more than one piece of the square. But this is the original partition of the square.) On the other hand, if we scale the above "maximal" partition of the plane so that each piece intersects the square, we get a partition of the square into 821 pieces.

63. Solution (C). For each node $X_{i,j}$ for $1 \leq i \leq 10$, there is a unique path with 10 moves from $X_{0,1}$ to $X_{i,j}$. For example the only path with 10 moves from $X_{0,1}$ to $X_{6,1}$ is

$$X_{0,1} \rightarrow X_{1,1} \rightarrow X_{1,2} \rightarrow X_{1,1} \rightarrow X_{1,2} \rightarrow X_{1,1} \rightarrow X_{2,1} \rightarrow X_{3,1} \rightarrow X_{4,1} \rightarrow X_{5,1} \rightarrow X_{6,1}$$

The number of nodes on levels 1 through 10 is $2 + 4 + \cdots + 2^{10} = 2(2^{10} - 1) = 2046$.

64. Solution (C). The expansion $(r + b)^6 = 1r^6 + 6r^5b + 15r^4b^2 + 20r^3b^3 + 15r^2b^4 + 6rb^5 + 1b^6$ provides an inventory of the 64 possible colorings of the cube. For example the coefficient 15 of the term r^4b^2 can be interpreted as saying that there are 15 ways to build the cube with four red faces and two blue faces. Among all 64 possible cubes, those with 6 red (only one of these), or with 5 red faces (there are 6 of these) must have such a vertex. Among the 15 with 4 red faces, 12 have such a vertex (they correspond to the 12 edges) and among the 20 with 3 red faces, exactly 8 (these correspond to the eight vertices) have such a vertex. So the number of 'good' cubes is 27, and the probability is $27/64$.

OR

There are six sides, so the number of ways the cube can be painted is $2^6 = 64$. To have no such vertex P is the same as having some pair of opposite sides both blue. Thus there are at most three patterns with more than three red sides but with no vertex P. There are exactly 20 ways to have exactly three red sides. So $(64 - 20)/2$ with more than three red sides. Thus $19 = 22 - 3$ is the number with at least one vertex P and more than three red sides. To have exactly three red sides and no vertex P, exactly one pair of opposite sides is red. There are three such pairs, and for each of these there are four choices for the other red side. So there are $8 = 20 - 12$ with a vertex P. Alternately, simply pick one of the eight vertices to be the one at the intersection of the three red sides. The total number with at least one vertex P is 27. So the probability is $27/64$.

65. Solution (C). Each of the 20 vertices is adjacent to three others, belongs to the same pentagonal face and is not adjacent to six others, and is not on a space diagonal with itself. So each vertex is the endpoint of 10 space diagonals. Since there are 20 vertices, there are 200 such space diagonal endpoints, and therefore 100 space diagonals. Alternatively, first note that there are 30 edges and there are $190 = \binom{20}{2}$ distinct line segments formed by connecting pairs of vertices. Thus the total number of diagonals is $160 = 190 - 30$. Each face has 5 edges and there are $10 = \binom{5}{1}$ distinct line segments connecting pairs of vertices in any one face. Thus each face has $5 = 10 - 5$ diagonals. These are the non-space diagonals. Since there are twelve faces, the number of non-space diagonals is $60 = 12 \cdot 5$. Therefore the number of space diagonals is $100 = 160 - 60$.

66. Solution (C). Let a represent the number of moves of type $(3, 8)$ (a negative a refers to moves of type $(-3, -8)$, etc.), b, the number of moves of type $(8, 3)$; c, the number of moves of type $(3, -8)$; and d, the number of moves of type $(8, -3)$. Then

$$a(3, 8) + b(8, 3) + c(3, -8) + d(8, -3) = (19, 0)$$

for any path from $(0, 0)$ to $(19, 0)$. Check to see that $a = 2, b = -3, c = -1$ and $d = 5$ works. Thus, 11 moves will suffice. On the other hand, we must have

$$3(a + c) + 8(b + d) = 19$$

and

$$8(a - c) + 3(b - d) = 0.$$

But $a + c$ is odd $\Rightarrow a - c$ is odd $\Rightarrow a - c \neq 0 \Rightarrow b - d \neq 0 \Rightarrow |b - d| \geq 8$ and $|a - c| \geq 3$ so that no sequence with fewer than 11 moves can work. The sequence

$$(0, 0), (3, 8), (11, 5), (19, 2), (16, 10), (8, 7), (0, 4), (8, 1), (11, 9), (19, 6), (11, 3), (19, 0)$$

is an example of an 11 move sequence that works.

35.2 Probability and Statistics

1. Solution (A). Exactly two of the four cards have primes, so just one of the six pairs consists of primes.

2. Solution (B). Each time you compare the new coin with the most heavy from the previous pair. You cannot do any better because the last coin might be the heaviest.

3. Solution (D). There are just $\binom{5}{3} = 10$ ways to the committee. One of these has three females and three have two males. The other all have two females and one males. Thus the probability is 0.6.

4. Solution (E). Consider the 'baskets' $\{1, 2, 3\}, \{4, 5, 6\}, \{7, 8, 9\}, \dots, \{19, 20, 21\}$. Put each of the 15 numbers into the set it helps to name. There are 7 subsets and 15 numbers so, by the Pigeonhole Principle, at least one of the baskets must have three members.

5. Solution (E). Put all the black unit cubes at the corners and along the edges of the big cube so they contribute either 3 or 2 square units to the surface. There are just enough black unit cubes to fill all eight corner positions and all the other $12 \times 2 = 24$ edge positions, so each face of the big cube is three-forths covered by black unit squares.

6. Solution (C). Since there is an odd number of coins, either the number of heads is odd and the number of tails is even, or the number of heads is even and the number of tails is odd. Since the coin is fair, half of the time the number of heads will be odd.

7. **Solution** (D). The probability of choosing a vowel written by an Englishman is $\frac{3}{6} \cdot 0.4 = 0.2$. The probability of choosing a vowel is

$$\frac{3}{6} \cdot 0.4 + \frac{2}{5} \cdot 0.6 = 0.44.$$

Therefore, the probability that the writer was English is $0.2/0.44 = 5/11$.

8. **Solution** (A). The probability of obtaining an ace on the first draw is $4/52 = 1/13$. If the first card drawn is an ace there are 3 aces remaining in the deck, which now consists of 51 cards. Thus, the probability of getting an ace on the second draw is $3/51 = 1/17$. The required probability is the product of the two, which is $1/221$.

9. **Solution** (E). The events "the second person has more tails than the first" and "the second person has more heads than the first" have the same probability and compliment each other. Hence, the probability of each event is $1/2$.

10. **Solution** (D). Since the total value of the three coins is exactly 0.35, the coins must be a quarter and two nickels. There are 6 ways to pick a quarter and two nickels out of a total of $\binom{8}{3} = 56$ ways to select three coins, so the probability is $6/56 = 3/28$.

11. **Solution** (B). The probability that a missile is **not** intercepted is $2/3$. The probability that any one missile hits its target is $P(hits) = (3/4) \cdot (2/3) = 1/2$. The three missiles are fired independently, so the desired probability is $1/2 \cdot 1/2 \cdot 1/2 = 1/8$.

12. **Solution** (B). There are $6 \cdot 4 = 24$ cubes with one face painted, and these show one painted face with probability $5/6$. There are $12 \cdot 2 = 24$ cubes with two painted faces and these show one painted face with probability $1/3$. The other 16 cubes show one painted face with probability 0. So the probability that one painted face shows is

$$p = 24/64 \cdot 5/6 + 24/64 \cdot 1/3 = 7/16.$$

13. **Solution** (A). There are 10^3 ways to pick three digits, and $6\binom{10}{3} = 720$ ways pick three *different* digits. In exactly one case out of six, the digits are in order, so the probability is $120/1000 = 3/25$.

14. **Solution** (D). There are $a = \binom{26}{13}$ ways to pick 13 of the 26 remaining cards. Since there are 5 hearts and 21 non-heart cards remaining, there are $b = \binom{5}{2}\binom{21}{11}$ ways to choose exactly 2 hearts. The probability b/a is $3527160/10400600 \approx 0.34$.

15. **Solution** (D). There are $6^3 = 216$ ordered triples of dice rolls. The product is prime precisely when two rolls are 1 and the third is a prime number. Since the prime (2,3, or 5) can appear in any of the three positions, there are 9 such triples, so the probability is $9/216 = 1/24$.

16. Solution (C). Each corner square touches 3 others. Each other edge square touches 5 others, and each interior square touches 8 others. Therefore there are 42 pairs of squares that touch one-another. There are $\binom{16}{2} = 120$ pairs of squares, so the probability is $\frac{42}{120} = \frac{7}{20}$.

17. Solution (D). There are four cubes with paint on three adjacent faces, 20 cubes with paint on two (adjacent) faces, and 28 cubes with only one painted face. The other 12 cubes have no painted faces. A cube with three painted faces has probability $1/2$ of landing so that two of the faces can be seen, and a cube with 2 painted faces has probability $2/3$ of landing so that two painted faces are showing. Thus the probability we seek is

$$(4/64) \cdot (1/2) + (20/64) \cdot (2/3) = 23/96.$$

18. Solution (C). The total number of the permutations of 6 is $6! = 6 \cdot 5 \cdot 4 \cdot 3 \cdot 2 \cdot 1$. There are $6 \cdot 4!$ outcomes with exactly two men seated between Sam and Peter. To see this note that the blanks can be filled in in four factorial ways: $S, \_, \_, P, \_, \_$ for each of the six possible positions of the S and P. The probability is therefore $6 \cdot 4!/6! = 1/5$.

OR

Note that once Peter has been seated, there is just one place out of the five remaining for Sam to sit so that the two are separated by exactly two men.

19. Solution (C). The probability of obtaining an ace on the first draw is $4/52 = 1/13$. If the first card drawn is an ace there are 3 aces remaining in the deck, which now consists of 51 cards. Thus, the probability of getting an ace on the second draw is $3/51 = 1/17$. The required probability is the product of the two, which is c) $1/221$. The problem can be solved by dividing:

$$\frac{\binom{4}{2}}{\binom{52}{2}} = \frac{4!/2!2!}{52!/50!2!} = 1/221.$$

20. Solution (B). Let x be the probability that Bill wins after a head is tossed and let y be the probability that he wins after a tail is tossed. After a head is tossed, there is a 60% chance of winning on the next toss. Since a tail occurs the remaining 40% of the time and the probability of winning after that is y, we must have $x = 0.6 + 0.4y$. Similarly, if a tail was tossed, Bill loses immediately 40% of the time. The remaining 60% of the time a head was tossed and the probability of winning is now x. Thus, $y = 0.6x$. Solving these equations, we find that $x = \frac{15}{19}$ and $y = \frac{9}{19}$. Since there is a 60% chance of a head on the first toss and a 40% chance of a tail, the probability of winning is $0.6x + 0.4y = \frac{63}{95}$.

OR

The winning toss sequences for Bill are HH, HTHH, HTHTHH, HTHTHTHH, ... THH, THTHH, THTHTHH, THTHTHTHH ... . Let h represent the probability of a head on one toss and t represent the probability of a tail. Summing the probabilities for each of these gives an "infinite" sum $h^2 + hth^2 + hthth^2 + hththth^2 + \cdots + th^2 + thth^2 + ththth^2 + thththth^2 + \cdots = h^2(1 + t)(1 + ht + (ht)^2 + (ht)^3 + \cdots) = h^2(1 + t)\dfrac{1}{1 - ht}$ (since $0 < ht < 1$). For $h = 0.6 = \dfrac{3}{5}$ and $t = 0.4 = \dfrac{2}{5}$, have $\left(\dfrac{3}{5}\right)^2 \cdot \dfrac{7}{5} \cdot \dfrac{1}{1 - \dfrac{6}{25}} = \dfrac{63}{95}$.

21. **Solution** (D). The first flip of heads comes from a two-headed coin with probability 0.8 and the flip comes from a normal coin with probability 0.2, so the probability that the second flip results in heads is $0.8 \cdot 1 + 0.2 \cdot .5 = 0.9$. Alternatively, the conditional probability of event A occurring given that event B equals the probability of both A and B occurring divided by the probability event B occurs. So first calculate the probability both tosses are heads, then divide by the probability the first toss is heads. The probability a particular coin is chosen is $1/4$. Thus the probability both tosses are heads is $(1/4) \cdot 1 + (1/4) \cdot 1 + (1/4) \cdot (1/4) + (1/4) \cdot 0 = 9/16$. For the first toss, the probability of heads is $(1/4) \cdot 1 + (1/4) \cdot 1 + (1/4) \cdot (1/2) = 5/8$. Thus the conditional probability is $(9/16)/(5/8) = 9/10$.

22. **Solution** (D). There are $\binom{39}{5}$ ways to draw five numbers altogether. You do not win if the set of five numbers drawn is disjoint of our selection or matches exactly one number. There are $\binom{34}{5}$ ways to select a disjoint set and $5\binom{34}{4}$ ways to draw five numbers that match exactly one of our numbers. Here 5 is the number of ways to select the matching number, and $\binom{34}{4}$ is the number of ways to select four numbers that do not match our selection. The probability that we do not win is

$$\frac{\binom{34}{5} + 5\binom{34}{4}}{\binom{39}{5}} = \frac{278,256 + 231,880}{575,757} \approx 0.89.$$

Thus the probability of winning is $1 - 0.89 = 0.11$, that is, 11%.

23. **Solution** (B). The five scores must be of the form $x, y, 92, 95, 95$ and have a sum of $5 \cdot 91 = 455$. Therefore $x + y = 173$.

24. **Solution** (E). The sum of the 27 scores was $27 \cdot 72 = 1944$, so the new mean must be $(1944 - 85)/26 = 71.5$.

25. **Solution** (B). Exactly $19 + 18 + 10 + 9 = 56$ percent of the families in the table have one or more children. Therefore, $n = 100 - 56 = 44$ and 44% of the families have no children. Finally, $.44 \times 5,000 = 2,200$.

26. **Solution** (C). Since the average of all 50 scores is 68, the sum of the scores is $50 \times 68 = 3400$, and the sum of the other 40 scores is $3400 - 1000 = 2400$. Therefore the average of these 40 scores is $2400/40 = 60$.

27. **Solution** (D). Translating the information into equations, where x represents the number of students who passed the final,

$$62(20 - x) + 92(x) = 20 \cdot 80 = 1600.$$

Solve this for x to get $x = 12$.

28. **Solution** (D). Denote the numbers by A, B, C, and D. Then

$$\frac{1}{2}(A + B) = 5,$$

$$\frac{1}{2}(B + C) = 4,$$

and

$$\frac{1}{2}(C + D) = 10.$$

Therefore

$$\frac{1}{2}(A + D) = \frac{1}{2}(A + B) - \frac{1}{2}(B + C) + \frac{1}{2}(C + D) = 11.$$

CHAPTER 36

NUMBER THEORY

36.1 Prime factors and Factorization

1. **Solution** (D). The prime factorization of 7! is $2^4 \cdot 3^2 \cdot 5 \cdot 7$. Each odd factor of 7! is a product of odd prime factors, and there are $3 \cdot 2 \cdot 2 = 12$ ways to choose the three exponents.

2. **Solution** (B). $2^{16} - 16 = 2^4(2^{12} - 1) = 2^4(2^6 + 1)(2^6 - 1) = 2^4 \cdot 63 \cdot 65 = 2^4 \cdot 9 \cdot 7 \cdot 5 \cdot 13$.

3. **Solution** (C). Note that $25y^2 = (5y)^2$ so $12x = 2^2 \cdot 3x$ must be a perfect square multiple of 5. The smallest integer $3x$ is $3^2 \cdot 5^2$, so $x = 3 \cdot 5^2 = 75$ and in this case $y = 6$. Thus $x + y = 81$.

4. **Solution** (B). Factor N into primes to get $N = 3^4 \cdot 5 \cdot 7 \cdot 13 \cdot 19$. The only way to divide these primes into three groups each with product less than 100 is $3^4, 5 \cdot 19$, and $7 \cdot 13$ since the 19 cannot be matched with a 3 ($3 \cdot 19 = 57$ and $57 \cdot 99 \cdot 99 < N$), so the sum is $81 + 95 + 91 = 267$.

5. **Solution** (E). We can write $K = 75 \cdot 74 \cdot 73 \cdot 72 \cdot 71! - 71! = 71!(T - 1)$, where T is the even number $75 \cdot 74 \cdot 73 \cdot 72$. So the answer we seek is the largest power of 2 that divides 71!. This is the number $\lfloor \frac{71}{2} \rfloor + \lfloor \frac{71}{4} \rfloor + \lfloor \frac{71}{8} \rfloor + \lfloor \frac{71}{16} \rfloor + \lfloor \frac{71}{32} \rfloor + \lfloor \frac{71}{64} \rfloor = 35 + 17 + 8 + 4 + 2 + 1 = 67$.

6. **Solution** (E). If the three integers are denoted $n - 1, n$, and $n + 1$, then their sum is $3n$ and their product is $n(n^2 - 1)$. Thus $n(n^2 - 1) = 33 \cdot 3n = 99n$, from which it follows that $n^2 - 1 = 99$ and so $n = 10$. The product is 990 and the sum of the digits is $9 + 9 + 0 = 18$.

7. **Solution** (D). Squaring the first equation, we get $a^2b^2 = a^{2b}$ so that $a^2b^2 = a/b$ or $a = 1/b^3$. Substituting this into the first equation, we find $1/b^2 = 1/b^{3b}$. This gives $2 = 3b$ so that $b = 2/3$. Then $a = 27/8$.

8. Solution (D). Since each divisor d of a number D can be paired with a divisor D/d, only the perfect squares can have an odd number of divisors. There are 6 perfect squares between 10 and 99.

9. Solution (B). Solve for a and c in terms of b to get $(6b + 24) \cdot b \cdot (b + 8) = 27{,}846$. Thus $b \cdot (b + 4) \cdot (b + 8) = 4641$. Factor the right side to get $4641 = 3 \cdot 7 \cdot 13 \cdot 17$, which can be written in the form $b \cdot (b + 4) \cdot (b + 8)$ only when $b = 13$. Thus $a + b + c = 102 + 13 + 21 = 136$.

10. Solution (D). If N is any odd integer, then all of $(N - 2), (N - 4), (N - 6),$ and $(N - 8)$ are odd, so their product is odd. Thus $(N - 2)(N - 4)(N - 6)(N - 8) - 1$ is even.

11. Solution (E). Let $\dfrac{n}{50 - n} = k^2$. Then $n = \dfrac{50k^2}{k^2 + 1}$. Because n is an integer and $k^2 + 1$ does not divide k^2 unless $k = 0$, we see that $k^2 + 1$ must divide $50 = 5 \cdot 5 \cdot 2$. This happens when $k = 0, 1, 2, 3$ or 7.

12. Solution (E). The prime factorization of N is $2^4 \cdot 5^3 \cdot 11$. Consequently there are $(4 + 1) \cdot (3 + 1)(1 + 1) = 40$ divisors.

13. Solution (B). Only $1.6 \times 10^{21} = 16 \times 10^{20}$ is a perfect fourth power of an integer.

14. Solution (A). $r = 11\sqrt{10!}$, $t = \sqrt{12 \cdot 11}\sqrt{10!}$ and $s = 10\sqrt{11}\sqrt{10!}$, so $r < t < s$.

15. Solution (C). Let N denote the largest such factor of $11! = 2^8 3^4 5^2 7^1 11^1$. Then N has a factorization which can include only the primes $2, 3, 5, 7$, and 11. But N must not be even and also N cannot be divisible by 3, because otherwise it would not be one bigger than a multiple of 6. Hence $N = 5^i 7^j 11^k$ where $i = 0, 1$ or 2, $j = 0$ or 1, and $k = 0$ or 1. Examine these numbers starting with the largest ($i = 2, j = 1, k = 1$), which leads to 1925 which is 5 larger than a multiple of 6. Next largest is the number obtained when $i = 1, j = 1$, and $k = 1$, which is 385, and this number is 1 larger than a multiple of 6.

16. Solution (B). Note that

$$10! = 2 \cdot 3 \cdot 4 \cdot 5 \cdot 6 \cdot 7 \cdot 8 \cdot 9 \cdot 10 = 2^8 3^4 5^2 7.$$

The sum $a + b + c + d$ is smallest if all four numbers are equal, in which case

$$a + b + c + d = 4\sqrt[4]{10!} = 2^2 \cdot 3 \cdot \sqrt{5} \cdot \sqrt[4]{7} \approx 174.58.$$

Hence

$$a + b + c + d \geq 175.$$

This optimum is obtained for $a = 40, b = 42, c = 45$, and $d = 48$.

17. **Solution** (C). From the factorization

$$9! = 2^7 \cdot 3^4 \cdot 5 \cdot 7,$$

we see that a, b and c must have prime factors among the primes $2, 3, 5,$ and 7. Because the cube root of $9!$ is close to 70, and $a \cdot b \cdot c = 9!$, it follows that $c - a$ is smallest when a and c are close to 70. Trying $2 \cdot 5 \cdot 7$ for one of the factors leaves $2^6 \cdot 3^4 = 72^2$. Thus $c - a \leq 72 - 70 = 2$. To see that no smaller difference is possible, note that $9!$ is not a perfect cube so $c - a$ cannot be zero. If $c - a = 1$, then there is an integer n such that either $n \cdot (n+1)^2 = 9!$ or $n^2 \cdot (n+1) = 9!$. But $n = 70$ makes these two left sides too small, and $n = 71$ makes them both too big.

18. **Solution** (B). If the numbers are denoted $k - 2, k - 1, k, k + 1$ and $k + 2$, then

$$(k-2)(k-1)k(k+1)(k+2) \div 5k = 100K$$

for some integer K. This can happen only when one of the integers $k - 2, k - 1, k + 1, k + 2$ is a multiple of 125 since only one of these numbers can be a multiple of 5. Let $k + 2 = 125$ to minimize the sum. Then $5 \cdot 123 = 615$.

19. **Solution** (D). From the condition (c), we examine pairs of prime numbers that differ by 6. Listing the primes gives

$$2, 3, 5, 7, 11, 13, 17, 19, 23, 29, 31, 37, 43, 47, 53, 59, 61, \ldots.$$

Now by checking we can see that the pair 53 and 59 is the first pair of primes that satisfies all conditions a, b, and c; therefore, $xy = 56$ and $x = 7, y = 2^3$.

OR

(i) start a list of the integers larger than 5 that are powers of primes, 7, 8, 9, 11, 13, 16, 17,...; (ii) note that the smallest product with $x < y$ is $7 \cdot 8$ which just happens to be exactly 3 more than the prime 53 and 3 less than the prime 59.

20. **Solution** (D). Suppose T and B are the top and bottom numbers on the first die and t and b for the second. Then the sum S equals $Tt + Tb + Bt + Bb = (T + B)(t + b) = 7 \cdot 7 = 49$.

21. **Solution** (E). Multiplying the given equality $a + b^{-1} = 13(b + a^{-1})$ by ab we obtain: $a(ab + 1) = 13b(ab + 1)$, or $(a - 13b)(ab + 1) = 0$. Since $ab + 1 > 0$, the given equation is equivalent to $a = 13b$. The inequality $a + b \leq 100$ means that $14b \leq 100$; therefore, the possible values of the positive integer b are $1, 2, \ldots, 7$, and there are 7 solutions: $(13, 1), (26, 2), \ldots, (91, 7)$.

OR

we can simplify the left hand side by multiplying top and bottom by ab. This yields $a(ab+1)/b(1+ab) = 13$, cancel the factor $ab+1$ to get $a/b = 13$. Since $100/13 = 7\frac{9}{13} > 7\frac{7}{13}$, there are 7 pairs.

22. **Solution** (B). Factoring the number yields $7, 11$, and 13 as factors. Since these are all prime, either they or one of their multiples must be among the k integers. Trying 13 itself, we write $10 \cdot 11 \cdot 12 \cdot 13 \cdot 14 = 240{,}240$.

To see that no other value besides $k = 5$ works, trying 26, the closest multiples of 11 are 22 and 33, but any product of consecutive integers including both 22 and 26 is too large, as is any such product containing both 26 and 33. The same reasoning holds for 39 and 52 (55 is the closest multiple of 11 and 53 is prime).

Alternatively, "slowly" factor 240,240 into smaller numbers – $240,240 = 10 \cdot 24,024 = 10 \cdot 12 \cdot 2002 = 10 \cdot 12 \cdot 2 \cdot 1001 = 10 \cdot 12 \cdot 2 \cdot 11 \cdot 91 = 10 \cdot 12 \cdot 2 \cdot 11 \cdot 7 \cdot 13$. Simply multiply the "2" and "7" together and reorder to have $240,240 = 10 \cdot 11 \cdot 12 \cdot 13 \cdot 14$. So $k = 5$ works.

23. **Solution** (C). To say the numbers have exactly four divisors means that they are each a product of two distinct prime numbers, say pq and rs. To say the sum of their divisors is the same means $1 + p + q + pq = 1 + r + s + rs$, and to say they are consecutive means $pq + 1 = rs$. Putting the two equations together gives $p + q = 1 + r + s$, which can happen only when one of the primes is even (hence 2). A little trial and error produces some primes that work: $p = 2, q = 7, r = 3$, and $s = 5$. So the two numbers could be 14 and 15, both with sum of factors equal to 24. The number of factors of $14 \cdot 15 = 2 \cdot 3 \cdot 5 \cdot 7$ is $2^4 = 16$.

Actually, we really only need to know that the conditions say that the four primes p, q, r, s are **distinct**. Suppose a factor in the first number is the same as a factor in the second number, say $p = r$. Then

$$rs - pq = 1 \Rightarrow r = \frac{1}{s - q}$$

which will not give a prime number (the only positive integer solution is $r = 1$). This means that the product of the two numbers is $pqrs$ with each prime distinct. The number of divisors is always 16.

36.2 Place value and Digits

1. **Solution** (E). The smallest such integer is 15789 by looking at the prime factorization of 2250. So the answer is 30.

2. **Solution** (D) The number N can be written in the form $N = 1000a + b$, where a is a digit and b is a three digit number. The condition implies that $9b = 1000a + b$, or equivalently, $8b = 1000a$.

The smallest nonzero digit a for which $1000a$ is a multiple of 8 is $a = 1$, which leads to $b = 125$. Hence $N = 1125$.

3. **Solution** (C). Note that both M and N must be two-digit numbers. Let $M = 10a + b$. Then $2(a + b) = 10a + b$ and it follows that $8a = b$ and $M = 18$. Similarly, we can show that $N = 27$, so their sum is 45.

4. **Solution** (D). Since $\text{aaaa}_{\text{ten}} + \text{bbb}_{\text{ten}} + \text{cc}_{\text{ten}} + \text{d}_{\text{ten}} = 1995$, it follows that $\text{a} = 1$ and $\text{bbb}_{\text{ten}} + \text{cc}_{\text{ten}} + \text{d}_{\text{ten}} = 884$. Since $13 = 11 + 2 \leq \text{cc}_{\text{ten}} + \text{d}_{\text{ten}} \leq 99 + 8 = 107$, we have $\text{b} = 7$, and then it follows that $\text{c} = 9$ and $\text{d} = 8$. Thus, $\text{a} \cdot \text{b} \cdot \text{c} \cdot \text{d} = 504$.

5. **Solution** (B). $54 = 1 \cdot 6^2 + 3 \cdot 6 + 0 \cdot 1$, so the answer is &@#.

6. **Solution** (E). The only sums of $even + odd$ that are not prime are $2 + 7$, $3 + 6$ and $4 + 5$. So N cannot be 7654321, but next best choice works–7652341. The remainder is 4.

7. **Solution** (C). The only numbers satisfying the condition are those permutations of the digits $1, 1, 2, 4$. There are 12 such permutations.

8. **Solution** (B). The numbers are $10, 12, 21, 23, 32, 34, \ldots, 78, 87, 89, 98$. So the sum of all of them can be split into two sums $10 + 21 + 32 + \cdots + 98 = 9(108)/2 = 486$ and $12 + 23 + \cdot + 89 = 8 \cdot 101/2 = 404$.

9. **Solution** (A). Note that

$$
\begin{aligned}
3b^2 + 2b + 4 &= 1(b + 2)^2 + 5(b + 2) + 5 \\
&= b^2 + 4b + 4 + 5b + 10 + 5 \\
&= b^2 + 9b + 19,
\end{aligned}
$$

so $2b^2 - 7b - 15 = 0$. Thus, $(b - 5)(2b + 3) = 0$. Only $b = 5$ makes sense.

10. **Solution** (C). There are just six such numbers. The six possible choices are $3 \cdot 3 \cdot 223 = 2007$, $3 \cdot 32 \cdot 23 = 2208$, $3 \cdot 322 \cdot 3 = 2898$, $33 \cdot 2 \cdot 23 = 1518$, $33 \cdot 22 \cdot 3 = 2178$, and $332 \cdot 2 \cdot 3 = 1992$. So the number 2007 is third in the list.

11. **Solution** (C). There are just two two-digit cubes, 27 and 64. Checking each one shows that 27 satisfies the requirements and 64 does not. The number 27 has 4 divisors.

12. **Solution** (B). The multiples of 9^3 between 9^4 and 9^6 are $10(9^3), 11(9^3), 12(9^3), \ldots, (9^3 - 1)(9^3)$. The sequence $10, 11, 12, \ldots, 9^3 - 1$ has 719 members.

Alternatively, use the fact that for integers $n \leq m$ there are $m - n + 1$ integers in the range $n \leq m$. So we have $(9^3 - 1) - 10 + 1 = 719$, since $10 \cdot 9^3$ is the smallest multiple of 9^3 that is larger than $9^4 = 9 \cdot 9^3$ and $(9^3 - 1) \cdot 9^3$ is the largest multiple of 9^3 that is smaller than $9^6 = 9^3 \cdot 9^3$.

13. Solution (D). Trying the smallest digits first, we are led to numbers of the form $12xy$, which in turn leads to 1236.

14. Solution (B). Note that

$$\frac{10^{27} + 2}{3} = \frac{10^{27} - 1}{3} + 1 = 333\ldots3 + 1 = 333\ldots34,$$

where the string has 27 digits. The sum of the digits is therefore $26 \cdot 3 + 4 = 82$.

15. Solution (B). A rule for divisibility by 11 is that the alternating sum of the digits should be a multiple of 11. Thus, divisibility by 11 implies that $2A + B - (B + 12) = 2A - 12$ is zero or ± 11. Since $2A - 12$ is even, it follows that $2A - 12 = 0$ and that $A = 6$. So the number $56BB76$ is a multiple of 3, and this implies that $B = 0, 3, 6$ or 9. So $A + B$ could be $6, 9, 12$, or 15 only. Alternatively, the sum of the digits is $12 + 2A + 2B$. If the number is divisible by 33, then it must be divisible by 3. Thus 3 divides $2A + 2B$. Since 2 and 3 are primes, 3 must divide $A + B$. For the choices given, the factor of 11 is not important as only one of the choices is a multiple of 3.

16. Solution (B). Construct a table.

House number	#of digits used
1-9	$9 \cdot 1 = 9$
10-99	$90 \cdot 2 = 180$
100-199	$100 \cdot 3 = 300$
200-299	$100 \cdot 3 = 300$
300-399	$100 \cdot 3 = 300$

The first four rows show that 789 of the 999 digits are used on the first 299 houses. That leaves 210 more digits to be used on the remaining houses. This is just enough for 70 more houses. So there are $299 + 70 = 369$ houses in the development.

17. Solution Note: For real x, $\lfloor x \rfloor$ is the largest integer that does not exceed x.
(D). Since $\dfrac{1}{13} = \overline{.076923}$ and a_i is the ith digit in the decimal expansion of $\dfrac{1}{13}$, the largest value of a_i must be 9.

18. Solution (C). Let N denote the five-digit number obtained by subtracting 100000 from the six-digit number in the problem. Then $3(100000+N) = 10N+1$, which is equivalent to $7N = 299999$ which yields $N = 42857$, so the sum of the digits of the six-digit number is

$$1 + 4 + 2 + 8 + 5 + 7 = 27.$$

19. Solution (C). The numbers are

$$1111, 111a, 111b, 1118, 11a1, 11aa, 11ab, 11a8.$$

The units digit of the sum

$$1 + a + b + 8 + 1 + a + b + 8 = 18 + 2(a + b)$$

is 4 and from this one deduces that the units digit of $2 \cdot (a + b)$ is 6. Thus the units digit of $a + b$ is either 8 or 3. Since $1 < a < b < 8$, the only possibilities are $a = 6, b = 7$; $a = 2, b = 6$; and $a = 3, b = 5$. Since there are only three choices, simply try them out to discover that $a = 3, b = 5$ is the one that works, so $a + b = 8$.

20. **Solution** (E). There are 2 in each decile, $10a + 4$ and $10a + 6$. The tens digit of $(10a + 4)^2 = 100a^2 + 80a + 16$ is the units digit of $8a + 1$, while the tens digit of $(10a + 6)^2 = 100a^2 + 120a + 36$ is the units digit of $2a + 3$, both of which are odd for any integer a. All the other tens digits of perfect squares are even: $(10a + b)^2 = 100a^2 + 20a + b^2$, the tens digit of which is the tens digit of $2a + b^2$, which is even if the tens digit of b^2 is even. But the tens digit of b^2 is even if $b \neq 4, b \neq 6$.

21. **Solution** (E). This is just base 6 enumeration in disguise. The base 6 representation of 2012 is 13152, so there are $1 + 3 + 1 + 5 + 2 = 12$ dots in the boxes after 2012 minutes.

22. **Solution** (C). Since the hundreds digits must be different, the smaller of the numbers N should have as large a tens digit as possible and the larger M should have as small a tens digit as possible. So we should choose 8 and 2 for these digits provided we can select the hundreds digits to be one apart. Given that the 2 and the 8 are not available, there are just two pairs that we can use, 3 and 4 and 6 and 7. Picking the tens digit of N to be as large as we can and the tens digit of M to be as small as possible leads to $M = 723$ and $N = 684$. Thus the answer we seek is $723 - 684 = 39$. Another possible pair is $M = 426$ and $N = 387$.

23. **Solution** (D). Symbolically, $9 \cdot abcd = dcba$. In other words,

$$9\left(1000a + 100b + 10c + d\right) = 1000d + 100c + 10b + a.$$

Thus

$$9000a + 900b + 90c + 9d = 1000d + 100c + 10b + a$$

and

$$8999a + 890b = 10c + 991d,$$

so we must have $a = 1$ so $d = 9$. Then

$$900b + 90c + 81 = 100c + 10b + 1,$$

so $890b + 80 = 10c$. Dividing by 10 yields $89b + 8 = c$ from which it follows that so $b = 0$ and $c = 8$. So the sum of the digits is $1 + 0 + 8 + 9 = 18$.

Alternatively, if the number is $N = \underline{abcd}$, then a must be 1, d must be 9 and b is at most 1 since $9 \cdot 1200 > 10,000$. Since $9 \cdot 9 = 81$, $b = 0$ implies the units digit of $9 \cdot c$ is 2 so $c = 8$, and $b = 1$ implies the units digit of $9 \cdot c$ is 3 so $c = 7$. Since $9 \cdot 1179$ is a five digit number, $N = 1089$. So the sum of the digits is 18 (the same as the sum for 1179).

Yet another alternative is to notice that the sum of the digits of N is the same as that of $\overline{N}$ and that since $\overline{N}$ is a multiple of 9, the sum of the digits $S(N) = S(\overline{N})$ must be a multiple of 9 as well. But 18 is the only multiple of 9 among the options.

24. **Solution** (D). If $n < 75$ then $S(n) \le S(69) = 15$ and $S(S(n)) \le 9$, so $n + S(n) + S(S(n)) < 99$. On the other hand, $75 + S(75) + S(S(75)) = 90$, $76 + S(76) + S(S(76)) = 93$, $77 + S(77) + S(S(77)) = 96$, and $78 + S(78) + S(S(78)) = 99$. So 78 is the least integer with the property and $S(78) = 15$.

Alternatively, let $T(N) = N + S(N) + S(S(N))$. Since neither $S(N)$ nor $S(S(N))$ can be zero, we must have $N \le 97$. Thus $S(N) \le 17$ (using $N = 89$) and $S(S(N)) \le 9$. This gives a crude lower bound for N of $99 - (17 + 9) = 73$. If $S(N + 1)$ has the same number of digits as $S(N)$, then $T(N + 1) = T(N) + 3$ since $S(N + 1)$ will be one more than $S(N)$ and $S(S(N + 1))$ will be one more than $S(N)$. Moreover, as long as $S(N + k)$ continues to have the same number of digits as $S(N)$, we have $T(N + k) = T(N) + 3k$. The first integer M greater than 73 such that $S(M)$ has only one digit is $M = 80$. Thus the formula will work from 73 up to 79, so for $1 \le k \le 6$. We have $T(73) = 73 + S(73) + S(S(73)) = 73 + 10 + 1 = 84$, and $99 - 84 = 15 = 3 \cdot 5$. Thus $99 = T(73) + 3 \cdot 5 = T(78)$. Therefore $N = 78$ is the smallest integer with $T(N) = 99$.

25. **Solution** (A). Let $x = \underline{abc}$ and $y = \underline{def}$. Then $1000y + x = 6(1000x + y)$, from which it follows that $857x = 142y$. Since 857 and 142 are relatively prime, 857 must divide y. Since y is a three-digit number, y must be 857, and $x = 142$. The sum of the digits of N is $8 + 5 + 7 + 1 + 4 + 2 = 27$.

26. **Solution** (C). Charlie's current age must be a two digit number. Suppose that Charlie's current age is $10a + b$ where a, b are one of the digits $0, \ldots, 9$ ($a \ne 0$). Then seven years ago, his age would be $10a + b - 7$. If $b - 7 \ge 0$ then the product of the digits can not be ab as it was. Therefore we can write $10a + b - 7 = 10(a - 1) + b + 3$ and the product of the digits $(a - 1)(b + 3)$ must equal ab. This gives $b = 3a - 3$. Since $0 \le b < 7$ we can only have $a = 2$ or 3 and current ages of 23 and 36. Only 23 is prime. The next time the product of the digits will be six is at age 32, which is 9 years later.

27. **Solution** (D). The distance cannot exceed 49 miles since otherwise the marker at 49 miles will have 49 on one side and a positive number on the other side and the sum of the digits would be greater than 13. Now look at the markers at one mile and ten miles. On the other side of these markers, we must have one of the numbers 48 and 39. Thus the first one must be 48 and the second

one 39. Thus, the distance can only be 49 miles. On the other hand, one can check that the distance 49 does give a solution.

28. **Solution** (E). The number can be written as

$$998999 + 996997 \cdot 10^6 + 994995 \cdot 10^1 2 + \cdots + 100101 \cdot 10^{6 \cdot 449}.$$

All of the powers 10^{6k} are 1 more than a multiple of 11 as are all of the six digit numbers. So N is $450 = 440 + 10$ larger than a multiple of 11. Thus the remainder is 10.

29. **Solution** (C). We want three digit numbers that can be written in the form $abc + cba$ i.e.

$$100(a + c) + 20b + c + a = 101(a + c) + 20b,$$

where a, b, c are non negative integers, $a \neq 0$, and $a + c \leq 9$. When $a + c = 1$, there are 10 values of b that work. This is true when $a + c = k$ for $k = 1 \ldots 8$. When $k = 9$, $101(a + c) + 20b$ will not be a three digit number unless $b \leq 4.5$. There are 5 such values of b $(0, 1, 2, 3, 4)$. Thus, there are $8 \cdot 10 + 5 = 85$ numbers that can be written as the sum of a three digit number and its reversal.

30. **Solution** (E). Consider all years $19xy$ with $9 \geq x \geq y \geq 0$ and subtract yx to see what year $19ab$ would be the birth year. Starting with $x = 1, y = 0$, a person who turned 01 in 1910 was born in 1909. That person will turn 12 in 1921, 23 in 1932, etc. To be 02 in 1920 means a birth year of 1918. The person who turned 03 in 1930 was born in 1927. Next is the one who turned 04 in 1940 and was born 1936. So the oldest child was born in 1945 and turned 05 in 1950. This child turned 16 in 1961, 27 in 1972, 38 in 1983 and 49 in 1994 for a total of five years where the child's age was yx in $19xy$. The next oldest turned 06 in 1960, so was born in 1954. The special birthdays are at 17 in 1971, 28 in 1982 and 39 in 1993. The middle child turned 07 in 1970, then was 18 in 1982 and 29 in 1992. Child number four turned 08 in 1980 and was 19 in 1991. Finally the youngest child turned 09 in 1990 and was born in 1981. There are $5 + 4 + 3 + 2 + 1 = 15$ years where a child turned yx in $19xy$.

31. **Solution** (D). An application of the distributive property yields that the sum of the nine numbers in the table is $(a + b + c)(c + d + e)$. Let $x = a + b + c$. Then $d + e + f = 63 - x$. The formula $y = x(63 - x)$ represents a parabola with vertex at $x = 63/2$ that opens down. So, the closer we choose x to $63/2$, the higher the point on the parabola. Given the values of the six numbers, the closest pairs are $a + b + c = 35, d + e + f = 28$ and $a + b + c = 28, d + e + f = 35$. Their product is $35 \cdot 28 = 980$.

32. **Solution** (A). Each Kaprekar number is the difference of two numbers with the same digit sum. These two numbers, $\underline{dcba}$ and $\underline{abcd}$ yield the same remainder when divided by 9. Hence their difference is a multiple of 9. Only 2936 is not a multiple of 9. On the other hand, $K(A) = B, K(B) = C, K(C) = D,$ and $K(D) = E$.

Rather involved alternate solution: Write the numbers as $1000d + 100c + 10b + a$ and $1000a + 100b + 10c + d$. Then the difference is $1000(d - a) + 100(c - b) + 10(b - c) + (a - d)$. $M = 1000w + 100x + 10y + z$ has four digits, so it must be that $a < d$. Thus $a - d < 0$. This means the units digit of M must be $z = 10 + a - d > 0$, the 10 must be deducted from $10(b - c)$.

If $b = c$, then the "tens" digit of M will be a 9, as will the "hundreds" digit and the "thousands" digit will be $d - a - 1$. The possibilities are $z = 1$, $y = x = 9$ and $w = 8$ (example: $a = 0 \leq b = c \leq d = 9$); $z = 2$, $y = x = 9$ and $w = 7$ ($a = 1 \leq b \leq c \leq d = 9$); $z = 3$, $y = x = 9$ and $w = 6$ ($a = 2 \leq b = c \leq d = 9$); $z = 4$, $y = x = 9$ and $w = 5$ ($a = 2 \leq b = c \leq d = 8$); $z = 5$, $y = x = 9$ and $w = 4$ ($a = 1 \leq b = c \leq d = 6$); $z = 6$, $y = x = 9$ and $w = 3$ ($a = 2 \leq b = c \leq d = 6$); $z = 7$, $y = x = 9$ and $w = 2$ ($a = 4 \leq b = c \leq d = 7$); $z = 8$, $y = x = 9$ and $w = 1$ ($a = 3 \leq b = c \leq d = 5$).

If $b < c$, then the 10 needed for $z = 10 + a - d > 0$, can come from $100 + 10(b - c)$ and the 100 needed for $100 + 10(b - c)$ can come from $100(c - b)$ without changing the thousands digit which will be $1000(d-a)$. Thus $y = 10+b-c-1$ and $x = c-b-1$. In this case the possible values of the pairs (w, z) and (x, y) are related. For (w, z) the possible pairs are $(1, 9)$, $(2, 8)$, $(3, 7)$, ..., $(8, 2)$, and $(9, 1)$. For (x, y) the possible pairs are $(0, 8)$, $(1, 7)$, $(2, 6)$, ..., $(7, 1)$ and $(8, 0)$. However, since $w = d - a$ and $a \leq b < c \leq d$, $x = c - b - 1 < c - b \leq d - a = w$. For a given w, any $x < w$ is OK, but no $x \geq w$ can occur. So $M = 3267$ can occur, but $M = 1269$ cannot.

As in the case that $x = y = 9$, we can use the above information to construct a number N where $K(N) = M$ with M a given number fitting the pattern (with $(x, y) \neq (9, 9)$). Start with $w \geq 1$ and $0 \leq x < w$. Next choose a pair $d > a$ with $d - a = w$ and set $c = d$ and $b = c - x - 1$. For $M = 6264$, take (for example) $d = 8$, $c = 8$, $b = 5$ and $a = 2$ [or $d = 9$ $c = 9$, $b = 6$, and $a = 3$]. Then $8852 - 2588 = 6264[= 9963 - 3699]$, as desired.

36.3 Modular arithmetic and Divisibility

1. **Solution** (B). The sequence of powers of 2 has the property that every term of the form 2^{4k} is one larger than a multiple of 5, that is, has the form $5m + 1$ for some integer m. To see this note that 2^4 is such a number and the product $(5k + 1)(5j + 1) = 25kj + 5k + 5j + 1$ of two numbers of this form also has this form. Hence $2^{100} = 5m + 1$ for some integer m.

2. **Solution** (B). Note that $3^4 = 81$, so the units digit of 3^{4k} is a 1 for each positive integer k. We can rewrite the product as $3 \cdot 3^{2012} = 3 \cdot (3^4)^{503}$. Thus the units digit is a 3.

3. **Solution** (B). Because 1995 is divisible by 3, all its higher powers are divisible by 9. The principle of *casting out nines* says that any positive integer differs from the sum of its digits by a multiple

of 9. Hence the sum of the digits of 1995^{10} must also be divisible by 9. This requires that N be replaced by 2.

4. **Solution** (E). The sum of the first k odd integers, $1 + 3 + 5 + \cdots + (2k - 1)$ is k^2. Therefore, n is the $\sqrt{9409}$th $= 97$th positive odd integer, which is $2 \cdot 97 - 1 = 193$.

5. **Solution** (A). There are two solutions, $x = 1, y = 4$, and $z = 7$; and $x = 2, y = 1$, and $z = 9$. In both cases, $z - 2x = 5$. Do you notice anything special about the numbers $28, 30, 31$, and 365?

OR

Using the notation of modular congruences,

$$
\begin{aligned}
28x + 30y + 31z &\equiv 30x + 30y + 30z - 2x + z \\
&\equiv 30(x + y + z) + z - 2x \\
&\equiv z - 2x \\
&\equiv 365 \\
&\equiv 5 \quad (\text{mod } 10).
\end{aligned}
$$

6. **Solution** (C). Let u be the middle number. Then the sum is $(u - 2)^2 + u^2 + (u + 2)^2 = 3u^2 + 8$. If the sum is S then $u = \sqrt{\frac{S-8}{3}}$ is an integer. This is only the case for option (C).

7. **Solution** (C). Since $(3n + 1)^2 = 9n^2 + 6n + 1$ and $(3n + 2)^2 = 9n^2 + 12n + 4$ are both one bigger than a multiple of 3, the sum $a^2 + b^2$ must be two bigger than a multiple 3.

8. **Solution** (D). The sum S of all the natural numbers less than 45 is $S = 44 \cdot (1 + 44)/2 = 22 \cdot 45 = 990$, so the sum we want is $S - T$ where

$$
T = 3 + 6 + 9 + \cdots + 42 = 3(1 + 2 + \cdots + 14) = 3(14/2)15 = 315.
$$

Thus $S - T = 990 - 315 = 675$.

9. **Solution** (C). The square of an odd number is one bigger than a multiple of 4: $(2k + 1)^2 = 4k^2 + 4k + 1$. Therefore the sum of squares of six odd integers is necessarily 2 bigger than a multiple of 4. Only 1998 satisfies this requirement. On the other hand, $1998 = 43^2 + 9^2 + 7^2 + 3^2 + 3^2 + 1^2$.

10. **Solution** (E). Suppose there are k twelves. Then the sum is $11(k + 3) + 12k = 23k + 33 = 23(k + 1) + 10$ so the sum must be 10 bigger than some multiple of 23. Only 240 of the options is such a number.

11. **Solution** (D). Let N denote the number. Since the sum

$$
1 + 2 + 3 + 4 + 5 + 6 = 21
$$

of the digits of N is a multiple of 3, N must be a multiple of 3. Exactly half the numbers formed from the digits 1 through 6 without replacement are even. A number is a multiple of 6 if it is both even and a multiple of 3. Thus, the probability that N is a multiple of 6 is $1/2$.

12. **Solution** (C). Two of the four consecutive integers must be even and at least one of these must be divisible by 4. Also, one of the four must be divisible by 3. Therefore, 4, 8, and 12 must be divisors of the product. Since 10 and 15 are not divisors of $1 \cdot 2 \cdot 3 \cdot 4$, the correct answer is 3.

13. **Solution** (A). If x and y are integers, then $3x + 6y = 3(x + 2y)$ is divisible by 3. Since 95 is not divisible by 3, there is no such solution.

14. **Solution** (C). Divide 839 by 19 to get $839 = 44 \cdot 19 + 3$. One pair of positive value is $q = 44$ and $r = 3$. If we try to make q any bigger, we are forced to make r negative.

15. **Solution** (C). The number d divides $341 = 11 \cdot 31$ and $713 = 23 \cdot 31$. The only common divisors are 1 and 31. Since we get nonzero remainders, $d = 31$.

16. **Solution** (A). Factor each of the numbers into primes: $50 = 2 \cdot 5^2$; $90 = 2 \cdot 3^2 \cdot 5$; $98 = 2 \cdot 7^2$; $270 = 2 \cdot 3^3 \cdot 5$; $686 = 2 \cdot 7^3$; $882 = 2 \cdot 3^2 \cdot 7^2$ and $1764 = 2^2 \cdot 3^2 \cdot 7^2$. The divisibility conditions imply that $N = k \cdot 2 \cdot 3^2 \cdot 5 \cdot 7^2$ and the non-divisibility conditions imply that none of $2, 3, 5$, and 7 are divisors of k. Since $9261000 = 2^3 \cdot 3^3 \cdot 5^3 \cdot 7^3$, it follows that only $k = 1$ works and $N = 4410$.

17. **Solution** (D). Suppose the number a has n digits each of which is d. Then a is divisible by 9 if and only if the sum $d + d + \cdots + d = nd$ of its digits is divisible by 9. Since $1 \le n \le 6$, one concludes that $d = 3, 6, 9$ are the only possibilities. With $d = 3$, we get the two numbers 333 and 333 333 divisible by 9. With $d = 6$, we get the two numbers 666 and 666 666 divisible by 9. With $d = 9$, we get the six numbers $9, 99, \ldots, 999\,999$ divisible by 9, so altogether there are $2 + 2 + 6 = 10$ solutions.

18. **Solution** (D). Note that $a_n \equiv \sum_1^n (2k - 1) \equiv n^2 \pmod 9$, and the thirtieth multiple of 9 among these is 90.

19. **Solution** (B). $Q(n) = \dfrac{4(n + 2)(n - 3)}{(n - 2)(n + 2)(n - 3)}$ is undefined for $n \in \{-2, 2, 3\}$ for all other values of n we have $Q(n) = \dfrac{4}{(n - 2)}$. We are looking looking for all values of n such that $n - 2$ divides 4 but $n \notin \{-2, 2, 3\}$. The set of divisors of 4 is $\{\pm 1, \pm 2, \pm 4\}$ so the set of possible values of n is $\{1, 3, 0, 4, -2, 6\} \setminus \{-2, 2, 3\}$, that is, $\{1, 0, 4, 6\}$. The sum of these elements is 11.

20. **Solution** (E). Lights numbered 1, 4, 9, 16, 25, 36, and 49 are on. A cord must be pulled an odd number of times for the light to be on. Consider light number x and factor x as a product of primes. There are 15 primes (2, 3, 5, ..., 47) that are less than 50. So

$$x = 2^{a_1} \cdot 3^{a_2} \cdots 47^{a_{15}}.$$

There are

$$(a_1 + 1)(a_2 + 1)...(a_{15} + 1)$$

factors of this number. The cord is pulled once for each factor. Since we need an odd number of pulls for the light to be on, a_1, a_2, ...a_{15} must all be even, hence x is a perfect square.

21. **Solution** (E). Using the formula

$$1^2 + 2^2 + \cdots + n^2 = \frac{n(n+1)(2n+1)}{6},$$

we have

$$1^2 + 2^2 + \cdots + 2012^2 = \frac{2012 \cdot 2013 \cdot 4025}{6} = 1006 \cdot 671 \cdot 4025 \equiv 0 \ (\text{mod } 10).$$

So the last digit is 0. Alternatively, using mod(10) relation,

$$1^2 + 2^2 + 3^2 + \cdots + 10^2 \equiv 5 \quad \text{mod } (10)$$

and this repeats for every 10 consecutive integers,

$$2012 = 201 \times 10 + 2, 2001^2 \equiv 1 \quad \text{mod } (10) \quad \text{and} \quad 2012^2 \equiv 4 \quad \text{mod } (10)$$

so

$$1^2 + 2^2 + \cdots + 2012^2 \equiv 201 \times 5 + 1 + 4 \equiv 0 \quad \text{mod } (10).$$

22. **Solution** (D). Let U denote the sum of the 90 units digits and T the sum of the 90 tens digits. Thus $U = 0 + 1 + 2 + \cdots + 9 + \cdots + 9 = 405$ and $T = 1 + 1 + \cdots + 1 + 2 + 2 + \cdots + 9 + 9 + \cdots + 9 = 450 = U + 45$. Since $99 = 9 \cdot 11$, we ask the two questions 'What is the remainder when N is divided by 11' and 'What is the remainder when N is divided by 9'. The divisibility test for 11 tells us that the remainder when a number N is divided by 11 is the same as when the alternating sum of the digits of N is divided by 11. That alternating sum is $U - T = -45$, and the remainder when $-45 = -5(11) + 10$ is divided by 11 is 10. The remainder when N is divided by 9 is the same as the remainder when the sum of digits of N is divided by 9. This sum of digits is $U + T = 9 \cdot 45 + 9 \cdot 50 = 9 \cdot 95$ which is a multiple of 9. So far we know that there are integers p and q such that $N = 9p$ (N is a multiple of 9) and $N = 11q + 10$ (when N is divided by 11, the remainder is 10). Let r be the remainder when N is divided by 99. Then there is an integer k such that $N = 99k + r$. Now $N/9 = 11k + r/9 = p$ shows r is a multiple of 9. Furthermore $N/11 = 9k + r/11 = q + 10/11$ shows $r = 11(q - 9k) + 10$ shows r is 10 more than a multiple of 11. The only number in the range 0 to 98 that is both a multiple of 9 and 10 bigger than a multiple of 11 is 54.

Alternatively, consider a base 100 expansion of a positive integer A with an even number of digits base 10, say $2n$ digits - -

$$A = a_n \cdot 100^n + a_{n-1} \cdot 100^{n-1} + \cdots + a_1 \cdot 100 + a_0.$$

Since $100 = 9 \cdot 11 + 1 = 99 + 1$, the remainder on dividing A by 99 is the same as the remainder on dividing the sum $a_n + a_{n-1} + \cdots + a_1 + a_0$ by 99. For the given integer, the corresponding sum is $10 + 11 + \cdots + 99 = (109 \cdot 90)/2$ (90 terms in an arithmetic sequence where the sum of the first and last terms is $109 = 10 + 99$). Now simply divide by 99 to get the remainder of 54 for both the sum and the given integer $101112\ldots99$.

36.4   Diophantine equations (and Euclidean algorithm)

1. **Solution** (D). We know x and y are integers and $x^2(x+1) = y^2$. Using the prime factorization of x and y, we see that $x + 1$ must be a perfect square. Since $1 \le x \le 100$, x can be $3, 8, 15, 24, 35, 48, 63, 80$, or 99. There are nine possible solutions.

2. **Solution** (A). Expand and cancel to get $4a^2 + 4ab + b^2 - a^2 - 4ab - 4b^2 = 9$, so $3a^2 - 3b^2 = 9$. This can happen only if $(a-b)(a+b) = 3$, which is true in case $a - b = 1$ and $a + b = 3$ since $a + b$ cannot be negative. It follows that $a = 2$ and $b = 1$, so $ab = 2$.

3. **Solution** (A). Rewrite the equation as

$$(a-7)^2 + (b-7)^2 + (c-7)^2 = 147$$

after completing the square three times. So we' re looking for three squares whose sum is 147. Note that all three are odd or just one is odd. Trying all three odd leads to

$$11^2 + 5^2 + 1^2 = 121 + 25 + 1 = 147.$$

This is the only solution.

4. **Solution** (D). Manipulate the equation to have equivalent forms (for nonzero integers) $6y - 12x = xy$ and $y(6-x) = 12x$. Since both x and y are positive integers, $y > 2x \ge 2$ and $6 > x \ge 1$. Also xy is an integer multiple of 6 and $12x/(6-x)$ is a positive integer. The latter restriction implies $x \ne 1$. But integer solutions for y exist when $x = 2, 3, 4, 5$: $x = 2, y = 6$; $x = 3, y = 12$; $x = 4, y = 24$; and $x = 5, y = 60$.

5. **Solution** (B). The integers cannot all be positive because the sum of the cubes would be greater than 405. Testing a few cubes, we see that

$$4^3 + 5^3 + 6^3 = 64 + 125 + 216 = 405$$

so the ten integers must be $-3, -2, -1, 0, 1, 2, 3, 4, 5, 6$, and their sum is 15.

6. **Solution** (E). Note that x has to be an integer since the left side is an integer. If $x < 0$ then $\lfloor x/2 \rfloor < x/2$ and $\lfloor 2x/3 \rfloor < 2x/3$, so $\lfloor \frac{x}{2} \rfloor + \lfloor \frac{2x}{3} \rfloor \le \frac{7x}{6} < x$. On the other hand, if $8 \le x$ then

$\lfloor x/2 \rfloor > 0.4x$ and $\lfloor 2x/3 \rfloor > 0.6x$, so there are no solutions except those in the interval $[0, 7]$ and there are six such integers, $0, 2, 3, 4, 5, 7$.

7. **Solution** (B). Every even multiple of 7 between 14 and $994 = 71 \cdot 14$ works. In other words $x = 1000 - 14k$ works for $k = 1, 2, \ldots, 71$.

8. **Solution** (D). The given equation is equivalent to $-888 < n^3 - 222 < 888$ which in turn is equivalent to $-666 < n^3 < 1110$, which has the integral solutions $-8, -7, -6, \ldots, 0, 1, 2, \ldots, 10$ of which there are 19.

9. **Solution** (C). Since $28^2 = a^3 - b^3$ is an even number, a and b must have the same parity (either both are odd or both are even). In either case, $a - b$ is even. Next note that $b \neq 1$ since $28^2 + 1^3$ is not a perfect cube. Therefore $b \geq 2$. Note that

$$28^2 = a^3 - b^3 = (a - b)(a^2 + ab + b^2).$$

Now $a - b < 8$ because $a - b \geq 8$ implies that

$$a^2 + ab + b^2 \leq 28^2/8 = 98,$$

but $a \geq b+8 \geq 10$ implies $a^2 + ab + b^2 \geq 100$. Thus $a - b$ is an even factor of 28^2 that is less than 8. The only possibilities are 2 and 4. Trying $a - b = 2$ yields the quadratic $a^2 + ab + b^2 = 196$ which reduces to

$$(b + 2)^2 + (b + 2)b + b^2 = 196,$$

which has no integer solutions. Trying $a - b = 4$ yields $a = b + 4$ and so

$$(b + 4)^2 + (4 + b)b + b^2 = 28^2 \div 4 = 196.$$

This reduces to $3b^2 + 12b - 180 = 0$, which can be factored to yield roots of $b = -10$ and $b = 6$. It follows that $a = 10$ and $a + b = 16$.

Alternatively, one could build a table with values $784 + n^3$ for integer values of n and see when a perfect cube comes up. Yet another alternative is to rewrite the equation as $28^2 + b^3 = a^3$. Since a and b are positive integers, $a^3 > 28^2 = 784$. Note that $1000 = 10^3$ is larger than 784 and $729 = 9^3$ is smaller. Thus the smallest integer that might possibly work for a is $a = 10$. This actually works since $1000 - 784 = 216 = 6^3$. So if there is a correct single answer, it must be that $a + b = 16$.

10. **Solution** (C). Divide every term by 2 to get

$$(x - y)^2 + 2y^2 = 27.$$

Note that $27 - 2y^2$ is an odd perfect square which could by 9 (for $y = 3$) or 25 (for $y = 1$). Both these lead to values of 6 for x. Other possible values of x are $\pm 4, 0$, and -6, but $x = 6$ is the only one for which both x and y are positive.

11. **Solution** (E). Note that $5m^2 = 2(1001 - n^2)$. This implies that m is even, so m^2 is a multiple of 4. Thus $1001 - n^2$ is a multiple of 10, and it follows that the units digit of n^2 is 1. The number $(10k + d)^2$ has a units digit of 1 only when the digit d is 1 or 9. Since 1 is not an option, the answer is 9. In fact one solution is $(16, 19)$.

12. **Solution** (A). Subtract the former from the latter to get $y^2 - x^2 - (y - x) = 114$, which factors, so that we can write $(y - x)(y + x - 1) = 2 \cdot 3 \cdot 19$. Trying various combinations yields $y - x = 6$ from which it follows that $x = 7$ and $y = 13$, so their sum is 20. Alternatively, replace y with $62 - x^2$ in the second equation and use the graphing calculator. OR, use $y - x = 6$ to conclude directly that $y + x - 1 = 19$ from the fact that $(y - x)(y + x - 1) = 6 \cdot 19$, and therefore $y + x = 20$.

13. **Solution** (E). Since $b^3 = a^2 - 1 = (a - 1)(a + 1)$, $a + 1$ could be the square of $a - 1$. This leads to the solution $a = 3$ and $b = 2$. By symmetry, the pair $(a, b) = (-3, 2)$ also works. It could also happen that one of $a - 1$ and $a + 1$ is zero. In these cases, we get $(1, 0), (-1, 0)$, and $(0, -1)$. In fact there are only five solutions, but the proof is well beyond the scope of this contest.

14. **Solution** (A). Let K denote this least number of cents Margaret could have and let n, d, and q denote the number of nickels, dimes, and quarters respectively Margaret has. Then $5n + 10d + 25q = K$ and $25n + 10q + 5d = K$. Subtracting one equation from the other yields $d = 4n - 3q$. We want to minimize q, so we try $q = 1$. Then $n = 1$ is not allowed so we try $n = 2$ and we get $d = 5$, so the value of K is 85. The other possible values can be easily eliminated.

15. **Solution** (A). From (1) it follows that

$$ab + a + b + 1 = (a + 1)(b + 1) = 323 = 17 \cdot 19$$

and from (2) it follows that

$$bc + b + c + 1 = (b + 1)(c + 1) = 399 = 19 \cdot 21.$$

Thus, $b = 18, a = 16$, and $c = 20$. Then $d = 8! \div (a \cdot b \cdot c) = 7$.

16. **Solution** (B). Note that $a^2 + b^2$ is even if both a and b have the same parity. Since $a^2 + b^2$ is odd, one of a and b is odd and the other is even. Suppose a is odd. Then

$$a^2 = (2k + 1)^2 = 4k^2 + 4k + 1$$

is one bigger than a multiple of 4. Also $b^2 = (2l)^2 = 4l^2$ is a multiple of 4.. Thus $a^2 + b^2$ is one bigger than a multiple of 4. Checking $201, 205, 209, 213, 217, 221$ and 225, we find that none are prime ($11|209, 7|217$ and $13|221$). Therefore 229 is the first viable candidate. And $229 = 2^2 + 15^2$, so $a = 2$ and $b = 15$ and $a + b = 17$.

17. Solution (D). Add the 2nd and 3rd equations together and factor to get

$$
\begin{aligned}
ac + ad + bc + bd &= a(c+d) + b(c+d) \\
&= (a+b)(c+d) \\
&= 77 = 7 \cdot 11
\end{aligned}
$$

It follows that

$$\{a+b, c+d\} = \{7, 11\}$$

and

$$a + b + c + d = 18.$$

The equations are satisfied if $a = 7, b = 4, c = 2$ and $d = 5$.

18. Solution (A). We can rewrite the equation as $x(y+8) = 83 - y$. The factor y + 8 is at least 9, so $1 \le x \le 9$. An alternate form for the equation is $y(x+1) = 83 - 8x$. The right hand side is odd, so both $x + 1$ and y are odd integers. Now simply try $x = 2, x = 4, x = 6, x = 8$. For $x = 2 : 3y = 83 - 16 = 67$, no integer solution. For $x = 4 : 5y = 83 - 40 = 43$, again no integer solution. For $x = 6, 7y = 83 - 48 = 35$, so $y = 5$. Finally, for $x = 8, 9y = 83 - 64 = 19$, no solution. An alternate way to solve is to add 8 to both sides of the original equation to get $xy + 8x + y + 8 = 91$. Factor to get $(x+1)(y+8) = 7 \cdot 13$. It must be that $x + 1 = 7$ and $y + 8 = 13$.

19. Solution (A). Factor the second equation and use the first equation. We have

$$m^3 + n^3 = (m+n)(m^2 - mn + n^2) = 3(m^2 - mn + n^2) = 117.$$

Therefore $39 = m^2 - mn + n^2 = (m+n)^2 - 3mn = 9 - 3mn$. Thus, $mn = -10$. Substituting this into $m^2 - mn + n^2 = 39$ we find that $m^2 + n^2 = 29$.

20. Solution (B). By multiplying each fraction by $5mn$, we can transform the equation into the equivalent $2mn = 5(m - n + 1)$. Notice that this implies that either m or n is a multiple of 5. This leads us to try various multiples of 5 for m. Alternatively, transform the equation further to get

$$2mn - 5m + 5n - 5 = 0.$$

Subtract $-15/2$ from both sides so that we can factor to get

$$m(2n - 5) + 5(2n - 5)/2 = -15/2.$$

Massage this to get $(2n - 5)(2m + 5) = -15$. Since m is positive, $2m + 5 > 0$. Therefore, $2n - 5 < 0$. The only values of n making $2n - 5 < 0$ are $n = 1$ and $n = 2$. The choice $n = 1$ requires $m = 0$. So the solution $m = 5, n = 2$ is unique. Alternatively, clear the fractions to get

$5(m+1) - 5n = 5m - 5n + 5 = 2mn$. Since m and n are positive integers, the same is true for $2mn$. Thus $m + 1 > n$ and $5 - 5n \leq 0$. If $n = 1$, we would have $5m = 2m$ which is impossible. Thus $n > 1$ and we have $5m > 2mn$ so $5 > 2n$. Therefore the only possible value for n is 2. Now simply solve for m: $5m - 10 + 5 = 4m$, so $m = 5$ and $mn = 10$.

21. **Solution** (B). Because the quotient has 5 digits but only 3 subtractions were performed, 2 digits of the quotient must be zero. These digits can not be the first or last. So $K = L = 0$. We see J times AB is NPQ, but 8 times AB is TU, so J must be 9. M times AB is YOZ, a three digit number, so M is also 9. We know AB times 8 has a two digit product, while AB times 9 has a three digit product. Therefore $AB = 12$. Since $CDEFGHI$ equals AB times $JK8LM$ plus one, we find the dividend to be 1089709.

CHAPTER 37

TRIGONOMETRY

1. Solution (D).

$$1.2^2 = (\sin x + \cos x)^2 = 1 + 2\sin x \cdot \cos x.$$

Hence $\sin 2x = 1.2^2 - 1 = 0.44$.

2. Solution (A). Note that

$$0.64 = (\sin\theta + \cos\theta)^2 = \sin^2\theta + \cos^2\theta + 2\sin\theta\cos\theta = 1 + 2\sin\theta\cos\theta = 1 + \sin(2\theta),$$

so $\sin(2\theta) = -0.36$.

3. Solution (A). We have

$$\log(\sin(2\theta)) = \log(2\sin(\theta)\cos(\theta)) = 0.$$

So we need $2\theta = 2n\pi + (\pi/2)$. However, we also need θ to be in the first quadrant as both $\sin(\theta)$ and $\cos(\theta)$ must be positive. So $\theta = 2k\pi + (\pi/4)$.

4. Solution (D). We have $f(x) = (\sin(x) - \cos(x) - 1)(\sin(x) + \cos(x) - 1) = (\sin(x) - 1)^2 - \cos^2(x) = \sin^2(x) - 2\sin(x) + 1 - \cos^2(x) = 2\sin^2(x) - 2\sin(x)$. Setting $z = \sin(x)$, we have a parabola $y = 2(z^2 - z)$ that opens up. The vertex is at $(1/2, -1/2)$.

5. Solution (C).

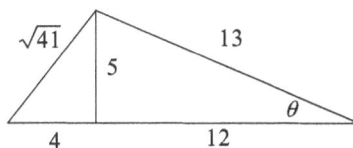

Use the diagram above or use the law of cosines: $41 = 16^2 + 13^2 - 2 \cdot 16 \cdot 13 \cos(\theta)$ to find $\cos(\theta) = \dfrac{12}{13}$. Since $\sin(\theta) = \sqrt{1 - \cos^2(\theta)} = \dfrac{5}{13}$, the area is $\dfrac{1}{2} \cdot 16 \cdot 5 = 40$.

6. **Solution** (E). By the Law of Cosines, $11^2 = 9^2 + 10^2 - 2 \cdot 9 \cdot 10 \cos(\angle ABC)$. Therefore, $\cos(\angle ABC) = (181 - 121) \div 180 = 1/3$.

7. **Solution** (C). Let h denote the distance from the landmark to the point on the road nearest the landmark. Then $\tan 52 = h/x$ and $\tan 28 = h/(x + 5)$. Solving each of these for x reveals that $h/\tan 52 = h/\tan 28 - 5$ which we can solve for h to get $h = 4.55$.

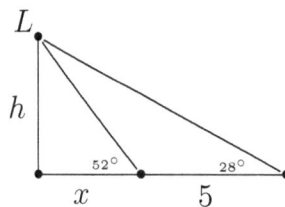

8. **Solution** (D). To keep the distance as short as possible, you must walk in a straight line. Because of your relative speeds, when you intercept Sandy, you must have walked twice as far as she has. In the diagram below, you walk $2x$ yards while Sandy walks x yards.

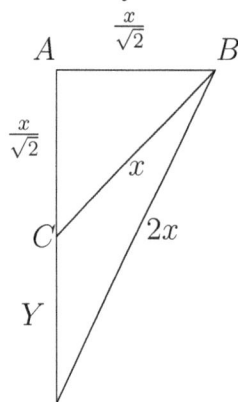

Triangle ABC is an isosceles right triangle with hypotenuse of length x. Using the Pythagorean theorem we find that $\left(Y + \dfrac{x}{\sqrt{2}}\right)^2 + \left(\dfrac{x}{\sqrt{2}}\right)^2 = (2x)^2$. This is equivalent to $3x^2 - \sqrt{2}\,xY - Y^2 = 0$. Using the quadratic formula and discarding the negative root, we find that $x = Y\left(\dfrac{\sqrt{2} + \sqrt{14}}{6}\right)$.

* This problem can also be solved using the law of cosines.

9. **Solution** (C). Use the law of cosines to get

$$(\sqrt{13})^2 = 5^2 + 6^2 - 2(5)(6)\cos\alpha.$$

Solve for $\cos\alpha$ to get $\cos\alpha = \dfrac{25 + 36 - 13}{60} = \dfrac{4}{5}$. Then note that $\sin\alpha = 3/5$, which means the altitude to the base of length 6 is 3, so the area is 9. Alternatively, build such a triangle from triangles we know more about. Start with a 3; 4; 5 triangle. Note that $\sqrt{13} = \sqrt{2^2 + 3^2}$ so we might try to build the triangle in the problem from a 2; 3; $\sqrt{13}$ and a 3; 4; 5. We can append them along the edge of length 3. Notice that since both triangles are right, we can append them so that the union is a triangular region with sides of length $4 + 2, 5$, and $\sqrt{13}$, whose area is easy to find. The altitude to the base 6 is 3, so the area is 9. Yet another approach would be to use Heron's formula.

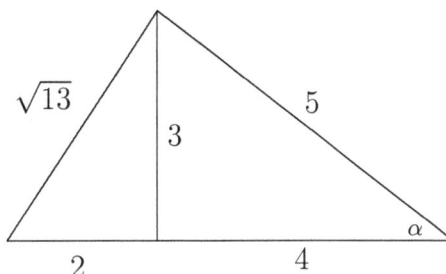

10. **Solution** (D). Let the reflection of points A and C be A' and C' respectively. Label point D such that $\overline{BD}$ is the bisector of angle ABC and D is on $\overline{AC}$. Since angles $ABD, A'BD, CBD,$ and $C'BD$ are all congruent, the reflection places C' on the segment $\overline{AB}$ and A' on the extension of $\overline{BC}$. The area of $\triangle ABC$ is found to be $[ABC] = \frac{1}{2}AB \cdot BC \sin \angle ABC = \frac{1}{2} \cdot 20 \cdot 5 \cdot \sin 60° = 25\sqrt{3}$. The requested area is twice the area of triangle ABD since $\overline{BD}$ lies on the line of reflection. Since $\overline{BD}$ is an angle bisector, we have

$$\frac{AD}{DC} = \frac{AB}{BC} = \frac{20}{5} = 4 \text{ and}$$
$$2[ABD] = 2 \cdot (AD/AC) \cdot [ABC] = 2 \cdot (4/5) \cdot [ABC]$$
$$= (8/5) \cdot 25\sqrt{3} = 40\sqrt{3}.$$

Apply the law of cosines to $\triangle ABC$ and obtain

$$x + y = \sqrt{20^2 + 5^2 - 2 \cdot 20 \cdot 5 \cdot \cos 60°} = 5\sqrt{13}.$$

Apply the law of sines to triangles BCD and BAD to get

$$\frac{5}{\sin \theta} = \frac{x}{\sin 30°}$$

and

$$\frac{20}{\sin(180 - \theta)} = \frac{y}{\sin 30°}.$$

Thus $x/y = 5/20$ and $y = \frac{4}{5}(x + y) = 4\sqrt{13}$. Applying the law of sines to $\triangle ABC$ yields $\frac{5}{\sin \alpha} = \frac{x+y}{\sin 60°}$ which implies that $\sin \alpha = \frac{\sqrt{3}}{2\sqrt{13}}$. Thus the desired area is $2 \cdot [ABD] = 2 \cdot \frac{1}{2} \cdot 20h = 20 \cdot y \sin \alpha = 40\sqrt{3}$.

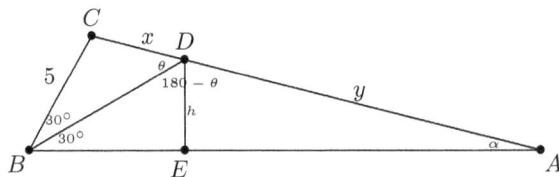

Solve for the areas of $\triangle ABA'$ and $\triangle ADA'$. Since $AB = A'B$ and $\angle ABA' = 60^{\circ}$, $\triangle ABA'$ is equilateral. So, $AA' = 20$ and the perpendicular from B to $\overline{AA'}$ has length $10\sqrt{3}$. $\triangle C'BC$ is similar to $\triangle ABA'$, so it is equilateral as well. Since $BC = 5$, so $C'C = 5$ and the length of the perpendicular from B to $\overline{C'C}$ has length $(5\sqrt{3})/2$ (one fourth the length of AA' and the perp from B to $\overline{AA'}$). Also triangles $\triangle ADA'$ and $\triangle CDC'$ are similar. So the length, z, of the perpendicular from D to $\overline{AA'}$ is four times the length, y, of the perp from D to $\overline{CC'}$. Thus $10\sqrt{3} = z + y + (5\sqrt{3})/2 = 5y + (5\sqrt{3})/2 \Rightarrow y = (3\sqrt{3})/2$ and $z = 6\sqrt{3}$. Therefore the area is $10(10\sqrt{3} - 6\sqrt{3}) = 40\sqrt{3}$.

CHAPTER 38

LOGARITHMS

1. Solution (A). Note that $2^{10x-1} = 1$ implies $10x = 1$, $x = 0.1$. So $\log x = -1$.

2. Solution (A). $\log_x 729 = \log_x 3^6 = 4 \Rightarrow x^4 = 3^6 = (3\sqrt{3})^4 \Rightarrow x = 3\sqrt{3}$.

3. Solution (B). The given equation can be transformed to get $5x = \dfrac{x^2}{14}$ which can be solved for x to yield $x = 70$. Since $\log_2 0$ is undefined, 0 is not a possible value of x.

4. Solution (E). Note that $\log x + \log(x+2) = \log(x(x+2)) = 3$ which implies that $x^2 + 2x = 10^3$ from which it follows that $x^2 + 2x - 10^3 = 0$. But one of the roots, $x = -1 - \sqrt{1 + 10^3}$ is not in the domain of $\log x$.

5. Solution (B). Since $\log_2 x(x+2) = 3$, we see that $x(x+2) = 2^3$. Solving this equation, we obtain $x = 2$ and $x = -4$. But the negative solution is not valid since $\log_2(-4)$ is not defined.

6. Solution (D). To find the zero, first solve

$$\log \sqrt{5x+5} + \frac{1}{2}\log(2x+1) = \log 15.$$

Then

$$\frac{1}{2}\{\log(5x+5)(2x+1)\} = \log 15.$$

Hence

$$\log(10x^2 + 15x + 5) = \log(15^2).$$

It follows that $10x^2 + 15x + 5 = 15^2 = 225$ which is equivalent to $2x^2 + 3x - 44 = 0$, the left side of which can be factored into $(2x + 11)(x - 4)$. Thus, $x = 4$ is the positive zero of f.

7. Solution (C). The number of decimal digits of an integer N is $\lfloor \log_{10} N \rfloor + 1$. Applying this to 3^{100}, we get $\lfloor \log_{10} 3^{100} \rfloor + 1 = \lfloor 100 \log_{10} 3 \rfloor + 1 = \lfloor 47.71 \rfloor + 1 = 48$.

8. Solution (B). The equation could be written as

$$\frac{b}{2} \log\left(\frac{b}{a}\right) - \frac{9a}{2} \log\left(\frac{b}{a}\right) = 1.$$

Since $b = ka$, then

$$\frac{ka}{2} \log(k) - \frac{9a}{2} \log(k) = 1.$$

So

$$\log k = \frac{2}{a(k-9)}$$

and

$$a = \frac{2}{(k-9)\log k}.$$

Since a is an integer, the only solution is when $k = 10$, so $a = 2$ and $b = 20$ and $b^2 - a^2 = 396$.

CHAPTER 39

MISCELLANEOUS

1. Solution (B). Donna's statement cannot be true so Donna and Eddie are lying. This makes Amy's statement and Ben's statement true. Therefore, Carrie is lying and the correct answer is (B).

2. Solution (C). If D's statement is false, then both A and B are also lying, which would mean that we have three liars, and that is impossible. So D's statement is true and therefore E's statement is false. Since C's statement is also false, it must be that A, B and D are honest. But A says it was B or C and B denies it, so only C is left.

3. Solution (A). The magic sum is 27 because the sum of the first nine odd positive integers is $9^2 = 81$. This means that we can fill in the 9 and the 17 as shown.

	1	
5	9	13
x	17	

We can quickly eliminate $x = 11$ and $x = 15$. If $x = 3$, then we would have to use the 7 in the bottom right square, which would require another 7 in the top right square. Thus $x = 7$. The complete square is shown below.

15	1	11
5	9	13
7	17	3

4. **Solution** (D). The solution is unique. The square with the '?' must be a 4. Begin in the cage in the upper left hand corner. (Note that the numbers 3 and 4 must occur in the cage in the upper right hand corner.) Then fill in the cage below it with sum 5. Eventually, you find the following solution:

5. **Solution** (B). Any two statements in the notebook contradict to each other. That means that there can be not more than one true statements. The remaining case – no true statements – is impossible: if this were true, the statement "This notebook contains 100 false statements" would be true, but this statement *is* in the notebook, so it would be false – a contradiction. Since there is one true statement, there are 99 are false statements. Such a statement exists in the notebook: "This notebook has exactly 99 false statements."

6. **Solution** (C). We want a number which satisfies the hypothesis but fails the conclusion. Seven is such a number.

7. **Solution** (D). Without loss of generality, let the left-most coordinate be 0. Since 19 is the largest difference, it follows that 19 is in the 5-element set. Let

$$S = \{0, a, b, c, 19\}$$

denote the set, with $a < b < c$. Then the set of differences of members of S is

$$\{a, b, c, 19 - a, 19 - b, 19 - c, 19, c - a, c - b, b - a\}$$

and the sum of all these is

$$4 \cdot 19 + 2c - 2a = 76 + 2(c - a).$$

On the other hand, the sum of the given distances is

$$2 + 4 + 5 + \cdots = 90 + k.$$

Therefore $76 + 2(c - a) = 90 + k$, and it follows that $2(c - a) = 14 + k$. This implies that k is even. Since $9 \leq k \leq 12$, we know that either $k = 10$ or $k = 12$. If $k = 10$, then $2(c - a) = 24$

and $c - a = 12$ which contradicts the listing of differences given. Therefore, $k = 12$ is the only possibility. Some trial and error leads to the two possible coordinate sets $\{0, 2, 7, 15, 19\}$ and $\{0, 4, 12, 17, 19\}$. The missing distance is $k = 12$.

8. **Solution** (C). It takes 11 trips back and forth. We can write this as follows: gg g w g gg g w g gg g gg where the first, third, fifth, etc symbol represents a trip from the starting bank to the destination bank, and the symbols in the even positions represent the return trips.

9. **Solution** (D). Note that the kth group ends with number $L = \dfrac{k(k+1)}{2}$. Since there are k numbers in the group, the group begins with number

$$F = \frac{k(k+1)}{2} - (k-1).$$

The average value of these numbers is

$$\frac{F+L}{2} = \frac{k^2+1}{2}$$

so the sum is

$$k\left(\frac{k^2+1}{2}\right).$$

When $k = 10$, the sum is 505.

10. **Solution** (B). Since

$$A = \{(x,y)|x^2 + y^2 \le 4\}$$
$$B = \{(x,y)|x + y \ge 2\}$$

the area of $A \cap B$ is one-fourth the area of the circle minus the area of the triangle in the first quadrant, which is

$$4\pi/4 - (2 \times 2)/2 = \pi - 2.$$

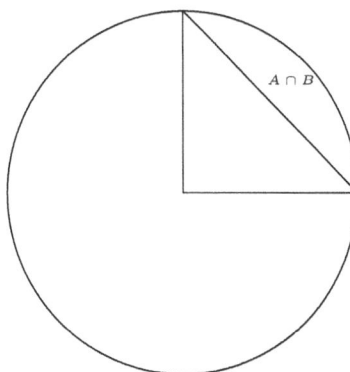

11. **Solution** (D). We must turn over every card with a non-prime number (to be sure it doesn't have a vowel on the other side), and every card with a vowel (to be sure it has a prime on the other side). Hence we must overturn five cards.

12. **Solution** (E). If K is the sum of the entries of each row, then

$$
\begin{aligned}
3K &= N + (N+3) + (N+6) + \ldots + (N+24) \\
&= 9N + 3 + 6 + \ldots + 24 = 9N + 8\left(\frac{3+24}{2}\right) \\
&= 9N + 4 \cdot 27,
\end{aligned}
$$

since the sum of the entries of an arithmetic sequence is the number of terms times the average of the first and last term. Hence,

$$
K = \frac{9N + 108}{3} = 3N + 36.
$$

13. **Solution** (D). If both $b = 0$ and $d = 0$, then $(ac)^2 = 36$ since $a = \pm 2$ and $c = \pm 3$. So all that is left is showing no larger value occurs. Since $(c, -d)$ is another point on the second circle, we may change $(ac - bd)^2$ to $(ac + bd)^2$ and further assume that all four numbers are nonnegative. For the points (a, b) and (c, d), using the trig identity $\cos(\alpha - \beta) = \cos(\alpha)\cos(\beta) + \sin(\alpha)\sin(\beta)$, the cosine of the angle between the rays to these points from the origin is given by $\cos(\theta) = (ac+bd)/6$ (where in this case:

$$
\cos(\alpha) = a/2, \sin(\alpha) = b/2
$$

and

$$
\cos(\beta) = c/3, \sin(\beta) = d/3
$$

- - for cosine it doesn't matter whether $\alpha \leq \beta$ or $\beta \leq \alpha$). We then have $ac + bd = 6 \cdot \cos(\theta)$. The largest value occurs when $\cos(\theta) = 1$, and thus when $\theta = 0°$ which puts the points (a, b) and (c, d) on the same ray from the origin. Any such pair (in the first quadrant) will work, and all give the same value of 6. Thus the largest value of $(ac - bd)^2$ is 36.

Alternatively, normalize the variables by putting $x = 2a$, $y = 2b$, $u = 3c$, $v = 3d$, so that

$$
a^2 + b^2 = c^2 + d^2 = 1.
$$

The problem is now to maximize $(ad - bc)^2$, or equivalently, to maximize $|ad - bc|$. Note that

$$
|ad| \leq \frac{a^2 + d^2}{2}, \ |bc| \leq \frac{b^2 + c^2}{2},
$$

so that

$$
|ad - bc| \leq |ad| + |bc| \leq \frac{a^2 + d^2 + b^2 + c^2}{2} = 1.
$$

The value 1 is attained, in particular, when $a = d$ and $b = -c$ (say, $a = d = 1$, $b = c = 0$). Therefore, $\max(ad - bc)^2 = 1$, and $\max(xv - yu)^2 = (2 \cdot 3)^2 = 36$.

14. **Solution** (C). Let r_1 and b_1 denote the number of red and black cards in pile A, so the numbers in pile B must be $r_2 = 26 - r_1$ and $b_2 = 26 - b_1$. Also, the numbers r_1 and b_1 satisfy $b_1 = 4r_1$ and $26 - r_1 = k(26 - b_1)$ for some integer k. Combining these equations and solving for r_1, we find that

$$r_1 = \frac{26(k-1)}{4k-1}.$$

Trying values of k starting at 2, we finally find that when $k = 10$, r_1 is an integer. Solving this we get $r_1 = \dfrac{9 \cdot 26}{39} = 6$ and thus $r_2 = 20$. Alternate Solution: Let r be the number of red cards in stack A and let b be the number of black cards in stack A. Then $4r = b > 0$ and $26 - r = p(26 - b)$ for some integer p. Since b is even and p is an integer, r must be even. Thus b is a positive integer multiple of 8. The only such numbers less than or equal to 26 are 8, 16 and 24. Having $b = 8$ requires $r = 2$. For this pair $24 = 26 - 2$ is not an integer multiple of $18 = 26 - 8$. Having $b = 16$ requires $r = 4$. For this pair $22 = 26 - 4$ is not an integer multiple of $10 = 26 - 16$. But for $b = 24$ we have $r = 6$ and

$$20 = 26 - 6 = 10(2) = 10(26 - 24).$$

For yet another approach, substitute $4r$ for b into the second equation and solve for r to get

$$r = 26(p - 1)/(4p - 1).$$

If the denominator is not an integer multiple of 13, then r will be. But that is impossible since $4r = b \leq 26$. Thus $4p - 1$ must be an odd integer multiple of 13. The smallest positive integer multiple of 13 that is one less than a multiple of 4 is 39. This occurs when $p = 10$. The corresponding value for r is 6. This yields $b = 24$. [Here is a formal proof that this is the only positive integer solution: If $0 < r < s$ are integer multiples of 13 that are both one less than a multiple of 4, then $s - r$ is both an integer multiple of 13 and an integer multiple of 4. Thus every positive integer multiple of 13 that is both one less than a multiple of 4 and strictly larger than 39 is of the form

$$39 + 4 \cdot 13k = 39 + 52k$$

for some positive integer k. Let $q = 10 + 13k$. Then $4q - 1 = 39 + 52k$. Substitute q in for p in the fraction $26(p - 1)/(4p - 1)$ and then replace q by $10 + 13k$. The resulting fraction is

$$26(9 + 13k)/(39 + 52k)$$

which reduces to

$$(18 + 26k)/(3 + 4k) = 6 + (2k/(3 + 4k)) > 6$$

since $k > 0$. Obviously, $2k/(3+4k)$ is never an integer when k is a positive. So the only solution is when $p = 10$.]

15. Solution (E). Factor the sum of the squares to get

$$1^2 + 4^2 + 7^2 + \cdots + 18^2 = 1056 = 2^5 \cdot 3 \cdot 11.$$

Notice that the only pair of dimensions that will accommodate the 18×18 square together with both the 14×14 and the 15×15 squares is 32×33. The four corners are unique. The only way to make room for the three largest squares is to put them in corners with the 14×14 square and the 15×15 square next to the 18×18 square. See the figure below. The only two squares that could fill the 3×15 gap left above the 15×15 square are the 7×7 and the 8×8 squares. Then the 1×1 must go in the tiny hole left. Finally the 10×10 and the 9×9 squares can be placed. So the sum of the areas of the four corner squares is 826.

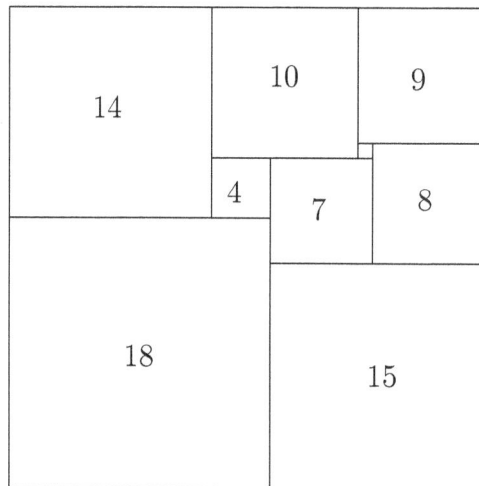

16. Solution (B). Note that

$$\log 6^{2004} = 2004 \log 6 = 1559.415\ldots$$

which means that 6^{2004} is a 1560 digit number that begins with the same digits as

$$10^{.415\ldots} = 2.6007\ldots,$$

so we have $6^{2004} = 26007\ldots$ and $D(6^{2004}) = 2$.

17. Solution (D). The solution is unique. Start with the 4 and build outwards.

	1		1	•	2		2	•	1
1	•	2		3	•	2	•	3	
	3	•	2	•	3		2	•	2
1	•	3		2	•	1		3	•
	3	•	2		1		1	•	2
2	•	4	•	1		1		2	
•	3	•	3		2	•	3	•	1
1		2	•	2	•	3	•	2	

Counting row by row, we see that there are 22 mines.

PART VI

SOLUTIONS TO PART III PAST CONTEST PROBLEMS

CHAPTER 40

SOLUTIONS TO 2016 AMC 8

40.1 2016 AMC 8 Solutions

1. **Solution** (C). There are 60 minutes in 1 hour, so 11 hours plus 5 minutes is equal to $11 \cdot 60 + 5 = 665$ minutes.

2. **Solution** (A). The area of $\triangle ACD$ is $\frac{1}{2} \cdot 8 \cdot 6 = 24$. The area of $\triangle MCD$ is $\frac{1}{2} \cdot 4 \cdot 6 = 12$. So the area of $\triangle AMC$ is $24 - 12 = 12$.

OR

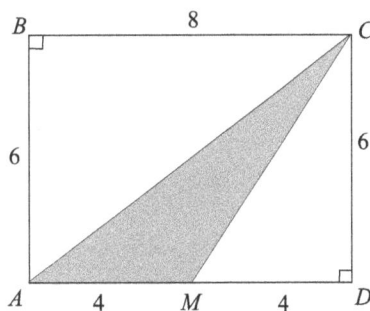

As seen in the diagram above, the altitude from C to the line of the base $\overline{AM}$ is $\overline{CD}$. Thus the area of the shaded $\triangle AMC$ is

$$\frac{1}{2} \cdot AM \cdot CD = \frac{1}{2} \cdot 4 \cdot 6 = 12.$$

3. **Solution** (A). The given scores $70, 80$, and 90 are a total of 30 above the stated average. Thus the remaining score is 30 points below the average, and $70 - 30 = 40$.

OR

Let x be the missing score. Then the sum $70 + 80 + 90 + x = 70 \cdot 4 = 280$. So x must be 40.

4. **Solution** (B). As a boy it took Cheenu 3 hours and 30 minutes, which is 210 minutes, to go 15 miles. That is a rate of $210 \div 15 = 14$ minutes per mile. As an old man it takes him 4 hours, or 240 minutes, to travel 10 miles. That is a rate of $240 \div 10 = 24$ minutes per mile. It takes him $24 - 14 = 10$ minutes more to walk a mile as an old man.

5. **Solution** (E). The two-digit numbers that leave a remainder of 3 when divided by 10 are: $13, 23, 33, 43, 53, 63, 73, 83, 93$. The two-digit numbers that leave a remainder of 1 when divided 9 are: $10, 19, 28, 37, 46, 55, 64, 73, 82, 91$. Among these two sets, 73 is the only common number. When 73 is divided by 11 the remainder is 7.

6. **Solution** (B).The 19 name lengths are $3, 3, 3, 3, 3, 3, 3, 4, 4, 4, 5, 6, 6, 6, 6, 7, 7, 7, 7$. The tenth value, 4, is the median.

7. **Solution** (B). The numbers 1^{2016}, 3^{2018}, 5^{2020} have even exponents and hence are squares. The number 2^{2017} is not a perfect square because it is twice a square $2\left(2^{2018}\right)^2$. Since $4^{2019} = \left(2^2\right)^{2019} 2^{4038}$, it is also a perfect square.

OR

A positive integer power of a square is again a square. This eliminates choices (A) and (D). An even power of any integer is a square. This eliminates choices (C) and (E). The only remaining choice is (B), and in fact, an odd power of a non-square cannot be a square.

8. **Solution** (C). Evaluate the expression by grouping as follows:

$$(100 - 98) + (96 - 94) + \cdots + (8 - 6) + (4 - 2) = 2 + 2 + \cdots + 2 + 2 = 2 \cdot 25 = 50.$$

9. **Solution** (B). The prime factorization of 2016 is: $2016 = \left(2^5\right)\left(3^2\right)(7)$, so the distinct prime divisors of 2016 are 2, 3, and 7, and their sum is $2 + 3 + 7 = 12$.

10. **Solution** (D). Since $2 * (5 * x) = 1$, it follows that $6 - (5 * x) = 1$, and so $5 * x = 5$. Applying the formula again, $15 - x = 5$, and therefore $x = 10$.

11. **Solution** (B). Let $\underline{ab}$ be the two digit number. Then $132 = (10a + b) + (10b + a) = 11(a + b)$. Thus $a + b = 12$. The possible numbers are: $39, 93, 48, 84, 57, 75$, and 66. There are seven two-digit numbers that meet this criterion.

12. **Solution** (B). Converting the given fractions to the same denominator, we see that $\frac{9}{12}$ of the girls and $\frac{8}{12}$ of the boys went on the trip. So the ratio of the number of girls to the number of boys was $9 : 8$, and it follows that $\frac{9}{17}$ of the students on the trip were girls.

OR

The number of boys and girls must be a common multiple of 4 and 3, the denominators of the fractions given in the problem. Suppose there are 12 boys and 12 girls in Jefferson Middle School. Then 9 girls and 8 boys went on the trip, for a total of 17 students. The fraction of girls on the trip is 9/17.

13. **Solution** (D). There are $6 \cdot 5 = 30$ possible pairs of numbers. For a product to be 0, either the first factor or the second factor must be 0, so there are $1 \cdot 5 + 5 \cdot 1 = 10$ such products. The desired probability is $10/30 = 1/3$.

14. **Solution** (A). In driving 350 miles, Karl used $\frac{350}{35} = 10$ gallons of gas, so he had $14 - 10 = 4$ gallons left in his tank. After buying 8 more gallons, he had $4 + 8 = 12$ gallons. When he arrived at his destination, he had $\frac{14}{2} = 7$ gallons left, so he used an additional $12 - 7 = 5$ gallons. This let him drive an additional $5 \cdot 35 = 175$ miles, so he drove a total of $350 + 175 = 525$ miles.

OR

Karl used $14 + 8 - \frac{14}{2} = 15$ gallons of gas on his trip, so he drove $15 \cdot 35 = 525$ miles.

15. **Solution** (C). Factor, using a difference of two squares:

$$
\begin{aligned}
13^4 - 11^4 &= \left(13^2 + 11^2\right)(13 + 11)(13 - 11) \\
&= 290 \cdot 24 \cdot 2 \\
&= 2 \cdot 145 \cdot 8 \cdot 3 \cdot 2 \\
&= 32 \cdot 435
\end{aligned}
$$

So the largest power of 2 that is a divisor of $13^4 - 11^4$ is 32.

16. **Solution** (D). Let N be the number of laps run by Annie when she passes Bonnie for the first time. The number of laps run by Bonnie is $N - 1$. Then $\frac{N}{N-1} = 1.25 = \frac{5}{4}$. so $N = 5$.

OR

For each lap Bonnie completes, Annie runs $1\frac{1}{4}$ laps, thus gaining $\frac{1}{4}$ of a lap on Bonnie during that time. Annie will pass Bonnie when Bonnie has run 4 laps, at which point Annie will have run 5.

17. **Solution** (D). If there were no restrictions, then 10^4 passwords would be possible. Among these, 10 passwords begin with 9 1 1, and have 10 options for the fourth digit. Thus $10^4 - 10 = 9990$ passwords satisfy the condition.

18. **Solution** (C). Divide the 216 sprinters into 36 groups of 6. Run 36 races to eliminate 180 sprinters, leaving 36 winners. Divide the 36 winners into 6 groups of 6, run 6 races to eliminate 30

sprinters, leaving 6 winners. Finally run the last race to determine the champion. The number of races run is $36 + 6 + 1 = 43$.

OR

When all the races have been run, 215 sprinters will have been eliminated. Since 5 sprinters are eliminated in each race, there are $\frac{215}{5} = 43$ races needed to determine the champion.

19. **Solution** (E). The average of the 25 even integers is $10000/25 = 400$. So 12 consecutive even integers will be larger than 400 and 12 consecutive even integers will be smaller than 400. The sum $376 + 378 + \cdots + 398 + 400 + 402 + \cdots + 422 + 424 = 10000$. The largest of these numbers is 424.

OR

The average of the 25 even integers is $\frac{10000}{25} = 400$. Since 12 consecutive even integers are larger than 400, the largest is $400 + 12 \cdot 2 = 424$.

20. **Solution** (A). If $b = 1$, then $a = 12$ and $c = 15$, and the least common multiple of a and c is 60. If $b > 1$, then any prime factor of b must also be a factor of both 12 and 15, and thus the only possible value is $b = 3$. In this case, a must be a multiple of 4 and a divisor of 12, so $a = 4$ or $a = 12$. Similarly, c must be a multiple of 5 and a divisor of 15, so $c = 5$ or $c = 15$. It follows that the least common multiple of a and c must be a multiple of 20. When $a = 4$, $b = 3$, and $c = 5$, the least common multiple of a and c is exactly 20.

21. **Solution** (B). Consider drawing all five chips and listing the 10 possible outcomes: RRRGG, RRGRG, RGRRG, GRRRG, GGRRR, GRGRR, RGGRR, GRRGR, RGRGR, RRGGR. All 10 of these outcomes are equally likely. The outcomes that end in G correspond to the outcomes where the 3 reds are drawn and the outcomes that end in R correspond to the outcomes where the 2 greens are drawn. The probability that the 3 reds are drawn is $\frac{4}{10} = \frac{2}{5}$.

22. **Solution** (C). The area of $\triangle BCE$ is $\frac{1}{2}(1)(4) = 2$. Triangles $\triangle CBH$ and $\triangle EFH$ are similar. Since $CB = \frac{1}{3}EF$, it follows that $IH = \frac{1}{3}GH = \frac{1}{4}IG = 1$. The area of $\triangle CBH$ is $\frac{1}{2}$, so the area of $\triangle ECH$ is $2 - \frac{1}{2} = \frac{3}{2}$. Thus the batwing's area is 3.

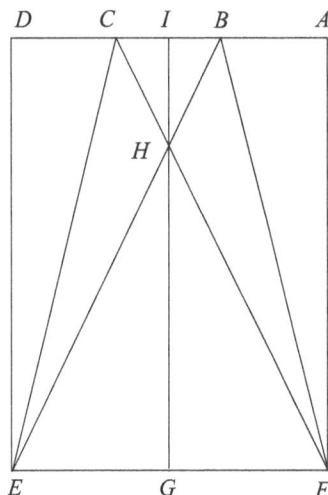

23. **Solution** (C). We know $\triangle AEB$ is equilateral since each of its sides is a radius of one of the congruent circles. Thus the measure of $\angle AEB$ is $60°$. Since $\overline{DB}$ is a diameter of circle A and $\overline{AC}$ is a diameter of circle B, it follows that $\angle DEB$ and $\angle AEC$ are both right angles. Therefore the degree measure of $\angle DEC$ is $90° + 90° - 60° = 120°$.

OR

We know $\triangle AEB$ is equilateral since each of its sides is a radius of one of the congruent circles. Thus the measures of $\angle AEB$ and $\angle EAB$ are both $60°$. Then the measure of $\angle DAE$ is $120°$, and since $\triangle DAE$ is isosceles, the measure of $\angle DEA$ is $30°$. Similarly, the measure of $\angle BEC$ is also $30°$. Therefore the degree measure of $\angle DEC$ is $30° + 60° + 30° = 120°$.

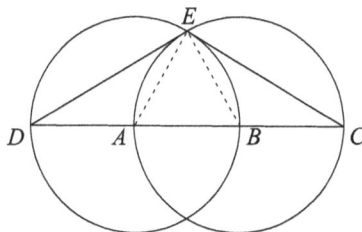

24. **Solution** (A). Since QRS is divisible by 5, we know that $S = 5$. Since PQR is divisible by 4, we know that QR is $12, 32$, or 24. So RST will be either $25T$ or $45T$ and divisible by 3. Using the available digits, 453 is the only number that is divisible by 3. So $T = 3$, $R = 4$, and $P = 1$.

25. **Solution** (B). Let O be the midpoint of base $\overline{AB}$ of $\triangle ABC$ and the center of the semicircle. Triangle $\triangle OBC$ is a right triangle with $OB = 8$ and $OC = 15$, and so, by the Pythagorean Theorem, $BC = 17$. Let E be the point where the semicircle intersects $\overline{BC}$, so radius $\overline{OE}$ is perpendicular to $\overline{BC}$. Then $\triangle OEB$ and $\triangle COB$ are similar, and therefore, $OE : CO = OB : CB$. Hence, $\frac{OE}{15} = \frac{8}{17}$ and so $OE = \frac{120}{17}$.

OR

Let O be the center of the semicircle, which is also the midpoint of base $\overline{AB}$. Since $OB = 8$ and $OC = 15$, then by the Pythagorean Theorem $BC = 17$. Let E be the point where the semicircle intersects $\overline{BC}$, so radius $\overline{OE}$ is perpendicular to $\overline{BC}$. Since the area of $\triangle OBC$ is $\frac{1}{2}(BC)(OE) = \frac{1}{2}(OB)(OC)$, then $\frac{1}{2}(17)(OE) = \frac{1}{2}(8)(15)$ and so $OE = 120/17$.

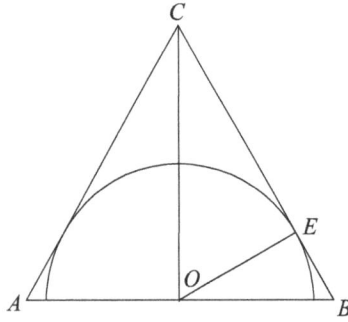

CHAPTER 41

SOLUTIONS TO 2016 AMC 10

41.1 2016 AMC 10A Solutions

1. **Solution** (B).
$$\frac{11! - 10!}{9!} = \frac{10! \cdot (11 - 1)}{9!} = \frac{10 \cdot 9! \cdot 10}{9!} = 100$$

2. **Solution** (C). The equation can be written $10^x \cdot (10^2)^{2x} = (10^3)^5$ or $10^x \cdot 10^{4x} = 10^{15}$. Thus $10^{5x} = 10^{15}$, so $5x = 15$ and $x = 3$.

3. **Solution** (C). Because $\$12.50 = 50 \cdot \$0.25$, Ben spent $\$50$. David spent $\$50 - \$12.50 = \$37.50$, and the two together paid $\$87.50$.

4. **Solution** (B).
$$\frac{3}{8} - \left(-\frac{2}{5}\right) \left\lfloor \frac{\frac{3}{8}}{-\frac{2}{5}} \right\rfloor = \frac{3}{8} + \frac{2}{5} \left\lfloor -\frac{15}{16} \right\rfloor = \frac{3}{8} + \frac{2}{5}(-1) = -\frac{1}{40}$$

5. **Solution** (D). Let the dimensions of the box be x, $3x$, and $4x$. Then the volume of the box is $12x^3$. Therefore the volume must be 12 times the cube of an integer. Among the choices, only $48 = 4 \cdot 12$, $96 = 8 \cdot 12$, and $144 = 12 \cdot 12$ are multiples of 12, and only for 96 is the other factor a perfect cube.

6. **Solution** (D). Each time Emilio replaces a 2 in the ones position by 1, Ximena's sum is decreased by 1. When Emilio replaces a 2 in the tens position by 1, Ximena's sum is decreased by 10. Ximena wrote 3 twos in the ones position $(2, 12, 22)$ and 10 twos in the tens position $(20, 21, 22, \ldots, 29)$. Thus Ximena's sum is greater than Emilio's sum by $3 \cdot 1 + 10 \cdot 10 = 103$.

7. **Solution** (D). The mean of the data values is

$$\frac{60 + 100 + x + 40 + 50 + 200 + 90}{7} = \frac{x + 540}{7} = x.$$

Solving this equation for x gives $x = 90$. Thus the data in nondecreasing order are $40, 50, 60, 90, 90, 100, 200$, so the median is 90 and the mode is 90, as required.

8. **Solution** (C). Working backwards, Fox must have approached the bridge for the third time with 20 coins in order to have no coins left after paying the toll. In the second crossing he must have started with 30 coins in order to have $20 + 40 = 60$ before paying the toll. So he must have started with 35 coins in order to have $30 + 40 = 70$ before paying the toll for the first crossing.

OR

Let c be the number of coins Fox had at the beginning. After three crossings he had $2(2(2c - 40) - 40) - 40 = 8c - 280$ coins. Setting this equal to 0 and solving gives $c = 35$.

9. **Solution** (D). There are

$$1 + 2 + \cdots + N = \frac{N(N+1)}{2}$$

coins in the array. Therefore $N(N+1) = 2 \cdot 2016 = 4032$. Because $N(N+1) \approx N^2$, it follows that $N \approx \sqrt{4032} \approx \sqrt{2^{12}} = 2^6 = 64$. Indeed, $63 \cdot 64 = 4032$, so $N = 63$ and the sum of the digits of N is 9.

10. **Solution** (B). Let the inner rectangle's length be x feet; then its area is x square feet. The middle region has area $3(x + 2) - x = 2x + 6$, so the difference in the arithmetic sequence is equal to $(2x + 6) - x = x + 6$. The outer region has area $5(x + 4) - 3(x + 2) = 2x + 14$, so the difference in the arithmetic sequence is also equal to $(2x + 14) - (2x + 6) = 8$. From $x + 6 = 8$, it follows that $x = 2$. The regions have areas 2, 10, and 18.

11. **Solution** (D). The diagonal of the rectangle from upper left to lower right divides the shaded region into four triangles. Two of them have a 1-unit horizontal base and altitude $\frac{1}{2} \cdot 5 = 2\frac{1}{2}$, and the other two have a 1-unit vertical base and altitude $\frac{1}{2} \cdot 8 = 4$. Therefore the total area is $2 \cdot \frac{1}{2} \cdot 1 \cdot 2\frac{1}{2} + 2 \cdot \frac{1}{2} \cdot 1 \cdot 4 = 6\frac{1}{2}$.

12. **Solution** (A). The product of three integers is odd if and only if all three integers are odd. There are 1008 odd integers among the 2016 integers in the given range. The probability that all the selected integers are odd is

$$p = \frac{1008}{2016} \cdot \frac{1007}{2015} \cdot \frac{1006}{2014}.$$

The first factor is $\frac{1}{2}$ and each of the other factors is less than $\frac{1}{2}$, so $p < \frac{1}{8}$.

13. **Solution** (B). The total number of seats moved to the right among the five friends must equal the total number of seats moved to the left. One of Dee and Edie moved some number of seats to the right, and the other moved the same number of seats to the left. Because Bea moved two seats to the right and Ceci moved one seat to the left, Ada must also move one seat to the left upon her return. Because her new seat is an end seat and its number cannot be 5, it must be seat 1. Therefore Ada occupied seat 2 before she got up. The order before moving was Bea-Ada-Ceci-Dee-Edie (or Bea-Ada-Ceci-Edie-Dee), and the order after moving was Ada-Ceci-Bea-Edie-Dee (or Ada-Ceci-Bea-Dee-Edie).

14. **Solution** (C). If the sum uses n twos and m threes, then $2n+3m = 2016$. Therefore $n = \frac{2016-3m}{2}$. Both m and n will be nonnegative integers if and only if m is an even integer from 0 to 672. Thus there are $\frac{672}{2} + 1 = 337$ ways to form the sum.

15. **Solution** (A). The circle of dough has radius 3 inches. The area of the remaining dough is $3^2 \cdot \pi - 7\pi = 2\pi$ in^2. Let r be the radius in inches of the scrap cookie; then $2\pi = \pi r^2$. Therefore $r = \sqrt{2}$ inches.

16. **Solution** (D). After reflection about the x-axis, the coordinates of the image are $A'(0, -2)$, $B'(-3, -2)$, and $C'(-3, 0)$. The counterclockwise $90°$-rotation around the origin maps this triangle to the triangle with vertices $A''(2, 0)$, $B''(2, -3)$, and $C''(0, -3)$. Notice that the final image can be mapped to the original triangle by interchanging the x- and y-coordinates, which corresponds to a reflection about the line $y = x$.

17. **Solution** (A). Let $N = 5k$, where k is a positive integer. There are $5k + 1$ equally likely possible positions for the red ball in the line of balls. Number these $0, 1, 2, 3, \ldots, 5k - 1, 5k$ from one end. The red ball will *not* divide the green balls so that at least $\frac{3}{5}$ of them are on the same side if it is in position $2k + 1, 2k + 2, \ldots, 3k - 1$. This includes $(3k - 1) - 2k = k - 1$ positions. The probability that $\frac{3}{5}$ or more of the green balls will be on the same side is therefore $1 - \frac{k-1}{5k+1} = \frac{4k+2}{5k+1}$. Solving the inequality $\frac{4k+2}{5k+1} < \frac{321}{400}$ for k yields $k > \frac{479}{5} = 95\frac{4}{5}$. The value of k corresponding to the required least value of N is therefore 96, so $N = 480$. The sum of the digits of N is 12.

18. **Solution** (C). The sum of the four numbers on the vertices of each face must be $\frac{1}{6} \cdot 3 \cdot (1 + 2 + \cdots + 8) = 18$. The only sets of four of the numbers that include 1 and have a sum of 18 are $\{1, 2, 7, 8\}$, $\{1, 3, 6, 8\}$, $\{1, 4, 5, 8\}$, and $\{1, 4, 6, 7\}$. Three of these sets contain both 1 and 8. Because two specific vertices can belong to at most two faces, the vertices of one face must be labeled with the numbers $1, 4, 6, 7$, and two of the faces must include vertices labeled 1 and 8. Thus 1 and 8 must mark two adjacent vertices. The cube can be rotated so that the vertex labeled 1 is at the lower left front, and the vertex labeled 8 is at the lower right front. The numbers 4, 6, and 7 must label vertices on the left face. There are $3! = 6$ ways to assign these three labels to the

three remaining vertices of the left face. Then the numbers 5, 3, and 2 must label the vertices of the right face adjacent to the vertices labeled 4, 6, and 7, respectively. Hence there are 6 possible arrangements.

19. **Solution** (E). Triangles APD and EPB are similar and $BE : DA = 1 : 3$, so $BP = \frac{1}{4}BD$. Triangles AQD and FQB are similar and $BF : DA = 2 : 3$, so $BQ = \frac{2}{5}BD$ and $QD = \frac{3}{5}BD$. Then $PQ = BQ - BP = (\frac{2}{5} - \frac{1}{4})BD = \frac{3}{20}BD$. Thus $BP : PQ : QD = \frac{1}{4} : \frac{3}{20} : \frac{3}{5} = 5 : 3 : 12$, and $r + s + t = 5 + 3 + 12 = 20$.

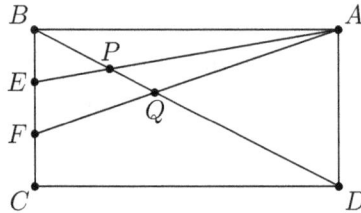

Note: The answer is independent of the dimensions of the original rectangle. Consider the figures below, showing the rectangle $ABCD$ with points E and F trisecting side $\overline{BC}$. Let G and H trisect $\overline{AD}$, and let M and N be the midpoints of $\overline{AB}$ and $\overline{CD}$. Then the segments $\overline{AE}$, $\overline{GF}$, and $\overline{HC}$ are equally spaced, implying that $BP = PR = RS = SD$ and showing that $BP : PD : BD = 1 : 3 : 4 = 5 : 15 : 20$. The segments $\overline{ME}$, $\overline{AF}$, $\overline{GC}$, and $\overline{HN}$ are also equally spaced, implying that $BT = TQ = QU = UV = VD$ and showing that $BQ : QD : BD = 2 : 3 : 5 = 8 : 12 : 20$. It then follows that $BP : PQ : QD = 5 : (15 - 12) : 12 = 5 : 3 : 12$.

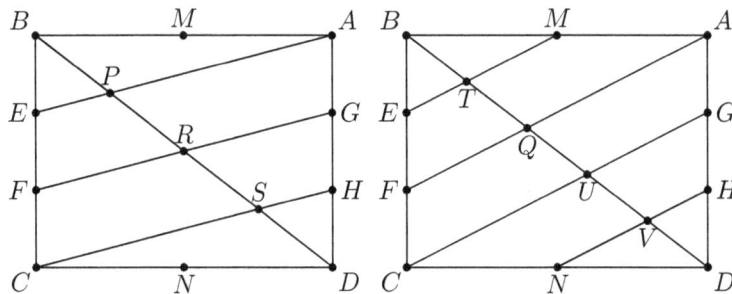

20. **Solution** (B). If a term contains all four variables a, b, c, and d, then it has the form $a^{i+1}b^{j+1}c^{k+1}d^{l+1}1^m$ for some nonnegative integers i, j, k, l, and m such that $(i + 1) + (j + 1) + (k + 1) + (l + 1) + m = N$ or $i + j + k + l + m = N - 4$. The number of terms can be counted using the stars and bars technique. The number of linear arrangements of $N - 4$ stars and 4 bars corresponds to the number of possible values of i, j, k, l, and m. Namely, in each arrangement the bars separate the stars into five groups (some of them can be empty) whose sizes are the values of i, j, k, l, and m. There are

$$\binom{N - 4 + 4}{4} = \binom{N}{4} = \frac{N(N - 1)(N - 2)(N - 3)}{4 \cdot 3 \cdot 2 \cdot 1} = 1001 = 7 \cdot 11 \cdot 13$$

such arrangements. So $N(N-1)(N-2)(N-3) = 4\cdot3\cdot2\cdot7\cdot11\cdot13 = 14\cdot13\cdot12\cdot11$. Thus the answer is $N = 14$.

21. **Solution** (D). Let X be the foot of the perpendicular from P to $\overline{QQ'}$, and let Y be the foot of the perpendicular from Q to $\overline{RR'}$. By the Pythagorean Theorem,

$$P'Q' = PX = \sqrt{(2+1)^2 - (2-1)^2} = \sqrt{8}$$

and

$$Q'R' = QY = \sqrt{(3+2)^2 - (3-2)^2} = \sqrt{24}.$$

The required area can be computed as the sum of the areas of the two smaller trapezoids, $PQQ'P'$ and $QRR'Q'$, minus the area of the large trapezoid, $PRR'P'$:

$$\frac{1+2}{2}\sqrt{8} + \frac{2+3}{2}\sqrt{24} - \frac{1+3}{2}\left(\sqrt{8}+\sqrt{24}\right) = \sqrt{6} - \sqrt{2}.$$

22. **Solution** (D). Let $110n^3 = p_1^{r_1}p_2^{r_2}\cdots p_k^{r_k}$, where the p_j are distinct primes and the r_j are positive integers. Then $\tau(110n^3)$, the number of positive integer divisors of $110n^3$, is given by

$$\tau(110n^3) = (r_1+1)(r_2+1)\cdots(r_k+1) = 110.$$

Because $110 = 2\cdot5\cdot11$, it follows that $k = 3$, $\{p_1, p_2, p_3\} = \{2, 5, 11\}$, and, without loss of generality, $r_1 = 1$, $r_2 = 4$, and $r_3 = 10$. Therefore

$$n^3 = \frac{p_1\cdot p_2^4\cdot p_3^{10}}{110} = p_2^3\cdot p_3^9, \quad \text{so} \quad n = p_2\cdot p_3^3.$$

It follows that $81n^4 = 3^4\cdot p_2^4\cdot p_3^{12}$, and because 3, p_2, and p_3 are distinct primes, $\tau(81n^4) = 5\cdot5\cdot13 = 325$.

23. **Solution** (A). From the given properties, $a\Diamond1 = a\Diamond(a\Diamond a) = (a\Diamond a)\cdot a = 1\cdot a = a$ for all nonzero a. Then for nonzero a and b, $a = a\Diamond1 = a\Diamond(b\Diamond b) = (a\Diamond b)\cdot b$. It follows that $a\Diamond b = \frac{a}{b}$. Thus

$$100 = 2016\Diamond(6\Diamond x) = 2016\Diamond\frac{6}{x} = \frac{2016}{\frac{6}{x}} = 336x,$$

so $x = \frac{100}{336} = \frac{25}{84}$. The requested sum is $25 + 84 = 109$.

24. **Solution** (E). Let $ABCD$ be the given quadrilateral inscribed in the circle centered at O, with $AB = BC = CD = 200$, as shown in the figure. Because the chords $\overline{AB}$, $\overline{BC}$, and $\overline{CD}$ are shorter than the radius, each of $\angle AOB$, $\angle BOC$, and $\angle COD$ is less than $60°$, so O is outside the quadrilateral $ABCD$. Let G and H be the intersections of $\overline{AD}$ with $\overline{OB}$ and $\overline{OC}$, respectively. Because $\overline{AD}$ and $\overline{BC}$ are parallel, and $\triangle OAB$ and $\triangle OBC$ are congruent and isosceles, it follows that $\angle ABO = \angle OBC = \angle OGH = \angle AGB$. Thus $\triangle ABG$, $\triangle OGH$, and $\triangle OBC$ are similar and isosceles with $\frac{AB}{BG} = \frac{OG}{GH} = \frac{OB}{BC} = \frac{200\sqrt{2}}{200} = \sqrt{2}$. Then $AG = AB = 200$, $BG = \frac{AB}{\sqrt{2}} = \frac{200}{\sqrt{2}} = 100\sqrt{2}$, and $GH = \frac{OG}{\sqrt{2}} = \frac{BO-BG}{\sqrt{2}} = \frac{200\sqrt{2}-100\sqrt{2}}{\sqrt{2}} = 100$. Therefore $AD = AG + GH + HD = 200 + 100 + 200 = 500$.

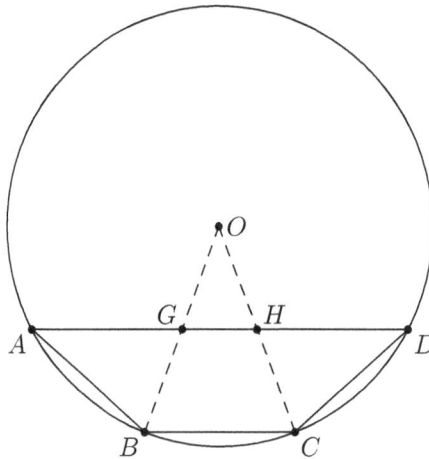

25. **Solution** (A). Because $\operatorname{lcm}(x,y) = 2^3 \cdot 3^2$ and $\operatorname{lcm}(x,z) = 2^3 \cdot 3 \cdot 5^2$, it follows that 5^2 divides z, but neither x nor y is divisible by 5. Furthermore, y is divisible by 3^2, and neither x nor z is divisible by 3^2, but at least one of x or z is divisible by 3. Finally, because $\operatorname{lcm}(y,z) = 2^2 \cdot 3^2 \cdot 5^2$, at least one of y or z is divisible by 2^2, but neither is divisible by 2^3. However, x must be divisible by 2^3. Thus $x = 2^3 \cdot 3^j$, $y = 2^k \cdot 3^2$, and $z = 2^m \cdot 3^n \cdot 5^2$, where $\max(j,n) = 1$ and $\max(k,m) = 2$. There are 3 choices for (j,n) and 5 choices for (k,m), so there are 15 possible ordered triples (x,y,z).

41.2 2016 AMC 10B Solutions

1. **Solution** (D).

$$\frac{2\left(\frac{1}{2}\right)^{-1} + \frac{\left(\frac{1}{2}\right)^{-1}}{2}}{\frac{1}{2}} = \left(2 \cdot 2 + \frac{2}{2}\right) \cdot 2 = 10$$

2. **Solution** (B).

$$\frac{2 \heartsuit 4}{4 \heartsuit 2} = \frac{2^3 \cdot 4^2}{4^3 \cdot 2^2} = \frac{2}{4} = \frac{1}{2}$$

3. **Solution** (D).

$$\left|\left|\left|\left|-2016\right|-(-2016)\right|-\left|-2016\right|\right|-(-2016)\right.$$

$$=\left|\left|2016+2016\right|-2016\right|+2016=2016+2016=4032$$

4. **Solution** (B). It took Zoey $1+2+3+\cdots+15=\frac{15\cdot16}{2}=120$ days to read the 15 books. Because $120=7\cdot17+1$, it follows that Zoey finished the 15th book on the same day of the week as the first, a Monday.

5. **Solution** (D). Because the mean is 8, it follows that the sum of the ages of all Amanda's cousins is $8\cdot4=32$. Because the median age is 5, the sum of the two middle ages is $5\cdot2=10$. Then the sum of the ages of Amanda's youngest and oldest cousins is $32-10=22$.

6. **Solution** (B). Because S has to be greater than 300, the digit sum has to be at least 4, and an example like $197+203=400$ shows that 4 is indeed the smallest possible value.

7. **Solution** (C). Let α and β be the measures of the angles, with $\alpha<\beta$. Then $\frac{\beta}{\alpha}=\frac{5}{4}$. Because $\alpha<\beta$, it follows that $90°-\beta<90°-\alpha$, so $90°-\alpha=2(90°-\beta)$. This leads to the system of linear equations $4\beta-5\alpha=0$ and $2\beta-\alpha=90°$. Solving the system gives $\alpha=60°$, $\beta=75°$. The requested sum is $\alpha+\beta=135°$.

8. **Solution** (A). Positive even powers of numbers ending in 5 end in 25. The tens digit of the difference is the tens digit of $25-17=08$, or 0.

9. **Solution** (C). Let the vertex of the triangle that lies in the first quadrant be (x,x^2). Then the base of the triangle is $2x$ and the height is x^2, so $\frac{1}{2}\cdot2x\cdot x^2=64$. Thus $x^3=64$, $x=4$, and $BC=2x=8$.

10. **Solution** (D). The weight of an object of uniform density is proportional to its volume. The volume of the triangular piece of wood of uniform thickness is proportional to the area of the triangle. The side length of the second piece is $\frac{5}{3}$ times the side length of the first piece, so the area of the second piece is $\left(\frac{5}{3}\right)^2$ times the area of the first piece. Therefore the weight is $12\cdot\left(\frac{5}{3}\right)^2=\frac{100}{3}\approx33.3$ ounces.

11. **Solution** (B). Let x be the number of posts along the shorter side; then there are $2x$ posts along the longer side. When counting the number of posts on all the sides of the garden, each corner post is counted twice, so $2x+2(2x)=20+4$. Solving this equation gives $x=4$. Thus the dimensions of the rectangle are $(4-1)\cdot4=12$ yards by $(8-1)\cdot4=28$ yards. The requested area is given by the product of these dimensions, $12\cdot28=336$ square yards.

12. **Solution** (D). The product of two integers is odd if and only if both integers are odd. Thus the probability that the product is odd is $\frac{3}{5} \cdot \frac{2}{4} = 0.3$, and the probability that the product is even is $1 - 0.3 = 0.7$.

13. **Solution** (D). Let x denote the number of sets of quadruplets. Then $1000 = 4 \cdot x + 3 \cdot (4x) + 2 \cdot (3 \cdot 4x) = 40x$. Thus $x = 25$, and the number of babies in sets of quadruplets is $4 \cdot 25 = 100$.

14. **Solution** (D). Note that $3 < \pi < 4$, $6 < 2\pi < 7$, $9 < 3\pi < 10$, and $12 < 4\pi < 13$. Therefore there are 3 1-by-1 squares of the desired type in the strip $1 \leq x \leq 2$, 6 1-by-1 squares in the strip $2 \leq x \leq 3$, 9 1-by-1 squares in the strip $3 \leq x \leq 4$, and 12 1-by-1 squares in the strip $4 \leq x \leq 5$. Furthermore there are 2 2-by-2 squares in the strip $1 \leq x \leq 3$, 5 2-by-2 squares in the strip $2 \leq x \leq 4$, and 8 2-by-2 squares in the strip $3 \leq x \leq 5$. There is 1 3-by-3 square in the strip $1 \leq x \leq 4$, and there are 4 3-by-3 squares in the strip $2 \leq x \leq 5$. There are no 4-by-4 or larger squares. Thus in all there are $3 + 6 + 9 + 12 + 2 + 5 + 8 + 1 + 4 = 50$ squares of the desired type within the given region.

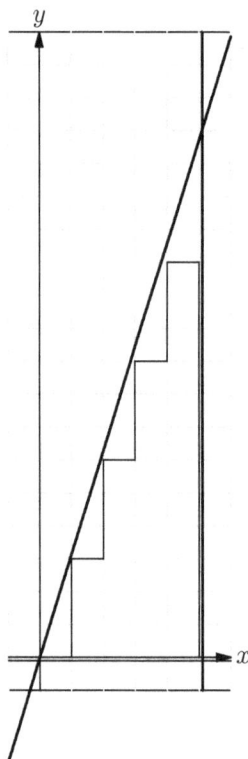

15. **Solution** (C). Shade the squares in a checkerboard pattern as shown in the first figure. Because consecutive numbers must be in adjacent squares, the shaded squares will contain either five odd numbers or five even numbers. Because there are only four even numbers available, the shaded squares contain the five odd numbers. Thus the sum of the numbers in all five shaded squares is $1 + 3 + 5 + 7 + 9 = 25$. Because all but the center add up to $18 = 25 - 7$, the center number must be 7. The situation described is actually possible, as the second figure demonstrates.

			3	4	5
			2	7	6
			1	8	9

16. **Solution** (E). Let r be the common ratio of the geometric series; then

$$S = \frac{1}{r} + 1 + r + r^2 + \cdots = \frac{\frac{1}{r}}{1-r} = \frac{1}{r - r^2}.$$

Because $S > 0$, the smallest value of S occurs when the value of $r - r^2$ is maximized. The graph of $f(r) = r - r^2$ is a downward-opening parabola with vertex $(\frac{1}{2}, \frac{1}{4})$, so the smallest possible value of S is $\frac{1}{(\frac{1}{4})} = 4$. The optimal series is $2, 1, \frac{1}{2}, \frac{1}{4}, \ldots$.

17. **Solution** (D). Suppose that one pair of opposite faces of the cube are assigned the numbers a and b, a second pair of opposite faces are assigned the numbers c and d, and the remaining pair of opposite faces are assigned the numbers e and f. Then the needed sum of products is $ace + acf + ade + adf + bce + bcf + bde + bdf = (a + b)(c + d)(e + f)$. The sum of these three factors is $2 + 3 + 4 + 5 + 6 + 7 = 27$. A product of positive numbers whose sum is fixed is maximized when the factors are all equal. Thus the greatest possible value occurs when $a + b = c + d = e + f = 9$, as in $(a, b, c, d, e, f) = (2, 7, 3, 6, 4, 5)$. This results in the value $9^3 = 729$.

18. **Solution** (E). A sum of consecutive integers is equal to the number of integers in the sum multiplied by their median. Note that $345 = 3 \cdot 5 \cdot 23$. If there are an odd number of integers in the sum, then the median and the number of integers must be complementary factors of 345. The only possibilities are 3 integers with median $5 \cdot 23 = 115$, 5 integers with median $3 \cdot 23 = 69$, $3 \cdot 5 = 15$ integers with median 23, and 23 integers with median $3 \cdot 5 = 15$. Having more integers in the sum would force some of the integers to be negative. If there are an even number of integers in the sum, say $2k$, then the median will be $\frac{j}{2}$, where k and j are complementary factors of 345. The possibilities are 2 integers with median $\frac{345}{2}$, 6 integers with median $\frac{115}{2}$, and 10 integers with median $\frac{69}{2}$. Again, having more integers in the sum would force some of the integers to be negative. This gives a total of 7 solutions.

19. **Solution** (D). Triangles AEP and CFP are similar and $FP : EP = CF : AE = 3 : 4$, so $FP = \frac{3}{7}EF$. Extend $\overline{AG}$ and $\overline{FC}$ to meet at point H; then $\triangle AEQ$ and $\triangle HFQ$ are similar.

Note that $\triangle HCG$ and $\triangle ABG$ are similar with sides in a ratio of $1:3$, so $CH = \frac{1}{3} \cdot 5$ and $FH = 3 + \frac{5}{3} = \frac{14}{3}$. Then $FQ : EQ = \frac{14}{3} : 4 = 7 : 6$, so $FQ = \frac{7}{13}FE$. Thus $PQ = FQ - FP = \left(\frac{7}{13} - \frac{3}{7}\right)FE = \frac{10}{91}FE$ and $\frac{PQ}{FE} = \frac{10}{91}$.

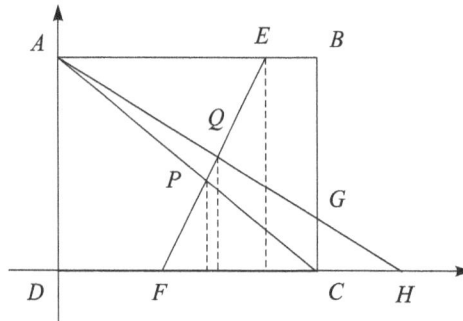

OR

Place the figure in the coordinate plane with D at the origin, A at $(0,4)$, and C at $(5,0)$. Then the equations of lines AC, AG, and EF are $y = -\frac{4}{5}x + 4$, $y = -\frac{3}{5}x + 4$, and $y = 2x - 4$, respectively. The intersections can be found by solving simultaneous linear equations: $P\left(\frac{20}{7}, \frac{12}{7}\right)$ and $Q\left(\frac{40}{13}, \frac{28}{13}\right)$. Because F, P, Q, and E are aligned, ratios of distances between these points are the same as ratios of the corresponding distances between their coordinates. Then

$$\frac{PQ}{FE} = \frac{\frac{40}{13} - \frac{20}{7}}{4 - 2} = \frac{10}{91}.$$

20. **Solution** (C). The scale factor for this transformation is $\frac{3}{2}$. The center of the dilation, D, must lie along ray $A'A$ (with A between A' and D), and its distance from A must be $\frac{2}{3}$ of its distance from A'. Because A is 3 units to the left of and 4 units below A', the center of the dilation must be 6 units to the left of and 8 units below A, placing it at $D(-4,-6)$. The origin is $\sqrt{(-4)^2 + (-6)^2} = 2\sqrt{13}$ units from D, so the dilation must move it half that far, or $\sqrt{13}$ units. Alternatively, note that the origin is 4 units to the right of and 6 units above D, so its image must be 6 units to the right of and 9 units above D; therefore it is located at $(2,3)$, a distance $\sqrt{2^2 + 3^2} = \sqrt{13}$ from the origin.

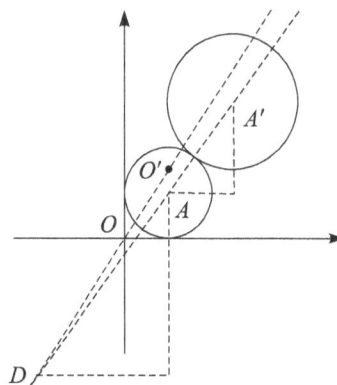

21. **Solution** (B). The graph of the equation is symmetric about both axes. In the first quadrant, the equation is equivalent to $x^2 + y^2 - x - y = 0$. Completing the square gives $(x - \frac{1}{2})^2 + (y - \frac{1}{2})^2 = \frac{1}{2}$, so the graph in the first quadrant is an arc of the circle that is centered at $C(\frac{1}{2}, \frac{1}{2})$ and contains the points $A(1, 0)$ and $B(0, 1)$. Because C is the midpoint of $\overline{AB}$, the arc is a semicircle. The region enclosed by the graph in the first quadrant is the union of isosceles right triangle AOB, where $O(0, 0)$ is the origin, and a semicircle with diameter $\overline{AB}$. The triangle and the semicircle have areas $\frac{1}{2}$ and $\frac{1}{2} \cdot \pi \left(\frac{\sqrt{2}}{2} \right)^2 = \frac{\pi}{4}$, respectively, so the area of the region enclosed by the graph in all quadrants is $4(\frac{1}{2} + \frac{\pi}{4}) = \pi + 2$.

22. **Solution** (A). There must have been $10 + 10 + 1 = 21$ teams, and therefore there were $\binom{21}{3} = \frac{21 \cdot 20 \cdot 19}{6} = 1330$ subsets $\{A, B, C\}$ of three teams. If such a subset does not satisfy the stated condition, then it consists of a team that beat both of the others. To count such subsets, note that there are 21 choices for the winning team and $\binom{10}{2} = 45$ choices for the other two teams in the subset. This gives $21 \cdot 45 = 945$ such subsets. The required answer is $1330 - 945 = 385$. To see that such a scenario is possible, arrange the teams in a circle, and let each team beat the 10 teams that follow it in clockwise order around the circle.

23. **Solution** (C). Extend sides $\overline{CB}$ and $\overline{FA}$ to meet at G. Note that $FC = 2AB$ and $ZW = \frac{5}{3}AB$. Then the areas of $\triangle BAG$, $\triangle WZG$, and $\triangle CFG$ are in the ratio $1^2 : (\frac{5}{3})^2 : 2^2 = 9 : 25 : 36$. Thus $\frac{[ZWCF]}{[ABCF]} = \frac{36 - 25}{36 - 9} = \frac{11}{27}$, and by symmetry, $\frac{[WCXYFZ]}{[ABCDEF]} = \frac{11}{27}$ also.

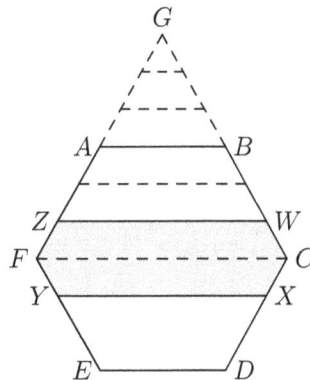

OR

Suppose that $AB = 1$; then $FZ = \frac{1}{3}$ and $FC = 2$. Trapezoid $WCFZ$, which is the upper half of hexagon $WCXYFZ$, can be tiled by 11 equilateral triangles of side length $\frac{1}{3}$, and the lower half similarly, making 22 such triangles. Hexagon $ABCDEF$ can be tiled by 6 equilateral triangles of side length 1, and each of these can be tiled by 9 equilateral triangles of side length $\frac{1}{3}$, making a total of $6 \cdot 9 = 54$ small triangles. The required ratio is $\frac{22}{54} = \frac{11}{27}$.

24. **Solution** (D). Let k be the common difference for the arithmetic sequence. If $b = c$ or $c = d$, then $k = bc - ab = cd - bc$ must be a multiple of 10, so $b = c = d$. However, the two-digit integers bc and cd are then equal, a contradiction. Therefore either (b, c, d) or $(b, c, d + 10)$ is an increasing arithmetic sequence.

> **Case 1:** (b, c, d) is an increasing arithmetic sequence. In this case the additions of k to ab and bc do not involve any carries, so (a, b, c) also forms an increasing arithmetic sequence, as does (a, b, c, d). Let $n = b - a$. If $n = 1$, the possible values of a are 1, 2, 3, 4, 5, and 6. If $n = 2$, the possible values of a are 1, 2, and 3. There are no possibilities with $n \geq 3$. Thus in this case there are 9 integers that have the required property: 1234, 2345, 3456, 4567, 5678, 6789, 1357, 2468, and 3579.

> **Case 2:** $(b, c, d + 10)$ is an increasing arithmetic sequence. In this case the addition of k to bc involves a carry, so $(a, b, c - 1)$ forms a nondecreasing arithmetic sequence, as does $(b, c - 1, (d + 10) - 2) = (b, c - 1, d + 8)$. Hence $(a, b, c - 1, d + 8)$ is a nondecreasing arithmetic sequence. Again letting $n = b - a$, note that $0 \leq c = d + (9 - n) \leq 9$ and $1 \leq a = d + (8 - 3n) \leq 9$. The only integers with the required properties are 8890 with $n = 0$; 5680 and 6791 with $n = 1$; 2470, 3581, and 4692 with $n = 2$; and 1482 and 2593 with $n = 3$. Thus in this case there are 8 integers that have the required property.

The total number of integers with the required property is $9 + 8 = 17$.

25. **Solution** (A). Note that for any x, $f(x + 1) = \sum_{k=2}^{10}(\lfloor kx + k \rfloor - k\lfloor x + 1 \rfloor) = \sum_{k=2}^{10}(\lfloor kx \rfloor + k - k\lfloor x \rfloor - k) = f(x)$. This implies that $f(x)$ is periodic with period 1. Thus the number of distinct values that $f(x)$ assumes is the same as the number of distinct values that $f(x)$ assumes for $0 \leq x < 1$. For these x, $\lfloor x \rfloor = 0$, so $f(x) = \sum_{k=2}^{10}\lfloor kx \rfloor$, which is a nondecreasing function of x. This function increases at exactly those values of x expressible as a fraction of positive integers with denominator between 2 and 10. There are 31 such values between 0 and 1. They are $\frac{1}{2}, \frac{1}{3}, \frac{2}{3}, \frac{1}{4}, \frac{3}{4}, \frac{1}{5}, \frac{2}{5}, \frac{3}{5}, \frac{4}{5}, \frac{1}{6}, \frac{5}{6}, \frac{1}{7}, \frac{2}{7}, \frac{3}{7}, \frac{4}{7}, \frac{5}{7}, \frac{6}{7}, \frac{1}{8}, \frac{3}{8}, \frac{5}{8}, \frac{7}{8}, \frac{1}{9}, \frac{2}{9}, \frac{4}{9}, \frac{5}{9}, \frac{7}{9}, \frac{8}{9}, \frac{1}{10}, \frac{3}{10}, \frac{7}{10}, \frac{9}{10}$. Thus $f(0) = 0$ and $f(x)$ increases 31 times for x between 0 and 1, showing that $f(x)$ assumes 32 distinct values.

CHAPTER 42

SOLUTIONS TO 2016 AMC 12

42.1 2016 AMC 12A Solutions

1. **Solution** (B).
$$\frac{11! - 10!}{9!} = \frac{10! \cdot (11 - 1)}{9!} = \frac{10 \cdot 9! \cdot 10}{9!} = 100$$

2. **Solution** (C). The equation can be written $10^x \cdot (10^2)^{2x} = (10^3)^5$ or $10^x \cdot 10^{4x} = 10^{15}$. Thus $10^{5x} = 10^{15}$, so $5x = 15$ and $x = 3$.

3. **Solution** (B).
$$\frac{3}{8} - \left(-\frac{2}{5}\right)\left\lfloor \frac{\frac{3}{8}}{-\frac{2}{5}} \right\rfloor = \frac{3}{8} + \frac{2}{5}\left\lfloor -\frac{15}{16} \right\rfloor = \frac{3}{8} + \frac{2}{5}(-1) = -\frac{1}{40}$$

4. **Solution** (D). The mean of the data values is
$$\frac{60 + 100 + x + 40 + 50 + 200 + 90}{7} = \frac{x + 540}{7} = x.$$

Solving this equation for x gives $x = 90$. Thus the data in nondecreasing order are $40, 50, 60, 90, 90, 100, 200$, so the median is 90 and the mode is 90, as required.

5. **Solution** (E). A counterexample must satisfy the hypothesis of being an even integer greater than 2 but fail to satisfy the conclusion that it can be written as the sum of two prime numbers.

6. **Solution** (D). There are
$$1 + 2 + \cdots + N = \frac{N(N+1)}{2}$$

Used with permission of the MAA.   **503**

coins in the array. Therefore $N(N + 1) = 2 \cdot 2016 = 4032$. Because $N(N + 1) \approx N^2$, it follows that $N \approx \sqrt{4032} \approx \sqrt{2^{12}} = 2^6 = 64$. Indeed, $63 \cdot 64 = 4032$, so $N = 63$ and the sum of the digits of N is 9.

7. **Solution** (D). The given equation is equivalent to $(x^2 - y^2)(x + y + 1) = 0$, which is in turn equivalent to $(x+y)(x-y)(x+y+1) = 0$. A product is 0 if and only if one of the factors is 0, so the graph is the union of the graphs of $x + y = 0$, $x - y = 0$, and $x + y + 1 = 0$. These are three straight lines, two of which intersect at the origin and the third of which does not pass through the origin. Therefore the graph consists of three lines that do not all pass through a common point.

8. **Solution** (D). The diagonal of the rectangle from upper left to lower right divides the shaded region into four triangles. Two of them have a 1-unit horizontal base and altitude $\frac{1}{2} \cdot 5 = 2\frac{1}{2}$, and the other two have a 1-unit vertical base and altitude $\frac{1}{2} \cdot 8 = 4$. Therefore the total area is $2 \cdot \frac{1}{2} \cdot 1 \cdot 2\frac{1}{2} + 2 \cdot \frac{1}{2} \cdot 1 \cdot 4 = 6\frac{1}{2}$.

9. **Solution** (E).

Let x be the common side length. Draw a diagonal between opposite corners of the unit square. The length of this diagonal is $\sqrt{2}$. The diagonal consists of two small-square diagonals and one small-square side length. Combining the previous two observations yields

$$2x\sqrt{2} + x = \sqrt{2}.$$

Solving this equation for x gives $x = \frac{4-\sqrt{2}}{7}$. The requested sum is $4 + 7 = 11$.

OR

Again let x be the common side length. Triangle ABC in the figure shown is a right triangle with sides $\frac{x}{2}$, $\frac{x}{2}$, and $1 - 2x$. By the Pythagorean Theorem,

$$\left(\frac{x}{2}\right)^2 + \left(\frac{x}{2}\right)^2 = (1 - 2x)^2.$$

Solving this equation and noting that $x < \frac{1}{2}$ yields $x = \frac{4-\sqrt{2}}{7}$, as above.

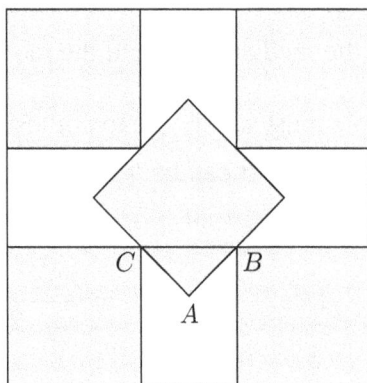

10. **Solution** (B). The total number of seats moved to the right among the five friends must equal the total number of seats moved to the left. One of Dee and Edie moved some number of seats to the right, and the other moved the same number of seats to the left. Because Bea moved two seats to the right and Ceci moved one seat to the left, Ada must also move one seat to the left upon her return. Because her new seat is an end seat and its number cannot be 5, it must be seat 1. Therefore Ada occupied seat 2 before she got up. The order before moving was Bea-Ada-Ceci-Dee-Edie (or Bea-Ada-Ceci-Edie-Dee), and the order after moving was Ada-Ceci-Bea-Edie-Dee (or Ada-Ceci-Bea-Dee-Edie).

11. **Solution** (E). Because 42 students cannot sing, $100 - 42 = 58$ can sing. Similarly, $100 - 65 = 35$ can dance, and $100 - 29 = 71$ can act. This gives a total of $58 + 35 + 71 = 164$. However, the students with two talents have been counted twice in this sum. Because there are 100 students in all, $164 - 100 = 64$ students must have been counted twice.

OR

Consider the three sets referred to in the problem: those who cannot sing, those who cannot dance, and those who cannot act. Students with one talent are in two of those sets, whereas students with two talents are in only one. Thus the total $42 + 65 + 29 = 136$ counts all students twice except those with two talents. The number of students with two talents is therefore $2 \cdot 100 - 136 = 64$.

12. **Solution** (C). Applying the Angle Bisector Theorem to $\triangle BAC$ gives $BD : DC = 6 : 8$, so $BD = \frac{6}{6+8} \cdot 7 = 3$. Then applying the Angle Bisector Theorem to $\triangle ABD$ gives $AF : FD = 6 : 3 = 2 : 1$.

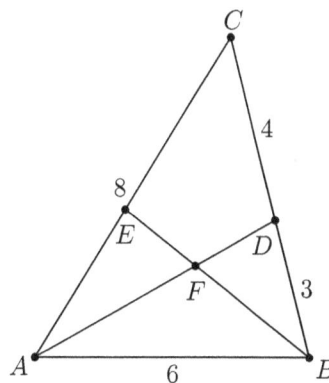

Note: More generally the ratio $AF : FD$ is $(AB + CA) : BC$, which equals $2 : 1$ whenever AB, BC, CA forms an arithmetic progression.

13. **Solution** (A). Let $N = 5k$, where k is a positive integer. There are $5k + 1$ equally likely possible positions for the red ball in the line of balls. Number these $0, 1, 2, 3, \ldots, 5k - 1, 5k$ from one end.

The red ball will *not* divide the green balls so that at least $\frac{3}{5}$ of them are on the same side if it is in position $2k + 1, 2k + 2, \ldots, 3k - 1$. This includes $(3k - 1) - 2k = k - 1$ positions. The probability that $\frac{3}{5}$ or more of the green balls will be on the same side is therefore $1 - \frac{k-1}{5k+1} = \frac{4k+2}{5k+1}$. Solving the inequality $\frac{4k+2}{5k+1} < \frac{321}{400}$ for k yields $k > \frac{479}{5} = 95\frac{4}{5}$. The value of k corresponding to the required least value of N is therefore 96, so $N = 480$. The sum of the digits of N is 12.

14. **Solution** (C). The sum of the four numbers on the vertices of each face must be $\frac{1}{6} \cdot 3 \cdot (1 + 2 + \cdots + 8) = 18$. The only sets of four of the numbers that include 1 and have a sum of 18 are $\{1, 2, 7, 8\}$, $\{1, 3, 6, 8\}$, $\{1, 4, 5, 8\}$, and $\{1, 4, 6, 7\}$. Three of these sets contain both 1 and 8. Because two specific vertices can belong to at most two faces, the vertices of one face must be labeled with the numbers $1, 4, 6, 7$, and two of the faces must include vertices labeled 1 and 8. Thus 1 and 8 must mark two adjacent vertices. The cube can be rotated so that the vertex labeled 1 is at the lower left front, and the vertex labeled 8 is at the lower right front. The numbers 4, 6, and 7 must label vertices on the left face. There are $3! = 6$ ways to assign these three labels to the three remaining vertices of the left face. Then the numbers 5, 3, and 2 must label the vertices of the right face adjacent to the vertices labeled 4, 6, and 7, respectively. Hence there are 6 possible arrangements.

15. **Solution** (D). Let X be the foot of the perpendicular from P to $\overline{QQ'}$, and let Y be the foot of the perpendicular from Q to $\overline{RR'}$. By the Pythagorean Theorem,

$$P'Q' = PX = \sqrt{(2+1)^2 - (2-1)^2} = \sqrt{8}$$

and

$$Q'R' = QY = \sqrt{(3+2)^2 - (3-2)^2} = \sqrt{24}.$$

The required area can be computed as the sum of the areas of the two smaller trapezoids, $PQQ'P'$ and $QRR'Q'$, minus the area of the large trapezoid, $PRR'P'$:

$$\frac{1+2}{2}\sqrt{8} + \frac{2+3}{2}\sqrt{24} - \frac{1+3}{2}\left(\sqrt{8} + \sqrt{24}\right) = \sqrt{6} - \sqrt{2}.$$

16. **Solution** (D). Let $u = \log_3 x$. Then $\log_x 3 = \frac{1}{u}$, $\log_{\frac{1}{3}} x = -u$, and $\log_x \frac{1}{3} = -\frac{1}{u}$. Thus each point at which two of the graphs of the given functions intersect in the (x, y)-plane corresponds

to a point at which two of the graphs of $y = u$, $y = \frac{1}{u}$, $y = -u$, and $y = -\frac{1}{u}$ intersect in the (u, y)-plane. There are 5 such points (u, y), namely $(0, 0)$, $(1, 1)$, $(-1, 1)$, $(1, -1)$, and $(-1, -1)$. The corresponding points of intersection on the graphs of the given functions are $(1, 0)$, $(3, 1)$, $(\frac{1}{3}, 1)$, $(3, -1)$, and $(\frac{1}{3}, -1)$.

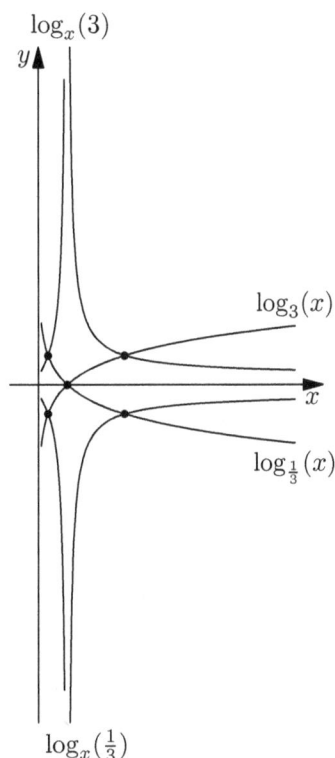

17. **Solution** (B). Without loss of generality, let the square and equilateral triangles have side length 6. Then the height of the equilateral triangles is $3\sqrt{3}$, and the distance of each of the triangle centers, E, F, G, and H, to the square $ABCD$ is $\sqrt{3}$. It follows that the diagonal of square $ABCD$ has length $6\sqrt{2}$, and the diagonal of square $EFGH$ has length equal to the side length of square $ABCD$ plus twice the distance from the center of an equilateral triangle to square $ABCD$ or $6 + 2\sqrt{3}$. The required ratio of the areas of the two squares is equal to the square of the ratio of the lengths of the diagonals of the two squares, or

$$\left(\frac{6 + 2\sqrt{3}}{6\sqrt{2}}\right)^2 = \left(\frac{3 + \sqrt{3}}{3\sqrt{2}}\right)^2 = \frac{12 + 6\sqrt{3}}{18} = \frac{2 + \sqrt{3}}{3}.$$

OR

Without loss of generality, place the square in the Cartesian plane with coordinates $A(-6, 0)$, $B(0, 0)$, $C(0, -6)$, and $D(-6, -6)$. The center of each equilateral triangle is the point at which the medians intersect, and this point is one third of the way from the midpoint of a side of the triangle to the opposite vertex. The height of an equilateral triangle with side 6 is $3\sqrt{3}$, so the centers are $\sqrt{3}$ units from the sides of the square. Therefore the coordinates are $E(-3, \sqrt{3})$, $F(\sqrt{3}, -3)$, $G(-3, -6 - \sqrt{3})$, and $H(-6 - \sqrt{3}, -3)$. The area of square $EFGH$ is half the product of the lengths of its diagonals, or $\frac{1}{2}(6 + 2\sqrt{3})^2 = 24 + 12\sqrt{3}$. Square $ABCD$ has area 36, so the desired ratio is $\frac{2+\sqrt{3}}{3}$.

18. **Solution** (D). Let $110n^3 = p_1^{r_1} p_2^{r_2} \cdots p_k^{r_k}$, where the p_j are distinct primes and the r_j are positive integers. Then $\tau(110n^3)$, the number of positive integer divisors of $110n^3$, is given by

$$\tau(110n^3) = (r_1 + 1)(r_2 + 1) \cdots (r_k + 1) = 110.$$

Because $110 = 2 \cdot 5 \cdot 11$, it follows that $k = 3$, $\{p_1, p_2, p_3\} = \{2, 5, 11\}$, and, without loss of generality, $r_1 = 1$, $r_2 = 4$, and $r_3 = 10$. Therefore

$$n^3 = \frac{p_1 \cdot p_2^4 \cdot p_3^{10}}{110} = p_2^3 \cdot p_3^9, \quad \text{so} \quad n = p_2 \cdot p_3^3.$$

It follows that $81n^4 = 3^4 \cdot p_2^4 \cdot p_3^{12}$, and because 3, p_2, and p_3 are distinct primes, $\tau(81n^4) = 5 \cdot 5 \cdot 13 = 325$.

19. **Solution** (B). Jerry arrives at 4 for the first time after an even number of tosses. Because Jerry tosses 8 coins, he arrives at 4 for the first time after either 4, 6, or 8 tosses. If Jerry arrives at 4 for the first time after 4 tosses, then he must have tossed HHHH. The probability of this occurring is $\frac{1}{16}$. If Jerry arrives at 4 for the first time after 6 tosses, he must have tossed 5 heads and 1 tail among the 6 tosses, and the 1 tail must have come among the first 4 tosses. Thus, there are 4 possible sequences of valid tosses, each with probability $\frac{1}{64}$, for a total of $\frac{1}{16}$. If Jerry arrives at 4 for the first time after 8 tosses, then he must have tossed 6 heads and 2 tails among the 8 tosses. Both tails must occur among the first 6 tosses; otherwise Jerry would have already reached 4 before the 8th toss. Further, at least 1 tail must occur in the first 4 tosses; otherwise Jerry would have already reached 4 after the 4th toss. Therefore there are $\binom{6}{2} - 1 = 14$ sequences for which Jerry first arrives at 4 after 8 tosses, each with probability $\frac{1}{256}$, for a total of $\frac{14}{256} = \frac{7}{128}$. Thus the probability that Jerry reaches 4 at some time during the process is $\frac{1}{16} + \frac{1}{16} + \frac{7}{128} = \frac{23}{128}$. The requested sum is $23 + 128 = 151$.

OR

Count the sequences of 8 heads or tails that result in Jerry arriving at 4. Any sequence with T appearing fewer than 3 times results in Jerry reaching 4. There are $\binom{8}{0} + \binom{8}{1} + \binom{8}{2} = 1 + 8 + 28 = 37$ such sequences. If Jerry's sequence contains exactly 3 Ts, then he reaches 4 only if he does so before getting his second T. As a result, Jerry can get at most one T in his first 5 tosses. This happens if the first 4 tosses are H and there is exactly one H in the last 4 tosses, or there is one T within the first 4 tosses followed by the remaining 5 Hs, accounting for $4 + 4 = 8$ ways for Jerry to get to 4 with exactly 3 Ts. Finally, the only way for Jerry to get to 4 by tossing exactly 4 Ts is HHHHTTTT. Jerry cannot get to 4 by tossing fewer than 4 Hs. Thus there are $37 + 8 + 1 = 46$ sequences where he reaches 4, out of $2^8 = 256$ equally likely ways to toss the coin 8 times. The required probability is $\frac{46}{256} = \frac{23}{128}$, and the requested sum is $23 + 128 = 151$.

20. **Solution** (A). From the given properties, $a \diamondsuit 1 = a \diamondsuit (a \diamondsuit a) = (a \diamondsuit a) \cdot a = 1 \cdot a = a$ for all nonzero a. Then for nonzero a and b, $a = a \diamondsuit 1 = a \diamondsuit (b \diamondsuit b) = (a \diamondsuit b) \cdot b$. It follows that $a \diamondsuit b = \frac{a}{b}$. Thus

$$100 = 2016 \diamondsuit (6 \diamondsuit x) = 2016 \diamondsuit \frac{6}{x} = \frac{2016}{\frac{6}{x}} = 336x,$$

so $x = \frac{100}{336} = \frac{25}{84}$. The requested sum is $25 + 84 = 109$.

21. **Solution** (E). Let $ABCD$ be the given cyclic quadrilateral with $AB = BC = CD = 200$, and let E and F be the feet of the perpendicular segments from B and C, respectively, to $\overline{AD}$, as shown in the figure. Let the center of the circle be O, and let $\angle AOB = \angle BOC = \angle COD = \theta$. Because inscribed $\angle BAD$ is half the size of central $\angle BOD = 2\theta$, it follows that $\angle BAD = \theta$. Let M be the midpoint of $\overline{AB}$. Then $\sin\left(\frac{\theta}{2}\right) = \frac{AM}{AO} = \frac{100}{200\sqrt{2}} = \frac{1}{2\sqrt{2}}$. Then $\cos\theta = 1 - 2\sin^2\left(\frac{\theta}{2}\right) = \frac{3}{4}$. Hence $AE = AB\cos\theta = 200 \cdot \frac{3}{4} = 150$, and $FD = 150$ as well. Because $EF = BC = 200$, the remaining side $AD = AE + EF + FD = 150 + 200 + 150 = 500$.

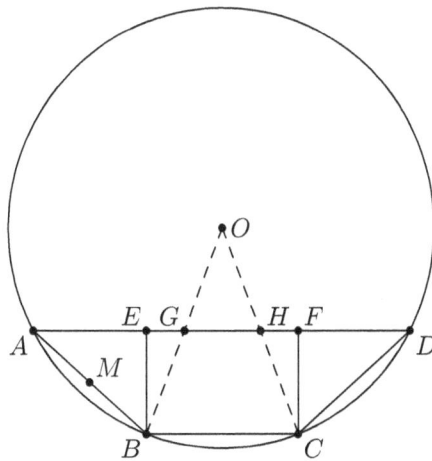

OR

Label the quadrilateral $ABCD$ and the center of the circle as in the first solution. Because the chords $\overline{AB}$, $\overline{BC}$, and $\overline{CD}$ are shorter than the radius, each of $\angle AOB$, $\angle BOC$, and $\angle COD$ is less than $60°$, so O is outside the quadrilateral $ABCD$. Let G and H be the intersections of $\overline{AD}$ with $\overline{OB}$ and $\overline{OC}$, respectively. Because $\overline{AD}$ and $\overline{BC}$ are parallel, and $\triangle OAB$ and $\triangle OBC$ are congruent and isosceles, it follows that $\angle ABO = \angle OBC = \angle OGH = \angle AGB$. Thus $\triangle ABG$, $\triangle OGH$, and $\triangle OBC$ are similar and isosceles with $\frac{AB}{BG} = \frac{OG}{GH} = \frac{OB}{BC} = \frac{200\sqrt{2}}{200} = \sqrt{2}$. Then

$AG = AB = 200$, $BG = \frac{AB}{\sqrt{2}} = \frac{200}{\sqrt{2}} = 100\sqrt{2}$, and $GH = \frac{OG}{\sqrt{2}} = \frac{BO-BG}{\sqrt{2}} = \frac{200\sqrt{2}-100\sqrt{2}}{\sqrt{2}} = 100$. Therefore $AD = AG + GH + HD = 200 + 100 + 200 = 500$.

OR

Let θ be the central angle that subtends the side of length 200. Then by the Law of Cosines, $\left(200\sqrt{2}\right)^2 + \left(200\sqrt{2}\right)^2 - 2\left(200\sqrt{2}\right)^2 \cos\theta = 200^2$, which gives $\cos\theta = \frac{3}{4}$. The Law of Cosines also gives the square of the fourth side of the quadrilateral as

$$\left(200\sqrt{2}\right)^2 + \left(200\sqrt{2}\right)^2 - 2\left(200\sqrt{2}\right)^2 \cos(3\theta)$$

$$= 160{,}000 - 160{,}000(4\cos^3\theta - 3\cos\theta) = 250{,}000.$$

Thus the fourth side has length $\sqrt{250{,}000} = 500$.

22. **Solution** (A). Because $\operatorname{lcm}(x,y) = 2^3 \cdot 3^2$ and $\operatorname{lcm}(x,z) = 2^3 \cdot 3 \cdot 5^2$, it follows that 5^2 divides z, but neither x nor y is divisible by 5. Furthermore, y is divisible by 3^2, and neither x nor z is divisible by 3^2, but at least one of x or z is divisible by 3. Finally, because $\operatorname{lcm}(y,z) = 2^2 \cdot 3^2 \cdot 5^2$, at least one of y or z is divisible by 2^2, but neither is divisible by 2^3. However, x must be divisible by 2^3. Thus $x = 2^3 \cdot 3^j$, $y = 2^k \cdot 3^2$, and $z = 2^m \cdot 3^n \cdot 5^2$, where $\max(j,n) = 1$ and $\max(k,m) = 2$. There are 3 choices for (j,n) and 5 choices for (k,m), so there are 15 possible ordered triples (x,y,z).

23. **Solution** (C). Let the chosen numbers be x, y, and z. The set of possible ordered triples (x,y,z) forms a solid unit cube, two of whose vertices are $(0,0,0)$ and $(1,1,1)$. The numbers fail to be the side lengths of a triangle with positive area if and only if one of the numbers is at least as great as the sum of the other two. The ordered triples that satisfy $z \geq x + y$ lie in the region on and above the plane $z = x + y$. The intersection of this region with the solid cube is a solid tetrahedron with vertices $(0,0,0)$, $(0,0,1)$, $(0,1,1)$, and $(1,0,1)$. The volume of this tetrahedron is $\frac{1}{6}$. The intersections of the solid cube with the regions defined by the inequalities $y \geq x + z$ and $x \geq y + z$ are solid tetrahedra with the same volume. Because at most one of the inequalities $z > x+y$, $y > x+z$, and $x > y+z$ can be true for any choice of x, y, and z, the three tetrahedra have disjoint interiors. Thus the required probability is $1 - 3 \cdot \frac{1}{6} = \frac{1}{2}$.

OR

As in the first solution, the set of possible ordered triples (x,y,z) forms a solid unit cube. First consider only the points for which $x > y$ and $x > z$. These points form a square pyramid whose vertex is $(0,0,0)$ and whose base has vertices at $(0,0,1)$, $(1,0,1)$, $(1,1,1)$, and $(0,1,1)$. Such an ordered triple corresponds to the side lengths of a triangle if and only if $z < x + y$. The plane $z = x + y$ passes through the vertex of the pyramid and bisects its base, so it bisects the volume of the pyramid. The probability of forming a triangle is the same as the probability of not forming a triangle. The same argument applies when y or z is the largest element in the triple. The probability of any two of x, y, and z being equal is 0, so this case can be ignored. Thus this event and its complement are equally likely; the probability is $\frac{1}{2}$.

24. **Solution** (B). Because a and b are positive, all the roots must be positive. Let the roots be $r, s,$ and t. Then

$$x^3 - ax^2 + bx - a = (x - r)(x - s)(x - t) = x^3 - (r + s + t)x^2 + (rs + st + tr)x - rst.$$

Therefore $r + s + t = a = rst$. The Arithmetic Mean–Geometric Mean Inequality implies that $27rst \le (r + s + t)^3 = (rst)^3$, from which $a = rst \ge 3\sqrt{3}$. Furthermore, equality is achieved if and only if $r = s = t = \sqrt{3}$. In this case $b = rs + st + tr = 9$.

25. **Solution** (E). Assume that $k = 2j \ge 2$ is even. The smallest perfect square with $k + 1$ digits is $10^k = (10^j)^2$. Thus the sequence of numbers written on the board after Silvia erases the last k digits of each number is the sequence

$$1 = \left\lfloor \frac{(10^j)^2}{10^k} \right\rfloor, \left\lfloor \frac{(10^j + 1)^2}{10^k} \right\rfloor, \ldots, \left\lfloor \frac{n^2}{10^k} \right\rfloor, \ldots$$

The sequence ends the first time that

$$\left\lfloor \frac{(n + 1)^2}{10^k} \right\rfloor - \left\lfloor \frac{n^2}{10^k} \right\rfloor \ge 2;$$

before that, every two consecutive terms are either equal or they differ by 1. Suppose that

$$\left\lfloor \frac{n^2}{10^k} \right\rfloor = a \quad \text{and} \quad \left\lfloor \frac{(n + 1)^2}{10^k} \right\rfloor \ge a + 2.$$

Then $n^2 < (a + 1)10^k$ and $(a + 2)10^k \le (n + 1)^2$. Thus

$$10^k = (a + 2)10^k - (a + 1)10^k < (n + 1)^2 - n^2 = 2n + 1.$$

It follows that $n = \frac{10^k}{2} + m$ for some positive integer m. Note that

$$\frac{n^2}{10^k} = \frac{1}{10^k}\left(\frac{10^k}{2} + m\right)^2 = \frac{1}{10^k}\left(\frac{10^{2k}}{4} + m \cdot 10^k + m^2\right) = \frac{10^k}{4} + m + \frac{m^2}{10^k}.$$

Because $k \ge 2$, it follows that 10^k is divisible by 4, and so

$$\left\lfloor \frac{n^2}{10^k} \right\rfloor = \frac{10^k}{4} + m + \left\lfloor \frac{m^2}{10^k} \right\rfloor \quad \text{and} \quad \left\lfloor \frac{(n + 1)^2}{10^k} \right\rfloor = \frac{10^k}{4} + m + 1 + \left\lfloor \frac{(m + 1)^2}{10^k} \right\rfloor.$$

The difference will be at least 2 for the first time when

$$\left\lfloor \frac{m^2}{10^k} \right\rfloor = 0 \quad \text{and} \quad \left\lfloor \frac{(m+1)^2}{10^k} \right\rfloor \geq 1,$$

that is, for m such that $m^2 < 10^k \leq (m+1)^2$, equivalently, $m < 10^j \leq m+1$. Thus $m = 10^j - 1$ and then

$$f(k) = f(2j) = a + 1 = \left\lfloor \frac{n^2}{10^k} \right\rfloor + 1 = \frac{10^k}{4} + m + 1 = \frac{10^{2j}}{4} + 10^j.$$

Therefore

$$\sum_{j=1}^{1008} f(2j) = \sum_{j=1}^{1008} \left(\frac{10^{2j}}{4} + 10^j \right) = 25 \sum_{j=0}^{1007} 10^{2j} + 10 \sum_{j=0}^{1007} 10^j$$

$$= \underbrace{252525\ldots 25}_{2016 \text{ digits}} + \underbrace{111\ldots 10}_{1009 \text{ digits}}.$$

Because there are no carries in the sum, the required sum of digits equals $1008 \cdot (2+5) + 1008 \cdot 1 = 1008 \cdot 8 = 8064$.

42.2   2016 AMC 12B Solutions

1. **Solution** (D).

$$\frac{2(\frac{1}{2})^{-1} + \frac{(\frac{1}{2})^{-1}}{2}}{\frac{1}{2}} = \left(2 \cdot 2 + \frac{2}{2} \right) \cdot 2 = 10$$

2. **Solution** (A). The harmonic mean of 1 and 2016 is

$$\frac{2 \cdot 1 \cdot 2016}{1 + 2016} = 2 \cdot \frac{2016}{2017} \approx 2 \cdot 1 = 2.$$

3. **Solution** (D).

$$\left| \left| \left| |-2016| - (-2016) \right| - |-2016| \right| - (-2016) \right.$$

$$= \left| \left| 2016 + 2016 \right| - 2016 \right| + 2016 = 2016 + 2016 = 4032$$

4. **Solution** (C). Let α and β be the measures of the angles, with $\alpha < \beta$. Then $\frac{\beta}{\alpha} = \frac{5}{4}$. Because $\alpha < \beta$, it follows that $90° - \beta < 90° - \alpha$, so $90° - \alpha = 2(90° - \beta)$. This leads to the system of linear equations $4\beta - 5\alpha = 0$ and $2\beta - \alpha = 90°$. Solving the system gives $\alpha = 60°$, $\beta = 75°$. The requested sum is $\alpha + \beta = 135°$.

5. **Solution** (B). Because $919 = 7 \cdot 131 + 2$, the war lasted 131 full weeks plus 2 days. Therefore it ended 2 days beyond Thursday, which is Saturday.

6. **Solution** (C). Let the vertex of the triangle that lies in the first quadrant be (x, x^2). Then the base of the triangle is $2x$ and the height is x^2, so $\frac{1}{2} \cdot 2x \cdot x^2 = 64$. Thus $x^3 = 64$, $x = 4$, and $BC = 2x = 8$.

7. **Solution** (D). In the first pass Josh marks out the odd numbers $1, 3, 5, 7, \ldots, 99$, leaving the multiples of 2: $2, 4, 6, 8, \ldots, 100$. In the second pass Josh marks out $2, 6, 10, \ldots, 98$, leaving the multiples of 4: $4, 8, 12, \ldots, 100$. Similarly, in the nth pass Josh marks out the numbers that are not multiples of 2^n, leaving the numbers that are multiples of 2^n. It follows that in the 6th pass Josh marks out the numbers that are multiples of 2^5 but not multiples of 2^6, namely 32 and 92. This leaves 64, the only number in his original list that is a multiple of 2^6. Thus the last number remaining is 64.

8. **Solution** (D). The weight of an object of uniform density is proportional to its volume. The volume of the triangular piece of wood of uniform thickness is proportional to the area of the triangle. The side length of the second piece is $\frac{5}{3}$ times the side length of the first piece, so the area of the second piece is $\left(\frac{5}{3}\right)^2$ times the area of the first piece. Therefore the weight is $12 \cdot \left(\frac{5}{3}\right)^2 = \frac{100}{3} \approx 33.3$ ounces.

9. **Solution** (B). Let x be the number of posts along the shorter side; then there are $2x$ posts along the longer side. When counting the number of posts on all the sides of the garden, each corner post is counted twice, so $2x + 2(2x) = 20 + 4$. Solving this equation gives $x = 4$. Thus the dimensions of the rectangle are $(4 - 1) \cdot 4 = 12$ yards by $(8 - 1) \cdot 4 = 28$ yards. The requested area is given by the product of these dimensions, $12 \cdot 28 = 336$ square yards.

10. **Solution** (A). The slopes of $\overline{PQ}$ and $\overline{RS}$ are -1, and the slopes of $\overline{QR}$ and $\overline{PS}$ are 1, so the figure is a rectangle. The side lengths are $PQ = (a - b)\sqrt{2}$ and $PS = (a + b)\sqrt{2}$, so the area is $2(a - b)(a + b) = 2(a^2 - b^2) = 16$. Therefore $a^2 - b^2 = 8$. The only perfect squares whose difference is 8 are 9 and 1, so $a = 3$, $b = 1$, and $a + b = 4$.

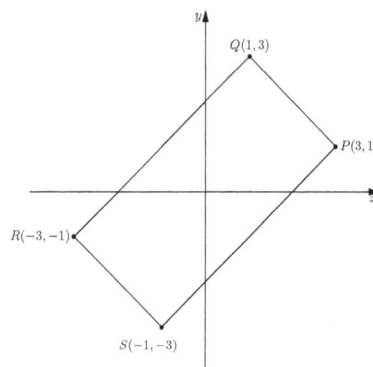

11. **Solution** (D). Note that $3 < \pi < 4$, $6 < 2\pi < 7$, $9 < 3\pi < 10$, and $12 < 4\pi < 13$. Therefore there are 3 1-by-1 squares of the desired type in the strip $1 \le x \le 2$, 6 1-by-1 squares in the

strip $2 \leq x \leq 3$, 9 1-by-1 squares in the strip $3 \leq x \leq 4$, and 12 1-by-1 squares in the strip $4 \leq x \leq 5$. Furthermore there are 2 2-by-2 squares in the strip $1 \leq x \leq 3$, 5 2-by-2 squares in the strip $2 \leq x \leq 4$, and 8 2-by-2 squares in the strip $3 \leq x \leq 5$. There is 1 3-by-3 square in the strip $1 \leq x \leq 4$, and there are 4 3-by-3 squares in the strip $2 \leq x \leq 5$. There are no 4-by-4 or larger squares. Thus in all there are $3 + 6 + 9 + 12 + 2 + 5 + 8 + 1 + 4 = 50$ squares of the desired type within the given region.

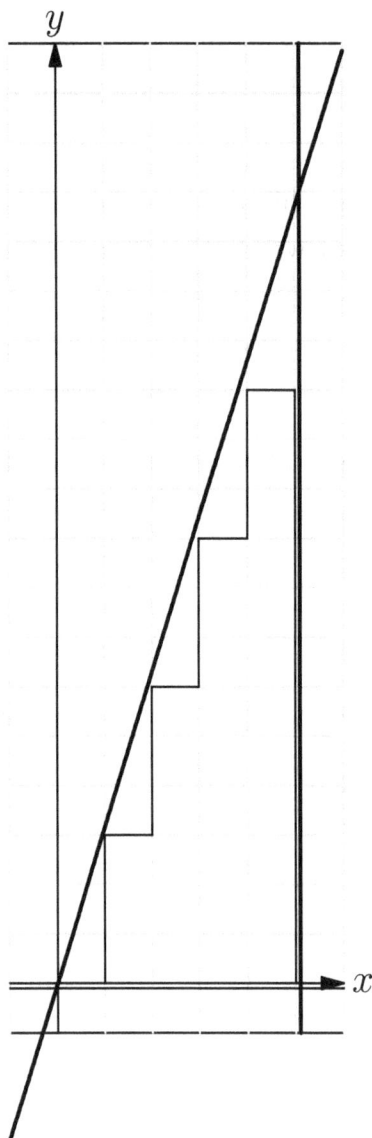

12. **Solution** (C). Shade the squares in a checkerboard pattern as shown in the first figure. Because consecutive numbers must be in adjacent squares, the shaded squares will contain either five odd numbers or five even numbers. Because there are only four even numbers available, the shaded squares contain the five odd numbers. Thus the sum of the numbers in all five shaded squares is

$1 + 3 + 5 + 7 + 9 = 25$. Because all but the center add up to $18 = 25 - 7$, the center number must be 7. The situation described is actually possible, as the second figure demonstrates.

13. **Solution** (E). Let Alice, Bob, and the airplane be located at points A, B, and C, respectively. Let D be the point on the ground directly beneath the airplane, and let h be the airplane's altitude, in miles. Then $\triangle ACD$ and $\triangle BCD$ are $30° - 60° - 90°$ right triangles with right angles at D, so $AD = \sqrt{3}h$ and $BD = \frac{h}{\sqrt{3}}$. Then by the Pythagorean Theorem applied to the right triangle on the ground,

$$100 = AB^2 = AD^2 + BD^2 = \left(\sqrt{3}h\right)^2 + \left(\frac{h}{\sqrt{3}}\right)^2 = \frac{10h^2}{3}.$$

Thus $h = \sqrt{30}$, and the closest of the given choices is 5.5.

14. **Solution** (E). Let r be the common ratio of the geometric series; then

$$S = \frac{1}{r} + 1 + r + r^2 + \cdots = \frac{\frac{1}{r}}{1 - r} = \frac{1}{r - r^2}.$$

Because $S > 0$, the smallest value of S occurs when the value of $r - r^2$ is maximized. The graph of $f(r) = r - r^2$ is a downward-opening parabola with vertex $(\frac{1}{2}, \frac{1}{4})$, so the smallest possible value of S is $\frac{1}{(\frac{1}{4})} = 4$. The optimal series is $2, 1, \frac{1}{2}, \frac{1}{4}, \ldots$.

15. **Solution** (D). Suppose that one pair of opposite faces of the cube are assigned the numbers a and b, a second pair of opposite faces are assigned the numbers c and d, and the remaining pair of opposite faces are assigned the numbers e and f. Then the needed sum of products is $ace + acf + ade + adf + bce + bcf + bde + bdf = (a + b)(c + d)(e + f)$. The sum of these three factors is $2 + 3 + 4 + 5 + 6 + 7 = 27$. A product of positive numbers whose sum is fixed is maximized when the factors are all equal. Thus the greatest possible value occurs when $a + b = c + d = e + f = 9$, as in $(a, b, c, d, e, f) = (2, 7, 3, 6, 4, 5)$. This results in the value $9^3 = 729$.

16. **Solution** (E). A sum of consecutive integers is equal to the number of integers in the sum multiplied by their median. Note that $345 = 3 \cdot 5 \cdot 23$. If there are an odd number of integers in the

sum, then the median and the number of integers must be complementary factors of 345. The only possibilities are 3 integers with median $5 \cdot 23 = 115$, 5 integers with median $3 \cdot 23 = 69$, $3 \cdot 5 = 15$ integers with median 23, and 23 integers with median $3 \cdot 5 = 15$. Having more integers in the sum would force some of the integers to be negative. If there are an even number of integers in the sum, say $2k$, then the median will be $\frac{j}{2}$, where k and j are complementary factors of 345. The possibilities are 2 integers with median $\frac{345}{2}$, 6 integers with median $\frac{115}{2}$, and 10 integers with median $\frac{69}{2}$. Again, having more integers in the sum would force some of the integers to be negative. This gives a total of 7 solutions.

17. **Solution** (D). Let $x = BH$. Then $CH = 8 - x$ and $AH^2 = 7^2 - x^2 = 9^2 - (8-x)^2$, so $x = 2$ and $AH = \sqrt{45}$. By the Angle Bisector Theorem in $\triangle ACH$, $\frac{AP}{PH} = \frac{CA}{CH} = \frac{9}{6}$, so $AP = \frac{3}{5}AH$. Similarly, by the Angle Bisector Theorem in $\triangle ABH$, $\frac{AQ}{QH} = \frac{BA}{BH} = \frac{7}{2}$, so $AQ = \frac{7}{9}AH$. Then $PQ = AQ - AP = (\frac{7}{9} - \frac{3}{5})AH = \frac{8}{45}\sqrt{45} = \frac{8}{15}\sqrt{5}$.

18. **Solution** (B). The graph of the equation is symmetric about both axes. In the first quadrant, the equation is equivalent to $x^2 + y^2 - x - y = 0$. Completing the square gives $(x - \frac{1}{2})^2 + (y - \frac{1}{2})^2 = \frac{1}{2}$, so the graph in the first quadrant is an arc of the circle that is centered at $C(\frac{1}{2}, \frac{1}{2})$ and contains the points $A(1, 0)$ and $B(0, 1)$. Because C is the midpoint of $\overline{AB}$, the arc is a semicircle. The region enclosed by the graph in the first quadrant is the union of isosceles right triangle AOB, where $O(0, 0)$ is the origin, and a semicircle with diameter $\overline{AB}$. The triangle and the semicircle have areas $\frac{1}{2}$ and $\frac{1}{2} \cdot \pi(\frac{\sqrt{2}}{2})^2 = \frac{\pi}{4}$, respectively, so the area of the region enclosed by the graph in all quadrants is $4(\frac{1}{2} + \frac{\pi}{4}) = \pi + 2$.

19. **Solution** (B). The probability that a flipper obtains his first head on the nth flip is $(\frac{1}{2})^n$, because the sequence of outcomes must be exactly TT ... TH, with $n - 1$ Ts. Therefore the probability that all of them obtain their first heads on the nth flip is $((\frac{1}{2})^n)^3 = (\frac{1}{8})^n$. The probability that all three flip their coins the same number of times is computed by summing an infinite geometric series:

$$\left(\frac{1}{8}\right)^1 + \left(\frac{1}{8}\right)^2 + \left(\frac{1}{8}\right)^3 + \cdots = \frac{\frac{1}{8}}{1 - \frac{1}{8}} = \frac{1}{7}.$$

20. **Solution** (A). There must have been $10 + 10 + 1 = 21$ teams, and therefore there were $\binom{21}{3} = \frac{21 \cdot 20 \cdot 19}{6} = 1330$ subsets $\{A, B, C\}$ of three teams. If such a subset does not satisfy the stated condition, then it consists of a team that beat both of the others. To count such subsets, note that there are 21 choices for the winning team and $\binom{10}{2} = 45$ choices for the other two teams in the subset. This gives $21 \cdot 45 = 945$ such subsets. The required answer is $1330 - 945 = 385$. To see that such a scenario is possible, arrange the teams in a circle, and let each team beat the 10 teams that follow it in clockwise order around the circle.

21. **Solution** (B). For any point P between B and D, let Q be the foot of the perpendicular from P to $\overline{CD}$, let P' be the intersection of $\overline{AQ}$ and $\overline{BD}$, and let Q' be the foot of the perpendicular from P' to $\overline{CD}$. Let $x = PQ$ and $y = P'Q'$. Because $\triangle PQD$ and $\triangle P'Q'D$ are isosceles right triangles, $DQ = x$ and $DQ' = y$. Because $\triangle ADQ$ is similar to $\triangle P'Q'Q$, $\frac{1}{x} = \frac{y}{x-y}$. Solving for y gives $y = \frac{x}{1+x}$.

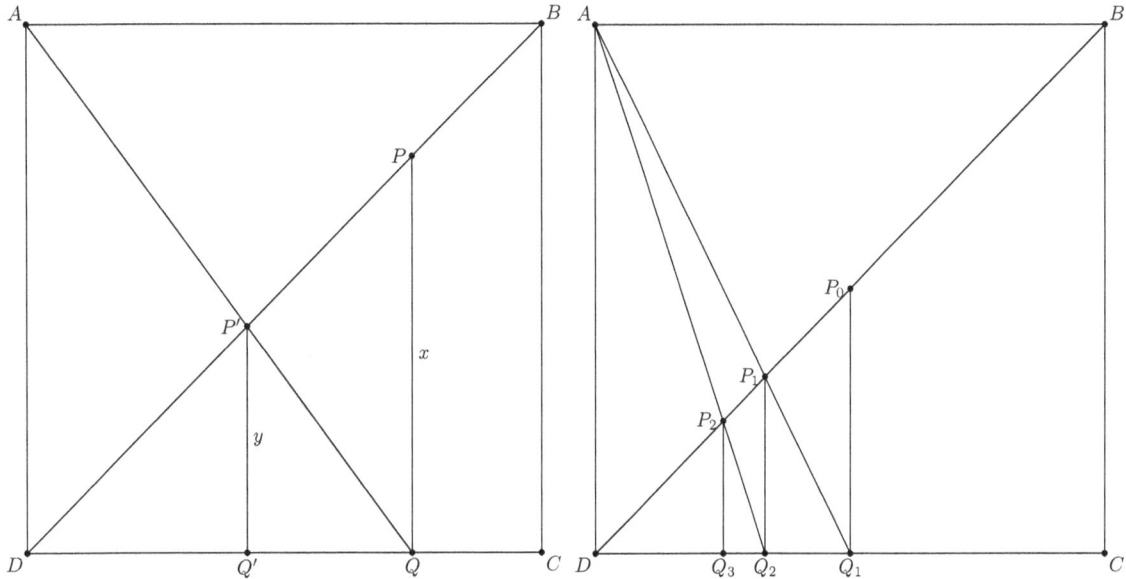

Now let P_0 be the midpoint of $\overline{BD}$. Then $P_0Q_1 = DQ_1 = \frac{1}{2}$. It follows from the analysis above that $P_1Q_2 = DQ_2 = \frac{1}{3}$, $P_2Q_3 = DQ_3 = \frac{1}{4}$, and in general $P_iQ_{i+1} = DQ_{i+1} = \frac{1}{i+2}$. The area of $\triangle DQ_iP_i$ is

$$\frac{1}{2} \cdot DQ_i \cdot P_iQ_{i+1} = \frac{1}{2} \cdot \frac{1}{i+1} \cdot \frac{1}{i+2} = \frac{1}{2}\left(\frac{1}{i+1} - \frac{1}{i+2}\right).$$

The requested infinite sum telescopes:

$$\sum_{i=1}^{\infty} \text{Area of } \triangle DQ_iP_i = \frac{1}{2}\left(\frac{1}{2} - \frac{1}{3} + \frac{1}{3} - \frac{1}{4} + \frac{1}{4} - \frac{1}{5} + \cdots\right).$$

Its value is $\frac{1}{2} \cdot \frac{1}{2} = \frac{1}{4}$.

22. **Solution** (B). Because $\frac{1}{n} = \frac{abcdef}{999999}$, it follows that n is a divisor of $10^6 - 1 = (10^3 - 1)(10^3 + 1) = 3^3 \cdot 7 \cdot 11 \cdot 13 \cdot 37$. Because $\frac{1}{n+6} = \frac{wxyz}{9999}$, it follows that $n + 6$ divides $10^4 - 1 = 3^2 \cdot 11 \cdot 101$. However, $n + 6$ does not divide $10^2 - 1 = 3^2 \cdot 11$, because otherwise the decimal representation of $\frac{1}{n+6}$ would have period 1 or 2. Thus $n = 101k - 6$, where $k = 1, 3, 9, 11, 33,$ or 99. Because $n < 1000$, the only possible values of k are $1, 3,$ and 9, and the corresponding values of n are $95, 297,$ and 903. Of these, only $297 = 3^3 \cdot 11$ divides $10^6 - 1$. Thus $n \in [201, 400]$. It may be checked that $\frac{1}{297} = 0.\overline{003367}$ and $\frac{1}{303} = 0.\overline{0033}$.

23. **Solution** (A). In the first octant, the first inequality reduces to $x + y + z \leq 1$, and the inequality defines the region under a plane that intersects the coordinate axes at $(1, 0, 0)$, $(0, 1, 0)$, and

$(0, 0, 1)$. By symmetry, the first inequality defines the region inside a regular octahedron centered at the origin and having internal diagonals of length 2. The upper half of this octahedron is a pyramid with altitude 1 and a square base of side length $\sqrt{2}$, so the volume of the octahedron is $2 \cdot \frac{1}{3} \cdot \left(\sqrt{2}\right)^2 \cdot 1 = \frac{4}{3}$. The second inequality defines the region obtained by translating the first region up 1 unit. The intersection of the two regions is bounded by another regular octahedron with internal diagonals of length 1. Because the linear dimensions of the third octahedron are half those of the first, its volume is $\frac{1}{8}$ that of the first, or $\frac{1}{6}$.

24. **Solution** (D). Note that $\gcd(a, b, c, d) = 77$ and $\operatorname{lcm}(a, b, c, d) = n$ if and only if $\gcd(\frac{a}{77}, \frac{b}{77}, \frac{c}{77}, \frac{d}{77}) = 1$ and $\operatorname{lcm}(\frac{a}{77}, \frac{b}{77}, \frac{c}{77}, \frac{d}{77}) = \frac{n}{77}$. Thus there are 77,000 ordered quadruples (a, b, c, d) such that $\gcd(a, b, c, d) = 1$ and $\operatorname{lcm}(a, b, c, d) = \frac{n}{77}$. Let $m = \frac{n}{77}$ and suppose that p is a prime that divides m. Let $A = A(p)$, $B = B(p)$, $C = C(p)$, $D = D(p)$, and $M = M(p) \geq 1$ be the exponents of p such that p^A, p^B, p^C, p^D, and p^M are the largest powers of p that divide a, b, c, d, and m, respectively. The gcd and lcm requirements are equivalent to $\min(A, B, C, D) = 0$ and $\max(A, B, C, D) = M$. For a fixed value of M, there are $(M + 1)^4$ quadruples (A, B, C, D) with each entry in $\{0, 1, \ldots, M\}$. There are M^4 of them for which $\min(A, B, C, D) \geq 1$, and also M^4 of them such that $\max(A, B, C, D) \leq M - 1$. Finally, there are $(M - 1)^4$ quadruples (A, B, C, D) such that $\min(A, B, C, D) \geq 1$ and $\max(A, B, C, D) \leq M - 1$. Thus the number of quadruples such that $\min(A, B, C, D) = 0$ and $\max(A, B, C, D) = M$ is equal to $(M + 1)^4 - 2M^4 + (M - 1)^4 = 12M^2 + 2 = 2(6M^2 + 1)$. Multiplying these quantities over all primes that divide m yields the total number of quadruples (a, b, c, d) with the required properties. Thus

$$77{,}000 = 2^3 \cdot 5^3 \cdot 7 \cdot 11 = \prod_{p \mid m} 2(6(M(p))^2 + 1).$$

Note that $6(M(p))^2 + 1$ is odd and this product must contain three factors of 2, so there must be exactly three primes that divide m. Let p_1, p_2, and p_3 be these primes. Note that $6 \cdot 1^2 + 1 = 7$, $6 \cdot 2^2 + 1 = 5^2$, and $6 \cdot 3^2 + 1 = 5 \cdot 11$. None of these could appear as a factor more than once because 77,000 is not divisible by 7^2, 5^4, or 11^2. Moreover, the product of these three is equal to $5^3 \cdot 7 \cdot 11$. All other factors of the form $6M^2 + 1$ are greater than these three, so without loss of generality the only solution is $M(p_1) = 1$, $M(p_2) = 2$, and $M(p_3) = 3$. It follows that $m = p_1^1 p_2^2 p_3^3$, and the smallest value of m occurs when $p_1 = 5$, $p_2 = 3$, and $p_3 = 2$. Therefore the smallest possible values of m and n are $5 \cdot 3^2 \cdot 2^3 = 360$ and $77(5 \cdot 3^2 \cdot 2^3) = 27{,}720$, respectively.

25. **Solution** (A). Express each term of the sequence (a_n) as $2^{\frac{b_n}{19}}$. (Equivalently, let b_n be the logarithm of a_n to the base $\sqrt[19]{2}$.) The recursive definition of the sequence (a_n) translates into $b_0 = 0$, $b_1 = 1$, and $b_n = b_{n-1} + 2b_{n-2}$ for $n \geq 2$. Then the product $a_1 a_2 \cdots a_k$ is an integer if and only if $\sum_{i=1}^k b_i$ is divisible by 19. Let $c_n = b_n \bmod 19$. It follows that $a_1 a_2 \cdots a_k$ is an integer if and only if $p_k = \sum_{i=1}^k c_i$ is divisible by 19. Let $q_k = p_k \bmod 19$. Because the largest answer choice is 21,

it suffices to compute c_k and q_k successively for k from 1 up to at most 21, until q_k first equals 0. The modular computations are straightforward from the definitions.

k	1	2	3	4	5	6	7	8	9	10	11	12	13	14	15	16	17
c_k	1	1	3	5	11	2	5	9	0	18	18	16	14	8	17	14	10
q_k	1	2	5	10	2	4	9	18	18	17	16	13	8	16	14	9	0

Thus the requested answer is 17.

OR

Using standard techniques, the recurrence relation for b_n can be solved to get $b_n = \frac{1}{3}(2^n - (-1)^n)$. Let $S_k = b_1 + b_2 + \cdots + b_k$. Then it is straightforward to show that $S_k = \frac{1}{3}(2^{k+1} - 1)$ for k odd, and $S_k = \frac{2}{3}(2^k - 1)$ for k even. Let $P_k = a_1 a_2 \cdots a_k$. It follows that, for k odd, P_k is an integer if and only if 19 divides $2^{k+1} - 1$; and, for k even, P_k is an integer if and only if 19 divides $2^k - 1$. A little computation shows that this first occurs at $k = 17$, when $2^{18} - 1 = 2^{18} - 1 = (2^9 - 1)(2^9 + 1) = 511 \cdot 513 = 511 \cdot 19 \cdot 27$. (In fact, one can show that P_k is an integer if and only if k is congruent to 0 or -1 mod 18.)

CHAPTER 43

SOLUTIONS TO 2016 AIME

43.1   2016 AIME I Solutions

1. **Solution** (336).

$$
\begin{aligned}
S(a) + S(-a) &= \frac{12}{1-a} + \frac{12}{1+a} = \frac{2}{12} \cdot \frac{12^2}{1-a^2} \\
&= \frac{1}{6} S(a) S(-a) = \frac{1}{6} \cdot 2016 = 336
\end{aligned}
$$

2. **Solution** (071). Because $1 + 2 + 3 + 4 + 5 + 6 = 21$, the probability of rolling a k with one of these dice is $\frac{k}{21}$. Then the probability of rolling a 7 with the pair of dice is

$$
\frac{1 \cdot 6 + 2 \cdot 5 + 3 \cdot 4 + 4 \cdot 3 + 5 \cdot 2 + 6 \cdot 1}{21^2} = \frac{56}{21^2} = \frac{8}{63}.
$$

   The requested sum is $8 + 63 = 71$.

3. **Solution** (810). There are 5 ways to start the path, then 9 ways to continue it on the upper pentagon (1, 2, 3, or 4 steps in either direction, or 0 steps). No matter where the path leaves the upper pentagon, there are 2 ways to move down to the lower pentagon, where it has 9 ways to continue horizontally before the final step down to the bottom vertex. There are $5 \cdot 9 \cdot 2 \cdot 9 = 810$ such paths.

4. **Solution** (108). Label the other vertices of the pyramid B, C, and D, where $\overline{AB}$ and $\overline{AC}$ are sides of a base of the prism. The dihedral angle of $60°$ is formed by faces $\triangle BAC$ and $\triangle BDC$. Let E be the midpoint of $\overline{BC}$. Then $\angle AED = 60°$. Because $\angle BAC = 120°$, it follows that

$\angle EAB = 60°$ and $AE = AB\cos 60° = 6$. Therefore in right $\triangle DAE$, $h = AE\tan 60° = 6\sqrt{3}$. Hence $h^2 = 36 \cdot 3 = 108$.

5. **Solution** (053). Let k represent the number of days it took Anh to finish the book. Then the total number pages read is $n + (n+1) + \cdots + (n+(k-1)) = \frac{k}{2}(2n+k-1)$, and the total number of minutes taken is $t + (t+1) + \cdots + (t+(k-1)) = \frac{k}{2}(2t+k-1)$. Thus $2 \cdot 374 = k(2n+k-1)$ and $2 \cdot 319 = k(2t+k-1)$. Note also that $55 = 374 - 319 = \frac{k}{2} \cdot 2(n-t) = k(n-t)$. Then k must be a common factor of 748, 638, and 55, so k must be a factor of $\gcd(748, 638, 55) = 11$. Because $k > 1$, it follows that $k = 11$. Then $2n + 10 = 68$ and $2t + 10 = 58$. Therefore $n = 29$ and $t = 24$. The requested sum is $29 + 24 = 53$.

6. **Solution** (013). Because I is the incenter of $\triangle ABC$, $\overline{AI}$ is the angle bisector of $\angle BAC$ and $\angle LAI = \angle IAC$. It is also clear that $\angle BAD = \frac{1}{2}\widehat{BD} = \angle BCD$. Thus $\angle AID = \angle DCA + \angle IAC = \angle DCB + \angle LAI = \angle BAD + \angle LAI = \angle DAI$ and $DA = DI = 5$. Because $\angle LAD = \angle BAD = \angle BCD = \angle ACD$ and $\angle ADL = \angle ADC$, $\triangle DAL$ is similar to $\triangle DCA$. Hence $\frac{DA}{DL} = \frac{DC}{DA}$, or $\frac{5}{3} = \frac{5+IC}{5}$ and $IC = \frac{10}{3}$. The requested sum is $10 + 3 = 13$.

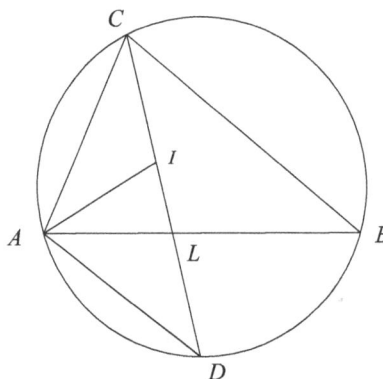

7. **Solution** (103). The complex number is real if either $\frac{\sqrt{ab+2016}}{ab+100}$ is real and $\frac{\sqrt{|a+b|}}{ab+100} = 0$ or $\frac{\sqrt{ab+2016}}{ab+100}$ is imaginary and equals $\frac{\sqrt{|a+b|}}{ab+100}i$. In the former case, $a = -b$, and $\frac{\sqrt{ab+2016}}{ab+100}$ must be real. This will occur if and only if $ab + 2016 \geq 0$ and $ab \neq -100$. So $a^2 \leq 2016$, and there are $2 \cdot \lfloor\sqrt{2016}\rfloor + 1 - 2 = 87$ possible values for a (including 0, and excluding ± 10). In the latter case, $\sqrt{ab+2016} = \sqrt{-|a+b|}$. If $a + b > 0$, then $ab + 2016 = -(a+b)$, which can be rewritten as $(a+1)(b+1) = -2015 = -5 \cdot 13 \cdot 31$. There are 8 integer solutions to this equation with $a + b > 0$. Similarly if $a + b < 0$, then $ab + 2016 = a + b$, which can be rewritten as $(a-1)(b-1) = -2015$. Again, there are 8 integer solutions to this equation with $a + b < 0$. This yields a total of $87 + 8 + 8 = 103$ possible ordered pairs.

8. **Solution** (162). For any permutation p

$$s(p) = 100(a_1 + a_4 + a_7) + 10(a_2 + a_5 + a_8) + (a_3 + a_6 + a_9).$$

Because the units digit of m is 0, it follows that $a_3 + a_6 + a_9 = 10$ or 20.

If $a_3 + a_6 + a_9 = 10$, then $a_1 + a_4 + a_7 + a_2 + a_5 + a_8 = 45 - 10 = 35$, and $s(p) = 90(a_1 + a_4 + a_7) + 360 \geq 90(1 + 2 + 3) + 360 = 900$.

If $a_3 + a_6 + a_9 = 20$, then $s(p) = 90(a_1 + a_4 + a_7) + 270 \geq 90 \cdot 6 + 270 = 810$, with equality if and only if $\{a_1, a_4, a_7\} = \{1, 2, 3\}$. The set $\{a_3, a_6, a_9\}$ must equal $\{4, 7, 9\}$, $\{5, 6, 9\}$, or $\{5, 7, 8\}$.

For each $\{a_i, a_{i+3}, a_{i+6}\}$, there are 3! ways to permute these numbers. Hence there are $n = 3 \cdot 3! \cdot 3! \cdot 3! = 648$ permutations p with $s(p) = m = 810$ and $|m - n| = |810 - 648| = 162$.

9. **Solution** (744). Let α, β, and γ denote the degree measures of $\angle BAC$, $\angle BAQ$, and $\angle CAS$, respectively, so that $\alpha + \beta + \gamma = 90°$. Then $AS = AC \cos \gamma = 31 \cos \gamma$ and $AQ = AB \cos \beta = 40 \cos \beta$. The area of rectangle $AQRS$ is

$$AS \cdot AQ = 31 \cdot 40 \cos \beta \cos \gamma$$
$$= 1240 \cdot \frac{1}{2} \left(\cos(\beta + \gamma) + \cos(\beta - \gamma) \right)$$
$$= 620 \left(\cos(90° - \alpha) + \cos(\beta - \gamma) \right)$$
$$\leq 620 (\sin \alpha + 1).$$

The extreme value is assumed when $\beta = \gamma = \frac{1}{2}(90° - \sin^{-1} \frac{1}{5}) \approx 39.23°$ giving the area $620(\frac{1}{5} + 1) = 744$.

10. **Solution** (504). The ratio $\frac{a_2}{a_1}$ must be rational, so let $a_2 = \frac{ba_1}{a}$, where a and b are relatively prime positive integers and $a < b$. Because $a_3 = \frac{b^2 a_1}{a^2}$ is also an integer, there is an integer c such that $a_1 = ca^2$ and $a_2 = cab$. Thus the sequence begins ca^2, cab, cb^2, $cb(2b - a)$. Examining a few terms of the sequence suggests that for $k \geq 1$

$$a_{2k-1} = c\left((k-1)b - (k-2)a\right)^2 \text{ and}$$
$$a_{2k} = c\left((k-1)b - (k-2)a\right)\left(kb - (k-1)a\right).$$

Then, because the sequence $a_{2k-1}, a_{2k}, a_{2k+1}$ is geometric, it would follow that

$$a_{2k+1} = \frac{a_{2k}^2}{a_{2k-1}} = \frac{c^2\left((k-1)b - (k-2)a\right)^2\left(kb - (k-1)a\right)^2}{c\left((k-1)b - (k-2)a\right)^2} = c\left(kb - (k-1)a\right)^2.$$

It would then follow that for $k \geq 1$ that

$$a_{2k} = c\left((k-1)b - (k-2)a\right)\left(kb - (k-1)a\right) \text{ and}$$
$$a_{2k+1} = c\left(kb - (k-1)a\right)^2.$$

Because $a_{2k}, a_{2k+1}, a_{2k+2}$ is arithmetic, it would follow that

$$
\begin{aligned}
a_{2k+2} &= 2a_{2k+1} - a_{2k} \\
&= 2c\left(kb - (k-1)a\right)^2 - c\left((k-1)b - (k-2)a\right)\left(kb - (k-1)a\right) \\
&= c\left(kb - (k-1)a\right)\left((k+1)b - ka\right).
\end{aligned}
$$

It can now be verified by mathematical induction that for all positive integers k

$$
\begin{aligned}
a_{2k} &= c\left((k-1)b - (k-2)a\right)\left(kb - (k-1)a\right) \text{ and} \\
a_{2k+1} &= c\left(kb - (k-1)a\right)^2.
\end{aligned}
$$

In particular, $a_{13} = c(6b - 5a)^2 = 2016 = 14 \cdot 12^2$. Therefore $6b - 5a$ is a factor of 12 and is also the seventh term in an arithmetic progression whose first two terms are a and b. Let $n = 6b - 5a$. Then $a < a + 6(b - a) = n$, and $6b = 5a + n \equiv n - a \pmod{6}$, implying that $n - a$ is a multiple of 6. Thus $6 < a + 6 \le n \le 12$, and the only solution for (a, b, n) in positive integers is $(6, 7, 12)$. The corresponding value of c is 14, and $a_1 = 14 \cdot 6^2 = 504$.

11. **Solution** (109). The given condition $(x-1)P(x+1) = (x+2)P(x)$ implies that $P(x)$ is divisible by $x - 1$. From this it follows that x divides $P(x + 1)$, and the given condition then implies that x divides $P(x)$. Substituting $x - 1$ in place of x in the given condition yields $(x - 2)P(x) = (x + 1)P(x - 1)$, which implies that $P(x)$ is divisible by $x + 1$. Thus there is a polynomial $L(x)$ such that $P(x) = x(x - 1)(x + 1)L(x)$. Substituting this for $P(x)$ in the given condition yields $(x - 1)(x + 1)x(x + 2)L(x + 1) = (x + 2)x(x - 1)(x + 1)L(x)$ or $L(x + 1) = L(x)$, implying that there is a constant c such that $L(x) = c$ for all x. It follows that $P(x) = c(x^3 - x)$. Now the condition $\left(P(2)\right)^2 = P(3)$ implies that $c = \frac{2}{3}$, so $P(\frac{7}{2}) = \frac{105}{4}$. The requested sum is $105 + 4 = 109$.

12. **Solution** (132). Considering the expression $e(m) = m^2 - m + 11$ modulo 2, 3, 5, and 7 shows that none of these primes can be a factor of $e(m)$. Thus the smallest possible value for $e(m)$ with four prime factors is 11^4, but there is no integer m for which $e(m) = 11^4$. The next candidate for the smallest value for $e(m)$ with four prime factors is $11^3 \cdot 13$. If there actually is an integer m such that $m^2 - m + 11 = 11^3 \cdot 13$, then because $m^2 - m + 11 = m(m - 1) + 11$ is a multiple of 11, there must be an integer k with either $m = 11k$ or $m = 11k + 1$. If $m = 11k$, the equation becomes $11k^2 - k + 1 = 11^2 \cdot 13$. It follows that $-k + 1$ must be divisible by 11. Clearly $k = 1$ does not work, but $k = 12$ satisfies the equation. If $m = 11k + 1$, there are no small values of k that satisfy the needed condition. Therefore the least positive integer m satisfying the needed condition is $m = 11 \cdot 12 = 132$.

13. **Solution** (273). Let $T(h)$ be the expected number of jumps it will take Freddy to reach the river when he is a distance h from it. The problem asks for the value of $T(3)$. Note that $T(0) = 0$, and

for each h with $1 \le h \le 23$ there is a probability of $\frac{1}{2}$ that Freddy will stay the same distance from the river, a probability of $\frac{1}{4}$ that he will get one jump closer to the river, and a probability of $\frac{1}{4}$ that he will get one jump farther away. Thus

$$T(h) = 1 + \frac{1}{4}T(h-1) + \frac{1}{4}T(h+1) + \frac{1}{2}T(h),$$

which simplifies to $2T(h) = 4 + T(h-1) + T(h+1)$. For the special case $h = 24$,

$$T(24) = 1 + \frac{1}{3}T(23) + \frac{2}{3}T(24),$$

which

simplifies to $T(24) = 3 + T(23)$. Summing the equations $2T(h) = 4 + T(h-1) + T(h+1)$ for $1 \le h \le 23$ yields

$$2\sum_{h=1}^{23} T(h) = 4 \cdot 23 + \sum_{h=0}^{22} T(h) + \sum_{h=2}^{24} T(h).$$

This simplifies to $T(1) + T(23) = 92 + T(24)$. Combining this with the equation $T(24) = 3 + T(23)$ yields $T(1) = 95$. From the recurrence $T(h+1) = 2T(h) - T(h-1) - 4$, it follows that $T(2) = 186$ and $T(3) = 273$.

14. **Solution** (574). Let $A = (0,0)$ and $B = (1001, 429) = (143 \cdot 7, 143 \cdot 3)$. Between the points $(0,0)$ and $(7,3)$, $\overline{AB}$ can intersect the square at lattice point (m,n) only if it passes between the upper-left and lower-right corners of the square. That is,

$$n + \frac{1}{10} \ge \frac{3}{7}\left(m - \frac{1}{10}\right) \text{ and } n - \frac{1}{10} \le \frac{3}{7}\left(m + \frac{1}{10}\right).$$

This implies $3m-1 \le 7n \le 3m+1$. The only lattice points (m,n) with $0 \le m \le 7$ which satisfy this requirement are $(0,0)$, $(2,1)$, $(5,2)$, and $(7,3)$. In the cases of $(0,0)$ and $(7,3)$, $\overline{AB}$ passes through the center of the square centered at that point, so it also intersects the circle centered at that point. In the cases of $(2,1)$ and $(5,2)$, $\overline{AB}$ passes through the lower-right and upper-left corners of the square centered at that point, respectively, so it does not interest the circle centered at that point. Altogether the segment joining $(0,0)$ to $(7,3)$ intersects 4 squares and 2 circles. The same conclusion applies to the segment joining $(7k-7, 3k-3)$ to $(7k, 3k)$ for each k with $1 \le k \le 143$. Because the points $(7k, 3k)$ belong to two of these segments for $1 \le k \le 142$, $\overline{AB}$ intersects $4 \cdot 143 - 142 = 430$ of the squares and $2 \cdot 143 - 142 = 144$ of the circles. The requested sum is $430 + 144 = 574$.

15. **Solution** (270). Note that lines AD, XY, BC are the radical axes of pairs of circles ω and ω_1, ω_1 and ω_2, ω_2 and ω, respectively. Therefore lines AD, XY, and BC are either parallel to each other or concurrent. In the former case, ω_1 and ω_2 would be the same size, and by symmetry,

$CX = DX$, contradicting the given condition. Hence it must be that lines AD, XY, and BC are concurrent at a point Z. Denote by M the intersection of segments $\overline{ZY}$ and $\overline{AB}$. By the Power of a Point Theorem it follows that $MA^2 = MX \cdot MY = MB^2$. In particular,

$$AB^2 = 4MA^2 = 4MX \cdot MY = 4MX(MX + XY)$$
$$= (2MX + XY)^2 - XY^2 = (MY + MX)^2 - XY^2.$$

Claim: $(MX + MY)^2 = CX \cdot DX$. The claim implies that $AB^2 = CX \cdot DX - XY^2 = 37 \cdot 67 - 47^2 = 270$.

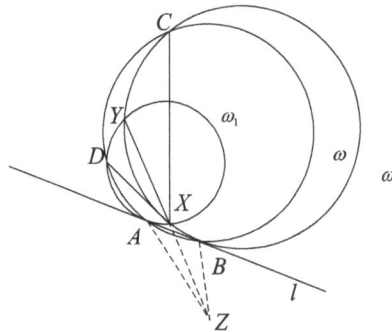

The proof of the claim is based on the following three observations: $ZAXB$ is cyclic; $\triangle XZC$ is similar to $\triangle XDZ$; and $ZAYB$ is a parallelogram.

Because $BCYX$ is cyclic, $\angle XBZ = \angle XYC$. Because $ADYX$ is cyclic, $\angle XAZ = \angle XYD$. Because C, Y, and D are collinear, $\angle XAZ + \angle XBZ = \angle XYD + \angle XYC = 180°$, from which it follows that $ZAXB$ is cyclic, establishing the first observation. This is also a direct consequence of Miquel's Theorem.

Because $BCYX$ and $ZAXB$ are cyclic, $\angle XCB = \angle XYB$ and $\angle ABX = \angle AZX$. Because $\overline{AB}$ is tangent to ω_2 at B, $\angle ABX = \angle XYB$. Combining the three equations yields $\angle XCZ = \angle XCB = \angle XYB = \angle ABX = \angle AZX = \angle DZX$. Likewise, $\angle XZC = \angle XDZ$. Hence $\triangle XZC$ is similar to $\triangle XDZ$, establishing the second observation.

As in the previous paragraph, $\angle XYB = \angle AZX$ or $\overline{BY} \parallel \overline{AZ}$. Similarly, $\overline{AY} \parallel \overline{BZ}$. Thus $ZAYB$ is a parallelogram, establishing the third observation.

Because $ZAYB$ is a parallelogram, $MY = MZ$ and $MX + MY = XM + MZ = XZ$. Because $\triangle XZC$ and $\triangle XDZ$ are similar, $\frac{XZ}{XC} = \frac{XD}{XZ}$ or $XZ^2 = XC \cdot XD$. Combining the last two equations yields $(MX + MY)^2 = XZ^2 = XC \cdot XD$, establishing the claim.

43.2   2016 AIME II Solutions

1. **Solution** (108). After the three people eat the peanuts, $444 - 5 - 9 - 25 = 405$ peanuts remain. Hence after eating the peanuts, for some positive integers b and d, Alex, Betty, and Charlie have

$b - d$, b, and $b + d$ peanuts, respectively. Then $(b - d) + b + (b + d) = 3b = 405$, and $b = 135$. Thus Betty originally had 144 peanuts. Because the initial numbers of peanuts were in geometric progression, for some $r > 1$, Alex originally had $\frac{144}{r}$ peanuts and Charlie originally had $144r$ peanuts. Because there was originally a total of 444 peanuts, it follows that $\frac{144}{r} + 144 + 144r = 444$. The only solution greater than 1 is $r = \frac{4}{3}$. Alex initially had $\frac{144}{r} = 108$ peanuts.

2. **Solution** (107). Let x be the probability that it rains on Saturday and not on Sunday, y be the probability that it rains on Sunday and not on Saturday, and z be the probability that it rains on both days. Then the conditions of the problem imply that $x + z = \frac{2}{5}$, $y + z = \frac{3}{10}$, and the conditional probabilities satisfy $\frac{z}{x+z} = 2 \cdot \frac{y}{1-(x+z)}$ or $\frac{5}{2}z = \frac{10}{3}y$. Thus $z = \frac{4}{3}y$, so $y + \frac{4}{3}y = \frac{3}{10}$ which gives $y = \frac{9}{70}$. The required probability is $x + y + z = \frac{2}{5} + \frac{9}{70} = \frac{37}{70}$. The requested sum is $37 + 70 = 107$.

3. **Solution** (265). The system is equivalent to

$$xyz + \log_5 x = 35$$
$$xyz + \log_5 y = 84$$
$$xyz + \log_5 z = 259.$$

Let $x = 5^\alpha$, $y = 5^\beta$, and $z = 5^\gamma$. Then

$$5^{\alpha+\beta+\gamma} + \alpha = 35$$
$$5^{\alpha+\beta+\gamma} + \beta = 84$$
$$5^{\alpha+\beta+\gamma} + \gamma = 259.$$

Adding these equations yields $3 \cdot 5^{\alpha+\beta+\gamma} + (\alpha+\beta+\gamma) = 378$. Let t be chosen so that $3t = \alpha+\beta+\gamma$. Then $3 \cdot 5^{3t} + 3t = 378$ and $125^t + t = 126$. Because $125^t + t$ is an increasing function of t, there is only one value of t satisfying this equation, and inspection shows that $t = 1$. Thus $\alpha+\beta+\gamma = 3$, implying that $5^3 + \alpha = 35$ and $\alpha = -90$; $5^3 + \beta = 84$ and $\beta = -41$; and $5^3 + \gamma = 259$ and $\gamma = 134$. The requested sum is $|\alpha| + |\beta| + |\gamma| = 90 + 41 + 134 = 265$.

4. **Solution** (180). Let the box contain r red cubes, g green cubes, and y yellow cubes. The given information implies that $r : g : y = 3 : 4 : 5$. Thus every $1 \times b \times c$ layer contains 15 yellow cubes, and each $a \times 1 \times c$ layer contains 15 red cubes. It follows that $bc = 36$ and $ac = 60$, so $abc^2 = 2160 = 2^4 \cdot 3^3 \cdot 5$. The value of abc is minimized when c is chosen to be as large as possible, that is, when $c = \gcd(36, 60) = 2^2 \cdot 3 = 12$. The corresponding values of a and b are 5 and 3, respectively, and the minimum volume of the box is 180. Note that this can be done if

each of the five $1 \times 3 \times 12$ layers is colored in the pattern

$$R \quad R \quad R \quad G \quad G \quad G \quad G \quad Y \quad Y \quad Y \quad Y \quad Y$$
$$R \quad R \quad R \quad G \quad G \quad G \quad G \quad Y \quad Y \quad Y \quad Y \quad Y \; .$$
$$R \quad R \quad R \quad G \quad G \quad G \quad G \quad Y \quad Y \quad Y \quad Y \quad Y$$

5. **Solution** (182). Let $a = BC_0$, $b = AC_0$, and $c = AB$. Then $C_0C_1 = \frac{ab}{c}$, and $\triangle BC_0C_1$ is similar to $\triangle BAC_0$ with ratio of similarity $\frac{BC_0}{AB} = \frac{a}{c}$. Furthermore, for $n \geq 2$, $\triangle BC_{n-1}C_n$ is similar to $\triangle BC_{n-2}C_{n-1}$ with the same ratio, so $C_{n-1}C_n = \frac{ba^n}{c^n}$. The sum of all lengths C_nC_{n-1} is

$$\sum_{n=1}^{\infty} \frac{ba^n}{c^n} = \frac{\frac{ab}{c}}{1 - \frac{a}{c}} = \frac{ab}{c - a}.$$

This is $6p$, so $ab = 6(c-a)(a+b+c) = 6(c^2 - a^2 + bc - ab) = 6(b^2 + bc - ab)$, from which $6c = 7a - 6b$. Squaring both sides gives $36c^2 = 36(a^2 + b^2) = 49a^2 - 84ab + 36b^2$, which implies that $13a - 84b = 0$. Because a and b are relatively prime, it follows that $a = 84$ and $b = 13$. Thus $c = 85$, and $p = 84 + 13 + 85 = 182$.

6. **Solution** (275). The polynomial $P(-x) = 1 + \frac{1}{3}x + \frac{1}{6}x^2$ has nonnegative coefficients equal in absolute value to the coefficients of $P(x)$. The coefficients of $Q(-x) = P(-x)P(-x^3)P(-x^5)P(-x^7)P(-x^9)$ are nonnegative as well because $Q(-x)$ is a product of five polynomials with nonnegative coefficients. Thus the sum of the absolute values of the coefficients of $Q(x)$ is equal to the sum of the coefficients of $Q(-x)$, which is $Q(-1) = P(-1)^5 = \left(\frac{3}{2}\right)^5 = \frac{243}{32}$. The requested sum is $243 + 32 = 275$.

7. **Solution** (840). Without loss of generality, let E and F be the vertices of $EFGH$ that are nearest A and B, respectively, and let I and J lie on $\overline{AB}$ and $\overline{BC}$, respectively. Because $\triangle EIF$ is similar to $\triangle JBI$, it follows that

$$\frac{EF}{IJ} = \frac{EF}{IF + FJ} = \frac{EF}{EI + IF} = \frac{IJ}{IB + BJ} = \frac{IJ}{AI + IB} = \frac{IJ}{AB},$$

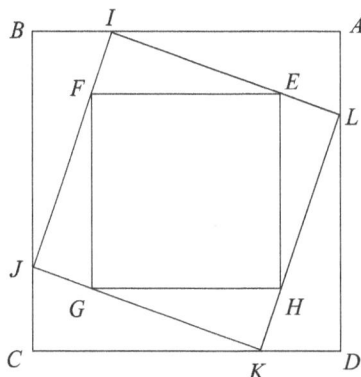

implying that AB, IJ, and EF, in that order, form a decreasing geometric sequence. Hence the three squares have areas $AB^2 = 2016$, $IJ^2 = 2016r$, and $EF^2 = 2016r^2$ for some $0 < r < 1$. For all areas to be integers, r must be rational, and when r^2 is written as a fraction in lowest terms, its denominator must divide $2016 = 2^5 \cdot 3^2 \cdot 7$. Thus the only possible denominators for r written in lowest terms are 2, 3, 4, 6, and 12. Note that if $x = BI$, then

$$IJ^2 = x^2 + (AB - x)^2 = 2x^2 - 2x \cdot AB + AB^2 = 2\left(x - \frac{1}{2}AB\right)^2 + \frac{1}{2}AB^2 \geq \frac{1}{2}AB^2,$$

which implies that $IJ \geq \frac{1}{\sqrt{2}}AB$. Similarly, $EF \geq \frac{1}{\sqrt{2}}IJ$, implying that $EF \geq \frac{1}{2}AB$. Thus the only possible values of r are $\frac{1}{2}, \frac{2}{3}, \frac{3}{4}, \frac{5}{6}, \frac{7}{12}$, and $\frac{11}{12}$. Therefore the difference between the largest and smallest possible values of the area of $IJKL$ is $\left(\frac{11}{12} - \frac{1}{2}\right)(2016) = 840$.

8. **Solution** (728). It is easier to count the ordered triples (a, b, c) of positive integers with the product abc equal to the product $11 \cdot 21 \cdot 31 \cdot 41 \cdot 51 \cdot 61 = 3^2 \cdot 7 \cdot 11 \cdot 17 \cdot 31 \cdot 41 \cdot 61$. Exactly one of a, b, and c is divisible by each of 7, 11, 17, 31, 41, and 61. Either one of a, b, and c is divisible by 9 or exactly two of a, b, and c are divisible by 3. Hence there are $3^6 \cdot (3 + 3) = 2 \cdot 3^7$ such ordered triples (a, b, c). In three of these $2 \cdot 3^7$ ordered triples, two of a, b, and c equal 1. In three of these $2 \cdot 3^7$ ordered triples, two of a, b, and c equal 3. In these six cases a, b, and c are not distinct. In the remaining $2 \cdot 3^7 - 6$ ordered triples, a, b, and c are distinct. Each of the required unordered triples is represented by $3! = 6$ of the $2 \cdot 3^7 - 6$ ordered triples (a, b, c). Therefore the requested number of sets is $\frac{2 \cdot 3^7 - 6}{6} = 3^6 - 1 = 728$.

9. **Solution** (262). Because $c_1 = 2 < 100$, it follows that $k - 1 \geq 2$, so $k \geq 3$. There are integers $d \geq 0$ and $r \geq 1$ such that $a_n = 1 + (n-1)d$ and $b_n = r^{n-1}$, so $100 = c_{k-1} = 1 + (k-2)d + r^{k-2}$ and $1000 = c_{k+1} = 1 + kd + r^k$. Subtracting these equations gives $900 = r^k - r^{k-2} + 2d = r^{k-3}(r-1)r(r+1) + 2d$. Because $(r-1)r(r+1)$ must be a multiple of 3, d is also a multiple of 3. Because $100 = 1 + (k-2)d + r^{k-2}$, it follows that r is also a multiple of 3. The restrictions $r^{k-2} \leq 99$ and $r^k \leq 999$ show that (r, k) must be one of $(3, 3), (3, 4), (3, 5), (3, 6), (6, 3)$, or $(9, 3)$. For the first five of these there is no integer value for d that satisfies all the required conditions, but if $r = 9$ and $k = 3$, then $d = 90$ does satisfy all the required conditions. In this case $c_k = 1 + (3 - 1)90 + 9^{3-1} = 262$.

10. **Solution** (043).

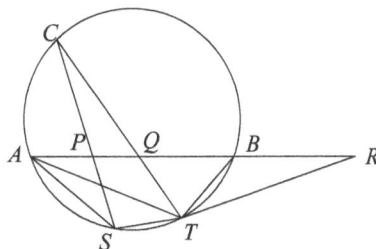

Extend $\overline{AB}$ through B to R so that $BR = 8$. Because $ACBT$ is cyclic, it follows by the Power of a Point Theorem that $CQ \cdot QT = AQ \cdot QB = 42$. Note that $PQ \cdot QR = 42 = CQ \cdot QT$. By the converse of the Power of a Point Theorem, it follows that $CPTR$ is cyclic. Because $CPTR$ and $ACTS$ are cyclic,

$$\angle BRT = \angle PRT = \angle PCT = \angle SCT = \angle SAT.$$

Because $ABTS$ is cyclic, it follows that $\angle AST = \angle RBT$. Hence $\triangle AST$ is similar to $\triangle RBT$, from which $\frac{AS}{ST} = \frac{RB}{BT}$ or $ST = AS \cdot \frac{BT}{RB} = \frac{35}{8}$. The requested sum is $35 + 8 = 43$.

11. **Solution** (749). Consider numbers of the form p^m, where p is a prime and m is a nonnegative integer. Then $(p^m)^k$ has $km + 1$ positive divisors. Thus all numbers N that are one more than a nonnegative multiple of k are k-nice. Conversely every k-nice number N must be one more than a nonnegative multiple of k. This is because if an integer a can be written in the form $p_1^{m_1} p_2^{m_2} \cdots p_i^{m_i}$, where $p_1, p_2, \ldots, p_i$ are distinct primes and $m_1, m_2, \ldots, m_i$ are nonnegative integers, then a^k has $N = (km_1 + 1)(km_2 + 1) \cdots (km_i + 1)$ positive divisors, and by considering each parenthesized factor of N modulo k, it is clear that $N \equiv 1 \pmod{k}$. It follows that there are $\left\lfloor \frac{999-1}{k} \right\rfloor + 1$ positive integers less than 1000 that are k-nice. Thus there are 143 positive integers less than 1000 that are 7-nice, and there are 125 positive integers less than 1000 that are 8-nice. Because 7 and 8 are relatively prime, there are $\left\lfloor \frac{998}{56} \right\rfloor + 1 = 18$ positive integers less than 1000 that are both 7-nice and 8-nice. Thus there are a total of $999 - (143 + 125 - 18) = 749$ positive integers less than 1000 that are neither 7-nice nor 8-nice.

12. **Solution** (732). Consider the problem of painting k regions in a row rather than in a ring. Let A_k be the number of ways to paint k regions in a row so that no two adjacent regions receive the same color and the first and last regions are painted different colors. Let B_k be the number of ways to paint the k regions so that no two adjacent regions receive the same color and the first and last regions are painted the same color. Then $A_1 = 0$ and $B_1 = 4$. Note that for $k \geq 1$, $A_{k+1} = 2A_k + 3B_k$ and $B_{k+1} = A_k$. Thus $A_1 = 0$, $A_2 = 12$, and for $k > 1$, $A_{k+1} = 2A_k + 3A_{k-1}$, which allow the calculation of $A_3 = 24$, $A_4 = 84$, $A_5 = 240$, and $A_6 = 732$. The requested count is equal to $A_6 = 732$. It is easy to verify that $A_k = 3^k + 3(-1)^k$ satisfies the required recursion.

13. **Solution** (371). There are $6! = 720$ configurations. The minimum possible score of 2 occurs when there is a rook on $(1, 1)$, and the maximum possible score of 7 occurs when all the rooks are arranged on the squares $(k, 7 - k)$ for $k = 1, 2, 3, 4, 5, 6$. Let a_n be the number of configurations whose score is exactly n, and let b_n be the number of configurations whose score is at least n. Then the total of all 720 scores is

$$2a_2 + 3a_3 + \cdots + 7a_7 = 2(a_2 + \cdots + a_7) + (a_3 + \cdots + a_7) + \cdots + (a_6 + a_7) + a_7$$
$$= 2b_2 + b_3 + b_4 + b_5 + b_6 + b_7.$$

In each case the rooks are placed in row order; that is, a square is chosen in row 1, then a square in row 2, and so forth.

Because every configuration has a score of at least 2, conclude that $b_2 = 720$. The configurations counted by b_3 do not have a rook in $(1,1)$, so there are 5 choices for where to put the first row's rook, and 5! positions for the remaining rooks, showing that $b_3 = 5 \cdot 5! = 600$. The configurations counted by b_4 can have a rook in any of 4 positions in row 1 and a rook in any of 4 positions in row 2, and there are 4! positions for the remaining rooks, showing that $b_4 = 4 \cdot 4 \cdot 4! = 384$. Using a similar line of reasoning, $b_5 = 3 \cdot 3 \cdot 3 \cdot 3! = 162$, $b_6 = 2 \cdot 2 \cdot 2 \cdot 2 \cdot 2! = 32$, and $b_7 = 1 \cdot 1 \cdot 1 \cdot 1 \cdot 1 \cdot 1! = 1$.

Then the total score is $2 \cdot 720 + 600 + 384 + 162 + 32 + 1 = 2619$. The average is $\frac{2619}{720} = \frac{291}{80}$. The requested sum is $291 + 80 = 371$.

14. **Solution** (450). Let H be the foot of the perpendicular from P to the plane of $\triangle ABC$. Because $PA = PB = PC$, it follows that $HA = HB = HC$; that is, H is the centroid of the equilateral $\triangle ABC$. Likewise H is also the foot of the perpendicular from Q to the plane of $\triangle ABC$. Hence O is the midpoint of $\overline{PQ}$ and $PQ = 2d$.

Let D be the midpoint of side $\overline{AB}$. Hence $\overline{PD} \perp \overline{AB}$ and $\overline{QD} \perp \overline{AB}$, from which it follows that $\angle PDQ$ is the angle formed by planes ABP and ABQ, and so $\angle PDQ = 120°$. Let $\angle PDH = x$ and $\angle QDH = y$. Then $\tan(x+y) = -\sqrt{3}$.

Set $AB = a$. Then $DC = \frac{\sqrt{3}a}{2}$, $DH = \frac{1}{3}DC = \frac{\sqrt{3}a}{6}$, and $PH = \tan x \cdot DH = \frac{\sqrt{3}a}{6}\tan x$. Likewise $QH = \frac{\sqrt{3}a}{6}\tan y$. Hence $2d = PQ = PH + HQ = \frac{\sqrt{3}a}{6}(\tan x + \tan y)$, or

$$\tan x + \tan y = \frac{4\sqrt{3}d}{a}.$$

Because $OP = OC = OQ$, conclude that O, the midpoint of $\overline{PQ}$, is the circumcenter of $\triangle PCQ$, from which it follows that $\angle PCQ = 90°$. Then $\overline{CH}$ is the altitude to the hypotenuse of right $\triangle CPQ$, implying that $CH^2 = PH \cdot QH$. Hence

$$CH^2 = \left(\frac{2}{3}DC\right)^2 = \frac{a^2}{3} = PH \cdot HQ = \frac{a^2 \tan x \tan y}{12},$$

implying that $\tan x \tan y = 4$.

By the Tangent Angle Addition Formula, $\tan(x+y) = \frac{\tan x + \tan y}{1 - \tan x \tan y}$ or

$$-\sqrt{3} = \frac{\frac{4\sqrt{3}d}{a}}{1-4},$$

implying that $d = \frac{3a}{4}$. Substituting $a = 600$ in the last equation yields $d = 450$.

15. **Solution** (863). For $1 \leq i \leq 216$, let $b_i = \sqrt{1 - a_i}$. Because $\sum_{i=1}^{216} a_i = 1$, it follows that $\sum_{i=1}^{216} b_i^2 = 215$. Also, if $\{x_i\}$ is a sequence of positive real numbers with $\sum_{i=1}^{216} x_i = 1$, then $\sum_{1 \leq i < j \leq 216} 2x_i x_j = \left(\sum_{i=1}^{216} x_i\right)^2 - \sum_{i=1}^{216} x_i^2 = 1 - \sum_{i=1}^{216} x_i^2$.

Observe that for each i, $\left(\frac{x_i}{b_i}\right)^2 - \frac{2x_i}{215} + \left(\frac{b_i}{215}\right)^2 = \left(\frac{x_i}{b_i} - \frac{b_i}{215}\right)^2 \geq 0$, and thus summing over $1 \leq i \leq 216$ yields $\sum_{i=1}^{216} \left(\left(\frac{x_i}{b_i}\right)^2 - \frac{2x_i}{215} + \left(\frac{b_i}{215}\right)^2\right) \geq 0$. Because $\sum_{i=1}^{216} b_i^2 = 215$ and $\sum_{i=1}^{216} x_i = 1$, it follows that $\sum_{i=1}^{216} \left(\frac{x_i}{b_i}\right)^2 - \frac{2}{215} + \frac{1}{215^2} \cdot 215 \geq 0$, which is equivalent to $\frac{1}{215} \leq \sum_{i=1}^{216} \left(\frac{x_i}{b_i}\right)^2 = \sum_{i=1}^{216} \frac{x_i^2}{1-a_i}$. Hence $\sum_{1 \leq i < j \leq 216} 2x_i x_j = 1 - \sum_{i=1}^{216} x_i^2 \leq \frac{214}{215} + \sum_{i=1}^{216} \frac{x_i^2}{1-a_i} - \sum_{i=1}^{216} x_i^2 = \frac{214}{215} + \sum_{i=1}^{216} \frac{a_i x_i^2}{1-a_i}$, so $\sum_{1 \leq i < j \leq 216} x_i x_j \leq \frac{107}{215} + \sum_{i=1}^{216} \frac{a_i x_i^2}{2(1-a_i)}$. Equality occurs in this inequality if and only if for each i, $\frac{x_i}{b_i} - \frac{b_i}{215} = 0$ or $x_i = \frac{1-a_i}{215}$. Therefore such a sequence $\{x_i\}$ is unique and $x_2 = \frac{3}{860}$. The requested sum is $3 + 860 = 863$.

CHAPTER 44

SOLUTIONS TO MATHCOUNTS PROBLEMS

1. Solution For fractions of the form $1/n$ where $10^{m-1} < n < 10^m$ the number $1/n$ can be represented as

$$\frac{1}{n} = \sum_{i=0}^{\infty} \frac{(10^m - n)^i}{10^{mi+m}}.$$

In the case $n = 98$, we have

$$\frac{1}{98} = \frac{2^0}{10^2} + \frac{2^1}{10^4} + \frac{2^2}{10^6} + \frac{2^3}{10^8} + \frac{2^4}{10^{10}} + \cdots.$$

So the tenth digit is 6. Check to be sure. A geometric series with first term $a = 1/10^m$ and ratio $r = \dfrac{10^m - n}{10^m}$ converges to the number $\dfrac{a}{1-r}$.

2. Solution Let x_0, x_1, x_2, x_3, and x_4 denote the five entries in the blanks. Since there are ten digits in the problem, we have $x_0 + x_1 + x_2 + x_3 + x_4 = 10$. Now $x_0 = 1$ because there is at least one 0 and there cannot be any other 0's because each of the digits appears at least once. Therefore there are at least two 1's. In case this is all the ones, we have $x_2 + x_3 + x_4 = 7$ and none of these are 1's. The only way for this to happen is that x_2, x_3, x_4 consists of two 2's and a 3. This would mean that there are four 2's (ie, $x_2 = 4$) in the problem, a contradiction. Therefore $x_1 \geq 3$. Now if $x_1 = 3$, we can easily find a solution.

digit	
0	1
1	3
2	2
3	3
4	1

Next consider $x_1 \geq 4$. To see that there is not solution with $x_1 = 4$ note that in this case two of the numbers x_2, x_3, and x_4 must be 1's. But $x_4 \geq 2$ so it must be $x_2 = x_3 = 1$. In this case we have two 4's, so there at least two 2's or at least two 3's. There is a solution for $x_1 = 5$. Its $x_0 = 1, x_1 = 5, x_2 = 1, x_3 = 1, x_4 = 1$.

3. (a) **Solution**

+	0	1	2	3
0	0	1	2	3
1	1	2	3	130
2	2	3	130	131
3	3	130	131	132

×	0	1	2	3
0	0	0	0	0
1	0	1	2	3
2	0	2	130	132
3	0	3	132	121

(b) **Solution** Omitting the -4 subscript, 1, 2, 3, 130, 131, 132, 133, 120, 121, 122, 123, 110, 111, 112.

(c) **Solution** Again omitting the -4 subscript, 13, 12, 11, 10, 23, 22, 21, 20, 33, 32, 31, 30, 1303, 1302.

(d) **Solution** If the leftmost digit is in the first, third, fifth, position, the number is positive. If the leftmost digit is in an even position, the number is negative.

(e) **Solution** If you add two positive numbers, their leftmost digits are in an odd position. If there is no carry, then the sum has a representation whose leftmost digit is in an odd position, and if there is carry, that carry is a two digit carry, and so, again the leftmost digit is in an odd position.

(f) **Solution** Repeated addition of positive numbers results in a positive number. The same argument works for the product of a positive number and a negative number.

Definition. If two symbols represent the integers that have a sum of zero, then the two integers are called *additive inverses* of each other. Note that from questions (1) and (2) we have that 1_{-4} and 13_{-4} represent additive inverses since $1 + 13 = 0$.

(g)

(h) **Solution** $123.32_{-4} = (-4)^2 + 2(-4)^1 + 3(-4)^0 + 3(-4)^{-1} + 2(-4)^{-2} = 11 - 0.75 + 0.125 = 83/8 = 10.375$.

(i) **Solution** Repeated division yields $99 = 13203_{-4}$.

(j) **Solution** Repeated division yields $17.5 = 102.2_{-4}$.

(k) **Solution** $1/2 = 1.2_{-4}$ by inspection, and $1/4 = 1.3_{-4}$. The number $1/16$ is easy to represent because $0.01_4 = 0.01_{-4}$.

(l) **Solution** Since multiplication by 16 just moves the radix point two places, we get an idea of the answer by finding the base -4 representation of $16/3$, which begins $132.....$ Thus we try the number $x = 1.\overline{32}_{-4}$. In the usual way, multiply by $(-4)^2 = 16$ and subtract to get $16x - x = 132.\overline{32} - 1.\overline{32} = 131$, which means that $15x = 5$. So $x = \frac{1}{3}$. Alternatively, write one third as a sum of powers of 4: $1/3 = (1/4) + (1/4)^2 + (1/4)^3 \cdots = 1.3 + .01 + .013 + .0001 + .00013 + .00001 + \cdots = 1.\overline{32}_{-4}$

(m) **Solution** Since $-4 \equiv 1 \pmod 5$, it follows that each base -4 numeral is congruent to the sum of its digits, modulo 5. Thus $d = 3$ works here.

(n) **Solution** The main idea is that of negative carry. Notice that in this problem, the $.1_{-4} + .3_{-4} = -1_{-4} = -1$, so that the addition looks a little like subtraction. $2312.12_{-4} + 13202.31_{-4} = 113.03 - 4$.

(o) **Solution** Recalling that one way to write -1 is 13, it seem reasonable to write zero as 14 for the moment. Note that both 14 and 28 can be thought of as zero. We may therefore consider the string of digits 1402814 as zero. Now place the number 1202313 directly below 1402814 and subtract. We get 200501 which we can rewrite as 213101. To check this, consider the sum $1202313 + 213101$. Now recall that carrying 13 is the same as subtracting 1 from the next column. Thus $1202313 + 213101 = 0$, proving that these two are additive inverses of each other.

(p) **Solution** Because of the last problem, subtraction of 1202313 is the same as adding its additive inverse 213101. So the number we want is

$$
\begin{array}{r}
1\ \ 1\ \ 3\ \ 2\ \ 0\ \ 0\ \ 3 \\
+ \qquad 2\ \ 1\ \ 3\ \ 1\ \ 0\ \ 1 \\
\hline
1\ \ 3\ \ 3\ \ 1\ \ 2\ \ 3\ \ 0
\end{array}
$$

(q) **Solution** $112.3_{-4} \times 33.2_{-4} = 2110.32_{-4}$.

4. (a) **Solution** 35. The value of $\dfrac{7!}{(7-3)!3!}$ is $\dfrac{7!}{4!3!} = \dfrac{7 \cdot 6 \cdot 5}{3 \cdot 2 \cdot 1} = 35$.

(b) **Solution** The number of subsets of a seven element set is the same as the number of ways to *select* three items from a set of seven where we do not distinguish two samples if and only if the order differs, and we sample without replacement. The typical notation for this number is $^{7}C_3$, or C_3^7, or $\binom{7}{3}$. There is a formula for C_r^n. It is $C_r^n = \dfrac{n!}{(n-r)!r!}$. In our special case, this is $\dfrac{7!}{(7-3)!3!} = \dfrac{7\cdot 6\cdot 5}{3\cdot 2\cdot 1} = 35.$

(c) **Solution** Consider the seven numbered blanks

$$\bar{1},\bar{2},\bar{3},\bar{4},\bar{5},\bar{6},\bar{7}$$

Let us agree to select four of these blanks in which to put 1's, and to put dots in the other three places. For example, $11\cdot 1\cdot\cdot 1$. Now this string is a coding for the solution $2,1,0,1)$ of the problem. How many solutions are there? The answer is that there is one solution for each way of selecting the positions in which to put the 1's, or equivalently, one solution for each way to select the three dots, which we know from the last problem is 35.

(d) **Solution** For each solution (a,b,c,d) of problem 2, $(a+1,b+1,c+1,d+1)$ is a solution to problem 3. Thus the answer is 35. Alternatively, consider the seven gaps determined by a string of eight 1's:

$$1\_1\_1\_1\_1\_1\_1\_1\_.$$

Select three of these seven positions in which to insert dots. For example

$$1\underset{.}{\cdot}1\_1\_1\_1\underset{.}{\cdot}1\_1\underset{.}{\cdot}1\_1$$

corresponds to the solution $(1,3,2,2)$.

(e) **Solution** Name the members of the club A,B,C,D,E,F,G, so selecting the three-person committee is the same as the first problem of selecting a three-member subset of a seven member set.

(f) **Solution** Every three points determines a triangle. For example Draw circle, label points, draw triangle.

(g) **Solution** Each path can be coded as a string of 4 u's and 3 r's where u means up and r means right. There are 35 such strings of us and r's.

(h) **Solution** Each point that belongs to two secant lines is determined by the four endpoints of the two secants. Since each set of four points determines exactly one such point, the answer we seek is just the number $\binom{7}{4} = 35$. One such point is shown below.

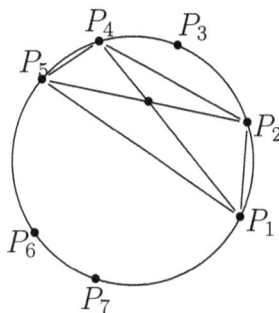

(i) **Solution** Of course this is just $\binom{7}{3} = 35$.

(j) **Solution** Just build Pascal's Triangle down to row 7. Again, we get 35.

CHAPTER 45

SOLUTIONS TO ARML CONTEST PROBLEMS: POWER QUESTION

45.1   Power Question 2009: Sign on the Label

Solutions to 2009 Power Question

1. (a) $(312, 123, 231)$

   (b) There are three: $41352, 42351, 51342$.

   (c) There are four: $352146, 362145, 452136, 562134$.

2. (a) We will prove this by contradiction. Suppose for some other 4-label L we have $S_3[L] = S_3[1234] = (123, 123)$. Write out L as a_1, a_2, a_3, a_4. From the first window of $S_3[L]$, we have $a_1 < a_2 < a_3$. From the second window, we have $a_2 < a_3 < a_4$. Connecting these inequalities gives $a_1 < a_2 < a_3 < a_4$, which forces $L = 1234$, a contradiction. Therefore, the 3-signature above is unique.

   (b) There are 11 others (12 in all, if we include $S_3[1234]$):

   $$S_3[1234] = (123, 123) \quad S_3[1243] = (123, 132) \quad S_3[1324] = (132, 213)$$
   $$S_3[1423] = (132, 213) \quad S_3[2134] = (213, 123) \quad S_3[2314] = (231, 213)$$
   $$S_3[3241] = (213, 231) \quad S_3[3421] = (231, 321) \quad S_3[4132] = (312, 132)$$
   $$S_3[4231] = (312, 231) \quad S_3[4312] = (321, 312) \quad S_3[4321] = (321, 321)$$

   (c) If $S_3[a_1, a_2, a_3, a_4] = (123, 321)$, then the first window forces $a_2 < a_3$, whereas the second window forces $a_2 > a_3$. This is impossible, so the 3-signature $(123, 321)$ is impossible.

(d) There are 18 impossible 3-signatures with two windows. In nine of these, the first window indicates that $a_2 < a_3$ (an increase), but the second window indicates that $a_2 > a_3$ (a decrease). In the other nine, the end of the first window indicates a decrease (that is, $a_2 > a_3$), but the beginning of the second window indicates an increase ($a_2 < a_3$). In general, for a 3-signature to be possible, the end of the first window and beginning of the second window must be consistent, indicating either an increase or a decrease. The impossible 3-signatures are:

$$(123, 321) \quad (123, 312) \quad (123, 213) \quad (132, 231) \quad (132, 132) \quad (132, 123)$$
$$(213, 321) \quad (213, 312) \quad (213, 213) \quad (231, 231) \quad (231, 132) \quad (231, 123)$$
$$(312, 321) \quad (312, 312) \quad (312, 213) \quad (321, 231) \quad (321, 132) \quad (321, 123)$$

3. (a) The first pair indicates an increase; the next three are decreases, and the last pair is an increase. So the 2-signature is $(12, 21, 21, 21, 12)$.

(b) There are several:

$$564312 \quad 564213 \quad 563214 \quad 465312 \quad 465213 \quad 463215$$
$$365412 \quad 365214 \quad 364215 \quad 265413 \quad 265314 \quad 264315$$
$$165413 \quad 165314 \quad 164315$$
$$453216 \quad 354216 \quad 254316 \quad 154326$$

4. In part (a), we can count by brute force, or use the formula from part (b) (with independent proof).

(a) For the case of 5-labels, brute force counting is tractable.

$$\boxed{12543, 13542, 14532, 23541, 24531, 34521}.$$

(b) The answer is $\boxed{\dbinom{2n}{n}}$. The shape of this signature is a wedge: n up steps followed by n down steps. The wedge for $n = 3$ is illustrated below:

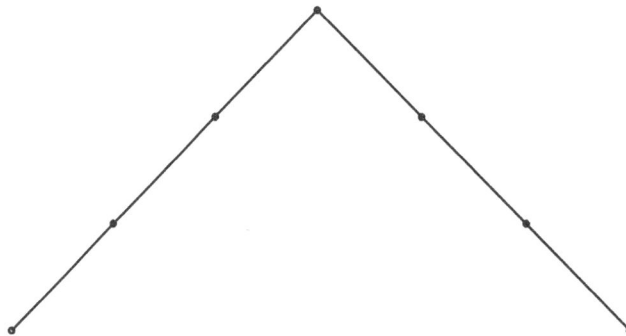

The largest number in the label, $2n + 1$, must be placed at the peak in the center. If we choose the numbers to put in the first n spaces, then they must be placed in increasing order. Likewise, the remaining n numbers must be placed in decreasing order on the downward sloping piece of the shape. Thus there are exactly $\binom{2n}{n}$ such labels.

5. (a) The answer is $\boxed{16}$. We have a shape with two peaks and a valley in the middle. The 5 must go on one of the two peaks, so we place it on the first peak. By the shape's symmetry, we will double our answer at the end to account for the 5-labels where the 5 is on the other peak.

The 4 can go to the left of the 5 or at the other peak. In the first case, shown below left, the 3 must go at the other peak and the 1 and 2 can go in either order. In the latter case, shown below right, the 1, 2, and 3 can go in any of 3! arrangements.

So there are $2! + 3! = 8$ possibilities. In all, there are 16 5-labels (including the ones where the 5 is at the other peak).

(b) The answer is $\boxed{7936}$. The shape of this 2-signature has four peaks and three intermediate valleys:

We will solve this problem by building up from smaller examples. Let f_n equal the number of $(2n + 1)$-labels whose 2-signature consists of n peaks and $n - 1$ intermediate valleys. In part (b) we showed that $f_2 = 16$. In the case where we have one peak, $f_1 = 2$. For the trivial case (no peaks), we get $f_0 = 1$. These cases are shown below.

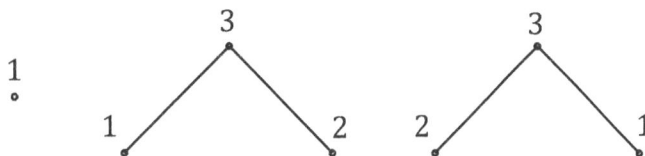

Suppose we know the peak on which the largest number, $2n+1$, is placed. Then that splits our picture into two shapes with fewer peaks. Once we choose which numbers from $1, 2, \ldots, 2n$

to place each shape, we can compute the number of arrangements of the numbers on each shape, and then take the product. For example, if we place the 9 at the second peak, as shown below, we get a 1-peak shape on the left and a 2-peak shape on the right.

For the above shape, there are $\binom{8}{3}$ ways to pick the three numbers to place on the left-hand side, $f_1 = 2$ ways to place them, and $f_2 = 16$ ways to place the remaining five numbers on the right.

This argument works for any $n > 1$, so we have shown the following:

$$f_n = \sum_{k=1}^{n} \binom{2n}{2k-1} f_{k-1} f_{n-k}.$$

So we have:

$$f_1 = \binom{2}{1} f_0^2 = 2$$

$$f_2 = \binom{4}{1} f_0 f_1 + \binom{4}{3} f_1 f_0 = 16$$

$$f_3 = \binom{6}{1} f_0 f_2 + \binom{6}{3} f_1^2 + \binom{6}{5} f_2 f_0 = 272$$

$$f_4 = \binom{8}{1} f_0 f_3 + \binom{8}{3} f_1 f_2 + \binom{8}{5} f_2 f_1 + \binom{8}{7} f_3 f_0 = \boxed{7936}.$$

6. (a) Signature (i) is possible, because it is the 3-signature of 12435. Signature (ii) is impossible. Let a 5-label be a_1, a_2, a_3, a_4, a_5. The second window of (ii) implies $a_3 < a_4$, whereas the third window implies $a_3 > a_4$, a contradiction.

   (b) There are $\boxed{p}$ such windows, regardless of ω_1.

   (c) Because the windows ω_1 and ω_2 overlap for $p - 1$ numbers in L, the first $p - 1$ integers in ω_2 must be in the same relative order as the last $p - 1$ integers in ω_1. Therefore, by choosing the last integer in ω_2 (there are p choices), the placement of the remaining $p - 1$ integers is determined. We only need to show that there exists some L such that $S_p[L] = (\omega_1, \omega_2)$.

   To do so, set $\omega_1 = w_1, w_2, \ldots, w_p$. Provisionally, let $L^{(k)} = w_1, w_2, \ldots, w_p, k + 0.5$ for $k = 0, 1, \ldots, p$. For example, if $p = 4$ and $\omega_1 = 3124$, then $L^{(2)} = 3, 1, 2, 4, 2.5$. First we show that $L^{(k)}$ has the required p-signature; then we can renumber the entries in $L^{(k)}$

in consecutive order to make them all integers, so for example $3, 1, 2, 4, 2.5$ would become $4, 1, 2, 5, 3$.

Even though the last entry in $L^{(k)}$ is not an integer, we can still compute $S_p[L^{(k)}]$, since all we require is that the entries are all distinct. In each $S_p[L^{(k)}]$, the first window is ω_1. When we compare $L^{(k)}$ to $L^{(k+1)}$, the last integer in ω_2 can increase by at most 1, because the last integer in the label jumps over at most one integer in positions 2 through p (those jumps are in boldface):

$$L^{(0)} = 3, 1, 2, 4, 0.5 \qquad S_4[L^{(0)}] = (3124, 234\underline{1})$$
$$L^{(1)} = 3, \mathbf{1}, 2, 4, 1.5 \qquad S_4[L^{(1)}] = (3124, 134\underline{2}) \qquad [\text{jumps over the } 1]$$
$$L^{(2)} = 3, 1, \mathbf{2}, 4, 2.5 \qquad S_4[L^{(2)}] = (3124, 124\underline{3}) \qquad [\text{jumps over the } 2]$$
$$L^{(3)} = 3, 1, 2, 4, 3.5 \qquad S_4[L^{(3)}] = (3124, 124\underline{3}) \qquad [\text{jumps over nothing}]$$
$$L^{(4)} = 3, 1, 2, \mathbf{4}, 4.5 \qquad S_4[L^{(4)}] = (3124, 123\underline{4}) \qquad [\text{jumps over the } 4]$$

The final entry of ω_2 in $S_p[L^{(0)}]$ is 1, because 0.5 is smaller than all other entries in $L^{(0)}$. Likewise, the final entry of ω_2 in $S_p[L^{(p)}]$ is p. Since the final entry increases in increments of 0 or 1 (as underlined above), we must see all p possibilities for ω_2. By replacing the numbers in each $L^{(k)}$ with the integers 1 through $p + 1$ (in the same relative order as the numbers in $L^{(k)}$), we have found the $(p+1)$-labels that yield all p possibilities.

7. (a) The answer is $\boxed{p! \cdot p^{n-p}}$. Call two consecutive windows in a p-signature **compatible** if the last $p - 1$ numbers in the first label and the first $p - 1$ numbers in the second label (their "overlap") describe the same ordering. For example, in the p-signature $(\ldots, 2143, 2431, \ldots)$, 2143 and 2431 are compatible. Notice that the last three digits of 2143 and the first three digits of 2431 can be described by the same 3-label, 132.

Theorem: A signature σ is possible if and only if every pair of consecutive windows is compatible.

Proof: ($\Rightarrow$) Consider a signature σ describing a p-label L. If some pair in σ is not compatible, then there is some string of $p - 1$ numbers in our label L that has two different $(p-1)$-signatures. This is impossible, since the p-signature is well-defined.

(Le *ftarrow*) Now suppose σ is a p-signature such that that every pair of consecutive windows is compatible. We need to show that there is at least one label L with $S_p[L] = \sigma$. We do so by induction on the number of windows in σ, using the results from 5(b).

Let $\sigma = \{\omega_1, \omega_2, \ldots, \omega_{k+1}\}$, and suppose $\omega_1 = a_1, a_2, \ldots, a_p$. Set $L_1 = \omega_1$.

Suppose that L_k is a $(p + k - 1)$-label such that $S_p[L_k] = \{\omega_1, \ldots, \omega_k\}$. We will construct L_{k+1} for which $S_p[L_{k+1}] = \{\omega_1, \ldots, \omega_{k+1}\}$.

As in 5(b), denote by $L_k^{(j)}$ the label L_k with a $j + 0.5$ appended; we will eventually renumber the elements in the label to make them all integers. Appending $j + 0.5$ does not affect any of the non-terminal windows of $S_p[L_k]$, and as j varies from 0 to $p - k + 1$ the final window of $S_p[L_k^{(j)}]$ varies over each of the p windows compatible with ω_k. Since ω_{k+1} is compatible with ω_k, there exists some j for which $S_p[L_k^{(j)}] = \{\omega_1, \ldots, \omega_{k+1}\}$. Now we renumber as follows: set $L_{k+1} = S_{k+p}[L_k^{(j)}]$, which replaces $L_k^{(j)}$ with the integers 1 through $k + p$ and preserves the relative order of all integers in the label.

By continuing this process, we conclude that the n-label L_{n-p+1} has p-signature σ, so σ is possible. $\qquad\qquad\square$

To count the number of possible p-signatures, we choose the first window ($p!$ choices), then choose each of the remaining $n - p$ compatible windows (p choices each). In all, there are $p! \cdot p^{n-p}$ possible p-signatures.

(b) The answer is $\boxed{n = 7, p = 5}$. Let P denote the probability that a randomly chosen p-signature is possible. We are given that $1 - P = 575$, so $P = \dfrac{1}{576}$. We want to find p and n for which

$$\frac{p! \cdot p^{n-p}}{(p!)^{n-p+1}} = \frac{1}{576}$$

$$\frac{p^{n-p}}{(p!)^{n-p}} = \frac{1}{576}$$

$$((p-1)!)^{n-p} = 576.$$

The only factorial that has 576 as an integer power is $4! = \sqrt{576}$. Thus $p = 5$ and $n - p = 2 \Rightarrow n = 7$.

8. (a) The p-signature is not unique because it equals both $S_3[625143]$ and $S_3[635142]$.

(b) Let $L = a_1, a_2, a_3, a_4, a_5, a_6$. We have $a_4 < a_6 < a_5$ (from window #4), $a_3 < a_1 < a_2$ (from window #1), and $a_2 < a_4$ (from window #2). Linking these inequalities, we get

$$a_3 < a_1 < a_2 < a_4 < a_6 < a_5 \quad \Rightarrow \quad L = 231465,$$

so $S_3[L]$ is unique.

(c) 12345 and 54321 are the only ones.

(d) $L = 645132$ will work. First, note that $S_3[645132] = (312, 231, 312, 132)$. Next, we need to show that $S_4[645132] = \{4231, 3412, 4132\}$ is unique. So let $L' = a_1, a_2, a_3, a_4, a_5, a_6$ be a 6-label such that $S_4[L'] = (4231, 3412, 4132)$.

We get $a_4 < a_6 < a_5$ (from window #3), $a_5 < a_2 < a_3$ (from window #2), and $a_3 < a_1$ (from window #1). Linking these inequalities, we get

$$a_4 < a_6 < a_5 < a_2 < a_3 < a_1 \quad \Rightarrow \quad L' = 645132,$$

so $L' = L$, which means $S_4[L]$ is unique.

9. (a) The n-labels with unique 2-signatures are $1, 2, \ldots, n$ and $n, n-1, \ldots, 1$, and their respective 2-signatures are $(12, 12, \ldots, 12)$ and $(21, 21, \ldots, 21)$.

Proof: (Not required for credit.) Let $L = a_1, a_2, \ldots, a_n$. The first signature above implies that $a_1 < a_2 < \cdots < a_n$, which forces $a_1 = 1, a_2 = 2$, and so on. Likewise, the second signature forces $a_1 = n, a_2 = n-1$, and so on.

To show that all other n-labels fail to have unique 2-signatures, we will show that in any other n-label L', there are two numbers k and $k+1$ that are not adjacent. By switching k and $k+1$ we get a label L'' for which $S_2[L'] = S_2[L'']$, since the differences between k and its neighbors in L' were at least 2 (and likewise for $k+1$).

To show that such a k and $k+1$ exist, we proceed by contradiction. Suppose that all such pairs are adjacent in L'. Then 1 and n must be at the ends of L' (or else some intermediate number k will fail to be adjacent to $k+1$ or $k-1$). But if $a_1 = 1$, then that forces $a_2 = 2$, $a_3 = 3, \ldots, a_n = n$. That is, we get the two labels we already covered. (We get the other label if $a_n = 1$.)

Therefore, none of the $n! - 2$ remaining labels has a unique 2-signature.

(b) $S_5[495138627]$ is unique. Let $L = a_1, \ldots, a_9$ and suppose $S_5[L] = S_5[495138627] = (\omega_1, \ldots, \omega_5)$. Then we get the following inequalities:

$a_4 < a_8$	[from ω_4]	$a_8 < a_5$	[from ω_4]
$a_5 < a_1$	[from ω_1]	$a_1 < a_3$	[from ω_1]
$a_3 < a_7$	[from ω_3]	$a_7 < a_9$	[from ω_5]
$a_9 < a_6$	[from ω_5]	$a_6 < a_2$	[from ω_2]

Combining, we get $a_4 < a_8 < a_5 < a_1 < a_3 < a_7 < a_9 < a_6 < a_2$, which forces $a_4 = 1, a_8 = 2, \ldots, a_2 = 9$. So the label L is forced and $S_5[495138627]$ is therefore unique.

(c) The answer is $\boxed{p = 16}$. To show this fact we will need to extend the idea from part 8(b) about "linking" inequalities forced by the various windows:

Theorem: A p-signature for an n-label L is unique if and only if for every $k < n$, k and $k+1$ are in at least one window together. That is, the **distance** between them in the n-label is less than p.

Proof: Suppose that for some k, the distance between k and $k+1$ is p or greater. Then the label L' obtained by swapping k and $k+1$ has the same p-signature, because there are no numbers between k and $k+1$ in any window and because the two numbers never appear in the same window.

If the distance between all such pairs is less than k, we need to show that $S_p[L]$ is unique. For $i = 1, 2, \ldots, n$, let r_i denote the position where i appears in L. For example, if $L = 4123$, then $r_1 = 2, r_2 = 3, r_3 = 4$, and $r_4 = 1$.

Let $L = a_1, a_2, \ldots, a_n$. Since 1 and 2 are in some window together, $a_{r_1} < a_{r_2}$. Similarly, for any k, since k and $k+1$ are in some window together, $a_{r_k} < a_{r_{k+1}}$. We then get a linked inequality $a_{r_1} < a_{r_2} < \cdots < a_{r_n}$, which can only be satisfied if $a_{r_1} = 1, a_{r_2} = 2, \ldots, a_{r_n} = n$. Therefore, $S_p[L]$ is unique. $\square$

From the proof above, we know that the signature is unique if and only if every pair of consecutive integers coexists in at least one window. Therefore, we seek the largest distance between consecutive integers in L. That distance is 15 (from 8 to 9, and from 17 to 18). Thus the smallest p is $\boxed{16}$.

10. Let s_k denote the number of such unique signatures. We proceed by induction with base case $k = 2$. From 8(c), a 2-signature for a label L is unique if and only if consecutive numbers in L appear together in some window. Because $k = 2$, the consecutive numbers must be adjacent in the label. The 2^2-labels 1234 and 4321 satisfy this condition,[1] so their 2^1-signatures are unique. Thus we have shown that $s_2 \geq 2 = 2^{2^2-3}$, and the base case is established.

Now suppose $s_k \geq 2^{2^k-3}$ for some $k \geq 2$. Let L_k be a 2^k-label with a unique 2^{k-1}-signature. Write $L_k = (a_1, a_2, \ldots, a_{2^k})$. We will *expand* L_k to form a 2^{k+1} label by replacing each a_i above with the numbers $2a_i - 1$ and $2a_i$ (in some order). This process produces a valid 2^{k+1}-label, because the numbers produced are all the integers from 1 to 2^{k+1}. Furthermore, different L_k's will produce different labels: if the starting labels differ at place i, then the new labels will differ at places $2i-1$ and $2i$. Therefore, each starting label produces 2^{2^k} distinct 2^{k+1}-labels through this process. Summarizing, each valid 2^k-label can be expanded to produce 2^{2^k} distinct 2^{k+1}-labels, none of which could be obtained by expanding any other 2^k-label.

It remains to be shown that the new label has a unique 2^{k-1}-signature. Because L_k has a unique 2^{k-1}-signature, for all $i \leq 2^k - 1$, both i and $i+1$ appeared in some 2^{k-1}-window. Therefore, there were fewer than $2^{k-1} - 1$ numbers between i and $i+1$. When the label is expanded, $2i$ and $2i-1$ are adjacent, $2i+1$ and $2i+2$ are adjacent, and $2i$ and $2i+1$ are fewer than $2 \cdot (2^{k-1} - 1) + 2 = 2^k$ places apart. Thus, every pair of adjacent integers is within some 2^k-window.

Since each pair of consecutive integers in our new 2^{k+1}-label coexists in some 2^k-window for every possible such expansion of L_k, that means all 2^{2^k} ways of expanding L_k to a 2^{k+1}-label

---

[1] The proof in 8(a) establishes that these are the only 2^2-labels with unique 2-signatures, thus giving $s_2 = 2$. That proof was not required for credit and is not needed here, since the inequality above is good enough for this problem.

result in labels with unique 2^k-signatures. We then get

$$
\begin{aligned}
s_{k+1} &\geq 2^{2^k} \cdot s_k \\
&\geq 2^{2^k} \cdot 2^{2^k-3} \\
&= 2^{2^{k+1}-3},
\end{aligned}
$$

which completes the induction.

45.2 Power Question 2010: Power of Circular Subdivisions

Solutions to 2010 Power Question

1a. By symmetry, P_1, P_2, the two plots sold on day 1, are centered on the y-axis, say at $(0, \pm y)$ with $y > 0$. Let these plots have radius r. Because P_1 is tangent to U, $y + r = 1$. Because P_1 is tangent to C, the distance from $(0, 1 - r)$ to $\left(\frac{1}{2}, 0\right)$ is $r + \frac{1}{2}$. Therefore

$$
\begin{aligned}
\left(\frac{1}{2}\right)^2 + (1 - r)^2 &= \left(r + \frac{1}{2}\right)^2 \\
1 - 2r &= r \\
r &= \frac{1}{3} \\
y = 1 - r &= \frac{2}{3}.
\end{aligned}
$$

Thus the plots are centered at $\left(0, \pm\frac{2}{3}\right)$ and have radius $\frac{1}{3}$.

1b. The four "removed" circles have radii $\frac{1}{2}, \frac{1}{2}, \frac{1}{3}, \frac{1}{3}$ so the combined area of the six remaining curvilinear territories is:

$$
\pi \left(1^2 - \left(\frac{1}{2}\right)^2 - \left(\frac{1}{2}\right)^2 - \left(\frac{1}{3}\right)^2 - \left(\frac{1}{3}\right)^2\right) = \frac{5\pi}{18}.
$$

2a. At the beginning of day 2, there are six c-triangles, so six incircles are sold, dividing each of the six territories into three smaller curvilinear triangles. So a total of 18 curvilinear triangles exist at the start of day 3, each of which is itself divided into three pieces that day (by the sale of a total of 18 regions bounded by the territories' incircles). Therefore there are 54 regions at the end of day 3.

2b. Each day, every curvilinear territory is divided into three smaller curvilinear territories. Let R_n be the number of regions at the end of day n. Then $R_0 = 2$, and $R_{n+1} = 3 \cdot R_n$. Thus R_n is a geometric sequence, so $R_n = 2 \cdot 3^n$. For $n > 0$, the number of plots sold on day n equals the number of territories existing at the end of day $n-1$, i.e., $X_n = R_{n-1}$, so $X_0 = X_1 = 2$, and for $n > 1$, $X_n = 2 \cdot 3^{n-1}$.

2c. The total number of plots sold up to and including day n is

$$
\begin{aligned}
2 + \sum_{k=1}^{n} X_k &= 2 + 2 \sum_{k=1}^{n} 3^{k-1} \\
&= 2 + 2 \cdot (1 + 3 + 3^2 + \ldots + 3^{n-1}) \\
&= 3^n + 1.
\end{aligned}
$$

Alternatively, proceed by induction: on day 0, there are $2 = 3^0 + 1$ plots sold, and for $n \geq 0$,

$$
\begin{aligned}
(3^n + 1) + X_{n+1} &= (3^n + 1) + 2 \cdot 3^n \\
&= 3 \cdot 3^n + 1 \\
&= 3^{n+1} + 1.
\end{aligned}
$$

3a. Use Descartes' Circle Formula with $a = b = 1$ and $c = \dfrac{3}{2}$ to solve for d:

$$
\begin{aligned}
2 \cdot \left(1^2 + 1^2 + \left(\frac{3}{2}\right)^2 + d^2 \right) &= \left(1 + 1 + \frac{3}{2} + d \right)^2 \\
\frac{17}{2} + 2d^2 &= \frac{49}{4} + 7d + d^2 \\
d^2 - 7d - \frac{15}{4} &= 0,
\end{aligned}
$$

from which $d = \dfrac{15}{2}$ or $d = -\dfrac{1}{2}$. These values correspond to radii of $\dfrac{2}{15}$, a small circle nestled between the other three, or 2, a large circle enclosing the other three.

Alternatively, start by scaling the kingdom with the first four circles removed to match the situation given. Thus the three given circles are internally tangent to a circle of radius $r = 2$ and curvature $d = -\dfrac{1}{2}$. Descartes' Circle Formula gives a quadratic equation for d, and the sum of the roots is $2 \cdot \left(1 + 1 + \dfrac{3}{2} \right) = 7$, so the second root is $7 + \dfrac{1}{2} = \dfrac{15}{2}$, corresponding to a circle of radius $r = \dfrac{2}{15}$.

3b. Apply Descartes' Circle Formula to yield

$$
(a + b + c + x)^2 = 2 \cdot (a^2 + b^2 + c^2 + x^2),
$$

a quadratic equation in x. Expanding and rewriting in standard form yields the equation

$$
x^2 - px + q = 0
$$

where $p = 2(a + b + c)$ and $q = 2(a^2 + b^2 + c^2) - (a + b + c)^2$.

The discriminant of this quadratic is

$$\begin{aligned}
p^2 - 4q &= 8(a+b+c)^2 - 8(a^2+b^2+c^2) \\
&= 16(ab + ac + bc).
\end{aligned}$$

This last expression is positive because it is given that $a, b, c > 0$. Therefore the quadratic has two distinct real roots, say d_1 and d_2. These usually correspond to two distinct radii, $r_1 = \dfrac{1}{|d_1|}$ and $r_2 = \dfrac{1}{|d_2|}$.

There are two possible exceptions. If $(a, b, c, 0)$ satisfies Descartes' Circle Formula, then one of the radii is undefined. The other case to consider is if $r_2 = r_1$, which would occur if $d_2 = -d_1$. This case can be ruled out because $d_1 + d_2 = p = 2(a+b+c)$, which must be positive if $a, b, c > 0$. (Notice too that this inequality rules out the possibility that both circles have negative curvature, so that there cannot be two distinct circles to which the given circles are internally tangent.)

When both roots d_1 and d_2 are positive, the three given circles are externally tangent to both fourth circles. When one is positive and one is negative, the three given circles are internally tangent to one circle and externally tangent to the other.

While the foregoing answers the question posed, it is interesting to examine the result from a geometric perspective: why are there normally two possible fourth circles? Consider the case when one of a, b, and c is negative (i.e., two circles are internally tangent to a third). Let A and B be circles internally tangent to C. Then A and B partition the remaining area of C into two c-triangles, each of which has an incircle, providing the two solutions.

If, as in the given problem, $a, b, c > 0$, then all three circles A, B, and C, are mutually externally tangent. In this case, the given circles bound a c-triangle, which has an incircle, corresponding to one of the two roots. The complementary arcs of the given circles bound an infinite region, and this region normally contains a second circle tangent to the given circles. To demonstrate this fact geometrically, consider shrink-wrapping the circles: the shrink-wrap is the border of the smallest convex region containing all three circles. (This region is called the *convex hull* of the circles). There are two cases to address. If only two circles are touched by the shrink-wrap, then one circle is wedged between two larger ones and completely enclosed by their common tangents. In such a case, a circle can be drawn so that it is tangent to all three circles as shown in the diagram below (shrink-wrap in bold; locations of fourth circle marked at D_1 and D_2).

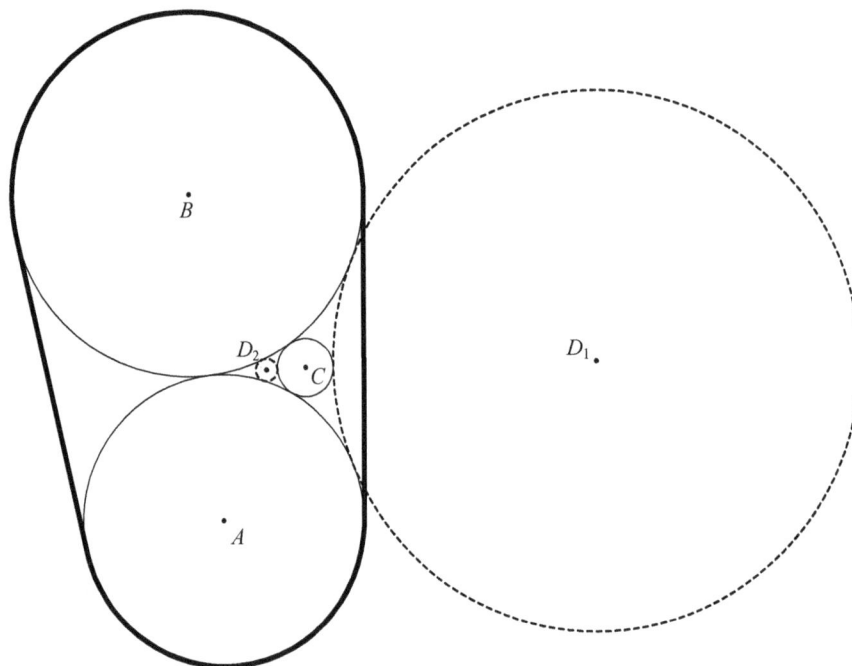

On the other hand, if the shrink-wrap touches all three circles, then it can be expanded to make a circle tangent to and containing A, B, and C, as shown below.

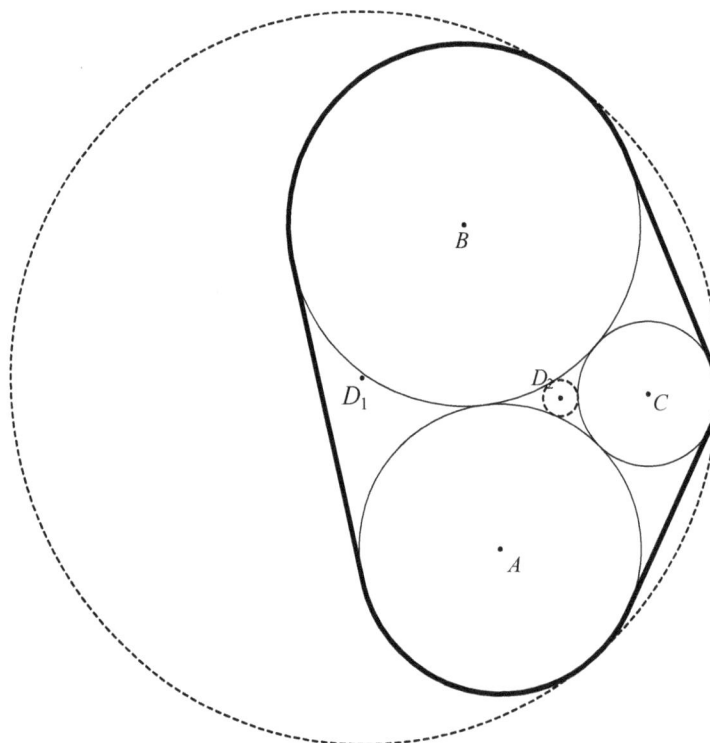

The degenerate case where $(a, b, c, 0)$ satisfies Descartes' Circle Formula is treated in 3c below.

One final question is left for the reader to investigate. Algebraically, it is possible that there is a double root if $p^2 - 4q = 0$. To what geometric situation does this correspond, and under what (geometric) conditions can it arise?

3c. In this case, the fourth "circle" is actually a line tangent to all three circles, as shown in the diagram below.

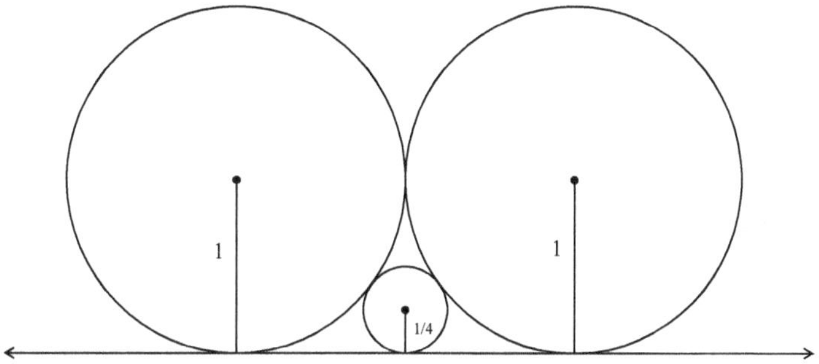

4a. Note that

$$\frac{1}{\rho} = \frac{\phi^2 - \phi}{\phi + \sqrt{\phi}}$$
$$= \phi - \sqrt{\phi}.$$

Therefore

$$\left(\rho - \frac{1}{\rho}\right)^2 = \left(2\sqrt{\phi}\right)^2 = 4\phi$$
$$= 2\left(\rho + \frac{1}{\rho}\right).$$

Multiplying both sides of the equation by ρ^2 gives $(\rho^2 - 1)^2 = 2(\rho^3 + \rho)$. Expand and isolate ρ^4 to obtain $\rho^4 = 2\rho^3 + 2\rho^2 + 2\rho - 1$.

Alternate Proof: Because $\phi^2 = \phi + 1$, any power of ρ can be expressed as an integer plus integer multiples of $\sqrt{\phi}$, ϕ, and $\phi\sqrt{\phi}$. In particular,

$$\rho^2 = \phi^2 + 2\phi\sqrt{\phi} + \phi$$
$$= 2\phi\sqrt{\phi} + 2\phi + 1,$$
$$\rho^3 = 4\phi\sqrt{\phi} + 5\phi + 3\sqrt{\phi} + 4, \text{ and}$$
$$\rho^4 = 12\phi\sqrt{\phi} + 16\phi + 8\sqrt{\phi} + 9.$$

Therefore

$$
\begin{aligned}
2\rho^3 + 2\rho^2 + 2\rho - 1 &= 2\left(4\phi\sqrt{\phi} + 5\phi + 3\sqrt{\phi} + 4\right) + 2\left(2\phi\sqrt{\phi} + 2\phi + 1\right) + 2\rho - 1 \\
&= 12\phi\sqrt{\phi} + 16\phi + 8\sqrt{\phi} + 9 \\
&= \rho^4.
\end{aligned}
$$

4b. If the radii are in geometric progression, then so are their reciprocals (i.e., curvatures). Without loss of generality, let $(a, b, c, d) = (a, ar, ar^2, ar^3)$ for $r > 1$. By Descartes' Circle Formula,

$$
\left(a + ar + ar^2 + ar^3\right)^2 = 2\left(a^2 + a^2 r^2 + a^2 r^4 + a^2 r^6\right).
$$

Cancel a^2 from both sides of the equation to obtain

$$
\left(1 + r + r^2 + r^3\right)^2 = 2\left(1 + r^2 + r^4 + r^6\right).
$$

Because $1 + r + r^2 + r^3 = (1 + r)(1 + r^2)$ and $1 + r^2 + r^4 + r^6 = (1 + r^2)(1 + r^4)$, the equation can be rewritten as follows:

$$
\begin{aligned}
(1 + r)^2 \left(1 + r^2\right)^2 &= 2\left(1 + r^2\right)\left(1 + r^4\right) \\
(1 + r)^2 \left(1 + r^2\right) &= 2\left(1 + r^4\right) \\
r^4 - 2r^3 - 2r^2 - 2r + 1 &= 0.
\end{aligned}
$$

Using the identity from 4a, $r = \rho$ is one solution; because the polynomial is palindromic, another real solution is $r = \rho^{-1} = \phi - \sqrt{\phi}$, but this value is less than 1. The product of the corresponding linear factors is $r^2 - 2\phi r + \phi^2 - \phi = r^2 - 2\phi r + 1$. Division verifies that the other quadratic factor of the polynomial is $r^2 + (2\phi - 2)r + 1$, which has no real roots because $(\sqrt{5} - 1)^2 < 1$.

5. The equation $(x + b + c + d)^2 = 2\left(x^2 + b^2 + c^2 + d^2\right)$ is quadratic with two solutions. Call them a and a'. These are the curvatures of the two circles which are tangent to circles with curvatures b, c, and d. Rewrite the equation in standard form to obtain $x^2 - 2\left(b + c + d\right)x + \ldots = 0$. Using the sum of the roots formula, $a + a' = 2(b + c + d) = 2(s - a)$. So $a' = 2s - 3a$, and therefore

$$
\begin{aligned}
s' &= a' + b + c + d \\
&= 2s - 3a + s - a \\
&= 3s - 4a.
\end{aligned}
$$

6. Day 1 starts with circles of curvature $-1, 2, 2$ bounding C and C'. The geometers mark off P, P' with curvature 3 yielding configurations $(-1, 2, 2, 3)$, $s = 6$. Then the king sells two plots of curvature 3.

a. To find plots sold on day 2 start with the configuration $(-1, 2, 2, 3)$ and compute the three distinct curvatures of circles conjugate to one of the circles in this configuration. Because $s = 6$, the curvatures are $2 \cdot 6 - 3(-1) = 15$, $2 \cdot 6 - 3 \cdot 2 = 6$, and $2 \cdot 6 - 3 \cdot 3 = 3$. However, if P is the circle of curvature 3 included in the orientation, then P' is the new conjugate circle of curvature 3, which was also marked off on day 2. Thus the only options are 15 and 6. In the first case, $s' = 3s - 4a = 3 \cdot 6 - 4(-1) = 22$; in the second, $s' = 3s - 4a = 3 \cdot 6 - 4 \cdot 2 = 10$.

b. On day 2, six plots are sold: two with curvature 15 from the configuration $(2, 2, 3, 15)$, and four with curvature 6 from the configuration $(-1, 2, 3, 6)$. The total area sold on day 2 is therefore

$$2 \cdot \frac{\pi}{15^2} + 4 \cdot \frac{\pi}{6^2} = \frac{3}{25}\pi,$$

which is exactly 12% of the unit circle.

c. Day 3 begins with two circles of curvature 15 from the configuration $(2, 2, 3, 15)$, and four circles of curvature 6 from the configuration $(-1, 2, 3, 6)$. Consider the following two cases:

Case 1: $(a, b, c, d) = (2, 2, 3, 15)$, $s = 22$

- $a = 2 : a' = 2s - 3a = \mathbf{38}$
- $b = 2 : b' = 2s - 3b = \mathbf{38}$
- $c = 3 : c' = 2s - 3c = \mathbf{35}$
- $d = 15 : d' = 2s - 3d = -1$, which is the configuration from day 1.

Case 2: $(a, b, c, d) = (-1, 2, 3, 6)$, $s = 10$

- $a = -1 : a' = 2s - 3a = \mathbf{23}$
- $b = 2 \ : b' = 2s - 3b = \mathbf{14}$
- $c = 3 \ : c' = 2s - 3c = \mathbf{11}$
- $d = 6 \ : d' = 2s - 3d = 2$, which is the configuration from day 1.

So the areas of the plots removed on day 3 are:

$$\frac{\pi}{38^2}, \ \frac{\pi}{35^2}, \ \frac{\pi}{23^2}, \ \frac{\pi}{14^2}, \ \text{and} \ \frac{\pi}{11^2}.$$

There are two circles with area $\dfrac{\pi}{35^2}$, and four circles with each of the other areas, for a total of 18 plots.

d. Because 18 plots were sold on day 3, the mean curvature is

$$\frac{2(38 + 38 + 35) + 4(23 + 14 + 11)}{18} = 23.$$

7. Proceed by induction. The base case, that all curvatures prior to day 2 are integers, was shown in problem 1a. Using the formula $a' = 2s - 3a$, if a, b, c, d, and s are integers on day n, then a', b', c', and d' are integer curvatures on day $n + 1$, proving inductively that all curvatures are integers.

8a. Notice that substituting az_A, bz_B, cz_C, dz_D for a, b, c, d respectively in the derivation of the formula in problem 4 leaves the algebra unchanged, so in general, $a'z_{A'} = 2\hat{s} - 3az_A$, and similarly for $b'z_{B'}, c'z_{C'}, d'z_{D'}$.

8b. It suffices to show that for each circle C with curvature c and center z_C (in the complex plane), cz_C is of the form $u + iv$ where u and v are integers. If this is the case, then each center is of the form $\left(\dfrac{u}{c}, \dfrac{v}{c}\right)$.

Proceed by induction. To check the base case, check the original kingdom and the first four plots: $U, C, C', P_1,$ and P_2. Circle U is centered at $(0,0)$, yielding $-1 \cdot z_U = 0 + 0i$. Circles C and C' are symmetric about the y-axis, so it suffices to check just one of them. Circle C has radius $\dfrac{1}{2}$ and therefore curvature 2. It is centered at $\left(\dfrac{1}{2}, \dfrac{0}{2}\right)$, yielding $2z_C = 2\left(\dfrac{1}{2} + 0i\right) = 1$. Circles P_1 and P_2 are symmetric about the x-axis, so it suffices to check just one of them. Circle P_1 has radius $\dfrac{1}{3}$ and therefore curvature 3. It is centered at $\left(\dfrac{0}{3}, \dfrac{2}{3}\right)$, yielding $3z_{P_3} = 0 + 2i$.

For the inductive step, suppose that $a\,z_A, b\,z_B, c\,z_C, d\,z_D$ have integer real and imaginary parts. Then by closure of addition and multiplication in the integers, $a'z_{A'} = 2\hat{s} - 3a\,z_A$ also has integer real and imaginary parts, and similarly for $b'z_{B'}, c'z_{C'}, d'z_{D'}$.

So for all plots A sold, $a\,z_A$ has integer real and imaginary parts, so each is centered at $\left(\dfrac{u}{c}, \dfrac{v}{c}\right)$ where u and v are integers, and c is the curvature.

9a. Let $s = a + b + c + d$. From problem 5, it follows that

$$\begin{aligned} \mathcal{C}(P^{(1)}) &= (b, c, d, a'), \\ \mathcal{C}(P^{(2)}) &= (a, c, d, b'), \\ \mathcal{C}(P^{(3)}) &= (a, b, d, c'), \end{aligned}$$

where $a' = 2s - 3a$, $b' = 2s - 3b$, and $c' = 2s - 3c$. Because $a \le b \le c$, it follows that $c' \le b' \le a'$. Therefore $P^{(3)}$ dominates $P^{(2)}$ and $P^{(2)}$ dominates $P^{(1)}$.

9b. Because $\mathcal{C}(P^{(1)}) = (b, c, d, a')$ and $\mathcal{C}(Q^{(1)}) = (x, y, z, w')$, it is enough to show that $a' \le w'$. As in the solution to problem 5, a and a' are the two roots of the quadratic given by Descartes' Circle Formula:

$$(X + b + c + d)^2 = 2(X^2 + b^2 + c^2 + d^2).$$

Solve by completing the square:

$$\begin{aligned} X^2 - 2(b+c+d)X + 2(b^2+c^2+d^2) &= (b+c+d)^2; \\ \left(X - (b+c+d)\right)^2 &= 2(b+c+d)^2 - 2(b^2+c^2+d^2) \\ &= 4(bc + bd + cd). \end{aligned}$$

Thus $a, a' = b + c + d \pm 2\sqrt{bc + bd + cd}$.

Because $a \leq b \leq c \leq d$, and only a can be less than zero, a must get the minus sign, and a' gets the plus sign:

$$a' = b + c + d + 2\sqrt{bc + bd + cd}.$$

Similarly,

$$w' = x + y + z + 2\sqrt{xy + xz + yz}.$$

Because P dominates Q, each term in the expression for a' is less than or equal to the corresponding term in the expression for w', thus $a' \leq w'$.

9c. Because $\mathcal{C}(P^{(3)}) = (a, b, d, c')$ and $\mathcal{C}(Q^{(3)}) = (w, x, z, y')$, it suffices to show that $c' \leq y'$. If $a \geq 0$, then the argument is exactly the same as in problem 9a, but if $a < 0$, then there is more to be done.

Arguing as in 9b, $c, c' = a + b + d \pm 2\sqrt{ab + ad + bd}$. If $a < 0$, then the other three circles are *internally* tangent to the circle of curvature a, so this circle has the largest radius. In particular, $\frac{1}{|a|} > \frac{1}{b}$. Thus $b > |a| = -a$, which shows that $a + b > 0$. Therefore c must get the minus sign, and c' gets the plus sign. The same argument applies to y and y'.

When $a < 0$, it is also worth considering whether the square roots are defined (and real). In fact, they are. Consider the diameters of the circles with curvatures b and d along the line through the centers of these circles. These two diameters form a single segment inside the circle with curvature a, so the sum of the diameters is at most the diameter of that circle: $\frac{2}{b} + \frac{2}{d} \leq \frac{2}{|a|}$. It follows that $-ad - ab = |a|d + |a|b \leq bd$, or $ab + ad + bd \geq 0$. This is the argument of the square root in the expressions for c and c'. An analogous argument shows that the radicands are nonnegative in the expressions for b and b'.

The foregoing shows that

$$c' = a + b + d + 2\sqrt{ab + ad + bd}$$

and, by an analogous argument for $w < 0$,

$$y' = w + x + z + 2\sqrt{wx + wz + xz}.$$

It remains to prove that $c' \leq y'$. Note that only a and w may be negative; b, c, d, x, y, and z are all positive. There are three cases.

(i) If $0 \leq a \leq w$, then $ab \leq wx$, $ad \leq wz$, and $bd \leq xz$, so $c' \leq y'$.

(ii) If $a < 0 \leq w$, then $ab + ad + bd \leq bd$, and $bd \leq xz \leq wx + wz + xz$, so $c' \leq y'$. (As noted above, both radicands are nonnegative.)

(iii) If $a \leq w < 0$, then it has already been established that $a + b$ is positive. Analogously, $a + d$, $w + x$, and $w + z$ are positive. Furthermore, $a^2 \geq w^2$. Thus $(a+b)(a+d) - a^2 \leq (w+x)(w+z) - w^2$, which establishes that $ab + ad + bd \leq wx + wz + xz$, so $c' \leq y'$.

10a. First, show by induction that every c-triangle on every day in the kingdom is proper. For the base case, both c-triangles at the end of day 0 have configuration $(-1, 2, 2, 3)$, so they are proper. For the inductive step, let T be a proper c-triangle. If $\mathcal{C}(T) = (a, b, c, d)$, then the three c-triangles obtained from T on the next day have configurations $\mathcal{C}(T^{(1)}) = (b, c, d, a')$, $\mathcal{C}(T^{(2)}) = (a, c, d, b')$, and $\mathcal{C}(T^{(3)}) = (a, b, d, c')$. According to problem 5, $c' = 2a + 2b - c + 2d = 2(a+b) + (d-c) + d$. Arguing as in the proof of 9c, $a + b \geq 0$. By the inductive hypothesis, T is proper, $d - c \geq 0$; therefore $c' \geq d$. Because $a' \geq b' \geq c'$, all three c-triangles are proper.

If $n = 0$, then both circles sold on day n have curvature 2, and this fits the formula: $n^2 + 2 = 0^2 + 2 = 2$.

Let P_0 be one of the c-triangles left at the end of day 0. For $m > 0$, let $P_m = P_{m-1}^{(3)}$. Use induction to prove the following two claims:

(i) $\mathcal{C}(P_m) = (-1, 2, m^2 + 2, (m+1)^2 + 2)$.

(ii) P_m dominates all c-triangles left at the end of day m.

For the moment, grant these two claims. Then (ii) implies that the incircle of P_{n-1} is at least as large as any plot sold on day n, and (i) shows that this incircle has curvature $n^2 + 2$.

For the base case, both c-triangles at the end of day 0 have associated circle configuration $\mathcal{C}(P_0) = (-1, 2, 2, 3) = (-1, 2, 0^2 + 2, 1^2 + 2)$, so either dominates the other.

For $m > 0$, assume inductively that $\mathcal{C}(P_{m-1}) = (-1, 2, (m-1)^2 + 2, m^2 + 2) = (a, b, c, d)$. Because $P_m = P_{m-1}^{(3)}$, $\mathcal{C}(P_m) = (a, b, d, c') = (-1, 2, m^2 + 2, c')$. Use algebra and the result of problem 5 to obtain $c' = 2(a + b + c + d) - 3c = (m+1)^2 + 2$. This completes the inductive step for (i).

Now let Q be any c-triangle left at the end of day $m - 1$, with $\mathcal{C}(Q) = (x, y, z, w)$. Any c-triangle left at the end of day m is of the form $Q^{(1)}$, $Q^{(2)}$, or $Q^{(3)}$ for some such Q. By the inductive hypothesis, P_{m-1} dominates Q. It has already been established that these c-triangles are both proper, so the results of problem 9 apply. By 9c, $P_m = P_{m-1}^{(3)}$ dominates $Q^{(3)}$, and by 9a, $Q^{(3)}$ dominates $Q^{(1)}$ and $Q^{(2)}$. This completes the inductive step for (ii).

10b. Let R_n be a c-triangle with configuration

$$\mathcal{C}(R_n) = (2\rho^{n-2}, 2\rho^{n-1}, 2\rho^n, 2\rho^{n+1}).$$

According to problem 4, these four numbers (a geometric progression with common ratio ρ) satisfy Descartes' Circle Formula, so there is such a c-triangle. The following inductive argument proves that for $n \geq 2$, each c-triangular plot remaining in the kingdom at the end of day n dominates R_n.

For the base case, problem 6 shows that the following are sufficient: $2 \leq 2\rho^0$, $3 \leq 2\rho^1$, $15 \leq 2\rho^2$, and $38 \leq 2\rho^3$. In fact, it is enough to calculate $\phi = \dfrac{1 + \sqrt{5}}{2} > 1.6$, $\sqrt{\phi} > 1.2$, $\rho = \phi + \sqrt{\phi} > 2.8$ to conclude that $2\rho > 5.6$, $2\rho^2 > 15.68$, and $2\rho^3 > 43.9$. These same calculations show that the main result is true for $n \leq 2$.

For the inductive step, let T be a c-triangle in the kingdom remaining at the end of day n. The inductive hypothesis is that T dominates R_n. By problems 9a and 9b, all the c-triangles obtained from T on day $n+1$ dominate $R_n^{(1)}$. All that remains is to show that $R_n^{(1)}$ has the right configuration, so that $R_{n+1} = R_n^{(1)}$.

One approach is to use the formulas from problems 4a and 5 to show that $2(2\rho^{n-2} + 2\rho^{n-1} + 2\rho^n + 2\rho^{n+1}) - 3 \cdot 2\rho^{n-2} = 2\rho^{n+2}$. A method that avoids calculation is to note that $\mathcal{C}(R_n^{(1)}) = (2\rho^{n-1}, 2\rho^n, 2\rho^{n+1}, x)$ for some x. Because Descartes' Circle Formula is quadratic in x, there are at most two possibilities. According to 4b, two solutions are given by $x = 2\rho^{n-2}$ and $x = 2\rho^{n+2}$, because both of these give geometric progressions with common ratio ρ. The first corresponds to R_n, so the second must correspond to $R_n^{(1)}$. This completes the induction.

If $n \leq 2$, then (as already noted) the solution to problem 6 shows that the curvature of the smallest plot sold on day n does not exceed $2\rho^n$. If $n > 2$, then this smallest plot is the incircle of some c-triangle T that remains at the end of day $n - 1$, with $n - 1 \geq 2$. Because T dominates R_{n-1}, the curvature of its incircle does not exceed that of R_{n-1}, which is $2\rho^n$.

45.3 Power Question 2011: Power of Triangles

Solutions to 2011 Power Question

1a.

$$
\begin{aligned}
\mathrm{Pa}(1,1) + \mathrm{Pa}(2,1) + \mathrm{Pa}(3,1) + \mathrm{Pa}(4,1) + \mathrm{Pa}(5,1) &= 1 + 2 + 3 + 4 + 5 &= \mathbf{15} \\
\mathrm{Pa}(2,2) + \mathrm{Pa}(3,2) + \mathrm{Pa}(4,2) + \mathrm{Pa}(5,2) + \mathrm{Pa}(6,2) &= 1 + 3 + 6 + 10 + 15 &= \mathbf{35} \\
\mathrm{Pa}(3,3) + \mathrm{Pa}(4,3) + \mathrm{Pa}(5,3) + \mathrm{Pa}(6,3) + \mathrm{Pa}(7,3) &= 1 + 4 + 10 + 20 + 35 &= \mathbf{70} \\
\mathrm{Pa}(4,4) + \mathrm{Pa}(5,4) + \mathrm{Pa}(6,4) + \mathrm{Pa}(7,4) + \mathrm{Pa}(8,4) &= 1 + 5 + 15 + 35 + 70 &= \mathbf{126}.
\end{aligned}
$$

1b. Notice that $\mathrm{Pa}(n, n) + \mathrm{Pa}(n+1, n) + \cdots + \mathrm{Pa}(n+k, n) = \mathrm{Pa}(n+k+1, n+1)$, so $m = n + k + 1$ and $j = n + 1$. (By symmetry, $j = k$ is also correct.) The equation is true for all n when $k = 0$, because the sum is simply $\mathrm{Pa}(n, n)$ and the right side is $\mathrm{Pa}(n + 1, n + 1)$, both of which are 1.

Proceed by induction on k. If $\text{Pa}(n,n)+\text{Pa}(n+1,n)+\cdots+\text{Pa}(n+k,n)=\text{Pa}(n+k+1,n+1)$, then adding $\text{Pa}(n+k+1,n)$ to both sides yields $\text{Pa}(n+k+1,n)+\text{Pa}(n+k+1,n+1)=\text{Pa}(n+k+2,n+1)$ by the recursive rule for Pa.

2a. By definition of Pa, $\text{Pa}(n,0)=\text{Pa}(n,n)=1$ for all nonnegative integers n, and this value is odd, so $\text{Pa}P(n,0)=\text{Pa}P(n,n)=1$ by definition.

2b.

$$
\begin{array}{ccccccccccccccccc}
& & & & & & & & 1 & & & & & & & & \\
& & & & & & & 1 & & 1 & & & & & & & \\
& & & & & & 1 & & 0 & & 1 & & & & & & \\
& & & & & 1 & & 1 & & 1 & & 1 & & & & & \\
& & & & 1 & & 0 & & 0 & & 0 & & 1 & & & & \\
& & & 1 & & 1 & & 0 & & 0 & & 1 & & 1 & & & \\
& & 1 & & 0 & & 1 & & 0 & & 1 & & 0 & & 1 & & \\
& 1 & & 1 & & 1 & & 1 & & 1 & & 1 & & 1 & & 1 & \\
1 & & 0 & & 0 & & 0 & & 0 & & 0 & & 0 & & 0 & & 1
\end{array}
$$

3a. Notice that

$$
\begin{aligned}
\text{Pa}(n,k) &= \frac{n!}{k!(n-k)!} \\
&= \frac{n}{k}\frac{(n-1)!}{(k-1)!((n-1)-(k-1))!} \\
&= \frac{n}{k}\cdot\text{Pa}(n-1,k-1).
\end{aligned}
$$

Examining the right hand side in the case where $n=2^j$ and $0<k<n$, the second factor, $\text{Pa}(n-1,k-1)$ is an integer, and the first factor has an even numerator, so $\text{Pa}(n,k)$ is even. Therefore $\text{Pa}P(n,k)=0$.

3b. Proceed by induction on j. It is easier to follow the logic when the computations are expressed modulo 2. The claim is that $\text{Pa}(n,k)\equiv\text{Pa}(n-2^j,k-2^j)+\text{Pa}(n-2^j,k)\bmod 2$. If $j=0$, so that $2^j=1$, then this congruence is exactly the recursive definition of $\text{Pa}(n,k)$ when $0<k<n$. For $k=0$ and $k=n$, the left-hand side is 1 and the right-hand side is either $0+1$ or $1+0$. For other values of k, all three terms are zero.

Now, assume $\mathrm{Pa}(n, k) \equiv \mathrm{Pa}(n - 2^j, k - 2^j) + \mathrm{Pa}(n - 2^j, k) \bmod 2$ for some $j \geq 0$, and let $n \geq 2^{j+1}$. For any k, apply the inductive hypothesis three times:

$$
\begin{aligned}
\mathrm{Pa}(n, k) &\equiv \mathrm{Pa}(n - 2^j, k - 2^j) + \mathrm{Pa}(n - 2^j, k) \bmod 2 \\
&\equiv \mathrm{Pa}(n - 2^j - 2^j, k - 2^j - 2^j) + \mathrm{Pa}(n - 2^j - 2^j, k - 2^j) \\
&\quad + \mathrm{Pa}(n - 2^j - 2^j, k - 2^j) + \mathrm{Pa}(n - 2^j - 2^j, k).
\end{aligned}
$$

But $u + u \equiv 0 \bmod 2$ for any integer u. Thus the two occurrences of $\mathrm{Pa}(n - 2^j - 2^j, k - 2^j)$ cancel each other out, and

$$
\begin{aligned}
\mathrm{Pa}(n, k) &\equiv \mathrm{Pa}(n - 2^j - 2^j, k - 2^j - 2^j) + \mathrm{Pa}(n - 2^j - 2^j, k) \\
&\equiv \mathrm{Pa}(n - 2^{j+1}, k - 2^{j+1}) + \mathrm{Pa}(n - 2^{j+1}, k) \bmod 2.
\end{aligned}
$$

3c. By part (b), $\mathrm{Pa}(n, k) \equiv \mathrm{Pa}(n - 2^j, k - 2^j) + \mathrm{Pa}(n - 2^j, k) \bmod 2$ when $j \geq 0$ and $n \geq 2^j$. If $2^j \leq n < 2^{j+1}$ and $0 \leq k < 2^j$, then $\mathrm{Pa}(n - 2^j, k - 2^j) = 0$, so

$$
\mathrm{Pa}(n, k) \equiv \mathrm{Pa}(n - 2^j, k - 2^j) + \mathrm{Pa}(n - 2^j, k) \equiv \mathrm{Pa}(n - 2^j, k) \bmod 2.
$$

Thus $\mathrm{Pa}P(n, k) = \mathrm{Pa}P(n - 2^j, k)$.

Alternate Solution: Problem 3a establishes the statement when $n = 2^j$. For $2^j < n < 2^{j+1}$, proceed by induction on n. Then $\mathrm{Pa}P(n, k) \equiv \mathrm{Pa}P(n - 1, k - 1) + \mathrm{Pa}P(n - 1, k) \bmod 2$, while $\mathrm{Pa}P(n - 1, k - 1) = \mathrm{Pa}P(n - 1 - 2^j, k - 1)$ and $\mathrm{Pa}P(n - 1, k) = \mathrm{Pa}P(n - 1 - 2^j, k)$. But $\mathrm{Pa}P(n - 1 - 2^j, k - 1) + \mathrm{Pa}P(n - 1 - 2^j, k) \equiv \mathrm{Pa}P(n - 2^j, k) \bmod 2$, establishing the statement.

4a.

	24		37		27		9		1		
		30		61		64		36		10	1
	36	91		125		100		46		11	1

4b. Using the given values yields the system of equations below.

$$
\begin{cases}
\mathrm{Cl}(1, 1) &= 1 &= a(1)^2 + b(1) + c \\
\mathrm{Cl}(2, 1) &= 7 &= a(2)^2 + b(2) + c \\
\mathrm{Cl}(3, 1) &= 19 &= a(3)^2 + b(3) + c
\end{cases}
$$

Solving this system, $a = 3, b = -3, c = 1$.

4c. Use induction on n. For $n = 1, 2, 3$, the values above demonstrate the theorem. If $\mathrm{Cl}(n, 1) = 3n^2 - 3n + 1$, then $\mathrm{Cl}(n + 1, 1) = \mathrm{Cl}(n, 0) + \mathrm{Cl}(n, 1) = 6n + (3n^2 - 3n + 1) = (3n^2 + 6n + 3) - (3n + 3) + 1 = 3(n + 1)^2 - 3(n + 1) + 1$.

5a. $\text{Cl}(11, 2) = 1000$.

5b. $\text{Cl}(n, 2) = (n - 1)^3$. Use induction on n. First, rewrite $\text{Cl}(n, 1) = 3n^2 - 3n + 1 = n^3 - (n - 1)^3$, and notice that $\text{Cl}(2, 2) = 1 = (2-1)^3$. Then if $\text{Cl}(n, 2) = (n-1)^3$, using the recursive definition, $\text{Cl}(n + 1, 2) = \text{Cl}(n, 2) + \text{Cl}(n, 1) = (n - 1)^3 + (n^3 - (n - 1)^3) = n^3$.

5c. $\text{Cl}(11, 3) = 2025$.

5d. Notice that $\text{Cl}(3, 3) = 1 = 1^3$, and then for $n > 3$, $\text{Cl}(n, 3) = \text{Cl}(n - 1, 2) + \text{Cl}(n - 1, 3)$; replacing $\text{Cl}(n - 1, 3)$ analogously on the right side yields the summation

$$\text{Cl}(n, 3) = \text{Cl}(n - 1, 2) + \text{Cl}(n - 2, 2) + \ldots + \text{Cl}(3, 2) + 1.$$

By 5b, $\text{Cl}(n - 1, 2) = (n - 2)^3$, so this formula is equivalent to

$$\text{Cl}(n, 3) = (n - 2)^3 + (n - 3)^3 + \ldots + 2^3 + 1.$$

Use the identity $1^3 + 2^3 + \cdots + m^3 = \dfrac{m^2(m + 1)^2}{4}$ and substitute $n - 2$ for m to obtain $\text{Cl}(n, 3) = \dfrac{(n - 2)^2(n - 1)^2}{4}$.

6. Notice that $\text{Cl}(n, 0) = 6 \cdot \text{Pa}(n, 1)$, and that for $n > 0$, $\text{Cl}(n, n) = \text{Pa}(n, n)$. From problem 4c, $\text{Cl}(n, 1) = 6\text{Pa}(n, 2) + 1$ (where $\text{Pa}(n, k) = 0$ if $k > n$).

For $k > 1$, repeated application of the formula $\text{Cl}(m, k) = \text{Cl}(m - 1, k - 1) + \text{Cl}(m - 1, k)$ allows each value of $\text{Cl}(n, k)$ to be written as a sum:

$$
\begin{aligned}
\text{Cl}(n, k) &= \text{Cl}(n - 1, k - 1) + \text{Cl}(n - 1, k) \\
&= \text{Cl}(n - 1, k - 1) + \text{Cl}(n - 2, k - 1) + \text{Cl}(n - 2, k), \text{ and eventually:} \\
&= \text{Cl}(n - 1, k - 1) + \text{Cl}(n - 2, k - 1) + \cdots + \text{Cl}(k, k - 1) + \text{Cl}(k, k) \\
&= \text{Cl}(n - 1, k - 1) + \text{Cl}(n - 2, k - 1) + \cdots + \text{Cl}(k - 1, k - 1),
\end{aligned}
$$

because $\text{Cl}(k, k) = \text{Cl}(k - 1, k - 1) = 1$. Then

$$
\begin{aligned}
\text{Cl}(n, 2) &= \text{Cl}(n - 1, 1) + \text{Cl}(n - 2, 1) + \cdots + \text{Cl}(1, 1) \\
&= (6\text{Pa}(n - 1, 2) + 1) + (6\text{Pa}(n - 2, 2) + 1) + \cdots + (6\text{Pa}(2, 2) + 1) + 1 \\
&= 6(\text{Pa}(n - 1, 2) + \text{Pa}(n - 2, 2) + \cdots + \text{Pa}(2, 2)) + (n - 1).
\end{aligned}
$$

By the identity from 1b, $\text{Pa}(n - 1, 2) + \text{Pa}(n - 2, 2) + \cdots + \text{Pa}(2, 2) = \text{Pa}(n, 3)$. Therefore $\text{Cl}(n, 2) = 6\text{Pa}(n, 3) + n - 1 = 6\text{Pa}(n, 3) + \text{Pa}(n - 1, 1)$.

The general formula is $\text{Cl}(n, k) = 6\text{Pa}(n, k + 1) + \text{Pa}(n - 1, k - 1)$. This formula follows by induction on n. If $n = k$, then $\text{Cl}(n, k) = 1$, and $6\text{Pa}(n, k+1) + \text{Pa}(n-1, k-1) = 6 \cdot 0 + 1 = 1$.

Then suppose for some $n \geq k$, $\text{Cl}(n, k) = 6\text{Pa}(n, k + 1) + \text{Pa}(n - 1, k - 1)$. It follows that

$$
\begin{aligned}
\text{Cl}(n + 1, k) &= \text{Cl}(n, k - 1) + \text{Cl}(n, k) \\
&= (6\text{Pa}(n, k) + \text{Pa}(n - 1, k - 2)) + (6\text{Pa}(n, k + 1) + \text{Pa}(n - 1, k - 1)) \\
&= 6(\text{Pa}(n, k) + \text{Pa}(n, k + 1)) + (\text{Pa}(n - 1, k - 2) + \text{Pa}(n - 1, k - 1)) \\
&= 6\text{Pa}(n + 1, k + 1) + \text{Pa}(n, k - 1).
\end{aligned}
$$

7a.

$$
\begin{array}{ccccccccccc}
& & & \dfrac{1}{5} & & \dfrac{1}{20} & & \dfrac{1}{30} & & \dfrac{1}{20} & & \dfrac{1}{5} & \\[2ex]
& \dfrac{1}{6} & & \dfrac{1}{30} & & \dfrac{1}{60} & & \dfrac{1}{60} & & \dfrac{1}{30} & & \dfrac{1}{6}
\end{array}
$$

7b. $\text{Le}(17, 1) = \text{Le}(16, 0) - \text{Le}(17, 0) = \dfrac{1}{17} - \dfrac{1}{18} = \dfrac{1}{306}$.

7c. $\text{Le}(17, 2) = \text{Le}(16, 1) - \text{Le}(17, 1) = \text{Le}(15, 0) - \text{Le}(16, 0) - \text{Le}(17, 1) = \dfrac{1}{2448}$.

8a. $\text{Le}(n, 1) = \text{Le}(n - 1, 0) - \text{Le}(n, 0) = \dfrac{1}{n} - \dfrac{1}{n + 1} = \dfrac{1}{n(n + 1)}$.

8b. Because $\text{Le}(n, 1) = \dfrac{1}{n} - \dfrac{1}{n + 1}$,

$$
\begin{aligned}
\sum_{i=1}^{2011} \text{Le}(i, 1) &= \sum_{i=1}^{2011} \left(\frac{1}{n} - \frac{1}{n + 1} \right) \\
&= \left(\frac{1}{1} - \frac{1}{2} \right) + \left(\frac{1}{2} - \frac{1}{3} \right) + \cdots + \left(\frac{1}{2010} - \frac{1}{2011} \right) + \left(\frac{1}{2011} - \frac{1}{2012} \right) \\
&= 1 - \frac{1}{2012} \\
&= \frac{2011}{2012}.
\end{aligned}
$$

8c. $\text{Le}(n, 2) = \text{Le}(n - 1, 1) - \text{Le}(n, 1) = \dfrac{1}{n(n - 1)} - \dfrac{1}{n(n + 1)} = \dfrac{2}{(n - 1)(n)(n + 1)}$. Note that this result appears in the table as a unit fraction because at least one of the integers $n - 1, n, n + 1$ is even.

9a. Extending the result of 8b gives

$$
\sum_{i=1}^{n} \text{Le}(i, 1) = \frac{1}{1} - \frac{1}{n},
$$

so as $n \to \infty$, $\sum_{i=1}^{n} \text{Le}(i, 1) \to 1$. This value appears as $\text{Le}(0, 0)$, so $n = k = 0$.

9b. $n = k = m - 1$.

9c. Because in general $\mathrm{Le}(i,m) = \mathrm{Le}(i-1,m-1) - \mathrm{Le}(i,m-1)$, a partial sum can be rewritten as follows:

$$
\begin{aligned}
\sum_{i=m}^{n} \mathrm{Le}(i,m) &= \sum_{i=m}^{n} \left(\mathrm{Le}(i-1,m-1) - \mathrm{Le}(i,m-1) \right) \\
&= \left(\mathrm{Le}(m-1,m-1) - \mathrm{Le}(m,m-1) \right) + \left(\mathrm{Le}(m,m-1) - \mathrm{Le}(m+1,m-1) \right) + \\
&\quad \cdots + \left(\mathrm{Le}(n-1,m-1) - \mathrm{Le}(n,m-1) \right) \\
&= \mathrm{Le}(m-1,m-1) - \mathrm{Le}(n,m-1).
\end{aligned}
$$

Because the values of $\mathrm{Le}(n,m-1)$ get arbitrarily small as n increases (proof: $\mathrm{Le}(i,j) < \mathrm{Le}(i-1,j-1)$ by construction, so $\mathrm{Le}(n,m-1) < \mathrm{Le}(n-m+1,0) = \dfrac{1}{n-m+1}$), the limit of these partial sums is $\mathrm{Le}(m-1,m-1)$. So $n = k = m-1$.

Note: This result can be extended even further. In fact, for every value of $k < n$,

$$
\mathrm{Le}(n,k) = \sum_{i=n+1}^{\infty} \mathrm{Le}(i,k+1).
$$

In other words, each entry in Leibniz's triangle is an infinite sum of the entries in the diagonal directly to its right, beginning with the entry below and to the right of the given one.

10. Note that $\dfrac{1}{3} = \mathrm{Le}(2,0) = \mathrm{Le}(3,0) + \mathrm{Le}(3,1)$. Also $\mathrm{Le}(3,0) = \mathrm{Le}(4,0) + \mathrm{Le}(4,1)$ and $\mathrm{Le}(4,0) = \mathrm{Le}(5,0) + \mathrm{Le}(5,1)$. So

$$
\begin{aligned}
\frac{1}{3} &= \mathrm{Le}(5,0) + \mathrm{Le}(5,1) + \mathrm{Le}(4,1) + \mathrm{Le}(3,1) \\
&= \frac{1}{6} + \frac{1}{30} + \frac{1}{20} + \frac{1}{12}.
\end{aligned}
$$

Therefore $a = 6, b = 12, c = 20, d = 30$ is a solution.

On the other hand, a similar analysis yields

$$
\begin{aligned}
\frac{1}{3} &= \mathrm{Le}(4,0) + \mathrm{Le}(5,1) + \mathrm{Le}(5,2) + \mathrm{Le}(3,1) \\
&= \frac{1}{5} + \frac{1}{30} + \frac{1}{60} + \frac{1}{12},
\end{aligned}
$$

so $a = 5, b = 12, c = 30, d = 60$ is another such solution.

Finally, notice that $\mathrm{Le}(3,1) = \dfrac{1}{12} = \mathrm{Le}(11,0)$, so that $\dfrac{1}{12}$ can be rewritten as $\mathrm{Le}(12,0) + \mathrm{Le}(12,1) = \dfrac{1}{13} + \dfrac{1}{156}$. Then

$$
\begin{aligned}
\frac{1}{3} &= \mathrm{Le}(4,0) + \mathrm{Le}(4,1) + \mathrm{Le}(3,1) \\
&= \mathrm{Le}(4,0) + \mathrm{Le}(4,1) + \mathrm{Le}(12,0) + \mathrm{Le}(12,1) \\
&= \frac{1}{5} + \frac{1}{20} + \frac{1}{13} + \frac{1}{156},
\end{aligned}
$$

yields $a = 5, b = 13, c = 20, d = 156.$

In general, once a triple a, b, c has been found, rewriting (for example)

$$\frac{1}{a} = \text{Le}(a-1, 0) = \text{Le}(a, 0) + \text{Le}(a, 1) = \frac{1}{a+1} + \frac{1}{a(a+1)}$$

creates an appropriate quadruple, although on occasion these values duplicate another value in the quadruple.

11. The formula is

$$\text{Le}(n, k) = \frac{1}{(n+1) \cdot \text{Pa}(n, k)},$$

or equivalently,

$$\frac{1}{(k+1) \cdot \text{Pa}(n+1, k+1)}.$$

Because $\text{Pa}(n, 0) = \text{Pa}(n, n) = 1$, when $k = 0$ or $k = n$, the formula is equivalent to the definition of $\text{Le}(n, 0) = \text{Le}(n, n) = \frac{1}{n+1}.$

To prove the formula for $1 \le k \le n - 1$, use induction on k. The base case $k = 0$ was proved above. If the formula holds for a particular value of $k < n$, then it can be extended to the case $k + 1$ using two identities:

$$\text{Le}(n+1, k+1) = \text{Le}(n, k) - \text{Le}(n+1, k) \text{ and}$$
$$(n+1)\text{Pa}(n, k) = \frac{(n+1)!}{k!(n-k)!}.$$

Using the first identity and the inductive hypothesis yields

$$\text{Le}(n+1, k+1) = \text{Le}(n, k) - \text{Le}(n+1, k)$$
$$= \frac{1}{(n+1) \cdot \text{Pa}(n, k)} - \frac{1}{(n+2) \cdot \text{Pa}(n+1, k)}.$$

Applying the second identity to the right side of the equation yields:

$$\text{Le}(n+1, k+1) = \frac{k!\,(n-k)!}{(n+1)!} - \frac{k!(n+1-k)!}{(n+2)!}$$
$$= \frac{k!\,(n-k)!}{(n+2)!}[(n+2) - (n+1-k)]$$
$$= \frac{k!\,(n-k)!}{(n+2)!}(k+1)$$
$$= \frac{(k+1)!\,((n+1)-(k+1))!}{(n+1)!\,(n+2)}$$
$$= \frac{1}{(n+2)\text{Pa}(n+1, k+1)}.$$

45.4 Power Question 2012: Cops and Robbers

Solutions to 2012 Power Question

1. a. Have the Cop stay at A for 2 days. If the Robber is not at A the first day, he must be at one of B_1–B_6, and because the Robber must move along an edge every night, he will be forced to go to A on day 2.

   b. The Cops should stay at $\{A_1, A_3, A_5\}$ for 2 days. If the Robber evades capture the first day, he must have been at an even-numbered hideout. Because he must move, he will be at an odd-numbered hideout the second day. Equivalently, the Cops could stay at $\{A_2, A_4, A_6\}$ for 2 days.

   c. Let $n = h(M)$. The following strategy will always catch a Robber within two days using $n-1$ Cops, which proves that $C(M) \leq n - 1$. Choose any subset S of $n - 1$ hideouts and position $n - 1$ Cops at the hideouts of S for 2 days. If the Robber is not caught on the first day, he must have been at the hideout not in S, and therefore must move to a hideout in S on the following day.

2. a. $C(M) = 6$. The result of 1c applies, so $C(M) \leq 6$. To see that 6 is minimal, note that it is possible for the Robber to go from any hideout to any other hideout in a single night. Suppose that on a given day, the Robber successfully evades capture. Without loss of generality, assume that the Robber was at hideout A_1. Then the only valid conclusion the Cops can draw is that on the next day, the Robber will *not* be at A_1. Thus the only hideout the Cops can afford to leave open is A_1 itself. If they leave another hideout open, the Robber might have chosen to hide there, and the process will repeat.

   b. The following strategy guarantees capture using three Cops for four consecutive days, so $C(M) \leq 3$. Position three Cops at $\{B, E, H\}$ for two days, which will catch any Robber who starts out at B, C, D, E, F, G, or H, because a Robber at C or D would have to move to either B or E, and a Robber at F or G would have to move to either E or H. So if the Robber is not yet caught, he must have been at I, J, K, L, or A for those two days. In this case, an analogous argument shows that placing Cops at B, H, K for two consecutive days will guarantee a capture.

   c. The result in 1c applies, so $C(\mathcal{K}_n) \leq n - 1$. It remains to show that $n - 1$ is minimal. As was argued in 2a, the Robber always has $n - 1$ choices of where to move the next day, so a strategy using less than $n - 1$ Cops can never guarantee capture. Thus $C(\mathcal{K}_n) > n - 2$.

3. The following argument shows that $C(\mathcal{P}_n) = 1$, and that capture occurs in at most $2n - 4$ days. It helps to draw the hideout map as in the following diagram, so that odd-numbered hideouts are

all on one level and even-numbered hideouts are all on another level; the case where n is odd is shown below. (A similar argument applies where n is even.)

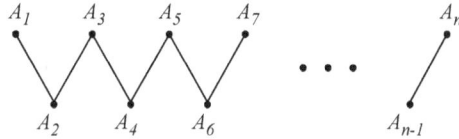

Each night, the Robber has the choice of moving diagonally left or diagonally right on this map, but he is required to move from top to bottom or vice-versa.

For the first $n - 2$ days, the Cop should search hideouts $A_2, A_3, \ldots, A_{n-1}$, in that order. For the next $n - 2$ days, the Cop should search hideouts $A_{n-1}, A_{n-2}, \ldots, A_2$, in that order.

The Cop always moves from top to bottom, or vice-versa, except that he searches A_{n-1} two days in a row. If the Cop is lucky, the Robber chose an even-numbered hideout on the first day. In this case, the Cop and the Robber are always on the same level (top or bottom) for the first half $(n - 2$ days) of the search. The Cop is searching from left to right, and the Robber started out to the right of the Cop (or at A_2, the first hideout searched), so eventually the Robber is caught (at A_{n-1}, if not earlier).

If the Robber chose an odd-numbered hideout on the first day, then the Cop and Robber will be on different levels for the first half of the search, but on the same level for the second half. For the second half of the search, the Robber is to the left of the Cop (or possibly at A_{n-1}, which would happen if the Robber moved unwisely or if he spent day $n - 2$ at hideout A_n). But now the Cop is searching from right to left, so again, the Cop will eventually catch the Robber.

The same strategy can be justified without using the zig-zag diagram above. Suppose that the Robber is at hideout A_R on a given day, and the Cop searches hideout A_C. Whenever the Cop moves to the next hideout or the preceding hideout, C changes by ± 1, and the Robber's constraint forces R to change by ± 1. Thus if the Cop uses the strategy above, on each of the first $n - 2$ days, either the difference $R - C$ stays the same or it decreases by 2. If on the first day $R - C$ is even, either $R - C$ is 0 (the Robber was at A_2 and is caught immediately) or is positive (because $C = 2$ and $R \geq 4$). Because the difference $R - C$ is even and decreases (if at all) by 2 each day, it cannot go from positive to negative without being zero. If on the first day $R - C$ is odd, then the Robber avoids capture through day $n - 2$ (because $R - C$ is still odd), but then on day $n - 1$, R changes (by ± 1) while C does not. So $R - C$ is now even (and is either 0 or negative), and henceforth either remains the same or increases by 2 each day, so again, $R - C$ must be zero at some point between day $n - 1$ and day $2n - 4$, inclusive.

4. a. $C(\mathcal{C}_n) = 2$. Position one Cop at A_1 and have the Cop search that hideout each day. This strategy reduces $\mathcal{C}_n$ to $\mathcal{P}_{n-1}$, because the Robber cannot get past A_1. Then the second Cop can pursue the strategy from problem 3 on hideouts $\{A_2, \ldots, A_n\}$ to guarantee capture. Thus $C(\mathcal{C}_n) \le 2$. On the other hand, one Cop is obviously insufficient because the Robber has two choices of where to go every day, so it is impossible for one Cop to cover both possible choices.

   b. If the second Cop makes it all the way to hideout A_n without capturing the Robber, then the second part of the strategy from part (a) can be improved by having the Cops walk back towards each other in a "pincers" move. The details are as follows. If n is odd, then when the second Cop reaches A_n, his and the first Cop's hideouts are both odd. By hypothesis, the Robber's hideout has the opposite (even) parity. (Otherwise the second Cop would have caught the Robber the first time around.) So the two Cops should each "wait" (that is, search A_1 and A_n again) for one day, after which they will have the same parity as the Robber. Then the first Cop should search hideouts in increasing order while the second Cop searches in decreasing order. This part of the search takes at most $\dfrac{n-1}{2}$ days, for a total of $(n-1) + 1 + \dfrac{n-1}{2}$ days, or $\dfrac{3n-1}{2}$ days. If n is even, then when the second Cop reaches hideout A_n, the Robber is on an odd hideout (but not A_1), and so the first Cop's hideout is *already* the same parity as the Robber's hideout. So on the next day, the first Cop should search A_2 while the second Cop should re-search A_n. Thereafter, the first Cop searches consecutively increasing hideouts while the second Cop searches consecutively decreasing ones as in the case of n odd. In this case, the search takes $(n-1) + 1 + \dfrac{n-2}{2} = \dfrac{3n-2}{2}$ days.

5. For the map M from 2a, $W(M) = 7$. The most efficient strategy is to use 7 Cops to blanket all the hideouts on the first day. Any strategy using fewer than 7 Cops would require 6 Cops on each of two consecutive days: given that any hideout can be reached from any other hideout, leaving more than one hideout unsearched on one day makes it is impossible to eliminate any hideouts the following day. So any other strategy would require a minimum of 12 Cop workdays.

   For the map M from 2b, $W(M) = 8$. The strategy outlined in 2b used three Cops for a maximum of four workdays, yielding 12 Cop workdays. The most efficient strategy is to use 4 Cops, positioned at $\{B, E, H, K\}$ for 2 days each. The following argument demonstrates that 8 Cop workdays is in fact minimal. First, notice that there is no advantage to searching one of the hideouts between vertices of the square (for example, C) without searching the other hideout between the same vertices (for example, D). The Robber can reach C on day n if and only if he is at either B or E on day $n - 1$, and in either case he could just as well go to D instead of C. So there is no situation in which the Robber is certain to be caught at C rather than at D. Additionally, the

Robber's possible locations on day $n + 1$ are the same whether he is at C or D on day n, so searching one rather than the other fails to rule out any locations for future days. So any successful strategy that involves searching C should also involve searching D on the same day, and similarly for F and G, I and J, and L and A. On the other hand, if the Robber must be at one of C and D on day n, then he must be at either B or E on day $n + 1$, because those are the only adjacent hideouts. So any strategy that involves searching both hideouts of one of the off-the-square pairs on day n is equivalent to a strategy that searches the adjacent on-the-square hideouts on day $n + 1$; the two strategies use the same number of Cops for the same number of workdays. Thus the optimal number of Cop workdays can be achieved using strategies that only search the "corner" hideouts B, E, H, K.

Restricting the search to only those strategies searching corner hideouts B, E, H, K, a total of 8 workdays can be achieved by searching all four hideouts on two consecutive days: if the Robber is at one of the other eight hideouts the first day, he must move to one of the two adjacent corner hideouts the second day. But each of these corner hideouts is adjacent to two other corner hideouts. So if only one hideout is searched, for no matter how many consecutive days, the following day, the Robber could either be back at the previously-searched hideout or be at any other hideout: no possibilities are ruled out. If two adjacent corners are searched, the Cops do no better, as the following argument shows. Suppose that B and E are both searched for two consecutive days. Then the Cops can rule out B, E, C, and D as possible locations, but if the Cops then switch to searching either H or K instead of B or E, the Robber can go back to C or D within two days. So searching two adjacent corner hideouts for two days is fruitless and costs four Cop workdays. Searching diagonally opposite corner hideouts is even less fruitful, because doing so rules out none of the other hideouts as possible Robber locations. Using three Cops each day, it is easy to imagine scenarios in which the Robber evades capture for three days before being caught: for example, if B, E, H are searched for two consecutive days, the Robber goes from I or J to K to L. Therefore if three Cops are used, four days are required for a total of 12 Cop workdays.

If there are more than four Cops, the preceding arguments show that the number of Cops must be even to produce optimal results (because there is no advantage to searching one hideout between vertices of the square without searching the other). Using six Cops with four at corner hideouts yields no improvement, because the following day, the Robber could get to any of the four corner hideouts, requiring at least four Cops the second day, for ten Cop workdays. If two or fewer Cops are at corner hideouts, the situation is even worse, because if the Robber is not caught that day, he has at least nine possible hideouts the following day (depending on whether the unsearched corners are adjacent or diagonally opposite to each other). Using eight Cops (with four at corner vertices) could eliminate one corner vertex as a possible location for the second day (if the non-corner hideouts searched are on adjacent sides of the square), but eight Cop workdays have already been used on the first day. So 8 Cop workdays is minimal.

6. A single Cop can only search one hideout in a day, so as long as M has two or more hideouts, there is no strategy that guarantees that a lone Cop captures the Robber the first day. Then either more than one Cop will have to search on the first day, or a lone Cop will have to search for at least 2 days; in either of these cases, $W(M) \geq 2$. On the other hand, n Cops can always guarantee a capture simply by searching all n hideouts on the first day, so $W(M) \leq n$.

To show that the lower bound cannot be improved, consider the *star on n hideouts*; that is, the map $\mathcal{S}_n$ with one central hideout connected to $n - 1$ outer hideouts, none of which is connected to any other hideout. (The map in 1a is $\mathcal{S}_7$.) Then $W(\mathcal{S}_n) = 2$, because a single Cop can search the central hideout for 2 days; if the Robber is at one of the outer hideouts on the first day, he must go to the central hideout on the second day. For the upper bound, the complete map $\mathcal{K}_n$ is an example of a map with $W(M) = n$. As argued above, the minimum number of Cops needed to guarantee a catch is $n - 1$, and using $n - 1$ Cops requires two days, for a total of $2n - 2$ Cop workdays. So the optimal strategy for $\mathcal{K}_n$ (see 2c) is to use n Cops for a single day.

7. a. Suppose M is bipartite, and let $\mathcal{A}$ and $\mathcal{B}$ be the sets of hideouts referenced in the definition. Because $\mathcal{A}$ and $\mathcal{B}$ are disjoint, either $|\mathcal{A}| \leq n/2$ or $|\mathcal{B}| \leq n/2$ or both. Without loss of generality, suppose that $|\mathcal{A}| \leq n/2$. Then position Cops at each hideout in $\mathcal{A}$ for two days. If the Robber was initially on a hideout in $\mathcal{B}$, he must move the following day, and because no hideout in $\mathcal{B}$ is connected to any other hideout in $\mathcal{B}$, his new hideout must be a hideout in $\mathcal{A}$.

    b. The given condition actually implies that the graph is bipartite. Let A_1 be a hideout in M, and let $\mathcal{A}$ be the set of all hideouts V such that all paths from A_1 to V have an even number of edges, as well as A_1 itself; let $\mathcal{B}$ be the set of all other hideouts in M. Notice that there are no edges from A_1 to any other element A_i in $\mathcal{A}$, because if there were, that edge would create a path from A_1 to A_i with an odd number of edges (namely 1). So A_1 has edges only to hideouts in $\mathcal{B}$. In fact, there can be no edge from any hideout A_i in $\mathcal{A}$ to any other hideout A_j in $\mathcal{A}$, because if there were, there would be a path with an odd number of edges from A_1 to A_j via A_i. Similarly, there can be no edge from any hideout B_i in $\mathcal{B}$ to any other hideout B_j in $\mathcal{B}$, because if there were such an edge, there would be a path from A_1 to B_j with an even number of edges via B_i. Thus hideouts in $\mathcal{B}$ are connected only to hideouts in $\mathcal{A}$ and vice versa; hence M is bipartite.

8. Because a Robber in $\mathcal{A}_i$ can only move to a hideout in $\mathcal{A}_{i-1}$ or $\mathcal{A}_{i+1}$, this map is essentially the same as the cyclic map $\mathcal{C}_k$. So the Cops should apply a similar strategy. First, position n/k Cops at the hideouts of set $\mathcal{A}_1$ and n/k Cops at the hideouts of set $\mathcal{A}_2$. On day 2, leave the first n/k

Cops at $\mathcal{A}_1$, but move the second group of Cops to $\mathcal{A}_3$. Continue until the second set of Cops is at $\mathcal{A}_k$; then "wait" one turn (search $\mathcal{A}_k$ again), and then search backward $\mathcal{A}_{k-1}$, $\mathcal{A}_{k-2}$, etc.

9. There are many examples. Perhaps the simplest to describe is the *complete bipartite map* on the hideouts $\mathcal{A} = \{A_1, A_2, \ldots, A_{17}\}$ and $\mathcal{B} = \{B_1, B_2, \ldots, B_{1995}\}$, which is often denoted $M = \mathcal{K}_{17,1995}$. That is, let M be the map with hideouts $\mathcal{A} \cup \mathcal{B}$ such that A_i is adjacent to B_j for all i and j, and that A_i is not adjacent to A_j, nor is B_i adjacent to B_j, for any i and j.

If 17 Cops search the 17 hideouts in $\mathcal{A}$ for two consecutive days, then they are guaranteed to catch the Robber. This shows that $C(M) \leq 17$ and that $W(M) \leq 34$. If fewer than 17 Cops search on a given day, then the Robber has at least one safe hideout: he has exactly one choice if the Cops search 16 of the hideouts A_i and the Robber was hiding at some B_j the previous day, 1995 choices if the Robber was hiding at some A_i the previous day. Thus $C(M) = 17$.

On the other hand, fewer than 34 Cop workdays cannot guarantee catching the Robber. Unless the Cops search every A_i (or every B_j) on a given day, they gain no information about where the Robber is (unless he is unlucky enough to be caught).

The complete bipartite map is not the simplest in terms of the number of edges. Try to find a map with 2012 hideouts and as few edges as possible that has a Cop number of 17 and a workday number of 34.

10. No such map exists. Let M be a map with at least four hideouts. The proof below shows that either $W(M) > 3$ or else M is a star (as in the solution to problem 6), in which case $W(M) = 2$.

First, suppose that there are four distinct hideouts A_1, A_2, B_1, and B_2 such that A_1 and A_2 are adjacent, as are B_1 and B_2. If the Cops make fewer than four searches, then they cannot search $\{A_1, A_2\}$ twice and also search $\{B_1, B_2\}$ twice. Without loss of generality, assume the Cops search $\{A_1, A_2\}$ at most once. Then the Robber can evade capture by moving from A_1 to A_2 and back again, provided that he is lucky enough to start off at the right one. Thus $W(M) > 3$ in this case. For the second case, assume that whenever A_1 is adjacent to A_2 and B_1 is adjacent to B_2, the four hideouts are *not* distinct. Start with two adjacent hideouts, and call them A_1 and A_2. Consider any hideout, say B, that is not one of these two. Then B must be adjacent to some hideout, say C. By assumption, A_1, A_2, B, and C are not distinct, so $C = A_1$ or $C = A_2$. That is, every hideout B is adjacent to A_1 or to A_2.

If there are hideouts B_1 and B_2, distinct from A_1 and A_2 and from each other, such that B_1 is adjacent to A_1 and B_2 is adjacent to A_2, then it clearly violates the assumption of this case. That is, either every hideout B, distinct from A_1 and A_2, is adjacent to A_1, or every such hideout is adjacent to A_2. Without loss of generality, assume the former.

Because every hideout is adjacent to A_1 (except for A_1 itself), the foregoing proves that the map M is a star. It remains to show that there are no "extra" edges in the map. For the sake of contradiction, suppose that B_1 and B_2 are adjacent hideouts, both distinct from A_1. Because the map has at least four hideouts, choose one distinct from these three and call it A_2. Then these four hideouts violate the assumption of this case.

45.5   Power Question 2013: Power of (Urban) Planning

Solutions to 2013 Power Question

1. a. Factoring, $6 = 2 \cdot 3^1$, $16 = 16 \cdot 3^0$, and $72 = 8 \cdot 3^2$, so $d(6) = 1/3$, $d(16) = 1$, and $d(72) = 1/9$.

    b. If $n = 3^k m$ where $3 \nmid m$, then $d(n) = 1/3^k$. So the smallest values of $d(n)$ occur when k is largest. The largest power of 3 less than 100 is $3^4 = 81$, so $d(81) = 1/3^4 = 1/81$ is minimal.

    c. If $d(n, 17) = 1/9$, then $9 \mid (n - 17)$ and $27 \nmid (n - 17)$. So $n = 17 + 9k$ where $k \in \mathbb{Z}$ is not divisible by 3. For $k = -1, 1, 2, 4$, this formula yields $n = 8, 26, 35$, and 53, respectively.

    d. There are many possible answers; the simplest is $h_n = 17 + 3^n$, yielding the sequence $\{20, 26, 44, 98, \ldots\}$.

2. a. Because $d(6) = 1/3$, $\mathcal{N}(6)$ is the set of all houses n such that $d(n) = 1/3$, equivalently, where $n = 3k$ and k is not divisible by 3. So $\mathcal{N}(6) = \{3, 6, 12, 15, 21, 24, \ldots\}$.

    b. Using similar logic to 2a, $\mathcal{N}(n) = \{m \in \mathbb{Z}^+ \mid 3 \nmid m\}$. So $\mathcal{N}(n) = \{1, 2, 4, 5, 7, 8, \ldots\}$.

    c. Here, $\mathcal{N}(n) = \{m \mid m = 27k, \text{ where } 3 \nmid k\}$. The ten smallest elements of $\mathcal{N}(n)$ are 27, 54, 108, 135, 189, 216, 270, 297, 351, and 378.

3. a. The statement is false. $d(n, m) < 1$ if $3 \mid (n - m)$, so for example, if $n = 2$ and $m = 5$.

    b. If $d(m) = d(n) = 1/3^k$, rewrite $m = 3^k m'$ and $n = 3^k n'$ where m' and n' are not divisible by 3. Then $m - n = 3^k(m' - n')$. If $3 \mid (m' - n')$, then $d(m, n) < 1/3^k$. If $3 \nmid (m' - n')$, then $d(m, n) = 1/3^k$. So the set of possible values of $d(m, n)$ is $\{3^{-q} \mid q \geq k, q \in \mathbb{Z}^+\}$.

    c. Because $d(17, m) = 1/81$, $17 - m = 81l$, where $l \in \mathbb{Z}$ and $3 \nmid l$. So $m = 17 - 81l$ and $m - 16 = 1 - 81l$. Hence $3 \nmid m - 16$, and $d(m, 16) = d(m - 16) = 1$.

4. a. Because $d(17, 34) = d(17) = 1$, $d(34, 51) = d(17) = 1$, and $d(17, 51) = d(34) = 1$, it follows that $d(17, 51) \leq d(17, 34) + d(34, 51)$. Because $d(17, 68) = d(51) = 1/3$, $d(68, 95) = d(27) = 1/27$, and $d(17, 95) = d(78) = 1/3$, it follows that $d(17, 95) \leq d(17, 68) + d(68, 95)$.

    b. Because $d(17, 68) = 1/3$, the condition implies that $3 \mid (c - 17)$ but $9 \nmid (c - 17)$. Thus either $c - 17 = 9k + 3$ or $c - 17 = 9k + 6$ for some $k \in \mathbb{Z}$. Solving for c yields $c = 9k + 20$ or $c = 9k + 23$, equivalently $c = 9k + 2$ or $c = 9k + 5$, $k \in \mathbb{Z}, k \geq 0$.

5. a. Write $a = 3^\alpha a_0$ and $b = 3^\beta b_0$, where $3 \nmid a_0$ and $3 \nmid b_0$. First consider the case $\alpha = \beta$. Then $a - b = 3^\alpha(a_0 - b_0)$. In this case, if $3 \mid (a_0 - b_0)$, then $3^\alpha(a_0 - b_0) = 3^\gamma c$, where $3 \nmid c$ and $\alpha < \gamma$; so $d(a - b) = 3^{-\gamma} < 3^{-\alpha} = d(a) = d(b)$. If $3 \nmid (a_0 - b_0)$, then $d(a, b) = 3^{-\alpha} = d(a) = d(b)$. If $\alpha \neq \beta$, suppose, without loss of generality, that $\alpha < \beta$ (so that $d(a) > d(b)$). Then $a - b = 3^\alpha(a_0 - 3^{\beta-\alpha}b_0)$. In this second factor, notice that the second term, $3^{\beta-\alpha}b_0$ is divisible by 3 but the first term a_0 is not, so their difference is not divisible by 3. Thus $d(a, b) = 3^{-\alpha} = d(a)$. Therefore $d(a, b) = d(a)$ when $d(a) > d(b)$, and similarly $d(a, b) = d(b)$ when $d(b) > d(a)$. Hence $d(a, b) \leq \max\{d(a), d(b)\}$.

   b. Note that $d(a, c) = d(a - c) = d((a - b) - (c - b)) = d(a - b, c - b)$. Then $d(a, c) \leq \max\{d(a, b), d(c, b)\}$ by part 5a, and the fact that $d(c, b) = d(b, c)$.

   c. Because $d(x, y) > 0$ for all $x \neq y$, $\max\{d(a, b), d(b, c)\} < d(a, b) + d(b, c)$ whenever $a \neq b \neq c$. Thus $d(a, c) < d(a, b) + d(b, c)$ when a, b, c are all different. If $a = b \neq c$, then $d(a, b) = 0$, $d(a, c) = d(b, c)$, so $d(a, c) \leq d(a, b) + d(b, c)$. And if $a = c \neq b$, then $d(a, c) = 0$ while $d(a, b) = d(b, c) > 0$, so $d(a, c) < d(a, b) + d(b, c)$.

   d. The foregoing shows that all ARMLopolitan triangles are isosceles! Examining the proof in 5a, note that if $d(a) \neq d(b)$, then $d(a, b) = \max\{d(a), d(b)\}$. Applying that observation to the proof in 5b, if $d(a, b) \neq d(b, c)$, then $d(a, c) = \max\{d(a, b), d(b, c)\}$. So either $d(a, b) = d(b, c)$ or, if not, then either $d(a, c) = d(a, b)$ or $d(a, c) = d(b, c)$. Thus all ARMLopolitan triangles are isosceles.

6. a. For example, $d(9) = 1/9$ and $d(9 + 18) = 1/27$. In general, if $n = 9 + 27k$, then $d(n) = 1/9$, but $d(n + 18) = d(27 + 27k) \leq 1/27$.

   b. Note that $d(m + 18, n + 18) = d((m + 18) - (n + 18)) = d(m - n) = d(m, n)$. In particular, if $m = n + 1$, then $d(m, n) = 1$, thus $d(m + 18, n + 18)$ is also 1.

7. a. No, they won't move. If $a \in \mathcal{N}(1)$, then $3 \nmid a$. Thus $3 \nmid (a + 18)$, and so $d(a + 18) = 1$.

   b. Yes, they will be split apart. If $a \in \mathcal{N}(9)$, then $9 \mid a$ but $27 \nmid a$, so $a = 27k + 9$ or $a = 27k + 18$. Adding 18 has different effects on these different sets of houses: $27k + 18 \mapsto 27k + 36$, which is still in $\mathcal{N}(9)$, while $27k + 9 \mapsto 27k + 27$, which is in $\mathcal{N}(27r)$ for some $r \in \mathbb{Z}^+$, depending on the exact value of a. For example, $a = 9 \mapsto 27$, while $a = 63 \mapsto 81$, and $d(81) = 3^{-4}$. That is, $d(a + 18) \leq 3^{-3}$ whenever $a = 27k + 9$, with equality unless $a \equiv 63 \bmod 81$.

   c. The question is whether it is possible that $x \notin \mathcal{N}(9)$ but $x + 18 \in \mathcal{N}(9)$. Because $d(9) = 1/9$, there are two cases to consider: $d(x) > 1/9$ and $d(x) < 1/9$. In the first case, $9 \nmid x$ (either x is not divisible by 3, or x is divisible by 3 and not 9). These houses do not move into $\mathcal{N}(9)$: $x \not\equiv 0 \bmod 9 \Rightarrow x + 18 \not\equiv 0 \bmod 9$. On the other hand, if $d(x) < 1/9$, then $27 \mid x$, i.e., $x = 27k$. Then $x + 18 = 27k + 18$ which is divisible by 9 but not by 27, so $d(x + 18) = 9$. So in fact **every** house x of ARMLopolis with $d(x) < 1/9$ will move into $\mathcal{N}(9)$!

d. As noted in 7c, all houses in $\mathcal{N}(3^k)$, where $k \geq 3$, will move into $\mathcal{N}(9)$. On the other hand, houses of the form $27k + 9$ will find that their distance from City Hall *decreases*, so each of those now-vacated neighborhoods will be filled by residents of the old $\mathcal{N}(9)$. None of $\mathcal{N}(1)$, $\mathcal{N}(2)$, or $\mathcal{N}(3)$ will be affected, however.

8. a. If $n = 3k+2$, then $d(23, n) = d(3k-21)$, and because $3k - 21 = 3(k-7)$, $d(3k-21) \leq 1/3$. On the other hand, if $d(23, n) \leq 1/3$, then $3 \mid (n - 23)$, so $n - 23 = 3k$, and $n = 3k + 23 = 3(k+7) + 2$.

   b. If $n = 3k + 2$, then $d(32, n) = d(3k - 30) \leq 1/3$. Similarly, if $d(32, n) \leq 1/3$, then $3 \mid n - 32 \Rightarrow n = 3k + 32 = 3(k + 10) + 2$.

   c. Show that $\mathcal{D}_r(x) \subseteq \mathcal{D}_r(z)$ and conversely. Suppose that $y \in \mathcal{D}_r(x)$. Then $d(x, y) \leq r$. Because $d(x, z) = r$, by 5b, $d(y, z) \leq \max\{d(x, y), d(x, z)\} = r$. So $y \in \mathcal{D}_r(z)$. Thus $\mathcal{D}_r(x) \subseteq \mathcal{D}_r(z)$. Similarly, $\mathcal{D}_r(z) \subseteq \mathcal{D}_r(x)$, so the two sets are equal.

9. a. The maximum possible distance $d(34, n_k)$ is $1/3$. This can be proved by induction on k: $d(n_1, 34) \leq 1/3$, and if both $d(n_{k-1}, 34) \leq 1/3$ and $d(n_{k-1}, n_k) \leq 1/3$, then $\max\{d(n_{k-1}, 34), d(n_{k-1}, n_k)\} \leq 1/3$ so by 5b, $d(34, n_k) \leq 1/3$.

   b. The sequence is contained in the disk for all rational values of r, by the same logic as in 9a.

10. a. $d(3/5) = 1/3$, $d(5/8) = 1$, and $d(7/18) = 9$.

   b. Because $d(4/3) = 1/3$, $\mathcal{N}(4/3) = \{p/q$ in lowest terms such that $3 \mid q$ and $9 \nmid q\}$. Thus the set of all possible (p, q) consists precisely of those ordered pairs $(p'r, q'r)$, where r is any non-zero integer, p' and q' are relatively prime integers, $3 \mid q'$, and $9 \nmid q'$.

   c. The houses in this sequence actually get arbitrarily far apart from each other, and so there's no single house which they approach. If

$$H_n = \frac{1}{0!} + \frac{1}{1!} + \cdots + \frac{1}{n!},$$

   then $d(H_n, H_{n-1}) = d(1/n!) = 3^k$, where $k = \lfloor n/3 \rfloor + \lfloor n/9 \rfloor + \lfloor n/27 \rfloor \dots$. In particular, if $n! = 3^k \cdot n_0$, where $k > 0$ and $3 \nmid n_0$, then $d(H_n, H_{n-1}) = 3^k$, which can be made arbitrarily large as n increases.

45.6 Power Question 2014: Power of Potlucks

Solutions to 2014 Power Question

1. a. There are $\binom{17}{2} = 136$ possible pairs of dishes, so $\mathcal{F}_{17}$ must have 136 people.

b. With d dishes there are $\binom{d}{2} = \dfrac{d^2 - d}{2}$ possible pairs, so $n = \dfrac{d^2 - d}{2}$. Then $2n = d^2 - d$, or $d^2 - d - 2n = 0$. Using the quadratic formula yields $d = \dfrac{1 + \sqrt{1 + 8n}}{2}$ (ignoring the negative value).

c. The town T' consists of all residents of T who do not know how to make D. Because T is full, every pair of dishes $\{d_i, d_j\}$ in $\mathrm{dish}(T)$ can be made by some resident r_{ij} in T. If $d_i \neq D$ and $d_j \neq D$, then $r_{ij} \in T'$. So every pair of dishes in $\mathrm{dish}(T) \setminus \{D\}$ can be made by some resident of T'. Hence T' is full.

2. a. Paul and Arnold cannot be in the same group, because they both make pie, and Arnold and Kelly cannot be in the same group, because they both make salad. Hence there must be at least two groups. But Paul and Kelly make none of the same dishes, so they can be in the same group. Thus a valid group assignment is

$$
\begin{aligned}
\text{Paul} &\mapsto 1 \\
\text{Kelly} &\mapsto 1 \\
\text{Arnold} &\mapsto 2.
\end{aligned}
$$

Hence $\mathrm{gr}(\text{ARMLton}) = 2$.

b. Sally and Ross both make calzones, Ross and David both make pancakes, and Sally and David both make steak. So no two of these people can be in the same group, and $\mathrm{gr}(\text{ARMLville}) = 3$.

3. a. Let the dishes be d_1, d_2, d_3, d_4 and let resident r_{ij} make dishes d_i and d_j, where $i < j$. There are six pairs of dishes, which can be divided into nonoverlapping pairs in three ways: $\{1, 2\}$ and $\{3, 4\}$, $\{1, 3\}$ and $\{2, 4\}$, and $\{1, 4\}$ and $\{2, 3\}$. Hence the assignment $r_{12}, r_{34} \mapsto 1$, $r_{13}, r_{24} \mapsto 2$, and $r_{14}, r_{23} \mapsto 3$ is valid, hence $\mathrm{gr}(\mathcal{F}_4) = 3$.

b. First, $\mathrm{gr}(\mathcal{F}_5) \geq 5$: there are $\binom{5}{2} = 10$ people in $\mathcal{F}_5$, and because each person cooks two different dishes, any valid group of three people would require there to be six different dishes—yet there are only five. So each group can have at most two people. A valid assignment using five groups is shown below.

Residents	Group
r_{12}, r_{35}	1
r_{13}, r_{45}	2
r_{14}, r_{23}	3
r_{15}, r_{24}	4
r_{25}, r_{34}	5

c. Now there are $\binom{6}{2} = 15$ people, but there are six different dishes, so it is possible (if done carefully) to place three people in a group. Because four people in a single group would require there to be eight different dishes, no group can have more than three people, and so $15/3 = 5$ groups is minimal. (Alternatively, there are five different residents who can cook dish d_1, and no two of these can be in the same group, so there must be at least five groups.) The assignment below attains that minimum.

Residents	Group
r_{12}, r_{34}, r_{56}	1
r_{13}, r_{25}, r_{46}	2
r_{14}, r_{26}, r_{35}	3
r_{15}, r_{24}, r_{36}	4
r_{16}, r_{23}, r_{45}	5

4. Pick some $n \geq 2$ and a full town $\mathcal{F}_n$ whose residents prepare dishes $d_1, \ldots, d_n$, and let $\mathrm{gr}(\mathcal{F}_n) = k$. Suppose that $f_n : \mathcal{F}_n \to \{1, 2, \ldots, k\}$ is a valid group assignment for $\mathcal{F}_n$. Then remove from $\mathcal{F}_n$ all residents who prepare dish d_n; by problem 1c, this operation yields the full town $\mathcal{F}_{n-1}$. Define $f_{n-1}(r) = f_n(r)$ for each remaining resident r in $\mathcal{F}_n$. If r and s are two (remaining) residents who prepare a common dish, then $f_n(r) \neq f_n(s)$, because f_n was a valid group assignment. Hence $f_{n-1}(r) \neq f_{n-1}(s)$ by construction of f_{n-1}. Therefore f_{n-1} is a valid group assignment on $\mathcal{F}_{n-1}$, and the set of groups to which the residents of $\mathcal{F}_{n-1}$ are assigned is a (not necessarily proper) subset of $\{1, 2, \ldots, k\}$. Thus $\mathrm{gr}(\mathcal{F}_{n-1})$ is at most k, which implies the desired result.

5. Because each chef knows how to prepare exactly two dishes, and no two chefs know how to prepare the same two dishes, each chef is counted exactly twice in the sum $\Sigma \, |\mathrm{chef}_T(D)|$. More formally, consider the set of "resident-dish pairs":

$$S = \{(r, D) \in T \times \mathrm{dish}(T) \mid r \text{ makes } D\}.$$

Count $|S|$ in two different ways. First, every dish D is made by $|\mathrm{chef}_T(D)|$ residents of T, so

$$|S| = \sum_{D \in \mathrm{dish}(T)} |\mathrm{chef}_T(D)|.$$

Second, each resident knows how to make exactly two dishes, so

$$|S| = \sum_{r \in T} 2 = 2\mathrm{pop}(T).$$

6. Let $D \in \text{dish}(T)$. Suppose that f is a valid group assignment on T. Then for $r, s \in \text{chef}_T(D)$, if $r \neq s$, it follows that $f(r) \neq f(s)$. Hence there must be at least $|\text{chef}_T(D)|$ distinct groups in the range of f, i.e., $\text{gr}(T) \geq |\text{chef}_T(D)|$.

7. For $n = 5$, this result is attained as follows:

Resident	Dishes
Amy	d_1, d_2
Benton	d_2, d_3
Carol	d_3, d_4
Devin	d_4, d_5
Emma	d_5, d_1

For each dish D, note that $\text{chef}_T(D) = 2$. But $\text{gr}(T) > 2$, because if T had at most two groups, at least one of them would contain three people, and choosing any three people will result in a common dish that two of them can cook. Hence T is heterogeneous.

For $n \geq 6$, it suffices to assign dishes to residents so that there are three people who must be in different groups and that no dish is cooked by more than two people, which guarantees that $\text{gr}(T) \geq 3$ and $\text{chef}_T(D) \leq 2$ for each dish D.

Resident	Dishes
Amy	d_1, d_2
Benton	d_1, d_3
Carol	d_2, d_3
Devin	d_4, d_5
Emma	d_5, d_6

Note that Devin's and Emma's dishes are actually irrelevant to the situation, so long as they do not cook any of d_1, d_2, d_3, which already have two chefs each. Thus we can adjust this setup for $n = 7$ by setting Devin's dishes as d_4, d_5 and Emma's dishes as d_6, d_7. (In this last case, Devin and Emma are extremely compatible: they can both be put in a group with anyone else in the town!)

8. a. Because the town is full, each pair of dishes is cooked by exactly one resident, so it is simplest to identify residents by the pairs of dishes they cook. Suppose the first resident cooks (d_1, d_2),

the second resident (d_2, d_3), the third resident (d_3, d_4), and so on, until the sixth resident, who cooks (d_6, d_1). Then there are 8 choices for d_1 and 7 choices for d_2. There are only 6 choices for d_3, because $d_3 \neq d_1$ (otherwise two residents would cook the same pair of dishes). For $k > 3$, the requirement that no two intermediate residents cook the same dishes implies that d_{k+1} cannot equal any of $d_1, \ldots, d_{k-1}$, and of course d_k and d_{k+1} must be distinct dishes. Hence there are $8 \cdot 7 \cdot 6 \cdot 5 \cdot 4 \cdot 3 = 20{,}160$ six-person resident cycles, not accounting for different starting points in the cycle and the two different directions to go around the cycle. Taking these into account, there are $20{,}160/(6 \cdot 2) = 1{,}680$ distinguishable resident cycles.

b. Using the logic from 8a, there are $d(d-1)\cdots(d-k+1)$ choices for $d_1, d_2, \ldots, d_k$. To account for indistinguishable cycles, divide by k possible starting points and 2 possible directions, yielding $\dfrac{d(d-1)\cdots(d-k+1)}{2k}$ or $\dfrac{d!}{2k(d-k)!}$ distinguishable resident cycles.

9. Note that for every $D \in \text{dish}(T), \text{chef}_T(D) \leq 2$, because otherwise, $r_1, r_2, \ldots, r_n$ could not be a resident cycle. Without loss of generality, assume the cycle is $r_1, r_2, \ldots, r_n$. If n is even, assign resident r_i to group 1 if i is odd, and to group 2 if i is even. This is a valid group assignment, because the only pairs of residents who cook the same dish are (r_i, r_{i+1}) for $i = 1, 2, \ldots, n-1$ and (r_n, r_1). In each case, the residents are assigned to different groups. This proves $\text{gr}(T) = 2$, so T is homogeneous.

On the other hand, if n is odd, suppose for the sake of contradiction that there are only two groups. Then either r_1 and r_n are in the same group, or for some i, r_i and r_{i+1} are in the same group. In either case, two residents in the same group share a dish, contradicting the requirement that no members of a group have a common dish. Hence $\text{gr}(T) \geq 3$ when n is odd, making T heterogeneous.

10. a. First note that the condition $|\text{chef}_T(D)| = 2$ for all D implies that $\text{pop}(T) = |\text{dish}(T)|$, using the equation from problem 5. So for the town in question, the population of the town equals the number of dishes in the town. Because no two chefs cook the same pair of dishes, it is impossible for such a town to have exactly two residents, and because each dish is cooked by exactly two chefs, it is impossible for such a town to have only one resident.

The claim is true for towns of three residents satisfying the conditions: such towns must have one resident who cooks dishes d_1 and d_2, one resident who cooks dishes d_2 and d_3, and one resident who cooks dishes d_3 and d_1, and those three residents form a cycle. So proceed by (modified) strong induction: assume that for some $n > 3$ and for all positive integers k such that $3 \leq k < n$, every town T with k residents and $|\text{chef}_T(D)| = 2$ for all $D \in \text{dish}(T)$ can be divided into a finite number of resident cycles such that each resident belongs to exactly one of the cycles. Let T_n be a town of n residents, and arbitrarily pick resident r_1 and dishes d_1

and d_2 cooked by r_1. Then there is exactly one other resident r_2 who also cooks d_2 (because $|\text{chef}_{T_n}(d_2)| = 2$). But r_2 also cooks another dish, d_3, which is cooked by another resident, r_3. Continuing in this fashion, there can be only two outcomes: either the process exhausts all the residents of T_n, or there exists some resident r_m, $m < n$, who cooks the same dishes as r_{m-1} and r_ℓ for $\ell < m - 1$.

In the former case, r_n cooks another dish; but every dish besides d_1 is already cooked by two chefs in T_n, so r_n must also cook d_1, closing the cycle. Because every resident is in this cycle, the statement to be proven is also true for T_n.

In the latter case, the same logic shows that r_m cooks d_1, also closing the cycle, but there are other residents of T_n who have yet to be accounted for. Let $C_1 = \{r_1, \ldots, r_m\}$, and consider the town T' whose residents are $T_n \setminus C_1$. Each of dishes $d_1, \ldots, d_m$ is cooked by two people in C_1, so no chef in T' cooks any of these dishes, and no dish in T' is cooked by any of the people in C_1 (because each person in C_1 already cooks two dishes in the set $\text{dish}(C_1)$). Thus $|\text{chef}_{T'}(D)| = 2$ for each D in $\text{dish}(T')$. It follows that $\text{pop}(T') < \text{pop}(T)$ but $\text{pop}(T') > 0$, so by the inductive hypothesis, the residents of T' can be divided into disjoint resident cycles. Thus the statement is proved by strong induction.

b. In order for T to be homogeneous, it must be possible to partition the residents into exactly two dining groups. First apply 10a to divide the town into finitely many resident cycles C_i, and assume towards a contradiction that such a group assignment $f : T \to \{1, 2\}$ exists. If $\text{pop}(T)$ is odd, then at least one of the cycles C_i must contain an odd number of residents; without loss of generality, suppose this cycle to be C_1, with residents $r_1, r_2, \ldots, r_{2k+1}$. (By the restrictions noted in part a, $k \geq 1$.) Now because r_i and r_{i+1} cook a dish in common, $f(r_i) \neq f(r_{i+1})$ for all i. Thus if $f(r_1) = 1$, it follows that $f(r_2) = 2$, and that $f(r_3) = 1$, etc. So $f(r_i) = f(r_1)$ if i is odd and $f(r_i) = f(r_2)$ if i is even; in particular, $f(r_{2k+1}) = f(1)$. But that equation would imply that r_1 and r_{2k+1} cook no dishes in common, which is impossible if they are the first and last residents in a resident cycle. So no such group assignment can exist, and $\text{gr}(T) \geq 3$. Hence T is heterogeneous.

11. a. In problem 5, it was shown that

$$2\text{pop}(T) = \sum_{D \in \text{dish}(T)} |\text{chef}_T(D)|.$$

Therefore $\sum_{D \in \text{dish}(T)} |\text{chef}_T(D)|$ is even. But if $|\text{chef}_T(D)| = 3$ for all $D \in \text{dish}(T)$, then the sum is simply $3\,|\text{dish}(T)|$, so $|\text{dish}(T)|$ must be even.

b. By problem 6, it must be the case that $\text{gr}(T) \geq 3$. Let $C = \{r_1, r_2, \ldots, r_n\}$ denote a resident cycle such that for every dish $D \in \text{dish}(T)$, there exists a chef in C that can prepare D. Each resident is a chef for two dishes, and every dish can be made by two residents in C (although by three in T). Thus the number of residents in the resident cycle C is equal to $|\text{dish}(T)|$, which was proved to be even in the previous part.

Define a group assignment by setting

$$f(r) = \begin{cases} 1 & \text{if } r \notin C \\ 2 & \text{if } r = r_i \text{ and } i \text{ is even} \\ 3 & \text{if } r = r_i \text{ and } i \text{ is odd.} \end{cases}$$

For any $D \in \text{dish}(T)$, there are exactly three D-chefs, and exactly two of them belong to the resident cycle C. Hence exactly one of the D-chefs r will have $f(r) = 1$. The remaining two D-chefs will be r_i and r_{i+1} for some i, or r_1 and r_n. In either case, the group assignment f will assign one of them to 2 and the other to 3. Thus any two residents who make a common dish will be assigned different groups by f, so f is a valid group assignment, proving that $\text{gr}(T) = 3$.

12. a. From problem 5,

$$2\text{pop}(T) = \sum_{D \in \text{dish}(T)} |\text{chef}_T(D)|.$$

Because $|\text{chef}_T(D)| = k$ for all $D \in \text{dish}(T)$, the sum is $k \cdot \text{dish}(T)$. Thus $2\text{pop}T = k \cdot \text{dish}(T)$, and so $k \cdot \text{dish}(T)$ must be even. By assumption, $|\text{dish}(T)|$ is odd, so k must be even.

b. Suppose for the sake of contradiction that there is some n for which the group $R = \{r \in T \mid f(r) = n\}$ has a D-chef for every dish D. Because f is a group assignment and f assigns every resident of R to group n, no two residents of R make the same dish. Thus for every $D \in \text{dish}(T)$, exactly one resident of R is a D-chef; and each D-chef cooks exactly one other dish, which itself is not cooked by anyone else in R. Thus the dishes come in pairs: for each dish D, there is another dish D' cooked by the D-chef in R and no one else in R. However, if the dishes can be paired off, there must be an even number of dishes, contradicting the assumption that $|\text{dish}(T)|$ is odd. Thus for every n, the set $\{r \in T \mid f(r) = n\}$ must be missing a D-chef for some dish D.

c. Let f be a group assignment for T, and let $R = \{r \in T \mid f(r) = 1\}$. From problem 12b, there must be some $D \in \text{dish}(T)$ with no D-chefs in R. Moreover, f cannot assign two D-chefs to the same group, so there must be at least k other groups besides R. Hence there are at least $1 + k$ different groups, so $\text{gr}(T) > k$.

13. a. Fix $D \in \text{dish}(\mathcal{F}_d)$. Then for every other dish $D' \in \text{dish}(\mathcal{F}_d)$, there is exactly one chef who makes both D and D', hence $|\text{chef}_{\mathcal{F}_d}(D)| = d - 1$, which is even because d is odd. Thus for

each $D \in \text{dish}(\mathcal{F}_d)$, $|\text{chef}_{\mathcal{F}_d}(D)|$ is even. Because $|\text{dish}(\mathcal{F}_d)| = d$ is odd and $|\text{chef}_{\mathcal{F}_d}(D)| = d - 1$ for every dish in $\mathcal{F}_d$, problem 12c applies, hence $\text{gr}(\mathcal{F}_d) > d - 1$.

Label the dishes $D_1, D_2, \ldots, D_d$, and label the residents $r_{i,j}$ for $1 \le i < j \le d$ so that $r_{i,j}$ is a D_i-chef and a D_j-chef. Define $f : \mathcal{F}_d \to \{0, 1, \ldots, d-1\}$ by letting $f(r_{i,j}) \equiv i + j \bmod d$.

Suppose that $f(r_{i,j}) = f(r_{k,\ell})$, so $i + j \equiv k + \ell \bmod d$. Then $r_{i,j}$ and $r_{k,\ell}$ are assigned to the same group, which is a problem if they are different residents but are chefs for the same dish. This overlap occurs if and only if one of i and j is equal to one of k and ℓ. If $i = k$, then $j \equiv \ell \bmod d$. As j and ℓ are both between 1 and d, the only way they could be congruent modulo d is if they were in fact equal. That is, $r_{i,j}$ is the same resident as $r_{k,\ell}$. The other three cases ($i = \ell$, $j = k$, and $j = \ell$) are analogous. Thus f is a valid group assignment, proving that $\text{gr}(\mathcal{F}_d) \le d$. Therefore $\text{gr}(\mathcal{F}_d) = d$.

b. In problem 4, it was shown that the sequence $\text{gr}(\mathcal{F}_2), \text{gr}(\mathcal{F}_3), \ldots$ is nondecreasing. If d is even, $\text{gr}(\mathcal{F}_d) \ge \text{gr}(\mathcal{F}_{d-1})$, and because $d - 1$ is odd, problem 13a applies: $\text{gr}(\mathcal{F}_{d-1}) = d - 1$. Hence $\text{gr}(\mathcal{F}_d) \ge d - 1$. Now it suffices to show that $\text{gr}(\mathcal{F}_d) \le d - 1$ by exhibiting a valid group assignment $f : \mathcal{F}_d \to \{1, 2, \ldots, d-1\}$.

Label the dishes $D_1, \ldots, D_d$, and label the residents $r_{i,j}$ for $1 \le i < j \le d$ so that $r_{i,j}$ is a D_i-chef and a D_j-chef. Let $R = \{r_{i,j} \mid i, j \ne d\}$. That is, R is the set of residents who are not D_d-chefs. Using 1c, R is a full town with $d - 1$ dishes, so from 12a, it has a group assignment $f : R \to \{1, 2, \ldots, d-1\}$. For each $D_i \in \text{dish}(\mathcal{F}_d)$, $i \ne d$, $|\text{chef}_R(D_i)| = d - 2$. Because there are $d - 1$ groups and $|\text{chef}_R(D_i)| = d - 2$, exactly one group n_i must not contain a D_i-chef for each dish D_i.

It cannot be the case that $n_i = n_j$ for $i \ne j$. Indeed, suppose for the sake of contradiction that $n_i = n_j$. Without loss of generality, assume that $n_i = n_j = 1$ (by perhaps relabeling the dishes). Then any resident $r \in R$ assigned to group 1 (that is, $f(r) = 1$) would be neither a D_i-chef nor a D_j-chef. The residents in R who are assigned to group 1 must all be chefs for the remaining $d - 3$ dishes. Because each resident cooks two dishes, and no two residents of group 1 can make a common dish,

$$|\{r \in R \mid f(r) = 1\}| \le \frac{d - 3}{2}.$$

For each of the other groups $2, 3, \ldots, d - 1$, the number of residents of R in that group is no more than $(d - 1)/2$, because there are $d - 1$ dishes in R, each resident cooks two dishes, and no two residents in the same group can make a common dish. However, because $d - 1$ is odd, the size of any group is actually no more than $(d - 2)/2$. Therefore

$$|R| = \sum_{k=1}^{d-1} |\{r \in R \mid f(r) = k\}|$$

$$= |\{r \in R \mid f(r) = 1\}| + \sum_{k=2}^{d-1} |\{r \in R \mid f(r) = k\}|$$

$$\leq \frac{d-3}{2} + \sum_{k=2}^{d-1} \frac{d-2}{2}$$

$$= \frac{d-3}{2} + \frac{(d-2)^2}{2}$$

$$= \frac{d^2 - 3d + 1}{2} < \frac{d^2 - 3d + 2}{2} = |R|.$$

This is a contradiction, so it must be that $n_i \neq n_j$ for all $i \neq j$, making f a valid group assignment on $\mathcal{F}_d$. Hence $\mathrm{gr}(\mathcal{F}_d) = d - 1$.

CHAPTER 46

SOLUTIONS TO UNC CHARLOTTE SUPER COMPETITION

46.1 2017 Comprehensive

1. **Solution** (C). The degree of the numerator is $7 \cdot 8 + 9$ and the degree of the denominator is 9, so the quotient has degree $65 - 9 = 56$.

2. **Solution** (A). Since the radius of the ball is 1, its center is moving in a smaller square S: $-2 \leq x \leq 2$, $-2 \leq y \leq 2$ (lifted to the height of 1 above the floor), getting reflected off its sides. Upon leaving the point $(2, 0)$, it attains the boundary of the square S successively at the points $(0, 2)$, $(-2, 0)$, $(0, -2)$, and then returns to the initial point $(2, 0)$ having traveled a distance of $L = (4)2\sqrt{2} = 8\sqrt{2}$. We have $11 < L < 12$ since $L^2 = 128$ is between 11^2 and 12^2.

3. **Solution** (C). The first equation of the system represents the circle U of radius 3 centered at the origin $(0, 0)$. The second equation represents the circle V_r of radius r centered at the point $(8, 6)$ whose distance from the origin equals 10. The uniqueness of the solution of the system means that the two circles are tangent to one another. This occurs for two values of r: $r = 10 - 3 = 7$ (V_r is tangent to U from outside) and $r = 10 + 3 = 13$ (U is tangent to V_r from inside). Hence $s = 7 + 13 = 20$.

4. **Solution** (D). Translate the center of the region from $(1, 1)$ to the origin and note that the resulting region described by $|x| + |y| \leq 2$ is a square with vertices $(0, \pm 2)$, $(\pm 2, 0)$, so its area is 8.

5. **Solution** (C). Number the points from 1 to 6 in clockwise fashion. Once 1's two vertex partners are selected, the figure is determined. But note that 1's partners can only be 2,3; 6,2; or 5,6.

6. Solution (E). Let x, y, z be the width, depth, and height of the box, respectively. Then $xy = 40$, $xz = 48$ and $yz = 30$. Hence

$$x^2 y^2 z^2 = (xy)(xz)(yz) = 40 \cdot 48 \cdot 30 = 57600.$$

Therefore, $xyz = 240$, and $z = (xyz)/(xy) = 240/40 = 6$.

7. Solution (E). The parabola, and hence the whole picture, is symmetric with respect to the y-axis. Therefore, the midpoint of the bottom side of the square is at the origin, so that the upper right vertex (x, y) of the square satisfies two equations $y = 2x$ and $y = 15 - x^2$. They imply that $2x = 15 - x^2$. This quadratic equation has a unique nonnegative root $x = 3$. Therefore, the sides of the square have length $2 \cdot 3 = 6$.

8. Solution (C). Suppose there are usually s students in the math class. Since exactly $1/4$ know how to play poker, s must be a multiple of 4. Similarly, $s - 3$ must be a multiple of 5. The only multiple of 4 between 15 and 40 that is 3 more than a multiple of 5 is $s = 28$. On Wednesday, there were $28 - 3 = 25$ students, and 5 knew how to play poker.

9. Solution (C). If all the 20 flowers were lilacs, then there would be a total of $20 \cdot 4 = 80$ petals. Hence some lilacs should be replaced with pansies. How many? Each such replacement adds one more petal. Therefore, $92 - 80 = 12$ lilacs should be replaced with pansies.

10. Solution (E). The number 8 appears in the ones place 85 times $(8, 18, 28, \ldots, 848)$, in the tens place 80 times $(80\ldots, 89;\ 180, \ldots, 189;\ 280\ldots, 289; \ldots; 780, \ldots, 789)$, and in the hundreds place 53 times $(800, 801, \ldots, 852)$. Hence it appears a total of $85 + 80 + 53 = 218$ times.

11. Solution (E). The number of ways four people can sit in a row is $4! = 24$. The number of favorable outcomes is four: (two possibilities for Jim and Paula) $\times$ (two ways to seat the second couple). Thus, the probability sought for equals $4/24 = 1/6$.

12. Solution (A). Let s be the radius of each small circle and let m be the radius of the medium size circle. We have $m + s = 10$. The centers of the four small circles are the vertices of a square with side length $2s$ and diagonal $2m$. So we have $s = m/\sqrt{2}$. Thus

$$m = \frac{10}{1 + 1/\sqrt{2}} = \frac{20}{2 + \sqrt{2}} = \frac{20(2 - \sqrt{2})}{4 - 2} = 10(2 - \sqrt{2}).$$

13. Solution (C). Set $x = 4x(1 - x)$, and solve to find $x = 3/4$.

14. Solution (D). First note that $a = 1$ is impossible since this would imply that $1/a + 1/b + 1/c > 1$. Similarly, $a \geq 3$ would imply that $1/a + 1/b + 1/c \leq 1/3 + 1/4 + 1/5 < 1$, which is not possible either. Therefore, $a = 2$. Now we need to find integers b and c such that $3 \leq b < c$ and

$1/b + 1/c = 1/2$. Again, $b \geq 4$ is ruled out since in this case $1/b + 1/c \leq 1/4 + 1/5 < 1/2$. Therefore, $b = 3$ and hence $c = 6$, so that $a + b + c = 2 + 3 + 6 = 11$.

15. **Solution** (D). Let the roots of the quadratic equation be u and $2u$. Then the equation can be rewritten in the form $(x - u)(x - 2u) = 0$, or, equivalently, $x^2 - 3ux + 2u^2 = 0$. It follows that $3u = 9$, so that $u = 3$ and hence $a = 2 \cdot 3^2 = 18$.

16. **Solution** (D). Let x_1 and x_2 be the roots of the equation. The latter can be rewritten as $(x - x_1)(x - x_2) = 0$ so that $x_1 + x_2 = 7$ and $x_1 x_2 = a$. Then $x_1^2 + x_2^2 = (x_1 + x_2)^2 - 2x_1 x_2 = 7^2 - 2a$. It follows that $49 - 2a = 39$ and hence $a = (49 - 39)/2 = 5$.

17. **Solution** (B). Let $a < b < c$ be the lengths of the legs and the hypotenuse of the triangle. Then $c - b = b - a = h$ for some $h > 0$. By the Pythagorean theorem, $(b - h)^2 + b^2 = (b + h)^2$ or $(b+h)^2 - (b-h)^2 = b^2$. Then $4bh = b^2$. It follows that $h = b/4$. The equality $(b-h) + b + (b+h) = 48$ implies that $b = 16$ so that $h = 4$ and the lengths of the legs are 12 and 16. Therefore, the area of the triangle equals $(1/2)12 \cdot 16 = 96$.

18. **Solution** (A). The lengths of the legs of the triangle are a and $\left(\dfrac{3}{2}\right)a$ for some $a > 0$. By the Pythagorean theorem, $a^2 + \left(\left(\dfrac{3}{2}\right)a\right)^2 = 52$, so that $\left(\dfrac{13}{4}\right)a^2 = 52$. It follows that the legs have lengths $a = 4$ and $\left(\dfrac{3}{2}\right)4 = 6$. Therefore, the area of the triangle is $\left(\dfrac{1}{2}\right)6 \cdot 4 = 12$.

19. **Solution** (B). The amount of non-watery part in 20 pounds of fresh cucumbers is $20 \cdot 0.1 = 2$ pounds. The same amount after a week makes 20% or $1/5$ of the total weight. Hence, the weight is $2 \cdot 5 = 10$.

20. **Solution** (D). We rewrite the equation as $\sqrt{a/9} = b/9$ or $9a = b^2$. This equation is satisfied by 4 pairs of digits: $(0,0)$, $(1,3)$, $(4,6)$, $(9,9)$.

46.2   2017 Level 3

1. **Solution** (B). The second equation, upon completing the squares, becomes $(x - 5)^2 + (y - 12)^2 = 64$. Therefore, both equations represent circles. Their centers are 13 units apart and the sum of their radii is also 13.

2. **Solution** (D). The sum of the coefficients of any polynomial $p(x)$ equals $p(1)$. Therefore, $s = h(1) = f(g(1))$. We have $g(1) = 0$ and hence $h(1) = f(0) = 27$.

3. **Solution** (B). The graph of the function $g(x) = |2x|$ on $[-10, 10]$ is a letter "V" with vertices at the points $(-10, 20)$, $(0, 0)$ and $(10, 20)$. The graph of $h(x) = g(x) - 10$ is the graph of $g(x)$ shifted downward by 10 units – the half of its height. The graph of $f(x) = |h(x)|$ is obtained from the

graph of $g(x)$ by reflecting its part lying below the x-axis (a smaller "V") across the x-axis, thus creating a letter "W".

4. **Solution** (D). The first equation $y + 2x = 0$ describes a line, while the second equation $(x - 3)^2 + (y - 6)^2 = r^2$ describes a circle of radius r centered at the point $(3, 6)$. The uniqueness of the simultaneous solution of the two equations means that the circle is tangent to the line, that is, r is the shortest distance from the circle's center to the line. To calculate r, consider the triangle with vertices $P(3, 6)$, $Q(-3, 6)$ and $O(0, 0)$. Considering the sides PQ or OQ as its bases, we obtain two expressions for the triangle's area which are, therefore, equal: $(1/2)6 \cdot 6 = (1/2)\sqrt{6^2 + 3^2} \cdot r$. Consequently, $r = (6 \cdot 6)/\sqrt{45} = 12/\sqrt{5}$. It follows that $5 < r < 6$ since $r^2 = \frac{144}{5} = 28\frac{4}{5}$ is between 5^2 and 6^2.

5. **Solution** (E). The centers are $(0, 0)$ $(4, 0)$ and $(7, 8)$, so the area of the triangle is $\frac{1}{2}(4)(8) = 16$

6. **Solution** (B). The second equation implies that $xy = z^2 + 1 \geq 1 > 0$, so that x and y are of the same sign. It follows then from the first equation that they are positive. Hence the Arithmetic Mean – Geometric Mean Inequality applies, which says that $(x+y)/2 \geq \sqrt{xy}$. In our case, $(x+y)/2 = 1$ while $\sqrt{xy} \geq 1$ so the inequality turns into equality; in this case it implies that $x = y$. Therefore, $x = y = 1$ and $z = 0$.

7. **Solution** (A). The numbers $x \geq 0$ for which the function $f(x) = \sqrt{a - \sqrt{a + x}} - x$ is defined form an interval $I = [0, b]$, where $b = a^2 - a$. In this interval the function $f(x)$ is continuous and decreasing from $\sqrt{a - \sqrt{a}} > 0$ to $-b < 0$. Therefore, there is exactly one value x_0 in this interval such that $f(x_0) = 0$. Since $f(0) > 0$ and $f(b) < 0$, we have $0 < x_0 < b$, so that x_0 is positive as required.

8. **Solution** (D). Let x, y, z be the width, depth, and height of the box, respectively. Then $xy = 40$, $xz = 48$ and $yz = 30$. Hence

$$x^2 y^2 z^2 = (xy)(xz)(yz) = 40 \cdot 48 \cdot 30 = 57600.$$

Therefore, $xyz = 240$, and $z = (xyz)/(xy) = 240/40 = 6$.

9. **Solution** (E). The parabola, and hence the whole picture, is symmetric with respect to the y-axis. Therefore, the midpoint of the bottom side of the square is at the origin, so that the upper right vertex (x, y) of the square satisfies two equations $y = 2x$ and $y = 15 - x^2$. They imply that $2x = 15 - x^2$. This quadratic equation has one nonnegative root $x = 3$. Therefore, the sides of the square have length $2 \cdot 3 = 6$.

10. **Solution** (C). If all the 20 flowers were lilacs, then there would be a total of $20 \cdot 4 = 80$ petals. Hence some lilacs should be replaced with pansies. How many? Each such replacement adds one more petal. Therefore, $92 - 80 = 12$ lilacs should be replaced with pansies.

11. Solution (D). The pair of equal digits $a = c$ can be selected in 9 ways $(1, 2, \ldots, 9)$ while, for any such selection, b can be chosen in 10 different ways $(0, 1, 2, \ldots, 9)$. Hence the number sought for is $9 \cdot 10 = 90$.

12. Solution (A). Let x be the number sought for. Then $2x = 3x^3$, or, equivalently, $x(2 - 3x^2) = 0$. Since $x > 0$, it follows that $x^2 = 2/3$ so that $x = \sqrt{2/3}$.

13. Solution (D). First note that $a = 1$ is impossible since this would imply that $1/a + 1/b + 1/c > 1$. Similarly, $a \geq 3$ would imply that $1/a + 1/b + 1/c \leq 1/3 + 1/4 + 1/5 < 1$, which is not possible either. Therefore, $a = 2$. Now we need to find integers b and c such that $3 \leq b < c$ and $1/b + 1/c = 1/2$. Again, $b \geq 4$ is ruled out since in this case $1/b + 1/c \leq 1/4 + 1/5 < 1/2$. Therefore, $b = 3$ and hence $c = 6$, so that $a + b + c = 2 + 3 + 6 = 11$.

14. Solution (D). Let the roots of the quadratic equation be u and $2u$. Then the equation can be rewritten in the form $(x - u)(x - 2u) = 0$, or, equivalently, $x^2 - 3ux + 2u^2 = 0$. It follows that $3u = 9$, so that $u = 3$ and hence $a = 2 \cdot 3^2 = 18$.

15. Solution (D). Let x_1 and x_2 be the roots of the equation. The latter can be rewritten as $(x - x_1)(x - x_2) = 0$ so that $x_1 + x_2 = 7$ and $x_1 x_2 = a$. Then $x_1^2 + x_2^2 = (x_1 + x_2)^2 - 2x_1 x_2 = 7^2 - 2a$. It follows that $49 - 2a = 39$ and hence $a = (49 - 39)/2 = 5$.

16. Solution (C). We have
$$S = \cot 1° \, \cot 2° \, \cot 3° \, \ldots \cot 89° \quad \text{and} \quad S = \cot 89° \cot 88° \cot 87° \ldots \cot 1°.$$
Note that $\cot(90° - x°) = \tan x°$ and hence $\cot(90° - x°) \cdot \cot x° = 1$. It follows that $S^2 = (\cot 1° \cot 89°)(\cot 2° \cot 88°) \ldots (\cot 89° \cot 1°) = 1$. Since $S > 0$, we also have $S = 1$.

17. Solution (B). Let $a < b < c$ be the lengths of the legs and the hypotenuse of the triangle. Then $c - b = b - a = h$ for some $h > 0$. By the Pythagorean theorem, $(b - h)^2 + b^2 = (b + h)^2$ or $(b+h)^2 - (b-h)^2 = b^2$. Then $4bh = b^2$. It follows that $h = b/4$. The equality $(b-h)+b+(b+h) = 48$ implies that $b = 16$ so that $h = 4$ and the lengths of the legs are 12 and 16. Therefore, the area of the triangle equals $(1/2)12 \cdot 16 = 96$.

18. Solution (A). The lengths of the legs of the triangle are a and $\left(\frac{3}{2}\right)a$ for some $a > 0$. By the Pythagorean theorem, $a^2 + \left(\left(\frac{3}{2}\right)a\right)^2 = 52$, so that $\left(\frac{13}{4}\right)a^2 = 52$. It follows that the legs have lengths $a = 4$ and $\left(\frac{3}{2}\right)a = 6$. Therefore, the area of the triangle is $\left(\frac{1}{2}\right)6 \cdot 4 = 12$.

19. Solution (D). We have $\sin \alpha = \cos\left(\frac{\pi}{2} - \alpha\right)$ so that
$$T_{10}(\sin \alpha) = T_{10}\left(\cos\left(\frac{\pi}{2} - \alpha\right)\right) = \cos\left(10\left(\frac{\pi}{2} - \alpha\right)\right)$$

$$= \cos(5\pi - 10\alpha) = -\cos(-10\alpha) = -\cos 10\alpha.$$

20. Solution (D). Let $u = \sqrt{1-x^2}$ and $v = \sqrt{1-y^2}$. Using the Arithmetic Mean – Geometric Mean Inequality, we obtain:

$$f(x,y) = xv + yu \le \left(\frac{1}{2}\right)(x^2 + v^2) + \left(\frac{1}{2}\right)(y^2 + u^2) = \left(\frac{1}{2}\right)(x^2 + u^2 + y^2 + v^2) = \left(\frac{1}{2}\right)(1+1) = 1.$$

Therefore, $f(x,y) \le 1$ for all x, y in the square Q. On the other hand, $f(1,0) = 1$.

CHAPTER 47

SOLUTIONS TO STATE MATH CONTEST OF NORTH CAROLINA PROBLEMS

47.1 2016 PART I: 20 MULTIPLE CHOICE PROBLEMS

1. **Solution** (B). Notice that $\left(1+\dfrac{1}{i}\right)\left(1-\dfrac{1}{i+1}\right) = 1$. Thus, the product is equal to $1+\dfrac{1}{n}$.

2. **Solution** (C). The equation $x^2 + y^2 = 1 - 2y$ is equivalent to $x^2 + (y+1)^2 = 2$ whose graph is a circle with center at $(0,-1)$ and radius $\sqrt{2}$. This circle intersect the x-axis at $(-1,0)$ and $(1,0)$. The area of the region below the circle and above the x-axis is equal to the area of one quarter of a circle with radius $\sqrt{2}$ minus the area of a right triangle with legs $\sqrt{2}$, i.e. $\dfrac{\pi}{2} - 1$.

3. **Solution** (C). Let $a = \sqrt{x}$. Then
$$a^2 - \frac{1}{a^2} = a + \frac{1}{a},$$
which is equivalent to
$$\left(a - \frac{1}{a}\right)\left(a + \frac{1}{a}\right) = a + \frac{1}{a}.$$
Since $x > 1$, we have $a > 1$; form the previous equation we get $a - \dfrac{1}{a} = 1$. Hence, $a^2 - 2 + \dfrac{1}{a^2} = 1$, which implies $a^2 + \dfrac{1}{a^2} = 3$. Therefore, $x + \dfrac{1}{x} = 3$.

4. **Solution** (A). Let A has coordinates (a,b). Then B has coordinates $(-a,-b)$. Both points are on the parabola; hence $b = 2a^2 + 4a - 2$ and $-b = 2a^2 - 4a - 2$. We get that $a = 1$ or $a = -1$. If $a = 1$, then $b = 4$; if $a = -1$, then $b = -4$. The length of the line segment joining A and B is $2\sqrt{17}$.

5. **Solution** (E). Let $\overline{abc}$ be a 3-digit number that is 34 times the sum of its digits. Then $100a + 10b + c = 34(a+b+c)$. Then $22a - 11c = 8b$, which implies that b is divisible by 11. Since b is a digit,

we get $b = 0$ and $c = 2a$. The 3-digit numbers that satisfy this property are: 102, 204, 306, and 408. Their sum is 1020.

6. **Solution** (D). Notice that the function f has the same value at points symmetric about 31. Since f has exactly three real roots, one of the roots must be 31, and the other two roots must be $31 - a$ and $31 + a$ for some nonzero real number a. The sum of the roots is 93.

7. **Solution** (A). Since a, b, and c are members of a geometric progression, in the given order, then $b^2 = ac$. Then

$$
\begin{aligned}
\frac{\log_b 3 \left(\log_{a^2} c - \log_c \sqrt{a}\right)}{\log_a 9 - 2\log_c 3} &= \frac{\frac{\log_3 3}{\log_3 b}\left(\frac{\log_3 c}{\log_3 a^2} - \frac{\log_3 \sqrt{a}}{\log_3 c}\right)}{\frac{\log_3 9}{\log_3 a} - 2\frac{\log_3 3}{\log_3 c}} \\
&= \frac{\frac{1}{\log_3 b}\left(\frac{\log_3 c}{2\log_3 a} - \frac{\frac{1}{2}\log_3 a}{\log_3 c}\right)}{\frac{2}{\log_3 a} - 2\frac{1}{\log_3 c}} \\
&= \frac{1}{4}\frac{1}{\log_3 b}(\log_3 c + \log_3 a) \\
&= \frac{1}{4}\frac{\log_3 ac}{\log_3 b} = \frac{\log_3 b^2}{4\log_3 b} \\
&= \frac{2\log_3 b}{4\log_3 b} = \frac{1}{2}.
\end{aligned}
$$

8. **Solution** (C). Let x_1, x_2, and x_3 be the integer roots of the given equation. Then $x_1 + x_2 + x_3 = 0$, $x_1 x_2 + x_2 x_3 + x_1 x_3 = -13$, and $x_1 x_2 x_3 = -a$. Then

$$x_1^2 + x_2^2 + x_3^2 = (x_1 + x_2 + x_3)^2 - 2(x_1 x_2 + x_2 x_3 + x_1 x_3) = 26.$$

Since x_1, x_2, x_3 are all integers, their squares are also integer numbers. We get that $\{x_1^2, x_2^2, x_3^2\} = \{0, 1, 25\}$ or $\{x_1^2, x_2^2, x_3^2\} = \{1, 9, 16\}$. If

$$\{x_1, x_2, x_3\} = \{\pm 0, \pm 1, \pm 5\},$$

then the equation $x_1 + x_2 + x_3 = 0$ is not satisfied. If

$$\{x_1, x_2, x_3\} = \{\pm 1, \pm 3, \pm 4\},$$

then there are two possibilities for the roots: $\{1, 3, -4\}$ and $\{-1, -3, 4\}$. Then $a = -12$ or $a = 12$.

9. **Solution** (A). If n is not divisible by neither 5 nor 7, Adam will finish when he moves forward $5n$ spaces and David will finish when he moves forward $7n$ spaces. Hence, if 5 does not divide n and 7 does not divide n, Adam will win.

If n is divisible by 5, but not by 7, then Adam will win; he will finish in $\dfrac{n}{5}$ moves and David will finish in $7n$ moves. If n is divisible by 7, by similar arguments as in the previous case, David will win. If n is divisible by both 5 and 7, then David will win. Hence, David wins if and only if n is divisible by 7. There are 13 2-digit numbers divisible by 7. Hence, Adam will win in all other cases, $90 - 13 = 77$. The probability that Adam will win is $\dfrac{77}{90}$.

10. Solution (B).

$$
\begin{aligned}
\sin^3 18° + \sin^2 18° &= \sin^2 18°(\sin 18° + 1) \\
&= \sin^2 18°(\sin 18° + \sin 90°) \\
&= 2\sin^2 18° \sin 54° \cos 36° \\
&= 2\sin^2 18° \cos^2 36° \\
&= \frac{2\sin^2 18° \cos^2 18° \cos^2 36°}{\cos^2 18°} \\
&= \frac{\sin^2 36° \cos^2 36°}{2\cos^2 18°} \\
&= \frac{\sin^2 72°}{8\cos^2 18°} \\
&= \frac{\cos^2 18°}{8\cos^2 18°} = \frac{1}{8}.
\end{aligned}
$$

11. Solution (B). Without loss of generality, assume that $\triangle AIB \sim \triangle ABC$. We first determine the pairs of equal angles of both triangles. The angles of $\triangle AIB$ are

$$
\frac{1}{2}\angle A, \ \frac{1}{2}\angle B, \ \text{and} \ \frac{\pi}{2} + \frac{1}{2}\angle C.
$$

It is clear that $\dfrac{1}{2}\angle A$ cannot be equal to $\angle A$, $\dfrac{1}{2}\angle B$ cannot be equal to $\angle B$, and $\dfrac{\pi}{2} + \dfrac{1}{2}\angle C$ cannot be equal to $\angle C$. Therefore either

$$
\angle A = \frac{1}{2}\angle B, \angle B = \frac{\pi}{2} + \frac{1}{2}\angle C, \angle C = \frac{1}{2}\angle A,
$$

or

$$
\angle A = \frac{\pi}{2} + \frac{1}{2}\angle C, \angle B = \frac{1}{2}\angle A, \angle C = \frac{1}{2}\angle B.
$$

In both cases we have that the largest angle is $\dfrac{4}{7}\pi$.

12. Solution (B). Let

$$
x = \arctan\frac{1}{2}, \ y = \arctan\frac{1}{4}, \ \text{and} \ z = \arctan\frac{1}{13}.
$$

Then

$$
\tan x = \frac{1}{2}, \ \tan y = \frac{1}{4}, \ \tan z = \frac{1}{13}, \ \text{and} \ x, y, z \in \left(0, \frac{\pi}{4}\right).
$$

Applying the identity

$$
\tan(\alpha + \beta) = \frac{\tan\alpha + \tan\beta}{1 - \tan\alpha\tan\beta}
$$

twice, we get

$$\tan\left(x + y + z\right) = 1.$$

Hence,

$$\arctan\frac{1}{2} + \arctan\frac{1}{4} + \arctan\frac{1}{13} = x + y + z = \arctan 1 = \frac{\pi}{4}.$$

13. Solution (C). Let D be a point on the circle such as AD is a diameter of the circle. Then

$$AD = 6, \ \angle ADB = 30°, \ \angle ABD = 90°, \ \text{and} \ \angle BAD = 60°.$$

Since $AD = 6$, we get $AB = 3$ and $BD = 3\sqrt{3}$. Since $\angle ACD = 90°$, by Pythagorean Theorem we get $CD = 4\sqrt{2}$. By Ptolemy's Theorem, we have $AB \cdot CD + BD \cdot AC = AD \cdot BC$. Hence, $BC = 2\sqrt{2} + \sqrt{3}$.

14. Solution (C). The first equation is equivalent to

$$x^2 + \frac{1}{2}y(1 + \cos 2\alpha) = \frac{1}{2}x \sin 2\alpha,$$

which is equivalent to

$$x \sin 2\alpha - y \cos 2\alpha = 2x^2 + y.$$

Now we get

$$\sin 2\alpha = \frac{x(2x^2 + y)}{x^2 + y^2}$$

and

$$\cos 2\alpha = -\frac{y(2x^2 + y)}{x^2 + y^2}.$$

Using the Pythagorean identity

$$\sin^2(2\alpha) + \cos^2(2\alpha) = 1$$

and simplifying, we get $\frac{(2x^2+y)^2}{x^2+y^2} = 1$. Hence,

$$4x^2 + 4y = 1.$$

15. Solution (C). It is clear that y must be a positive integer and x could be positive or negative integer. Since $6^{12} = 2^{12}3^{12}$ it follows that $x = \pm 2^a 3^b$ and $y = 2^c 3^d$ for some nonnegative integers a, b, c, d. Then

$$x^2 y^3 = 2^{2a+3c}3^{2b+3d} = 2^{12}3^{12}.$$

Hence, $2a + 3c = 12$ and $2b + 3d = 12$. The solutions of these equations are:

$$(a, c), (b, d) \in \{(0, 4), (3, 2), (6, 0)\}.$$

Any of the values for a can be paired with any of the values for b; then the values of c and d are determined. Hence, there 9 pairs of positive integers (x, y) that satisfied the given equation. Since x can be negative as well, the total number of solutions is 18.

16. Solution (D). Applying Manelaus' Theorem to $\triangle ACM$ and the line BP we get

$$\frac{AQ}{QC} \cdot \frac{CP}{PM} \cdot \frac{MB}{BA} = 1.$$

Sine $\dfrac{CP}{PM} = \dfrac{1}{2}$ and $\dfrac{MB}{BA} = \dfrac{1}{2}$, we get $AQ = 4QC$. This implies $QC = \dfrac{AC}{5}$. Since $AN = NC = \dfrac{AC}{2}$, we have $NQ = \dfrac{3AC}{10}$. Therefore, $\dfrac{CQ + AN}{NQ} = \dfrac{\frac{7AC}{10}}{\frac{3AC}{10}} = \dfrac{7}{3}$.

17. Solution (C). $f(x,y) = (x + y + 1)^2 + 2(y + 1)^2 + 1$. Hence, the minimum of f is 1 and it is achieved for $x = 0$ and $y = -1$.

18. Solution (A).

$$a + b + c + abc = 1 + ab + bc + ca - (1 - a)(1 - b)(1 - c) = 2 - (1 - a)(1 - b)(1 - c) \le 2.$$

However, 2 is achieved when at least one of a, b, or c is 1, which is not possible. Hence, $a + b + c + abc$ does not have a maximum when $a, b, c \in (0, 1)$ and $ab + bc + ca = 1$.

19. Solution (C). Every prime greater than 3 is of form $6k + 1$ or $6k - 1$ for some positive integer k. Since $p + 6$ is prime, we get $p \ne 2$ and $p \ne 3$. Since $p + 10$ is prime, we get $p \ne 6k - 1$. Hence, $p = 6k + 1$ for some positive integer k. If $q > 6$, then q cannot have the form $6s - 1$ ($q + 10$ must be prime); hence, if $q > 6$, then $q = 6s + 1$ for some positive integer s. However, in this case we would get $p + q + 1 = 3(2k + 2s + 1)$, which is not prime. Hence, $q \le 5$. Clearly $q \ne 2$ and $q \ne 5$. Therefore, $q = 3$. To find p, we are looking for prime numbers of form $6k + 1$ such that $p + 6$, $p + 10$, and $p + 4$ are all primes and $1 \le k \le 16$. We get four such numbers: $7, 13, 37, 97$. Therefore, there four ordered pairs of primes that satisfy the given requirements: $(7, 3)$, $(13, 3)$, $(37, 3)$, and $(97, 3)$.

20. Solution (A). Since $\sin x < x$ on $\left(0, \dfrac{\pi}{2}\right)$ and since $f(x) = \cos x$ is decreasing on

$$\left(0, \frac{\pi}{2}\right),$$

we have

$$\sin(\cos x) < \cos x < \cos(\sin x).$$

Then

$$b = \sin(\cos b) < \cos b, \ \cos c < \cos(\sin c) = c.$$

Hence,

$$\cos b - b > 0 = \cos a - a > \cos c - c.$$

Since the function

$$g(x) = \cos x - x$$

is decreasing on $\left(0, \dfrac{\pi}{2}\right)$ and

$$g(b) > g(a) > g(c),$$

we conclude $b < a < c$.

47.2   2016 PART II: 10 INTEGER ANSWER PROBLEMS

1. **Solution** (621). Let $100a + 10b + c$ be the largest number we can make with the cards, $a > b > c$. The second largest number is $100a + 10c + b$. The sum of the largest and the second largest number is $200a + 11(b + c) = 1233$. Hence, $a \le 6$. If $a = 6$, then $b + c = 3$. This implies $b = 2$ and $c = 1$. Therefore, the largest number is 621.

2. **Solution** (3456). Let r and h be the radius and the height of the cylinder. Then $V = \pi r^2 h$ and $A = 2\pi r^2 + 2\pi rh$. From $\pi r^2 h = 2\pi r^2 + 2\pi rh$ we get $rh = 8r + 8h$, which is equivalent to $(r - 8)(h - 8) = 64$. Since $r \ge 1$ and $h \ge 1$, we get $r - 8 > -8$ and $h - 8 > 8$. Hence,

$$(r, h) \in \{(9, 72), (10, 40), (12, 24), (16, 16), (24, 12), (40, 10), (72, 9)\}.$$

The smallest volume of 3456π is obtained for $r = 12$, $h = 24$.

3. **Solution** (6098). Assume $n + 2002 = x^2$ and $n - 2002 = y^2$ for some nonnegative integer numbers x and y. Then $n = \dfrac{1}{2}(x^2 + y^2)$ and $x^2 - y^2 = 4004$. The last equality implies $(x - y)(x + y) = 2^2 \cdot 7 \cdot 11 \cdot 13$. Notice that $0 < x - y \le x + y$ and $x - y$ and $x + y$ are both even. Hence, $(x - y, x + y) \in \{(2, 2002), (14, 286), (22, 182), (26, 154)\}$, which implies $(x, y) \in \{(1002, 1000), (150, 136), (102, 80), (90, 64)\}$. The values of n are: 1002002, 20498, 8402, 6098. The smallest value of n is 6098.

4. **Solution** (80). The given inequality is equivalent to $\dfrac{n}{5} \le i \le \dfrac{n}{2}$; hence, we are looking for the smallest integer n such that there are exactly 25 integers i that satisfy $\dfrac{n}{5} \le i \le \dfrac{n}{2}$. Therefore, $\dfrac{n}{2} - \dfrac{n}{5} \ge 24$; it is 24 because it is possible that both fractions on the left side of the inequality to be integer numbers. Hence, $n \ge 80$. If $n = 80$, then $\dfrac{80}{5} = 16$, $\dfrac{80}{2} = 40$, and there are exactly 25 integer numbers i such that $16 \le i \le 40$.

5. **Solution** (20201). The equation $|x| + |y| = 0$ has one solution. Consider the equations $|x| + |y| = k$ for all $1 \le k \le 100$. We will consider four cases:

   (a) Assume $x \ge 0$ and $y \ge 0$. Then the equation $x + y = k$ has $k + 1$ solutions:$(0, k), (1, k - 1), \dots, (k - 1, 1), (k, 0)$.

   (b) Assume $x \ge 0$ and $y < 0$. Then the equation $x - y = k$ has k solutions:$(0, -k), (1, -(k - 1)), \dots, (k - 2, -2), (k - 1, -1)$.

(c) Assume $x < 0$ and $y \geq 0$. Then the equation $-x + y = k$ has k solutions: $(-1, k-1), (-2, k-2), \ldots, (-(k-2), 2), (-(k-1), 1), (-k, 0)$.

(d) Assume $x < 0$ and $y < 0$. Then the equation $-x - y = k$ has $k - 1$ solutions: $(-1, -(k-1)), (-2, -(k-2)), \ldots, (-(k-2), -2), (-(k-1), -1)$.

Hence, for each integer $k > 0$, the equation $|x| + |y| = k$ has $4k$ solutions. Therefore, the number of lattice points inside the region $|x| + |y| \leq 100$ is

$$1 + \sum_{k=1}^{100} 4k = 1 + 4 \cdot \frac{100 \cdot 101}{2} = 20,201.$$

6. **Solution** (50). For any positive integers i and j, if $j^2 \leq i < (j+1)^2$, then $\lfloor \sqrt{i} \rfloor = j$. Hence,

$$\sum_{i=j^2}^{(j+1)^2-1} \lfloor \sqrt{i} \rfloor = \sum_{i=j^2}^{(j+1)^2-1} j = j((j+1)^2 - j^2) = j(2j+1).$$

Then

$$\sum_{i=1}^{48} \lfloor \sqrt{i} \rfloor = \sum_{i=1}^{3} \lfloor \sqrt{i} \rfloor + \sum_{i=4}^{8} \lfloor \sqrt{i} \rfloor + \sum_{i=9}^{15} \lfloor \sqrt{i} \rfloor + \sum_{i=16}^{24} \lfloor \sqrt{i} \rfloor + \sum_{i=25}^{35} \lfloor \sqrt{i} \rfloor + \sum_{i=36}^{48} \lfloor \sqrt{i} \rfloor = 203.$$

Since $217 - 203 = 14 = 2 \cdot 7$, we get

$$\sum_{i=1}^{50} \lfloor \sqrt{i} \rfloor = \sum_{i=1}^{48} \lfloor \sqrt{i} \rfloor + \sum_{i=49}^{50} \lfloor \sqrt{i} \rfloor = 203 + 14 = 217.$$

Therefore, $n = 50$.

7. **Solution** (12). Let $f(a, b, c, d) = (a-b)(b-c)(c-d)(d-a)(a-c)(b-d)$. Since $f(0, 1, 2, 3) = -12$, the greatest integer that divides the product $(a-b)(b-c)(c-d)(d-a)(a-c)(b-d)$ must be a divisor of 12. Between any four integers at least two are congruent to each other modulo 3, which implies $f(a, b, c, d) \equiv 0 \pmod 3$. We also have that between four integers either two are even and two are odd or at least three have the same parity; either way, $f(a, b, c, d) \equiv 0 \pmod 4$. Hence, $f(a, b, c, d)$ is always divisible by 12. Therefore, the greatest integer which divides $(a-b)(b-c)(c-d)(d-a)(a-c)(b-d)$ is 12.

8. **Solution** (203). Denote the vertex of the angle by O. Let $r = OA_0$, $a = A_0A_1$, $s = OB_0$, $b = B_0B_1$, $x_0 = Area(OA_0B_0)$, and $x_n = Area(A_{n-1}A_nB_nB_{n-1})$, $1 \leq n \leq 100$. Denote $r_0 = \frac{a}{r}$ and $s_0 = \frac{b}{s}$. Then

$$\frac{x_n}{x_0} = \frac{Area(OA_nB_n)}{x_0} - \frac{Area(OA_{n-1}B_{n-1})}{x_0} = \frac{(r+na)(s+nb)}{rs} - \frac{(r+(n-1)a)(s+(n-1)b)}{rs}.$$

Hence,

$$\frac{x_n}{x_0} = (1 + nr_0)(1 + ns_0) - (1 + (n-1)r_0)(1 + (n-1)s_0) = r_0 + s_0 + (2n-1)r_0s_0.$$

This implies $x_n = (r_0 + s_0 + (2n-1)r_0 s_0)x_0$ and $x_{n+1} - x_n = 2r_0 s_0 x_0$. Hence, the sequence $\{x_n\}$ is an arithmetic progression. Since $x_1 = 5, x_2 = 7$, we get that $x_{100} = 5 + 99 \cdot 2 = 203$

9. **Solution** (976). It is clear that n has exactly 10 positive divisors. If the prime factorization of n is $p_1^{\alpha_1} p_2^{\alpha_2} \cdots p_k^{\alpha_k}$, then the number of divisors of n is $(\alpha_1 + 1)(\alpha_2 + 1)\cdots(\alpha_k + 1)$. Since 10 can be written as a product of 1 and 10, or 2 and 5, we get that $n = p^9$ or $n = p_1 p_2^4$ where p, p_1 and p_2 are prime numbers. The n has form p^9, then the largest possible value is $n = 2^9 = 512$. If n has form $p_1 p_2^4$, then $p_2 < 5$ since $2 \cdot 5^4 = 1250$. If $p_2 = 2$, then the largest n is $61 \cdot 2^4 = 976$. If $p_2 = 3$, then the largest n is $11 \cdot 3^4 = 891$. Hence, the largest values for n that satisfies the properties describes in the question is 976.

10. **Solution** (31). Notice that $2^{10} = 1024 > 10^3$. Then

$$2^{100} = (2^{10})^{10} > (10^3)^10 = 10^{30}.$$

On the other hand, $2^{13} = 8192 < 10^4$. Then

$$2^{100} = 2^9 \cdot (2^{13})^7 < 512 \cdot (10^4)^7 = 512 \cdot 10^{28} = 5.12 \cdot 10^{30}.$$

Hence, $10^{30} < 2^{100} < 5.12 \cdot 10^{30}$. Since both 10^{30} and $5.12 \cdot 10^{30}$ have 31 digits, we conclude that 2^{100} has 31 digits.

The following problem, will be used only as part of a tie-breaking procedure. Do not work on it until you have completed the rest of the test.

TIE BREAKER PROBLEM

Solution (74). Let x and y be any integer numbers. Using the identity (i) we get

$$1 \circ y = 1 \circ (1 + (y-1)) = (1 \circ 1) - (y-1) = 1 - y + 1 = 2 - y.$$

Using the identity (ii) and the previous identity, we get

$$x \circ y = (1 + (x-1)) \circ y = (1 \circ y) + 2(x-1) = 2 - y + 2x - 2 = 2x - y.$$

Hence, $47 * 20 = 2 \cdot 47 - 20 = 74$.

47.3 2017 PART I: 20 MULTIPLE CHOICE PROBLEMS

1. **Solution** (C). The actual mean is $\frac{1}{100}(45,000 \cdot 100 + 53,000 - 35,000) = 45,180$.

2. **Solution** (D). Let a, b, and c be the lengths of the sides on the rectangular parallelepiped such that $ab = 18$, $bc = 40$, and $ac = 80$. If we multiply the last three equations we get $a^2 b^2 c^2 = 57600$. Hence, the volume of the parallelepiped is $V = abc = \sqrt{a^2 b^2 c^2} = 240$.

3. **Solution** (B). The number $10^{2017} - 2017$ has form $99\ldots97983$ where the number of 9s before the digit 7 is 2013. The sum of the digits of $10^{2017} - 2017$ is $9 \cdot 2013 + 7 + 9 + 8 + 3 = 18,144$.

4. **Solution** (A). Let Y be the event that the two balls are yellow, and S be the event that the two balls are the same color. Then $P(Y) = \binom{4}{2} = 6$ and $P(S) = \binom{4}{2} + \binom{3}{2} + \binom{6}{2} = 6 + 3 + 15 = 24$. Hence the probability that both balls are yellow provided that they are the same color is $P(Y|S) = \frac{P(Y)}{P(S)} = \frac{1}{4}$.

5. **Solution** (D). We will use the property $\log_{a^n} b^n = \log_a b$. Then $(\log_3 5 + \log_9 25 + \log_{27} 125 + \cdots + \log_{3^n} 5^n)\log_{25} \sqrt[2n]{27} = n(\log_3 5)(\log_{25} \sqrt[2n]{27}) = \frac{n\log 5}{\log 3} \cdot \frac{3\log 3}{4n\log 5} = \frac{3}{4}$.

6. **Solution** (D). Let q be the ratio of the geometric progression x, y, and z, and let d be the difference of the arithmetic progression x, $2y$, and $3z$. Then $y = xq$, $z = xq^2$, and $2y - x = 3z - 2y$. By substituting the first two equations in the third one we get $x(3q^2 - 4q + 1) = 0$. Since $x \neq 0$, we get $q = 1$ or $q = \frac{1}{3}$. Thus, the sum of all possible ratios of the geometric progression is $\frac{4}{3}$.

7. **Solution** (A). The quadratic equation has only one real solution if its discriminant is 0, i.e. $D = (a+5)^2 - 56 = 0$. Then $a^2 + 10a - 31 = 0$. By Viete's formulas, the sum of the roots of the last equation is -10.

8. **Solution** (B). Let a, b, and c be the lengths of the sides of the triangle and let $h_a = 12$, $h_b = 15$, and $h_c = 20$ be the lengths of its altitudes. From $A = \frac{ah_a}{2} = \frac{bh_b}{2} = \frac{ch_c}{2}$ we get $a : b = h_b : h_a$, $b : c = h_c : h_b$, and $a : c = h_c : h_a$. Then $b = \frac{4}{5}a$ and $c = \frac{3}{5}a$. The area of the triangle is $A = \frac{ah_a}{2} = 6a$. Let $s = \frac{1}{2}(a + b + c) = \frac{6}{5}a$. Using Heron's formula we get $A = \sqrt{s(s-a)(s-b(s-c)} = \frac{6a^2}{25}$. Thus $6a = \frac{6a^2}{25}$. Since $a > 0$, we get $a = 25$ and $A = 150$.

9. **Solution** (D). If zero 1s are used, then there are $2^6 = 64$ numbers. If one 1 is used, then there are $\binom{6}{1}2^5 = 192$ such numbers. If two 2s are used, then we have $2^4 = 16$ ways for four of the digits that are not 1s; there five position where to put the two 1s; hence there are $16\binom{5}{2} = 160$ numbers in this case. If three 1s are used, we have $2^3 = 8$ ways for the other three digits (non 1s), and $\binom{4}{3} = 4$ for the three 1s; thus there are 32 such numbers. We cannot use four 1s. Therefore, there are $64 + 192 + 160 + 32 = 448$ such numbers.

10. **Solution** (C). The given equation is equivalent to $(1 + x)(1 + x^2 - x^4) = 0$. Hence $x = -1$ or $1 + x^2 - x^4 = 0$. From the second equation we get $x^2 = \frac{1 \pm \sqrt{5}}{2}$. Since $\frac{1-\sqrt{5}}{2} < 0$, the real solutions of $1 + x^2 - x^4 = 0$ are $\pm\sqrt{\frac{1+\sqrt{5}}{2}}$. The product of all real solutions of the given equation is $\frac{1+\sqrt{5}}{2}$.

11. Solution (A).

$$\left(1 - \frac{4}{1}\right)\left(1 - \frac{4}{3^2}\right)\left(1 - \frac{4}{5^2}\right)\left(1 - \frac{4}{7^2}\right) \cdots \left(1 - \frac{4}{99^2}\right)$$

$$= \left(1 - \frac{2}{1}\right)\left(1 + \frac{2}{1}\right)\left(1 - \frac{2}{3}\right)\left(1 + \frac{2}{3}\right) \cdots \left(1 - \frac{2}{99}\right)\left(1 + \frac{2}{99}\right)$$

$$= -1 \cdot 3 \cdot \frac{1}{3} \cdot \frac{5}{3} \cdot \frac{3}{5} \cdot \frac{7}{5} \cdot \frac{5}{7} \cdots \frac{97}{95} \cdot \frac{95}{97} \cdot \frac{99}{97} \cdot \frac{97}{99} \cdot \frac{101}{99}$$

$$= -\frac{101}{99}$$

12. Solution (A). Let u and v be the length of the legs of the triangle ABC and w be the length of its hypotenuse. Then $u + v = -\frac{b}{a}$ and $uv = \frac{c}{a}$. Now we have

$$w^2 = u^2 + v^2 = (u + v)^2 - 2uv = \left(\frac{b}{a}\right)^2 - 2\frac{c}{a} = \frac{b^2 - 2ac}{a^2}.$$

Thus $\left(\frac{w}{2}\right)^2 = \frac{b^2 - 2ac}{4a^2}$. Since the radius if the circumscribed circle is $R = \frac{w}{2}$, we have $A = \pi R^2 = \pi \frac{b^2 - 2ac}{4a^2}$.

13. Solution (B). Since $\log_{2^k} a = \frac{1}{k} \log_2 a$ for every $k = 1, 2, \ldots, n$, we get

$$\log_{2^{k-1}} a \log_{2^k} a = \frac{1}{(k-1)k} \log_2^2 a \text{ for } k = 2, 3, \ldots, n.$$

Then

$$\sum_{k=2}^{n} \log_{2^{k-1}} a \log_{2^k} a = \sum_{k=2}^{n} \frac{\log_2^2 a}{(k-1)k} = \log_2^2 a \sum_{k=2}^{n} \frac{1}{(k-1)k} = \log_2^2 a \sum_{k=2}^{n} \left(\frac{1}{k-1} - \frac{1}{k}\right)$$

$$= \log_2^2 a \left(1 - \frac{1}{n}\right).$$

14. Solution (B). We first have

$$\frac{a\sqrt{3} + b}{b\sqrt{3} + c} = \frac{a\sqrt{3} + b}{b\sqrt{3} + c} \cdot \frac{b\sqrt{3} - c}{b\sqrt{3} - c} = \frac{(3ab - bc) + (b^2 - ac)\sqrt{3}}{3b^2 - c}.$$

Since $\frac{a\sqrt{3} + b}{b\sqrt{3} + c}$ is a rational number, we have $b^2 = ac$. Then

$$a^2 + b^2 + c^2 = a^2 + ac + c^2 = (a + c)^2 - ac = (a + c)^2 - b^2 = (a + c - b)(a + c + b).$$

Hence $\frac{a^2 + b^2 + c^2}{a + b + c} = a + c - b$.

15. Solution (D). Let F be a point on the side $\overline{AB}$ such that $\overline{CF} \parallel \overline{AD}$. Since the triangle BCD is isosceles, $BD \perp AD$, and $\overline{CF} \parallel \overline{AD}$, we get that $\overline{CF}$ is a bisects the angle BCD. Now we have that $\angle BCF = \angle FCD = \angle BFC$ which implies that th triangle BCF is isosceles, i.e. $FB = BC = 7$. Since $AF = CD = 7$, we get $AB = 14$.

16. **Solution** (B). Subtracting the first equation from the second equation we get $x + y^2 - x^2 - y = 24$, which implies that $(y - x)(y + x - 1) = 24$. Since $y - x < y + x - 1$, we have the following systems: $y - x = 1, y + x - 1 = 24$; $y - x = 2, y + x - 1 = 12$; $y - x = 3, y + x - 1 = 8$; and $y - x = 4, y + x - 1 = 6$. Only the first and the third system have integer solutions: $(3, 6)$ and $(12, 13)$. The pair $(3, 6)$ implies that z is negative. Hence, the only solution of the system is the triple $(12, 13, 57)$.

17. **Solution** (D). Let the regular pentagon $ABCDE$ be inscribed in a circle of radius 1 and center O, let a and d be the lengths of its side and diagonal, respectively. Applying the Law of Cosine for the triangle AOB we get $a^2 = 1^2 + 1^2 - 2\cos 72° = 2 - 2\cos 72°$. Applying the Law of Cosine for the triangle AOC we get $d^2 = 1^2 + 1^2 - 2\cos 144° = 2 + 2\cos 36°$. Then

$$a^2 + d^2 = 4 + 2(\cos 36° - \cos 72°) = 4 + 4\sin 18° \sin 54° = 4 + 2\frac{2\cos 18° \sin 18° \sin 54°}{\cos 18°} =$$

$$= 4 + \frac{2\sin 36° \cos 36°}{\cos 18°} = 4 + \frac{\sin 72°}{\cos 18°} = 4 + 1 = 5.$$

18. **Solution** (E).

$$(1 - \cot 1°)(1 - \cot 2°)(1 - \cot 3°) \cdots (1 - \cot 44°)$$
$$= \left(1 - \frac{\cos 1°}{\sin 1°}\right)\left(1 - \frac{\cos 2°}{\sin 2°}\right) \cdots \left(1 - \frac{\cos 44°}{\sin 44°}\right)$$
$$= \left(\frac{\sin 1° - \cos 1°}{\sin 1°}\right)\left(\frac{\sin 2° - \cos 2°}{\sin 2°}\right) \cdots \left(\frac{\sin 44° - \cos 44°}{\sin 44°}\right).$$

Using the identity
$$\sin \alpha - \cos \alpha = \sqrt{2}\sin(\alpha - 45°),$$

we get

$$\left(\frac{\sin 1° - \cos 1°}{\sin 1°}\right)\left(\frac{\sin 2° - \cos 2°}{\sin 2°}\right) \cdots \left(\frac{\sin 44° - \cos 44°}{\sin 44°}\right)$$
$$= \frac{\sqrt{2}\sin(1° - 45°)\sqrt{2}\sin(2° - 45°) \cdots \sqrt{2}\sin(44° - 45°)}{\sin 1° \sin 2° \cdots \sin 44°}$$
$$= \frac{(\sqrt{2})^{44}(-1)^{44}\sin 44° \sin 43° \cdots \sin 1°}{\sin 1° \sin 2° \cdots \sin 44°} = 2^{22}.$$

19. **Solution** (B). From $a_{n+1} = \frac{n+1}{n}a_n + 1$ we get $na_{n+1} - (n + 1)a_n = n$. By substituting $n = 1, 2, 3, \ldots, 2016$ in the last equation we get: $1a_2 - 2a_1 = 1$, $2a_3 - 3a_2 = 2$, $3a_4 - 4a_3 = 3$, ..., $2016a_{2017} - 2017a_{2016} = 2016$. Adding all these equations gives $-2(a_1 + a_2 + \cdots + a_{2016}) + 2016a_{2017} = 1 + 2 + 3 + \cdots + 2016 = \frac{2016 \cdot 2017}{2}$. Hence $a_1 + a_2 + \cdots + a_{2016} = -\frac{1}{2}\left(\frac{2016 \cdot 2017}{2} - 2016 \cdot 2017\right) = 504 \cdot 2017.$

20. Solution (C). We will use Cauchy's Inequality: Let $a_1, a_2, \cdots, a_n$ and $b_1, b_2, \cdots, b_n$ be real numbers. Then

$$(a_1 b_1 + a_2 b_2 + \cdot + a_n b_n)^2 \leq (a_1^2 + a_2^2 + \cdots + a_n^2)(b_1^2 + b_2^2 + \cdots + b_n^2)$$

and equality holds if and only if $b_i = 0$ for all $i = 1, 2, \ldots, n$ or $a_i = k b_i$ for all $i = 1, 2, \ldots, n$ and some constant k.

$$(2x + y)^2 = \left(\frac{2}{\sqrt{3}} \cdot \sqrt{3}x + \frac{1}{\sqrt{2}} \cdot \sqrt{2}y \right)^2 \leq \left(\left(\frac{2}{\sqrt{3}} \right)^2 + \left(\frac{1}{\sqrt{2}} \right)^2 \right) ((\sqrt{3}x)^2 + (\sqrt{2}y)^2).$$

Hence, $(2x+y)^2 \leq \frac{11}{6}(3x^2 + 2y^2)$. Since $3x^2 + 2y^2 \leq 6$, we get $(2x+y)^2 \leq 11$, i.e. $2x + y \leq \sqrt{11}$. The equality holds if $\frac{2}{\sqrt{3}} \cdot \sqrt{2}y = \frac{1}{\sqrt{2}} \cdot \sqrt{3}x$, i.e $3x = 4y$. Then $x = \frac{4}{\sqrt{11}}$ and $y = \frac{3}{\sqrt{11}}$. Therefore, the greatest value is $\sqrt{11}$.

47.4 2017 PART II: 10 INTEGER ANSWER PROBLEMS

1. Solution (1497). Let n be a number that satisfies the properties stated in the problem. Then $n + 5$ is divisible by 5, 7, and 12, which implies it is divisible by the least common multiple of 5, 7, and 12. Since $\text{lcm}(5, 7, 12) = 252$, we get that $n + 5 = 252k$ where $k = 1, 2, 3$ since n must be a three digit number. Hence n could be 247, 499, or 751. Their sum is 1497.

2. Solution (65). Setting $x = 1$ in $f(x+y) = 3^y f(x) + 2^x f(y)$, we get $f(1+y) = 3^y f(1) + 2f(y) = 3^y + 2f(y)$ for all $y \in \mathbb{R}$. Setting $y = 1$ in $f(x+y) = 3^y f(x) + 2^x f(y)$, we get $f(x+1) = 3f(x) + 2^x f(1) = 3f(x) + 2^x$ for all $x \in \mathbb{R}$. By renaming the variable y to x, we get that $f(x+1) = 3^x + 2f(x)$ and $f(x+1) = 3f(x) + 2^x$. Then $f(x) = 3^x - 2^x$ and $f(4) = 65$.

3. Solution (19). The given equation can be written as $3^x(1 + 3^{y-x} + 3^{z-x}) = 27 \cdot 811$. Since 3^x and 811 are relatively prime, and $1 + 3^{y-x} + 3^{z-x}$ and 27 are also relatively prime, we get $3^x = 27$ and $1 + 3^{y-x} + 3^{z-x} = 811$. Thus $x = 3$. Then the second equation becomes $3^{y-3} + 3^{z-3} = 810$, i.e. $3^{y-3}(1 + 3^{z-y}) = 81 \cdot 10$. Since 3^{y-3} and 10 are relatively prime, and $1 + 3^{z-y}$ and 81 are also relatively prime, we get $3^{y-3} = 81$ and $1 + 3^{z-y} = 10$. The second equation fives $y = 7$, and from the third equation we get $z = 9$. Thus $x + y + z = 19$.

4. Solution (2). Denote $x = \sqrt{a-5}$ and $y = a$. It is clear that $y = x^2 + 5$. Since $x \geq 0$, we see that the points z_1 lie on the half parabola $y = x^2 + 5, x \geq 0$. The points z_2 lie on the ellipse $\frac{x^2}{4} + \frac{y^2}{9} = 1$. The closest two points from both sets are the points $(0, 5)$ and $(0, 3)$. Thus the minimum of $|z_1 - z_2|$ is 2.

5. Solution (4). It is an easy observation that an integer number is congruent to the sum of its digits modulo 9. The sum of the digits $0 + 1 + 2 + \cdots 9 = 45$ is divisible by 9 and $2^{29} \equiv 2^2(2^3)^9 \equiv$

$2^2(-1)^9 \equiv -4 \equiv 5(\bmod\ 9)$. Hence 2^{29} gives a reminder 5 when divided by 9, which implies that the missing digit is 4.

6. **Solution** (2). Let K be a point on $\overline{BE}$ such that $\overline{DK}\|\overline{AC}$. The triangles BCE and BDK are similar. Hence $\frac{BK}{BE} = \frac{DK}{CE} = \frac{BD}{BC} = \frac{1}{5}$; thus $BK = \frac{1}{5}BE$ and $DK = \frac{1}{5}CE$. The triangles DKF and AEF are also similar, and this implies $\frac{KF}{FE} = \frac{KD}{AE} = \frac{KD}{CE} = \frac{1}{5}$. Let $KF = k$. Then $EF = 5k$. From the equality $BE = BK + KF + FE$ we get $BK = \frac{3}{2}k$. From $BF = BK + KF = \frac{5}{2}k$. Thus $\frac{EF}{BF} = 2$.

7. **Solution** (9). Since $x^2 - 6x = x(x-6)$, we will consider the given equation on the intervals $(-\infty, 0]$, $(0, 6)$, and $[6, \infty)$. On the intervals $(-\infty, 0]$ and $[6, \infty)$ the given equation has the form $x^2 - 6x - m = 0$ and its solutions are $x = 3 \pm \sqrt{9+m}$. If $m > 0$, then $x = 3 + \sqrt{9+m} \in [6, \infty)$ and $x = 3 - \sqrt{9+m} \in (-\infty, 0]$. If $m < 0$ then $x = 3 \pm \sqrt{9+m} \in (0, 6)$ which is not the interval we are considering. If $m = 0$, then the equation has two zeros. On the interval $(0, 6)$ the given equation has the form $x^2 - 6x + m = 0$. The given equation will have exactly three zeros if $m > 0$ and the equation $x^2 - 6x + m = 0$ has only one zero on $(0, 6)$. Hence, the discriminant of $x^2 - 6x + m = 0$ must be 0, i.e. $36 - 4m = 0$. Therefore the given equation has exactly three zeros for $m = 9$.

8. **Solution** (11). The given equation is equivalent to $\log(1 + a^2) + \log(100 + b^2) = \log a + \log b + 2\log 2 + 1$. Using properties of logarithmic function we get $\log(1+a^2)(100+b^2) = \log(40ab)$ which implies $(1+a^2)(100+b^2) = 40ab$. Since a and b are positive numbers, using the inequality between arithmetic mean and geometric mean we get $1+a^2 \geq 2a$ and $100+b^2 \geq 20b$ and equality holds when $a = 1$ and $b = 10$. Hence, if $a \neq 1$ and $b \neq 10$, $(1+a^2)(100+b^2) > 40ab$. Therefore, $a = 1$, $b = 10$, and $a + b = 11$.

9. **Solution** (99). If $\min\{a, b, c\} \geq 3$, then $\left(1+\frac{1}{a}\right)\left(2+\frac{1}{b}\right)\left(3+\frac{1}{c}\right) \leq \frac{280}{27} < \frac{91}{8}$. Otherwise $\min\{a, b, c\} \leq 2$ and we consider three cases. If $\min\{a, b, c\} = c = 2$, then $a \geq 3$ and $b \geq 3$ and

$$\left(1+\frac{1}{a}\right)\left(2+\frac{1}{b}\right)\left(3+\frac{1}{2}\right) \leq \left(1+\frac{1}{3}\right)\left(2+\frac{1}{3}\right)\left(3+\frac{1}{2}\right) = \frac{98}{9} < \frac{91}{8}.$$

If $\min\{a, b, c\} = b = 2$, then $a \geq 3$ and $c \geq 3$ and $\left(1+\frac{1}{a}\right)\left(2+\frac{1}{2}\right)\left(3+\frac{1}{c}\right) \leq \left(1+\frac{1}{3}\right)\left(2+\frac{1}{2}\right)\left(3+\frac{1}{3}\right) = \frac{100}{9} < \frac{91}{8}$. If $\min\{a, b, c\} = a = 2$, then $b \geq 3$ and $c \geq 4$ or $b \geq 4$ and $c \geq 3$ since b and c are distinct (we could have considered subcases similar to this in the previous cases, but it was not necessary). Then

$$\left(1+\frac{1}{a}\right)\left(2+\frac{1}{b}\right)\left(3+\frac{1}{c}\right) \leq \left(1+\frac{1}{2}\right)\left(2+\frac{1}{3}\right)\left(3+\frac{1}{4}\right) = \frac{91}{8},$$

or

$$\left(1+\frac{1}{a}\right)\left(2+\frac{1}{b}\right)\left(3+\frac{1}{c}\right) \leq \left(1+\frac{1}{2}\right)\left(2+\frac{1}{4}\right)\left(3+\frac{1}{3}\right) = \frac{90}{8} < \frac{91}{8}.$$

Therefore, the maximum of $\left(1 + \frac{1}{a}\right)\left(2 + \frac{1}{b}\right)\left(3 + \frac{1}{c}\right)$ is $\frac{91}{8}$ which is achieved for $a = 2$, $b = 3$, and $c = 4$. Hence $m = 91$, $n = 8$ and $m + n = 99$.

10. **Solution** (8).

$$x^{256} - 256^{32} = x^{(2^8)} - (2^8)^{32} = x^{(2^8)} - 2^{(2^8)}$$
$$= \left(x^{(2^7)} + 2^{(2^7)}\right)\left(x^{(2^7)} - 2^{(2^7)}\right)$$
$$= \left(x^{(2^7)} + 2^{(2^7)}\right)\left(x^{(2^6)} + 2^{(2^6)}\right)\left(x^{(2^6)} - 2^{(2^6)}\right)$$
$$= \cdots$$
$$= \left(x^{(2^7)} + 2^{(2^7)}\right)\left(x^{(2^6)} + 2^{(2^6)}\right)\cdots(x + 2)(x - 2).$$

Hence, the sum of the squares of the real roots of the equation is 8.

TIE BREAKER PROBLEM

Solution (130). If n is odd, then all of its divisors are odd. Then $d_1^2 + d_2^2 + d_3^2 + d_4^2 \equiv 0 (\mathrm{mod}\ 4)$, which is not possible since we assumed n is odd. Hence n is even, and $d_1 = 1$, $d_2 = 2$. This implies $n \equiv 1 + 0 + d_3^2 + d_4^2 \pmod{4}$. If both d_3 and d_4 are even, then $n \equiv 1 (\mathrm{mod}\ 4)$; if both d_3 and d_4 are odd, then $n \equiv 3 (\mathrm{mod}\ 4)$; and if one of d_3 and d_4 is even and the other one is odd, then $n \equiv 2 (\mathrm{mod}\ 4)$. In any case, $n \not\equiv 0 \pmod{4}$. Hence $4 \nmid n$ and $(d_3, d_4) = (p, q)$ or $(d_3, d_4) = (p, 2p)$ for some odd primes p and q. If $(d_3, d_4) = (p, q)$, then $n \equiv 3 (\mathrm{mod}\ 4)$, which is not possible since n is an even integer. If $(d_3, d_4) = (p, 2p)$, then $n = 5(1 + p^2)$, and thus $5 \mid n$. Hence $d_3 = 5$ and $d_4 = 10$. Therefore $n = 130$.

CHAPTER 48

SOLUTIONS TO BAMO PROBLEMS

A. Solution There is just one coloring; it's forced: Without loss of generality, suppose that the corners are colored in order, R, G, B, P, as shown below.

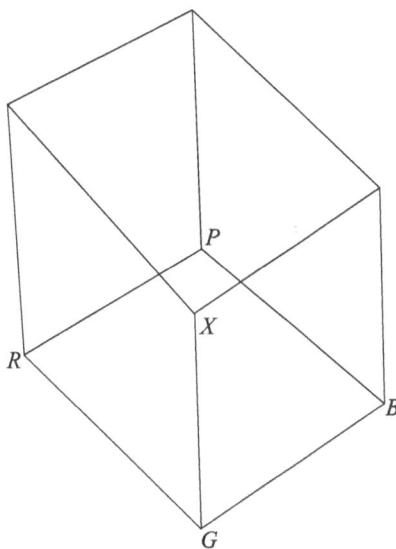

Notice that vertex X cannot be R, B, or G; it must therefore be colored P. Likewise, every other vertex on the top has only one possible color, since each top vertex is part of a two faces which collectively use three different colors on the bottom level. So there is only one solution, namely, starting from vertex X and going counterclockwise (as seen from above): P, R, G, B.

B. Solution (a) Here is a sketch that shows why there is no solution. After the first move, without loss of generality, there will be two empty locations next to one another, and after the next move, there will be two pegs left, separated by one empty space on one side and by two empty spaces on the other side. After this, it is impossible to proceed, so we are stuck with two pegs.

(b) For even n it is pretty easy to construct a solution:

Suppose the holes are numbered $0, 1, 2, ..., n-1$ with hole 0 initially empty. Then do the following steps: Jump 2 into 0, 4 into 2, 6 into 4, 8 into 6, ... $n-2$ into $n-4$. Now you're left with a peg in $n-1$ next to 0, and holes $2, 4, 6, ..., n-4$ are filled, so just jump the peg in $n-1$ over all the even pegs like a successive capture in checkers, and you're done.

C. Solution Let's write the number of cookies in the red and blue plate, respectively, as an ordered pair (x, y), so that the legal moves are to $(x-2, y)$ or $(x, y-2)$ or $(x-1, y+1)$. Thus the only positions with no legal move are $(0,0)$ and $(0,1)$, and since cookies are eaten in pairs, the final position is determined by the original number of cookies.

Starting from (20,14), we know that eventually all the cookies will be eaten, so there are exactly $(20+14)/2 = 17$ cookie-eating moves. There may also be some number of moves from the first pile to the second pile, but since an even number of cookies are eaten from each pile, there must be an even number of such moves. Thus, the total number of moves in the game is odd, and the first player gets the last legal move.

For general starting positions, there are a few cases to examine depending on whether the total number of cookies is even and the number of cookies in pile 2 is even, but the logic is similar.

D. Solution Triangle CBQ is isosceles, so the perpendicular bisector of side BQ is angle bisector of angle C. Similarly for BP and A. The intersection of these two bisectors is the circumcenter of BPQ and the incenter of ABC.

3. Solution Move the y's to the other side by dividing both sides by $y + \sqrt{1+y^2}$:

$$x + \sqrt{1+x^2} = \frac{1}{y + \sqrt{1+y^2}}.$$

Then get rid of the denominator on the right hand side by multiplying top and bottom of the fraction by $-y + \sqrt{1+y^2}$:

$$x + \sqrt{1+x^2} = \frac{\sqrt{1+y^2)} - y}{(y+\sqrt{1+y^2})(\sqrt{1+y^2}-y)} = \frac{\sqrt{1+y^2)} - y}{(1+y^2) - y^2} = \sqrt{1+y^2} - y.$$

Now move back y to the left and the radical $\sqrt{1+x^2}$ to the right:

$$x + y = \sqrt{1+y^2} - \sqrt{1+x^2}.$$

Now, again "rationalize" the right hand side by multiplying it and dividing by $\sqrt{1+y^2}+\sqrt{1+x^2}$:

$$
\begin{aligned}
x+y &= \frac{(\sqrt{1+y^2}-\sqrt{1+x^2})(\sqrt{1+y^2}+\sqrt{1+x^2})}{\sqrt{1+y^2}+\sqrt{1+x^2}} \\
&= \frac{(1+y^2)-(1+x^2)}{\sqrt{1+y^2}+\sqrt{1+x^2}} \\
&= \frac{y^2-x^2}{\sqrt{1+y^2}+\sqrt{1+x^2}}.
\end{aligned}
$$

Move the denominator over the left hand side and factor $y^2 - x^2 = (y-x)(y+x)$ on the right hand side:

$$(x+y)(\sqrt{1+y^2}+\sqrt{1+x^2}) = (x+y)(x-y).$$

Finally, move the right hand side to the left and factor out $(x+y)$:

$$(x+y)(\sqrt{1+y^2}+\sqrt{1+x^2}-(x-y)) = 0,$$

or, equivalently,

$$(x+y)((\sqrt{1+y^2}+y)+(\sqrt{1+x^2}-x)) = 0.$$

Regardless of what x and y are, we always have $1+y^2 > y^2$ and $1+x^2 > x^2$, and hence $\sqrt{1+y^2} > |y|$ and $\sqrt{1+x^2} > |x|$, which implies $\sqrt{1+y^2}+y > 0$ and $\sqrt{1+x^2}-x > 0$. Thus, the second factor in the product above, $(\sqrt{1+y^2}+y)+(\sqrt{1+x^2}-x)$ is always positive. So the product can equal 0 only if $x+y=0$.

OR

Suppose

$$(x+\sqrt{1+x^2})(y+\sqrt{1+y^2}) = 1.$$

Multiply both sides by $-y+\sqrt{1+y^2}$ to get

$$x+\sqrt{1+x^2} = -y+\sqrt{1+y^2}.$$

Rearrange to get

$$x+y = \sqrt{1+y^2}-\sqrt{1+x^2}. \tag{48.1}$$

By a symmetrical argument, we also have

$$x+y = \sqrt{1+x^2}-\sqrt{1+y^2}. \tag{48.2}$$

Average (48.1) and (48.2) to conclude that $x+y=0$.

OR

We note that $y = -x$ gives $(x + \sqrt{1 + x^2})(y + \sqrt{1 + y^2} = 1 + x^2 - x^2 = 1$. Further the function $x + \sqrt{1 + x^2}$ is monotone increasing (easy with calculus) so for a given x there is at most one solution to $(x + \sqrt{1 + x^2})(y + \sqrt{1 + y^2} = 1$, hence that solution always is $-x$.

OR

The equality can't hold if x and y are of the same sign. If $x = 0$ then $y = 0$ and vice versa. So WLOG, assume $x > 0 \; y < 0$. There are angles X and Y in $(-\pi/2, \pi/2)$ such that $\cot X = x$ and $\cot Y = y$. Then $\sqrt{1 + x^2} = \dfrac{1}{\sin X}$ and $\sqrt{1 + y^2} = -\dfrac{1}{\sin Y}$,

$$x + \sqrt{1 + x^2} = \frac{\cos X + 1}{\sin X} = \cot X/2,$$

$$y + \sqrt{1 + y^2} = \frac{\cos y - 1}{\sin y} = \tan -Y/2,$$

so $\tan(-Y/2)\cot(X/2) = 1$, so $\tan X/2 = \tan -Y/2$, so $X = -Y$.

4. **Solution** Yes! in fact, $F_{54} = 86267571272$ is a multiple of 2014 (although you are not required to find this index), and every 54th Fibonacci number thereafter will be a multiple of 2014. To see why, we write the sequence (mod 2014): our goal is to show that it equals zero eventually. Although conventionally, the Fibonacci sequence starts with $F_1 = F_2 = 1$, we can extend it *backwards*—this is the crux idea—with $F_0 = 0$. , and the recurrence formula forces $F_{-1} = 1, F_2 = -1, F_{-3} = 2, F_{-4} = 3$, etc., so the "bi-infinite" Fibonacci sequence looks like this:

$$\ldots, -8, 5, -3, 2, -1, 1, 0, 1, 1, 2, 3, 5, 8, \ldots$$

Clearly, it is symmetrical, although the signs alternate for the values to the left of F_0.

Next, we can show that the sequence is eventually *periodic*: There are only 2014 different values (mod 2014), and thus 2014^2 possible distinct consecutive pairs of numbers. By the pigeonhole principle, eventually, after at most $2014^2 + 1$ steps, we will see the same consecutive pair repeated, and this will then determine the rest of the sequence, with repeating blocks of the same numbers, *ad infinitum*.

We will be done if the periodic block begins with $F_0 = 0$, since this would imply infinitely many zeros. But perhaps the periodic block didn't start at the beginning. We will use the "extending backwards" idea to show that, in fact, periodicity must start at the beginning of the sequence (F_0).

Suppose that a periodic block starts at $F_M = a, F_{M+1} = b$, and has length L, in other words, ends at F_{M+L-1}, and suppose that $M > 0$. Notice that by going backwards, we can compute $F_{M+L-1} = b - a$, since the next periodic block starts at index $M + L$, and $F_{M+L} = a$ and $F_{M+L+1} = b$. Likewise, we can keep going backwards from F_M to deduce that $F_{M-1} = F_{M+L-1}$

and $F_{M-2} = F_{M+L-2}$, etc., so eventually we will get $F_0 = F_L$. So the periodicity starts at F_0 and we are guaranteed to see zeros every L steps.

Remark: This proof shows that the length of the period is at most $2014^2 + 1$, when in fact it was much smaller (namely, 54). It can be proven that for a prime p, the maximum period for divisibility (mod p) is $p + 1$.

5. **Solution** We suppose the desired conclusion is false, and seek a contradiction. Refer to the players as $P_0, P_1, \ldots, P_{2n}$ in increasing order of their rating. Call a game an *upset* if the lower-rated player beat the higher-rated player.

Let $r = \lfloor \sqrt{2k} \rfloor$. By assumption, for each value $i = 1, 2, \ldots, r$, player P_{n-i} won either less than $n - r$ or more than $n + r$ games. In the first case, player P_{n-i} must have lost more than $r - i$ games against lower-rated players; in the second case, player P_{n-i} must have won more than $r + i$ games against higher-rated players. Either way, P_{n-i} participated in at least $r - i + 1$ upsets.

By similar arguments, for each $i = 1, 2, \ldots, r$, player P_{n+i} also participated in at least $r - i + 1$ upsets, and player P_n participated in at least $r + 1$ upsets. Thus, altogether, we have at least

$$[r + (r-1) + (r-2) + \cdots + 1] + [r + (r-1) + (r-2) + \cdots + 1] + (r+1) = (r+1)^2$$

participations in upsets. Every upset involved precisely two players, so this requires a total of at least $(r+1)^2/2$ upsets.

However, $\sqrt{2k} < r + 1$, so $k < (r+1)^2/2$. So the number of upsets is less than $(r+1)^2/2$, a contradiction.

CHAPTER 49

SOLUTIONS TO PURPLE COMET MATH MEET

---

1. **Solution** (8). The equation simplifies to $(6 + 8)x + (7 + 9) = (10 + 12) + (11 + 13)x$ or $14x + 16 = 22 + 24x$. Thus, $-10x = 6$ and $x = -\frac{3}{5}$. The requested sum is $3 + 5 = 8$.

2. **Solution** (19). $12! = 12 \cdot 11 \cdot 10 \cdot 9 \cdot 8 \cdot 7 \cdot 6 \cdot 5 \cdot 4 \cdot 3 \cdot 2 \cdot 1 = (2 \cdot 2 \cdot 3) \cdot 11 \cdot (2 \cdot 5) \cdot (3 \cdot 3) \cdot (2 \cdot 2 \cdot 2) \cdot 7 \cdot (2 \cdot 3) \cdot 5 \cdot (2 \cdot 2) \cdot 3 \cdot 2 \cdot 1$ which is a total for 19 prime factors.

3. **Solution** (66). The entire grid forms a square with side length 10, so the grid has area 100. The figure shown, thus, has area 100 minus the areas of two trapezoids and four right triangles shaded in the diagram shown below. Both trapezoids have bases of length 2 and 8 with height 2, so each has area $\frac{2+8}{2}2 = 10$. The largest right triangle has legs with lengths 2 and 6, so it has area $\frac{2 \cdot 6}{2} = 6$. Two small right triangles each have two legs of length 2, so each has area $\frac{2 \cdot 2}{2} = 2$. The last right triangle has a hypotenuse of length 4 and an altitude to the hypotenuse of length 2, so its area is $\frac{2 \cdot 4}{2} = 4$. Therefore, the area of the ten sided figure is $100 - (10 + 10 + 6 + 2 + 2 + 4) = 100 - 34 = 66$.

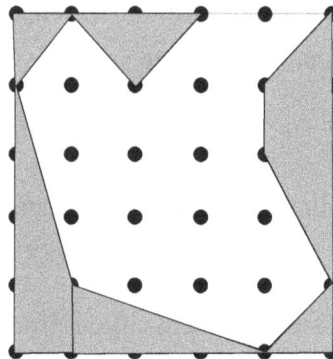

4. **Solution** (34825). Let $y = \dfrac{x}{100}$. Sally's 2009 salary is given by

$$37000(1 + y)(1 + y)(1 - 2y) = 34825.$$

If the salary increases and decreases are reversed, her 2009 salary would be

$$37000(1 - 2y)(1 + y)(1 + y)$$

which is still equal to 34825 since multiplication is a commutative operation. This happens if x is approximately 13.4144 percent.

5. **Solution** (98). The number 2400 has 36 positive integer divisors. One can check all 18 pairs of these divisors, although the minimum sum will occur when the pair of divisors are closest to $\sqrt{2400}$ which is between 48 and 49. The pairs of divisors closest to this is 48 and 50 whose sum is 98 which is the least possible.

6. **Solution** (7245). Grouping the numbers in the sum in groups of three gives

$$
\begin{aligned}
& 1 + 2 - 3 + 4 + 5 - 6 + 7 + 8 - 9 + \cdots + 208 + 209 - 210 \\
=\ & (1 + 2 - 3) + (4 + 5 - 6) + (7 + 8 - 9) + \cdots + (208 + 209 - 210) \\
=\ & 0 + 3 + 6 + \cdots + 207 \\
=\ & 3(0 + 1 + 2 + \cdots + 69).
\end{aligned}
$$

Since the sum of the first n positive integers is given by $\dfrac{n(n+1)}{2}$, the desired sum is $3 \cdot \dfrac{69 \cdot 70}{2} = 7245$.

7. **Solution** (7).
$$5^{2010} \cdot 16^{502} = 5^{2010} \cdot \left(2^4\right)^{502} = 5^{2010} \cdot 2^{2008} = 5^2 \cdot 10^{2008}$$

which is the number 25 followed by 2008 zeros. Thus, the sum of its digits is $2 + 5 = 7$.

8. **Solution** (288). Each of the small squares has side length $\sqrt{3}$ and diagonal length $\sqrt{3}\sqrt{2} = \sqrt{6}$. Since the large square has sides equal to the sum of the lengths of two of the diagonals and one of the sides of the small squares, the area of the large square is

$$\left(\sqrt{3} + 2\sqrt{6}\right)^2 = 3 + 4 \cdot \sqrt{3}\sqrt{6} + 4 \cdot 6 = 27 + 12\sqrt{2}.$$

Each small square has area 3, and there are nine small squares with a total area of 27. Thus, the desired area is $12\sqrt{2} = \sqrt{288}$. The requested answer is 288.

9. **Solution** (64). If the two numbers are reciprocals, their product is

$$1 = \frac{80 - 6\sqrt{n}}{n} \cdot \frac{80 + 6\sqrt{n}}{n} = \frac{80^2 - 6^2 n}{n^2}$$

or

$$n^2 + 36n - 80^2 = 0$$

which has solutions

$$n = \frac{-36 \pm \sqrt{36^2 + 4 \cdot 80^2}}{2} = \frac{-36 \pm 4\sqrt{9^2 + 40^2}}{2} = -18 \pm 2 \cdot 41.$$

Thus, the positive solution is $2 \cdot 41 - 18 = 64$.

10. **Solution** (182). Let the median be m. The sum of all the numbers in S is $9 \cdot 202$. The sum of the five smallest numbers in S is $5 \cdot 100$. The sum of the five largest numbers in S is $5 \cdot 300$. Thus, $m = 5 \cdot 100 + 5 \cdot 300 - 9 \cdot 202 = 500 + 1500 - 1818 = 182$.

11. **Solution** (11). There are ten marbles in all, so there are $\binom{10}{2} = 45$ equally likely ways to select two marbles. There is no way to select two white marbles, $\binom{2}{2} = 1$ way to select two blue marbles, $\binom{3}{2} = 3$ ways to select two red marbles, and $\binom{4}{2} = 6$ ways to select two green marbles. Thus, the probability of selecting two marbles of the same color is

$$\frac{0 + 1 + 3 + 6}{45} = \frac{10}{45} = \frac{2}{9}.$$

The requested sum is $2 + 9 = 11$.

12. **Solution** (51). The answer is the least positive integer d such that

$$\left(\frac{d}{2}\right)^2 (\pi - 3.14) \geq 1.$$

That is,

$$d^2 \geq \frac{4}{\pi - 3.14} \approx \frac{4}{0.00159} \approx 2511.$$

The value of d is thus 51.

13. **Solution** (630). An arithmetic progression containing the terms 4 and 10 is completely determined by the positions in the progression of the 4 and 10. For each

$$(a_1, a_2, a_3, a_4, a_5, a_6, a_7, a_8, a_9, a_{10}) \in S$$

there is exactly one

$$(b_1, b_2, b_3, b_4, b_5, b_6, b_7, b_8, b_9, b_{10}) \in S$$

where the positions of the 4 and the 10 are reversed. For example, the sequence

$$(-2, 1, 4, 7, 10, 13, 16, 19, 22, 25)$$

and the sequence

$$(16, 13, 10, 7, 4, 1, -2, -5, -8, -11)$$

have their 4 and 10 terms in opposite positions. Then

$$(a_1 + b_1, a_2 + b_2, a_3 + b_3, a_4 + b_4, a_5 + b_5, a_6 + b_6, a_7 + b_7, a_8 + b_8, a_9 + b_9, a_{10} + b_{10})$$

is also an arithmetic progression, but since two of its terms are $4 + 10 = 14$, it is a constant progression with all of its terms equal to 14. In particular, $a_{10} + b_{10} = 14$. It follows that mean of all tenth terms of progressions in S is $\frac{14}{2} = 7$, and the requested sum of all tenth terms is 7 times the number of elements in S. Since an element of S is completely determined by the placement of the terms 4 and 10, and there are ten possible placements of the term 4, and nine possible placements of the term 10 once the 4 has been placed, the size of S is $10 \cdot 9 = 90$. Thus, the requested sum is $7 \cdot 90 = 630$.

14. **Solution** (126). The quadratic formula says that the two roots of the polynomial are $\frac{-b + \sqrt{b^2 - 8c}}{4}$ and $\frac{-b - \sqrt{b^2 - 8c}}{4}$. These two roots differ by $\frac{\sqrt{b^2 - 8c}}{2} = 30$ or $b^2 - 8c = 3600$. Since c is positive, the smallest value of b and c occur when $3600 + 8c$ is the smallest perfect square multiple of 8 exceeding 60^2. This is $64^2 = 4096$ which occurs when $b = 64$ and $c = \frac{4096 - 3600}{8} = 62$. Thus, the least possible value of $b + c$ is $64 + 62 = 126$.

15. **Solution** (47). Since a and b are not relatively prime, there must be a prime number p_1 that divides both a and b. Similarly, there must be a prime number p_2 that divides both b and c, but since a and c are relatively prime, p_1 and p_2 are distinct. Similarly, there is a third prime p_3 that divides both c and d, and a fourth prime p_4 that divides both d and e, with all four primes being distinct. Thus, the smallest possible value for the sum

$$a + b + c + d + e$$

is

$$p_1 + p_1 \cdot p_2 + p_2 \cdot p_3 + p_3 \cdot p_4 + p_4.$$

It follows that the four primes should be the smallest four prime numbers: 2, 3, 5, 7. It is also clear that p_1 and p_4 should be the largest of those four prime numbers. An assignment that minimizes the sum is $p_1 = 5, p_2 = 3, p_3 = 2, p_4 = 7$ resulting in the sum

$$5 + 5 \cdot 3 + 3 \cdot 2 + 2 \cdot 7 + 7 = 5 + 15 + 6 + 14 + 7 = 47.$$

16. **Solution** (21). More generally, let X be a point on median $\overline{AD}$, and let line $\overline{BX}$ intersect side $\overline{CA}$ at point Y as shown below. Let Z be a point on side $\overline{BC}$ so that $\overline{YZ}$ is parallel to $\overline{XD}$. Then

triangles BDX and BZY are similar, and triangles YZC and ADC are similar. This implies that $\dfrac{BZ}{YX} = \dfrac{BD}{DX}$ and $\dfrac{CZ}{YZ} = \dfrac{CD}{AD}$, so

$$\frac{BD}{DX} + \frac{CD}{AD} = \frac{BZ + CZ}{YZ} = \frac{BC}{YZ}.$$

Also, $\dfrac{YC}{YZ} = \dfrac{AC}{AD}$, so

$$YC = YZ\frac{AC}{AD} = \frac{BC}{\frac{BD}{DX} + \frac{CD}{AD}} \cdot \frac{AC}{AD} = \frac{AC}{\frac{BD}{BC}\frac{AD}{DX} + \frac{CD}{BC}} = \frac{2}{\frac{AD}{DX} + 1} \cdot AC.$$

Applying this when $DX = \dfrac{1}{3} \cdot AD$ shows that $GC = \dfrac{2}{3+1} \cdot 70 = 35$. Applying it when $DX = \frac{2}{3} \cdot AD$ shows that $HC = \dfrac{2}{\frac{3}{2}+1} \cdot 70 = 56$. Thus, $GH = 56 - 35 = 21$. Note that the solution does not depend on the lengths of $\overline{AB}$ and $\overline{BC}$.

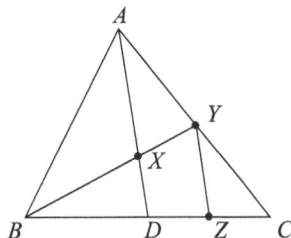

Alternatively, drop perpendiculars from A, G and E to $\overline{BH}$ and denote the feet of these perpendiculars by K, L and M, respectively. Triangles AFB and BFE have equal areas since $\overline{AF} \equiv \overline{FE}$, and the corresponding heights are the same. These triangles share side $\overline{BF}$ so the corresponding heights must be equal, $AK = EM$.

Triangles BEM and BGL are similar because they have two congruent angles, and thus $\dfrac{GL}{EM} = \dfrac{BG}{BE}$. Since E is one third the distance from D to A along a median, E is the centroid of triangle ABC, and $\dfrac{BG}{BE} = \dfrac{3}{2}$. Then the ratio of similarity must be the same for triangles LHG and KHA, thus $HG = \dfrac{3}{2}AH$. Since E is the centroid, $\overline{BG}$ is a median, and $AG = 35$, $HG = \dfrac{3}{5} \cdot 35 = 21$.

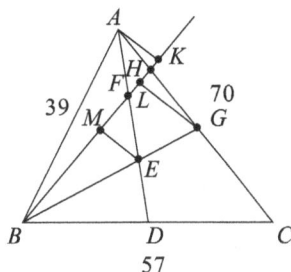

17. **Solution** (2521). Nine people can sit in a circle in 8! equally likely arrangements. For the arrangement to have members of the same team not sitting next to each other, Greg, who is on both

the swim team and the tennis team, must sit next to both Barb and Cory. Similarly, Alan, who is on both the golf team and the tennis team, must sit next to both Emma and Fran, and Doug, who is on both the golf team and the swim team, must sit next to both Hope and Inga. For each of these groups of three people: {Barb,Greg,Cory}, {Emma,Alan,Fran}, {Hope,Doug,Inga}, there are two arrangements of the people in the group (for example, Barb,Greg,Cory and Cory,Greg,Barb). There are also $2! = 2$ arrangements of the three groups around the circle. It follows that there are $2 \cdot 2 \cdot 2 \cdot 2 = 16$ arrangements of the people so that no two members of the same team sit next to each other. The requested probability is

$$\frac{16}{8!} = \frac{16}{8 \cdot 7 \cdot 6 \cdot 5 \cdot 4 \cdot 3 \cdot 2 \cdot 1} = \frac{1}{7 \cdot 6 \cdot 5 \cdot 4 \cdot 3} = \frac{1}{2520}.$$

The requested sum is $1 + 2520 = 2521$.

18. **Solution** (122). Because $3^2 + 4^2 = 5^2$, there is an angle ϕ so that $\cos\phi = \frac{4}{5}$ and $\sin\phi = \frac{3}{5}$. Recall the angle sum formula $\cos(\phi + \theta) = \cos\phi \cdot \cos\theta - \sin\phi \cdot \sin\theta$ and the double angle formulas $\cos 2\theta = 2\cos^2\theta - 1$ and $\sin 2\theta = 2\sin\theta\cos\theta$. Thus,

$$\cos 2\phi = 2\cos^2\phi - 1 = 2\left(\frac{4}{5}\right)^2 - 1 = \frac{7}{25}$$

and

$$\sin 2\phi = 2\sin\phi\cos\phi = 2\frac{4}{5}\frac{3}{5} = \frac{24}{25}$$

. From the given information one sees that

$$\cos(\phi + \theta) = \frac{4\cos\theta - 3\sin\theta}{5} = \frac{13}{15}.$$

Then,

$$\cos 2(\phi + \theta) = 2\cos^2(\phi + \theta) - 1 = 2\left(\frac{13}{15}\right)^2 - 1 = \frac{338}{225} - 1 = \frac{113}{225}.$$

Also,

$$\cos 2(\phi + \theta) = \cos 2\phi \cdot \cos 2\theta - \sin 2\phi \cdot \sin 2\theta = \frac{7}{25}\cos 2\theta - \frac{24}{25}\sin 2\theta = \frac{1}{25}\frac{m}{n}.$$

It follows that $\frac{m}{n} = 25\frac{113}{225} = \frac{113}{9}$. The requested sum is $113 + 9 = 122$.

Alternatively, set $x = \cos\theta$ and $y = \sin\theta$. Then it is known that $4x - 3y = \frac{13}{3}$ and $x^2 + y^2 = 1$, and, from the double angle formulas for $\sin\theta$ and $\cos\theta$, the requested value is

$$7\cos 2\theta - 24\sin 2\theta = 7(2x^2 - 1) - 24(2xy).$$

But then

$$7(2x^2 - 1) - 24(2xy) = 14x^2 - 7 - 48xy$$
$$= -18(x^2 + y^2) + 2(4x - 3y)^2 - 7$$
$$= -18(1) + 2\left(\frac{13}{3}\right)^2 - 7 = \frac{113}{9}.$$

19. Solution (147).

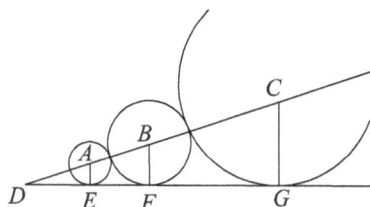

Let circles A, B, and C have centers A, B, C and radii a, b, and c, and be tangent to their common tangent at points E, F, and G, respectively. Let D be the intersection of the common tangent and the line containing the centers of the circles. Then triangles ADE, BDF, and CDG are similar. It follows that

$$\frac{a}{AD} = \frac{b}{BD} = \frac{b}{AD + a + b}.$$

This yields

$$\frac{AD}{a} = \frac{b + a}{b - a},$$

so by similar triangles

$$\frac{AD}{a} = \frac{CD}{c} = \frac{AD + a + 2b + c}{c} = \frac{a\frac{b+a}{b-a} + a + 2b + c}{c}.$$

Solving for c gives

$$c = \frac{a\frac{b+a}{b-a} + a + 2b}{\frac{b+a}{b-a} - 1} = \frac{a(b+a) + (a + 2b)(b - a)}{(b+a) - (b-a)} = \frac{b^2}{a} = \frac{42^2}{12} = 147.$$

Alternatively, note that there is a dilation centered at D that maps circle A to circle B. The constant of this dilation is $\frac{42}{12} = \frac{7}{2}$. The same dilation moves circle B to circle C, thus the radius of circle C is $\frac{7}{2} \cdot 42 = 147$.

20. Solution (162). Begin with six open positions. To get an arrangement of digits having the desired property, choose a position to place the digit 6. The 6 can be placed in any of the 3 rightmost positions. Once the position of the 6 has been chosen, choose a position to place the digit 5. The 5 can be placed in any of the 3 rightmost positions not occupied by the 6. Continuing this way, the digits 6, 5, 4, and 3 can all be placed in one of 3 positions, the 3 rightmost positions which are left open. Finally, there will be 2 available positions to place the 2, and only one position left for the 1. Thus, the number of ways to choose an arrangement with the desired property is $3^4 \cdot 2 = 162$.

21. **Solution** (891). The list contains 100 numbers. The sum is

$$
\begin{aligned}
a &= 0.9 \times (99 + 999 + 9999 + \cdots + 999 \cdots 9) \\
&= 0.9 \times [(100 - 1) + (1000 - 1) + (10000 - 1) + \cdots + (1000 \cdots 0 - 1)] \\
&= 0.9 \times \left[\sum_{n=0}^{101} (10^n - 1) - 9 \right] \\
&= \frac{9}{10} \cdot \left[\sum_{n=0}^{101} 10^n - 111 \right] \\
&= \frac{9}{10} \cdot \left[\frac{10^{102} - 1}{10 - 1} - 111 \right] \\
&= \frac{1}{10} \left[(10^{102} - 1) - 999 \right] \\
&= 10^{101} - 100.
\end{aligned}
$$

The decimal representation of this number is 101 digits long and consists of 99 nines followed by 2 zeros. The sum of the digits of this number is $99 \cdot 9 = 891$.

22. **Solution** (42). More generally, let a_n be the number of ways one can draw n non-intersecting line segments between the pairings of $2n$ points on a circle. Clearly, $a_1 = 1$. If a_1 through a_n are known, a_{n+1} can be calculated as follows. Number the points in the circle in order from 1 to $2n + 2$. Point 1 will need to be paired with one of the n points labeled with an even number so that there will be an even number of points lying on each side of the line segment drawn from point 1. If point 1 is paired with point $2k$, then there are $2k - 2$ points left on one side of the segment, and $2n - 2k$ points left on the other side. It follows that there are $a_{k-1} \cdot a_{n-k}$ ways to pair the remaining points. This gives the recursive formula

$$
a_{n+1} = \sum_{k=1}^{n+1} a_{k-1} \cdot a_{n-k}
$$

assuming that a_0 is defined to be 1. Thus,

$$
a_2 = a_0 \cdot a_1 + a_1 \cdot a_0 = 2,
$$

$$
a_3 = a_0 \cdot a_2 + a_1 \cdot a_1 + a_2 \cdot a_0 = 5,
$$

$$
a_4 = a_0 \cdot a_3 + a_1 \cdot a_2 + a_2 \cdot a_1 + a_3 \cdot a_0 = 14,
$$

$$
a_5 = a_0 \cdot a_4 + a_1 \cdot a_3 + a_2 \cdot a_2 + a_3 \cdot a_1 + a_4 \cdot a_0 = 42.
$$

Thus, the answer is 42. These numbers are also the Catalan numbers

$$
a_n = C(n) = \frac{\binom{2n}{n}}{n + 1}.
$$

23. Solution (19).

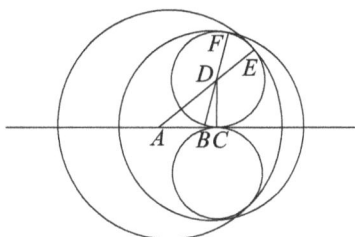

Let A be the center of the circle with radius 10, and B be the center of the circle with radius 8. The two small circles are tangent to each other and the line AB at point C. Let D be the center of one of the small circles, E be the point where that small circle is tangent to the circle centered at A, and F be the point where that small circle is tangent to the circle centered at B. The triangles ACD and BCD are right triangles with right angles at C. Let r be the radius of the small circles. Then $CD = r$. Since $AE = 10$, it follows from the Pythagorean Theorem that

$$AC = \sqrt{(10 - r)^2 - r^2}.$$

Since $BF = 8$, it follows from the Pythagorean Theorem that

$$BC = \sqrt{(8 - r)^2 - r^2}.$$

But $AC = BC + 3$, so

$$\sqrt{(10 - r)^2 - r^2} = \sqrt{(8 - r)^2 - r^2} + 3.$$

Square both sides to get

$$100 - 20r = 64 - 16r + 6\sqrt{(8 - r)^2 - r^2} + 9.$$

Solving for the radical term and squaring again yields

$$6\sqrt{(8 - r)^2 - r^2} = 27 - 4r$$

and

$$36\,(64 - 16r) = 27^2 - 216r + 16r^2.$$

This simplifies to

$$16r^2 + 360r - 1575 = 0$$

which has one positive solution at

$$r = \frac{-360 + \sqrt{360^2 + 4 \cdot 16 \cdot 1575}}{2 \cdot 16} = \frac{-360 + 24\sqrt{15^2 + 175}}{32} = \frac{-360 + 24 \cdot 20}{32} = \frac{15}{4}.$$

The requested sum is $15 + 4 = 19$.

24. Solution (24). Rewrite the condition as

$$mn - 20m + 10n = 0$$

and then

$$mn - 20m + 10n - 200 = -200$$

which factors as

$$(m + 10)(n - 20) = -200.$$

Thus, (m, n) satisfies the desired condition if the product of $m + 10$ and $n - 20$ is -200. The 12 positive integer divisors of 200 are 1, 2, 4, 5, 8, 10, 20, 25, 40, 50, 100, and 200. If $m + 10$ is any one of these positive integer divisors of 200 or the negative of any one of them, then setting $n = \dfrac{-200}{m + 10} + 20$ lets (m, n) satisfy the required condition. Thus, there are 24 ordered pairs satisfying the condition.

25. Solution (10). First note that since

$$x^3 + 3x + 1 = (x - x_1)(x - x_2)(x - x_3),$$

we get

$$x_1 + x_2 + x_3 = 0, \ x_1 x_2 + x_1 x_3 + x_2 x_3 = 3, \ \text{and} \ x_1 x_2 x_3 = -1.$$

It also follows that

$$0 = (x_1 + x_2 + x_3)^2 = x_1^2 + x_2^2 + x_3^2 + 2(x_1 x_2 + x_1 x_3 + x_2 x_3)$$

and

$$x_1^2 + x_2^2 + x_3^2 = -6.$$

Finally, since each x_j satisfies $x^3 + 3x + 1 = 0$, one gets that $x_1^3 + x_2^3 + x_3^3 = -3(x_1 + x_2 + x_3) - 3 = -3$. Now,

$$
\begin{aligned}
&= \frac{x_1^2}{(5x_2 + 1)(5x_3 + 1)} + \frac{x_2^2}{(5x_1 + 1)(5x_3 + 1)} + \frac{x_3^2}{(5x_1 + 1)(5x_2 + 1)} \\
&= \frac{x_1^2(5x_1 + 1) + x_2^2(5x_2 + 1) + x_3^2(5x_3 + 1)}{(5x_1 + 1)(5x_2 + 1)(5x_3 + 1)} \\
&= \frac{5(x_1^3 + x_2^3 + x_3^3) + (x_1^2 + x_2^2 + x_3^2)}{5^3(x_1 x_2 x_3) + 5^2(x_1 x_2 + x_1 x_3 + x_2 x_3) + 5(x_1 + x_2 + x_3) + 1} \\
&= \frac{5(-3) - 6}{125(-1) + 25(3) + 5(0) + 1} \\
&= \frac{-21}{-125 + 75 + 1} = \frac{21}{49} = \frac{3}{7}.
\end{aligned}
$$

The requested sum is $3 + 7 = 10$.

26. Solution (71). First consider the parabola with equation $y = x^2$ and two parallel lines with equations $y = mx + b_1$ and $y = mx + b_2$ for constants m, $b_1 > 0$, and $b_2 > 0$. Note that the axis of symmetry of the parabola is the y axis. The first line intersects the parabola at two points with x coordinates

$$\frac{m \pm \sqrt{m^2 + 4b_1}}{2},$$

and the second line at x coordinates

$$\frac{m \pm \sqrt{m^2 + 4b_2}}{2}.$$

It follows that the difference between the x coordinates of the intersection points to the left of the y axis is

$$\frac{\sqrt{m^2 + 4b_1} - \sqrt{m^2 + 4b_2}}{2}$$

which is also the difference between the x coordinates of the intersection points to the right of the y axis. Since all parabolas are geometrically similar, this shows that when two parallel lines intersect a parabola, the difference in the distances from the axis of symmetry for two intersection points on one side of the axis is the same as the difference in distances from the axis of symmetry for the two intersection points on the other side of the axis.

Because the slope of the axis of symmetry will not change if the parabola is translated seven units in the negative x direction and 6 units in the negative y direction, it can be assumed that the given parabola passes through the four points $(0,0)$, $(0,6)$, $(11,13)$, and $(11,42)$. Assume that the axis of symmetry of the parabola has slope m. Then there are constants a and b so that the line parallel to the axis which passes through $(11,13)$ has equation $y = mx + a$, and the line parallel to the axis which passes through $(11,42)$ has equation $y = mx + b$. It follows that $13 = 11m + a$ and $42 = 11m + b$ which implies that

$$b - a = 42 - 13 = 29.$$

The result of the previous paragraph says that the lines $y = mx + b$ and $y = mx + 6$ must be the same distance apart as the lines $y = mx$ and $y = mx + a$. Thus, $b - 6 = 0 - a$ or $a + b = 6$. This lets us solve for b to get $b = \dfrac{35}{2}$. So, the line parallel to the axis of symmetry which passes through the point $(11,42)$ also passes through the point $\left(0, b = \dfrac{35}{2}\right)$. Conclude that the slope m of this line, and thus the slope of the axis of symmetry, is

$$\frac{42 - \dfrac{35}{2}}{11} = \frac{49}{22}.$$

The requested sum is $49 + 22 = 71$.

27. **Solution** (35). Apply the Cauchy-Schwartz inequality to the vectors

$$(2(\sin a + \cos a), 1)$$

and $(\sin b, \cos b)$ to get that

$$\left[4\left(\sin a + \cos a\right)^2 + 1\right]\left(\sin^2 b + \cos^2 b\right) \geq \left[2(\sin a + \cos a)\sin b + \cos b\right]^2 = 3^2$$

which simplifies to

$$\left[4(\sin^2 a + 2\sin a \cos a + \cos^2 a) + 1\right]\left(\sin^2 b + \cos^2 b\right) = 5 + 4\sin 2a \geq 9.$$

Thus, $\sin 2a \geq 1$, and it follows that $\sin 2a = 1$. Use the identity

$$\tan \frac{x}{2} = \frac{\sin x}{1 + \cos x}$$

to conclude that

$$\tan a = \frac{1}{1 + 0} = 1.$$

Since equality occurs in the Cauchy-Schwartz inequality only when the vectors are parallel, conclude that

$$\frac{\sin b}{\cos b} = \frac{2(\sin a + \cos a)}{1}$$

and

$$\tan^2 b = 4\left(\sin^2 a + 2\sin a \cos a + \cos^2 a\right) = 4\left(1 + \sin 2a\right) = 8.$$

Now

$$3\tan^2 a + 4\tan^2 b = 3(1) + 4(8) = 35.$$

28. **Solution** (97). The sum

$$\sum_{n=3}^{\infty} \frac{1}{n^5 - 5n^3 + 4n} = \sum_{n=3}^{\infty} \frac{1}{(n-2)(n-1)n(n+1)(n+2)}$$

$$= \frac{1}{4}\sum_{n=3}^{\infty} \frac{(n+2) - (n-2)}{(n-2)(n-1)n(n+1)(n+2)}$$

$$= \frac{1}{4}\sum_{n=3}^{\infty} \frac{1}{(n-2)(n-1)n(n+1)} - \frac{1}{(n-1)n(n+1)(n+2)}.$$

This last summation telescopes leaving

$$\frac{1}{4}\frac{1}{(3-2)(3-1)3(3+1)} = \frac{1}{4 \cdot 1 \cdot 2 \cdot 3 \cdot 4} = \frac{1}{96}.$$

The requested sum is $1 + 96 = 97$.

29. **Solution** (175). Without loss of generality, let the square have side length 18. Set up coordinate axes so that $\overline{BC}$ is the x-axis and $\overline{AB}$ is the y-axis. Then the coordinates of point $A = (0, 18)$,

$B = (0,0)$, $C = (18,0)$, and $D = (18,18)$. Let the coordinates of point E be $(0,y)$. Then the coordinates of point $F = (y,0)$, $G = (18-y,18)$, $H = \left(\frac{y}{2}, \frac{y}{2}\right)$, $I = \left(18 - \frac{y}{2}, 9\right)$, and $J = \left(9, \frac{18+y}{4}\right)$. The line $\overline{GJ}$ is perpendicular to the line $\overline{HI}$, so the slope of $\overline{GJ} = \dfrac{\frac{54-y}{4}}{9-y}$ must be the negative reciprocal of the slope of $\overline{HI} = \dfrac{9 - \frac{y}{2}}{18 - y} = \frac{1}{2}$. It follows that $\dfrac{54-y}{4(9-y)} = -2$ so $y = 14$.

The areas of pentagon $AEHJG$ and quadrilateral $CIHF$ can be calculated from the coordinates of their vertices $A = (0,18)$, $E = (0,14)$, $H = (7,7)$, $J = (9,8)$, $G = (4,18)$, $C = (18,0)$, $I = (11,9)$, and $F = (14,0)$. This can be done in a number of ways including by calculating half the pseudo-determinants

$$\frac{1}{2}\begin{vmatrix} 0 & 18 \\ 0 & 14 \\ 7 & 7 \\ 9 & 8 \\ 4 & 18 \\ 0 & 18 \end{vmatrix} = \frac{1}{2}\left[\begin{array}{c}(0 \cdot 14 + 0 \cdot 7 + 7 \cdot 8 + 9 \cdot 18 + 4 \cdot 18) - \\ (0 \cdot 18 + 4 \cdot 8 + 9 \cdot 7 + 7 \cdot 14 + 0 \cdot 18)\end{array} \right] = \frac{97}{2}$$

and

$$\frac{1}{2}\begin{vmatrix} 18 & 0 \\ 11 & 9 \\ 7 & 7 \\ 14 & 0 \\ 18 & 0 \end{vmatrix} = \frac{1}{2}\left[\begin{array}{c}(18 \cdot 9 + 11 \cdot 7 + 7 \cdot 0 + 14 \cdot 0) - \\ (18 \cdot 0 + 14 \cdot 7 + 7 \cdot 9 = 11 \cdot 0)\end{array} \right] = 39.$$

The desired ratio of areas is $\dfrac{\frac{97}{2}}{39} = \dfrac{97}{78}$, and the requested sum is $97 + 78 = 175$.

The Purple Comet! Math Meet staff apologizes and regrets that we all missed the typographical error in the original presentation of Problem 30. The problem originally asked for $4x^2 + 8y^3$ instead of the correct $8x^2 + 4y^3$. Since the original problem does not have a non-negative integer answer, this problem had to be eliminated from the competition.

30. **Solution** (20). First note that the second equation is not the same as the first equation with the variables x and y interchanged. This shows that when the equations are satisfied, x and y are two

distinct values. Rewrite the two equations as

$$x^4 - (x-1)^2 = 2y^3 - 4\sqrt{5} - 2$$

$$y^4 - (y-1)^2 = 2x^3 + 4\sqrt{5} - 2$$

and add the two equations to get

$$\left(x^4 - x^2 + 2x + 1 - 2x^3\right) + \left(y^4 - y^2 + 2y + 1 - 2y^3\right) = 0$$

and

$$(x^2 - x - 1)^2 + (y^2 - y - 1)^2 = 0.$$

It follows that both $x^2 - x - 1 = 0$ and $y^2 - y - 1 = 0$ so x and y must be the two roots of $x^2 - x - 1$. The two values that satisfy the original equation are $x = \dfrac{1-\sqrt{5}}{2}$ and $y = \dfrac{1+\sqrt{5}}{2}$. Finally,

$$8x^2 + 4y^3 = 8\frac{1 - 2\sqrt{5} + (\sqrt{5})^2}{2^2} + 4\frac{1 + 3\sqrt{5} + 3(\sqrt{5})^2 + (\sqrt{5})^3}{2^3} = (12 - 4\sqrt{5}) + (8 + 4\sqrt{5}) = 20.$$

CHAPTER 50

SOLUTIONS TO WISCONSIN MATH TALENT SEARCH

1. Solution If $t = (a, b, c)$, let $s(t) = a + b + c$ be the sum of the three real numbers which make up the triple. Notice that

$$s(t') = a' + b' + c' = (a + b) + (b + c) + (c + a) = 2(a + b + c) = 2s(t).$$

Now let us start with $t_0 = (a_0, b_0, c_0)$, let $t_1 = t'_0$ let $t_2 = t'_1$ and so on. Then the above formula implies that

$$s(t_1) = 2s(t_0), \ s(t_2) = 2s(t_1) = 2^2 s(t_0), \ s(t_3) = 2s(t_2) = 2^3 s(t_0),$$

and in general that $s(t_n) = 2^n s(t_0)$. In particular, if $t_n = t_0$ for some $n \geq 1$, then $s(t_0) = s(t_n) = 2^n s(t_0)$, and hence $s(t_0) = 0$. We have shown that if $t_n = t_0$ for some $n \geq 1$, then $s(t_0) = 0$. We now show that $s(t_0) = 0$ implies that we return to the original triple in six steps. Since $0 = s(t_0) = a_0 + b_0 + c_0$, we have

$$t_1 = (a_0 + b_0, b_0 + c_0, c_0 + a_0)$$

and

$$t_2 = (a_0 + b_0 + b_0 + c_0, b_0 + c_0 + c_0 + a_0, c_0 + a_0 + a_0 + b_0) = (b_0, c_0, a_0).$$

Similarly, we get

$$t_4 = (c_0, a_0, b_0)$$

and then

$$t_6 = (a_0, b_0, c_0) = t_0.$$

Reprinted by permission of the Wisconsin Mathematics, Engineering and Science Talent Search.

2. **Solution** Suppose $m = abc = 100a + 10b + c$. If we rearrange the digits of m in all six possible ways, then each of $a, b,$ and c will occur precisely twice in the 1's place, twice in the 10's place, and twice in the 100's place. Thus the sum of the six rearrangements is

$$2(a + b + c)(100 + 10 + 1) = 222(a + b + c)$$

and hence the arithmetic mean of these six numbers is $37(a + b + c)$. By assumption,

$$100a + 10b + c = m = 37(a + b + c),$$

so

$$7a = 3b + 4c \quad \text{or} \quad 7(a - c) = 3(b - c).$$

Now $-9 \leq b - c \leq 9$ and 7 divides $b - c$, so there are just three possibilities. If $b - c = 0$, then $a = b = c$ and we obtain the numbers aaa with $a = 1, 2, \ldots, 9$. If $b - c = 7$, then $a - c = 3$, so $b = c + 7, a = c + 3$ and there are only three possibilities for c namely $0, 1, 2$. These yield $m = 370, 481,$ and 592. Finally, if $b - c = -7$, then $a - c = -3$, so $b = c - 7, a = c - 3$. Thus $c = 7, 8, 9$ and these yield $m = 407, 518,$ and 629. Of course, one could do this all by computer, but that would be less of a challenge.

3. **Solution** We will show that there are no such functions. Suppose the opposite. Then with $x = y = -1/2$ we would get $f(-1/2)g(-1/2) = 0$. Thus $f(-1/2)$ or $g(-1/2)$ must be equal to 0. But if $f(-1/2) = 0$ then $x = -1/2, y = 1/2$ would give

$$f(-1/2)g(1/2) = 0 \neq -1/2 + 1/2 + 1 = 1.$$

We get into a similar contradiction when $g(-1/2) = 0$.

4. **Solution** If there are g green and r red balls in one box, then we would pick a green ball with probability $g/(g + r)$ from that box. The other box would have $5 - g$ green and $7 - r$ red balls, so picking a ball randomly would produce a green ball with probability $\dfrac{5 - g}{12 - (r + g)}$. Since the boxes are chosen with equal probability, the probability of choosing a green ball with this arrangement of balls will be the average of the two probabilities: $\dfrac{1}{2}\left(\dfrac{g}{g + r} + \dfrac{5 - g}{12 - (r + g)}\right)$.

We have to figure out how to maximize this quantity for the possible values of r and g. These values are $0 \leq g \leq 5$ and $0 \leq r \leq 7$, with the $(g, r) = (0, 0)$ and $(5, 7)$ cases excluded. Suppose that we put a single green ball in one of the boxes and the remaining 4 green and 7 red balls in the other box. Then the probability of winning is $\dfrac{1}{2}\left(\dfrac{1}{1} + \dfrac{4}{11}\right)$. We will show that we cannot get a higher winning probability than that. If one of the boxes contains only red balls, then the probability of winning is at most $1/2$ (since we immediately lose if we pick the all-red box), which is less than $\dfrac{15}{22}$. If one of the boxes contains only green balls and more than one, then moving one of them to the other box will increase the probability of winning. This is because the probability of picking a green ball increased in the 'mixed' box (by the additional green), and it did not change for the all-green box. So among configurations with an all green box, the example above with a box with just one green ball maximizes your probability of winning. Now assume that there are red and green balls in both boxes. We assume that the first box has at most as many balls as the second one, meaning that $r + g \leq 12 - (r + g)$. (The proof will work the same way in the opposite case.) Now take a green ball from the second box, a red from the first, and switch them. Then the probability of winning changed by $\frac{1}{2}\left(\frac{g}{g+r} + \frac{5-g}{12-(r+g)}\right)$. which means that it did not decrease. Repeating this step a couple of more times we either get all the greens in the first box or all the reds in the second box. But we have already checked that those cases have a winning probability at most $\frac{15}{22}$, so this must be true for our starting configuration as well. This shows that the probability of winning is at most $\frac{15}{22}$ for all possible arrangements.

5. **Solution** Clearly, each number in the sequence is positive, and the sequence is increasing. By squaring $a_{k+1} = a_k + 2/a_k$, we get $a_{k+1}^2 = a_k^2 + 4 + 4/a_k^2$ for $k \geq 0$. Using this identity repeatedly:

$$
\begin{aligned}
a_{2015}^2 &= a_{2014}^2 + 4 + \frac{4}{a_{2014}^2} \\
&= \left(a_{2013}^2 + 4 + \frac{4}{a_{2013}^2}\right) + 4 + \frac{4}{a_{2014}^2} \\
&= a_{2013}^2 + 2\cdot 4 + \frac{4}{a_{2014}^2} + \frac{4}{a_{2013}^2} \\
&= \left(a_{2012}^2 + 4 + \frac{4}{a_{2012}^2}\right) + 2\cdot 4 + \frac{4}{a_{2014}^2} + \frac{4}{a_{2013}^2} \\
&= a_{2012}^2 + 3\cdot 4 + \frac{4}{a_{2014}^2} + \frac{4}{a_{2013}^2} + \frac{4}{a_{2012}^2} \\
&\;\;\vdots \\
&= a_0^2 + 2015\cdot 4 + \frac{4}{a_{2014}^2} + \frac{4}{a_{2013}^2} + \frac{4}{a_{2012}^2} + \cdots + \frac{4}{a_0^2}
\end{aligned}
$$

The last expression is at least as big as $a_0^2 + 2015\cdot 4 + 4/a_0^2 = 8065 > 89^2$ which proves $a_{2015} > 89$.

CHAPTER 51

SOLUTIONS TO USA MATHEMATICAL TALENT
SEARCH(2015-2016)

51.1 USAMTS-Round 1 Problems

1. Solution Consider the following 2×2 square

a	b
c	d

and suppose $a < c$. By hypothesis, we must have either $a + d = b + c$ or $ad = bc$. Either way, it follows that $b < d$. Since $9 < 11$, repeatedly applying this observation tells us that every number in the top row is less than the number directly below it. Similarly, every other pair of consecutive rows must satisfy the same constraint, and we conclude that every column is strictly increasing from top to bottom.

Given the same 2×2 configuration, if a positive integer n divides c but not ad, then we must have $a + d = b + c$.

Finally, consider the following 3×2 configuration:

a	e	c
b	f	d

Reprinted by permission of USA Mathematical Talent Search, a Program of the Art of Problem Solving Initiative.

625

In this case, we claim that if $\gcd(a, b) = 1$ and $b - a > d - c$, then we must have $a + f = b + e$. To see this, let $b = a + k$ and suppose $af = be$. This can be rearranged to $a(f - e) = ke$, or

$$\frac{a}{b - a} = \frac{e}{f - e}.$$

Since a and $b - a$ are relatively prime, we conclude that $f - e \geq b - a$ and $e \geq a$. This implies that we must have $fc = ed$. Rewriting e as $\frac{af}{b}$ yields $ad = bc$, and by the same argument we have $d - c \geq b - a$, a contradiction. So we must have $a + f = b + e$.

Combining these observations with a little trial-and-error, one can construct the following unique solution.

3	9	12	6	1
5	11	14	7	2
10	22	28	21	16
15	27	33	26	21
20	36	44	37	32

2. **Solution** By adding the two equations we have $ab + ac + bc = 1110$. So $a, b,$ and c are the roots of the polynomial

$$f(x) = (x - a)(x - b)(x - c) = x^3 - dx^2 + 1110x - 1000,$$

where $d > 0$. We compare this to the polynomial

$$g(x) = (x - 1)(x - 10)(x - 100) = x^3 - 111x^2 + 1110x - 1000.$$

Since $a < 1$, we know that $g(a) < 0$. Notice that $g(x) - f(x) = (d - 111)x^2$. Since $g(a) = g(a) - f(a) < 0$, we conclude that $g(x) - f(x) = (d - 111)x^2$ is negative for all $x \neq 0$. Therefore,

$$g(b) = g(b) - f(b) < 0$$

and

$$g(c) = g(c) - f(c) < 0.$$

This means that b and c are in $(0, 1) \cup (10, 100)$. Since $abc = 1000$, we see that $bc > 1000$, which implies that b and c are both greater than 10. Thus, $10 < b, c < 100$, as desired.

Note: This proof can be generalized. Given any two cubic polynomials differing in exactly one coefficient and having six distinct positive roots, one can show that there are only two possible orderings of the six roots.

3. Solution Define a function $f : P \to \mathbb{R}$ by setting

$$f(Q) = [QV_1V_2] - [QV_2V_3] + \cdots + (-1)^{n-1}[QV_nV_1].$$

That is, $f(Q)$ is the area of the blue regions minus the area of the green regions given a fixed point Q. We examine what happens to f as we move Q along a fixed line. To do this, we look at the area of the triangle with base V_1V_2.

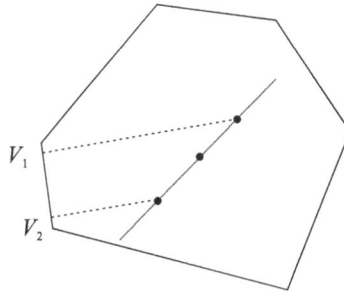

The base stays fixed and the height changes linearly as Q moves along this line. The same is true for all other sides of P, which means that $f(Q)$ changes linearly as Q moves along a fixed line inside P.

Now suppose A is not a vertex and $f(A) = 0$. We'll show that P contains at least one more balancing point. In order to show this, we will consider two cases: (1) A is in the interior of P and (2) A is on an edge of P.

Suppose first that A is in the interior of P. Take two segments ℓ_1 and ℓ_2 containing A in their interiors inside P. If either segment contains another balancing point, we're done. So suppose neither does. Since f changes linearly as we move along either segment, we know that we can find two points B and C on ℓ_1 and ℓ_2, respectively, such that $f(B)$ and $f(C)$ have opposite signs.

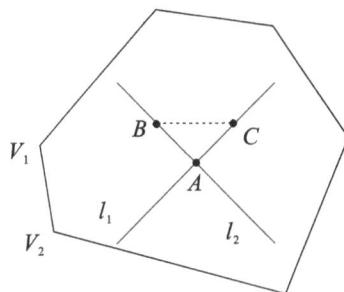

By convexity, we know that the segment connecting B and C is completely contained in P. Since f changes linearly along this segment, there must be some D between B and C for which $f(D) = 0$. Therefore, we have found another balancing point.

Now suppose that A is on an edge, V_iV_{i+1}. Pick some point R inside P and consider the segment QR. Again, if either V_iV_{i+1} or QR contains another balancing point we're done, so suppose neither does. Since f changes linearly as we move along V_iV_{i+1} and QR, we know that we can find two points B' and C' on V_iV_{i+1} and QR, respectively, such that $f(B')$ and $f(C')$ have opposite signs. By convexity, we know that the segment connecting B' and C' is completely contained in P. Since f changes linearly on this segment, there must be some D' between B' and C' for which $f(D') = 0$, and once again we have found another balancing point.

Combining both of these cases, we see that if P has exactly one balancing point, it must be a vertex.

Note: A hexagon with vertices at $(0,0), (0,2), (1,10), (3,10), (4,14/3)$, and $(4,8/3)$ has exactly one balancing point at $x = 0$.

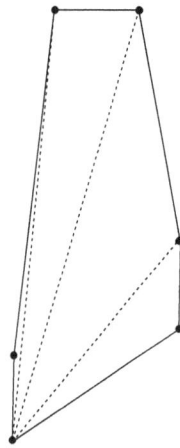

One can calculate the areas of the four triangles (from left to right) to be 1, 10, 13, and 4, and we see that $(0,0)$ is indeed a balancing point.

4. **Solution** We interpret players as points in the plane with coordinates $(x, y) = $ (weight, height) and interpret the requirements graphically. Each player is a lattice point on or within the right triangle defined by the lines $y = x + 1$, $x = 190$, and $y = 197$.

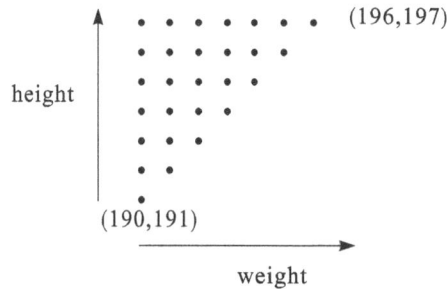

Requirement (i) states that if a player (a, b) is on the team, then so is everyone on or within the rectangle above and to the left of (a, b). An example for player $(192, 194)$ is shown below.

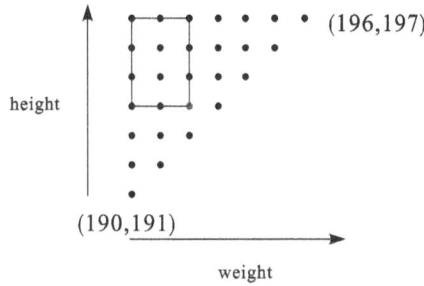

Requirement (ii) states that if a player (a, b) is on the team, then no one on the line $x = b$ can be on the team. Combined with requirement (i), this additionally implies that no one to the right of the line $x = b$ can be on the team if a player (a, b) is on the team. An example for player $(190, 193)$ is shown below.

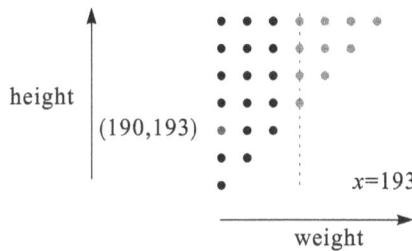

Notice that the red line is determined by the shortest player selected for the team on the line $x = 190$. Call this player p_0, and let his height be $190 + m$. Then requirement (ii) tells us that all players selected for the team are on the lines $x = 190, x = 191, \ldots, x = 190 + (m - 1)$. Requirement (i) implies that the players selected for the team on the line $x = 190 + k$ are determined by the shortest player selected for the team on this line. Let p_k be the shortest player on the line $x = 190 + k$. Then it suffices to count the number of ways to select $p_0, \ldots, p_{m-1}$. (If there are no players selected for the team on the line $x = 190 + k$, we simply ignore $p_k, \ldots, p_{m-1}$.)

To do this, draw a path containing only steps up and to the right starting at p_0 and ending at one of the m dots in the top row to the left of the line $x = 190 + m$, such that the lowest point on the path with x-coordinate $190 + k$ is p_k (if there are no players with x-coordinate $190 + k$, our path will stay completely to the left of the line $x = 190 + k$). An example path is shown below with $p_0 = (190, 193), p_1 = (191, 194)$, and $p_2 = (192, 195)$.

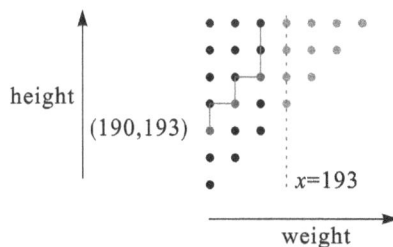

These paths correspond to teams because they determine $p_0, \ldots, p_{m-1}$. Notice that these paths are in one to one correspondence with paths starting at p_0 that end at $(189 + m, 198)$: we simply remove the portion of any given path above the line $y = 197$. Each path from our initial player to $(189+m, 198)$ must have $m-1$ steps to the right and $7 - (m-1)$ steps up. Thus the total number of paths from our initial player to $(189 + m, 198)$ is $\binom{7}{m-1}$. Summing over all $m \le 7$, we get

$$\sum_{m=1}^{7} \binom{7}{m-1} = 2^7 - 1$$

paths. This corresponds to the number of teams with at least one player, so in total the number of possible teams is $2^7 = 128$.

Challenge: Can you give a bijective proof mapping possible teams to binary strings of 0's and 1's of length 7?

5. **Solution** $n = 6$ works by taking $d_1 = 1$, $d_2 = 2$, and $d_3 = 3$. Similarly, $n = 6m$ works by taking $d_1 = m$, $d_2 = 2m$, and $d_3 = 3m$. We will show that these are the only possible values of n.

We'll assume without loss of generality that $d_1 < d_2 < \cdots < d_k$. Then, by dividing out by any common factor, we'll assume that $d_1, d_2, \ldots, d_k$ share no (non-trivial) common divisor, making our goal to show that $n = 6$.

Suppose two consecutive terms d_i and d_{i+1} have a common divisor r. Then $r \mid (d_{i+1} - d_i)$, which is the common difference of the arithmetic sequence. This implies that r divides all d_i, but we assumed that $d_1, d_2, \ldots, d_k$ shared no common divisor, so $r = 1$.

Since d_k and d_{k-1} are relatively prime and are both factors of n, we have $n \ge d_k d_{k-1}$. Then, $d_k \ge k$, so $n \ge k d_{k-1}$. For $k \ge 3$, d_{k-1} is greater than or equal to the average of the d_i, which means that n is at least k times the average of d_i. That is,

$$n \ge d_1 + d_2 + \cdots + d_k.$$

In order for equality to hold, we need $n = k d_{k-1} = d_k d_{k-1}$. The first equality implies that d_{k-1} is the average of the k-term arithmetic sequence, so $k = 3$. The second equality then tells us that $d_3 = 3$. That is, $d_1 = 1$, $d_2 = 2$, $d_3 = 3$, and $n = 6$, as desired.

To conclude it now suffices to eliminate $k = 2$. In this case, we have $n = d_1 + d_2$ and $d_1 \ne d_2$, which means that $d_2 > \frac{n}{2}$, contradicting the fact that d_2 is a divisor of n. Thus $k = 2$ is impossible.

51.2   USAMTS-Round 2 Problems

1. **Solution** In this solution, we use the notation $R_i C_j$ to denote the square in row i and column j. For example, the given 5 is in $R_4 C_2$.

The larger numbers give us the strongest restrictions, so we examine them first.

11				6			**12**
	10	1					
			2				8
	5			2			
					7	9	
12				3		**11**	**10**

The only pairs of squares that are a distance of 12 apart are the two pairs of diagonally opposite corners (R_1C_1, R_6C_8) and (R_1C_8, R_6C_1). However, the only square a distance of 10 from the given 10 is R_6C_8. So the 10 must be placed there. Thus, we can place the 12's in the corners (R_1C_8, R_6C_1).

For the pair of 11's to be a distance of 11 apart, one of them must be in a corner. Since three corners are already taken, we place an 11 in the remaining corner, R_1C_1. The two squares that are a distance of 11 from this are R_5C_8 and R_6C_7. Since column 8 already contains 12, 8, and 10, we cannot place any more numbers in that column and we must place the 11 in R_6C_7.

11		**7**		6			12
9	10	1					
			2				8
	5			2			
	8				7	9	
12				3		11	10

Next we resolve the 9, 8, and 7. There are two possible positions for the 7: it can be either in R_3C_1 or R_1C_3. Similarly, the remaining 8 can be placed in R_4C_1, R_6C_3, or R_5C_2 and the remaining 9 can be placed in R_2C_1 or R_1C_2. The 8 cannot be placed in R_6C_3 because row 6 already contains 12, 3, 11, and 10. So all possible positions for the non-given 8 and 9 are in columns 1 and 2. This means that columns 1 and 2 will each contain 3 numbers after the 8 and 9 are placed. So we cannot place the 7 in column 1, as this would force column 1 to contain 4 numbers. So we place the 7 in R_1C_3. This means that row 1 contains four numbers and we must place the 9 in R_2C_1. Once the 9 is placed, column 1 contains three numbers and we must place the 8 in R_5C_2.

Now we are forced to place the 6 in R_5C_3. This forces us to place the 1 in R_2C_4. The only way to place two more numbers in row 3 is to place the 5 in R_3C_6 and the 4 in R_3C_7. We conclude by placing the remaining 4 in R_4C_4 and the remaining 3 in R_4C_6.

11		7		6		12
9	10	1	**1**			
			2	**5**	**4**	8
	5		**4**	2	3	
	8	**6**			7	9
12				3	11	10

2. Solution The answer is yes and there are many possible solutions. We present one here.

We start with a cube and glue square pyramids on two of its faces. There are two distinct ways to do this: the two pyramids can replace either adjacent faces or opposite faces. We show that we can find nets for each of these polyhedra that produce the same pair of polygons.

In the diagram below, we replace two opposite faces in the cube net with square pyramids, then cut the resulting net along the red edge to produce two polygons.

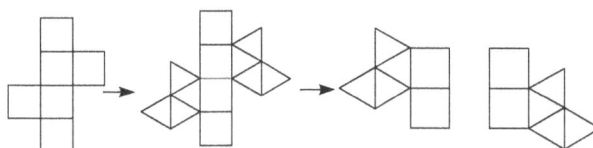

In the diagram below, we replace two adjacent faces in the cube net with square pyramids, then cut the resulting net along the red edge to produce two polygons.

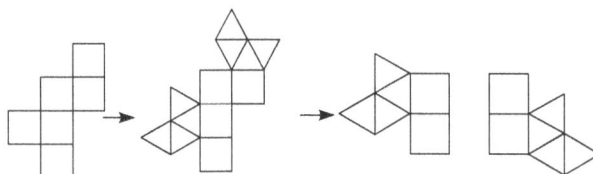

3. Solution Fix n. Let

$$S_r = \sum_{k=n-r}^{n} \frac{k \cdot k! \cdot \binom{n}{k}}{n^k}$$

be the sum of the last $r + 1$ terms of the given sum. We claim that

$$S_r = \frac{n!}{r! n^{n-r-1}}.$$

For $r = 0$, notice that we can write the last term in the sum as

$$\frac{n \cdot n! \cdot \binom{n}{n}}{n^n} = \frac{n \cdot n!}{n^n} = \frac{n!}{0! n^{n-1}}.$$

So the claim holds for $r = 0$. Now suppose the claim holds for $r = m - 1$ and we will show that it's true for $r = m$ (assuming $m < n$).

The sum of the last $m + 1$ terms is the sum of the last m terms, plus the $(m + 1)$st term from the end. That is,

$$S_m = S_{m-1} + \frac{(n-m)n!/m!}{n^{n-m}}.$$

By the inductive hypothesis, the sum of the last m terms is

$$S_{m-1} = \frac{n!}{(m-1)!n^{n-m}}.$$

So the sum of the last $m + 1$ terms is

$$S_m = \frac{n!}{(m-1)!n^{n-m}} + \frac{(n-m)n!/m!}{n^{n-m}} = \frac{n!(n-m) + n! \cdot m}{(m)!n^{n-m}}$$

$$= \frac{n! \cdot n}{m!n^{n-m}}$$

$$= \frac{n!}{m!n^{n-m-1}}.$$

Therefore, for all $r < n$, the sum of the last $r + 1$ terms is $\frac{n!}{r!n^{n-r-1}}$. Substituting $r = n - 1$, we have

$$\sum_{k=1}^{n} \frac{k \cdot k! \cdot \binom{n}{k}}{n^k} = \frac{n!}{(n-1)!n^{n-(n-1)-1}}$$

$$= \frac{n}{n^0} = n.$$

Dividing through by n gives the result.

OR

Suppose we roll an n-sided die repeatedly until we roll any number for the second time. We calculate the probability that the first repeat is on the $(k + 1)$st roll.

In order for the $(k + 1)$st roll to be the first repeat, the previous k rolls must have all been distinct. The number of sequences of k distinct rolls is $k!\binom{n}{k}$, because we can choose any k numbers from our die to show up and place them in any order. Since there are n^k total sequences of k rolls, the probability that the first k rolls are distinct is $\frac{k!\binom{n}{k}}{n^k}$. Since the $(k + 1)$st roll is a repeat, it must be one of the k numbers already seen. The probability of this happening is $\frac{k}{n}$. In total, the probability that our $(k + 1)$st roll is the first repeat is

$$\frac{k}{n} \cdot \frac{k!\binom{n}{k}}{n^k} = \frac{k \cdot k! \cdot \binom{n}{k}}{n^{k+1}}.$$

Summing over all $k \leq n$, we find that the probability that we have at least one repeat after $n + 1$ rolls is

$$\sum_{k=1}^{n} \frac{k \cdot k! \cdot \binom{n}{k}}{n^{k+1}}.$$

Since we're rolling an n-sided die, we're guaranteed to have a repeat by the $(n+1)$st roll. In particular,

$$\sum_{k=1}^{n} \frac{k \cdot k! \cdot \binom{n}{k}}{n^{k+1}} = 1.$$

Factoring an n out of the denominator, we have

$$\frac{1}{n} \sum_{k=1}^{n} \frac{k \cdot k! \cdot \binom{n}{k}}{n^{k}} = 1.$$

4. **Solution** Fix an integer b and write $H(x) = P(x+b) - P(b)$. If H is identically 0, then $P(x)$ is a constant function. Substituting $P(x) = c$, the condition says that c divides 0, which is true. So constant functions work. For the remainder, we will assume that $P(x)$ is non-constant. In that case, $H(x)$ and $P(x)$ have the same leading term, so we can write

$$H(x) = P(x) + r(x),$$

where $r(x)$ is either identically 0 or $\deg r(x) < \deg P(x)$. For any integer a, we have that $H(a) = P(a) + r(a)$ is a multiple $P(a)$. In particular, for any integer a, $r(a)$ is a multiple of $P(a)$.

The degree of $r(x)$ is less than the degree of $P(x)$ (or $r(x)$ is identically 0), so we can choose some M such that for all integers $a \geq M$, $|P(a)| > |r(a)|$. Since $r(a)$ is a multiple of $P(a)$ for each such a, this implies that $r(a) = 0$ for all integers $a \geq M$. Therefore, $r(x)$ is a polynomial with infinitely many zeros, and must be identically 0. Hence $H(x) = P(x)$ and

$$P(x+b) - P(b) = P(x).$$

Since b was an arbitrary integer, this equation holds for all integers b and real numbers x. Plugging in $x = b = 0$ yields $P(0) = 0$. Plugging in $x = 1, b = 1$ gives us $P(2) = 2P(1)$. Then plugging in $x = 2, b = 1$ gives us $P(3) = 2P(1) + P(1) = 3P(1)$. Continuing this way by induction, plugging in $x = k, b = 1$ gives us

$$P(k+1) = kP(1) + P(1) = (k+1)P(1).$$

Therefore, $P(x) - xP(1)$ is a polynomial with infinitely many zeros, and must be identically 0. So if $P(x)$ is not constant, $P(x) = xP(1)$ is a linear function with no constant term.

Therefore, the only non-constant solution is $P(x) = cx$ for some constant c. Substituting $P(x) = cx$, the given condition says that $ca + cb - cb = ca$ is a multiple of ca, which is true. So both of these classes of functions work and our solution is $\boxed{P(x) = cx \text{ or } P(x) = c}$ for some integer c.

Note We ended up deriving an equation of the form $P(x+b) = P(x) + P(b)$. This is a version of Cauchy's functional equation. It turns out that if a function $f : \mathbb{Q} \to \mathbb{Q}$ satisfies $f(x+y) =$

$f(x) + f(y)$ for all x and y, then it must be of the form $f(x) = cx$. The proof is similar to what we did above, except we can no longer take advantage of the fact that we know f is a polynomial.

5. **Solution** The answer is $\frac{3}{2}n^2 - 2$. Let $n = 2k$ and we will show that the minimum number of elements in S is $6k^2 - 2$. We begin with a Lemma.

Lemma: Let k be a positive integer. If we have a $(2k+1) \times (2k+1)$ grid of squares, with some squares shaded so that all shaded squares are connected by side, as in condition (i) in the problem, and no 2×2 subgrid is entirely unshaded, then the number of shaded squares is at least $2k^2 - 1$.

Proof: Place the $(2k+1) \times (2k+1)$ grid on the coordinate plane with each square having area 1. The shaded squares together form a polygon whose area is equal to the total number of shaded squares. Since no 2×2 subgrid is unshaded, all of the interior vertices of the grid are inside or on the boundary of the polygon; there are $4k^2$ such vertices. By Pick's Theorem, we have that the area is $\frac{B}{2} + I - 1$ where B is the number of boundary vertices and I is the number of interior vertices. As $B + I \geq 4k^2$, we have

$$\frac{B}{2} + I - 1 \geq \frac{B+I}{2} - 1 \geq \frac{4k^2}{2} - 1 = 2k^2 - 1,$$

as desired.

Let S be a valid subset of the original grid, and we will show that it contains at least $6k^2 - 2$ squares. Consider the set of vertices of the n^2 blocks of 2×2 squares.

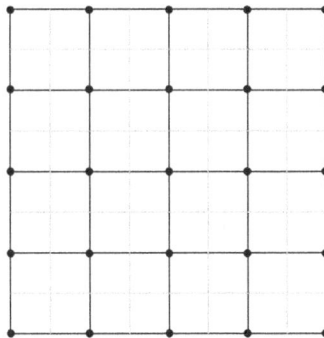

There are $(2k+1)^2$ such vertices. Construct a $(2k+1) \times (2k+1)$ grid G of squares so that each square corresponds to one of these $(2k+1)^2$ vertices. Shade a square in G if the corresponding vertex touches a square of S. An example of such a grid and shading is shown below.

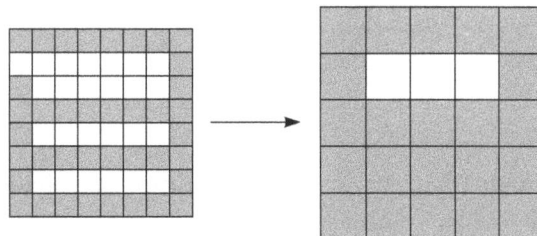

By (i), the shaded squares in G are connected by side, and by (ii) no 2×2 block of squares in G can be fully unshaded. Therefore, by the lemma we have at least $2k^2 - 1$ shaded squares in G.

Next, construct a graph G' as follows: the vertices correspond to the shaded squares in G. Whenever we have two adjacent squares in the same 2×2 block both in S, we draw an edge in G' connecting the two vertices that the two squares in S touch. An example graph is shown below.

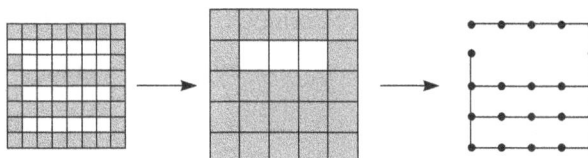

By (i), G' is a connected graph. If we arbitrarily break the cycles in G', it has $2k^2 - 2$ edges. Let $S(G')$ be a spanning tree of G'. Since G' has at least $2k^2 - 1$ vertices, we know that $S(G')$ has at least $2k^2 - 2$ edges.

Notice that each edge in $S(G')$ corresponds to two unit squares in the same 2×2 block both being shaded. Since $S(G')$ contains no cycles, each edge corresponds to a unique such pair of unit squares. We use this to place a lower bound on the total number of shaded squares in the original grid.

In the original grid, we require at least one shaded square in each of the $4k^2$ blocks. Also, an additional shaded square in the original grid is required for each edge of $S(G')$. There are at least $2k^2 - 2$ edges in $S(G')$, so in total we have at least $6k^2 - 2$ shaded squares.

We now exhibit a construction of $6k^2 - 2$ shaded squares. In the accompanying diagrams, we demonstrate the construction for $n = 4$. Number the rows and columns from 1 to $4k$, and denote the square in row i column j as (i, j). Shade in the $4k - 2$ squares $(i, 2)$ for $2 \leq i \leq 4k - 1$.

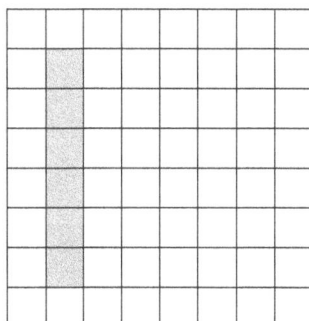

Next, for the k values of i with $1 \leq i \leq 4k$ and $i \equiv 2 \pmod 4$, shade in $4k - 3$ squares (i, j) for $3 \leq j \leq 4k - 1$.

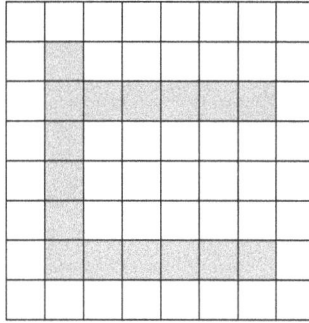

Finally, shade in $k(2k-1)$ squares of the form $(2i-1, 4j-1)$, for $2 \leq i \leq 2k$ and $1 \leq j \leq k$.

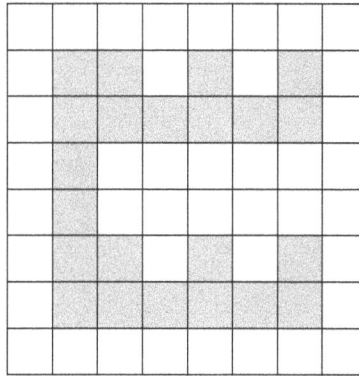

The total number of shaded squares is

$$4k - 2 + k(4k - 3) + k(2k - 1) = 4k^2 - 3k + 4k - 2 + 2k^2 - k = 6k^2 - 2 = \frac{3}{2}n^2 - 2,$$

as desired.

CHAPTER 52

SOLUTIONS TO HARVARD-MIT MATHEMATICS TOURNAMENT

52.1 Harvard-MIT Mathematics Tournament 2015

52.1.1 HMMT 2015 - Individual

1. **Solution** We can write $a + ab + abc = a(1 + b + bc)$. Since 11 is prime, $a = 11$ or $a = 1$. But since b, c are both positive integers, we cannot have $a = 11$, and so $a = 1$. Then

$$1 + b + bc = 11 \implies b + bc = 10 \implies b(c + 1) = 10,$$

and since c is a positive integer, only $b = 1, 2, 5$ are possible. This gives the $\boxed{3}$ triples $(a, b, c) = (1, 1, 9), (1, 2, 4), (1, 5, 1)$.

2. **Solution** $\left(\dfrac{\pi - 2}{4}\right)$. We require $a + b > 1$ and $a^2 + b^2 < 1$. Geometrically, this is the area enclosed in the quarter-circle centered at the origin with radius 1, not including the area enclosed by $a + b < 1$ (an isosceles right triangle with side length 1). As a result, our desired probability is $\dfrac{\pi - 2}{4}$.

3. **Solution** (13). Suppose instead Neo started at a weight of 2015 pounds, instead had green pills, which halve his weight, and purple pills, which increase his weight by a pound, and he wished to reduce his weight to one pound. It is clear that, if Neo were able to find such a sequence of pills in the case where he goes from 2015 pounds to 1 pound, he can perform the sequence in reverse (replacing green pills with red pills and purple pills with blue pills) to achieve the desired weight, so this problem is equivalent to the original.

Reprinted by permission of Harvard-MIT Math Tournament.

Suppose at some point, Neo were to take two purple pills followed by a green pill; this changes his weight from $2k$ to $k+1$. However, the same effect could be achieved using less pills by first taking a green pill and then taking a purple pill, so the optimal sequence will never contain consecutive purple pills. As a result, there is only one optimal sequence for Neo if he is trying to lose weight: take a purple pill when his weight is odd, and a green pill when his weight is even. His weight thus becomes

$$2015 \rightarrow 2016 \rightarrow 1008 \rightarrow 504 \rightarrow 252 \rightarrow 126 \rightarrow 63$$

$$\rightarrow 64 \rightarrow 32 \rightarrow 16 \rightarrow 8 \rightarrow 4 \rightarrow 2 \rightarrow 1$$

which requires a total of $\boxed{13}$ pills. Reversing this sequence solves the original problem directly.

4. **Solution** (130π). Let O be the center of the circle and let M be the midpoint of segment AB and let N be the midpoint of segment CD. Since quadrilateral $OMPN$ is a rectangle we have that $ON = MP = AM - AP = 3$ so

$$OC = \sqrt{ON^2 + NC^2} = \sqrt{9 + 121} = \sqrt{130}$$

Hence the desired area is $\boxed{130\pi}$.

5. **Solution** (672). From each of the sets $\{1,2,3\},\{4,5,6\},\{7,8,9\},\ldots$ at most 1 element can be in S. This leads to an upper bound of $\left\lceil \frac{2015}{3} \right\rceil = \boxed{672}$ which we can obtain with the set $\{1,4,7,\ldots,2014\}$.

6. **Solution** (-35). For almost all integers x, $f(x) \neq -x - 20$. If $f(x) = -x - 20$, then

$$f(-x - 20 + 2x + 20) = 15 \implies -x - 20 = 15 \implies x = -35.$$

Now it suffices to prove that the $f(-35)$ can take any value.

$f(-35) = 15$ in the function $f(x) \equiv 15$. Otherwise, set $f(-35) = c$, and $f(x) = 15$ for all other x. It is easy to check that these functions all work.

7. **Solution** (252). Note that $\angle MIA = \angle BAI = \angle CAI$, so $MI = MA$. Similarly, $NI = NB$. As a result,

$$CM + MN + NC = CM + MI + NI + NC = CM + MA + NB + NC = AC + BC = 48.$$

Furthermore, $AC^2 + BC^2 = 36^2$. As a result, we have $AC^2 + 2AC \cdot BC + BC^2 = 48^2$, so $2AC \cdot BC = 48^2 - 36^2 = 12 \cdot 84$, and so

$$\frac{AC \cdot BC}{2} = 3 \cdot 84 = \boxed{252}.$$

8. Solution $\left(\dfrac{35}{36}\right)$. The I-altitudes of triangles AIB and CID are both equal to the radius of ω, hence have equal length. Therefore $\dfrac{[AIB]}{[CID]} = \dfrac{AB}{CD}$. Also note that $[AIB] = IA \cdot IB \cdot \sin AIB$ and $[CID] = IC \cdot ID \cdot \sin CID$, but since lines IA, IB, IC, ID bisect angles $\angle DAB, \angle ABC, \angle BCD, \angle CDA$ respectively we have that $\angle AIB + \angle CID = (180° - \angle IAB - \angle IBA) + (180° - \angle ICD - \angle IDC) = 180°$. So, $\sin AIB = \sin CID$. Therefore $\dfrac{[AIB]}{[CID]} = \dfrac{IA \cdot IB}{IC \cdot ID}$. Hence

$$\frac{AB}{CD} = \frac{IA \cdot IB}{IC \cdot ID} = \boxed{\frac{35}{36}}.$$

9. Solution $\left(\dfrac{1}{2015}\right)$. Write $k = \frac{m}{n}$, for relatively prime integers m, n. For the property not to hold, there must exist integers a and b for which

$$\frac{a}{b} < \frac{m}{n} < \frac{a+1}{b+1}$$

(i.e. at some point, Guildenstern must "jump over" k with a single win) $\iff$ $an + n - m > bm > an$ hence there must exist a multiple of m strictly between an and $an + n - m$.

If $n - m = 1$, then the property holds as there is no integer between an and $an + n - m = an + 1$. We now show that if $n - m \neq 1$, then the property does not hold. By Bézout's Theorem, as n and m are relatively prime, there exist a and x such that $an = mx - 1$, where $0 < a < m$. Then $an + n - m \geq an + 2 = mx + 1$, so $b = x$ satisfies the conditions. As a result, the only possible k are those in the form $\dfrac{n}{n+1}$.

We know that Rosencrantz played at most 2015 games, so the largest non-perfect winrate he could possibly have is $\dfrac{2014}{2015}$. Therefore, $k \in \left\{\dfrac{1}{2}, \dfrac{2}{3}, \dots, \dfrac{2014}{2015}\right\}$, the product of which is $\boxed{\dfrac{1}{2015}}$.

10. Solution (43). For convenience, let $n = 101$. Compute the number of functions such that $f^n(1) = 1$. Since n is a prime, there are 2 cases: the order of 1 is either 1 or n. The first case gives n^{n-1} functions, and the second case gives $(n-1)!$ functions. By symmetry, the number of ways for $f^n(1) = 2$ is

$$\frac{1}{n-1} \cdot (n^n - n^{n-1} - (n-1)!) = n^{n-1} - (n-2)!.$$

Plugging in $n = 101$, we need to find

$$101^{100} - 99! \equiv (-2)^{-2} - \frac{101!}{6}$$

$$= 1/4 - 1/6 = 1/12 = \boxed{43} \pmod{103}.$$

52.1.2 HMMT 2015 - Guts round

1. **Solution** (2014). After k minutes, the diseased plants are the ones with taxicab distance at most k from the center. The plants on the corner are the farthest from the center and have taxicab distance 2014 from the center, so all the plants will be diseased after 2014 minutes.

2. **Solution** $\left(\dfrac{1+\sqrt{5}}{2}\right)$. Let the shorter leg have length ℓ, and the common ratio of the geometric sequence be $r > 1$. Then the length of the other leg is ℓr, and the length of the hypotenuse is ℓr^2. Hence,

$$\ell^2 + (\ell r)^2 = (\ell r^2)^2$$

$$\implies \ell^2(r^2 + 1) = \ell^2 r^4$$

$$\implies r^2 + 1 = r^4$$

Hence, $r^4 - r^2 - 1 = 0$, and therefore $r^2 = \dfrac{1 \pm \sqrt{5}}{2}$. As $r > 1$, we have $r^2 = \boxed{\dfrac{1+\sqrt{5}}{2}}$, completing the problem as the ratio of the hypotenuse to the shorter side is $\dfrac{\ell r^2}{\ell} = r^2$.

3. **Solution** (1). The area of the parallelogram can be made arbitrarily small, so the smallest positive integer area is 1.

4. **Solution** (60%). We see there are a total of $100 + 3 \times 100 + 300 = 700$ points, and he needs $70\% \times 700 = 490$ of them. He has $100 + 60 + 70 + 80 = 310$ points before the final, so he needs 180 points out of 300 on the final, which is 60%.

5. **Solution** $(-20, 28, 38)$. Let x, y, z be the integers. We have

$$\frac{x+y}{2} + z = 42$$
$$\frac{y+z}{2} + x = 13$$
$$\frac{x+z}{2} + y = 37$$

Adding these three equations yields $2(x+y+z) = 92$, so $\frac{x+y}{2} + z = 23 + \frac{z}{2} = 42$ so $z = 38$. Similarly, $x = -20$ and $y = 28$.

6. **Solution** $\left(\dfrac{1}{3}\right)$. Let X be the midpoint of segment AM. Note that $OM \perp MX$ and that $MX = \dfrac{1}{2}$ and $OX = \dfrac{1}{2} + r$ and $OM = 1 - r$. Therefore by the Pythagorean theorem, we have

$$OM^2 + MX^2 = OX^2 \implies (1-r)^2 + \frac{1}{2^2} = \left(\frac{1}{2} + r\right)^2$$

which we can easily solve to find that $r = \boxed{\dfrac{1}{3}}$.

7. **Solution** (116). Note that $2015 = 5 \times 13 \times 31$ and that $N = 2^{30} \cdot 3^{12} \cdot 5^4$ has exactly 2015 positive factors. We claim this is the smallest such integer. Note that $N < 2^{66}$.

   If n has 3 distinct prime factors, it must be of the form $p^{30}q^{12}r^4$ for some primes p, q, r, so $n \geq 2^{30} \cdot 3^{12} \cdot 5^4$.

   If n has 2 distinct prime factors, it must be of the form $p^e q^f > 2^{e+f}$ where $(e+1)(f+1) = 2015$. It is easy to see that this means $e + f > 66$ so $n > 2^{66} > N$.

   If n has only 1 prime factor, we have $n \geq 2^{2014} > N$.

   So N is the smallest such integer, and the sum of its prime factors is $2 \cdot 30 + 3 \cdot 12 + 5 \cdot 4 = 116$.

8. **Solution** (8060). We need both c/d and d/c to be integers, which is equivalent to $|c| = |d|$, or $d = \pm c$. So there are 4030 ways to pick c and 2 ways to pick d, for a total of 8060 pairs.

9. **Solution** (3). Since $|z| = |\bar{z}|$ we may divide by $|z|$ and assume that $|z| = 1$. Then $\bar{z} = \dfrac{1}{z}$, so we are looking for the smallest positive integer n such that there is a $2n^{\text{th}}$ root of unity in the first quadrant. Clearly there is a sixth root of unity in the first quadrant but no fourth or second roots of unity, so $n = \boxed{3}$ is the smallest.

10. **Solution** (2160). Note that T, E, A are used an odd number of times. Therefore, one must go in the middle spot and the other pair must match up. There are are

$$3 \cdot 2 \left(\frac{6!}{2!} \right) = 2160$$

ways to fill in the first six spots with the letters T, H, E, M, M and a pair of different letters. The factor of 3 accounts for which letter goes in the middle.

11. **Solution** ($\pm 1, \pm 2$). By Fermat's Last Theorem, we know $n < 3$. Suppose $n \leq -3$. Then $a^n + b^n = c^n \implies (bc)^{-n} + (ac)^{-n} = (ab)^{-n}$, but since $-n \geq 3$, this is also impossible by Fermat's Last Theorem. As a result, $|n| < 3$.

   Furthermore, $n \neq 0$, as $a^0 + b^0 = c^0 \implies 1 + 1 = 1$, which is false. We now just need to find constructions for $n = -2, -1, 1, 2$. When $n = 1$, $(a, b, c) = (1, 2, 3)$ suffices, and when $n = 2$, $(a, b, c) = (3, 4, 5)$ works nicely. When $n = -1$, $(a, b, c) = (6, 3, 2)$ works, and when $n = -2$, $(a, b, c) = (20, 15, 12)$ is one example. Therefore, the working values are $n = \boxed{\pm 1, \pm 2}$.

12. **Solution** ($2\sqrt{2}$). Let $ABCD$ be a square with $A = (0, 0), B = (1, 0), C = (1, 1), D = (0, 1)$, and P be a point in the same plane as $ABCD$. Then the desired expression is equivalent to $AP + BP + CP + DP$. By the triangle inequality, $AP + CP \geq AC$ and $BP + DP \geq BD$, so

the minimum possible value is $AC + BD = 2\sqrt{2}$. This is achievable when $a = b = \dfrac{1}{2}$, so we are done.

13. **Solution** (4). Piet can change the colors of at most 5 squares per minute, so as there are 16 squares, it will take him at least four minutes to change the colors of every square. Some experimentation yields that it is indeed possible to make the entire grid blue after 4 minutes; one example is shown below:

Here, jumping on the squares marked with an X provides the desired all-blue grid.

14. **Solution** (50). Let x be the length of BH. Note that quadrilateral $ABDE$ is cyclic, so by Power of a Point, $x(56 - x) = 20 \cdot 15 = 300$. Solving for x, we get $x = 50$ or 6. We must have $BH > HD$ so $x = 50$ is the correct length.

15. **Solution** (7). We have $1111_b = b^3 + b^2 + b + 1 = (b^2 + 1)(b + 1)$. Note that

$$\gcd(b^2 + 1, b + 1) = \gcd(b^2 + 1 - (b + 1)(b - 1), b + 1) = \gcd(2, b + 1),$$

which is either 1 or 2. If the gcd is 1, then there is no solution as this implies $b^2 + 1$ is a perfect square, which is impossible for positive b. Hence the gcd is 2, and $b^2 + 1, b + 1$ are both twice perfect squares.

Let $b + 1 = 2a^2$. Then

$$b^2 + 1 = (2a^2 - 1)^2 + 1 = 4a^4 - 4a^2 + 2 = 2(2a^4 - 2a^2 + 1),$$

so

$$2a^4 - 2a^2 + 1 = (a^2 - 1)^2 + (a^2)^2$$

must be a perfect square. This first occurs when $a^2 - 1 = 3, a^2 = 4 \implies a = 2$, and thus $b = \boxed{7}$. Indeed, $1111_7 = 20^2$.

16. **Solution** (4061). If none are of x, y, z are zero, then there are $4 \cdot 10^3 = 4000$ ways, since xyz must be positive. Indeed, $(abc)^2 = xyz$. So an even number of them are negative, and the ways to choose an even number of 3 variables to be negative is 4 ways. If one of x, y, z is 0, then one of a, b, c is zero at least. So at least two of x, y, z must be 0. If all 3 are zero, this gives 1 more

solution. If exactly 2 are negative, then this gives $3 \cdot 20$ more solutions. This comes from choosing one of x, y, z to be nonzero, and choosing its value in 20 ways.

Our final answer is $4000 + 60 + 1 = 4061$.

17. **Solution** $\left(\dfrac{2 - \sqrt{3}}{4}\right)$. Note that $AE = BF = CG = DH = 1$ at all times. Suppose that the squares have rotated θ radians. Then

$$\angle O_1 O_2 H = \frac{\pi}{4} - \theta = \angle O_1 DH,$$

so

$$\angle HDC = \frac{\pi}{4} - \angle O_1 DH = \theta.$$

Let P be the intersection of AB and EH and Q be the intersection of BC and GH. Then $PH \parallel BQ$ and $HQ \parallel PB$, and $\angle PHG = \dfrac{\pi}{2}$, so $PBQH$ - our desired intersection - is a rectangle. We have

$$BQ = 1 - QC = 1 - \sin\theta$$

and $HQ = 1 - \cos\theta$, so our desired area is $(1 - \cos\theta)(1 - \sin\theta)$. After 5 minutes, we have $\theta = \dfrac{2\pi}{12} = \dfrac{\pi}{6}$, so our answer is $\boxed{\dfrac{2 - \sqrt{3}}{4}}$.

18. **Solution** (101). Firstly, $f(100, 100) = 101$.

To see this is maximal, note that $f(x, y) \le \max\{x, y\} + 1$, say by induction on $x + y$.

19. **Solution** (634). Let a_n denote the number of ways to color a $2 \times n$ grid subject only to the given constraint, and b_n denote the number of ways to color a $2 \times n$ grid subject to the given constraint, but with the added restriction that the first column cannot be colored black-black.

Consider the first column of a $2 \times n$ grid that is not subject to the additional constraint. It can be colored black-white or white-black, in which case the leftmost 2x2 square is guaranteed not to be monochromatic, and so the remaining $2 \times (n-1)$ subgrid can be colored in a_{n-1} ways. Otherwise, it is colored white-white or black-black; WLOG, assume that it's colored black-black. Then the remaining $2 \times (n-1)$ subgrid is subject to both constraints, so there are b_{n-1} ways to color the remaining subgrid. Hence $a_n = 2a_{n-1} + 2b_{n-1}$.

Now consider the first column of a $2 \times n$ grid that is subject to the additional constraint. The first column cannot be colored black-black, and if it is colored white-black or black-white, there are a_{n-1} ways to color the remaining subgrid by similar logic to the previous case. If it is colored white-white, then there are b_{n-1} ways to color the remaining subgrid, again by similar logic to the previous case. Hence $b_n = 2a_{n-1} + b_{n-1}$.

Therefore, we have $b_n = 2a_{n-1} + \frac{1}{2}(a_n - 2a_{n-1})$, and so $a_n = 2a_{n-1} + 2b_{n-1} = 2a_{n-1} + 2(2a_{n-2} + \frac{1}{2}(a_{n-1} - 2a_{n-2})) = 3a_{n-1} + 2a_{n-2}$. Finally, we have $a_0 = 1$ (as the only possibility is to, well, do nothing) and $a_1 = 4$ (as any 2×1 coloring is admissible), so $a_2 = 14, a_3 = 50, a_4 = 178, a_5 = \boxed{634}$.

20. **Solution** (5994). Let $n = \overline{abc}$, and assume without loss of generality that $a \geq b \geq c$. We have $k \mid 100a + 10b + c$ and $k \mid 100a + 10c + b$, so $k \mid 9(b - c)$. Analogously, $k \mid 9(a - c)$ and $k \mid 9(a - b)$. Note that if $9 \mid n$, then 9 also divides any permutation of ns digits, so $9 \mid f(n)$ as well; ergo, $f(n) \geq 9$, implying that $k \geq 9$. If k is not a multiple of 3, then we have

$$k \mid c - a \implies k \leq c - a < 9,$$

contradiction, so $3 \mid k$.

Let

$$x = \min(a - b, b - c, a - c).$$

If $x = 1$, then we have $k \mid 9$, implying $k = 9$ - irrelevant to our investigation. So we can assume $x \geq 2$. Note also that $x \leq 4$, as

$$2x \leq (a - b) + (b - c) = a - c \leq 9 - 1,$$

and if $x = 4$ we have $n = 951 \implies f(n) = 3$. If $x = 3$, then since

$$3 \mid k \mid 100a + 10b + c \implies 3 \mid a + b + c,$$

we have

$$a \equiv b \equiv c \pmod 3$$

(e.g. if $b - c = 3$, then $b \equiv c \pmod 3$, so $a \equiv b \equiv c \pmod 3$ - the other cases are analogous). This gives us the possibilites $n = 147, 258, 369$, which give $f(n) = 3, 3, 9$ respectively.

Hence we can conclude that $x = 2$; therefore $k \mid 18$. We know also that $k \geq 9$, so either $k = 9$ or $k = 18$. If $k = 18$, then all the digits of n must be even, and n must be a multiple of 9; it is clear that these are sufficient criteria. As n's digits are all even, the sum of them is also even, and hence their sum is 18. Since $a \geq b \geq c$, we have

$$a + b + c = 18 \leq 3a \implies a \geq 6,$$

but if $a = 6$ then $a = b = c = 6$, contradicting the problem statement. Thus $a = 8$, and this gives us the solutions $n = 882, 864$ along with their permutations.

It remains to calculate the sum of the permutations of these solutions. In the $n = 882$ case, each digit is either 8, 8, or 2 (one time each), and in the $n = 864$ case, each digit is either 8, 6, or 4 (twice each). Hence the desired sum is

$$111(8 + 8 + 2) + 111(8 \cdot 2 + 6 \cdot 2 + 4 \cdot 2) = 111(54) = \boxed{5994}.$$

21. Solution (2530). This solution will be presented in the general case with n colors. Our problem asks for $n = 10$.

We isolate three cases:

Case 1: Every unit square has the same color

In this case there are clearly n ways to color the square.

Case 2: Two non-adjacent squares are the same color, and the other two squares are also the same color (but not all four squares are the same color).

In this case there are clearly $\binom{n}{2} = \dfrac{n(n-1)}{2}$ ways to color the square.

Case 3: Every other case

Since without the "rotation" condition there would be n^4 colorings, we have that in this case by complementary counting there are $\dfrac{n^4 - n(n-1) - n}{4}$ ways to color the square.

Therefore the answer is

$$n + \frac{n^2 - n}{2} + \frac{n^4 - n^2}{4} = \frac{n^4 + n^2 + 2n}{4} = \boxed{2530}$$

22. Solution $\left(1, \dfrac{3 + \sqrt{3}i}{2}, \dfrac{1 - \sqrt{3}i}{2}, \dfrac{3 - \sqrt{3}i}{2}, \dfrac{1 + \sqrt{3}i}{2}\right)$. The $x^5 - 5x^4$ at the beginning of the polynomial motivates us to write it as

$$(x - 1)^5 + x^3 - 3x^2 + 4x - 2$$

and again the presence of the $x^3 - 3x^2$ motivates writing the polynomial in the form

$$(x - 1)^5 + (x - 1)^3 + (x - 1).$$

Let a and b be the roots of the polynomial $x^2 + x + 1$. It's clear that the roots of our polynomial are given by 1 and the roots of the polynomials $(x - 1)^2 = a$ and $(x - 1)^2 = b$. The quadratic formula shows that WLOG

$$a = \frac{-1 + \sqrt{3}i}{2} \quad \text{and} \quad b = \frac{-1 - \sqrt{3}i}{2}$$

so we find that either

$$x - 1 = \pm\frac{1 + \sqrt{3}i}{2} \quad \text{or } x - 1 = \pm\frac{1 - \sqrt{3}i}{2}.$$

Hence our roots are

$$\boxed{1, \frac{3 + \sqrt{3}i}{2}, \frac{1 - \sqrt{3}i}{2}, \frac{3 - \sqrt{3}i}{2}, \frac{1 + \sqrt{3}i}{2}}.$$

23. **Solution** (4097). Let $n = a^4 + b$ where a, b are integers and $0 < b < 4a^3 + 6a^2 + 4a + 1$. Then

$$\sqrt[4]{n} - \lfloor \sqrt[4]{n} \rfloor < \frac{1}{2015}$$

$$\sqrt[4]{a^4 + b} - a < \frac{1}{2015}$$

$$\sqrt[4]{a^4 + b} < a + \frac{1}{2015}$$

$$a^4 + b < \left(a + \frac{1}{2015}\right)^4$$

$$a^4 + b < a^4 + \frac{4a^3}{2015} + \frac{6a^2}{2015^2} + \frac{4a}{2015^3} + \frac{1}{2015^4}$$

To minimize $n = a^4 + b$, we clearly should minimize b, which occurs at $b = 1$. Then

$$1 < \frac{4a^3}{2015} + \frac{6a^2}{2015^2} + \frac{4a}{2015^3} + \frac{1}{2015^4}.$$

If $a = 7$, then

$$\frac{6a^2}{2015^2}, \frac{4a}{2015^3}, \frac{1}{2015^4} < \frac{1}{2015},$$

so

$$\frac{4a^3}{2015} + \frac{6a^2}{2015^2} + \frac{4a}{2015^3} + \frac{1}{2015^4} < \frac{4 \cdot 7^3 + 3}{2015} < 1,$$

so $a \geq 8$. When $a = 8$, we have

$$\frac{4a^3}{2015} = \frac{2048}{2015} > 1,$$

so $a = 8$ is the minimum.

Hence, the minimum n is $8^4 + 1 = \boxed{4097}$.

24. **Solution** $\left(\frac{1}{16}\right)$. At every second, each ant can travel to any of the three vertices they are not currently on. Given that, at one second, the three ants are on different vertices, the probability of them all going to the same vertex is $\frac{1}{27}$ and the probability of them all going to different vertices is $\frac{11}{27}$, so the probability of the three ants all meeting for the first time on the nth step is $\left(\frac{11}{27}\right)^{n-1} \times \frac{1}{27}$. Then the probability the three ants all meet at the same time is $\sum_{i=0}^{\infty} \left(\frac{11}{27}\right)^i \times \frac{1}{27} = \frac{\frac{1}{27}}{1 - \frac{11}{27}} = \frac{1}{16}$.

25. **Solution** $\left(\frac{4225}{64}\pi\right)$. By the properties of reflection, the circumradius of $P_A P_B P_C$ equals the circumradius of ABC. Therefore, the circumcircle of $P_A P_B P_C$ must be externally tangent to the circumcircle of ABC. Now it's easy to see that the midpoint of the 2 centers of ABC and $P_A P_B P_C$ lies on the circumcircle of ABC. So the locus of P is simply the circumcircle of ABC. Since $[ABC] = \frac{abc}{4R}$, we find the circumradius is $R = \frac{13 \cdot 14 \cdot 15}{84 \cdot 4} = \frac{65}{8}$, so the enclosed region has area $\frac{4225}{64}\pi$.

26. Solution $(\log_2 2015 - 1)$. Let $g(x) = f(x) + 1$. Substituting g into the functional equation, we get that

$$g(xy) - 1 = g(x) - 1 + g(y) - 1 + 1$$
$$g(xy) = g(x) + g(y).$$

Also, $g(2) = 1$. Now substitute $x = e^{x'}, y = e^{y'}$, which is possible because $x, y \in \mathbb{R}^+$. Then set $h(x) = g(e^x)$. This gives us that

$$g(e^{x'+y'}) = g(e^{x'}) + g(e^{y'}) \implies h(x' + y') = h(x') + h(y')$$

for al $x', y' \in \mathbb{R}$. Also h is continuous. Therefore, by Cauchy's functional equation, $h(x) = cx$ for a real number c. Going all the way back to g, we can get that $g(x) = c \log x$. Since $g(2) = 1, c = \dfrac{1}{\log 2}$. Therefore,

$$g(2015) = c \log 2015 = \frac{\log 2015}{\log 2} = \log_2 2015.$$

Finally,

$$f(2015) = g(2015) - 1 = \log_2 2015 - 1.$$

27. Solution $\left(16\sqrt{17} + 8\sqrt{5}\right)$. By the triangle inequality, $AP + CP \geq AC$ and $BP + DP \geq BD$. So P should be on AC and BD; i.e. it should be the intersection of the two diagonals. Then

$$AP + BP + CP + DP = AC + BD,$$

which is easily computed to be $16\sqrt{17} + 8\sqrt{5}$ by the Pythagorean theorem.

Note that we require the intersection of the diagonals to actually *exist* for this proof to work, but $ABCD$ is convex and this is not an issue.

28. Solution $\left(\dfrac{5\sqrt{3}}{3}\right)$. First we find the direction of a line perpendicular to both of these lines. By taking the cross product

$$(2, 3, 1) \times (-1, 1, 2) = (5, -5, 5)$$

we find that the plane $x - y + z + 3 = 0$ contains the first line and is parallel to the second. Now we take a point on the second line, say the point $(3, 0, -1)$ and find the distance between this point and the plane. This comes out to

$$\frac{|3 - 0 + (-1) + 3|}{\sqrt{1^2 + 1^2 + 1^2}} = \frac{5}{\sqrt{3}} = \boxed{\frac{5\sqrt{3}}{3}}.$$

29. Solution $\left(\dfrac{1}{2}\right)$. For $k > \dfrac{1}{2}$, I claim that the second sequence must converge. The proof is as follows: by the Cauchy-Schwarz inequality,

$$\left(\sum_{n \geq 1} \frac{\sqrt{a_n}}{n^k}\right)^2 \leq \left(\sum_{n \geq 1} a_n\right)\left(\sum_{n \geq 1} \frac{1}{n^{2k}}\right)$$

Since for $k > \dfrac{1}{2}, \sum_{n \geq 1} \dfrac{1}{n^{2k}}$ converges, the right hand side converges. Therefore, the left hand side must also converge.

For $k \leq \dfrac{1}{2}$, the following construction surprisingly works: $a_n = \dfrac{1}{n \log^2 n}$. It can be easily verified that $\sum_{n \geq 1} a_n$ converges, while $\sum\limits_{n \geq 1} \dfrac{\sqrt{a_n}}{n^{\frac{1}{2}}} = \sum\limits_{n \geq 1} \dfrac{1}{n \log n}$ does not converge.

30. **Solution** (7). We can obtain $n = 7$ in the following way: Consider a rhombus $ABCD$ made up of two equilateral triangles of side length 1, where $\angle DAB = 60°$. Rotate the rhombus clockwise about A to obtain a new rhombus $AB'C'D'$ such that $DD' = 1$. Then one can verify that the seven points A, B, C, D, B', C', D' satisfy the problem condition.

To prove that $n = 8$ points is unobtainable, one interprets the problem in terms of graph theory. Consider a graph on 8 vertices, with an edge drawn between two vertices if and only if the vertices are at distance 1 apart. Assume for the sake of contradiction that this graph has no three points, no two of which are at distance 1 apart (in terms of graph theory, this means the graph has no independent set of size 3).

First, note that this graph cannot contain a complete graph of size 4 (it's clear that there can't exist four points in the plane with any two having the same pairwise distance).

I claim that every vertex has degree 4. It is easy to see that if a vertex has degree 5 or higher, then there exists an independent set of size 3 among its neighbors, contradiction (one can see this by drawing the 5 neighbors on a circle of radius 1 centered at our initial vertex and considering their pairwise distances). Moreover, if a vertex has degree 3 or lower then there are at least four vertices that are not at distance 1 from that vertex, and since not all four of these vertices can be at distance 1 from one another, there exists an independent set of of size 3, contradiction.

Now, we consider the complement of our graph. Every vertex of this new graph has degree 3 and by our observations, contains no independent set of size 4. Moreover, by assumption this graph contains no triangle (a complete graph on three vertices). But we can check by hand that there are only six distinct graphs on eight vertices with each vertex having degree 3 (up to isomorphism), and five of these graphs contain a triangle, and the remaining graph contains an independent set of size 4, contradiction!

Hence the answer is $\boxed{n = 7}$.

31. **Solution** $\left(\dfrac{27}{35}\right)$. We interpret the problem with geometric probability. Let the three segments have lengths $x, y, 1 - x - y$ and assume WLOG that $x \geq y \geq 1 - x - y$. The every possible (x, y) can be found in the triangle determined by the points $\left(\dfrac{1}{3}, \dfrac{1}{3}\right), (\dfrac{1}{2}, \dfrac{1}{2}), (1, 0)$ in $\mathbb{R}^2$, which has area $\dfrac{1}{12}$.

The line $x = 3(1 - x - y)$ intersects the lines $x = y$ and $y = 1 - x - y$ at the points

$$\left(\frac{3}{7}, \frac{3}{7}\right) \text{ and } \left(\frac{3}{5}, \frac{1}{5}\right)$$

Hence $x \leq 3(1 - x - y)$ if (x, y) is in the triangle determined by points

$$\left(\frac{1}{3}, \frac{1}{3}\right), \left(\frac{3}{7}, \frac{3}{7}\right), \left(\frac{3}{5}, \frac{1}{5}\right)$$

which by shoelace has area $\dfrac{2}{105}$. Hence the desired probability is given by

$$\frac{\frac{1}{12} - \frac{2}{105}}{\frac{1}{12}} = \boxed{\frac{27}{35}}$$

32. **Solution** (2029906). Lemma: n is expressible as

$$\left\lceil \frac{x}{2} \right\rceil + y + xy$$

iff $2n + 1$ is not a Fermat Prime.

Proof: Suppose n is expressible. If $x = 2k$, then

$$2n + 1 = (2k + 1)(2y + 1),$$

and if $x = 2k - 1$, then $n = k(2y + 1)$. Thus, if $2n + 1$ isn't prime, we can factor $2n + 1$ as the product of two odd integers $2x + 1$, $2y + 1$ both greater than 1, resulting in positive integer values for x and y. Also, if n has an odd factor greater than 1, then we factor out its largest odd factor as $2y + 1$, giving a positive integer value for x and y. Thus n is expressible iff $2n + 1$ is not prime or n is not a power of 2. That leaves only the n such that $2n + 1$ is a prime one more than a power of two. These are well-known, and are called the Fermat primes.

It's a well-known fact that the only Fermat primes ≤ 2015 are 3, 5, 17, 257, which correspond to $n = 1, 2, 8, 128$. Thus the sum of all expressible numbers is

$$\frac{2015 \cdot 2016}{2} - (1 + 2 + 8 + 128) = 2029906.$$

33. **Solution** (6). Let A, B, C, D be the four points. There are 6 pairwise distances, so at least three of them must be equal.

Case 1: There is no equilateral triangle. Then WLOG we have $AB = BC = CD = 1$.

- Subcase 1.1: $AD = 1$ as well. Then $AC = BD \neq 1$, so $ABCD$ is a square.

- Subcase 1.2: $AD \neq 1$. Then $AC = BD = AD$, so A, B, C, D are four points of a regular pentagon.

Case 2: There is an equilateral triangle, say ABC, of side length 1.

- Subcase 2.1: There are no more pairs of distance 1. Then D must be the center of the triangle.

- Subcase 2.2: There is one more pair of distance 1, say AD. Then D can be either of the two intersections of the unit circle centered at A with the perpendicular bisector of BC. This gives us 2 kites.

- Subcase 2.3: Both $AD = BD = 1$. Then $ABCD$ is a rhombus with a $60°$ angle.

This gives us 6 configurations total.

34. **Solution** (4104). A computer search yields that the second smallest number is 4104. Indeed, $4104 = 9^3 + 15^3 = 2^3 + 16^3$

35. **Solution** $\left(2^{2015} + \left\lfloor \left(\frac{3}{2}\right)^{2015} \right\rfloor - 2\right)$. In general, if $k \leq 471600000$, then any integer can be expressed as the sum of $2^k + \left\lfloor \left(\frac{3}{2}\right)^k \right\rfloor - 2$ integer kth powers. This bound is optimal.

The problem asking for the minimum number of k-th powers needed to add to any positive integer is called Waring's problem.

36. **Solution** (ACGPRES). The smallest counterexamples are:

- Polya's conjecture: 906,150,257

- Euler's sum of powers: 31,858,749,840,007,945,920,321

- Cyclotomic polynomials: 105

- Prime race: 23,338,590,792

- Seventeen   conjecture:
8,424,432,925,592,889,329,288,197,322,308,900,672,459,420,460,792,433

- Goldbach's other conjecture: 5777

- Average square: 44

52.1.3   HMMT 2015 - Team round

1. **Solution** (20). There are three possible triangles: either $\angle ABC = \angle BCA$, in which case $\angle BAC = 180 - 2x$, $\angle ABC = \angle BAC$, in which case $\angle BAC = x$, or $\angle BAC = \angle BCA$, in which case $\angle BAC = \frac{180 - x}{2}$. These sum to $\frac{540 - 3x}{2}$, so we have $\frac{540 - 3x}{2} = 240 \implies x = \boxed{20}$.

2. **Solution** (7 yellow coins). Let r, y, b denote the numbers of red, yellow, and blue coins respectively. Note that each of the three possible exchanges do not change the parities of $y - r$, $b - y$, or $b - r$, and eventually one of these differences becomes zero. Since $b - r$ is the only one of these

differences that is originally even, it must be the one that becomes zero, and so Bassanio will end with some number of yellow coins. Furthermore, Bassanio loses a coin in each exchange, and he requires at least five exchanges to rid himself of the blue coins, so he will have at most $12 - 5 = 7$ yellow coins at the end of his trading.

It remains to construct a sequence of trades that result in seven yellow coins. First, Bassanio will exchange one yellow and one blue coin for one red coin, leaving him with four red coins, three yellow coins, and four blue coins. He then converts the red and blue coins into yellow coins, resulting in $\boxed{7 \text{ yellow coins}}$, as desired.

3. **Solution** $\left(\dfrac{2014}{2015}\right)$. Note that $n = \lfloor n \rfloor + \{n\}$, so

$$\frac{\lfloor n \rfloor}{n} = \frac{\lfloor n \rfloor}{\lfloor n \rfloor + \{n\}}$$
$$= \frac{2015}{2016}$$
$$\implies 2016\lfloor n \rfloor = 2015\lfloor n \rfloor + 2015\{n\}$$
$$\implies \lfloor n \rfloor = 2015\{n\}$$

Hence, $n = \lfloor n \rfloor + \{n\} = \dfrac{2016}{2015}\lfloor n \rfloor$, and so n is maximized when $\lfloor n \rfloor$ is also maximized. As $\lfloor n \rfloor$ is an integer, and $\{n\} < 1$, the maximum possible value of $\lfloor n \rfloor$ is 2014. Therefore, $\{n\} = \dfrac{\lfloor n \rfloor}{2015} = \boxed{\dfrac{2014}{2015}}$.

4. **Solution** (65536). First, we claim that the set $\{2, 4, 8, 256, 65536\}$ is not good. Assume the contrary and say $2 \in S$. Then since $2^2 = 4$, we have $4 \in T$. And since $4^4 = 256$, we have $256 \in S$. Then since $256^2 = 65536$, we have $65536 \in T$. Now, note that we cannot place 8 in either S or T, contradiction.

Hence $n \le 65536$. And the partition $S = \{2, 3\} \cup \{256, 257, \ldots, 65535\}$ and $T = \{4, 5, \ldots, 255\}$ shows that $n \ge 65536$. Therefore $n = \boxed{65536}$.

5. **Solution** (176). Kelvin needs (at most) $i(10 - i)$ hops to determine the ith lilypad he should jump to, then an additional 11 hops to actually get across the river. Thus he requires $\sum_{i=1}^{10} i(10-i)+11 = \boxed{176}$ hops to guarantee success.

6. **Solution** (52). Denote friendship between two people a and b by $a \sim b$. Then, assuming everyone is friends with themselves, the following conditions are satisfied:

- $a \sim a$

- If $a \sim b$, then $b \sim a$

- If $a \sim b$ and $b \sim c$, then $a \sim c$

Thus we can separate the five people into a few groups (possibly one group), such that people are friends within each group, but two people are enemies when they are in different groups. Here comes the calculation. Since the number of group(s) can be 1, 2, 3, 4, or 5, we calculate for each of those cases. When there's only one group, then we only have 1 possibility that we have a group of 5, and the total number of friendship assignments in this case is $\binom{5}{5} = 1$; when there are two groups, we have $5 = 1 + 4 = 2 + 3$ are all possible numbers of the two groups, with a total of $\binom{5}{1} + \binom{5}{2} = 15$ choices; when there are three groups, then we have $5 = 1+1+3 = 1+2+2$, with $\binom{5}{3} + \dfrac{\binom{5}{1}\binom{5}{2}}{2} = 25$ possibilities; when there are four of them, then we have $5 = 1 + 1 + 1 + 2$ be its only possibility, with $\binom{5}{2} = 10$ separations; when there are 5 groups, obviously we have 1 possibility. Hence, we have a total of $1 + 15 + 25 + 10 + 1 = \boxed{52}$ possibilities.

Alternatively, we can also solve the problem recursively. Let B_n be the number of friendship graphs with n people, and consider an arbitrary group. If this group has size k, then there are $\binom{n}{k}$ possible such groups, and B_{n-k} friendship graphs on the remaining $n - k$ people. Therefore, we have the recursion

$$B_n = \sum_{k=0}^{n} \binom{n}{k} B_{n-k}$$

with the initial condition $B_1 = 1$. Calculating routinely gives $B_5 = 52$ as before.

7. **Solution** $(49 + 20\sqrt{6})$. Note that $\angle APB = 180° - \angle BPC = \angle CPD = 180° - \angle DPA$ so $4[BPC][DPA] = (PB \cdot PC \cdot \sin BPC)(PD \cdot PA \cdot \sin DPA) = (PA \cdot PB \cdot \sin APB)(PC \cdot PD \cdot \sin CPD) = 4[APB][CPD] = 2400 \implies [BPC][DPA] = 600$. Hence by AM-GM we have that

$$[BPC] + [DPA] \geq 2\sqrt{[BPC][DPA]} = 20\sqrt{6}$$

so the minimum area of quadrilateral $ABCD$ is $\boxed{49 + 20\sqrt{6}}$.

8. **Solution** $((a, b, c, d) = (128, 32, 16, 4)$ or $(a, b, c, d) = (160, 16, 8, 4))$. It's easy to guess that there are solutions such that a, b, c, d are in the form of n^x, where n is a rather small number. After a few attempts, we can see that we obtain simple equations when $n = 2$ or $n = 3$: for $n = 2$, the equation becomes in the form of

$$2^t + 2^t + 2^{t+1} = 2^{t+2}$$

for some non-negative integer t; for $n = 3$, the equation becomes in the form of

$$3^t + 3^t + 3^t = 3^{t+1}$$

for some non-negative integer t. In the first case, we hope that t is a multiple of two of $3, 4, 5$, that $t + 1$ is a multiple of the last one, and that $t + 2$ is a multiple of 11. Therefore,

$$t \equiv 15, 20, 24 \pmod{60}$$

and

$$t \equiv 9 \pmod{11}.$$

It's easy to check that the only solution that satisfies the given inequality is the solution with $t = 20$, and $(a, b, c, d) = (128, 32, 16, 4)$. In the case where $n = 3$, we must have that t is a multiple of 60, which obviously doesn't satisfy the inequality restriction. Remark: By programming, we find that the only two solutions are

$$(a, b, c, d) = (128, 32, 16, 4)$$

and

$$(a, b, c, d) = (160, 16, 8, 4),$$

with the the former being the intended solution.

9. **Solution** $\left(\dfrac{507}{16384} \text{ or } \dfrac{2^{10} - 10}{2^{15}} \text{ or } \dfrac{2^9 - 5}{2^{14}} \right)$. First, we find the probability that all vertices have even degree. Arbitrarily number the vertices 1, 2, 3, 4, 5, 6. Flip the coin for all the edges out of vertex 1; this vertex ends up with even degree with probability $\dfrac{1}{2}$. Next we flip for all the remaining edges out of vertex 2; regardless of previous edges, vertex 2 ends up with even degree with probability $\dfrac{1}{2}$, and so on through vertex 5. Finally, if vertices 1 through 5 all have even degree, vertex 6 must also have even degree. So all vertices have even degree with probability $\dfrac{1}{2^5} = \dfrac{1}{32}$. There are $\binom{6}{2} = 15$ edges total, so there are 2^{15} total possible graphs, of which 2^{10} have all vertices with even degree. Observe that exactly 10 of these latter graphs are not good, namely, the $\dfrac{1}{2}\binom{6}{3}$ graphs composed of two separate triangles. So $2^{10} - 10$ of our graphs are good, and the probability that a graph is good is $\dfrac{2^{10} - 10}{2^{15}}$.

10. **Solution** (25). Call a number *good* if it is not *bad*. We claim all good numbers are products of distinct primes, none of which are equivalent to 1 modulo another.

We first show that all such numbers are *good*. Consider $n = p_1 p_2 \ldots p_k$, and let x be a number satisfying

$$x \equiv c \pmod{p_1 p_2 \ldots p_k}$$

and

$$x \equiv 1 \pmod{(p_1 - 1)(p_2 - 1) \ldots (p_k - 1)}.$$

Since, by assumption, $p_1 p_2 \ldots p_k$ and $(p_1 - 1)(p_2 - 1) \ldots (p_k - 1)$ are relatively prime, such an x must exist by CRT. Then $x^x \equiv c^1 = c \pmod{n}$, for any c, as desired.

We now show that all other numbers are *bad*. Suppose that there exist some $p_1, p_2 \mid n$ such that $\gcd(p_1, p_2 - 1) \neq 1$ (which must hold for some two primes by assumption), and hence $\gcd(p_1, p_2 - 1) = p_1$. Consider some c for which $p_1 c$ is not a p_1th power modulo p_2, which must exist as $p_1 c$ can take any value modulo p_2 (as p_1, p_2 are relatively prime). We then claim that $x^x \equiv p_1 c \pmod{n}$ is not solvable.

Since $p_1 p_2 \mid n$, we have $x^x \equiv p_1 c \pmod{p_1 p_2}$, hence $p_1 \mid x$. But then $x^x \equiv p_1 c$ is a p_1th power modulo p_2 as $p_1 \mid x$, contradicting our choice of c. As a result, all such numbers are *bad*.

Finally, it is easy to see that n is *bad* if it is not squarefree. If p_1 divides n twice, then letting $c = p_1$ makes the given equivalence unsolvable.

Hence, there are 16 numbers (13 primes: 2, 3, 5, 7, 11, 13, 17, 19, 23, 29, 31, 37, 41; and 3 semiprimes: $3 \cdot 5 = 15$, $3 \cdot 11 = 33$, $5 \cdot 7 = 35$) that are *good*, which means that $41 - 16 = \boxed{25}$ numbers are *bad*.

52.1.4   HMMT 2015 - Theme round

1. **Solution** (3). Let line BE hit line DA at Q. It's clear that triangles AQP and FBP are similar so

$$\frac{AP}{PF} = \frac{AQ}{BF} = \frac{2AD}{\frac{2}{3}BC} = \boxed{3}$$

2. **Solution** (88). Let the four numbers be a, b, c, d, so that the other four numbers are ab, ad, bc, bd. The sum of these eight numbers is

$$a + b + c + d + ab + ad + bc + bd = (a + c) + (b + d) + (a + c)(b + d) = 2015,$$

and so $(a + c + 1)(b + d + 1) = 2016$. Since we seek to minimize $a + b + c + d$, we need to find the two factors of 2016 that are closest to each other, which is easily calculated to be $42 \cdot 48 = 2016$; this makes $a + b + c + d = \boxed{88}$.

3. **Solution** $\left(\dfrac{5\sqrt{2} - 3}{6}\right)$. Let A be the center of the square in the lower left corner, let B be the center of the square in the middle of the top row, and let C be the center of the rightmost square in the middle row. It's clear that O is the circumcenter of triangle ABC - hence, the desired radius is merely the circumradius of triangle ABC minus $\dfrac{1}{2}$. Now note that by the Pythagorean theorem, $BC = \sqrt{2}$ and $AB = AC = \sqrt{5}$ so we easily find that the altitude from A in triangle ABC has length $\dfrac{3\sqrt{2}}{2}$. Therefore the area of triangle ABC is $\dfrac{3}{2}$. Hence the circumradius of triangle ABC is

given by

$$\frac{BC \cdot CA \cdot AB}{4 \cdot \frac{3}{2}} = \frac{5\sqrt{2}}{6}$$

and so the answer is $\dfrac{5\sqrt{2}}{6} - \dfrac{1}{2} = \boxed{\dfrac{5\sqrt{2} - 3}{6}}$.

4. **Solution** (6). We claim that the answer is 6.

On Aziraphale's first two turns, it is always possible for him to take 2 adjacent squares from the central four; without loss of generality, suppose they are the squares at $(1,1)$ and $(1,2)$. If allowed, Aziraphale's next turn will be to take one of the remaining squares in the center, at which point there will be seven squares adjacent to a red square, and so Aziraphale can guarantee at least two more adjacent red squares. After that, since the number of blue squares is always at most the number of red squares, Aziraphale can guarantee another adjacent red square, making his score at least 6.

If, however, Crowley does not allow Aziraphale to attain another central red square – i.e. coloring the other two central squares blue – then Aziraphale will continue to take squares from the second row, WLOG $(1,3)$. If Aziraphale is also allowed to take $(1,0)$, he will clearly attain at least 6 adjacent red squares as each red square in this row has two adjacent squares to it, and otherwise (if Crowley takes $(1,0)$), Aziraphale will take $(0,1)$ and guarantee a score of at least $4 + \frac{4}{2} = 6$ as there are 4 uncolored squares adjacent to a red one.

Therefore, the end score will be at least 6. We now show that this is the best possible for Aziraphale; i.e. Crowley can always limit the score to 6. Crowley can play by the following strategy: if Aziraphale colors a square in the second row, Crowley will color the square below it, if Aziraphale colors a square in the third row, Crowley will color the square above it. Otherwise, if Aziraphale colors a square in the first or fourth rows, Crowley will color an arbitrary square in the same row. It is clear that the two "halves" of the board cannot be connected by red squares, and so the largest contiguous red region will occur entirely in one half of the grid, but then the maximum score is $4 + \frac{4}{2} = 6$.

The optimal score is thus both at least 6 and at most 6, so it must be 6 as desired.

5. **Solution** (12). We claim that the answer is 12. We first show that if 13 squares are colored red, then some four form an axis-parallel rectangle. Note that we can swap both columns and rows without affecting whether four squares form a rectangle, so we may assume without loss of generality that the top row has the most red squares colored; suppose it has k squares colored. We may further suppose that, without loss of generality, these k red squares are the first k squares in the top row from the left.

Consider the $k \times 5$ rectangle formed by the first k columns. In this rectangle, no more than 1 square per row can be red (excluding the top one), so there are a maximum of $k + 4$ squares colored red. In the remaining $(5 - k) \times 5$ rectangle, at most $4(5 - k)$ squares are colored red (as the top row of this rectangle has no red squares), so there are a maximum of

$$(k + 4) + 4(5 - k) = 24 - 3k$$

squares colored red in the 5×5 grid. By assumption, at least 13 squares are colored red, so we have

$$13 \leq 24 - 3k \iff k \leq 3.$$

Hence there are at most 3 red squares in any row. As there are at least 13 squares colored red, this implies that at least 3 rows have 3 red squares colored. Consider the 3×5 rectangle formed by these three rows. Suppose without loss of generality that the leftmost three squares in the top row are colored red, which forces the rightmost three squares in the second row to be colored red. But then, by the Pigeonhole Principle, some 2 of the 3 leftmost squares or some 2 of the 3 rightmost squares in the bottom row will be colored red, leading to an axis-parallel rectangle – a contradiction.

Hence there are most 12 squares colored red. It remains to show that there exists some coloring where exactly 12 squares are colored red, one example of which is illustrated below:

	R	R	R	R
R	R			
R		R		
R			R	
R				R

The maximum number of red squares, therefore, is $\boxed{12}$.

6. Solution $\left(\dfrac{1}{561} \text{ or } \dfrac{105}{\binom{36}{4}} \right)$. Firstly, there are $\binom{36}{4}$ possible combinations of points. Call a square *proper* if its sides are parallel to the coordinate axes and *improper* otherwise. Note that every *improper* square can be inscribed in a unique *proper* square. Hence, an $n \times n$ *proper* square represents a total of n squares: 1 proper and $n - 1$ improper.

There are thus a total of

$$\sum_{i=1}^{6} i(6-i)^2 = \sum_{i=1}^{6}(i^3 - 12i^2 + 36i)$$

$$= \sum_{i=1}^{6} i^3 - 12\sum_{i=1}^{6} i^2 + 36\sum i = 1^6 i$$

$$= 441 - 12(91) + 36(21)$$

$$= 441 - 1092 + 756$$

$$= 105$$

squares on the grid. Our desired probability is thus $\dfrac{105}{\binom{36}{4}} = \boxed{\dfrac{1}{561}}$.

7. **Solution** (1470). Consider the directed graph with $1,2,3,4,5,6,7$ as vertices, and there is an edge from i to j if and only if $f(i) = j$. Since the bottom row is equivalent to the top one, we have $f^6(x) = x$. Therefore, the graph must decompose into cycles of length 6, 3, 2, or 1. Furthermore, since no other row is equivalent to the top one, the least common multiple of the cycle lengths must be 6. The only partitions of 7 satisfying these constraints are $7 = 6 + 1$, $7 = 3 + 2 + 2$, and $7 = 3 + 2 + 1 + 1$.

If we have a cycle of length 6 and a cycle of length 1, there are 7 ways to choose which six vertices will be in the cycle of length 6, and there are $5! = 120$ ways to determine the values of f within this cycle (to see this, pick an arbitrary vertex in the cycle: the edge from it can connect to any of the remaining 5 vertices, which can connect to any of the remaining 4 vertices, etc.). Hence, there are $7 \cdot 120 = 840$ possible functions f in this case.

If we have a cycle of length 3 and two cycles of length 2, there are $\dfrac{\binom{7}{2}\binom{5}{2}}{2} = 105$ possible ways to assign which vertices will belong to which cycle (we divide by two to avoid double-counting the cycles of length 2). As before, there are $2! \cdot 1! \cdot 1! = 2$ assignments of f within the cycles, so there are a total of 210 possible functions f in this case.

Finally, if we have a cycle of length 3, a cycle of length 2, and two cycles of length 1, there are $\binom{7}{3}\binom{4}{2} = 210$ possible ways to assign the cycles, and $2! \cdot 1! \cdot 0! \cdot 0! = 2$ ways to arrange the edges within the cycles, so there are a total of 420 possible functions f in this case.

Hence, there are a total of $840 + 210 + 420 = \boxed{1470}$ possible f.

8. **Solution** (70). Let the expected number of minutes it will take the rook to reach the upper right corner from the top or right edges be E_e, and let the expected number of minutes it will take the rook to reach the upper right corner from any other square be E_c. Note that this is justified because the expected time from any square on the top or right edges is the same, as is the expected

time from any other square (this is because swapping any two rows or columns doesn't affect the movement of the rook). This gives us two linear equations:

$$E_c = \frac{2}{14}(E_e + 1) + \frac{12}{14}(E_c + 1)$$

$$E_e = \frac{1}{14}(1) + \frac{6}{14}(E_e + 1) + \frac{7}{14}(E_c + 1)$$

which gives the solution $E_e = 63$, $E_c = \boxed{70}$.

9. **Solution** (0). Without loss of generality, suppose that the top left corner contains a 1, and examine the top left 3×4:

1	x	x	x
x	x	x	*
x	x	x	*

There cannot be another 1 in any of the cells marked with an x, but the 3×3 on the right must contain a 1, so one of the cells marked with a * must be a 1. Similarly, looking at the top left 4×3:

1	x	x
x	x	x
x	x	x
x	*	*

One of the cells marked with a * must also contain a 1. But then the 3×3 square diagonally below the top left one:

1	x	x	x
x	x	x	*
x	x	x	*
x	*	*	?

must contain multiple 1s, which is a contradiction. Hence no such supersudokus exist.

10. **Solution** (71.8). Label the squares using coordinates, letting the top left corner be $(0,0)$. The burrito will end up in 10 (not necessarily different) squares. Call them $p_1 = (x_1, y_1) = (0,0), p_2 = (x_2, y_2), \ldots, p_{10} = (x_{10}, y_{10})$. p_2 through p_{10} are uniformly distributed throughout the square. Let $d_i = |x_{i+1} - x_i| + |y_{i+1} - y_i|$, the taxicab distance between p_i and p_{i+1}.

After 1 minute, the pigeon will eat 10% of the burrito. Note that if, after eating the burrito, the pigeon throws it to a square taxicab distance d from the square it's currently in, it will take exactly d minutes for it to reach that square, regardless of the path it takes, and another minute for it to eat 10% of the burrito.

Hence, the expected number of minutes it takes for the pigeon to eat the whole burrito is

$$1 + E\left(\sum_{i=1}^{9}(d_i + 1)\right) = 1 + E\left(\sum_{i=1}^{9} 1 + |x_{i+1} - x_i| + |y_{i+1} - y_i|\right)$$

$$= 10 + 2 \cdot E\left(\sum_{i=1}^{9}|x_{i+1} - x_i|\right)$$

$$= 10 + 2 \cdot \left(E(|x_2|) + E(\sum_{i=2}^{9}|x_{i+1} - x_i|)\right)$$

$$= 10 + 2 \cdot (E(|x_2|) + 8 \cdot E(|x_{i+1} - x_i|))$$

$$= 10 + 2 \cdot \left(4.5 + 8 \cdot \frac{1}{100} \cdot \sum_{k=1}^{9} k(20 - 2k)\right)$$

$$= 10 + 2 \cdot (4.5 + 8 \cdot 3.3)$$

$$= 71.8$$

52.2 HMMT 2016

52.2.1 HMMT 2016 - Algebra

1. **Solution** $\left(\dfrac{20}{29}\right)$. From the problem, let A denote the point z on the unit circle, B denote the point 1.45 on the real axis, and O the origin. Let AH be the height of the triangle OAH and H lies on the segment OB. The real part of z is OH. Now we have $OA = 1, OB = 1.45$, and $AB = 1.05$. Thus

$$OH = OA \cos \angle AOB = \cos \angle AOB = \frac{1^2 + 1.45^2 - 1.05^2}{2 \cdot 1 \cdot 1.45} = \frac{20}{29}.$$

2. **Solution** (1,2,4). $n = 1$ works. If n has an odd prime factor, you can factor, and this is simulated also by $n = 8$:

$$a^{2k+1} + 1 = (a + 1)(\sum_{i=0}^{2k}(-a)^i)$$

with both parts larger than one when $a > 1$ and $k > 0$. So it remains to check 2 and 4, which work. Thus the answers are 1,2,4.

3. **Solution** (7294927). From the given conditions, we want to calculate

$$\sum_{i=0}^{3}\sum_{j=i}^{3}(10^i + 10^j)^2.$$

By observing the formula, we notice that each term is an exponent of 10. 10^6 shows up 7 times, 10^5 shows up 2 times, 10^4 shows up 9 times, 10^3 shows up 4 times, 10^2 shows up 9 times, 10 shows 2 times, 1 shows up 7 times. Thus the answer is 7294927.

4. **Solution** (14). Let r_i denote the remainder when 2^i is divided by 25. Note that because $2^{\phi(25)} \equiv 2^{20} \equiv 1 \pmod{25}$, r is periodic with length 20. In addition, we find that 20 is the order of 2 mod 25. Since 2^i is never a multiple of 5, all possible integers from 1 to 24 are represented by $r_1, r_2, ..., r_{20}$ with the exceptions of 5, 10 ,15, and 20. Hence, $\sum_{i=1}^{20} r_i = \sum_{i=1}^{24} i - (5+10+15+20) = 250$.

We also have

$$\sum_{i=0}^{2015} \left\lfloor \frac{2^i}{25} \right\rfloor = \sum_{i=0}^{2015} \frac{2^i - r_i}{25}$$

$$= \sum_{i=0}^{2015} \frac{2^i}{25} - \sum_{i=0}^{2015} \frac{r_i}{25}$$

$$= \frac{2^{2016} - 1}{25} - \sum_{i=0}^{1999} \frac{r_i}{25} - \sum_{i=0}^{15} \frac{r_i}{25}$$

$$= \frac{2^{2016} - 1}{25} - 100 \left(\frac{250}{25} \right) - \sum_{i=0}^{15} \frac{r_i}{25}$$

$$\equiv \frac{2^{2016} - 1}{25} - \sum_{i=0}^{15} \frac{r_i}{25} \pmod{100}$$

We can calculate $\sum_{i=0}^{15} r_i = 185$, so

$$\sum_{i=0}^{2015} \left\lfloor \frac{2^i}{25} \right\rfloor \equiv \frac{2^{2016} - 186}{25} \pmod{100}$$

Now $2^{\phi(625)} \equiv 2^{500} \equiv 1 \pmod{625}$, so $2^{2016} \equiv 2^{16} \equiv 536 \pmod{625}$. Hence $2^{2016} - 186 \equiv 350 \pmod{625}$, and $2^{2016} - 186 \equiv 2 \pmod 4$. This implies that $2^{2016} - 186 \equiv 350 \pmod{2500}$, and so $\dfrac{2^{2016} - 186}{25} \equiv \boxed{14} \pmod{100}$.

5. **Solution** (3). A quick telescope gives that

$$a_1 + \cdots + a_n = 2a_1 + a_3 + a_{n-1} - a_{n-2}$$

for all $n \geq 3$:

$$\sum_{k=1}^{n} a_k = a_1 + a_2 + a_3 + \sum_{k=1}^{n-3} (a_k - 2a_{k+1} + 2a_{k+2})$$

$$= a_1 + a_2 + a_3 + \sum_{k=1}^{n-3} a_k - 2 \sum_{k=2}^{n-2} a_k + \sum_{k=3}^{n-1} a_k$$

$$= 2a_1 + a_3 - a_{n-2} + a_{n-1}.$$

Putting $n = 100$ gives the answer.

One actual value of a_2 which yields the sequence is $a_2 = \dfrac{742745601954}{597303450449}$.

6. **Solution** (50). We claim that all odd numbers are special, and the only special even number is 2. For any even $N > 2$, the numbers relatively prime to N must be odd. When we consider $k = 3$, we see that N can't be expressed as a sum of 3 odd numbers.

Now suppose that N is odd, and we look at the binary decomposition of N, so write $N = 2^{a_1} + 2^{a_2} + ... + 2^{a_j}$ as a sum of distinct powers of 2. Note that all these numbers only have factors of 2 and are therefore relatively prime to N. We see that $j < \log_2 N + 1$.

We claim that for any $k \geq j$, we can write N as a sum of k powers of 2. Suppose that we have N written as

$$N = 2^{a_1} + 2^{a_2} + ... + 2^{a_k}.$$

Suppose we have at least one of these powers of 2 even, say 2^{a_1}. We can then write

$$N = 2^{a_1 - 1} + 2^{a_1 - 1} + 2^{a_2} + ... + 2^{a_k},$$

which is $k + 1$ powers of 2. The only way this process cannot be carried out is if we write N as a sum of ones, which corresponds to $k = N$. Therefore, this gives us all $k > \log_2 N$.

Now we consider the case $k = 2$. Let 2^a be the largest power of 2 such that $2^a < N$. We can write

$$N = 2^a + (N - 2^a).$$

Note that since 2^a and N are relatively prime, so are $N - 2^a$ and N. Note that $a < \log_2 N$. Now similar to the previous argument, we can write 2^a as a sum of k powers of 2 for $1 < k < 2^a$, and since $2^a > \dfrac{N}{2}$, we can achieve all k such that $2 \leq k < \dfrac{N}{2} + 1$.

Putting these together, we see that since $\dfrac{N}{2} + 1 > \log_2 N$ for $N \geq 3$, we can achieve all k from 2 through N, where N is odd.

7. **Solution** (14). Note that

$$2^{10n} = 1024^n = 1.024^n \times 10^{3n}.$$

So 2^{10n} has roughly $3n + 1$ digits for relatively small n's. (Actually we have that for $0 < x < 1$,

$$(1 + x)^2 = 1 + 2x + x^2 < 1 + 3x.$$

Therefore, $1.024^2 < 1.03^2 < 1.09$, $1.09^2 < 1.27$, $1.27^2 < 1.81 < 2$, and $2^2 = 4$, so $1.024^{16} < 4$. Thus the conclusion holds for $n \leq 16$.)

For any positive integer $n \leq 16$,

$$A = \sum_{i=1}^{n} 2^{10i} \times 10^{\sum_{j=i+1}^{n}(3j+1)}.$$

Let

$$A_i = 2^{10i} \times 10^{\sum_{j=i+1}^{n}(3j+1)}$$

for $1 \leq i \leq n$, then we know that

$$A - 2^{10n} = \sum_{i=1}^{n-1} A_i$$

and

$$A_i = 2^{10i+\sum_{j=i+1}^{n}(3j+1)} \times 5^{\sum_{j=i+1}^{n}(3j+1)} = 2^{u_i} \times 5^{v_i}$$

where

$$u_i = 10i + \sum_{j=i+1}^{n}(3j+1),$$

$$v_i = \sum_{j=i+1}^{n}(3j+1).$$

We have that

$$u_i - u_{i-1} = 10 - (3i+1) = 3(3-i).$$

Thus, for $1 \leq i \leq n-1$, u_i is minimized when $i = 1$ or $i = n-1$, with

$$u_1 = \frac{3n^2 + 5n + 12}{2}$$

and $u_{n-1} = 13n - 9$. When $n = 5$,

$$A - 2^{10n} = A_1 + A_2 + A_3 + A_4 = 2^{10} \times 10^{46} + 2^{20} \times 10^{39} + 2^{30} \times 10^{29} + 2^{40} \times 10^{16}$$

is at most divisible by 2^{57} instead of 2^{170}. For all other n's, we have that $u_1 \neq u_{n-1}$, so we should have that both $170 \leq u_1$ and $170 \leq u_{n-1}$. Therefore, since $170 \leq u_{n-1}$, we have that $14 \leq n$. We can see that $u_1 > 170$ and $14 < 16$ in this case. Therefore, the minimum of n is $\boxed{14}$.

8. Solution (30). Note that, if k is relatively prime to n, there exists a unique $0 < k^{-1} < n$ such that $kk^{-1} \equiv 1 \pmod{n}$. Hence, if $k^2 \not\equiv 1 \pmod{n}$, we can pair k with its inverse to get a product of 1.

If $k^2 \equiv 1 \pmod{n}$, then $(n-k)^2 \equiv 1 \pmod{n}$ as well, and $k(n-k) \equiv -k^2 \equiv -1 \pmod{n}$. Hence these k can be paired up as well, giving products of -1. When $n \neq 2$, there is no k such that $k^2 \equiv 1 \pmod{n}$ and $k \equiv n - k \pmod{n}$, so the total product $\pmod{n}$ is $(-1)^{\frac{m}{2}}$, where m is the number of k such that $k^2 \equiv 1 \pmod{n}$.

For prime p and positive integer i, the number of solutions to $k^2 \equiv 1 \pmod{p^i}$ is 2 if p is odd, 4 if $p = 2$ and $i \geq 3$, and 2 if $p = i = 2$. So, by Chinese remainder theorem, if we want the product to be -1, we need $n = p^k, 2p^k$, or 4. We can also manually check the $n = 2$ case to work.

Counting the number of integers in the allowed range that are of one of these forms (or, easier, doing complementary counting), we get an answer of 30.

(Note that this complicated argument basically reduces to wanting a primitive root.)

9. **Solution** (995×2^{998}). Define an $n \times n$ matrix $A_n(x)$ with entries $a_{i,j} = x$ if $i \equiv j \pm 1 \pmod{n}$ and 1 otherwise. Let $F(x) = \sum_{\pi \in S_n} (-1)^{f(\pi)} x^{g(\pi)}$ (here $(-1)^{f(\pi)}$ gives the sign $\prod \dfrac{\pi(u) - \pi(v)}{u - v}$ of the permutation π). Note by construction that $F(x) = \det(A_n(x))$.

We find that the eigenvalues of $A_n(x)$ are $2x + n - 2$ (eigenvector of all ones) and $(x-1)(\omega_j + \omega_j^{-1})$, where $\omega_j = e^{\frac{2\pi ji}{n}}$, for $1 \leq j \leq n - 1$. Since the determinant is the product of the eigenvalues,

$$F(x) = (2x + n - 2)2^{n-1}(x - 1)^{n-1} \prod_{k=1}^{n-1} \cos\left(\frac{2\pi k}{n}\right).$$

Evaluate the product and plug in $x = -1$ to finish. (As an aside, this approach also tells us that the sum is 0 whenever n is a multiple of 4.)

10. **Solution** (52). Consider a triangle ABC with Fermat point P such that $AP = a, BP = b, CP = c$. Then

$$AB^2 = AP^2 + BP^2 - 2AP \cdot BP \cos(120°)$$

by the Law of Cosines, which comes

$$AB^2 = a^2 + ab + b^2$$

and hence $AB = 3$. Similarly, $BC = \sqrt{52}$ and $AC = 7$.

Furthermore, we have

$$BC^2 = 52 = AB^2 + BC^2 - 2AB \cdot BC \cos\angle BAC$$
$$= 3^2 + 7^2 - 2 \cdot 3 \cdot 7 \cos\angle BAC$$
$$= 58 - 42\cos\angle BAC$$

And so $\cos\angle BAC = \dfrac{1}{7}$.

Invert about A with arbitrary radius r. Let B', P', C' be the images of B, P, C respectively. Since $\angle APB = \angle AB'P' = 120°$ and $\angle APC = \angle AC'P' = 120°$, we note that $\angle B'P'C' = 120° - \angle BAC$, and so

$$\cos\angle B'P'C' = \cos(120° - \angle BAC)$$
$$= \cos 120° \cos\angle BAC - \sin 120° \sin\angle BAC$$
$$= \frac{1}{2}\left(\frac{1}{7}\right) - \frac{\sqrt{3}}{2}\left(\frac{4\sqrt{3}}{7}\right)$$
$$= -\frac{13}{14}$$

Furthermore, using the well-known result

$$B'C' = \frac{r^2 BC}{AB \cdot AC}$$

for an inversion about A, we have

$$B'P' = \frac{BPr^2}{AB \cdot AP}$$
$$= \frac{br^2}{a \cdot 3}$$
$$= \frac{br^2}{3a}$$

and similarly $O'C' = \dfrac{cr^2}{7a}$, $B'C' = \dfrac{r^2\sqrt{52}}{21}$. Applying the Law of Cosines to $B'P'C'$ gives us

$$B'C'^2 = B'P'^2 + P'C'^2 - 2B'P' \cdot P'C' \cos(120° - \angle BAC)$$
$$\implies \frac{52r^4}{21^2} = \frac{b^2 r^4}{9a^2} + \frac{c^2 r^4}{49a^2} + \frac{13bcr^4}{147a^2}$$
$$\implies \frac{52}{21^2} = \frac{b^2}{9a^2} + \frac{c^2}{49a^2} + \frac{13bc}{147a^2}$$
$$\implies \frac{52}{21^2} = \frac{49b^2 + 39bc + 9c^2}{21^2 a^2}$$

and so $\dfrac{49b^2 + 39bc + 9c^2}{a^2} = \boxed{52}$.

Motivation: the desired sum looks suspiciously like the result of some Law of Cosines, so we should try building a triangle with sides $\dfrac{7b}{a}$ and $\dfrac{3c}{a}$. Getting the $\dfrac{39bc}{a}$ term is then a matter of setting $\cos\theta = -\dfrac{13}{14}$. Now there are two possible leaps: noticing that $\cos\theta = \cos(120 - \angle BAC)$, or realizing that it's pretty difficult to contrive a side of $\dfrac{7b}{a}$ – but it's much easier to contrive a side of $\dfrac{b}{3a}$. Either way leads to the natural inversion idea, and the rest is a matter of computation.

52.2.2   HMMT 2016 - Combinatorics

1. **Solution** (4). For S_3, either all three lines are parallel (4 regions), exactly two are parallel (6 regions), or none are parallel (6 or seven regions, depending on whether they all meet at one point), so $|S_3| = 3$. Then, for S_4, either all lines are parallel (5 regions), exactly three are parallel (8 regions), there are two sets of parallel pairs (9 regions), exactly two are parallel (9 or 10 regions), or none are parallel (8, 9, 10, or 11 regions), so $|S_4| = 4$.

2. **Solution** (6). Let E be the expected value of the resulting string. Starting from the empty string,

   - We have a $\dfrac{1}{2}$ chance of not selecting the letter M; from here the length of the resulting string is $1 + E$.

- We have a $\frac{1}{4}$ chance of selecting the letter M followed by a letter other than M, which gives a string of length $2 + E$.

- We have a $\frac{1}{4}$ chance of selecting M twice, for a string of length 2.

Thus, $E = \frac{1}{2}(1 + E) + \frac{1}{4}(2 + E) + \frac{1}{4}(2)$. Solving gives $E = 6$.

3. **Solution** (2520). First consider the case $a, b > 0$. We have $720 = 2^4 \cdot 3^2 \cdot 5$, so the number of divisors of 720 is $5 * 3 * 2 = 30$. We consider the number of ways to select an ordered pair (a, b) such that a, b, ab all divide 720. Using the balls and urns method on each of the prime factors, we find the number of ways to distribute the factors of 2 across a and b is $\binom{6}{2}$, the factors of 3 is $\binom{4}{2}$, the factors of 5 is $\binom{3}{2}$. So the total number of ways to select (a, b) with a, b, ab all dividing 720 is $15 * 6 * 3 = 270$. The number of ways to select any (a, b) with a and b dividing 720 is $30 * 30 = 900$, so there are $900 - 270 = 630$ ways to select a and b such that a, b divide 720 but ab doesn't.

Now, each $a, b > 0$ corresponds to four solutions $(\pm a, \pm b)$ giving the final answer of 2520. (Note that $ab \neq 0$.)

4. **Solution** (81). We break this into cases. First, if the middle edge is not included, then there are $6 * 5 = 30$ ways to choose two distinct points for the figure to begin and end at. We could also allow the figure to include all or none of the six remaining edges, for a total of 32 connected figures not including the middle edge. Now let's assume we are including the middle edge. Of the three edges to the left of the middle edge, there are 7 possible subsets we can include (8 total subsets, but we subtract off the subset consisting of only the edge parallel to the middle edge since it's not connected). Similarly, of the three edges to the right of the middle edge, there are 7 possible subsets we can include. In total, there are 49 possible connected figures that include the middle edge. Therefore, there are $32 + 49 = 81$ possible connected figures.

5. **Solution** $\left(\dfrac{63}{2}\right)$. Consider M in binary. Assume we start with $M = 0$, then add a to M, then add $2b$ to M, then add $4c$ to M, and so on. After the first addition, the first bit (defined as the rightmost bit) of M is toggled with probability $\frac{1}{2}$. After the second addition, the second bit of M is toggled with probability $\frac{1}{2}$. After the third addition, the third bit is toggled with probability $\frac{1}{2}$, and so on for the remaining three additions. As such, the six bits of M are each toggled with probability $\frac{1}{2}$ - specifically, the kth bit is toggled with probability $\frac{1}{2}$ at the kth addition, and is never toggled afterwards. Therefore, each residue from 0 to 63 has probability $\frac{1}{64}$ of occurring, so they are all equally likely. The expected value is then just $\frac{63}{2}$.

6. Solution Let $N = n + r$, and $M = n$. Then $r = N - M$, and $s = a_N - a_M$, and

$$d = r + s = (a_N + N) - (a_M + M).$$

So we are trying to find the number of possible values of

$$(a_N + N) - (a_M + M),$$

subject to $N \geq M$ and $a_N \geq a_M$.

Divide the a_i into the following "blocks":

- $a_1 = 1, a_2 = 0,$

- $a_3 = 1, a_4 = 0,$

- $a_5 = 3, a_6 = 2, a_7 = 1, a_8 = 0,$

- $a_9 = 7, a_{10} = 6, \ldots, a_{16} = 0,$

and so on. The kth block contains a_i for $2^{k-1} < i \leq 2^k$. It's easy to see by induction that $a_{2^k} = 0$ and thus $a_{2^k+1} = 2^k - 1$ for all $k \geq 1$. Within each block, the value $a_n + n$ is constant, and for the kth block ($k \geq 1$) it equals 2^k. Therefore,

$$d = (a_N + N) - (a_M + M)$$

is the difference of two powers of 2, say $2^n - 2^m$. For any $n \geq 1$, it is clear there exists an N such that $a_N + N = 2^n$ (consider the nth block). We can guarantee $a_N \geq a_M$ by setting $M = 2^m$. Therefore, we are searching for the number of integers between 1 and 2016 that can be written as $2^n - 2^m$ with $n \geq m \geq 1$. The pairs (n, m) with $n > m \geq 1$ and $n \leq 10$ all satisfy $1 \leq 2^n - 2^m \leq 2016$ (45 possibilities). In the case that $n = 11$, we have that $2^n - 2^m \leq 2016$ so $2^m \geq 32$, so $m \geq 5$ (6 possibilities). There are therefore $45 + 6 = 51$ jetlagged numbers between 1 and 2016.

7. Solution (24, 28, 32). *Ed. note: I'm probably horribly abusing notation*

Define the *generating function* of an event A as the polynomial

$$g(A, x) = \sum p_i x^i$$

where p_i denotes the probability that i occurs during event A. We note that the generating is multiplicative; i.e.

$$g(A \text{ AND } B, x) = g(A)g(B) = \sum p_i q_j x^{i+j}$$

where q_j denotes the probability that j occurs during event B.

In our case, events A and B are the rolling of the first and second dice, respectively, so the generating functions are the same:

$$g(\text{die}, x) = \frac{1}{8}x^1 + \frac{1}{8}x^2 + \frac{1}{8}x^3 + \frac{1}{8}x^4 + \frac{1}{8}x^5 + \frac{1}{8}x^6 + \frac{1}{8}x^7 + \frac{1}{8}x^8$$

and so

$$g(\text{both dice rolled}, x) = g(\text{die}, x)^2 = \frac{1}{64}(x^1 + x^2 + x^3 + x^4 + x^5 + x^6 + x^7 + x^8)^2$$

where the coefficient of x^i denotes the probability of rolling a sum of i.

We wish to find two alternate dice, C and D, satisfying the following conditions:

- C and D are both 8-sided dice; i.e. the sum of the coefficients of $g(C, x)$ and $g(D, x)$ are both 8 (or $g(C, 1) = g(D, 1) = 8$).

- The faces of C and D are all labeled with a positive integer; i.e. the powers of each term of $g(C, x)$ and $g(D, x)$ are positive integer (or $g(C, 0) = g(D, 0) = 0$).

- The probability of rolling any given sum upon rolling C and D is equal to the probability of rolling any given sum upon rolling A and B; i.e. $g(C, x)g(D, x) = g(A, x)g(B, x)$.

Because the dice are "fair" – i.e. the probability of rolling any face is $\frac{1}{8}$ – we can multiply $g(A, x), g(B, x), g(C, x)$ and $g(D, x)$ by 8 to get integer polynomials; as this does not affect any of the conditions, we can assume $g(C, x)$ and $g(D, x)$ are integer polynomials multiplying to $(x^1 + x^2 + \ldots + x^8)^2$ (and subject to the other two conditions as well). Since $\mathbb{Z}$ is a UFD (i.e. integer polynomials can be expressed as the product of integer polynomials in exactly one way, up to order and scaling by a constant), all factors of $g(C, x)$ and $g(D, x)$ must also be factors of $x^1 + x^2 + \ldots + x^8$. Hence it is useful to factor $x^1 + x^2 + \ldots + x^8 = x(x + 1)(x^2 + 1)(x^4 + 1)$.

We thus have $g(C, x)g(D, x) = x^2(x+1)^2(x^2+1)^2(x^4+1)^2$. We know that $g(C, 0) = g(D, 0) = 0$, so $x \mid g(C, x), g(D, x)$. It remains to distribute the remaining term $(x + 1)^2(x^2 + 1)^2(x^4 + 1)^2$; we can view each of these 6 factors as being "assigned" to either C or D. Note that since $g(C, 1) = g(D, 1) = 8$, and each of the factors $x + 1, x^2 + 1, x^4 + 1$ evaluates to 2 when $x = 1$, exactly three factors must be assigned to C and exactly three to D. Finally, assigning $x + 1, x^2 + 1$, and $x^4 + 1$ to C results in the standard die, with $a = b = 28$.. This gives us the three cases (and their permutations):

- $g(C, x) = x(x + 1)^2(x^2 + 1)$, $g(D, x) = x(x^2 + 1)(x^4 + 1)^2$. In this case we get $g(C, x) = x^5 + 2x^4 + 2x^3 + 2x^2 + x$ and $g(D, x) = x^{11} + x^9 + 2x^7 + 2x^5 + x^3 + x$, so the "smaller" die has faces $5, 4, 4, 3, 3, 2, 2$, and 1 which sum to 24.

- $g(C, x) = x(x+1)(x^2+1)^2$, $g(D, x) = x(x+1)(x^4+1)^2$. In this case we have $g(C, x) = x^6 + x^5 + 2x^4 + 2x^3 + x^2 + x$ and $g(D, x) = x^{10} + x^9 + 2x^6 + 2x^5 + x^2 + x$, so the "smaller" die has faces $6, 5, 4, 4, 3, 3, 2$ and 1 which sum to 28.

- $g(C, x) = x(x^2+1)^2(x^4+1)$, $g(D, x) = x(x+1)^2(x^4+1)$. In this case we have $g(C, x) = x^9 + 2x^7 + 2x^5 + 2x^3 + x$ and $g(D, x) = x^7 + 2x^6 + x^5 + x^3 + 2x^2 + x$, so the "smaller die" has faces $7, 6, 6, 5, 3, 2, 2, 1$ which sum to 32.

Therefore, $\min\{a, b\}$ is equal to $\boxed{24, 28, \text{ or } 32}$.

8. **Solution** $(2 \cdot (3^{2017} - 2^{2017}))$. For each $f, g \in X$, we define

$$d(f, g) := \min_{0 \le i \le 2016} \left(\max(f(i), g(i)) \right) - \max_{0 \le i \le 2016} \left(\min(f(i), g(i)) \right)$$

Thus we desire $\max_{g \in X} d(f, g) = 2015$.

First, we count the number of functions $f \in X$ such that

$$\exists g : \min_i \max\{f(i), g(i)\} \ge 2015 \text{ and } \exists g : \min_i \max\{f(i), g(i)\} = 0.$$

That means for every value of i, either $f(i) = 0$ (then we pick $g(i) = 2015$) or $f(i) \ge 2015$ (then we pick $g(i) = 0$). So there are $A = 3^{2017}$ functions in this case.

Similarly, the number of functions such that

$$\exists g : \min_i \max\{f(i), g(i)\} = 2016 \text{ and } \exists g : \min_i \max\{f(i), g(i)\} \le 1$$

is also $B = 3^{2017}$.

Finally, the number of functions such that

$$\exists g : \min_i \max\{f(i), g(i)\} = 2016 \text{ and } \exists g : \min_i \max\{f(i), g(i)\} = 0$$

is $C = 2^{2017}$.

Now $A + B - C$ counts the number of functions with $\max_{g \in X} d(f, g) \ge 2015$ and C counts the number of functions with $\max_{g \in X} d(f, g) \ge 2016$, so the answer is $A + B - 2C = 2 \cdot (3^{2017} - 2^{2017})$.

9. **Solution** (30212). We decompose into cycle types of σ. Note that within each cycle, all vertices have the same degree; also note that the tree has total degree 14 across its vertices (by all its seven edges).

For any permutation that has a 1 in its cycle type (i.e it has a fixed point), let $1 \le a \le 8$ be a fixed point. Consider the tree that consists of the seven edges from a to the seven other vertices - this permutation (with a as a fixed point) is an automorphism of this tree.

For any permutation that has cycle type $2 + 6$, let a and b be the two elements in the 2-cycle. If the 6-cycle consists of c, d, e, f, g, h in that order, consider the tree with edges between a and b, c, e, g and between b and d, f, h. It's easy to see σ is an automorphism of this tree.

For any permutation that has cycle type $2 + 2 + 4$, let a and b be the two elements of the first two-cycle. Let the other two cycle consist of c and d, and the four cycle be e, f, g, h in that order. Then consider the tree with edges between a and b, a and c, b and d, a and e, b and f, a and g, b and h. It's easy to see σ is an automorphism of this tree.

For any permutation that has cycle type $2 + 3 + 3$, let a and b be the vertices in the 2-cycle. One of a and b must be connected to a vertex distinct from a, b (follows from connectedness), so there must be an edge between a vertex in the 2-cycle and a vertex in a 3-cycle. Repeatedly applying σ to this edge leads to a cycle of length 4 in the tree, which is impossible (a tree has no cycles). Therefore, these permutations cannot be automorphisms of any tree.

For any permutation that has cycle type $3 + 5$, similarly, there must be an edge between a vertex in the 3-cycle and a vertex in the 5-cycle. Repeatedly applying σ to this edge once again leads to a cycle in the tree, which is not possible. So these permutations cannot be automorphisms of any tree.

The only remaining possible cycle types of σ are $4 + 4$ and 8. In the first case, if we let x and y be the degrees of the vertices in each of the cycles, then $4x + 4y = 14$, which is impossible for integer x, y. In the second case, if we let x be the degree of the vertices in the 8-cycle, then $8x = 14$, which is not possible either.

So we are looking for the number of permutations whose cycle type is not

$$2 + 2 + 3, \ 8, 4 + 4, \ 3 + 5.$$

The number of permutations with cycle type $2 + 2 + 3$ is $\binom{8}{2}\frac{1}{2}\binom{6}{3}(2!)^2 = 1120$, with cycle type 8 is $7! = 5040$, with cycle type $4 + 4$ is $\frac{1}{2}\binom{8}{4}(3!)^2 = 1260$, with cycle type $3 + 5$ is $\binom{8}{3}(2!)(4!) = 2688$. Therefore, by complementary counting, the number of permutations that ARE automorphisms of some tree is $8! - 1120 - 1260 - 2688 - 5040 = 30212$.

10. **Solution** (18). The answer is 18.

First, we will show that Kristoff must carry at least 18 ice blocks. Let

$$0 < x_1 \le x_2 \le \cdots \le x_n$$

be the weights of ice blocks he carries which satisfy the condition that for any $p, q \in \mathbb{Z}_{\ge 0}$ such that $p + q \le 2016$, there are disjoint subsets I, J of $\{1, \ldots, n\}$ such that $\sum_{\alpha \in I} x_\alpha = p$ and $\sum_{\alpha \in J} x_\alpha = q$.

Claim: For any i, if $x_1 + \cdots + x_i \leq 2014$, then

$$x_{i+1} \leq \left\lfloor \frac{x_1 + \cdots + x_i}{2} \right\rfloor + 1.$$

Proof: Suppose to the contrary that

$$x_{i+1} \geq \left\lfloor \frac{x_1 + \cdots + x_i}{2} \right\rfloor + 2.$$

Consider when Anna and Elsa both demand $\left\lfloor \frac{x_1 + \cdots + x_i}{2} \right\rfloor + 1$ kilograms of ice (which is possible as $2 \times \left(\lfloor \frac{x_1 + \cdots + x_i}{2} \rfloor + 1 \right) \leq x_1 + \cdots + x_i + 2 \leq 2016$). Kristoff cannot give any ice x_j with $j \geq i+1$ (which is too heavy), so he has to use from $x_1, \ldots, x_i$. Since he is always able to satisfy Anna's and Elsa's demands,

$$x_1 + \cdots + x_i \geq 2 \times \left(\lfloor \frac{x_1 + \cdots + x_i}{2} \rfloor + 1 \right) \geq x_1 + \cdots + x_i + 1.$$

A contradiction.

It is easy to see $x_1 = 1$, so by hand we compute obtain the inequalities $x_2 \leq 1$, $x_3 \leq 2$, $x_4 \leq 3$, $x_5 \leq 4$, $x_6 \leq 6$, $x_7 \leq 9$, $x_8 \leq 14$, $x_9 \leq 21$, $x_{10} \leq 31$, $x_{11} \leq 47$, $x_{12} \leq 70$, $x_{13} \leq 105$, $x_{14} \leq 158$, $x_{15} \leq 237$, $x_{16} \leq 355$, $x_{17} \leq 533$, $x_{18} \leq 799$. And we know $n \geq 18$; otherwise the sum $x_1 + \cdots + x_n$ would not reach 2016.

Now we will prove that $n = 18$ works. Consider the 18 numbers named above, say $a_1 = 1$, $a_2 = 1$, $a_3 = 2$, $a_4 = 3$, $\ldots$, $a_{18} = 799$. We claim that with $a_1, \ldots, a_k$, for any $p, q \in \mathbb{Z}_{\geq 0}$ such that $p + q \leq a_1 + \cdots + a_k$, there are two disjoint subsets I, J of $\{1, \ldots, k\}$ such that $\sum_{\alpha \in I} x_\alpha = p$ and $\sum_{\alpha \in J} x_\alpha = q$. We prove this by induction on k. It is clear for small $k = 1, 2, 3$. Now suppose this is true for a certain k, and we add in a_{k+1}. When Kristoff meets Anna first and she demands p kilograms of ice, there are two cases.

Case I: if $p \geq a_{k+1}$, then Kristoff gives the a_{k+1} block to Anna first, then he consider $p' = p - a_{k+1}$ and the same unknown q. Now $p' + q \leq a_1 + \cdots + a_k$ and he has $a_1, \ldots, a_k$, so by induction he can successfully complete his task.

Case II: if $p < a_{k+1}$, regardless of the value of q, he uses the same strategy as if $p + q \leq a_1 + \cdots + a_k$ and he uses ice from $a_1, \ldots, a_k$ without touching a_{k+1}. Then, when he meets Elsa, if $q \leq a_1 + \cdots + a_k - p$, he is safe. If $q \geq a_1 + \cdots + a_k - p + 1$, we know

$$q - a_{k+1} \geq a_1 + \cdots + a_k - p + 1 - \left(\lfloor \frac{a_1 + \cdots + a_k}{2} \rfloor + 1 \right) \geq 0.$$

So he can give the a_{k+1} to Elsa first then do as if $q' = q - a_{k+1}$ is the new demand by Elsa. He can now supply the ice to Elsa because $p + q' \leq a_1 + \cdots + a_k$. Thus, we finish our induction.

Therefore, Kristoff can carry those 18 blocks of ice and be certain that for any $p + q \leq a_1 + \cdots + a_{18} = 2396$, there are two disjoint subsets $I, J \subseteq \{1, \ldots, 18\}$ such that $\sum_{\alpha \in I} a_\alpha = p$ and $\sum_{\alpha \in J} a_\alpha = q$. In other words, he can deliver the amount of ice both Anna and Elsa demand.

52.2.3 HMMT 2016 - Geometry

1. **Solution** (2, 6). The dodecagon has to be a "plus shape" of area 20, then just try the three non-congruent possibilities.

2. **Solution** (14). Let H_B be the reflection of H over AC and let H_C be the reflection of H over AB. The reflections of H over AB, AC lie on the circumcircle of triangle ABC. Since the circumcenters of triangles AH_CB, AH_BC are both O, the circumcenters of AHB, AHC are reflections of O over AB, AC respectively. Moreover, the lines from O to the circumcenters in question are the perpendicular bisectors of AB and AC. Now we see that the distance between the two circumcenters is simply twice the length of the midline of triangle ABC that is parallel to BC, meaning the distance is $2(\frac{1}{2}BC) = 14$.

3. **Solution** $\left(\dfrac{25\sqrt{3}}{12} - \dfrac{4\pi}{3} \right)$. Let $K(P)$ denote the area of P. Note that $K(T) - K(\omega) = 3(X - Y)$, which gives our answer.

4. **Solution** $\left(\dfrac{7\sqrt{3}}{3} \right)$. We note that D is the circumcenter O of ABC, since $2\angle C = \angle ATB = \angle AOB$. So we are merely looking for the circumradius of triangle ABC. By Heron's Formula, the area of the triangle is $\sqrt{9 \cdot 6 \cdot 1 \cdot 2} = 6\sqrt{3}$, so using the formula $\dfrac{abc}{4R} = K$, we get an answer of $\dfrac{3 \cdot 8 \cdot 7}{4 \cdot 6\sqrt{3}} = \dfrac{7\sqrt{3}}{3}$. Alternatively, one can compute the circumradius using trigonometric methods or the fact that $\angle A = 60°$.

5. **Solution** (462). The lines in question are the radical axes of the 9 circles. Three circles with noncollinear centers have a radical center where their three pairwise radical axes concur, but all other intersections between two of the $\binom{9}{2}$ lines can be made to be distinct. So the answer is

$$\binom{\binom{9}{2}}{2} - 2\binom{9}{3} = 462$$

by just counting pairs of lines, and then subtracting off double counts due to radical centers (each counted three times).

6. **Solution** $\left(\dfrac{21\sqrt{3}}{8} \right)$. Let segments AI and EF meet at K. Extending AK to meet the circumcircle again at Y, we see that X and Y are diametrically opposite, and it follows that AX and EF are

parallel. Therefore the height from X to $\overline{UV}$ is merely AK. Observe that $AE = AF$, so $\triangle AEF$ is equilateral; since MN, MP are parallel to AF, AE respectively, it follows that

$$\triangle MVU, \triangle UEN, \triangle FPV$$

are equilateral as well. Then

$$MV = MP - PV = \frac{1}{2}AC - FP = \frac{1}{2}AC - AF + AP = \frac{1}{2}AC - AF + \frac{1}{2}AB = \frac{1}{2}BC,$$

since E, F are the tangency points of the incircle. Since $\triangle MVU$ is equilateral, we have

$$UV = MU = MV = \frac{1}{2}BC.$$

Now we can compute $BC = 7$, whence $UV = \frac{7}{2}$ and

$$AK = \frac{AB + AC - BC}{2} \cdot \cos 30° = \frac{3\sqrt{3}}{2}.$$

Hence, the answer is $\frac{21\sqrt{3}}{8}$.

7. **Solution** (1021). Note that the triangle is a right triangle with right angle at A. Therefore,

$$R^2 = \frac{(7-2)^2 + (11-6)^2}{4} = \frac{25}{2} = (25)(2^{-1}) \equiv 1021 \pmod{2017}.$$

(An equivalent approach works for general triangles; the fact that the triangle is right simply makes the circumradius slightly easier to compute.)

8. **Solution** $(2 + 4\ln(2))$. A vertex P_i is part of the convex hull if and only if it is not contained in the triangle formed by the origin and the two adjacent vertices. Let the probability that a given vertex is contained in the aforementioned triangle be p. By linearity of expectation, our answer is simply $6(1 - p)$. Say $|P_0| = a, |P_2| = b$. Stewart's Theorem and the Law of Cosines give that p is equal to the probability that

$$|P_1| < \sqrt{ab - ab\frac{a^2 + b^2 + ab}{(a+b)^2}} = \frac{ab}{a+b};$$

alternatively this is easy to derive using coordinate methods. The corresponding double integral evaluates to

$$p = \frac{2}{3}(1 - \ln(2)),$$

thus telling us our answer.

9. **Solution** $\left(\frac{28}{5}\sqrt{69}\right)$. Let O be circumcenter, R the circumradius and r the common inradius. We have

$$IO^2 = JO^2 = R(R - 2r)$$

by a result of Euler; denote x for the common value of IO and JO. Additionally, we know $AJ = AB = AD = 49$ (angle chase to find that $\angle BJA = \angle JBA$). Since A is the midpoint of the arc $\overset{\frown}{BD}$ not containing C, both J and A lie on the angle bisector of angle $\angle BCD$, so C, J, A are collinear. So by Power of a Point we have

$$R^2 - x^2 = 2Rr = AJ \cdot JC = 49 \cdot 24.$$

Next, observe that the angle bisector of angle BAD contains both I and O, so A, I, O are collinear. Let M be the midpoint of IJ, lying on $\overline{BD}$. Let K be the intersection of IO and BD. Observing that the right triangles $\triangle IMO$ and $\triangle IKM$ are similar, we find

$$IM^2 = IK \cdot IO = rx,$$

so

$$IJ^2 = 4rx.$$

Now apply Stewart's Theorem to $\triangle AOJ$ to derive

$$R\left(x(R-x) + 4rx\right) = 49^2 x + x^2(R-x).$$

Eliminating the common factor of x and rearranging gives

$$49^2 - (R-x)^2 = 4Rr = 48 \cdot 49$$

so $R - x = 7$. Hence

$$R + x = \frac{49 \cdot 24}{7} = 168,$$

and thus

$$2R = 175, \ 2x = 161.$$

Thus

$$r = \frac{49 \cdot 24}{175} = \frac{168}{25}.$$

Finally,

$$IJ = 2\sqrt{rx} = 2\sqrt{\frac{84 \cdot 161}{25}} = \frac{28\sqrt{69}}{5}.$$

10. **Solution** $\left(\sqrt{\frac{1}{3}(7 + 2\sqrt{13})}\right)$. Let the B-mixtilinear incircle ω_B touch Γ at T_B, BA at B_1 and BC at B_2. Define $T_C \in \Gamma$, $C_1 \in CB$, $C_2 \in CA$, and ω_C similarly. Call I the incenter of triangle ABC, and γ the incircle.

We first identify two points on the radical axis of the B and C mixtilinear incircles:

- The midpoint M of arc BC of the circumcircle of ABC. This follows from the fact that M, B_1, T_B are collinear with

$$MB^2 = MC^2 = MB_1 \cdot MT_B$$

and similarly for C.

- The midpoint N of ID. To see this, first recall that I is the midpoint of segments $B_1 B_2$ and $C_1 C_2$. From this, we can see that the radical axis of ω_B and γ contains N (since it is the line through the midpoints of the common external tangents of ω_B, γ). A similar argument for C shows that the midpoint of ID is actually the radical center of the $\omega_B, \omega_C, \gamma$.

Now consider a homothety with ratio 2 at I. It sends line MN to the line through D and the A-excenter I_A (since M is the midpoint of II_A, by "Fact 5"). Since DH was supposed to be parallel to line MN, it follows that line DH passes through I_A; however a homothety at D implies that this occurs only if H is the midpoint of the A-altitude.

Let $a = BC$, $b = CA = 2$ and $c = AB = \sqrt{3}$. So, we have to just find the value of a such that the orthocenter of ABC lies on the midpoint of the A-altitude. This is a direct computation with the Law of Cosines, but a more elegant solution is possible using the fact that H has barycentric coordinates

$$(S_B S_C : S_C S_A : S_A S_B),$$

where

$$S_A = \frac{1}{2}(b^2 + c^2 - a^2)$$

and so on. Indeed, as H is on the A-midline we deduce directly that

$$S_B S_C = S_A(S_B + S_C) = a^2 S_A \implies \frac{1}{4}(a^2 - 1)(a^2 + 1) = \frac{1}{2}a^2(7 - a^2).$$

Solving as a quadratic in a^2 and taking the square roots gives

$$3a^4 - 14a^2 - 1 = 0 \implies a = \sqrt{\frac{1}{3}(7 + 2\sqrt{13})}$$

as desired.

52.2.4 HMMT 2016 - Guts

1. Solution $\left(\sqrt{10}\right)$. We have

$$(x - y)^2 + (x + y)^2 = 2(x^2 + y^2),$$

so $(x - y)^2 = 10$, hence $|x - y| = \sqrt{10}$.

2. Solution $\left(1 - \left(\frac{13}{16}\right)^5\right)$. During any given minute, the probability that Sherry doesn't catch the train is $\frac{1}{4} + \left(\frac{3}{4}\right)^2 = \frac{13}{16}$. The desired probability is thus one minus the probability that she doesn't catch the train for the next five minutes: $1 - \left(\frac{13}{16}\right)^5$.

3. **Solution** ($\sqrt{2}$). If W is the center of the circle then I is the incenter of $\triangle RWZ$. Moreover, $PRIZ$ is a rhombus. It follows that PI is twice the inradius of a 1-1-$\sqrt{2}$ triangle, hence the answer of $2 - \sqrt{2}$. So $LI = \sqrt{2}$.

Alternatively, one can show (note, really) that the triangle OIL is isosceles.

4. **Solution** $\left(\dfrac{8}{9}\right)$. Short version: third player doesn't matter; against 1 opponent, by symmetry, you'd both play the same strategy. Type A beats B, B beats C, and C beats A all with probability $5/9$. It can be determined that choosing each die with probability $1/3$ is the best strategy. Then, whatever you pick, there is a $1/3$ of dominating, a $1/3$ chance of getting dominated, and a $1/3$ chance of picking the same die (which gives a $1/3 \cdot 2/3 + 1/3 \cdot 1/3 = 1/3$ chance of rolling a higher number). Fix your selection; then the expected payout is then

$$1/3 \cdot 5/9 + 1/3 \cdot 4/9 + 1/3 \cdot 1/3 = 1/3 + 1/9 = 4/9.$$

Against 2 players, your EV is just $E(p1) + E(p2) = 2E(p1) = 8/9$.

5. **Solution** (15).

$$\left\lfloor \left(\frac{10}{4}\right)^3 \right\rfloor = \left\lfloor \frac{125}{8} \right\rfloor = 15.$$

6. **Solution** $\left(\dbinom{4030}{2015}\right)$. The general answer is $\dbinom{2(n-1)}{n-1}$: Simply note that the points in each column must be taken in order, and anything satisfying this avoids intersections, so just choose the steps during which to be in the first column.

7. **Solution** (28). For $0 \le k \le 6$, to obtain a score that is $k \pmod 6$ exactly k problems must get a score of 1. The remaining $6 - k$ problems can generate any multiple of 7 from 0 to $7(6-k)$, of which there are $7 - k$. So the total number of possible scores is $\sum_{k=0}^{6} (7 - k) = 28$.

8. **Solution** (875). For any n, we have

$$W(n, 1) = W(W(n, 0), 0) = (n^n)^{n^n} = n^{n^{n+1}}.$$

Thus,

$$W(555, 1) = 555^{555^{556}}.$$

Let $N = W(555, 1)$ for brevity, and note that $N \equiv 0 \pmod{125}$, and $N \equiv 3 \pmod 8$. Then,

$$W(555, 2) = W(N, 1) = N^{N^{N+1}}$$

is 0 $\pmod{125}$ and 3 $\pmod 8$.

From this we can conclude (by the Chinese Remainder Theorem) that the answer is 875.

9. **Solution** $\left(\dfrac{1999008}{1999012}\right)$. There are $\dbinom{2000}{2} + 8\dbinom{2}{2} = 1999008$ ways to get socks which are matching colors, and four extra ways to get a red-green pair, hence the answer.

10. **Solution** $\left(\dfrac{91}{6}\right)$. Let S, T be the intersections of the tangents to the circumcircle of ABC at A, C and at A, B respectively. Note that $ASCO$ is cyclic with diameter SO, so the circumcenter of AOC is the midpoint of OS, and similarly for the other side. So the length we want is $\dfrac{1}{2}ST$. The circumradius R of ABC can be computed by Heron's formula and $K = \dfrac{abc}{4R}$, giving $R = \dfrac{65}{8}$. A few applications of the Pythagorean theorem and similar triangles gives $AT = \dfrac{65}{6}, AS = \dfrac{39}{2}$, so the answer is $\boxed{\dfrac{91}{6}}$.

11. **Solution** (12). First, $\phi^!(n)$ is even for all odd n, so it vanishes modulo 2.

    To compute the remainder modulo 25, we first evaluate $\phi^!(3) + \phi^!(7) + \phi^!(9) \equiv 2 + 5 \cdot 4 + 5 \cdot 3 \equiv 12$ (mod 25). Now, for $n \geq 11$ the contribution modulo 25 vanishes as long as $5 \nmid n$.

    We conclude the answer is 12.

12. **Solution** (61). We have two cases, depending on whether we choose the middle edge. If so, then either all the remaining edges are either to the left of or to the right of this edge, or there are edges on both sides, or neither; in the first two cases there are 6 ways each, in the third there are $16 + 1 = 17$ ways, and in the last there is 1 way. Meanwhile, if we do not choose the middle edge, then we have to choose a beginning and endpoint, plus the case where we have a loop, for a total of $6 \cdot 5 + 1 = 31$ cases. This gives a total of $6 + 6 + 17 + 1 + 31 = 61$ possible cases.

13. **Solution** $(48 + \sqrt{2016})$. There are no integer solutions to $a^2 + b^2 = 2016$ due to the presence of the prime 7 on the right-hand side (by Fermat's Christmas Theorem). Assuming $a < b$, the minimal solution $(a, b) = (3, 45)$ which gives the answer above.

14. **Solution** $\left(\dfrac{462}{5}\right)$. Note that $AEF \sim ABC$. Let the vertices of the triangle whose area we wish to compute be P, Q, R, opposite A, E, F respectively. Since H, O are isogonal conjugates, line AH passes through the circumcenter of AEF, so $QR \parallel BC$.

    Let M be the midpoint of BC. We claim that $M = P$. This can be seen by angle chasing at E, F to find that $\angle PFB = \angle ABC, \angle PEC = \angle ACB$, and noting that M is the circumcenter of $BFEC$. So, the height from P to QR is the height from A to BC, and thus if K is the area of ABC, the area we want is $\dfrac{QR}{BC}K$.

    Heron's formula gives $K = 84$, and similar triangles QAF, MBF and RAE, MCE give

    $$QA = \frac{BC \tan B}{2 \ \tan A}, \ RA = \frac{BC \tan C}{2 \ \tan A},$$

so that

$$\frac{QR}{BC} = \frac{\tan B + \tan C}{2 \tan A} = \frac{\tan B \tan C - 1}{2} = \frac{11}{10},$$

since the height from A to BC is 12. So our answer is $\boxed{\dfrac{462}{5}}$.

15. **Solution** $(\sqrt{7})$. Consider the polynomial $P(z) = z^7 - 1$. Let $z = e^{ix} = \cos x + i \sin x$. Then

$$z^7 - 1 = \left(\cos^7 x - \binom{7}{2} \cos^5 x \sin^2 x + \binom{7}{4} \cos^3 x \sin^4 x - \binom{7}{6} \cos x \sin^6 x - 1 \right)$$
$$+ i \left(-\sin^7 x + \binom{7}{2} \sin^5 x \cos^2 x - \binom{7}{4} \sin^3 x \cos^4 x + \binom{7}{6} \sin x \cos 6x \right)$$

Consider the real part of this equation. We may simplify it to $64 \cos^7 x - \ldots - 1$, where the middle terms are irrelevant. The roots of P are $x = \frac{2\pi}{7}, \frac{4\pi}{7}, \ldots$, so $\prod_{k=1}^{7} \cos\left(\frac{2\pi k}{7}\right) = \frac{1}{64}$. But

$$\prod_{k=1}^{7} \cos\left(\frac{2\pi k}{7}\right) = \left(\prod_{k=1}^{3} \cos\left(\frac{k\pi}{7}\right) \right)^2$$

so $\prod_{k=1}^{3} \cos\left(\frac{k\pi}{7}\right) = \frac{1}{8}$.

Now consider the imaginary part of this equation. We may simplify it to $-64 \sin^{11} x + \ldots + 7 \sin x$, where again the middle terms are irrelevant. We can factor out $\sin x$ to get $-64 \sin^{10} x + \ldots + 7$, and this polynomial has roots $x = \frac{2\pi}{7}, \ldots, \frac{12\pi}{7}$ (but not 0). Hence $\prod_{k=1}^{6} \sin\left(\frac{2\pi k}{7}\right) = -\frac{7}{64}$. But, like before, we have

$$\prod_{k=1}^{6} \sin\left(\frac{2\pi k}{7}\right) = -\left(\prod_{k=1}^{3} \sin\left(\frac{2\pi k}{7}\right) \right)^2$$

hence $\prod_{k=1}^{3} \sin\left(\frac{k\pi}{7}\right) = \frac{\sqrt{7}}{8}$. As a result, our final answer is $\frac{\frac{\sqrt{7}}{8}}{\frac{1}{8}} = \boxed{\sqrt{7}}$.

16. **Solution** (9). Only $n \equiv 1 \pmod{210}$ work. Proof: we require $\gcd(n, 210) = 1$. Note that $\forall p \le 7$ the order of $n \pmod{p}$ divides $p - 1$, hence is relatively prime to any $p \le 7$. So $n^n \equiv 1 \pmod{p} \iff n \equiv 1 \pmod{p}$ for each of these p.

17. **Solution** (20). Odd a fail for parity reasons and $a \equiv 2 \pmod 3$ fail for mod 3 reasons. This leaves $a \in \{4, 6, 10\}$. It is easy to construct p and q for each of these, take $(p, q) = (3, 5), (5, 11), (3, 7)$, respectively.

18. **Solution** $\left(\frac{1}{2}\right)$. Let the points be $0, \ldots, 7 \pmod 8$, and view Alice's reveal as revealing the three possible locations of the apple. If Alice always picks $0, 2, 4$ and puts the apple randomly at 0 or 4, by symmetry Bob cannot achieve more than $\frac{1}{2}$. Here's a proof that $\frac{1}{2}$ is always possible.

Among the three revealed indices a, b, c, positioned on a circle, two must (in the direction in which they're adjacent) have distance at least 3, so without loss of generality the three are $0, b, c$ where $1 \leq b < c \leq 5$. Modulo reflection and rotation, the cases are: $(0, 1, 2)$: Bob places at 1 and wins. $(0, 1, 3)$: Bob places at 1 half the time and 3 half the time, so wherever the apple is Bob wins with probability $\frac{1}{2}$. $(0, 1, 4)$: Bob places at 1 or 4, same as above. $(0, 2, 4)$: Bob places at 1 or 3, same as above. $(0, 2, 5)$: Bob places at 1 or 5, same as above.

These cover all cases, so we're done.

19. **Solution** (1). Note

$$\sum_{i=0}^{2016} (-1)^i \cdot \frac{\binom{n}{i}\binom{n}{i+2}}{\binom{n}{i+1}^2} = \sum_{i=0}^{2016} (-1)^i \cdot \frac{(i+1)(n-i-1)}{(i+2)(n-i)},$$

So

$$\lim_{n\to\infty} \sum_{i=0}^{2016} (-1)^i \cdot \frac{\binom{n}{i}\binom{n}{i+2}}{\binom{n}{i+1}^2} = \sum_{i=0}^{2016} (-1)^i \cdot \frac{(i+1)}{(i+2)} = 1 - \sum_{i=2}^{2016} \frac{(-1)^i}{i} \approx \ln(2).$$

Then $\frac{1}{A} \approx \frac{1}{\ln(2)} \approx 1.44$, so the answer is 1.

20. **Solution** $\left(\frac{1170\sqrt{37}}{1379}\right)$. Observe that BG is the B-symmedian, and thus $\frac{AG}{GC} = \frac{c^2}{a^2}$. Stewart's theorem gives us

$$BG = \sqrt{\frac{2a^2c^2b}{b(a^2+c^2)} - \frac{a^2b^2c^2}{a^2+c^2}} = \frac{ac}{a^2+c^2}\sqrt{2(a^2+c^2)-b^2} = \frac{390\sqrt{37}}{197}.$$

Then by similar triangles,

$$ZW = HY\frac{ZA}{HA} = BG\frac{YC}{GC}\frac{ZA}{HA} = BG\frac{1}{2}\frac{6}{7} = \boxed{\frac{1170\sqrt{37}}{1379}}$$

where $\frac{ZA}{HA}$ is found with mass points or Ceva.

21. **Solution** $\left(\left(\frac{2}{3}, \frac{2}{9}\right)\right)$. We have the recurrence

$$E_n = \frac{1}{2}(E_{n-1}+1) + \frac{1}{2}(E_{n-2}+1),$$

or

$$E_n = 1 + \frac{1}{2}(E_{n-1} + E_{n-2}),$$

for $n \geq 2$.

Let $F_n = E_n - \frac{2}{3}n$. By directly plugging this into the recurrence for E_n, we get the recurrence

$$F_n = \frac{1}{2}(F_{n-1} + F_{n-1}).$$

The roots of the characteristic polynomial of this recurrence are 1 and $-\frac{1}{2}$, so

$$F_n = A + B(-\frac{1}{2})^n$$

for some A and B depending on the initial conditions. But clearly we have $E_0 = 0$ and $E_1 = 1$ so $F_0 = 0$ and $F_1 = \frac{1}{3}$ so $A = \frac{2}{9}$ and $B = -\frac{2}{9}$.

Hence,

$$E_n = \frac{2}{3}n + \frac{2}{9} - \frac{2}{9}(-\frac{1}{2})^n,$$

so

$$\lim_{n\to\infty}(E_n - \frac{2}{3}n - \frac{2}{9}) = 0.$$

Hence $\left(\frac{2}{3}, \frac{2}{9}\right)$ is the desired pair.

22. **Solution** (10). The square of the radius of a nice circle is the sum of the square of two integers.

The nice circle of radius r intersects (the open segment) $\overline{AB}$ if and only if a point on $\overline{AB}$ is a distance r from the origin. $\overline{AB}$ consists of the points $(20, t)$ where t ranges over $(15, 16)$. The distance from the origin is $\sqrt{20^2 + t^2} = \sqrt{400 + t^2}$. As t ranges over $(15, 16)$, $\sqrt{400 + t^2}$ ranges over $(\sqrt{625}, \sqrt{656})$, so the nice circle of radius r intersects $\overline{AB}$ if and only if $625 < r^2 < 656$.

The possible values of r^2 are those in this range that are the sum of two perfect squares, and each such value corresponds to a unique nice circle. By Fermat's Christmas theorem, an integer is the sum of two squares if an only if in its prime factorization, each prime that is 3 mod 4 appears with an even exponent (possibly 0.) In addition, since squares are 0, 1, or 4 mod 8, we can quickly eliminate integers that are 3, 6, or 7 mod 8.

Now I will list all the integers that aren't 3, 6, or 7 mod 8 in the range and either supply the bad prime factor or write "nice" with the prime factorization.

626: nice $(2 \cdot 313)$

628: nice $(2^2 \cdot 157)$

629: nice $(17 \cdot 37)$

632: 79

633: 3

634: nice $(2 \cdot 317)$

636: 3

637: nice $(7^2 \cdot 13)$

640: nice $(2^7 \cdot 5)$

641: nice (641)

642: 3

644: 7

645: 3

648: nice $(2^3 \cdot 3^4)$

649: 11

650: nice $(2 \cdot 5^2 \cdot 13)$

652: 163

653: nice (653). There are 10 nice circles that intersect $\overline{AB}$.

23. **Solution** $\left(1 - \left(\dfrac{1}{2}\right)^{2016}\right)$. Let $q = 1 - p$. Then

$$
\sum_{k=1}^{\infty} \left(1 - \sum_{n=0}^{k-1} \frac{e^{-t}t^n}{n!}\right) q^{k-1}p = \sum_{k=1}^{\infty} q^{k-1}p - \sum_{k=1}^{\infty}\sum_{n=0}^{k-1} \frac{e^{-t}t^n}{n!} q^{k-1}p
$$

$$
= 1 - \sum_{k=1}^{\infty}\sum_{n=0}^{k-1} \frac{e^{-t}t^n}{n!} q^{k-1}p
$$

$$
= 1 - \sum_{n=0}^{\infty}\sum_{k=n+1}^{\infty} \frac{e^{-t}t^n}{n!} q^{k-1}p
$$

$$
= 1 - \sum_{n=0}^{\infty} \frac{e^{-t}t^n}{n!} q^{n}
$$

$$
= 1 - \sum_{n=0}^{\infty} \frac{e^{-t}(qt)^n}{n!}
$$

$$
= 1 - e^{-t}e^{qt} = 1 - e^{-pt}.
$$

Thus the answer is $1 - \left(\dfrac{1}{2}\right)^{2016}$.

24. **Solution** (4030). We claim that Γ_2 is the incircle of $\triangle B_1 A_2 C$. This is because $\triangle B_1 A_2 C$ is similar to $A_1 B_1 C$ with dilation factor $\sqrt{5}-2$, and by simple trigonometry, one can prove that Γ_2 is similar to Γ_1 with the same dilation factor. By similarities, we can see that for every k, the incircle of $\triangle A_k B_k C$ is Γ_{2k-1}, and the incircle of $\triangle B_k A_{k+1} C$ is Γ_{2k}. Therefore, $A_1 B_{2016}$ intersects all $\Gamma_1, \ldots, \Gamma_{4030}$ but not Γ_k for any $k \geq 4031$.

25. **Solution** $\left(\dfrac{3^{8068} - 81}{80}\right)$. Let E_0 be the expected number of flips needed. Let E_1 be the expected number more of flips needed if the first flip landed on H. Let E_2 be the expected number more if the first two landed on HM. In general, let E_k be the expected number more of flips needed if the first k flips landed on the first k values of the sequence HMMTHMMT...HMMT.

We have

$$E_i = \begin{cases} 1 + \dfrac{1}{3}E_{i+1} + \dfrac{1}{3}E_1 + \dfrac{1}{3}E_0 & i \not\equiv 0 \pmod 4 \\ 1 + \dfrac{1}{3}E_{i+1} + \dfrac{2}{3}E_0 & i \equiv 0 \pmod 4 \end{cases}$$

Using this relation for $i = 0$ gives us $E_1 = E_0 - 3$. Let $F_i = \dfrac{1}{3^i}E_i$. By simple algebraic manipulations we have

$$F_{i+1} - F_i = \begin{cases} -\dfrac{2}{3^{i+1}} \cdot E_0 & i \not\equiv 0 \pmod 4 \\ -\dfrac{1}{3^i} - \dfrac{2}{3^{i+1}} \cdot E_0 & i \equiv 0 \pmod 4 \end{cases}$$

We clearly have $F_{2016\cdot 4} = 0$ and $F_0 = E_0$. So adding up the above relations for $i = 0$ to $i = 2016 \cdot 4 - 1$ gives

$$-E_0 = -2E_0 \sum_{i=1}^{2016\cdot 4} \frac{1}{3^i} - \sum_{k=0}^{2015} \frac{1}{3^{4k}}$$

$$= E_0 \left(\frac{1}{3^{2016\cdot 4}} - 1 \right) - \frac{1 - \frac{1}{3^{2016\cdot 4}}}{\frac{80}{81}}$$

so $E_0 = \dfrac{3^{8068} - 81}{80}$.

26. **Solution** (283). We see that the smallest such n must be a prime power, because if two numbers are distinct mod n, they must be distinct mod at least one of the prime powers that divide n. For $k \geq 2$, if $a \uparrow\uparrow k$ and $a \uparrow\uparrow (k+1)$ are distinct mod p^r, then $a \uparrow\uparrow (k-1)$ and $a \uparrow\uparrow k$ must be distinct mod $\phi(p^r)$. In fact they need to be distinct mod $\dfrac{\phi(p^r)}{2}$ if $p = 2$ and $r \geq 3$ because then there are no primitive roots mod p^r.

Using this, for $1 \leq k \leq 5$ we find the smallest prime p such that there exists a such that $a \uparrow\uparrow k$ and $a \uparrow\uparrow (k+1)$ are distinct mod p. The list is: $3, 5, 11, 23, 47$. We can easily check that the next largest prime for $k = 5$ is 139, and also any prime power other than 121 for which $a \uparrow\uparrow 5$ and $a \uparrow\uparrow 6$ are distinct is also larger than 139.

Now if $a \uparrow\uparrow 6$ and $a \uparrow\uparrow 7$ are distinct mod p, then $p - 1$ must be a multiple of 47 or something that is either 121 or at least 139. It is easy to see that 283 is the smallest prime that satisfies this.

If n is a prime power less than 283 such that $a \uparrow\uparrow 6$ and $a \uparrow\uparrow 7$ are distinct mod n, then the prime can be at most 13 and clearly this doesn't work because $\phi(p^r) = p^{r-1}(p-1)$.

To show that 283 works, choose a so that a is a primitive root mod $283, 47, 23, 11, 5$ and 3. This is possible by the Chinese Remainder theorem, and it is easy to see that this a works by induction.

27. **Solution** $\left(\dfrac{56\pi\sqrt{3}}{9}\right)$. Let Γ be an ellipse passing through $A = (2,0), B = (0,3), C = (0,7), D =$
$(6,0)$, and let $P = (0,0)$ be the intersection of AD and BC. $\dfrac{\text{Area of } \Gamma}{\text{Area of } ABCD}$ is unchanged under
an affine transformation, so we just have to minimize this quantity over situations where Γ is a
circle and $\dfrac{PA}{PD} = \dfrac{1}{3}$ and $\dfrac{PB}{BC} = \dfrac{3}{7}$. In fact, we may assume that $PA = \sqrt{7}, PB = 3, PC =$
$7, PD = 3\sqrt{7}$. If $\angle P = \theta$, then we can compute lengths to get

$$r = \frac{\text{Area of } \Gamma}{\text{Area of } ABCD} = \pi\frac{32 - 20\sqrt{7}\cos\theta + 21\cos^2\theta}{9\sqrt{7}\cdot\sin^3\theta}$$

Let $x = \cos\theta$. Then if we treat r as a function of x,

$$0 = \frac{r'}{r} = \frac{3x}{1 - x^2} + \frac{42x - 20\sqrt{7}}{32 - 20x\sqrt{7} + 21x^2}$$

which means that

$$21x^3 - 40x\sqrt{7} + 138x - 20\sqrt{7} = 0.$$

Letting $y = x\sqrt{7}$ gives

$$0 = 3y^3 - 40y^2 + 138y - 140 = (y - 2)(3y^2 - 34y + 70)$$

The other quadratic has roots that are greater than $\sqrt{7}$, which means that the minimum ratio is
attained when

$$\cos\theta = x = \frac{y}{\sqrt{7}} = \frac{2}{\sqrt{7}}.$$

Plugging that back in gives that the optimum $\dfrac{\text{Area of } \Gamma}{\text{Area of } ABCD}$ is $\dfrac{28\pi\sqrt{3}}{81}$, so putting this back into
the original configuration gives Area of $\Gamma \geq \dfrac{56\pi\sqrt{3}}{9}$. If you want to check on Geogebra, this
minimum occurs when the center of Γ is $\left(\dfrac{8}{3}, \dfrac{7}{3}\right)$.

28. **Solution** $\left(\dfrac{2^{135} - 2^{128} + 1}{2^{119}\cdot 129}\right)$. Let $n = 8$.

First, consider any given student S and an antigen a foreign to him/her. Assuming S has been
bitten, we claim the probability S will suffer due to a is

$$1 - \frac{2^{2^{n-1}+1} - 1}{2^{2^{n-1}}(2^{n-1} + 1)}.$$

Indeed, let $N = 2^{n-1}$ denote the number of students with a. So considering just these students and
summing over the number bitten, we obtain a probability

$$\frac{1}{2^N}\sum_{t=0}^{N}\binom{N}{t}\binom{N}{t}\frac{t}{t+1} = \frac{1}{2^N}\frac{2^N N - 2^N + 1}{N + 1}.$$

We now use linearity over all pairs (S, a) of students S and antigens a foreign to them. Noting that each student is bitten with probability $\frac{1}{2}$, and retaining the notation $N = 2^{n-1}$, we get

$$\frac{1}{2} \sum_{k=0}^{n} \left[\binom{n}{k} \cdot k \left(\frac{2^N N - 2^N + 1}{2^N(N+1)} \right) \right] = \frac{nN(2^N N - 2^N + 1)}{2^{N+1}(N+1)}.$$

Finally, setting $n = 8 = 2^3$ and $N = 2^{n-1} = 2^7 = 128$, we get the claimed answer.

29. **Solution** $(1 + \log(2016))$. Letting $f(x)$ be the expected number of cuts if the initial length of the string is x, we get the integral equation $f(x) = 1 + \frac{1}{x} \int_1^x f(y)dy$. Letting $g(x) = \int_1^x f(y)dy$, we get $dg/dx = 1 + \frac{1}{x}g(x)$. Using integrating factors, we see that this has as its solution $g(x) = x\log(x)$, and thus $f(x) = 1 + \log(x)$.

30. **Solution** (22). First consider when $n \geq m$, so let $n = m + d$ where $d \geq 0$. Then we have

$$2^m(m + d - 2^d m) = 2^m(m(1 - 2^d) + d),$$

which is non-positive unless $m = 0$. So our first set of solutions is $m = 0, n = 2^j$.

Now, we can assume that $m > n$, so let $m = n + d$ where $d > 0$. Rewrite

$$2^m n - 2^n m = 2^{n+d} n - 2^n(n + d) = 2^n((2^d - 1)n - d).$$

In order for this to be a power of 2, $(2^d - 1)n - d$ must be a power of 2. This implies that for some j, $2^j \equiv -d \pmod{2^d - 1}$. But notice that the powers of 2 $\pmod{2^d - 1}$ are $1, 2, 4, \ldots, 2^{d-1}$ ($2^d \equiv 1$ so the cycle repeats).

In order for the residues to match, we need $2^j + d = c(2^d - 1)$, where $0 \leq j \leq d - 1$ and $c \geq 1$. In order for this to be true, we must have

$$2^{d-1} + d \geq 2^d - 1 \iff d + 1 \geq 2^{d-1}.$$

This inequality is only true for $d = 1, 2, 3$. We plug each of these into the original expression $(2^d - 1)n - d$.

For $d = 1$: $n - 1$ is a power of 2. This yields the set of solutions $(2^j + 2, 2^j + 1)$ for $j \geq 0$.

For $d = 2$: $3n - 2$ is a power of 2. Note that powers of 2 are $-2 \pmod 3$ if and only if it is an even power, so $n = \dfrac{2^{2j} + 2}{3}$. This yields the solution set $\left(\dfrac{2^{2j} + 8}{3}, \dfrac{2^{2j} + 2}{3} \right), j \geq 0$.

For $d = 3$: $7n - 3$ is a power of 2. Powers of 2 have a period of 3 when taken $\pmod 7$, so inspection tells us $7n - 3 = 2^{3j+2}$, yielding the solution set $\left(\dfrac{2^{3j+2} + 24}{7}, \dfrac{2^{3j+2} + 3}{7} \right), j \geq 0$.

Therefore, all the solutions are of the form

$$(m, n) = (0, 2^j), (2^j + 2, 2^j + 1), (\dfrac{2^{2j} + 8}{3}, \dfrac{2^{2j} + 2}{3}), (\dfrac{2^{3j+2} + 24}{7}, \dfrac{2^{3j+2} + 3}{7})$$

for $j \geq 0$. Restricting this family to $m, n \leq 100$ gives $7 + 7 + 5 + 3 = 22$.

31. **Solution** (172). We claim that $44, 56, 72$ are the only good numbers. It is easy to check that these numbers work.

Now we prove none others work. First, remark that as $n = 1, 2$ fail so we have $\varphi(n)$ is even, thus n is even. This gives us $\varphi(n) \leq n/2$. Now remark that $\tau(n) < 2\sqrt{n}$, so it follows we need

$$n/2 + 8\sqrt{n} > n \implies n \leq 256.$$

This gives us a preliminary bound. Note that in addition we have $8\tau(n) > n$.

Now, it is easy to see that powers of 2 fail. Thus let $n = 2^a p_1^b$ where p_1 is an odd prime. From $8\tau(n) > n$ we get $8(a+1)(b+1) > 2^a p_1^b \geq 2^a 3^b$ from which we get that (a, b) is one of

$$(1,1), (1,2), (1,3), (2,1), (2,2), (3,1), (3,2), (4,1).$$

Remark that $p_1 \leq \sqrt[b]{\dfrac{8(a+1)(b+1)}{2^a}}$. From this we can perform some casework:

- If $a = 1, b = 1$ then $p_1 - 1 + 16 = 2p_1$ but then $p = 15$, absurd.
- If $a = 1, b = 2$ then we have $p_1 \leq 5$ which is obviously impossible.
- If $a = 1, b = 3$ then $p_1 \leq 4$ which is impossible.
- If $a = 2, b = 1$ then $p_1 \leq 12$ and it is easy to check that $p_1 = 11$ and thus $n = 44$ is the only solution.
- If $a = 2, b = 2$ then $p_1 \leq 4$ which is impossible.
- If $a = 3, b = 1$ then $p_1 \leq 8$ and only $p_1 = 7$ or $n = 56$ works.
- If $a = 3, b = 2$ then $p_1 \leq 3$ and $p_1 = 3, n = 72$ works.
- If $a = 4, b = 1$ then $p_1 \leq 1$ which is absurd.

Now suppose n is the product of 3 distinct primes, so $n = 2^a p_1^b p_2^c$ so we have

$$8(a+1)(b+1)(c+1) > 2^a 3^b 5^c$$

then we must have (a, b, c) equal to one of

$$(1,1,1), (1,2,1), (2,1,1), (3,1,1).$$

Again, we can do some casework:

- If $a = b = c = 1$ then $8\tau(n) = 64 > 2p_1 p_2$ but then $p_1 = 3, p_2 = 5$ or $p_1 = 3, p_2 = 7$ is forced neither of which work.
- If $a = 1, b = 2, c = 1$ then $8\tau(n) = 96 > 2p_1^2 p_2$ but then $p_1 = 3, p_2 = 5$ is forced which does not work.

- If $a = 2, b = 1, c = 1$ then $8\tau(n) = 96 > 4p_1p_2$ forces $p_1 = 3, p_2 = 5$ or $p_1 = 3, p_2 = 7$ neither of which work.

- If $a = 3, b = 1, c = 1$ then $8\tau(n) = 108 > 8p_1p_2$ which has no solutions for p_1, p_2.

Finally, take the case where n is the product of at least 4 distinct primes. But then $n \geq 2 \cdot 3 \cdot 5 \cdot 7 = 210$ and as $2 \cdot 3 \cdot 5 \cdot 11 > 256$, it suffices to check only the case of 210. But 210 clearly fails, so it follows that $44, 56, 72$ are the only good numbers so we are done.

32. **Solution** (216). We perform casework on the point three vertices away from $(0,0)$. By inspection, that point can be $(\pm 8, \pm 3), (\pm 7, \pm 2), (\pm 4, \pm 3), (\pm 3, \pm 2), (\pm 2, \pm 1)$ or their reflections across the line $y = x$. The cases are as follows:

If the third vertex is at any of $(\pm 8, \pm 3)$ or $(\pm 3, \pm 8)$, then there are 7 possible hexagons. There are 8 points of this form, contributing 56 hexagons.

If the third vertex is at any of $(\pm 7, \pm 2)$ or $(\pm 2, \pm 7)$, there are 6 possible hexagons, contributing 48 hexagons.

If the third vertex is at any of $(\pm 4, \pm 3)$ or $(\pm 3, \pm 4)$, there are again 6 possible hexagons, contributing 48 more hexagons.

If the third vertex is at any of $(\pm 3, \pm 2)$ or $(\pm 2, \pm 3)$, then there are again 6 possible hexagons, contributing 48 more hexagons.

Finally, if the third vertex is at any of $(\pm 2, \pm 1)$, then there are 2 possible hexagons only, contributing 16 hexagons.

Adding up, we get our answer of $\boxed{216}$.

33. **Solution** (1984).

```
lucas_ones n = length . filter (elem '1') $ take (n + 1) lucas_strs
    where
        lucas = 2 : 1 : zipWith (+) lucas (tail lucas)
        lucas_strs = map show lucas


main = putStrLn . show $ lucas_ones 2016
```

34. **Solution** (1416528).

```
# 1 = on ground, 0 = raised, 2 = back on ground
cache = {}


def pangzi(legs):
```

```
if legs == (2,2,2,2,2,2): return 1
elif legs.count(0) > 3: return 0
elif legs[0] + legs[1] + legs[2] == 0: return 0
elif legs[3] + legs[4] + legs[5] == 0: return 0
elif cache.has_key(legs): return cache[legs]

cache[legs] = 0
for i in xrange(6): # raise a leg
  if legs[i] == 1:
    new = list(legs)
    new[i] = 0
    cache[legs] += pangzi(tuple(new))
  elif legs[i] == 0: # lower a leg
    new = list(legs)
    new[i] = 2
    cache[legs] += pangzi(tuple(new))
return cache[legs]

print pangzi((1,1,1,1,1,1))
```

35. **Solution** $(327680 \cdot 2^{16})$. This is Hadamard's maximal determinant problem. There's an upper bound of $n^{\frac{1}{2}n}$ which empirically seems to give reasonably good estimates, but in fact this is open for general n.

36. **Solution** (49). The first main insight is that all the cubics pass through the points A, B, C, H (orthocenter), O, and the incenter and three excenters. Since two cubics intersect in at most nine points, this is all the intersections of a cubic with a cubic.

On the other hand, it is easy to see that among intersections of circles with circles, there are exactly 3 points; the incircle is tangent to the nine-point circle at the Feuerbach point while being contained completely in the circumcircle; on the other hand for this obtuse triangle the nine-point circle and the circumcircle intersect exactly twice.

All computations up until now are exact, so it remains to estimate:
- Intersection of the circumcircle with cubics. Each cubic intersects the circumcircle at an even number of points, and moreover we already know that A, B, C are among these, so the number of additional intersections contributed is either 1 or 3; it is the former only for the Neuberg cubic which has a "loop". Hence the actual answer in this case is $1 + 3 + 3 + 3 + 3 = 13$ (but an estimate of $3 \cdot 5 = 15$ is very reasonable).

- Intersection of the incircle with cubics. Since $\angle A$ is large the incircle is small, but on the other hand we know I lies on each cubic. Hence it's very likely that each cubic intersects the incircle twice (once "coming in" and once "coming out"). This is the case, giving $2 \cdot 5 = 10$ new points.

- Intersection of the nine-point with cubics. We guess this is close to the 10 points of the incircle, as we know the nine-point circle and the incircle are tangent to each other. In fact, the exact count is 14 points; just two additional branches appear.

In total, $N = 9 + 3 + 13 + 10 + 14 = 49$.

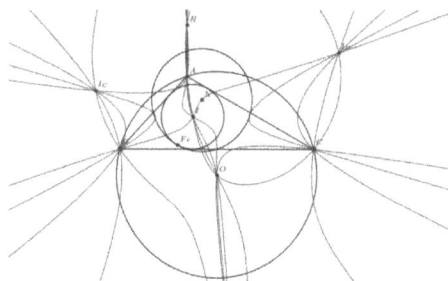

52.2.5   HMMT 2016 - Team

1. **Solution** Since cubes are 0 or ± 1 modulo 9, by inspection we see that we must have $a^3 \equiv b^3 \equiv 0$ (mod 3) for this to be possible. Thus a, b are divisible by 3. But then we get $3^3 \mid 2016$, which is a contradiction.

   One can also solve the problem in the same manner by taking modulo 7 exist, since all cubes are 0 or ± 1 modulo 7. The proof can be copied literally, noting that $7 \mid 2016$ but $7^3 \nmid 2016$.

2. **Solution** (329). In order for $c_n \neq 0$, we must have $\gcd(n, 210) = 1$, so we need only consider such n. The number $n^{c_n} - 1$ is divisible by 210 iff it is divisible by each of 2, 3, 5, and 7, and we can consider the order of n modulo each modulus separately; c_n will simply be the LCM of these orders. We can ignore the modulus 2 because order is always 1. For the other moduli, the sets of orders are

$$a \in \{1, 2\} \quad \bmod 3$$
$$b \in \{1, 2, 4, 4\} \quad \bmod 5$$
$$c \in \{1, 2, 3, 3, 6, 6\} \quad \bmod 7.$$

By the Chinese Remainder Theorem, each triplet of choices from these three multisets occurs for exactly one n in the range $\{1, 2, \ldots, 210\}$, so the answer we seek is the sum of $\operatorname{lcm}(a, b, c)$ over a,

b, c in the Cartesian product of these multisets. For $a = 1$ this table of LCMs is as follows:

	1	2	3	3	6	6
1	1	2	3	3	6	6
2	2	2	6	6	6	6
4	4	4	12	12	12	12
4	4	4	12	12	12	12

which has a sum of $21 + 56 + 28 + 56 = 161$. The table for $a = 2$ is identical except for the top row, where $1, 3, 3$ are replaced by $2, 6, 6$, and thus has a total sum of 7 more, or 168. So our answer is $161 + 168 = \boxed{329}$.

This can also be computed by counting how many times each LCM occurs:

- 12 appears 16 times when $b = 4$ and $c \in \{3, 6\}$, for a contribution of $12 \times 16 = 192$;

- 6 appears 14 times, 8 times when $c = 6$ and $b \leq 2$ and 6 times when $c = 3$ and $(a, b) \in \{(1, 2), (2, 1), (2, 2)\}$, for a contribution of $6 \times 14 = 84$;

- 4 appears 8 times when $b = 4$ and $a, c \in \{1, 2\}$, for a contribution of $4 \times 8 = 32$;

- 3 appears 2 times when $c = 3$ and $a = b = 1$, for a contribution of $3 \times 2 = 6$;

- 2 appears 7 times when $a, b, c \in \{1, 2\}$ and $(a, b, c) \neq (1, 1, 1)$, for a contribution of $2 \times 7 = 14$;

- 1 appears 1 time when $a = b = c = 1$, for a contribution of $1 \times 1 = 1$.

The result is again $192 + 84 + 32 + 6 + 14 + 1 = 329$.

3. **Solution** (14040). We present five different solutions and outline a sixth and seventh one. In what follows, let $a = BC$, $b = CA$, $c = AB$ as usual, and denote by r and R the inradius and circumradius. Let $s = \frac{1}{2}(a + b + c)$.

In the first five solutions we will only prove that

$$\angle AIO = 90° \implies b + c = 2a.$$

Let us see how this solves the problem. This lemma implies that $s = 216$. If we let E be the foot of I on AB, then $AE = s - BC = 72$, consequently the inradius is $r = \sqrt{97^2 - 72^2} = 65$. Finally, the area is $sr = 216 \cdot 65 = \boxed{14040}$.

Proof (First Solution): Since $OI \perp DA$, $AI = DI$. Now, it is a well-known fact that $DI = DB = DC$ (this is occasionally called "Fact 5"). Then by Ptolemy's Theorem,

$$DB \cdot AC + DC \cdot AB = DA \cdot BC \implies AC + AB = 2BC.$$

Proof (Second Solution): As before note that I is the midpoint of AD. Let M and N be the midpoints of AB and AC, and let the reflection of M across BI be P; thus $BM = BP$. Also, $MI = PI$, but we know $MI = NI$ as I lies on the circumcircle of triangle AMN. Consequently, we get $PI = NI$; moreover by angle chasing we have

$$\angle INC = \angle AMI = 180° - \angle BPI = \angle IPC.$$

Thus triangles INC and PIC are congruent (CI is a bisector) so we deduce $PC = NC$. Thus,

$$BC = BP + PC = BM + CN = \frac{1}{2}(AB + AC).$$

Proof (Third Solution): We appeal to Euler's Theorem, which states that $IO^2 = R(R - 2r)$. Thus by the Pythagorean Theorem on $\triangle AIO$ (or by Power of a Point) we may write

$$(s - a)^2 + r^2 = AI^2 = R^2 - IO^2 = 2Rr = \frac{abc}{2s}$$

with the same notations as before. Thus, we derive that

$$\begin{aligned}
abc &= 2s\left((s - a)^2 + r^2\right) \\
&= 2(s - a)\left(s(s - a) + (s - b)(s - c)\right) \\
&= \frac{1}{2}(s - a)\left((b + c)^2 - a^2 + a^2 - (b - c)^2\right) \\
&= 2bc(s - a).
\end{aligned}$$

From this we deduce that $2a = b + c$, and we can proceed as in the previous solution.

Proof (Fourth Solution): From Fact 5 again ($DB = DI = DC$), drop perpendicular from I to AB at E; call M the midpoint of BC. Then, by AAS congruency on AIE and CDM, we immediately get that $CM = AE$. As $AE = \frac{1}{2}(AB + AC - BC)$, this gives the desired conclusion.

Proof (Fifth Solution): This solution avoids angle-chasing and using the fact that BI and CI are angle-bisectors. Recall the perpendicularity lemma, where

$$WX \perp YZ \iff WY^2 - WZ^2 = XY^2 - XZ^2.$$

Let B' be on the extension of ray CA such that $AB' = AB$. Of course, as in the proof of the angle bisector theorem, $BB' \parallel AI$, meaning that $BB' \perp IO$. Let I' be the reflection of I across A; of course, I' is then the incenter of triangle $AB'C'$. Now, we have $B'I^2 - BI^2 = B'O^2 - BO^2$ by the perpendicularity and by power of a point $B'O^2 - BO^2 = B'A \cdot B'C$. Moreover

$BI^2 + B'I^2 = BI^2 + BI'^2 = 2BA^2 + 2AI^2$ by the median formula. Subtracting, we get $BI^2 = AI^2 + \frac{1}{2}(AB)(AB - AC)$. We have a similar expression for CI, and subtracting the two results in $BI^2 - CI^2 = \frac{1}{2}(AB^2 - AC^2)$. Finally,

$$BI^2 - CI^2 = \frac{1}{4}[(BC + AB - AC)^2 - (BC - AB + AC)^2]$$

from which again, the result $2BC = AB + AC$ follows.

Proof: [Sixth Solution, outline] Use complex numbers, setting $I = ab + bc + ca$, $A = -a^2$, etc. on the unit circle (scale the picture to fit in a unit circle; we calculate scaling factor later). Set $a = 1$, and let $u = b + c$ and $v = bc$. Write every condition in terms of u and v, and the area in terms of u and v too. There should be two equations relating u and v: $2u + v + 1 = 0$ and $u^2 = \frac{130}{97}^2 v$ from the right angle and the 144 to 97 ratio, respectively. The square area can be computed in terms of u and v, because the area itself is antisymmetric so squaring it suffices. Use the first condition to homogenize (not coincidentally the factor $(1 - b^2)(1 - c^2) = (1 + bc)^2 - (b + c)^2 = (1 + v)^2 - u^2$ from the area homogenizes perfectly... because $AB \cdot AC = AI \cdot AI_A$, where I_A is the A-excenter, and of course the way the problem is set up $AI_A = 3AI$.), and then we find the area of the scaled down version. To find the scaling factor simply determine $|b - c|$ by squaring it, writing in terms again of u and v, and comparing this to the value of 144.

Proof: [Seventh Solution, outline] Trigonometric solutions are also possible. One can you write everything in terms of the angles and solve the equations; for instance, the $\angle AIO = 90°$ condition can be rewritten as $\frac{1}{2}\cos\frac{B - C}{2} = 2\sin\frac{B}{2}\sin\frac{C}{2}$ and the 97 to 144 ratio condition can be rewritten as $\dfrac{2\sin\frac{B}{2}2\sin\frac{C}{2}}{\sin A} = \dfrac{97}{144}$. The first equation implies $\sin\frac{A}{2} = 2\sin\frac{B}{2}\sin\frac{C}{2}$, which we can plug into the second equation to get $\cos\frac{A}{2}$.

4. **Solution** (3,5). Constructions for $n = 3$ and $n = 5$ are easy. For $n > 5$, color the odd rows black and the even rows white. If the squares can be paired in the way desired, each pair we choose must have one black cell and one white cell, so the numbers of black cells and white cells are the same. The number of black cells is $\frac{n + 1}{2}n - 4$ or $\frac{n + 1}{2}n - 5$ depending on whether the removed center cell is in an odd row. The number of white cells is $\frac{n - 1}{2}n$ or $\frac{n - 1}{2}n - 1$. But

$$\left(\frac{n + 1}{2}n - 5\right) - \frac{n - 1}{2}n = n - 5$$

so for $n > 5$ this pairing is impossible. Thus the answer is $n = 3$ and $n = 5$.

5. **Solution** ($p = 2$ and $p \equiv 3 \pmod 4$). Clearly $p = 2$ works with solutions $(0, 0)$ and $(1, 1)$ and not $(0, 1)$ or $(1, 0)$.

If $p \equiv 3 \pmod 4$ then -1 is not a quadratic residue, so for $x^3 + 4x \neq 0$, exactly one of $x^3 + 4x$ and $-x^3 - 4x$ is a square and gives two solutions (for positive and negative y), so there's exactly two solutions for each such pair $\{x, -x\}$. If x is such that $x^3 + 4x = 0$, there's exactly one solution.

If $p \equiv 1 \pmod 4$, let i be a square root of $-1 \pmod p$. The right hand side factors as $x(x + 2i)(x - 2i)$. For $x = 0, 2i, -2i$ this is zero, there is one choice of y, namely zero. Otherwise, the right hand side is nonzero. For any fixed x, there are either 0 or 2 choices for y. Replacing x by $-x$ negates the right hand side, again producing two choices for y since -1 is a quadratic residue. So the total number of solutions (x, y) is $3 \pmod 4$, and thus there cannot be exactly p solutions.

Remark This is a conductor 36 elliptic curve with complex multiplication, and the exact formula for the number of solutions is given in *http://www.mathcs.emory.edu/ ono/publications-cv/pdfs/026.pdf.*

6. Solution $\left(\binom{43}{21} - 1\right)$. Let a_n be the number of well-filled subsets whose maximum element is n (setting $a_0 = 1$). Then it's easy to see that

$$a_{2k+1} = a_{2k} + a_{2k-1} + \cdots + a_0$$
$$a_{2k+2} = (a_{2k+1} - C_k) + a_{2k} + \cdots + a_0.$$

where C_k is the number of well-filled subsets of size $k + 1$ with maximal element $2k + 1$.

We proceed to compute C_k. One can think of such a subset as a sequence of numbers

$$1 \le s_1 < \cdots < s_{k+1} \le 2k + 1$$

such that $s_i \ge 2i - 1$ for every $1 \le i \le k + 1$. Equivalently, letting $s_i = i + 1 + t_i$ it's the number of sequences

$$0 \le t_1 \le \cdots \le t_{k+1} \le k + 1$$

such that $t_i \ge i$ for every i. This gives the list of x-coordinates of steps up in a Catalan path from $(0, 0)$ to $(k + 1, k + 1)$, so

$$C_k = \frac{1}{k+2}\binom{2(k+1)}{(k+1)}$$

is equal to the $(k+1)$th Catalan number.

From this we can solve the above recursion to derive that

$$a_n = \binom{n}{\lfloor (n-1)/2 \rfloor}.$$

Consequently, for even n,

$$a_0 + \cdots + a_n = a_{n+1} = \binom{n+1}{\lfloor n/2 \rfloor}.$$

Putting $n = 42$ gives the answer, after subtracting off the empty set (counted in a_0).

7. Solution $\left(\dfrac{2^{2017}+1}{3}\right)$. Define $g(x) = 2x^2 - 1$, so that $q(x) = -\dfrac{1}{2} + g\left(x + \dfrac{1}{2}\right)$. Thus

$$q^N(x) = 0 \iff \frac{1}{2} = g^N\left(x + \frac{1}{2}\right)$$

where $N = 2016$.

But, viewed as function $g : [-1,1] \to [-1,1]$ we have that $g(x) = \cos(2\arccos(x))$. Thus, the equation $q^N(x) = 0$ is equivalent to

$$\cos\left(2^{2016}\arccos\left(x + \frac{1}{2}\right)\right) = \frac{1}{2}.$$

Thus, the solutions for x are

$$x = -\frac{1}{2} + \cos\left(\frac{\pi/3 + 2\pi n}{2^{2016}}\right) \quad n = 0, 1, \ldots, 2^{2016} - 1.$$

So, the roots are negative for the values of n such that

$$\frac{1}{3}\pi < \frac{\pi/3 + 2\pi n}{2^{2016}} < \frac{5}{3}\pi$$

which is to say

$$\frac{1}{6}(2^{2016} - 1) < n < \frac{1}{6}(5 \cdot 2^{2016} - 1).$$

The number of values of n that fall in this range is

$$\frac{1}{6}(5 \cdot 2^{2016} - 2) - \frac{1}{6}(2^{2016} + 2) + 1 = \frac{1}{6}(4 \cdot 2^{2016} + 2) = \frac{1}{3}(2^{2017} + 1).$$

8. Solution $\left(\dfrac{3\pi}{13} - \dfrac{4}{13}\log\dfrac{3}{2}\right)$. We have

$$\int_0^\pi \frac{2\sin\theta + 3\cos\theta - 3}{13\cos\theta - 5}\,d\theta = 2\int_0^{\pi/2} \frac{2\sin 2x + 3\cos 2x - 3}{13\cos 2x - 5}\,dx$$

$$= 2\int_0^{\pi/2} \frac{4\sin x \cos x - 6\sin^2 x}{8\cos^2 x - 18\sin^2 x}\,dx$$

$$= 2\int_0^{\pi/2} \frac{\sin x(2\cos x - 3\sin x)}{(2\cos x + 3\sin x)(2\cos x - 3\sin x)}\,dx$$

$$= 2\int_0^{\pi/2} \frac{\sin x}{2\cos x + 3\sin x}.$$

To compute the above integral we want to write $\sin x$ as a linear combination of the denominator and its derivative:

$$
\begin{aligned}
2\int_0^{\pi/2} \frac{\sin x}{2\cos x + 3\sin x} &= 2\int_0^{\pi/2} \frac{-\frac{1}{13}[-3(2\cos x + 3\sin x) + 2(3\cos x - 2\sin x)]}{2\cos x + 3\sin x} \\
&= -\frac{2}{13}\left[\int_0^{\pi/2}(-3) + 2\int_0^{\pi}\frac{-2\sin x + 3\cos x}{2\cos x + 3\sin x}\right] \\
&= -\frac{2}{13}\left[-\frac{3\pi}{2} + 2\log(3\sin x + 2\cos x)\,\big|_0^{\pi/2}\right] \\
&= -\frac{2}{13}\left[-\frac{3\pi}{2} + 2\log\frac{3}{2}\right] \\
&= \frac{3\pi}{13} - \frac{4}{13}\log\frac{3}{2}.
\end{aligned}
$$

9. **Solution** Say a set S *s-meets* F if it shares at least s elements with each set in F. Suppose no such set of size (at most) $r-1$ exists. (Each $S \in F$ s-meets F by the problem hypothesis.)

Let T be a maximal set such that $T \subseteq S$ for infinitely many $S \in F$, which form $F' \subseteq F$ (such T exists, since the empty set works). Clearly $|T| < r$, so by assumption, T does not s-meet F, and there exists $U \in F$ with $|U \cap T| \le s-1$. But U s-meets F', so by pigeonhole, there must exist $u \in U \setminus T$ belonging to infinitely many $S \in F'$, contradicting the maximality of T.

Comment. Let X be an infinite set, and $a_1, \ldots, a_{2r-2-s}$ elements not in X. Then $F = \{B \cup \{x\} : B \subseteq \{a_1, \ldots, a_{2r-2-s}\}, |B| = r-1, x \in X\}$ shows we cannot replace $r-1$ with any smaller number.

OR

We can also use a more indirect approach (where the use of contradiction is actually essential).

Fix $S \in F$ and $a \in S$. By assumption, $S \setminus \{a\}$ does not s-meet F, so there exists $S' \in F$ such that S' contains at most $s-1$ elements of $S \setminus \{a\}$, whence $S \cap S'$ is an s-set containing a. We will derive a contradiction from the following lemma:

Lemma. Let F, G be families of r-sets such that any $f \in F$ and $g \in G$ share at least s elements. Then there exists a finite set H such that for any $f \in F$ and $g \in G$, $|f \cap g \cap H| \ge s$.

Proof. Suppose not, and take a counterexample with $r+s$ minimal; then F, G must be infinite and $r > s > 0$.

Take arbitrary $f_0 \in F$ and $g_0 \in G$; then the finite set $X = f_0 \cup g_0$ meets F, G. For every subset $Y \subseteq X$, let $F_Y = \{S \in F : S \cap X = Y\}$; analogously define G_Y. Then the F_Y, G_Y partition F, G, respectively. For any F_Y and $y \in Y$, define $F_Y(y) = \{S \setminus \{y\} : S \in F_Y\}$.

Now fix subsets $Y, Z \subseteq X$. If one of F_Y, G_Z is empty, define $H_{Y,Z} = \emptyset$.

Otherwise, if Y, Z are disjoint, take arbitrary $y \in Y$, $z \in Z$. By the minimality assumption, there exists finite $H_{Y,Z}$ such that for any $f \in F_Y(y)$ and $g \in G_Z(z)$, $|f \cap g \cap H_{Y,Z}| \geq s$.

If Y, Z share an element a, and $s = 1$, take $H_{Y,Z} = \{a\}$. Otherwise, if $s \geq 2$, we find again by minimality a finite $H_{Y,Z}(a)$ such that for $f \in F_Y(a)$ and $g \in G_Z(a)$, $|f \cap g \cap H_{Y,Z}| \geq s - 1$; then take $H_{Y,Z} = H_{Y,Z}(a) \cup \{a\}$.

Finally, we see that $H = \bigcup_{Y,Z \subseteq X} H_{Y,Z}$ shares at least s elements with each $f \cap g$ (by construction), contradicting our assumption.

10. Solution

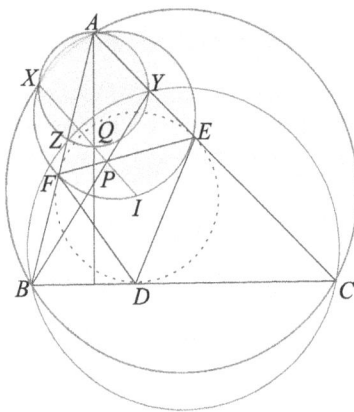

The proof proceeds through a series of seven lemmas.

Lemma 1. Lines DP and EF are the internal and external angle bisectors of $\angle BPC$.

Proof: Since DEF the cevian triangle of ABC with respect to its Gregonne point, we have that

$$-1 = \left(\overline{EF} \cap \overline{BC}, D; B, C\right).$$

Then since $\angle DPF = 90°$ we see P is on the Apollonian circle of BC through D. So the conclusion follows.

Lemma 2. Triangles BPF and CEP are similar.

Proof: Invoking the angle bisector theorem with the previous lemma gives

$$\frac{BP}{BF} = \frac{BP}{BD} = \frac{CP}{CD} = \frac{CP}{CE}.$$

But $\angle BFP = \angle CEP$, so $\triangle BFP \sim \triangle CEP$.

Lemma 3. Quadrilateral $BZYC$ is cyclic; in particular, line YZ is the antiparallel of line BC through $\angle BAC$.

Proof: Remark that $\angle YBZ = \angle PBF = \angle ECP = \angle YCZ$.

Lemma 4. The circumcircles of triangles AYZ, AEF, ABC are concurrent at a point X such that $\triangle XBF \sim \triangle XCE$.

Proof: Note that line EF is the angle bisector of $\angle BPZ = \angle CPY$. Thus

$$\frac{ZF}{FB} = \frac{ZP}{PB} = \frac{YP}{PC} = \frac{YE}{EC}.$$

Then, if we let X be the Miquel point of quadrilateral $ZYCB$, it follows that the spiral similarity mapping segment BZ to segment CY maps E to F; therefore the circumcircle of $\triangle AEF$ must pass through X too.

Lemma 5. Ray XP bisects $\angle FXE$.

Proof: The assertion amounts to

$$\frac{XF}{XE} = \frac{BF}{EC} = \frac{FP}{PE}.$$

The first equality follows from the spiral similarity $\triangle BFX \sim \triangle CEX$, while the second is from $\triangle BFP \sim \triangle CEP$. So the proof is complete by the converse of angle bisector theorem.

Lemma 6. Points X, P, I are collinear.

Proof: On one hand, $\angle FXI = \angle FAI = \frac{1}{2}\angle A$. On the other hand, $\angle FXP = \frac{1}{2}\angle FXE = \frac{1}{2}\angle A$. Hence, X, Y, I collinear.

Lemma 7. Points X, Q, I are collinear.

Proof: On one hand, $\angle AXQ = 90°$, because we established earlier that line YZ was antiparallel to line BC through $\angle A$, hence $AQ \perp BC$ means exactly that $\angle AZQ = AYQ = 90°$. On the other hand, $\angle AXI = 90°$ according to the fact that X lies on the circle with diameter AI. This completes the proof of the lemma.

Finally, combining the final two lemmas solves the problem.

52.2.6 HMMT 2016 - Invitational Competition

1. Solution The path Theseus traces out is a closed, non-self-intersecting path in the plane. Since each move is along a line segment, the path forms the boundary of a polygon in the plane. WLOG this polygon is traversed counterclockwise (i.e. with the interior of the polygon always on Theseus's left); the argument in the clockwise case is analogous, with a sign flip.

Let $i = 0$ correspond to the direction east, $i = 1$ correspond to north, and so forth. For $0 \le i \le 3$ and $j = i \pm 1$ let X_{ij} be the number of times Theseus switches from going in direction i to going in direction j (where indices are taken mod 4). So, $X = X_{1,2}, Y = X_{2,1}$. We have the following relations among the X_{ij}:

The sum $n = \sum X_{i,j}$ is the number of vertices of the polygon. The sum of the interior angles of the polygon is thus $\pi(n-2)$. Each counterclockwise turn contributes $\frac{\pi}{2}$, and each clockwise turn contributes $\frac{3\pi}{2}$ to this sum. The number of the former is counted by $\sum X_{i,i+1}$, and the latter by $\sum X_{i+1,i}$. So, $\pi(n-2) = \frac{\pi}{2} \sum X_{i,i+1} + \frac{3\pi}{2} \sum X_{i+1,i} = \pi n + \frac{\pi}{2} \sum (X_{i+1,i} - X_{i,i+1})$, which gives $\sum(X_{i,i+1} - X_{i+1,i}) = 4$.

Finally, observe by starting from the middle of any edge that since we start and end in the same direction, the number of times we turn into direction i must equal the number of times we turn out of it. So $X_{i,i+1} + X_{i,i-1} = X_{i+1,i} + X_{i-1,i}$, which rearranges to give $X_{i,i+1} - X_{i+1,i} = X_{i-1,i} - X_{i-1,i}$. So $4 = \sum(X_{i,i+1} - X_{i+1,i}) = 4(X_{1,2} - X_{2,1}) = 4(X - Y)$.

So $X - Y = 1$ as desired.

2. **Solution** Let R be the circumradius of $\triangle ABC$. Recall that the circumcircle of $\triangle HBC$ has radius equal to R. By a homothety at H with factor 2 at H, it follows that $\triangle HMN$ has circumradius $R/2$. Thus ω has circumradius $R/2$, and since O lies on ω the conclusion follows.

Remark This problem is a slight modification of a proposal by *Dhroova Aiylam*.

3. **Solution** If $f(z_1) = f(z_2)$, then plugging in (w, x, y, z_1) and (w, x, y, z_2) yields $z_1 = z_2$. Thus, f is injective. Substitution of (z, w) with $(1, f(f(1)))$ in the main equation yields

$$f\Big(f(f(1))xf(yf(1))\Big) = f(xf(y))$$

Because f is injective, we have

$$f(yf(1)) \cdot f(f(1)) = f(y), \forall y \in \mathbb{N} \qquad (*)$$

We can easily show by induction that

$$f(y) = f(yf(1)^n)f(f(1))^n, \forall n \in \mathbb{N}$$

Thus, $f(f(1))^n | f(y), \forall n \in \mathbb{N}$ which implies that $f(f(1)) = 1$. Using this equality in $(*)$ yields $f(1) = 1$. Substitution of (y, z) with $(1, 1)$ in the main equation yields $f(xy) = f(x)f(y)$.

So,

$$f(n!) = f\Big(\prod_{i=1}^{n} i\Big) = \prod_{i=1}^{n} f(i).$$

By injectivity, each $f(i)$ in this product is a distinct positive integer, so their product is at least $\prod_{i=1}^{n} i = n!$, as desired.

Remark The equation condition of f in the problem is actually equivalent to the following two conditions combined:

$$\text{(C1) } f(xy) = f(x)f(y), \forall x, y \in \mathbb{N}, \text{ and}$$
$$\text{(C2) } f(f(f(x))) = x, \forall x \in \mathbb{N}$$

4. **Solution** Let n be the (odd) degree of P. Suppose, for contradiction, that P has no integer roots, and let

$$Q(x) = xP(x) = a_n x^{n+1} + a_{n-1}x^n + \cdots + a_0 x^1.$$

WLOG $a_n > 0$, so there exists $M > 0$ such that $Q(M) < Q(M+1) < \cdots$ and $Q(-M) < Q(-M-1) < \cdots$ (as $n+1$ is even).

There exist finitely many solutions (t, y) for any fixed t, so infinitely many solutions (x, y) with $|x|, |y| \geq M$. By definition of M, there WLOG exist infinitely many solutions with $|x| \geq M$ and $y \geq -x > 0$ (otherwise replace $P(t)$ with $P(-t)$).

Equivalently, there exist infinitely many (t, k) with $t \geq M$ and $k \geq 0$ such that

$$0 = Q(t+k) - Q(-t) = a_n[(t+k)^{n+1} - (-t)^{n+1}] + \cdots + a_0[(t+k)^1 - (-t)^1].$$

If $k \neq 0$, then $Q(x+k) - Q(-x)$ has constant term $Q(k) - Q(0) = kP(k) \neq 0$, and if $k = 0$, then $Q(x+k) - Q(-x) = 2(a_0 x^1 + a_2 x^3 + \cdots + a_{n-1}x^n)$ has x^1 coefficient nonzero (otherwise $P(0) = 0$), so no *fixed* k yields infinitely many solutions.

Hence for any $N \geq 0$, there exist $k = k_N \geq N$ and $t = t_N \geq M$ such that $0 = Q(t+k) - Q(-t)$. But then the triangle inequality yields

$$\underbrace{a_n[(t+k)^{n+1} - (-t)^{n+1}]}_{=\sum_{i=0}^{n-1} -a_i[(t+k)^{i+1}-(-t)^i]} \leq 2(t+k)^n(|a_{n-1}| + \cdots + |a_0|). \tag{52.1}$$

But

$$(t+k)^{n+1} - t^{n+1} \geq (t+k)^{n+1} - t \cdot (t+k)^n = k \cdot (t+k)^n,$$

[1] so $a_n k_N \leq 2(|a_{n-1}| + \cdots + |a_0|)$ for all N. Since $k_N \geq N$, this gives a contradiction for sufficiently large N.

Remark Another way to finish after the triangle inequality estimate (52.1) is to observe that $(t+k)^{n+1} - (-t)^{n+1} = (t+k)^{n+1} - t^{n+1} \geq t^n \cdot k + k^{n+1}$ (by binomial expansion), so $\dfrac{k_N(t_N^n + k_N^n)}{(t_N + k_N)^n}$

---

[1] However, the mean value theorem estimate $(t+k)^{n+1} - t^{n+1} \geq k \cdot (n+1)t^n$ will not suffice a priori, if k happens to be much larger than t. Of course, it is easy to prove that k cannot be too much larger than t (even before finishing the problem); the execution would just take slightly longer.

(which by Holder's inequality is at least $\frac{k_N}{2^{n-1}} \geq \frac{N}{2^{n-1}}$) is bounded above (independently of N), contradiction. (Here we use Holder's inequality to deduce $(1+1)^{n-1}(a^n + b^n) \geq (a+b)^n$ for positive reals a, b.)

Remark The statement is vacuous for even-degree polynomials. The case of degree 3 polynomials was given as ISL 2002 A3.

5. **Solution** (N/A). In general we can assume that each $a_i > 1$, since replacing $a_i = 1$ by some large integer a creates a set T containing the original T as a subset (by setting $e_i = 0$).

We proceed by induction on n. For the base case $n = 1$, an arithmetic progression of length at least 3 would give $a_1^{e_1} + a_1^{e_3} = 2a_1^{e_2}$ where $e_3 > e_2 > e_1$. But $a_1^{e_3} \geq 2a_1^{e_2}$, so this is impossible. Thus our result holds for $n = 1$.

Assume the result is true for some $n-1$, and let A_{n-1} be a number such that the longest progression when $|S| = n - 1$ has length less than A_{n-1}. Let M be a large integer that we will choose later. Take an arithmetic progression of length $2M + 1$, calling the terms $b_1, b_2, \ldots, b_{2M+1}$. Note that $b_{2M+1} \leq 2b_{M+1}$, since b is a sequence of positive integers. For each term from b_{M+1} to b_{2M+1} assign to it the maximum power that is part of the sum. Call this value $c(b_i)$. More explicitly, if $b_i = \sum_{j=1}^{n} a_j^{e_j}$, then $c(b_i) = \max(a_1^{e_1}, a_2^{e_2}, \ldots, a_n^{e_n})$.

Since

$$\frac{b_{M+1}}{n} \leq c(b_i) \leq b_{2M+1} \leq 2b_{M+1}$$

for $M + 1 \leq i \leq 2M + 1$, there are at most

$$\sum_{j=1}^{n}(\log_{a_i} 2n + 1) \leq n(\log_2 2n + 1)$$

different values of $c(b_i)$. By Van der Waerden's Theorem, there exists a value of M such that coloring an arithmetic progression of length M with $n(\log_2 2n + 1)$ colors yields a monochromatic arithmetic progression of length A_{n-1}. In particular, we can take

$$M = W(A_{n-1}, n(\log_2 2n + 1)),$$

where $W(n, k)$ denotes the Van der Waerden number. So, we color $b_{M+1}, \ldots, b_{2M+1}$ by their $c(b_i)$. Subtracting the common perfect power from each term of the monochromatic arithmetic progression obtained gives an arithmetic progression of A_{n-1} integers expressible as the sum of perfect powers of distinct numbers in $S \backslash a_j$. By the inductive hypothesis, this is a contradiction. So no arithmetic progression of length $2M + 1$ can be contained in T, and we can take $A_n = 2M + 1$.

By induction, we thus have such an A_n for all n.